Jerusalem Studies in Philosophy and History of Science

Series Editor
Orly Shenker, The Hebrew University of Jerusalem, The Sidney M. Edelstein Center for the History and Philosophy of Science, Technology and Medicine, Jerusalem, Israel

Jerusalem Studies in Philosophy and History of Science sets out to present state of the art research in a variety of thematic issues related to the fields of Philosophy of Science, History of Science, and Philosophy of Language and Linguistics in their relation to science, stemming from research activities in Israel and the near region and especially the fruits of collaborations between Israeli, regional and visiting scholars.

More information about this series at http://www.springer.com/series/16087

Meir Hemmo • Orly Shenker
Editors

Quantum, Probability, Logic

The Work and Influence of Itamar Pitowsky

Springer

Sidney M. Edelstein Center
for the History and Philosophy
of Science, Technology and Medicine

Editors
Meir Hemmo
University of Haifa
Haifa, Israel

Orly Shenker
The Hebrew University of Jerusalem
Jerusalem, Israel

ISSN 2524-4248 ISSN 2524-4256 (electronic)
Jerusalem Studies in Philosophy and History of Science
ISBN 978-3-030-34315-6 ISBN 978-3-030-34316-3 (eBook)
https://doi.org/10.1007/978-3-030-34316-3

This Springer imprint is published by the registered company Springer Nature Switzerland AG.
The registered company address is: Gewerbestrasse 11, 6330 Cham, Switzerland

Dedicated to the memory of Itamar Pitowsky

Photograph by: Lev Vaidman

Preface

Itamar Pitowsky (1950–2010) was an outstanding philosopher of science, especially physics, and a prominent and loved member of the small community in this field. This volume marks a decade since Pitowsky's untimely death. The essays in it address his work and influence on the contemporary scene in the philosophy of physics. They have been written by leading philosophers of physics and mathematical physicists especially for this volume, engaging in various ways with Pitowsky's work. The topics of the essays range from the interpretation of quantum mechanics and the measurement problem, quantum nonlocality and contextuality, the nature of the quantum probability, probability logic, quantum computation, the computational complexity of physical systems, and the mathematical aspect of physics. Thus, the essays provide a broad perspective on the state of the art in philosophy of physics as well as its evolution during the past decade, touching upon open questions and issues that are under debates in contemporary thinking.[1] The result is an overview of the depth and profoundness of Pitowsky's work.

Pitowsky's unique approach to the foundations of modern physics, computation, and proof theory combined mathematical and physical skills with philosophical sensitivity. While, in his publications, he cautiously refrained from committing himself explicitly to broad metaphysical and epistemological views, those were clearly present in his work and expressed in discussions in the many lectures and workshops in which he was present (the last one held in his honor in Jerusalem in 2008; the lectures given there, including Pitowsky's very last paper, appeared in the volume *Probability in Physics* edited by Yemima Ben-Menahem and Meir Hemmo Springer, 2012). If we can point to one central idea, which is the gist of many of his contributions, it is a down-to-earth emphasis on empirical data as the central, and possibly only, compelling basis on which one ought to form one's scientific and philosophical creed. This point of departure can be gleaned from his views on the foundations of quantum mechanics, statistical mechanics, computation and its relation to physics, logic and probability, and general philosophy of science.

[1]The essays in the volume are arranged in alphabetical order of the authors' surnames.

Needless to say, he combined all of these with original mathematical insights, a combination that is manifest in all of his publications, beginning with his early work on the nonclassical nature of quantum probability presented in his influential and widely cited book *Quantum Probability – Quantum Logic* (Springer 1989) and developed jointly with his former teacher and collaborator Jeffrey Bub, in a paper published in his last year (Bub and Pitowsky 2010; a lecture on which can be seen online at https://vimeo.com/4845944). In the words of Jeffrey Bub and William Demopoulos (2010), Itamar Pitowsky's work "profoundly influenced our understanding of what is characteristically nonclassical about quantum probabilities and quantum logic."[2]

A list of Pitowsky's publications is provided on p. xiii of this volume; and his intellectual genealogy as it appeared in his website is on p. ix.

We wish to thank the authors who wrote their articles especially for this volume, Jeff Bub for helping us in preparing it, and the reviewers who have made significant contributions to the articles they have read.

Haifa, Israel — Meir Hemmo
Jerusalem, Israel — Orly Shenker

[2]Bub, J., Demopoulos, W: Itamar Pitowsky 1950–2010. *Studies in History and Philosophy of Modern Physics* 41, 85 (2010).

Itamar Pitowsky: Academic Genealogy and the Arrow of Time

As you go backward in your family tree, the number of ancestors grows exponentially with the number of generations, which means that it is practically impossible to list all of them. The solution is easy enough: from the large number of ancestors, you choose one with some distinction. Then you list the people in the branch leading from this person to you while ignoring the rest. The rest include a multitude of anonymous people, a few crooks, and perhaps some outright criminals. (If you go back far enough, you are likely to find them). This is how your *yichus* is manufactured.

Academic genealogy, on the other hand, consists of a single branch or at most a few. You simply list your PhD supervisor, then his/her PhD supervisor, and so on. It turns out that there is a web site devoted to academic genealogy of mathematicians from which some of the information below is taken and can be continued further down the generations to the seventeenth century. So here is part of it:

1. Jeffrey Bub, PhD from London University, 1966

2. David Joseph Bohm, PhD from the University of California, Berkeley, 1943

(© Mark Edwards)

3. J. Robert Oppenheimer, PhD from the University of Göttingen, 1927

(Los Alamos National Laboratory)

4. Max Born, PhD from the University of Göttingen, 1907

5. Carl David Tolmé Runge, PhD from the University of Berlin, 1880

6a. Supervisor #1: Ernst Eduard Kummer, PhD from Martin-Luther-Universität Halle-Wittenberg, 1831

6b. Supervisor #2: Karl Theodor Wilhelm Weierstrass, PhD from the University of Königsberg, 1854

And so on . . .

Going forward in time changes one's perspective completely. The great Weierstrass has tens of thousands of academic descendants, so academic genealogy is not a very big deal either. However, academic trees, unlike family trees, grow at different rates backward and forward. As more PhD's graduate, the entropy increases.

Itamar Pitowsky: Selected Publications

Itamar Pitowsky: Resolution of the Einstein-Podolsky-Rosen and Bell paradoxes. Physical Review Letters, 48, 1299–1302 (1982)

Pitowsky, Itamar: Substitution and truth in quantum logic. Philosophy of Science 49(3), 380–401 (1982)

Pitowsky, Itamar: Where the Theory of Probability Fails. PSA: Proceedings of the Biennial Meeting of the Philosophy of Science Association, pp. 616–623 (1982)

Pitowsky, Itamar: The Logic of Fundamental Processes: Nonmeasurable Sets and Quantum Mechanics. Ph.D. Dissertation, The University of Western Ontario, Canada (1983)

Pitowsky, Itamar: Deterministic model of spin and statistics. Physical Review D 27, 2316–2326 (1983)

Pitowsky, Itamar: Quantum mechanics and value definiteness. Philosophy of Science 52(1), 154–156 (1985)

Pitowsky, Itamar: On the status of statistical inferences. Synthese 63(2), 233–247 (1985)

Pitowsky, Itamar: Unified field theory and the conventionality of geometry. Philosophy of Science 51(4), 685–689 (1984)

Bub, Jeffrey, Pitowsky, Itamar: Critical notice on Popper's postscript to the Logic of Scientific Discovery. *Canadian Journal of Philosophy* 15(3), 539–552 (1986)

Itamar Pitowsky: The range of quantum probabilities. Journal of Mathematical Physics 27, 1556–1565 (1986)

Itamar Pitowsky: Quantum logic and the "conspiracy" of nature. Iyyun, The Jerusalem Philosophical Quarterly 36 (3/4), 204–234 (1987) (in Hebrew)

Pitowsky, Itamar: Quantum Probability – Quantum Logic. Lecture Notes in Physics 321, Springer-Verlag (1989)

Itamar Pitowsky: From George Boole to John Bell: The origin of Bell's inequality. In: Kafatos, M. (ed.) Bell's Theorem, Quantum Theory and the Conceptions of the Universe, pp. 37–49, Kluwer, Dordrecht (1989)

Itamar Pitowsky: The physical Church thesis and physical computational complexity. lyyun, The Jerusalem Philosophical Quarterly 39, 81–99 (1990)

Pitowsky, Itamar: Bohm's quantum potentials and quantum gravity. Foundations of Physics 21(3), 343–352 (1991)

Itamar Pitowsky: Correlation polytopes their geometry and complexity. Mathematical Programming, 50, 395–414 (1991)

Pitowsky, Itamar: Why does physics need mathematics? A comment. In: Edna Ullmann-Margalit (ed.) The Scientific Enterprise, pp. 163–167, Kluwer Academic Publishers (1992)

Clifton, Robert, Pagonis, Constantine, Pitowsky, Itamar: Relativity, quantum mechanics and EPR. PSA: Proceedings of the Biennial Meeting of the Philosophy of Science Association, pp. 114–128 (1992)

Pitowsky, Itamar: George Boole's 'conditions of possible experience' and the quantum puzzle. British Journal for the Philosophy of Science 45(1), 95–125 (1994)

Pitowsky, Itamar: Laplace's demon consults an oracle: The computational complexity of prediction. Studies in History and Philosophy of Modern Physics 27(2), 161–180 (1996)

Pitowsky, Itamar, Shoresh, Noam: Locality, factorizability, and the Maxwell-Boltzmann distribution. Foundations of Physics 26(9), 1231–1242 (1996)

Itamar Pitowsky: Infinite and finite Gleason's theorems and the logic of indeterminacy. Journal of Mathematical Physics 39, 218–228 (1998)

Pitowsky, Itamar: Local fluctuations and local observers in equilibrium statistical mechanics. Studies in History and Philosophy of Modern Physics 32(4), 595–607 (2001)

Pitowsky, Itamar, Svozil, Karl: Optimal tests of quantum nonlocality. Physical Review A 64, 014102 (2001)

Pitowsky, Itamar: Range theorems for quantum probability and entanglement. arXiv:quant-ph/0112068 (2001)

Ben-Menahem, Yemima, Pitowsky, Itamar (2001). Introduction. Studies in History and Philosophy of Modern Physics 32(4), 503–510 (2001)

Pitowsky, Itamar: Quantum speed-up of computations. Philosophy of Science 69 (S3), S168–S177 (2002)

Pitowsky, Itamar: Generalizations of Kochen and Specker's theorem and the effectiveness of Gleason's theorem. Studies in History and Philosophy of Modern Physics 35(2), 177–194 (2003)

Pitowsky, Itamar: Betting on the outcomes of measurements: A Bayesian theory of quantum probability. Studies in History and Philosophy of Modern Physics 34 (3), 395–414 (2003)

Hemmo, Meir, Pitowsky, Itamar: Probability and nonlocality in many minds interpretations of quantum mechanics. British Journal for the Philosophy of Science 54(2), 225–243 (2003)

Shagrir, Oron, Pitowsky, Itamar: Physical hypercomputation and the church–turing thesis. Minds and Machines 13(1), 87–101 (2003)

Pitowsky, Itamar: Macroscopic objects in quantum mechanics: A combinatorial approach. Physical Review A 70, 022103 (2004)

Pitowsky, Itamar: Random witnesses and the classical character of macroscopic objects. philsci-archive.pitt.edu/id/eprint/1703 (2004)

Hrushovski, Ehud, Pitowsky, Itamar: Generalizations of Kochen and Specker's theorem and the effectiveness of Gleason's theorem. Studies in History and Philosophy of Modern Physics 35(2), 177–194 (2004)

Pitowsky, Itamar: On the definition of equilibrium. Studies in History and Philosophy of Modern Physics 37(3), 431–438 (2006)

Dolev, Shahar, Pitowsky, Itamar, Tamir Boaz: Grover's quantum search algorithm and Diophantine approximation. Physical Review A 73, 022308 (2006)

Pitowsky, Itamar: Quantum mechanics as a theory of probability. In: Demopoulos, William, Pitowsky, Itamar (eds.) Physical Theory and Its Interpretation: Essays in Honor of Jeffrey Bub. Western Ontario series in philosophy of science (72), pp. 213–240, Springer (2007)

Pitowsky, Itamar: From logic to physics: How the meaning of computation changed over time. In: Cooper, B.S., Löwe, B., Sorbi, A. (eds.) *Computation and Logic in the Real World,* pp. 621–631, Springer, Berlin Heidelberg (2007)

Pitowsky, Itamar: On Kuhn's The Structure of Scientific Revolutions. Iyyun: The Jerusalem Philosophical Quarterly 56, 119 (2007)

Hemmo, Meir, Pitowsky, Itamar: Quantum probability and many worlds. Studies in History and Philosophy of Modern Physics 38(2), 333–350 (2007)

Pitowsky, Itamar: Geometry of quantum correlations. Physical Review A 77, 062109 (2008)

Pitowsky, Itamar: New Bell inequalities for the singlet state: Going beyond the Grothendieck bound. Journal of Mathematical Physics 49, 012101 (2008)

Bub, Jeffrey, Pitowsky, Itamar (2010). Two dogmas about quantum mechanics. In: Saunders, Simon, Barrett, Jonathan, Kent, Adrian, Wallace, David (eds.) Many Worlds? Everett, Quantum Theory, & Reality, pp. 433–459, Oxford University Press (2010).

Pitowsky, Itamar: Typicality and the role of the Lebesgue measure in statistical mechanics. In: Ben-Menahem, Yemima, Hemmo Meir (eds.) Probability in Physics, pp. 41–58, Springer-Verlag Berlin Heidelberg (2012)

Contents

Contributors

Samson Abramsky Department of Computer Science, University of Oxford, Oxford, UK

Valia Allori Department of Philosophy, Northern Illinois University, Dekalb, IL, USA

Guido Bacciagaluppi Descartes Centre for the History and Philosophy of the Sciences and the Humanities, Utrecht University, Utrecht, Netherlands

Veronika Baumann Vienna Center for Quantum Science and Technology (VCQ), Faculty of Physics, University of Vienna, Vienna, Austria
Faculty of Informatics, Università della Svizzera italiana, Lugano, Switzerland

Yemima Ben-Menahem Department of Philosophy, The Hebrew University of Jerusalem, Jerusalem, Israel

Joseph Berkovitz Institute for the History and Philosophy of Science and Technology, University of Toronto, Victoria College, Toronto, ON, Canada

Harvey R. Brown Faculty of Philosophy, University of Oxford, Oxford, UK

Časlav Brukner Vienna Center for Quantum Science and Technology (VCQ), Faculty of Physics, University of Vienna, Vienna, Austria
Institute of Quantum Optics and Quantum Information (IQOQI), Austrian Academy of Sciences, Vienna, Austria

Jeffrey Bub Department of Philosophy, Institute for Physical Science and Technology, Joint Center for Quantum Information and Computer Science, University of Maryland, College Park, MD, USA

B. Jack Copeland Department of Philosophy, University of Canterbury, Christchurch, New Zealand

Mauro Dorato Department of Philosophy, University of Rome Tre, Rome, Italy

John Earman Department of History and Philosophy of Science, University of Pittsburgh, Pittsburgh, PA, USA

Laura Felline Department of Philosophy, University of Roma Tre, Rome, Italy

Amit Hagar Department of History and Philosophy of Science and Medicine, Indiana University, Bloomington, IN, USA

Richard Healey Philosophy Department, University of Arizona, Tucson, AZ, USA

Meir Hemmo Department of Philosophy, University of Haifa, Haifa, Israel

Gábor Hofer-Szabó Research Center for the Humanities, Budapest, Hungary

Ehud Hrushovski Mathematical Institute, Oxford, UK

Gil Kalai Einstein Institute of Mathematics, The Hebrew University of Jerusalem, Jerusalem, Israel
Efi Arazy School of Computer Science, Interdisciplinary Center Herzliya, Herzliya, Israel

Avi Levy Department of Philosophy, University of Haifa, Haifa, Israel

Wayne C. Myrvold Department of Philosophy, The University of Western Ontario, London, ON, Canada

Gal Ben Porath Department of History and Philosophy of Science, University of Pittsburgh, Pittsburgh, PA, USA

Michael Redhead Centre for Philosophy of Natural and Social Science, London School of Economics and Political Science, London, UK

Laura Ruetsche Department of Philosophy, University of Michigan, Ann Arbor, MI, USA

Simon Saunders Faculty of Philosophy, University of Oxford, Oxford, UK

Oron Shagrir Department of Philosophy, The Hebrew University of Jerusalem, Jerusalem, Israel

Orly Shenker Edelstein Center for History and Philosophy of Science, The Hebrew University of Jerusalem, Jerusalem, Israel

Andrew W. Simmons Department of Physics, Imperial College London, London, UK

Karl Svozil Institute for Theoretical Physics, Vienna University of Technology, Vienna, Austria

Jos Uffink Philosophy Department and Program in History of Science, Technology and Medicine, University of Minnesota, Minneapolis, MN, USA

Lev Vaidman School of Physics and Astronomy, Tel-Aviv University, Tel Aviv, Israel

Alexander Wilce Department of Mathematical Sciences, Susquehanna University, Selinsgrove, PA, USA

Chapter 1
Classical Logic, Classical Probability, and Quantum Mechanics

Samson Abramsky

Abstract We give an overview and conceptual discussion of some of our results on contextuality and non-locality. We focus in particular on connections with the work of Itamar Pitowsky on correlation polytopes, Bell inequalities, and Boole's "conditions of possible experience".

Keywords Logic · Probability · Correlation polytopes · Bell inequalities · Contextual fraction · Bell's theorem

1.1 Introduction

One of Itamar Pitowsky's most celebrated contributions to the foundations of quantum mechanics was his work on *correlation polytopes* Pitowsky (1989, 1991, 1994), as a general perspective on Bell inequalities. He related these to Boole's "conditions of possible experience" Boole (1862), and emphasized the links between correlation polytopes and classical logic.

My own work on the sheaf-theoretic approach to non-locality and contextuality with Adam Brandenburger Abramsky and Brandenburger (2011), on logical Bell inequalities with Lucien Hardy Abramsky and Hardy (2012), on robust constraint satisfaction with Georg Gottlob and Phokion Kolaitis Abramsky et al. (2013), and on the contextual fraction with Rui Soares Barbosa and Shane Mansfield Abramsky et al. (2017), is very much in a kindred spirit with Pitowsky's pioneering contributions. I will survey some of this work, making a number of comparisons and contrasts with Pitowsky's ideas.

S. Abramsky (✉)
Department of Computer Science, University of Oxford, Oxford, UK
e-mail: samson.abramsky@cs.ox.ac.uk

M. Hemmo, O. Shenker (eds.), *Quantum, Probability, Logic*,
Jerusalem Studies in Philosophy and History of Science,
https://doi.org/10.1007/978-3-030-34316-3_1

1.2 Boole's "Conditions of Possible Experience"

We quote Pitowsky's pellucid summary (Pitowsky 1994, p. 100):

> Boole's problem is simple: we are given rational numbers which indicate the relative frequencies of certain events. If no logical relations obtain among the events, then the only constraints imposed on these numbers are that they each be non-negative and less than one. If however, the events are logically interconnected, there are further equalities or inequalities that obtain among the numbers. The problem thus is to determine the numerical relations among frequencies, in terms of equalities and inequalities, which are induced by a set of logical relations among the events. The equalities and inequalities are called "conditions of possible experience".

More formally, we are given some basic events $E_1, \ldots, E_n$, and some boolean functions $\varphi_1, \ldots, \varphi_m$ of these events. Such a function φ can be described by a propositional formula in the variables $E_1, \ldots, E_n$.

Suppose further that we are given probabilities $p(E_i)$, $p(\varphi_j)$ of these events.

> Question: What numerical relationships between the probabilities can we infer from the logical relationships between the events?

Pitowksy's approach is to define a *correlation polytope* $c(\vec{E}, \vec{\varphi})$ induced by the given events, and to characterize the "conditions of possible experience" as the facet-defining inequalities for this polytope. He emphasizes the computational difficulty of obtaining these inequalities, and gives no direct characterization other than a brute-force computational approach. Indeed, one of his main results is the NP-completeness of an associated problem, of determining membership of the polytope. We shall return to these ideas later.

The point we wish to make now is that it is possible to give a very direct answer to Boole's question, which explicitly relates logical consistency conditions to arithmetical relationships. This insight is the key observation in Abramsky and Hardy (2012).

With notation as above, suppose that the formulas φ_j are not simultaneously satisfiable. This means that $\bigwedge_{i=1}^{m-1} \varphi_i \rightarrow \neg\varphi_m$, or by contraposition and De Morgan's law:

$$\varphi_m \rightarrow \bigvee_{i=1}^{m-1} \neg\varphi_i.$$

Passing to probabilities, we infer:

$$p(\varphi_m) \leq p(\bigvee_{i=1}^{m-1} \neg\varphi_i) \leq \sum_{i=1}^{m-1} p(\neg\varphi_i) = \sum_{i=1}^{m-1}(1-p(\varphi_i)) = (m-1)-\sum_{i=1}^{m-1} p(\varphi_i).$$

The first inequality is the monotonicity of probability, i.e. $E \subseteq E'$ implies $p(E) \leq p(E')$, while the second is the subadditivity of probability measures.[1] Collecting terms, we obtain:

$$\sum_{i=1}^{m} p(\varphi_i) \leq m - 1. \tag{1.1}$$

Thus a consistency condition on the formulas φ_i implies a non-trivial arithmetical inequality on the probabilities $p(\varphi_i)$.

1.2.1 An Example: The Bell Table

We can directly apply the above inequality to deduce a version of Bell's theorem Bell (1964). Consider the following table.

A B	(0, 0)	(1, 0)	(0, 1)	(1, 1)
a b	1/2	0	0	1/2
a b'	3/8	1/8	1/8	3/8
a' b	3/8	1/8	1/8	3/8
a' b'	1/8	3/8	3/8	1/8

Here we have two agents, Alice and Bob. Alice can choose from the measurement settings a or a', and Bob can choose from b or b'. These choices correspond to the rows of the table. The columns correspond to the joint outcomes for a given choice of settings by Alice and Bob, the two possible outcomes for each individual measurement being represented by 0 and 1. The numbers along each row specify a probability distribution on these joint outcomes. Thus for example, the entry in row 2, column 2 of the table says that when Alice chooses setting a and Bob chooses setting b', then with probability $1/8$, Alice obtains a value of 1, and Bob obtains a value of 0.

A standard version of Bell's theorem uses the probability table given above. This table can be realized in quantum mechanics, e.g. by a Bell state

$$\frac{|00\rangle + |11\rangle}{\sqrt{2}},$$

subjected to spin measurements in the XY-plane of the Bloch sphere, at a relative angle of $\pi/3$. See the supplemental material in Abramsky et al. (2017) for details.

[1] Also known as Boole's inequality (Weisstein 2019).

1.2.1.1 Logical Analysis of the Bell Table

We now pick out a subset of the elements of each row of the table, as indicated in the following table.

A	B	(0,0)	(1,0)	(0,1)	(1,1)
a	b	**1/2**	0	0	**1/2**
a	b'	**3/8**	1/8	1/8	**3/8**
a'	b	**3/8**	1/8	1/8	**3/8**
a'	b'	1/8	**3/8**	**3/8**	1/8

We can think of basic events E_a, $E_{a'}$, E_b, $E_{b'}$, where e.g. E_a is the event that the quantity measured by a has the value 0. Note that, by our assumption that each measurement has two possible outcomes,[2] $\neg E_a$ is the event that a has the value 1. Writing simply a for E_a etc., the highlighted positions in the table are represented by the following propositions:

$$\begin{aligned}
\varphi_1 &= \quad a \wedge b \quad \vee \neg a \wedge \neg b = a \leftrightarrow b \\
\varphi_2 &= \quad a \wedge b' \quad \vee \neg a \wedge \neg b' = a \leftrightarrow b' \\
\varphi_3 &= \quad a' \wedge b \quad \vee \neg a' \wedge \neg b = a' \leftrightarrow b \\
\varphi_4 &= \neg a' \wedge b' \vee \; a' \wedge \neg b' = a' \oplus b'.
\end{aligned}$$

The first three propositions pick out the correlated outcomes for the variables they refer to; the fourth the anticorrelated outcomes. These propositions are easily seen to be contradictory. Indeed, starting with φ_4, we can replace a' with b using φ_3, b with a using φ_1, and a with b' using φ_2, to obtain $b' \oplus b'$, which is obviously unsatisfiable.

We see from the table that $p(\varphi_1) = 1$, $p(\varphi_i) = 6/8$ for $i = 2, 3, 4$. Hence the violation of the Bell inequality (1.1) is $1/4$.

We may note that the logical pattern shown by this jointly contradictory family of propositions underlies the familiar CHSH correlation function.

1.2.2 The General Form

We can generalize the inequality (1.1). Given a family of propositions $\Phi = \{\varphi_i\}$, we say it is *K-consistent* if the size of the largest consistent subfamily of Φ is K.[3] If a family $\{\varphi_i\}_{i=1}^m$ is not simultaneously satisfiable, then it must be K-consistent for some $K < m$.

---

[2] And, implicitly, the assumption that each quantity has a definite value, whether we measure it or not.

[3] The size of a family $\{\varphi_j\}_{j \in J}$ is the cardinality of J. Note that repetitions are allowed in the family.

Theorem 1 *Suppose that we have a K-consistent family $\{\varphi_i\}_{i=1}^m$ over the basic events $E_1, \ldots, E_n$. For any probability distribution on the set $\mathbf{2}^{\vec{E}}$ of truth-value assignments to the E_i, with induced probabilities $p(\varphi_i)$ for the events φ_i, we have:*

$$\sum_{i=1}^{m} p(\varphi_i) \leq K. \tag{1.2}$$

See Abramsky and Hardy (2012) for the (straightforward) proof. Note that the basic inequality (1.1) is a simple consequence of this result.

We thus have a large class of inequalities arising directly from logical consistency conditions. As shown in Abramsky and Hardy (2012), even the basic form (1.1) is sufficient to derive the main no-go results in the literature, including the Hardy, GHZ and Kochen-Specker "paradoxes".

More remarkably, as we shall now go on to see, this set of inequalities is *complete*. That is, *every* facet-defining inequality for the non-local, or more generally non-contextual polytopes, is equivalent to one of the form (1.2). In this sense, we have given a complete answer to Boole's question.

The following quotation from Pitowsky suggests that he may have envisaged the possibility of such a result (Pitowsky 1991, p. 413):

> In fact, all facet inequalities for $c(n)$ should follow from "Venn diagrams", that is, the possible relations among n events in a probability space.

1.3 From Correlation Polytopes to Non-contextual Polytopes

We continue with the notation $E_1, \ldots, E_n$ for basic events, and $\varphi_1, \ldots, \varphi_m$ for boolean combinations of these events. We write $\mathbf{2} := \{0, 1\}$.

Pitowsky defines the correlation polytope $c(\vec{E}, \vec{\varphi}) \subseteq \mathbb{R}^{n+m}$ as follows. Each assignment $s \in \mathbf{2}^{\vec{E}}$ of 0 or 1 to the basic events E_i extends to a truth assignment for the formulae φ_j by the usual truth-table method, and hence determines a 0/1-vector v_s in $\mathbb{R}^{n+m}$. The polytope $c(\vec{E}, \vec{\varphi})$ is defined to be the convex hull of the set of vectors v_s, for $s \in \mathbf{2}^{\vec{E}}$.

In particular, Pitowsky focusses almost exclusively on the case where the formulae are pairwise conjunctions $\varphi_{ij} := E_i \wedge E_j$. One effect of restricting to this case is that m is bounded quadratically by n. We shall return to this point when we discuss complexity issues.

The main computational problem that Pitowsky focusses on is the following:

Instance: a rational vector $v \in \mathbf{Q}^{n+m}$.
Question: is v in $c(\vec{E}, \vec{\varphi})$?

For the case where $\vec{\varphi}$ is the set of all pairwise conjunctions φ_{ij} of basic events, Pitowsky shows that this problem is NP-complete.

We can see that the (quite standard) version of Bell's theorem described in Sect. 1.2.1 amounts to the statement that the vector

$$v = [p_a, p_{a'}, p_b, p_{b'}, 1, 6/8, 6/8, 6/8]^T$$

is not in the correlation polytope $c(E_a, E_{a'}, E_b, E_{b'}, \varphi_1, \varphi_2, \varphi_3, \varphi_4)$. Here p_a is the probability of the event E_a. This can be calculated as a marginal from the first or second rows of the Bell table, as $p_a = 1/2$. The fact that we get the same answer, whether we use the first or second row, is guaranteed by the *no-signalling condition* (Jordan 1983; Ghirardi et al. 1980), which will hold for all such tables which can be generated using quantum states and observables (Abramsky and Brandenburger 2011). Similarly for the other basic events.

We make a number of comments:

- Firstly, notice that the Bell table as given in Sect. 1.2.1 corresponds to the data arising directly from a Bell experiment. This and similar tables should be seen as setting the standard for what we are trying to describe.
- From this point of view, we notice firstly that to get a natural correspondence between these tables and correlation polytopes, we need to consider propositions beyond pairwise conjunctions. This makes a huge difference as regards the size of problem instances. While there are only $O(n^2)$ pairwise conjunctions on n variables, there are 2^{2^n} distinct boolean functions.
- At the same time, there is redundancy in the correlation polytope representation, since thanks to the no-signalling condition, the probabilities of the basic events can be calculated as marginals from the probabilities of the joint outcomes.
- The most important point is that there is a *structure* inherent in the Bell table and its generalizations, which is "flattened out" by the correlation polytope representation. Capturing this structure explicitly can lead to deeper insights into the non-classical phenomena arising in quantum mechanics.

Taking up the last point, we quote again from Pitowsky's admirably clear exposition (Pitowsky 1994, p. 111–112):

> One of the major purposes of quantum mechanics is to organize and predict the relative frequencies of events observable in various experiments — in particular the cases where Boole's conditions are violated. For that purpose a mathematical formalism has been invented which is essentially a new kind of probability theory. It uses no concept of 'population' but rather a primitive concept of 'event' or more generally 'observable' (which is the equivalent of the classical 'random variable'). In addition, to every particular physical system (which can be one 'thing' — an electron, for example — or consists of a few 'things') the theory assigns a state. The state determines the probabilities for the events or, more generally, the expectations of the observables. What this means operationally is that if we have a source of physical systems, all in the same state, then the relative frequency of a given event will approach the value of the probability, which is theoretically determined by the state.

> For certain families of events the theory stipulates that they are commeasurable. This means that, in every state, the relative frequencies of all these events can be measured on one single sample. For such families of events, the rules of classical probability — Boole's conditions in particular — are valid. Other families of events are not commeasurable, so their frequencies must be measured in more than one sample. The events in such families nevertheless exhibit logical relations (given, usually, in terms of algebraic relations among observables). But for some states, the probabilities assigned to the events violate one or more of Boole's conditions associated with those logical relations.

The point we would like to emphasize is that tables such as the Bell table in Sect. 1.2.1 can—and do—arise from experimental data, without presupposing any particular physical theory. This data *does* clearly involve concepts from classical probability. In particular, for each set C of "commeasurable events", there is a sample space, namely the set of possible joint outcomes of measuring all the observables in C. Moreover, there is a well-defined probability distribution on this sample space.

Taking the Bell table for illustration, the "commeasurable sets", or *contexts*, are the sets of measurements labelling the rows of the table. The sample space for a given row, with measurement α by Alice and β by Bob, is the set of joint outcomes $(\alpha = x, \beta = y)$ with $x, y \in \{0, 1\}$. The probabilities assigned to these joint outcomes in the table form a (perfectly classical) probability distribution on this sample space.

Thus the structure of the table as a whole is a family of probability distributions, each defined on a different sample space. However, these sample spaces are not completely unrelated. They overlap in sharing common observables, and they "agree on overlaps" in the sense that they have consistent marginals. This is exactly the force of the no-signalling condition.

The nature of the "non-classicality" of the data tables generated by quantum mechanics (and also elsewhere, see e.g. Abramsky 2013), is that there is no distribution over the global sample space of outcomes for *all* observables, which recovers the empirically accessible data in the table by marginalization. This is completely equivalent to a statement of the more traditional form: "there is no local or non-contextual hidden variable model" (or, in currently fashionable terminology: there is no "ontological model" of a certain form).

Our preferred slogan for this state of affairs is:

> Contextuality (with non-locality as a special case) arises when we have a family of data which is locally consistent, but globally inconsistent.

1.3.1 Formalization

We briefly summarise the framework introduced in Abramsky and Brandenburger (2011), and extensively developed subsequently. The main objects of study are *empirical models*: tables of data, specifying probability distributions over the joint outcomes of specified sets of compatible measurements. These can be thought of as statistical data obtained from some experiment or as the observations predicted by some theory.

A *measurement scenario* is an abstract description of a particular experimental setup. It consists of a triple $\langle X, \mathcal{M}, O\rangle$ where: X is a finite set of measurements; O is a finite set of outcome values for each measurement; and $\mathcal{M}$ is a set of subsets of X. Each $C \in \mathcal{M}$ is called a *measurement context*, and represents a set of measurements that can be performed together.

Examples of measurement scenarios include multipartite Bell-type scenarios familiar from discussions of nonlocality, Kochen–Specker configurations, measurement scenarios associated with qudit stabiliser quantum mechanics, and more. For example, the Bell scenario from Sect. 1.2.1, where two experimenters, Alice and Bob, can each choose between performing one of two different measurements, a or a' for Alice and b or b' for Bob, obtaining one of two possible outcomes, is represented as follows:

$$X = \{a, a', b, b\} \qquad\qquad O = \{0, 1\}$$
$$\mathcal{M} = \{\{a, b\}, \{a, b'\}, \{a', b\}, \{a', b'\}\}.$$

Given this description of the experimental setup, then either performing repeated runs of such experiments with varying choices of measurement context and recording the frequencies of the various outcome events, or calculating theoretical predictions for the probabilities of these outcomes, results in a probability table like that in Sect. 1.2.1.

Such data is formalised as an *empirical model* for the given measurement scenario $\langle X, \mathcal{M}, O\rangle$. For each valid choice of measurement context, it specifies the probabilities of obtaining the corresponding joint outcomes. That is, it is a family $\{e_C\}_{C\in\mathcal{M}}$ where each e_C is a probability distribution on the set O^C of functions assigning an outcome in O to each measurement in C (the rows of the probability table).

We require that the marginals of these distributions agree whenever contexts overlap, i.e.

$$\forall C, C' \in \mathcal{M}.\ e_C|_{C\cap C'} = e_{C'}|_{C\cap C'},$$

where the notation $e_C|_U$ with $U \subseteq C$ stands for marginalisation of probability distributions (to 'forget' the outcomes of some measurements): for $t \in O^U$, $e_C|_U(t) := \sum_{s\in O^C, s|_U=t} e_C(t)$. The requirement of *compatibility of marginals* is a generalisation of the usual *no-signalling* condition, and is satisfied in particular by all empirical models arising from quantum predictions (Abramsky and Brandenburger 2011).

An empirical model is said to be *non-contextual* if this family of distributions can be obtained as the marginals of a single probability distribution on global assignments of outcomes to all measurements, i.e. a distribution d on O^X

(where O^X acts as a canonical set of deterministic hidden variables) such that $\forall C \in \mathcal{M}.\ d|_C = e_C$. Otherwise, it is said to be *contextual*. Equivalently Abramsky and Brandenburger (2011), contextual empirical models are those which have no realisation by factorisable hidden variable models; thus for Bell-type measurement scenarios contextuality specialises to the usual notion of *nonlocality*.

Noncontextuality characterizes classical behaviours. One way to understand this is that it reflects a situation in which the physical system being measured exists at all times in a definite state assigning outcome values to all properties that can be measured. Probabilistic behaviour may still arise, but only via stochastic mixtures or distributions on these global assignments. This may reflect an averaged or aggregate behaviour, or an epistemic limitation on our knowledge of the underlying global assignment.

1.3.2 The Non-contextual Polytope

Suppose we are given a measurement scenario $\langle X, \mathcal{M}, O \rangle$. Each global assignment $t \in O^X$ induces a deterministic empirical model δ^t:

$$\delta^t_C(s) = \begin{cases} 1, & t|_C = s \\ 0 & \text{otherwise.} \end{cases}$$

Note that $t|_C$ is the function t restricted to C, which is a subset of its domain X.

We have the following result from (Abramsky and Brandenburger 2011, Theorem 8.1):

Theorem 2 *An empirical model $\{e_C\}$ is non-contextual if and only if it can be written as a convex combination $\sum_{j \in J} \mu_j \delta^{t_j}$ where $t_j \in O^X$ for each $j \in J$. This means that for each $C \in \mathcal{M}$,*

$$e_C = \sum_j \mu_j \delta^{t_j}_C.$$

Given a scenario $\langle X, \mathcal{M}, O \rangle$, define $m := \sum_{C \in \mathcal{M}} |O^C| = |\{\langle C, s \rangle \mid C \in \mathcal{M}, s \in O^C\}|$ to be the number of joint outcomes as we range over contexts. For example, in the case of the Bell table there are four contexts, each with four possible outcomes, so $m = 16$. We can regard an empirical model $\{e_C\}$ over $\langle X, \mathcal{M}, O \rangle$ as a real vector $\mathbf{v}^e \in \mathbb{R}^m$, with $\mathbf{v}^e[\langle C, s \rangle] = e_C(s)$.

We define the *non-contextual polytope* for the scenario $\langle X, \mathcal{M}, O \rangle$ to be the convex hull of the set of deterministic models δ^t, where t ranges over O^X. By the preceding theorem, this is exactly the set of non-contextual models.

Thus we have captured the question as to whether an empirical model is contextual in terms of membership of a polytope. The facet inequalities for this family of polytopes, as we range over measurement scenarios, give a general notion of Bell inequalities.

As explained in Abramsky and Hardy (2012), a complete finite set of rational facet inequalities for the contextual polytope over a measurement scenario can be computed using Fourier-Motzkin elimination. This procedure is doubly-exponential in the worst case, but standard optimizations reduce this to a single exponential. Despite this high complexity, Fourier-Motzkin elimination is widely used in computer-assisted verification and polyhedral computation (Strichman 2002; Christof et al. 1997).

1.3.3 Completeness of Logical Bell Inequalities

Suppose we are given a scenario $\langle X, \mathcal{M}, O \rangle$. A rational inequality is given by a rational vector $\mathbf{r}$ and a rational number r. An empirical model $\mathbf{v}$ satisfies this inequality if $\mathbf{r} \cdot \mathbf{v} \leq r$. Two inequalities are equivalent if they are satisfied by the same empirical models.

Theorem 3 *A rational inequality is satisfied by all non-contextual empirical models over* $\langle X, \mathcal{M}, O \rangle$ *if and only if it is equivalent to a logical Bell inequality of the form* (1.2).

For the proof, see Abramsky and Hardy (2012). Combining this result with the previous observations, we obtain:

Theorem 4 *The polytope of non-contextual empirical models over any scenario* $\langle X, \mathcal{M}, O \rangle$ *is determined by a finite set of logical Bell inequalities. Moreover, these inequalities can be obtained effectively from the scenario. Thus an empirical model over any scenario is contextual if and only if it violates one of finitely many logical Bell inequalities.*

It is worth reflecting on the conceptual import of these results. They are saying that the scope for non-classical correlations available to quantum mechanics, or any other physical theory, arises entirely from families of events which are locally consistent, but globally inconsistent. Thus non-classical probabilistic behaviour rests on this kind of logical structure.

1.4 The Contextual Fraction

We now turn to computational aspects. Fix a measurement scenario $\langle X, \mathcal{M}, O \rangle$. Let $n := |O^X|$ be the number of global assignments g, and $m := |\{\langle C, s \rangle \mid C \in \mathcal{M}, s \in O^C\}|$ be the number of local assignments ranging over contexts. The *incidence matrix* Abramsky and Brandenburger (2011) $\mathbf{M}$ is an $m \times n$ (0, 1)-matrix that records the restriction relation between global and local assignments:

$$\mathbf{M}[\langle C, s\rangle, g] := \begin{cases} 1 \text{ if } g|_C = s; \\ 0 \text{ otherwise.} \end{cases}$$

As already explained, an empirical model e can be represented as a vector $\mathbf{v}^e \in \mathbb{R}^m$, with the component $\mathbf{v}^e[\langle C, s\rangle]$ recording the probability given by the model to the assignment s at the measurement context C, $e_C(s)$. This vector is a flattened version of the table used to represent the empirical model. The columns of the incidence matrix, $\mathbf{M}[-, g]$, are the vectors corresponding to the (non-contextual) deterministic models δ^g obtained from global assignments $g \in O^X$. Recall that every non-contextual model can be written as a convex combination of these. A probability distribution on global assignments can be represented as a vector $\mathbf{d} \in \mathbb{R}^n$ with non-negative components, and then the corresponding non-contextual model is represented by the vector $\mathbf{M}\,\mathbf{d}$. So a model e is non-contextual if and only if there exists $\mathbf{d} \in \mathbb{R}^n$ such that:

$$\mathbf{M}\,\mathbf{d} = \mathbf{v}^e \qquad \text{and} \qquad \mathbf{d} \geq \mathbf{0}.$$

It is also natural to consider a relaxed version of this question, which leads us to the contextual fraction.

Given two empirical models e and e' on the same measurement scenario and $\lambda \in [0, 1]$, we define the empirical model $\lambda e + (1 - \lambda)e'$ by taking the convex sum of probability distributions at each context. Compatibility is preserved by this convex sum, hence it yields a well-defined empirical model.

A natural question to ask is: what fraction of a given empirical model e admits a non-contextual explanation? This approach enables a refinement of the binary notion of contextuality *vs* non-contextuality into a quantitative grading. Instead of asking for a probability distribution on global assignments that marginalises to the empirical distributions at each context, we ask only for a subprobability distribution[4] b on global assignments O^X that marginalises at each context to a subdistribution of the empirical data, thus explaining a fraction of the events, i.e. $\forall C \in \mathcal{M}.\ b|_C \leq e_C$. Equivalently, we ask for a convex decomposition

$$e = \lambda e^{NC} + (1 - \lambda)e' \tag{1.3}$$

where e^{NC} is a non-contextual model and e' is another (no-signalling) empirical model. The maximum weight of such a global subprobability distribution, or the maximum value of λ in such a decomposition,[5] is called the *non-contextual fraction* of e, by analogy with the *local fraction* previously introduced for models on Bell-

[4]A subprobability distribution on a set S is a map $b\colon S \longrightarrow \mathbb{R}_{\geq 0}$ with finite support and $w(b) \leq 1$, where $w(b) := \sum_{s \in S} b(s)$ is called its weight. The set of subprobability distributions on S is ordered pointwise: b' is a subdistribution of b (written $b' \leq b$) whenever $\forall s \in S.\ b'(s) \leq b(s)$.

[5]Note that such a maximum exists, i.e. that the supremum is attained. This follows from the Heine–Borel and extreme value theorems since the set of such λ is bounded and closed.

type scenarios Elitzur et al. (1992). We denote it by $\mathsf{NCF}(e)$, and the contextual fraction by $\mathsf{CF}(e) := 1 - \mathsf{NCF}(e)$.

The notion of contextual fraction in general scenarios was introduced in Abramsky and Brandenburger (2011), where it was proved that a model is non-contextual if and only if its non-contextual fraction is 1.

A global subprobability distribution is represented by a vector $\mathbf{b} \in \mathbb{R}^n$ with non-negative components, its weight being given by the dot product $\mathbf{1} \cdot \mathbf{b}$, where $\mathbf{1} \in \mathbb{R}^n$ is the vector whose n components are each 1. The following LP thus calculates the non-contextual fraction of an empirical model e:

$$\begin{array}{ll} \texttt{Find} & \mathbf{b} \in \mathbb{R}^n \\ \texttt{maximising} & \mathbf{1} \cdot \mathbf{b} \\ \texttt{subjectto} & \mathbf{M}\,\mathbf{b} \leq \mathbf{v}^e \\ \texttt{and} & \mathbf{b} \geq \mathbf{0} \quad . \end{array} \tag{1.4}$$

An *inequality* for a scenario $\langle X, \mathcal{M}, O\rangle$ is given by a vector $\mathbf{a} \in \mathbb{R}^m$ of real coefficients indexed by local assignments $\langle C, s\rangle$, and a bound R. For a model e, the inequality reads $\mathbf{a} \cdot \mathbf{v}^e \leq R$, where

$$\mathbf{a} \cdot \mathbf{v}^e = \sum_{C \in \mathcal{M}, s \in O^C} \mathbf{a}[\langle C, s\rangle]\, e_C(s).$$

Without loss of generality, we can take R to be non-negative (in fact, even $R = 0$) as any inequality is equivalent to one of this form. We call it a *Bell inequality* if it is satisfied by every non-contextual model. If, moreover, it is saturated by some non-contextual model, the Bell inequality is said to be *tight*. A Bell inequality establishes a bound for the value of $\mathbf{a} \cdot \mathbf{v}^e$ amongst non-contextual models e. For more general models, this quantity is limited only by the algebraic bound[6]

$$\|\mathbf{a}\| := \sum_{C \in \mathcal{M}} \max \left\{ \mathbf{a}[\langle C, s\rangle] \mid s \in O^C \right\}.$$

The *violation* of a Bell inequality $\langle \mathbf{a}, R\rangle$ by a model e is $\max\{0, \mathbf{a} \cdot \mathbf{v}^e - R\}$. However, it is useful to normalise this value by the maximum possible violation in order to give a better idea of the *extent* to which the model violates the inequality. The *normalised violation* of the Bell inequality by the model e is

$$\frac{\max\{0, \mathbf{a} \cdot \mathbf{v}^e - R\}}{\|\mathbf{a}\| - R}.$$

[6]We will consider only inequalities satisfying $R < \|\mathbf{a}\|$, which excludes inequalities trivially satisfied by all models, and avoids cluttering the presentation with special caveats about division by 0.

Theorem 5 *Let e be an empirical model.*

1. *The normalised violation by e of any Bell inequality is at most* $\mathsf{CF}(e)$*;*
2. *if* $\mathsf{CF}(e) > 0$*, this bound is attained, i.e. there exists a Bell inequality whose normalised violation by e is* $\mathsf{CF}(e)$*;*
3. *moreover, for any decomposition of the form* $e = \mathsf{NCF}(e)e^{NC} + \mathsf{CF}(e)e^{SC}$*, this Bell inequality is tight at the non-contextual model* e^{NC} *(provided* $\mathsf{NCF}(e) > 0$*) and maximally violated at the strongly contextual model* e^{SC}*.*

The proof of this result is based on the Strong Duality theorem of linear programming (Dantzig and Thapa 2003). This provides an LP method of calculating a witnessing Bell inequality for any empirical model e. For details, see the supplemental material in Abramsky et al. (2017).

1.5 Remarks on Complexity

We now return to the issue of the complexity of deciding whether a given empirical model is contextual.

It will be useful to consider the class of $(n, k, 2)$ Bell scenarios. In these scenarios there are n agents, each of whom has a choice of k measurement settings, and all measurements have 2 possible outcomes. This gives rise to a measurement scenario $\langle X, \mathcal{M}, O\rangle$ where $|X| = nk$. Each $C \in \mathcal{M}$ consists of n measurements, one chosen by each agent. Thus $|\mathcal{M}| = k^n$. For each context C, there are 2^n possible outcomes.

An empirical model for an $(n, k, 2)$ Bell scenario is thus given by a vector of $k^n 2^n$ probabilities. Thus the size of instances is exponential in the natural parameter n. This is the real obstacle to tractable computation as we increase the number of agents.

Given an empirical model $\mathbf{v}^e$ as an instance, we can use the linear program (1.4) given in the previous section to determine if it is contextual. The size of the linear program is determined by the incidence matrix, which has dimensions $p \times q$, where p is the dimension of $\mathbf{v}^e$, and $q = 2^{nk}$.

If we treat n as the complexity parameter, and keep k fixed, then $q = O(s^k)$, where s is the size of the instance. Thus the linear program has size polynomial in the size of the instance, and membership of the non-contextual polytope can be decided in time polynomial in the size of the instance.

There is an interesting contrast with Pitowsky's results on the NP-completeness of deciding membership in the correlation polytope of all binary conjunctions of basic events. In Pitowsky's case, the size of instances for these special forms of correlation polytope is polynomial in the natural parameter, which is the number of basic events. If we consider the correlation polytopes which would correspond directly to empirical models, the same argument as given above would apply: the instances would have exponential size in the natural parameter, while membership in the polytope could be decided by linear programming in time polynomial in the instance size.

We can also consider the situation where we fix n, and treat k as the complexity parameter. In this case, note that the size of instances are polynomial in k, while the size of the incidence matrix is exponential in k. Thus linear programming does not help. In fact, this seems to be the case that Pitowsky primarily had in mind. With $n = 2$, the restriction to binary conjunctions makes good sense, and all the examples he discusses are of this kind.

We also mention the results obtained in Abramsky and Hardy (2012); Abramsky et al. (2013); Abramsky (2017); Abramsky et al. (2017), which study the analogous problem with respect to *possibilistic contextuality*, that is, whether the supports of the probability distributions in the empirical model can be obtained from a set of global assignments. In that case, the complexity of deciding contextuality is shown to be NP-complete in the complexity parameter k in general; a precise delineation is given of the tractability boundary in terms of the values of the parameters. Moreover, as Rui Soares Barbosa has pointed out,[7] if we take n as the complexity parameter, there is a simple algorithm for detecting possibilistic contextuality which is polynomial in the size of the instance. Thus the complexity of detecting possibilistic contextuality runs completely in parallel with the probabilistic case.

1.6 The "Edge of Logical Contradiction" *vs.* the "Boundary of Paradox"

We give a final quotation from (Pitowsky 1994, p. 113):

> A violation of Boole's conditions of possible experience cannot be encountered when all the frequencies concerned have been measured on a single sample. Such a violation simply entails a logical contradiction; 'observing' it would be like 'observing' a round square. We expect Boole's conditions to hold even when the frequencies are measured on distinct large random samples. But they are systematically violated, and there is no easy way out (see below). We thus live 'on the edge of a logical contradiction'. An interpretation of quantum mechanics, an attempt to answer the WHY question, is thus an effort to save logic.

In my view, this states the extent of the challenge posed to classical logic by quantum mechanics too strongly. As we have discussed, the observational data predicted by quantum mechanics and confirmed by actual experiments consists of families of probability distributions, each defined on different sample spaces, corresponding to the different contexts. Since the contexts overlap, there are relationships between the sample spaces, which are reflected in coherent relationships between the distributions, in the form of consistent marginals. But there is no "global" distribution, defined on a sample space containing all the observable quantities, which accounts for all the empirically observable data. This *does* pose a challenge to the understanding of quantum mechanics as a physical theory, since it implies

[7] Personal communication.

that we cannot ascribe definite values to the physical quantities being measured, independent of whether or in what context they are measured. It does not, however, challenge classical logic and probability, which can be used to describe exactly this situation.

For this reason, I prefer to speak of contextuality as living "on the boundary of paradox" Abramsky (2017), as a signature of non-classicality, and one for which there is increasing evidence that it plays a fundamental rôle in quantum advantage in information-processing tasks (Anders and Browne 2009; Raussendorf 2013; Howard et al. 2014). So this boundary seems likely to prove a fruitful place to be. But we never actually cross the boundary, for exactly the reasons vividly expressed by Pitowsky in the opening two sentences of the above quotation.

There is much more to be said about the connections between logic and contextuality. In particular:

- There are notions of possibilistic and strong contextuality, and of All-versus-Nothing contextuality, which give a hierarchy of strengths of contextuality, and which can be described in purely logical terms, without reference to probabilities (Abramsky and Brandenburger 2011).
- These "possibilistic" forms of contextuality can be connected with logical paradoxes in the traditional sense. For example, the set of contradictory propositions used in Sect. 1.2.1 to derive Bell's theorem form a "Liar cycle" (Abramsky et al. 2015).
- There is also a topological perspective on these ideas. The four propositions from Sect. 1.2.1 form a discrete version of the Möbius strip. Sheaf cohomology can be used to detect contextuality (Abramsky et al. 2015).
- The logical perspective on contextuality leads to the recognition that the same structures arise in many non-quantum areas, including databases (Abramsky 2013), constraint satisfaction (Abramsky et al. 2013), and generic inference (Abramsky and Carù 2019).
- In Abramsky et al. (2017), an inequality of the form $p_F \geq \mathsf{NCF}(e)d$ is derived for several different settings involving information-processing tasks. Here p_F is the failure probability, $\mathsf{NCF}(e)$ is the non-contextual fraction of an empirical model viewed as a resource, and d is a parameter measuring the "difficulty" of the task. Thus the inequality shows the necessity of increasing the amount of contextuality in the resource in order to increase the success probability. In several cases, including certain games, communication complexity, and shallow circuits, the parameter d is $\frac{n-K}{n}$, where K is the K-consistency we encountered in formulating logical Bell inequalities (1.2).

We refer to the reader to the papers cited above for additional information on these topics.

1.7 Concluding Remarks

Itamar Pitowsky's work on quantum foundations combines lucid analysis, conceptual insights and mathematically sophisticated and elegant results. We have discussed some recent and ongoing work, and related it to his contributions, which remain a continuing source of inspiration.

Acknowledgements My thanks to Meir Hemmo and Orly Shenker for giving me the opportunity to contribute to this volume in honour of Itamar Pitowsky. I had the pleasure of meeting Itamar on several occasions when he visited Oxford.

I would also like to thank my collaborators in the work I have described in this paper: Adam Brandenburger, Rui Soares Barbosa, Shane Mansfield, Kohei Kishida, Ray Lal, Giovanni Carù, Lucien Hardy, Phokion Kolaitis and Georg Gottlob. My thanks also to Ehtibar Dzhafarov, whose Contextuality-by-Default theory (Dzhafarov et al. 2015) has much in common with my own approach, for our ongoing discussions.

References

Abramsky, S. (2013). Relational databases and Bell's theorem. In *In search of elegance in the theory and practice of computation* (pp. 13–35). Berlin: Springer.

Abramsky, S. (2017). Contextuality: At the borders of paradox. In E. Landry (Ed.), *Categories for the working philosopher*. Oxford: Oxford University Press.

Abramsky, S., & Brandenburger, A. (2011). The sheaf-theoretic structure of non-locality and contextuality. *New Journal of Physics, 13*(11), 113036.

Abramsky, S., & Carù, G. (2019). Non-locality, contextuality and valuation algebras: a general theory of disagreement. *Philosophical Transactions of the Royal Society A, 377*(2157), 20190036.

Abramsky, S., & Hardy, L. (2012). Logical Bell inequalities. *Physical Review A, 85*(6), 062114.

Abramsky, S., Gottlob, G., & Kolaitis, P. (2013). Robust constraint satisfaction and local hidden variables in quantum mechanics. In *Twenty-Third International Joint Conference on Artificial Intelligence*.

Abramsky, S., Barbosa, R. S., Kishida, K., Lal, R., & Mansfield, S. (2015). Contextuality, cohomology and paradox. In S. Kreutzer (Ed.), *24th EACSL Annual Conference on Computer Science Logic (CSL 2015), volume 41 of Leibniz International Proceedings in Informatics (LIPIcs)* (pp. 211–228). Dagstuhl, Schloss Dagstuhl–Leibniz-Zentrum für Informatik.

Abramsky, S., Barbosa, R. S., & Mansfield, S. (2017). Contextual fraction as a measure of contextuality. *Physical Review Letters, 119*(5), 050504.

Anders, J., & Browne, D. E. (2009). Computational power of correlations. *Physical Review Letters, 102*(5), 050502.

Bell, J. S. (1964). On the Einstein-Podolsky-Rosen paradox. *Physics, 1*(3), 195–200.

Boole, G. (1862). On the theory of probabilities. *Philosophical Transactions of the Royal Society of London, 152*, 225–252.

Christof, T., Löbel, A., & Stoer, M. (1997). PORTA-POlyhedron Representation Transformation Algorithm. Publicly available via ftp://ftp.zib.de/pub/Packages/mathprog/polyth/porta.

Dantzig, G. B., & Thapa, M. N. (2003). *Linear programming 2: Theory and extenstions* (Springer series in operations research and financial engineering). New York: Springer.

Dzhafarov, E. N., Kujala, J. V., & Cervantes, V. H. (2015). Contextuality-by-default: A brief overview of ideas, concepts, and terminology. In *International Symposium on Quantum Interaction* (pp. 12–23). Heidelberg: Springer.

Elitzur, A. C., Popescu, S., & Rohrlich, D. (1992). Quantum nonlocality for each pair in an ensemble. *Physics Letters A, 162*(1), 25–28.
Ghirardi, G. C., Rimini, A., & Weber, T. (1980). A general argument against superluminal transmission through the quantum mechanical measurement process. *Lettere Al Nuovo Cimento (1971–1985), 27*(10), 293–298.
Howard, M., Wallman, J., Veitch, V., & Emerson, J. (2014). Contextuality supplies the 'magic' for quantum computation. *Nature, 510*(7505), 351–355.
Jordan, T. F. (1983). Quantum correlations do not transmit signals. *Physics Letters A, 94*(6–7), 264.
Pitowsky, I. (1989). *Quantum probability, quantum logic* (Lecture notes in physics, Vol. 321). Berlin: Springer.
Pitowsky, I. (1991). Correlation polytopes: Their geometry and complexity. *Mathematical Programming, 50*(1–3), 395–414.
Pitowsky, I. (1994). George Boole's "Conditions of possible experience" and the quantum puzzle. *The British Journal for the Philosophy of Science, 45*(1), 95–125.
Raussendorf, R. (2013). Contextuality in measurement-based quantum computation. *Physical Review A, 88*(2), 022322.
Strichman, O. (2002). On solving Presburger and linear arithmetic with SAT. In *Formal methods in computer-aided design* (pp. 160–170). Heidelberg: Springer.
Weisstein, E. W. (2019). Bonferroni inequalities. http://mathworld.wolfram.com/BonferroniInequalities.html. MathWorld–A Wolfram Web Resource.

Chapter 2
Why Scientific Realists Should Reject the Second Dogma of Quantum Mechanics

Valia Allori

Abstract The information-theoretic approach to quantum mechanics, proposed by Bub and Pitowsky, is a realist approach to quantum theory which rejects the "two dogmas" of quantum mechanics: in this theory measurement results are not analysed in terms of something more fundamental, and the quantum state does not represent physical entities. Bub and Pitowsky's approach has been criticized because of their argument that kinematic explanations are more satisfactory than dynamical ones. Moreover, some have discussed the difficulties the information-theoretic interpretation faces in making sense of the quantum state as epistemic. The aim of this paper is twofold. First, I argue that a realist should reject the second dogma without relying on the alleged explanatory superiority of kinematical explanation over dynamical ones, thereby providing Bub and Pitowsky with a way to avoid the first set of objections to their view. Then I propose a functionalist account of the wavefunction as a non-material entity which does not fall prey of the objections to the epistemic account or the other non-material accounts such as the nomological view, and therefore I supply the proponents of the information-theoretic interpretation with a new tool to overcome the second set of criticisms.

Keywords Two dogmas · Quantum information · Wave-function ontology · Primitive ontology · Functionalism · Epistemic interpretation of the wave-function · Nomological interpretation of the wave-function

2.1 Introduction

Bub and Pitowsky (2010) have proposed the so-called information-theoretic (IT) approach to quantum mechanics. It is a realist approach to quantum theory which rejects what Bub and Pitowsky call the 'two dogmas' of quantum mechanics.

V. Allori (✉)
Department of Philosophy, Northern Illinois University, Dekalb, IL, USA
e-mail: vallori@niu.edu

M. Hemmo, O. Shenker (eds.), *Quantum, Probability, Logic*, Jerusalem Studies in Philosophy and History of Science,
https://doi.org/10.1007/978-3-030-34316-3_2

These dogmas are as follows: (D1) measurement results should not be taken as unanalysable primitives; and (D2) the wavefunction represents some physical reality. In contrast, in the IT approach measurements are not analysed in terms of something more fundamental, and the wavefunction does not represent physical entities. Bub and Pitowsky reject the first dogma arguing that their approach is purely kinematic, and that dynamical explanations (used to account for experimental outcomes) add no further insight. Their approach has been criticized by Brown and Timpson (2006) and by Timpson (2010), who argue that kinematic theories should not be a model for realist quantum theories. The second dogma is rejected as a consequence of the rejection of the first. Bub and Pitowsky assume that their rejection of the wavefunction as representing physical entities commits them to an epistemic view in which the wavefunction reflects is a reflection of our ignorance, and as such they have been criticized by some authors, such as Leifer (2014) and Gao (2017).

Little has been done to investigate whether it is possible to reject the second dogma without appealing to the alleged superiority of dynamical explanations, and whether one could think of the wavefunction as not representing matter without falling prey to the objections to the epistemic view (or to the so-called nomological view). The aim of this paper is to explore such issues. Therefore, first I argue that Bub and Pitowsky, as well as scientific realists in general, should reject the second dogma to start with, *independently of whether they decide to accept or deny the first dogma.* This argument supports the IT account to motivate their view, but it also stands alone as an argument to reject the wavefunction as a material entity in general. Then, in connection with this, I propose a functional account of the wavefunction which avoids the objections of the alternative views. This nicely fits in the IT interpretation by dissolving the objections against it which relies on the wavefunction being epistemic, but it is also helpful in answering general questions about the nature of the wavefunction.

Here's the scheme of the paper. In Sect. 2.2 I present the IT approach of Bub and Pitowsky. Then in Sect. 2.3 I discuss the distinction between dynamical and kinematical explanations, as well as the principle/constructive theory distinction, and how they were used by Bub and Pitwosky to motivate their view. Section 2.4 concludes this part of the paper by presenting the objections to the view connected with the explanatory superiority of kinematic theories. In Sect. 2.5 I move to the most prominent realist view which accepts dogma 2, namely wavefunction realism. Then in Sect. 2.6 I provide alternative reasons why a realist should reject the second dogma by introducing the primitive ontology approach. I connect the rejection of the first dogma to scientific realism in Sect. 2.7, showing how a realist may or may not consistently reject dogma 1. Then I move to the second set of objections to the IT account based on the epistemic wavefunction. First, in Sect. 2.8 I review the objections to the ontic accounts, while in Sect. 2.9 I discuss the ones to the epistemic account. Finally in Sect. 2.10 I propose an account of how the wavefunction based on functionalism, and I argue that it is better than the alternatives. In the last section I summarize my conclusions.

2.2 The IT Account, the Two Dogmas, and the Measurement Problem

The IT approach to quantum theory is a realist interpretation of the quantum formalism according to which quantum mechanics is fundamentally about measurement outcomes, whose statistics are recovered by the constraints that Hilbert space structure imposes on the quantum dynamics, and in which the wavefunction has the role of describing an agent's preferences (or expectations or degrees of belief) about measurement outcomes. The approach comes at the end of two lines of research first developed independently and then together.[1]

There are different components to the view, the first of which is given by the CBH theorem (Clifton et al. 2003), which offers an understanding of the quantum formalism in terms of simple physical principles,[2] which can be suitably translated into information-theoretic constraints on the probability of experimental outcomes. In light of this, Bub (2004, 2005) argues that quantum mechanics is about 'information,' taken as primitive. In this way, there is a sense in which the CBH constraints make quantum mechanics a kinematic theory (see Sect. 2.3): experimental outcomes are fundamental, and the theory poses constraints on what these outcomes are supposed to be. Bub also claims that further dynamical explanations are not necessary and therefore quantum mechanics does not need any additional fundamental ontology, even if is possible to provide one.[3]

The second ingredient of the IT account is given by Pitowsky's idea that quantum mechanics provides us with a new probability theory (Pitowsky 2007). Like classical probability theory consists of a space of possible events and a measure over it, here the space of possible events is the lattice of the closed subspaces of Hilbert space. Thus, the event – space defines the possible experiment outcomes (which are the fundamental elements of the theory) and the wavefunction encodes the experimenter's beliefs about them. According to Pitowsky, the advantage of this approach is that it can 'dissolve' the measurement problem. First he argues that there are two measurement problems:

- *The "big" measurement problem:* the problem of providing a dynamical explanation of why particular experiments have the outcomes they do; and

[1]See, respectively Clifton et al. (2003), Bub (2004, 2005, 2007), Pitowsky (2007), Bub and Pitowsky (2010).

[2]Here they are: the impossibility of superluminal information transfer between two physical systems by performing measurements on one of them, the impossibility of broadcasting the information contained in an unknown physical state, and the impossibility of unconditionally secure bit commitment.

[3]He writes: "you can, if you like, tell a story along Bohmian, or similar, lines [. . .] but, given the information-theoretic constraints, such a story can, in principle, have no excess empirical content over quantum mechanics" (Bub 2005, p. 542).

- *The "small" measurement problem*: the problem of explaining how an effectively classical probability space of macroscopic measurement outcomes arises from a quantum measurement process.

And then he argues that the 'big' one, which is what is usually called the measurement problem, is actually a pseudo-problem, given that in his approach one cannot talk about one single experimental results, just only the statistics.

One immediate reaction to this interpretation is that it is not truly realist, and therefore it is not in direct competition with the 'traditional' realist quantum theories such as the pilot-wave theory (Bohm 1952), the spontaneous collapse theory (Ghirardi et al. 1986) and Everettian mechanics (Everett 1957). However, Bub and Pitowsky argue that there is a robust sense in which their intent is realist, and that people who deny this implicitly accepts on two 'dogmas,' which they think have little justification:

D1: measurement should never be included as an unanalyzable primitive in a fundamental physical theory, and
D2: the wavefunction represents physical reality.

Bub and Pitowsky claim that it is usually thought that realists have to accept them to solve the measurement problem. However, they say, this is true only if they have in mind the 'big' problem, but in order to be realist one merely needs to solve the 'small' problem. In fact, they point out, take the Everett interpretation: it is considered a realist theory even if it only solves the 'small' problem. In Everett's theory, in fact, measurements have no definite outcomes and the experimental statistics are obtained in terms of a many-worlds structure. This structure acts classically due to decoherence plus the idea of considering probabilities as an agent's preferences (Wallace 2002). For Bub and Pitowsky, the difference between their approach and Everett is that while Everettians accept the two dogmas, they do not. However, they have in common that they still solve the 'small' measurement problem, given that in the IT account measurements are taken to be primitive, which is readily solved by the constraints imposed by the Hilbert space structure. This, they argue, shows that one could be realist even if rejecting the two dogmas of quantum mechanics. (In Sect. 2.5 I will provide another example of a realist quantum theory which rejects dogma 2).

Moreover, building on what Bub previously did, Bub and Pitowsky also argue that one should reject both dogmas. They do this by appealing to the distinction between dynamical and kinematic explanation, connected with the notions of principle and constructive theories which was originally introduced by Einstein (1919). They argue that kinematic theories are more satisfactory that dynamical ones, and that, given that the IT approach is the only realist interpretation that could be thought as a kinematic theory, it is to be preferred to the alternatives. Since this theory rejects the two dogmas, so should the realist. Let's discuss this argument in the next section.

2.3 The IT Approach as a Kinematic Theory

Bub and Pitowsky argue that the IT interpretation is a kinematic theory, and that this kind of theories are to be preferred, as dynamical theories add no further insight. Originally, the argument has been presented by Bub in terms of the principle/constructive theories distinction, so let's start with that. Principle theories are formulated in terms principles, which are used as constraints on physically possible processes, as in thermodynamics ('no perpetual motion machines'). Instead, constructive theories involve the dynamical reduction of macroscopic objects in terms of their microscopic constituents, as in the kinetic theory (which reduces the behavior of gases to the motion of atoms).[4] Einstein introduced this distinction when discussing his 1905 theory of relativity, which he regarded as a principle theory, as it was formulated in terms of the two principles of equivalence of inertial frames for all physical laws, and constancy of the velocity of light (in vacuum for all inertial frames). This theory explains relativistic effects (such as length contraction and time dilation) as the physical phenomena compatible with the theory's principles. By contrast, since Lorentz's theory (1909) derives the relativistic transformations and the relativistic effects from the electromagnetic properties of the ether and its interactions with matter, is a constructive theory.[5]

According to Bub and Pitowsky, Lorentz's constructive theory came first, then only later Einstein formulated his (principle theory of) special relativity: Minkowski provided kinematic constraints which relativistic dynamics has to obey, and then Einstein came up with an interpretation for special relativity. Bub and Pitowsky also claim that we should take the fact that Einstein's kinematic theory has been preferred over Lorentz's dynamical theory as evidence that such type of theories are to be preferred in general. In addition, they argue that the fact that Lorentz's theory can constructively explain Lorentz invariance justified a realist interpretation of special relativity as a principle theory. But after this initial use, the constructive counterpart is *no longer necessary*. Similarly, Bub and Pitowsky argue that IT is kinematic theory in that it provides constraints on the phenomena without

[4]Here's how Balashov and Janssen put it:"In a theory of principle [. . .] one explains the phenomena by showing that they necessarily occur in a world in accordance with the postulates. Whereas theories of principle are about the *phenomena*, constructive theories aim to get at the underlying *reality*. In a constructive theory one proposes a (set of) model(s) for some part of physical reality [. . .]. One explains the phenomena by showing that the theory provides a model that gives an empirically adequate description of the salient features of reality" (Balashov and Janssen 2003).

[5]Again in the worlds of Balashov and Janssen: "Consider the phenomenon of length contraction. Understood purely as a theory of principle, SR explains this phenomenon if it can be shown that the phenomenon necessarily occurs in any world that is in accordance with the relativity postulate and the light postulate. By its very nature such a theory-of-principle explanation will have nothing to say about the reality behind the phenomenon. A constructive version of the theory, by contrast, explains length contraction if the theory provides an empirically adequate model of the relevant features of a world in accordance with the two postulates. Such constructive-theory explanations do tell us how to conceive of the reality behind the phenomenon" (Balashov and Janssen 2003).

explaining them dynamically. That is, Hilbert space should be recognized as a "*the kinematic framework for the physics of an indeterministic universe*, just as Minkowski space-time provides the kinematic framework for the physics of a non-Newtonian, relativistic universe" (Bub and Pitowsky 2010, emphasis in the original text). Because of this, no other explanation for the experimental results is necessary.[6] More in detail, "the information-theoretic view of quantum probabilities as 'uniquely given from the start' by the structure of Hilbert space as a kinematic framework for an indeterministic physics is the proposal to interpret Hilbert space as a constructive theory of information-theoretic structure or probabilistic structure" (Bub and Pitowsky 2010).

As a consequence, we should reject the first dogma to obtain a kinematic theory, not a dynamical one. According to Bub and Pitowsky, the 'big' measurement problem is a dynamical problem, and as such a pseudo-problem when considering kinematic theories like theirs: as relativistic effects, such as length contraction and time dilation, are a problem for Newtonian mechanics but ceases to be such when looking at the geometry of Minkowski space-time and thus at the kinematics, here the 'big' problem dissolves when looking at the constraints that the Hilbert space structure imposes on physical events, given by experimental outcomes.

The next step is to observe that by rejecting dogma 1, the wavefunction can only be connected to experimenters' degrees of belief, rather than representing something real. Because of this, therefore, it is argued that we should also reject dogma 2. In other words, the rejection of the first dogma allows to make quantum theory a kinematic theory, which in turns dissolves the 'big' measurement problem, while the role of the rejection of the second dogma is to find a place for the wavefunction in the interpretation. That is, it allows to answer the question: if measurements are primitive, what is the wavefunction?

Be that as it may, in the next section we will start discussing some criticism raised against the IT account.

2.4 Objections to the Explanatory Superiority of Kinematic Theories

The IT framework has been criticized for mainly two reasons: on the one hand, objections have been raised against the alleged superiority of kinematic theories, therefore undermining the rejection of dogma 1; and on the other hand, people have pointed out the difficulties of considering the wavefunction as epistemic, as a consequence of the rejection of dogma 2. In this section, we will review the first

[6]"There is no deeper explanation for the quantum phenomena of interference and entanglement than that provided by the structure of Hilbert space, just as there is no deeper explanation for the relativistic phenomena of Lorentz contraction and time dilation than that provided by the structure of Minkowski space-time" (Bub and Pitowsky 2010).

kind of criticisms, while in Sect. 2.9 we will discuss the ones based on the nature of the wavefunction.

First of all, Timpson (2010) argues that the dogmas Bub and Pitowsky consider are not dogmas at all, as they can be derived from more general principles like realism. In fact, regarding dogma 1, a realist recognizes measurement results and apparatuses as merely complicated physical systems without any special physical status. Regarding dogma 2, we just proceed as we did in classical theories: there's a formalism, there's an evolution equation, then one is realist about what the equation is about, which, in the case of quantum mechanics, is the wavefunction. However, I think Timpson misses the point. The dogmas are dogmas *for the realist*: the realist is convinced that she has to accept them to solve the measurement problem. What Bub and Pitowsky are arguing is that realist does not have to accept these dogmas: Bub and Pitowsky in fact provide a realist alternative which does not rely on these dogmas. Timpson is on target in that a realist will likely endorse dogma 1 (what's special about measurements?), but Bub and Pitowsky argue that that is not the only option. Namely, one can recognize dogma 1 as being a dogma and therefore may want to recognize the possibility of rejecting it. One advantage of doing so, as we saw, is that in this way quantum theory becomes a kinematic theory, if one believes in the explanatory superiority of these type of theories. This leads one to recognize dogma 2 also as a dogma, and then to recognize the possibility of rejecting it as well.

Moreover, let me notice that Timpson also points out that Bub and Pitowsky provide us with no positive reason why one should be unhappy with such dogmas. I will present in Sect. 2.9 reasons for being extremely suspicious of dogma 2. Indeed, as I will argue later, one can restate the situation as follows: a realist, based on the reasons Timpson points out, *at first* is likely to accept *both* dogmas. However, upon further reflection, after realizing that by accepting dogma 2 one needs to face extremely hard problems, the realist should reject it.

Brown and Timpson (2006) point out that the first dogma was not a dogma at all for Einstein.[7] In this way, they criticize the historical reconstruction of Bub and Pitowsky for their argument for the superiority of kinematic theories. In the view of Brown and Timpson, Einstein's 1905 theory was seen by Einstein himself as a temporary theory, to be discarded once a deeper theory would make itself

[7]In fact, discussing his theory of special relativity he wrote: "One is struck that the theory [...] introduces two kinds of physical things, i.e., (1) measuring rods and clocks, (2) all other things, e.g., the electromagnetic field, the material point, etc. This, in a certain sense, is inconsistent; strictly speaking measuring rods and clocks would have to be represented as solutions of the basic equations (objects consisting of moving atomic configurations), not, as it were, as theoretically self-sufficient entities. However, the procedure justifies itself because it was clear from the very beginning that the postulates of the theory are not strong enough to deduce from them sufficiently complete equations [...] in order to base upon such a foundation a theory of measuring rods and clocks. [...] But one must not legalize the mentioned sin so far as to imagine that intervals are physical entities of a special type, intrinsically different from other variables ('reducing physics to geometry', etc.)" (Einstein 1949b).

available.[8] Indeed, they point out, Einstein introduced the principle/constructive distinction to express *his own dissatisfaction* of the theory at the time. According to them, it was Einstein's view that kinematic theories are typically employed when dynamical theories are either unavailable or too difficult to build.[9] In this way, they argue that it is too much of a stretch to argue that Einstein would have encouraged a kinematic theory interpretation of quantum mechanics, as Bub and Pitowsky propose. However, even if I think that Brown and Timpson are correct, it is not clear why this should be an issue for Bub and Pitowsky. In fact, the charge at this stage is merely that Einstein did not consider dogma 1 as such, and that he did not like kinematic explanations. And to this Bub and Pitowsky can simply reply that they disagree with Einstein, and insist that there is value in kinematic theories, contrarily to what Einstein himself thought.

Therefore, let's move to the more challenging charge. Brown and Timpson argue against the explanatory superiority of kinematic theories by pointing out that only dynamical theories provide insight of the reality underlying the phenomena.[10] That is, regularities and constraints over the possible experimental findings lack explanatory power because we still do not know why these constraints obtain. A reason is provided only by a dynamical, constructive theory, in which one is told the mechanism, or the microscopic story, that gives rise to the observed behavior. For instance, it has been argued that the economy of thermodynamic reasoning is trumped by the insight that statistical mechanics provides (Bell 1976; Brown and Timpson 2006).

Whether dynamical theories are actually more explanatory than kinematic theories remains a controversial issue, and certainly more should be written in this regard.[11] Nonetheless, even if one were to accept that in some contexts (such as in special relativity or thermodynamics) that dynamical theories are better, Bub (2004) has argued that a kinematic quantum theory is superior to any dynamical account. In fact, while Einstein's theory of Brownian motion, which led to the acceptance of the atomic theory of matter as more than a useful fiction, showed the *empirical* limits of thermodynamics and thus the superiority of the dynamical theory, no such advantage exists in the quantum domain. Bub in fact argues that such

[8]"The methodology of Einstein's 1905 theory represents a victory of pragmatism over explanatory depth; and that its adoption only made sense in the context of the chaotic state of physics at the start of the 20th century" (Brown and Timpson 2006).

[9]For according to Einstein, "when we say we have succeeded in understanding a group of natural processes, we invariably mean that a constructive theory has been found which covers the processes in question" (Brown and Timpson 2006).

[10]See also Brown (2005), and Brown and Pooley (2004).

[11]See Flores (1999), who argues that principle theories, in setting out the kinematic structure, are more properly viewed as explaining by unification, in the Friedman (1974)/Kitcher (1989) sense. Moreover, see Felline (2011) for an argument against Brown's view in the framework of special relativity. See also van Camp (2011) who argues that it is not the kinematic feature of Einsteinian relativity that makes it preferred to the Lorenzian dynamics but the constructive component, in virtue of which the theory possesses a conceptual structure.

an advantage is given *only* when a dynamical theory can provide some 'excess' in empirical content. Since dynamical constructive quantum theories (such as the pilot-wave theory) are empirically equivalent to a kinematic quantum theory, then (unlike kinetic theory) no dynamical quantum theory can ever provide more insight than a kinematic one. Critics however could point out immediately that there are dynamical quantum theories such as the spontaneous collapse theory, which are not empirically equivalent to quantum mechanics. Moreover, it is not obvious at all that the only reason to prefer dynamical theories is because they can point to new data.

Therefore, it seems safe to conclude that an important step in Bub and Pitowsky's argument for their view rests on controversial issues connected with whether a kinematic explanation is better than a dynamical one. Since the issue has not been settled, I think it constitutes a serious shortcoming of the IT interpretation as it stands now, and it would be nice if instead it could be grounded on a less controversial basis. Indeed, this is what I aim to do in Sect. 2.6. Before doing this, however, I will present in the next section the most popular realist approach which accepts the second dogma.

2.5 Accepting the Second Dogma: Wavefunction Realism

Part of Bub and Pitowsky's reasoning is based on their willingness to dissolve the 'big' measurement problem. So let's take a step back and reconsider this problem again. As is widely known, quantum theory has always been regarded as puzzling, at best. As recalled by Timpson (2010), within classical mechanics the scientific realist can read out the ontology from the formalism of the theory: the theory is about the temporal evolution of a point in three-dimensional space, and therefore the world is made of point-like particles. Similarly, upon opening books on quantum theory, a mathematical entity and an evolution equation for it immediately stand out: the wavefunction and the Schrödinger equation. So, the natural thought is to take this object as describing the fundamental ontology. Even if we have just seen how it can be readily justified, as we have seen, Bub and Pitowsky think it is a dogma, in the sense that it is something that a realist has the option of rejecting. Here I want to reinforce the argument that this is indeed a dogma, and that this dogma should be rejected.

As emphasized by Schrödinger when discussing his cat paradox to present the measurement problem (1935b), assuming the second dogma makes the theory empirically inadequate. In fact, superposition states, namely states in which 'stuff' is both here and there at the same time, are mathematically possible solutions of the Schrödinger equation. This feature is typical of waves, as the wavefunction is in virtue of obeying Schrödinger's equation (which is a wave equation). However, if the wavefunction provides a complete description of the world, superpositions may exist for everything, including cats in boxes being alive and dead at the same time. The problem is that we never observe such macroscopic superpositions, and this is, schematically, the measurement problem: if we open the box and we measure the state of the cat, we find the cat either dead or alive. The realist, keeping the second dogma true, traditionally identified their next task in finding theories to solve this problem. That is, the realist concluded that, in Bell's words, "either the wavefunction, as given by the Schrödinger equation, is not everything, or it is not right" (Bell 1987, 201). Maudlin (1995) has further elaborated the available options by stating that the following three claims are mutually inconsistent: (A) the wavefunction of a system is complete; (B) the wavefunction always evolves to the Schrödinger equation; and (C) measurements have determinate outcomes. Therefore, the most popular realist quantum theories are traditionally seen as follows: the pilot-wave theory (Bohm 1952) rejects A, the spontaneous collapse theory (Ghirardi et al. 1986) denies B, while the many-world theory (Everett 1957) denies C.

Notice that it is surprising that even if dogma 2 has been challenged or resisted in the 1920s by de Broglie, Heisenberg, Einstein and Lorentz (see next section),[12] the physicists proposing the realist quantum theories mentioned above did not do so. Indeed, the view which accepts the second dogma has been the accepted view among the realists for a very long time. In this sense it was a dogma, as Bub and Pitowsky rightly call it, until it was recently recognized to be an assumption and the name of *wavefunction realism* was given to the view which accepts dogma 2.

Even if it is not often discussed, most wavefunction realists acknowledge that the measurement problem leads to another problem, the so-called *configuration space* problem. The wavefunction by construction is an object that lives on configuration space. This is classically defined as the space of the configurations of all particles, which, by definition, is a space with a very large number of dimensions. So, given that for dogma 2 the ontology is given by the wavefunction, if physical space is the space in which the ontology lives, then physical space is configuration space. Accordingly, material objects are represented by a field in configuration space. That is, the arena in which physical phenomena take place seems to be fundamentally three-dimensional, but fundamentally it is not; and physical objects seem to be three-dimensional, while fundamentally they are not. Thus, the challenge for the wavefunction realist is to account for the fact that the world appears so different from what it actually is.[13] According to this view, the realist theories mentioned above are all taken to be theories about the wavefunction, aside from the pilot-wave theory which had both particles and waves. In contrast, since in all the other theories everything is 'made of' wavefunctions, particles and particle-like behaviors are best seen as emergent, one way or another.[14]

Be that as it may, the configuration space problem, after it was ignored since the 1920s, started to be recognized and discussed only in the 1990s, and at the moment it

[12]See Bacciagaluppi and Valentini (2009) for an interesting discussion of the various positions about this issue and others at the 1927 Solvay Congress.

[13]See most notably Albert (1996, 2013, 2015), Lewis (2004, 2005, 2006, 2013), Ney (2012, 2013, 2015, 2017, forthcoming), North (2013).

[14]See Albert (2015) and Ney (2017, forthcoming) for two different approaches to this.

is acknowledged that it has *no universally accepted solution.*[15] Regardless, strangely enough, most people keep identifying the measurement problem (who has been solved) and not the configuration space problem (which has not) as the problem for the realist.[16] In the next section I will present a proposal to eliminate this problem once for all.

2.6 Why Realists Should Reject the Second Dogma

In the last two decades or so, some authors have started to express resistance to the characterization of quantum theories in terms of having a wavefunction ontology.[17] They propose that in all quantum theories material entities are not represented by the wavefunction but by some other mathematical entity in three-dimensional space (or four-dimensional space-time) which they dubbed the primitive ontology (PO) of the theory. The idea is that the ontology of a fundamental physical theory should always be represented by a mathematical object in three-dimensional space, and quantum mechanics is not an exception. Given that the wavefunction is a field in configuration space, it cannot represent the ontology of quantum theory.

This view constitutes a realist approach, just like wavefunction realism. In contrast with it, however, the configuration space problem never arises because physical objects are not described by the wavefunction. In this way, not only the pilot-wave theory but also the spontaneous collapse and many-worlds are 'hidden variable' theories, in the sense that matter needs to be described by something else (in three-dimensional space) and not by the wavefunction. For instance, primitive ontologists talk about GRWm and GRWf as spontaneous localization theories with a matter density m and an event ('flash') ontology f, as opposed to what they call GRW0, namely the spontaneous localization theory with merely the wavefunction (Allori et al. 2008).[18] The proponents of this view therefore, contrarily to Bub and Pitowsky, do not deny the first dogma: measurement processes are 'regular' physical processes analyzable in terms of their microscopic constituents (the PO).

---

[15] With this I do not mean that there are no proposals but simply that these proposals are work-in-progress, rather than full-blown solutions.

[16] One may think that the fact that there is more than one solution to the measurement problem implies that it has not really been solved. That is, one may think that the measurement problem would be solved only if we had a unique answer to it. While this is certainly true, the sense in which it has been solved (albeit not uncontroversially so) is that all its possible solutions are fully developed, mature accounts, in contrast with the proposals to solve the configuration space problem, which are only in their infancy. This is therefore the sense in which the configuration space problem is more serious than the measurement problem.

[17] Dürr et al. (1992), Allori et al. (2008), Allori (2013a, b).

[18] Notice however, that they are hidden variable theories only in the sense that one cannot read their ontology in the formalism of quantum mechanics, which is, in this approach, fundamentally incomplete.

Nevertheless, like the IT interpretation, they reject the second dogma: they deny the wavefunction representing physical objects. As we will see in Sect. 2.8, however, they do not take the wavefunction as epistemic: this view rejects the second dogma in the sense that the wavefunction is not material but not in the sense that it does not represent some objective feature of the world.

In this framework, one of the motivations for the rejection of dogma 2 is the conviction that the configuration space problem is a *more serious problem* than the measurement problem. The measurement problem is the problem of making sense of unobserved macroscopic superpositions which are consequences of the formalism of the theory. I wish to argue now that one important reason why one should drop the second dogma comes from the recognition that, even if the realist solves the measurement problem, the corresponding realist theories would still face the configuration space problem, which is arguably much harder to solve and which can be entirely avoided if one rejects the second dogma.

First, take the wavefunction-only theories, like the spontaneous localization and many-worlds theories. If one could keep the wavefunction in three-dimensional space, then both these theories could identify a 'particle' as a localized three-dimensional field given by the wavefunction (the superposition wavefunction spontaneously localizes in the spontaneous localization theory, and decoherence separates the different braches in Everett's theory). This would solve the measurement problem. However, these theories would need a wavefunction in configuration space to be empirically adequate, and therefore one ends up facing the configuration space problem. So, wavefunction realists solve one problem, but they have a new one. In the case of the pilot-wave theory, moreover, the argument for the rejection of the second dogma is even clearer, because the theory does not even solve the measurement problem from the point of view of the wavefunction realist. In fact, in this view matter would be made by both particles and waves. That is, the cat is both made of waves and of particles, and because of its wave-like nature, she can be in unobserved macroscopic superposition. The measurement problem is solved only when one considers the theory a theory of particles only, without the wave-counterpart representing matter. This is how indeed the theory is often discussed: there are particles (which make up the cat), which are pushed around by the wave. This is however still misleading because it assumes that some wave physically exists, and this becomes problematical as soon as one realizes that this wave is in configuration space: What kind of physical entity is it? How does it interact with the particles? A better way of formulating the theory avoiding all this is to say that cats are made of particles and the wavefunction is not material, therefore rejecting the second dogma.[19]

Regardless, what's so bad about the configuration space problem? According to the proponents of the PO approach, it is extremely hard to solve, if not unsolvable.[20] Let us briefly see some reasons. One difficulty was recognized already

[19]This is implicit in Dürr et al. (1992).

[20]Allori et al. (2008), Allori (2013a, b).

by de Broglie, who noticed that, since for the wavefunction realist there are no particles at the fundamental level, "it seems a little paradoxical to construct a configuration space with the coordinates of points which do not exist" (de Broglie 1927). Interestingly, this argument can be tracked down historically also to Heisenberg, who very vividly said to Bloch (1976), referring to configuration space realism: "Nonsense, [. . .] space is blue and birds fly through it", to express the ultimate unacceptability of building a theory in which there was no fundamental three-dimensional space.[21] Also, as already noticed, in wavefunction realism the traditional explanatory schema of the behavior of macroscopic entities in terms of microscopic ones needs to be heavily revised: contrarily to the classical case, tables and chairs are not macroscopic three-dimensional objects composed of microscopic three-dimensional particles; rather, as already noticed, they are macroscopic three-dimensional objects 'emerging from' high-dimensional wavefunction. And this problem has to be solved in absence of a truly compelling argument to do so. In fact, what is the positive argument for dogma 2? The one we already saw is that we should do what we did in Newtonian mechanics, namely extract the ontology from the evolution equation of the theory. However here the situation is different. Newton started off with an *a priori* metaphysical hypothesis and built his theory around it, so there is no surprise that one has no trouble figuring out the ontology of the theory. In contrasts, quantum mechanics was not developed like that. Rather, as it is known, different formulations have been proposed to account for the experimental data (Heisenberg matrix mechanics and Schrödinger wave mechanics) without allowing ideas about ontology to take precedence. Just like Bub and Pitowsky claim, before the solution of the measurement problem, the theory is a kinematic theory: just like thermodynamics, it put constraints on possible experimental outcomes *without* investigating how and why these outcomes come about. The main difference between the PO and the IT approaches is that the former provides a framework of constructive quantum theories based on the microscopic PO, which can be seen as the dynamical counterpart of the latter, in which measurements are fundamental.[22]

Recently, a more compelling argument for wavefunction realism has been defended, namely that wavefunction realism is the only picture in which the world is local and separable.[23] The idea is that, since fundamentally all that exists is the

[21] Similar concerns have been expressed by Lorentz, who in a 1926 letter to Schrödinger wrote: "If I had to choose now between your wave mechanics and the matrix mechanics, I would give the preference to the former, because of its greater intuitive clarity, so long as one only has to deal with the three coordinates *x, y, z*. If, however, there are more degrees of freedom, then I cannot interpret the waves and vibrations physically, and I must therefore decide in favor of matrix mechanics" (Przibram 1967). Similarly, Schrödinger wrote: "The direct interpretation of this wave function of six variables in three-dimensional space meets, at any rate initially, with difficulties of an abstract nature" (1926, p. 39). Again: "I am long past the stage where I thought that one can consider the w-function as somehow a direct description of reality" (1935a). This is also a concern heartfelt by Einstein, who expressed this view in many letters, e.g.: "The field in a many-dimensional coordinate space does not smell like something real" (Einstein 1926).

[22] For more on the comparison between the PO approach and the IT framework, see Dunlap (2015).

[23] See Loewer (1996) and Ney (forthcoming) and references therein.

high dimensional wavefunction, there are no other facts that fail to be determined by local facts about the wavefunction. First, let me notice that this is true also for the IT account, in which what is real are the (three-dimensional) experimental outcomes. However, while a more thorough analysis of this should be done somewhere else, let me point out that it is an open question whether the notions of locality and separability used in this context are the ones which have been of interest in the literature. Einstein first proposed the EPR experiment to discard quantum mechanics because it would show three-dimensional nonlocality. Arguably,[24] Bell instead proved that the locality assumption in EPR was false and that nonlocality is a fact of nature. In other words, to show that the world is separable and local in configuration space seems to give little comfort to people like Einstein who wanted a world which is local and separable in three dimensional space.

Be that as it may, a further reason to reject the second dogma is that, if we accept it and somehow we manage to solve the configuration space problem, still we would lose important symmetries.[25] However, this is not the case if we drop the second dogma. In fact, the law of evolution of the wavefunction should no longer be regarded as playing a central role in determining the symmetries of the theory. Indeed, they are determined by the ontology, whatever it is (the PO in the PO approaches, and measurement outcomes in the IT framework). If one assumes that the wavefunction does not represent matter, and wants to keep the symmetry, then one can allow the wavefunction to transform as to allow for this.[26]

Another reason to reject the second dogma, even assuming the configuration space problem is solved, is the so-called problem of *empirical incoherence*. A theory is said to be empirically incoherent when its truth undermines our empirical justification for believing it to be true. Arguably, any theory gets confirmation by spatiotemporal observations. However, wavefunction realism rejects space-time as fundamental so that its fundamental entities are not spatiotemporal. Because of this, our observations are not spatiotemporal observations, and thus provide no evidence for the theory in the first place.[27]

Moreover, it seems to me that insisting on holding on to dogma 2 at all costs would ultimately undermines the original realist motivation. In fact, the realist, in

---

[24]See Bell and Gao (2016) and references therein for a discussion of Bell's theorem and its implications.

[25]See Allori et al. (2008) and Allori (2018, 2019). For instance, quantum mechanics with wavefunction ontology turns out not Galieli invariant. In fact the wavefunction is a scalar field in configuration space, and as such will remain the same under a Galilean transformation: $\psi(r) \rightarrow \psi(r - vt)$. In contrast, invariance would require a more complicated transformation: $(r, t) \rightarrow e^{\frac{i}{\hbar}\left[mvr - \frac{1}{2}mv^2 t\right]}\psi\left(r - vt, t\right)$. A similar argument can be drawn to argue that if one accepts the second dogma then the theory is also not time reversal invariant.

[26]See respectively Allori (2018, 2019) for arguments based on Galilean symmetry and time reversal symmetry that the wavefunction is best seen as a projective ray in Hilbert space rather than a physical field.

[27]Barrett (1999), Healey (2002), Maudlin (2007). For responses, see Huggett and Wüthrich (2013) and Ney (2015).

trying to make sense of the puzzles and paradoxes of quantum mechanics, as we have seen, would try to find the ontology with methods that worked in the past. That is, she will look at the evolution equation and she will identify the wavefunction as the ontology of the theory. This move allows us to see the measurement problem. To solve this problem the wavefunction realist develops theories with have an evolution equation for an object in configuration space. Thus, she needs also to solve the configuration space problem. Assuming a solution is possible, it will be radically different from the familiar constructive theories since it would involve emergence of the three-dimensional world from the multidimensional configuration space, rather that the dynamical explanation of the three-dimensional macroscopic data in terms of the three-dimensional microscopic ontology. However, the quest for a dynamical explanation, typical of proponents of constructive theories, in which macroscopic objects are thought of as composed of microscopic entities, motivated the whole realist approach to quantum theory to start with. To put it in another way, the wavefunction realist assumes that realist quantum theories (which solve the measurement problem) are satisfactory as they are, and looks to their fundamental equation to find the ontology. However the cost of assuming this amounts to ending up with a theory which will certainly not involve a dynamical explanation of the phenomena, which however is arguably the idea that motivated the realist to solve the measurement problem to start with. In contrast the primitive ontologist recognizes the disanalogy with classical mechanics (as mentioned earlier in this section), in which the strategy of looking at the evolution equation to find the ontology worked because Newton had already an ontology in mind when he constructed the theory (in contrast with quantum theory). Thus, she rejects dogma 2 by dropping the idea that the various realist theories are satisfactory as they are[28]: the wavefunction never describes matter, and therefore *never* provides the complete description of a physical system, *contra* Bell's characterization of the measurement problem. Physical objects have a different ontology, which (to avoid the configuration space problem) is postulated to be in three-dimensional space. In this way, one can have a dynamical explanation of the phenomena in terms of the fundamental ontology. In this sense, the PO approach provides with constructive dynamical theories corresponding to the kinematic theory given by the IT interpretation.

Before concluding this section, let me make some remarks. First, it is interesting to note how the discussion above, in which we have motivated the rejection of the second dogma with multiple arguments, provides a response to Timpson, who complained that Bub and Pitowsky provide no reason why one would be unhappy with dogma 2.

In addition, let me mention another theory which accepts the second dogma, as well as the first, which has been proposed to avoid the configurations space problem. The idea is to think as the wavefunction as a multi-field in three-dimensional

[28]Notice that there is nothing strange in this: the wavefunction realist did not accept quantum mechanics as it is because it suffered from the measurement problem, and in order to solve it, she finds it acceptable to, for instance, change the theory by modifying the Schrödinger equation.

field. That is, the wavefunction is a 'poly-wave' on three-dimensional space, a generalization of an ordinary classical field: as a classical field specifies a definite field value for each location in three-dimensional space, the multi-field, given an N-particle system, specifies a value for the N-tuple of points in three-dimensional space.[29] Even if it has the advantage of having no configuration space problem without dropping the second dogma, however this approach is problematical because it loses important symmetry properties, given that the multi-field transforms differently from how a classical field would (Belot 2012).[30]

On a different note, observe that the reasons I provided for the rejection of dogma 2 are not necessarily connected with dogma 1. On one hand, one can reject it with dogma 2, as in the IT approach, and have a macroscopic ontology of measurement results without having the measurement problem. The existence of the IT interpretation, which I think is best seen as a refinement of Bohr's original account, is thus the counterexample to the claim that a scientific realist cannot solve the measurement problem by rejecting both dogmas. On the other hand, PO theories solve the measurement problem by rejecting only the second dogma. Criticisms to theories which reject the first dogma has been put forward. Most notably Bell (1987) argues that by taking measurements as primitive one introduces a fundamental vagueness in the theory connected with characterizing what a measurement is. This, it is argued, makes the theory hardly satisfactory. Moreover, Egg (2018) argues that rejecting the first dogma is incompatible with standard accounts of scientific realism, according to which reality is more than what we can observe or measure. In the next section we will discuss in a little more detail the different options open to the realist, and the factors which are likely to be relevant in making a choice among the alternatives.

2.7 The Dogmas, Scientific Realism, and the Role of Explanation

Let's summarize some of the points made so far, and draw some partial conclusions. A scientific realist thinks it is possible to explain the manifest image of our senses in terms of the scientific image given to us by our best theories (Sellars 1962). We have different possibilities:

1. Reject dogma 2, thus assume a microscopic ontology in three-dimensions, as advised by the PO approach;

[29]Forrest (1988), Belot (2012), Hubert and Romano (2018).

[30]Moreover, a new formulation of the pilot-wave theory has been proposed (Norsen 2010) in which to each particle is associated a three-dimensional filed given by a conditional wave-function, namely a function ψ defined by the wave-function of the universe Ψ once the positions of all the other particles in the universe $Y(t)$ are fixed: $\psi_t(x) := \Psi(x, Y(t))$. However, to recover the correct trajectories one needs to add infinitely many fields, and this makes the theory hardly satisfactory.

2. Reject both dogmas, thus assume a macroscopic ontology in three-dimensions, as advised by the IT interpretation; and
3. Accept both dogmas, thus assume a non-three-dimensional ontology, as in wavefunction realism.[31]

In the first case, one can provide a constructive and dynamical explanation of the manifest image, both of the experimental results and the properties of macroscopic objects. In the second case, instead, the only thing to explain is why certain statistical patterns are observed, and this is done using kinematical constrain on the possible physical events, not in a dynamical way. In the third case the situation is more complicated: one needs to explain both the empirical statistics and the 'emergence' of macroscopic objects in a way which is neither (obviously) kinematic nor dynamics, neither constructive nor using principles. No such account has been proven so far as being uncontroversial or fully developed. Because of this, and the other problems discussed in the previous section, all other things being equal, at this stage one can argue that the first two alternatives are to be preferred over the third.

So, which of the first two should we chose? This is going to depend on whether one prefers dynamical to kinematic explanations. That is, Bub and Pitowsky find option 1 unnecessary, while the proponents of the PO approach find 2 lacking explanatory power. If so, we have found a new role for the kinematic/dynamic distinction. In the original argument by Bub and Pitowsky the distinction came in the first step of the argument to reject the first dogma on the basis of which dogma 2 was later rejected. Because the superiority of kinetic theories is a controversial matter, however, I think this argument is not likely to convince anyone already inclined to prefer dynamical theories to drop dogma 2. In contrast, in the argument I proposed, the reasoning against dogma 2 is independent of this distinction, which only plays a role in selecting which of the alternatives theories which *already reject* dogma 2.

2.8 On the Status of the Wavefunction

Before discussing the objections to the IT interpretation based on the epistemic account of the wavefunction, let us take another step back. The different accounts of the nature of the wavefunction are usually classified as either '*ontic*' or '*epistemic*'. Let's see what this means, and review some of the main objections to the ontic

[31] Interestingly, there's logically a fourth option, namely a theory which rejects dogma 1 without rejecting dogma 2. This would be a theory in measurement outcomes are left unanalyzed and the wavefunction represents physical objects. I am not sure who would hold such a view, but it seems to me that such a theory would presumably not appeal many people: while it solves the measurement problem (by rejecting dogma 1) it does not solve the configuration space problem (since it accepts dogma 2). Moreover, this theory leaves a mystery what the relation between the wavefunction and measurement processes is. Regardless, it nicely shows the relation between the two dogmas, with dogma 2 being in my opinion the one responsible for the problems for the quantum realist.

accounts, while we will leave the criticism to the epistemic accounts to the next section.

An account is said to be ontic when the wavefunction (or the density matrix which may also define the quantum state) represents some sort of objective feature of the world. In contrast, an account is epistemic if the wavefunction (or the density matrix) has to do with our knowledge of the physical system. Each view has their own sub-classification. As we have seen, wavefunction realism and the wavefunction as multi-field are ontic approaches in that they regard the wavefunction as a physical field. However, there are other approaches that could be labelled as ontic which do not have this feature. One of these ontic interpretations is the view according to which the wavefunction is a law of nature. This is a peculiar account because it has been formulated in the PO framework, which denies the second dogma. Roughly put, the idea is that since the wavefunction does not represent matter but has the role of governing the behavior of matter, then the wavefunction has a nomological character and thus is best understood as a law of nature.[32] Notice that this account of the wavefunction is still ontic (the wavefunction expresses some nomic feature of the world) even if the second dogma is rejected (the wavefunction does not represent material entities), and thus one can dub this view as 'ψ-non-material' rather than 'ψ-ontic', which does not seem sufficiently specific (another non-material view is the one according to which the wavefunction is a property, see below). Perhaps the most serious problem[33] for this approach is that one usually thinks of laws as time-independent entities, while the wavefunction evolves itself in time.[34] Another challenge for the view is that while laws are taken to be unique, the wavefunction is not. Replies have been provided, relying on a future quantum cosmology in which the wave function would be static.[35] However, they have been received with skepticism, given that one would want to get a handle on the nature of the wavefunction in the current situation, without having to wait for a future theory of quantum cosmology.

Perhaps a better way of capturing this idea can be found in a Humean framework, regarding the wavefunction as part of the Humean mosaic.[36] Nonetheless, many still find the account wanting, most prominently because they find Humeanism with respect to laws misguided for other reasons. A different but connected approach that can be thought as nomological is to think of the wave function as a dispositional property or more generally as a property of the ontology.[37] Not surprisingly, even if these approaches do not have the time-dependence problem, they nevertheless are not immune to criticisms, given that the dispositional properties are notoriously a

[32]Dürr et al. (1992, 1997), Goldstein and Zanghì (2013).

[33]For more objections, see Belot (2012), Callender (2015), Esfeld et al. (2014), Suárez (2015).

[34]Brown and Wallace (2005).

[35]Goldstein and Teufel (2001).

[36]Miller (2014), Esfeld (2014), Callender (2015), and Bhogal and Perry (2017).

[37]Suàrez (2007, 2015), Monton (2013), Esfeld et al. (2014), and Gao (2014).

tough nut to crack and moreover the wavefunction does not behave like a regular property.[38]

Moving on to the epistemic approaches, they have in common the idea that the wavefunction should be understood as describing our incomplete knowledge of the physical state, rather than the physical state itself. One of the motivations for the epistemic approaches is that they immediately eliminate the measurement problem altogether. In fact, schematically, in a Schrödinger cat type of experiment, the cat is never in a superposition state; rather, it is the experimenter who does not know what its state is. When I open the box and find the cat alive, I simply have updated my knowledge about the status of the cat, who now I know to always have been alive during the experiment.

The epistemic approaches can be distinguished into the 'purely epistemic,' or neo-Copenhagen, accounts, and the ones that invoke hidden variables, which one can dub 'realist-epistemic' approaches. The prototype of the latter type is Einstein's original 'ignorance' or 'statistical' interpretation of the wavefunction (Einstein 1949a), according to which the current theory is fundamentally incomplete: there are some hidden variables which describe the reality under the phenomena, and whose behavior is statistically well described by the wavefunction.[39] For a review of these epistemic approaches, see Ballentine (1970).[40] In contrast, the neo-Copenhagen approaches reject that the above mentioned hidden variables exist, or are needed. The wavefunction thus describes our knowledge of measurement outcomes, rather than of an underlying reality. This view has a long story that goes back to some of the founding fathers of the Copenhagen school as one can see, for instance, in Heisenberg (1958) and in Peierls (1991), who writes that the wavefunction "represents our knowledge of the system we are trying to describe." Similarly, Bub and Pitowsky write: "quantum state is a credence function, a bookkeeping device for keeping track of probabilities".[41]

A crucial difference between the ontic and the epistemic approaches is that while in the former the description provided by the wavefunction is objective, in the latter it is not. This is due to the fact that different agents, or observers, may have different information about the same physical system. Thus, many epistemic states may correspond to the same ontic state.

[38]For an assessment, see Suàrez (2015).

[39]Einstein was convinced that "assuming the success of efforts to accomplish a complete physical description, the statistical quantum theory would, within the framework of future physics, take an approximately analogous position to the statistical mechanics within the framework of classical mechanics. I am rather firmly convinced that the development of theoretical physics will be of this type; but the path will be lengthy and difficult" (Einstein 1949a, b).

[40]For a more modern approach, see Spekkens (2007).

[41]Other neo-Copenhagen approaches include Bayesian approaches (Fuchs and Peres 2000; Fuchs 2002; Fuchs and Schack 2009, 2010, Caves et al. 2002a, 2002b, 2007), pragmatist approaches (Healey 2012, 2015, 2017; Friederich 2015), and relational ones (Rovelli 1996).

2.9 Objections Based on the Wavefunction Being Epistemic

As we anticipated, since the IT interpretation rejects the first dogma and takes measurements as primitive and fundamental, one is left with the question about what the wavefunction is. According to Bub and Pitowsky, the wavefunction (or the density matrix) is not fundamental. Thus, they spend not too much time in discussing what the nature of the wavefunction is, and Timpson (2010) finds this troublesome. However, it seems natural for them to consider the wavefunction as epistemic, given their overall approach: the wavefunction adds no empirical content to the theory, and thus is best seen not as a description of quantum systems but rather as reflecting the assigning agents' epistemic relations to the systems.

One epistemic approach which shows important similarity with Bub and Pitowsky's view is the one put forward by Healey (2012, 2015, 2017). According to him, physical situations are *not* described by the wavefunction itself. Rather the wavefunction prescribes how our epistemic attitudes towards claims about physical situations should be. For instance, in a particles interference experiment, an experimenter uses the Born rule to generate the probability that the particle is located in some region. The wavefunction therefore prescribes what the experimenter should believe about particle location. This view is epistemic in the sense that the wavefunction provides a guide to the experimenter's beliefs. Moreover, it is pragmatic in the sense that concerns about explanation are taken to be prior to representational ones. In this sense, the wavefunction has an explanatory role. This explanatory role is also shared by Bub and Pitowsky's account, since in their account the wavefunction allows the definition of the principles of the theory which constrain the possible physical phenomena, including interference.

Anyway, because of its epistemic approach, the IT interpretation has received many criticisms, and the situation remains extremely controversial.[42] Let me review what the main criticisms are. First of all, it has been argued that the epistemic approaches cannot explain the interference phenomena in the two-slit experiment for particles: how can our ignorance about certain details can make some phenomenon such as interference happen? It seems that some objective feature of the world needs to exist in order to explain the interference fringes, and the obvious candidate is the wavefunction (Leifer 2014; Norsen 2017). Moreover, in particle experiments with the Mach-Zender interferometer outcomes change depending on whether we know or not which path the particle has taken. This seems incompatible with the epistemic view: if the wavefunction represent our ignorance about which path has been taken, then coming to know that would change the wavefunction but would not produce any physical effect (Bricmont 2016).

Moreover, the so-called no-go theorems for hidden variables (starting from the ones of von Neumann 1932, and Bell 1987) have been claimed to show that the realist-epistemic view is untenable, so that they can be considered no-go theorems

[42] See Gao (2017) for a review of the objections, and Leifer (2014) for a defense.

for this type of approach. In fact, if the wavefunction represents our ignorance of an underlying reality in which matter has some property A, say, spin in a given direction, then the wavefunction provides the statistical distribution of $v(A)$, the values of A. Indeed, if this interpretation is true, $v(A)$ exists for more than one property, since it would be arbitrary to say that spin exists only on certain directions. However, the no-hidden variable theorems show that mathematically no $v(A)$ can agree with the quantum predictions. That is, one cannot introduce hidden variables to *all* observables at once.[43]

Recently, a new theorem called PBR theorem, from the initials of its proponents (Pusey et al. 2012) has been proposed as a no-go theorem for the epistemic theories in general. This theorem is supposed to show that, in line of the strategy used by the other no-go theorems, epistemic approaches are in conflict with the statistical predictions of quantum mechanics by showing that if the wavefunction is epistemic, then it requires the existence of certain relations which are mathematically impossible. Therefore, if quantum mechanics is empirically adequate, then the wavefunction must be ontic (either material, or not).

This objections have been subject to an extensive examination, and proponents of the epistemic views have proposed several replies, which are however in the eyes of many still tentative.[44] Bub and Pitowsky respond that they can explain interference as a phenomenon constrained by the structure of Hilbert space, more specifically by the principle of 'no universal broadcasting.' Moreover, PBR's conclusion, as well as the conclusions of the other no-go theorems, applies only to theories that assume an underlying reality, which they reject. However, these answers appear unsatisfactory to someone who would want to know why these phenomena obtain, rather than having a mere description. So, without entering too much in the merit of these objections, in the view of many they pose a serious challenge to the epistemic views, and, accordingly, to the IT interpretation. Therefore, one may wonder whether it is possible to deny the second dogma while staying away from the epistemic views without endorsing the non-material interpretations discussed in Sect. 2.8, given that they suffer from serious objections too. This is the enterprise I will take on in the next section.

2.10 The Wavefunction Is as the Wavefunction Does

We have reviewed in the previous sections the various non-material accounts of the wavefunction (the nomological approach, which is non-material but ontic, and the epistemic ones) and their criticisms. In this section, I propose a new account of the wavefunction based on functionalism, which I argue is better than the alternatives and could be adopted both by the proponents of the PO approach and by Bub and Pitowsky.

[43]Notice that the pilot-wave theory circumvents this theorem because some, but not all, experiments reveal pre-existing properties (for more on this, see Bricmont 2016; Norsen 2017).

[44]Again, see Leifer (2014), Gao (2017), and references therein.

Functionalism, broadly speaking, is the view that certain entities can be 'reduced' to their functional role. To use a powerful slogan, 'a table is as a table does.' Strategies with a functionalist core have been used in the philosophy of mind for a long time (starting from Putnam 1960), and they have been recently used also in philosophy of physics.

For instance Knox (2013, 2014) argues that spacetime can be functionalized in the classical theory of gravity. That is, spacetime can be thought of as non-fundamental (emergent) in virtue of the fact that spacetime plays the role of defining inertial frames. Interestingly, Albert (2015) defends wavefunction realism using his own brand of functionalism. He in fact argues that ordinary three-dimensional objects are first functionalized in terms of their causal roles, and the wavefunction dynamically uses these relations in a way which gives rise to the empirical evidence.[45] Lam and Wüthrich (2017) use functionalist strategies in quantum gravity. There are many proposals for how to articulate a quantum theory of gravity, however it has been argued that in all these theories spacetime is not fundamental, but rather it is emergent from some non-spatiotemporal structures (Huggett and Wuthrich forthcoming). They argue that in order for spacetime to emerge it suffices to recover only those features which are functionally relevant in producing observations. However, not surprisingly, this approach faces the same objections of wavefunction realism. An objection to this approach is that it is unclear how space and time can indeed emerge from a fundamental non-spatiotemporal ontology: first, it is unclear what a non-spatiotemporal fundamental could be; second, it is doubtful whether the right notion of emergence is the one proposed (Lam and Esfeld 2013). In addition, there is the already mentioned problem of empirical incoherence, namely that the theory undermines its own confirmation. Arguably, any theory gets confirmation by spatiotemporal observations. However, a theory in which spacetime is functionalized is a theory whose fundamental entities are not spatiotemporal. Because of this, our observations are not spatiotemporal observations, and thus provide no evidence for the theory in the first place.[46]

As anticipated, my idea is that one should functionally define the wavefunction, similarly as how people wish to functionalize space, space-time and material objects, and define it in terms of the role it plays in the theory. That is, *the wavefunction is as the wavefunction does*. I am going to show that this view captures the appeals of the other accounts without falling prey to their objections. In other words, by thinking of the wavefunction functionally one can capture the intuitions that the wavefunction is law-like (as proposed by the nomological approach), and that it is explanatory rather than representational (as proposed by the epistemic and pragmatic approaches) avoiding the problem of time-dependence, non-uniqueness, and no-go theorems.

The wavefunction plays different but interconnected roles in realist quantum theories which reject the second dogma, which however all contribute to the same

[45] Ney (2017) provides a criticism and discusses her own alternative. See also Ney (forthcoming).

[46] For responses, see Huggett and Wüthrich (2013), and Ney (2015, forthcoming).

goal: to *reproduce the empirical data*. First, the wavefunction plays a nomological role. This is straightforward in the case of PO theories, where it is one of the ingredients necessary to write the dynamics for the PO, which in turn constructively accounts for the measurement outcomes.[47] In the case of the IT interpretation, the dynamics plays no role, but the wavefunction can be seen as nomological in the sense that allows the principles or the kinematic constraints of the theory to be expressed, which in turn limit the physically possible phenomena and thus explain the data.

Connected to this, the wavefunction does not represent matter but it has an explanatory role: it is a necessary ingredient to account for the experimental results. As we have seen, the PO approach and the IT interpretation differ in what counts as explanatory, the former favoring a dynamical account, and the latter a kinematic one. Accordingly, the explanatory role of the wavefunction in these theories is different. In the PO approach the wavefunction helps defining the dynamics for the PO, and in this way accounts for the empirical results constructively as macroscopically produced by the 'trajectories' of the microscopic PO. In the IT interpretation the wavefunction provides the statistics of the measurement outcomes via the Born rule, therefore explaining the probabilistic data by imposing kinematical constraints.

Notice that this explanatory role allows to be pragmatic about the wavefunction: it is a useful object to adequately recover the empirical predictions. Pragmatically, both in the PO approach and in the IT framework one uses the wavefunction to generate the probability outcomes given by the Born rule, but one *could have chosen another object*, like a density matrix, or the same object evolving according to another evolution equation.[48] The fact that quantum mechanics is formulated in terms of the wavefunction is contingent to other super-empirical considerations like for example simplicity and explanatory power: it is the simplest and most explanatory choice one could have made.

This pragmatic take is compatible also with the nomological role of the wavefunction. In fact, as we have seen, the wavefunction is not a law of nature, strictly speaking, rather it is merely an ingredient in the law. Because of this the wavefunction does not have to be unique, like potentials are not unique, and can be dispensable: other entities, such as the density matrix, can play its role. In this way, the wavefunction plays a nomological role, but avoiding the usual objections to the nomological view.

Moreover, notice that this approach, even if pragmatic, is not epistemic: the wavefunction does not have to do with our knowledge of the system (*contra* Bub and Pitowsky's original proposal). As a consequence, it does not suffer from the usual problems of the epistemic approaches. Moreover, in the case of the IT approach, this

[47] In the context of the pilot-wave theory, the wavefunction can be taken as a force or a potential, but one should not think of it as material (even if Belousek 2003 argues otherwise).

[48] In the PO framework, for instance, see Allori et al. (2008) for unfamiliar formulations of the pilot-wave theory and the spontaneous collapse theory respectively with a 'collapsing' wave function and a non-collapsing one respectively.

view remains true to the ideas which motivated the interpretation, namely to reject both dogmas, and fits well with their ideas of quantum probability.

Also, I wish to remark that, since all the theories we are considering reject the second dogma, this view also avoids the objection of empirical incoherence of other functionalist approaches. In fact the problem originates only if the fundamental entities of the theory are not spatiotemporal, and this is not the case in the IT frameworks (in which measurements outcomes are fundamental and are in three-dimensional space) and in the PO approach (in which the PO is a spatiotemporal entity).

To summarize, the wavefunction plays distinctive but interconnected roles which act together and allow to recover the empirical data. First, the wavefunction is an ingredient in the laws of nature, and as such contributes in determining the trajectories of matter. As such, it has an explanatory role in accounting of the experimental results, even if it is not representational one. Because of this, the wavefunction can be thought in pragmatic terms without considering it connected to our knowledge of the system. Formulated in this way, the view has the advantages of capturing the main motivations for the other views without falling prey of their main objections:

1. The wavefunction has a nomological role, without being a law; so it does not matter whether it is time-evolving or unique;
2. The wavefunction has an explanatory role without necessarily being uniquely defined or connected to the notion of dynamical explanation;
3. The wavefunction has a pragmatic role in recovering the empirical data without being epistemic, thereby avoiding the interference problem and the no-go theorems linked with the epistemic view.

These roles combine together to define the wavefunction functionally: the wavefunction is whatever function it plays, namely to recover the experimental results. This role can be accomplished either directly, in the IT framework, or indirectly, through the 'trajectories' of the PO, but always remaining in spacetime, therefore bypassing the configuration space problem and the worry of empirical incoherence. This account of the wavefunction thus is arguably better than the other taken individually, and accordingly is the one that the proponents of realist theories which reject the second dogma should adopt. In particular, it can be adopted by the proponents of the PO approach, as well as by Bub and Pitowsky to strengthen their views.

2.11 Conclusion

As we have seen, the IT interpretation rejects the two dogmas of quantum mechanics. First, it rejects dogma 1. Using the CBH theorem, which kinematically constrains measurements outcomes, they see quantum theory as a theory with measurements as primitive. Then, what is the wavefunction? The IT interpretation naturally interpret it as epistemic, therefore rejecting the second dogma. It is argued that both dogmas give rise to the measurement problem, and that by rejecting them both it becomes a pseudo-problem.

Because of this sequence of arguments the IT interpretation has the following two features that could be seen as shortcomings:

1. The rejection of dogma1 is based on the superiority of kinematic theories, which is controversial;
2. The rejection of dogma 2 makes them endorse the epistemic view, which is also considered problematical.

My aim in this paper was:

A. To show that any realist should reject dogma 2, even if she may or may not reject dogma 1, and
B. To provide a functionalist approach to the wavefunction which captures some of the motivations of the other accounts, including IT, while avoiding their main objections.

Aside from standing for themselves, in doing A, I also have provided the IT interpretation with tools to avoid the first set of objections; while in doing B, a way to reject the second. In fact, first I have argued that Bub and Pitowsky should have rejected dogma 2 *independently* of dogma 1 (on the basis of the configuration space problem, the loss of symmetries, and empirical incoherence), thereby completely bypassing the connection with the alleged superiority of kinematic theories. Indeed, my reasons for rejecting dogma 2 are less contentious in the sense that even those who endorse dogma 2 (such as the wavefunction realists) recognize them as problems for the tenability of dogma 2. After rejecting dogma 2 for these reasons, someone may also drop dogma 1 to make quantum mechanics a kinematic framework (if they prefer them for independent reasons). However, one may decide otherwise: a realist who prefers constructive theories based on a dynamical explanation could still keep dogma 1. Thus, if the IT proponents follow the route proposed here (drop dogma 2 first) then they can avoid their first type of challenges. As we have seen, the second set of objections comes from them endorsing the epistemic view. However, my functionalist approach to the wavefunction, which has merits in its own right, is compatible with dropping both dogmas without committing us to an epistemic view of the wavefunction. Because of this, it can be adopted by Bub and Pitowsky to avoid the second set of problems.

Acknowledgements Thank you to Meir Hemmo and Orly Shenker for inviting me to contribute to this volume. Also, I am very grateful to Laura Felline for her constructive comments on an earlier version of this paper, and to the participants of the International Workshop on "The Meaning of the Wave Function", Shanxi University, Taiyuan, China, held in October 2018, for important feedback on the second part of this paper.

References

Albert, D. Z. (1996). Elementary quantum metaphysics. In J. T. Cushing, A. Fine, & S. Goldstein (Eds.), *Bohmian mechanics and quantum theory: An appraisal* (pp. 277–284). Dordrecht: Kluwer.

Albert, D. Z. (2013). Wave function realism. In D. Z. Albert & A. Ney (Eds.), *The wave function: Essay on the metaphysics of quantum mechanics* (pp. 52–57). New York: Oxford University Press.

Albert, D. Z. (2015). *After physics*. Cambridge: Harvard University Press.

Allori, V. (2013a). On the metaphysics of quantum mechanics. In S. Lebihan (Ed.), *Precis de la Philosophie de la Physique*. Paris: Vuibert: 116–151.

Allori, V. (2013b). Primitive ontology and the structure of fundamental physical theories. In D. Z. Albert & A. Ney (Eds.), *The wave-function: Essays in the metaphysics of quantum mechanics* (pp. 58–75). New York: Oxford University Press.

Allori, V. (2018). A new argument for the Nomological interpretation of the wave function: The Galilean group and the classical limit of nonrelativistic quantum mechanics. *International Studies in the Philosophy of Science, 31*(2), 177–188.

Allori, V. (2019). Quantum mechanics, time and ontology. *Studies in History and Philosophy of Modern Physics, 66*, 145–154.

Allori, V., Goldstein, S., Tumulka, R., & Zanghì, N. (2008). On the common structure of Bohmian mechanics and the Ghirardi-Rimini-Weber Theory. *The British Journal for the Philosophy of Science, 59*(3), 353–389.

Bacciagaluppi, G., & Valentini, A. (2009). *Quantum theory at the crossroads: Reconsidering the 1927 Solvay conference*. Cambridge: Cambridge University Press.

Balashov, Y., & Janssen, M. (2003). Critical notice: Presentism and relativity. *The British Journal for the Philosophy of Science, 54*, 327–346.

Ballentine, L. E. (1970). The statistical interpretation of quantum mechanics. *Reviews of Modern Physics, 42*(4), 358–381.

Barrett, J. A. (1999). *The quantum mechanics of minds and worlds*. New York: Oxford University Press.

Bell, J. S. (1976). How to teach special relativity. *Progress in Scientific Culture*, 1. Reprinted in Bell (1987), pp. 67–80.

Bell, J. S. (1987). *Speakable and unspeakable in quantum mechanics*. Cambridge: Cambridge University Press.

Bell, M., & Gao, S. (2016). *Quantum nonlocality and reality, 50 years of Bell's theorem*. Cambridge: Cambridge University Press.

Belot, G. (2012). Quantum states for primitive ontologists. *European Journal for Philosophy of Science, 2*(1), 67–83.

Belousek, D. W. (2003). Formalism, ontology and methodology in Bohmian mechanics. *Foundations of Science, 8*(2), 109–172.

Bhogal, H., & Perry, Z. (2017). What the Humean should say about entanglement. *Noûs, 1*, 74–94.

Bloch, F. (1976). Heisenberg and the early days of quantum mechanics. *Physics Today*.

Bohm, D. (1952). A suggested interpretation of the quantum theory in terms of 'hidden variables' I & II. *Physical Review, 85*(2), 166–179. 180–193.

Bricmont, J. (2016). *Making sense of quantum mechanics*. Cham: Springer.

Brown, H. R. (2005). *Physical relativity: Space-time structure from a dynamical perspective* (forthcoming). Oxford: Oxford University Press.

Brown, H. R., & Pooley, O. (2004). Minkowski space-time: A glorious non-entity. In D. Dieks (Ed.), *The ontology of spacetime* (pp. 67–89). Amsterdam: Elsevier.

Brown, H. R., & Timpson, C. (2006). Why special relativity should not be a template for a fundamental reformulation of quantum mechanics. In W. Demopoulos & I. Pitowsky (Eds.), *Physical theory and its interpretation: Essays in honor of Jeffrey Bub* (pp. 29–42). Dordrecht: Springer.

Brown, H. R., & Wallace, D. (2005). Solving the measurement problem: de Broglie-Bohm loses out to Everett. *Foundations of Physics, 35*, 517–540.
Bub, J. (2004). Why the quantum? *Studies in the History and Philosophy of Modern Physics, 35*, 241–266.
Bub, J. (2005). Quantum mechanics is about quantum information. *Foundations of Physics, 35*(4), 541–560.
Bub, J. (2007). Quantum probabilities as degrees of belief. *Studies in History and Philosophy of Modern Physics, 39*, 232–254.
Bub, J., & Pitowsky, I. (2010). Two dogmas about quantum mechanics. In S. Saunders, J. Barrett, A. Kent, & D. Wallace (Eds.), *Many worlds?: Everett, quantum theory, and reality* (pp. 433–459). Oxford: Oxford University Press.
Callender, C. (2015). One world, one Beable. *Synthese, 192*, 3153–3177.
Caves, C. M., Fuchs, C. A., & Schack, R. (2002a). Quantum probabilities as Bayesian probabilities. *Physical Review A, 65*, 022305.
Caves, C. M., Fuchs, C. A., & Schack, R. (2002b). Unknown quantum states: The quantum de Finetti representation. *Journal of Mathematical Physics, 44*, 4537–4559.
Caves, C. M., Fuchs, C. A., & Schack, R. (2007). Subjective probability and quantum certainty. *Studies in History and Philosophy of Modern Physics, 38*(2), 255–274.
Clifton, R., Bub, J., & Halvorson, H. (2003). Characterizing quantum theory in terms of information theoretic constraints. *Foundations of Physics, 33*(11), 1561.
de Broglie, L. (1927). La Nouvelle Dynamique des Quanta. In: *Solvay conference, Electrons et Photons*, translated in G. Bacciagaluppi and A. Valentini (2009) *Quantum theory at the crossroads: Reconsidering the* 1927 *Solvay conference* (pp. 341–371). Cambridge: Cambridge University Press.
Dunlap, L. (2015). On the common structure of the primitive ontology approach and information-theoretic interpretation of quantum theory. *Topoi, 34*(2), 359–367.
Dürr, D., Goldstein, S., & Zanghí, N. (1992). Quantum equilibrium and the origin of absolute uncertainty. *Journal of Statistical Physics, 67*, 843–907.
Dürr, D., Goldstein, S., & Zanghì, N. (1997). Bohmian mechanics and the meaning of the wave function. In R. S. Cohen, M. Horne, & J. Stachel (Eds.), *Experimental metaphysics — Quantum mechanical studies for Abner Shimony* (Vol. 1: Boston Studies in the Philosophy of Science) (Vol. 193, pp. 25–38). Boston: Kluwer Academic Publishers.
Egg, M. (2018). Dissolving the measurement problem is not an option for realists. *Studies in History and Philosophy of Science Part B: Studies in History and Philosophy of Modern Physics, 66*, 62–68.
Einstein, A. (1919). What is the theory of relativity? *The London Times*. Reprinted in Einstein, A. (1982). *Ideas and opinions* (pp. 227–232). New York: Crown Publishers, Inc.
Einstein, A. (1926). Einstein to Paul Ehrenfest, 18 June, 1926, EA 10–138. Translated in Howard (1990), p. 83.
Einstein, A. (1949a). Autobiographical notes. In P. A. Schilpp (Ed.), *Albert Einstein: Philosopher-scientist*. Evanston, IL: The Library of Living Philosophers.
Einstein, A. (1949b). Reply to criticisms. In P. A. Schilpp (Ed.), *Albert Einstein: Philosopher-scientist*. Evanston, IL, The Library of Living Philosophers.
Esfeld, M. A. (2014). Quantum Humeanism, or: Physicalism without properties. *The Philosophical Quarterly, 64*, 453–470.
Esfeld, M. A., Lazarovici, D., Hubert, M., & Dürr, D. (2014). The ontology of Bohmian mechanics. *The British Journal for the Philosophy of Science, 65*, 773–796.
Everett, H., III. (1957). 'Relative state' formulation of quantum mechanics. *Reviews of Modern Physics, 29*, 454–462.
Felline, L. (2011). Scientific explanation between principle and constructive theories. *Philosophy of Science, 78*(5), 989–1000.
Flores, F. (1999). Einstein's theory of theories and types of theoretical explanation. *International Studies in the Philosophy of Science, 13*(2), 123–134.
Forrest, P. (1988). *Quantum metaphysics*. Oxford: Blackwell.

Friederich, S. (2015). *Interpreting quantum theory: A therapeutic approach*. Basingstoke: Palgrave Macmillan.
Friedman, M. (1974). Explanation and scientific understanding. *Journal of Philosophy, 71*, 5–19.
Fuchs, C. A. (2002). Quantum mechanics as quantum information (and only a little more). In A. Khrennikov (Ed.), *Quantum theory: Reconsideration of foundations*. Växjö: Vaxjo University Press.
Fuchs, C. A., & Peres, A. (2000). Quantum theory needs no 'interpretation'. *Physics Today, 53*(3), 70–71.
Fuchs, C. A., & Schack, R. (2009). Quantum-Bayesian coherence. *Reviews of Modern Physics, 85*, 1693.
Fuchs, C. A., & Schack, R. (2010). A quantum-Bayesian route to quantum-state space. *Foundations of Physics, 41*(3).
Gao, S. (2014). Reality and meaning of the wave function. In S. Gao (Ed.), *Protective measurement and quantum reality: Toward a new understanding of quantum mechanics* (pp. 211–229). Cambridge: Cambridge University Press.
Gao, S. (2017). *The meaning of the wave function: In search of the ontology of quantum mechanics*. Cambridge: Cambridge University Press.
Ghirardi, G. C., Rimini, A., & Weber, T. (1986). Unified dynamics for microscopic and macroscopic systems. *Physical Review D, 34*, 470.
Goldstein, S., & Teufel, S. (2001). Quantum Spacetime without observers: Ontological clarity and the conceptual foundations of quantum gravity. In C. Callender & N. Huggett (Eds.), *Physics meets philosophy at the Planck scale*. Cambridge: Cambridge University Press.
Goldstein, S., & Zanghì, N. (2013). Reality and the role of the wave function in quantum theory. In D. Albert & A. Ney (Eds.), *The wave-function: Essays in the metaphysics of quantum mechanics* (pp. 91–109). New York: Oxford University Press.
Healey, R. (2002). Can physics coherently deny the reality of time? *Royal Institute of Philosophy Supplement, 50*, 293–316.
Healey, R. (2012). Quantum theory: A pragmatist approach. *The British Journal for the Philosophy of Science, 63*(4), 729–771.
Healey, R. (2015). How quantum theory helps us explain. *The British Journal for the Philosophy of Science, 66*, 1–43.
Healey, R. (2017). *The quantum revolution in philosophy*. Oxford: Oxford University Press.
Heisenberg, W. (1958). *Physics and philosophy: The revolution in modern science*. London: George Allen & Unwin.
Howard, D. (1990). Nicht sein kann was nicht sein darf,' or the prehistory of EPR, 1909–1935: Einstein's early worries about the quantum mechanics of composite systems. In A. I. Miller (Ed.), *Sixty-two years of uncertainty: Historical, philosophical, and physical inquiries into the foundations of quantum mechanics* (pp. 61–111). New York/London: Plenum.
Hubert, M., & Romano, D. (2018). The wave-function as a multi-field. *European Journal for Philosophy of Science*, 1–17.
Huggett, N., & Wüthrich, C. (2013). Emergent spacetime and empirical (in) coherence. *Studies in Histories and Philosophy of Science, B44*, 276–285.
Huggett, N., & C. Wüthrich (Eds.) (forthcoming), *Out of nowhere: The emergence of spacetime in quantum theories of gravity*. Oxford: Oxford University Press.
Kitcher, P. (1989). Explanatory unification and the causal structure of the world. In P. Kitcher & W. Salmon (Eds.), *Minnesota studies in the philosophy of science* (Vol. 13, pp. 410–503). Minnesota: University of Minnesota Press.
Knox, E. (2013). Effective Spacetime geometry. *Studies in the History and Philosophy of Modern Physics, 44*, 346–356.
Knox, E. (2014). *Spacetime structuralism or Spacetime functionalism?* Manuscript.
Lam, V., & Esfeld, M. A. (2013). A dilemma for the emergence of Spacetime in canonical quantum gravity. *Studies in History and Philosophy of Modern Physics, 44*, 286–293.
Lam, V., & Wüthrich, C. (2017). *Spacetime is as Spacetime does*. Manuscript.

Leifer, M. S. (2014). Is the quantum state real? An extended review of ψ-ontology theorems. *Quanta, 3*, 67–155.

Lewis, P. J. (2004). Life in configuration space. *The British Journal for the Philosophy of Science, 55*, 713–729.

Lewis, P. J. (2005). Interpreting spontaneous collapse theories. *Studies in History and Philosophy of Modern Physics, 36*, 165–180.

Lewis, P. J. (2006). GRW: A case study in quantum ontology. *Philosophy Compass, 1*, 224–244.

Lewis, P. J. (2013). Dimension and illusion. In D. Albert & A. Ney (Eds.), *The wave-function: Essays in the metaphysics of quantum mechanics* (pp. 110–125). New York: Oxford University Press.

Loewer, B. (1996). Humean Supervenience. *Philosophical Topics, 24*, 101–127.

Lorentz, H. A. (1909). *The theory of electrons*. New York: Columbia University Press.

Maudlin, T. (1995). Three measurement problems. *Topoi, 14*, 7–15.

Maudlin, T. (2007). Completeness, supervenience and ontology. *Journal of Physics A, 40*, 3151–3171.

Miller, E. (2014). Quantum entanglement, bohmian mechanics, and humean supervenience. *Australasian Journal of Philosophy, 92*(3), 567–583.

Monton, B. (2013). Against 3N-dimensional space. In D. Z. Albert & A. Ney (Eds.), *The Wave-function: Essays in the metaphysics of quantum mechanics* (pp. 154–167). New York: Oxford University Press.

Ney, A. (2012). The status of our ordinary three-dimensions in a quantum universe. *Noûs, 46*, 525–560.

Ney, A. (2013). Ontological reduction and the wave-function ontology. In D. Albert & A. Ney (Eds.), *The wave-function: Essays in the metaphysics of quantum mechanics* (pp. 168–183). New York: Oxford University Press.

Ney, A. (2015). Fundamental physical ontologies and the constraint of empirical coherence. *Synthese, 192*(10), 3105–3124.

Ney, A. (2017). Finding the world in the wave-function: Some strategies for solving the macro-object problem. *Synthese*, 1–23.

Ney, A. (forthcoming). *Finding the world in the wave function*. Oxford: Oxford University Press.

Norsen, T. (2010). The theory of (exclusively) local Beables. *Foundations of Physics, 40*(12), 1858–1884.

Norsen, T. (2017). *Foundations of quantum mechanics: An exploration of the physical meaning of quantum theory*. Cham: Springer.

North, J. (2013). The structure of the quantum world. In D. Z. Albert & A. Ney (Eds.), *The wave-function: Essays in the metaphysics of quantum mechanics* (pp. 184–202). New York: Oxford University Press.

Peierls, R. (1991). In defence of 'measurement'. *Physics World*, January, pp. 19–20.

Pitowsky, I. (2007). Quantum mechanics as a theory of probability. In W. Demopoulos & I. Pitowsky (Eds.), *Physical theory and its interpretation: Essays in honor of Jeffrey Bub*. Berlin: Springer.

Przibram, K (1967). *Letters on wave mechanics* (trans. Martin Klein). New York: Philosophical Library.

Pusey, M., Barrett, J., & Rudolph, T. (2012). On the reality of the quantum state. *Nature Physics, 8*, 475–478.

Putnam, H. (1960). *Minds and machines*, reprinted in Putnam (1975b), pp. 362–385.

Putnam, H. (1975). *Mind, language, and reality*. Cambridge: Cambridge University Press.

Rovelli, C. (1996). Relational quantum mechanics. *International Journal of Theoretical Physics, 35*(8), 1637–1678.

Schrödinger, E. (1926). Quantisierung als Eigenwertproblem (Zweite Mitteilung). *Annalen der Physik, 79*, 489–527. English translation: Quantisation as a Problem of Proper Values. Part II.

Schrödinger, E. (1935a). *Schrödinger to Einstein*, 19 August 1935. Translated in Fine, A. (1996). *The Shaky game: Einstein realism and the quantum theory* (p. 82). Chicago: University of Chicago Press.

Schrödinger, E. (1935b). Die gegenwärtige Situation in der Quantenmechanik. *Die Naturwissenschaften*, *23*, 807–812, 823–828, 844–849.

Sellars, W. (1962). Philosophy and the scientific image of man. In R. Colodny (Ed.), *Frontiers of science and philosophy* (pp. 35–78). Pittsburgh, PA: University of Pittsburgh Press.

Spekkens, R. W. (2007). Evidence for the epistemic view of quantum states: A toy theory. *Physical Review A, 75*, 032110.

Suárez, M. (2007). Quantum Propensities. *Studies in the History and Philosophy of Modern Physics, 38*, 418–438.

Suarez, M. (2015). Bohmian dispositions. *Synthese, 192*(10), 3203–3228.

Timpson, C. (2010). Rabid dogma? Comments on Bub and Pitowsky. In S. Saunders, J. Barrett, A. Kent, & D. Wallace (Eds.), *Many worlds? Everett, quantum theory, and reality* (pp. 460–466). Oxford: Oxford University Press.

Van Camp, W. (2011). On kinematic versus dynamic approaches to special relativity. *Philosophy of Science, 78*(5), 1097–1107.

von Neuman, J. (1932). *Mathematische Grundlagen der Quantenmechanik.* Springer. Translated by R. T. Beyer as: Mathematical Foundations of Quantum Mechanics. Princeton: Princeton University, 1955.

Wallace, D. M. (2002). Everettian rationality: Defending Deutsch's Approach to probability in the Everett interpretation. *Studies in History and Philosophy of Science Part B: Studies in History and Philosophy of Modern Physics, 34*(3), 415–439.

Chapter 3
Unscrambling Subjective and Epistemic Probabilities

Guido Bacciagaluppi

Abstract There are two notions in the philosophy of probability that are often used interchangeably: that of subjective probabilities and that of epistemic probabilities. This paper suggests they should be kept apart. Specifically, it suggests that the distinction between subjective and objective probabilities refers to what probabilities *are*, while the distinction between epistemic and ontic probabilities refers to what probabilities are *about*. After arguing that there are *bona fide* examples of subjective ontic probabilities and of epistemic objective probabilities, I propose a systematic way of drawing these distinctions in order to take this into account. In doing so, I modify Lewis's notion of chances, and extend his Principal Principle in what I argue is a very natural way (which in fact makes chances fundamentally conditional). I conclude with some remarks on time symmetry, on the quantum state, and with some more general remarks about how this proposal fits into an overall Humean (but not quite neo-Humean) framework.

Keywords Subjective probabilities · Epistemic probabilities · Principal principle · Quantum state · Humeanism

I believe I first met Itamar Pitowsky in what I think was the spring of 1993, when he came to spend a substantial sabbatical period at Wolfson College, Cambridge, while writing his paper on George Boole's 'Conditions of Possible Experience' (Pitowsky 1994). At that time, the Cambridge group was by far one of the largest research groups in philosophy of physics worldwide, and Wolfson had the largest share of the group. Among others, that included two further Israelis, my friends and fellow PhD-students Meir Hemmo, who is co-editing this volume, and Jossi Berkovitz, who had worked under Itamar's supervision on de Finetti's probabilistic subjectivism and its

---

G. Bacciagaluppi (✉)
Descartes Centre for the History and Philosophy of the Sciences and the Humanities, Utrecht University, Utrecht, Netherlands
e-mail: g.bacciagaluppi@uu.nl

M. Hemmo, O. Shenker (eds.), *Quantum, Probability, Logic*,
Jerusalem Studies in Philosophy and History of Science,
https://doi.org/10.1007/978-3-030-34316-3_3

application to quantum mechanics (published in English as Berkovitz 2012). We were all lunch regulars at Wolfson, and discussions on the philosophy of physics, of probability, and of mathematics were thus not limited to the setting of the Friday seminars in the HPS department (or the immediately preceding lunches at the Eraina Taverna).

My earliest introduction to de Finetti's subjectivism was in fact through Itamar and Jossi, and Itamar's seminal work on quantum probability (Pitowsky 1989, 1994) looms large within my formative years. The present paper relates most closely to Itamar's more recent work on Bayesian quantum probabilities (Pitowsky 2003, 2007). I shall not be agreeing with Itamar on everything, but that fits into the wider spirit of friendship, co-operation, integrity, and pursuit of knowledge that makes philosophers of physics an ideal model of an academic community, and Itamar an ideal model of a philosoper of physics. I dedicate this paper to his memory.

3.1 Subjective or Epistemic Probabilities?

This paper is about the notion of subjective probabilities and that of epistemic probabilities, and how to keep them apart. Since terminology varies, I should briefly sketch which uses of the terms I intend. (A fuller explication will be given in Sects. 3.2 and 3.3.)

I take the use of the term 'subjective probability' to be fairly uniform in the literature, and to be fairly well captured by the idea of *degrees of belief* (and as such 'assigned by agents' and presumably 'living in our heads'), as opposed to some appropriate notion of objective probabilities ('out there'), sometimes cashed out in terms of frequencies or propensities. I shall take the standard explication of objective probabilities as that given by David Lewis with his notion of chance (Lewis 1980).

Along these lines, all probability assignments can be taken as subjective in the first place, but some such assignments may correspond to objective probabilities 'out there' in the world, determined by actual matters of fact. Probability assignments that do not correspond to objective probabilities shall be termed 'purely' or 'merely' subjective in the following. In this terminology, the subjective probabilities of a subjectivist like de Finetti, for whom objective probabilities do not exist, are the prime example of purely subjective probabilities, even though usually they will be firmly if pragmatically rooted in matters of actual fact (as further discussed in Sects. 3.4 and 3.10).

The term 'epistemic probabilities' has a wider variety of meanings, of which two are probably dominant, depending on whether one emphasises the fact that we lack knowledge of something or whether one emphasises the knowledge that we do have. The one I shall focus on is the former, that of *ignorance-interpretable* probabilities (i.e. of probabilities attaching to propositions expressing a matter of fact, but whose truth is unknown), which is perhaps most common in the philosophy of physics. The other refers to probability assignments whose *values* reflect our state of knowledge. The classical notion of probability based on the principle of indifference or an objective Bayesian view based on the principle of maximum entropy (to which we shall presently return) can thus also be termed epistemic.

This second notion of 'epistemic probabilities', which is perhaps most common in the philosophy of probability, belongs somewhere along the axis between subjective and objective probabilities. It is often contrasted with subjective probabilities when these are understood as merely 'prudential' or 'pragmatic' or 'instrumental'. This distinction is challenged by Berkovitz (2019), who convincingly argues that de Finetti's instrumental probabilities are also epistemic in this sense. Indeed, when talking about subjective probabilities in Sects. 3.4, 3.6 and 3.10, I shall always implicitly take them to be so. This second sense of 'epistemic probabilities' is not the topic of this paper, and will be discussed no further.

Epistemic probabilities in the sense of ignorance-interpretable ones instead are liable to be assimilated to subjective probabilities, but, as I shall argue, they ought not to, because the distinction between probabilities that are epistemic in this sense and those that are not (we shall call them 'ontic') is conceptually orthogonal to the subjective-objective distinction.

Take for example Popper's discussion in Vol. III of the *Postscript* (Popper 1982).[1] After treating extensively of his propensity interpretation in Vol. II, Popper starts Vol. III (on quantum mechanics) with a discussion of subjective vs objective probabilities, complaining that physicists conflate the two (especially in quantum mechanics). Popper's contrast for the objective view of probabilities is what he throughout calls the subjective view, which for him is the view that probabilistic statements are about our (partial) *ignorance*. In the case of classical statistical mechanics, he complains that '[t]his interpretation leads to the absurd result that the molecules escape from our bottle because we do not know all about them' (Popper 1982, p. 109).

True, we care about whether our coffee goes cold, and the coffee does not care about our feelings. And, at least in the sense in which Popper took his own notion of propensities to be explanatory of observed frequencies (cf. Bub and Pitowsky 1985, p. 541), it seems reasonable to say that while the behaviour of the coffee is a major factor in explaining why we come to have certain subjective expectations, it is only objective probabilities that can explain the behaviour of the coffee. But there is also a sense in which the probabilities of (classical) statistical mechanics are *clearly* epistemic: we do believe the coffee or the gas in the bottle to be in a definite microstate, but we do not know it. Also the statistical mechanical description of the gas presupposes there to be a particular matter of fact about the state of the molecules, which is however left out of the description. If we read Popper's complaint as being that epistemic probabilities *must* be subjective, or rather purely

[1]Incidentally, reviewed by Itamar when it came out (Bub and Pitowsky 1985). Note that my examples of possible assimilation or confusion between epistemic and subjective probabilities (in the senses just sketched) may well prove to be straw-mannish under further scrutiny, but I just wish to make a *prima facie* case that these two notions can indeed be confused. The rest of the paper will make the extended case for why we should distinguish them and how.

subjective (i.e. unrelated to objective probabilities), then it seems that *no* explanation of the kind Popper envisages is possible at all in statistical mechanics.

Or take Jaynes, e.g. in 'Some Random Observations' (1985). For Jaynes all probabilities are fundamentally subjective in the sense that 'a *probability* is something that we *assign* in order to represent a state of knowledge', as opposed to an actual frequency, which is a 'factual property of the real world' (p. 120). While for Jaynes the probabilities of classical statistical mechanics are thus of course epistemic, unlike for Popper they are meant to be (at least partially?) *objective*, in the sense that there are (rational?) principles, first and foremost the principle of maximum entropy, that provide a way for choosing our probability assignments based on those facts we do know.[2]

Part of Jaynes's motivation came from the complaint that the mathematics of quantum theory 'describes in part physical law, in part the process of human inference, all scrambled together in such a way that nobody has seen how to separate them' (p. 123). He continues:

> Many years ago I became convinced that this unscrambling would require that probability theory itself be reformulated so that it recognizes explicitly the role of human information and thus restores the distinction between reality and our knowledge of reality, that has been lost in present quantum theory.[3]

Jaynes' position, however, faces a different problem precisely in the case of quantum mechanics. Whether or not probabilities are purely subjective, for Jaynes they are always epistemic: '[p]robabilities in present quantum theory express the incompleteness of human knowledge just as truly as did those in classical statistical mechanics' (1985, p. 122). And this ominously leads him to question the possibility of fundamental indeterminism.

---

[2]It is unclear to what extent this 'objective Bayesian' strategy indeed yields objective probabilities. Consider for instance situations in which information about the physical situation is in principle readily available but is glibly ignored. In this case, even if we use principles such as maximum entropy, one might complain that there is little or no connection between the physical situation and the probabilities we assign. On the other hand, there may be situations in which the physical situation itself appears to systematically lead to situations of ignorance, e.g. because of something like mixing properties of the dynamics. In this case, there may arguably be a systematic link between certain objective physical situations and certain ways of assigning probabilities, which may go some way towards justifying the claim that these probabilities are objective. (Cf. discussions about whether high-level probabilities are objective, e.g. Glynn (2009), Lyon (2011), Emery (2013) and references therein.) If so, note that this will be the case whether or not there is *anyone* to assign those probabilities. Many thanks to Jossi Berkovitz for discussion of related points.

[3]For an example of what Jaynes has in mind, see Heisenberg's 'The Copenhagen interpretation of quantum theory' (Heisenberg 1958, Chap. 3). In Sect. 3.9 below, I shall allude to what I think are the origins of these aspects of Heisenberg's views, but I agree with Jaynes that (at least in the form they are usually presented) they are hard to make sense of. Note also that the 'Copenhagen interpretation' as publicised there appears to have been largely a (very successful) public relations exercise by Heisenberg (cf. Howard 2004). In reality, physicists like Bohr, Heisenberg, Born and Pauli held related but distinct views about how quantum mechanics should be understood.

Indeed, as Jaynes sees it, in the case of quantum mechanics human knowledge is limited by dogmatic adherence to indeterminism in the Copenhagen interpretation. But in fact, even though the 'Copenhagen interpretation' may have been dogmatic, indeterminism was the least worrying of its dogmas: a wide spectrum of non-dogmatic approaches to the foundations of quantum mechanics today embrace indeterminism in some form or other. Thus, if (as Jaynes seems to think) subjective probabilities *must* be epistemic, then they seem to be inappicable to many if not most fundamental approaches to quantum mechanics.[4]

These problems with both Popper's and Jaynes' positions seem to me to stem from a scrambling together of subjective and epistemic probabilities. I shall take it as uncontroversial that we often form degrees of belief about facts we do not know (so that such probabilities are both subjective and epistemic), but this in itself does not show that all epistemic probabilities should be subjective, or that all subjective probabilities should be epistemic. Indeed, being a degree of belief is conceptually distinct from being a probability about something we ignore: the characterisation of a probability as subjective is in terms of what that probability *is*, that of a probability as epistemic is in terms of what that probability is *about*.

Admittedly, even if we can conceptually distinguish between subjective and epistemic probabilities, it could be that the two categories are coextensive. And if all epistemic probabilities are merely subjective, or all subjective probabilities are also epistemic, then Popper's or Jaynes's worries become legitimate again. As a matter of fact, typical cases of epistemic probabilities are arguably subjective ('Is Beethoven's birthday the 16th of December?'[5]), and, say, probabilities in plausibly indeterministic scenarios are typically thought to be objective ('Will this atom decay within the next minute?'). And, indeed, while there is an entrenched terminology for probabilities that are not purely subjective (namely 'objective probabilities'), non-epistemic probabilities do not even seem to have a separate standard name (we shall call them 'ontic probabilities'). It is indeed tempting to think that all epistemic probabilties are subjective and all ontic probabilities are objective.

I urge the reader to resist this temptation. I believe that we can make good use of the logical space that lies between these distinctions. As the Popper and Jaynes quotations above are meant to suggest, objectivism about probability will be

[4]Spontaneous collapse approaches are invariably indeterministic (unless one counts among them also non-linear modifications of the Schrödinger equation, which however have not proved successful), as is Nelson's stochastic mechanics. Everett's theory vindicates indeterminism at the emergent level of worlds. And even many pilot-wave theories are stochastic, namely those modelled on Bell's (1984) theory based on fermion number density. Thus, among the main fundamental approaches to quantum mechanics, only theories modelled on de Broglie and Bohm's original pilot-wave theory provide fully deterministic underpinnings for it. (For a handy reference to philosophical issues in quantum mechanics, see Myrvold (2018b) and references therein. For more on Nelson see Footnote 40.)

[5]16 December is the traditional date for Beethoven's birthday, but we know from documentary evidence only that he was baptised on 17 December 1770 – Beethoven was *probably* born on 16 December.

self-limiting if it takes itself to apply only to ontic probabilities, as will subjectivism about probability if it takes itself to apply only to epistemic probabilities. The landscape of the philosophy of probability will be both richer and more easily surveyable if we unscramble subjective and epistemic probabilities in a way that allows also for (purely) subjective ontic probabilities as well as for epistemic objective probabilities. While this is surely not the first paper to make some of these claims (indeed, Jaynes is obviously proposing that epistemic probabilities can be objective[6]), I believe the way it approaches these questions will be original.

In what follows, I shall first recall the standard discussion of subjective and objective probabilities given by David Lewis (Sect. 3.2), then try and spell out what one typically means by the distinction between epistemic and ontic probabilities (Sect. 3.3). This will provide enough background to argue that there are *bona fide* cases of probabilities that are both purely subjective and ontic (Sect. 3.4), and of probabilities that are both epistemic and objective (Sect. 3.5). This will lead to an extension of Lewis's definition of objective probabilities and of his Principal Principle in which chances become fundamentally conditional, thus redrawing the subjective-objective distinction (Sect. 3.6), and to a more systematic discussion also of the epistemic-ontic distinction, clarifying some further ambiguity (Sect. 3.7). I conclude the paper with some additional remarks on time symmetry and on the quantum state (Sects. 3.8 and 3.9),[7] and with more general remarks on how the proposed framework for conceptualising probabilities fits into an overall Humean framework – at the same time distancing myself from some typical neo-Humean views (Sect. 3.10).

3.2 Subjective vs Objective Probabilities

Recall the standard distinction between *subjective* and *objective* probabilities: subjective probabilities are coherent degrees of belief, generally taken to live in our head, while objective probabilities are meant to live out there in the world, properties of external objects or reduced to such properties. I shall follow (half of) the literature in taking subjective probabilities as well-understood, and objective probabilities (if they exist) as being the more mysterious concept. While the literature on subjective probabilities is considerable and varied, I shall follow what is nowadays often considered the standard analysis of how subjective and objective probabilities relate – and which in fact provides an implicit definition of objective probabilities given an

[6]With the qualifications mentioned in Footnote 2. Thanks to both Jos Uffink and Ronnie Hermens for making me include more about Jaynes.

[7]These (especially the latter) can be skipped by readers who are not particularly familiar with debates on the metaphysics of time or, respectively, on the nature of the quantum state and of collapse in quantum mechanics.

understanding of subjective probabilities – David Lewis's 'A Subjectivist's Guide to Objective Chance' (Lewis 1980).[8]

Importantly for our later discussion, Lewis takes objective probabilities (also called 'chances') to be functions of propositions and times. Assuming for the sake of example that there are chances for rolls of dice, before I roll a pair of dice we have the intuition that the chance of rolling 11 is 1/18. Once I have rolled the first die, the chance of 11 either goes down to 0 and stays there (if I have rolled one of the numbers 1 to 4), or it goes up to 1/6 (if I have rolled 5 or 6). Then when I roll the second die the chance of 11 goes definitively either down to 0 or up to 1.

The idea behind Lewis's analysis is that objective probabilities, if they exist, are the kind of thing that – in a sense to be made precise – would rationally constrain our subjective probabilities (also called 'credences'[9]) *completely*. Credences of course are standardly taken to be rationally constrained both synchronically by the probability calculus and diachronically by Bayesian conditionalisation.[10] Few other requirements are widely recognised as able to constrain their values further (maybe only the requirement that credences should be non-dogmatic, i. e. not take the values 0 or 1, is one commanding near-universal support – but see the remarks below in Sect. 3.4). Thus, the criterion that Lewis proposes for objective probabilities is very stringent.

What Lewis has in mind is the following. The chance at t of A is x iff there is a proposition $\mathrm{ch}_t(A) = x$ such that (subject to a few provisos to be spelled out) conditionalising on $\mathrm{ch}_t(A) = x$ will always yield the *unique* value x for our credence,

$$\mathrm{cr}_t\left(A \,\middle|\, \mathrm{ch}_t(A) = x\right) = x, \tag{3.1}$$

whatever the form of our credence function at the time t. Of course, if we further believe at t that the chance at t of A is in fact x, i. e. if $\mathrm{cr}_t(\mathrm{ch}_t(A) = x) = 1$, then our credence at t in A is also x (by the probability calculus), or if we learn at t that $\mathrm{ch}_t(A) = x$, our credence in A becomes x (by Bayesian conditionalisation).

The stringency of Lewis's requirements on what is to count as rationally compelling is such that there are few candidates in the literature for satisfying them, none of which are uncontroversial. And because Lewis himself believes in 'Humean

[8]While in a Cambridge HPS mood, I cannot refrain from thanking Jeremy Butterfield for (among so many other things!) introducing me to this wonderful paper and all issues around objective probabilities.

[9]Again, terminology varies somewhat, but I shall take 'credences' to be synonymous with 'subjective probabilities'.

[10]Bayesian conditionalisation is often taken as the unique rational way in which one can change one's credences. But there may be other considerations competing with it, in particular affecting background theoretical knowledge. This will generally be implicit (but not needed) in much of this paper. Thanks to Jossi Berkovitz for pointing out to me that de Finetti himself was not committed to conditionalisation as the only way for updating probabilities, believing instead that agents may have reasons for otherwise changing their minds (cf. Berkovitz 2012, p. 16).

supervenience', for him chances have to supervene on 'particular matters of fact', making it especially hard to find plausible explicit candidates for objective chances. This problem is quaintly known in the literature as the 'big bad bug'.[11]

The provisos are, first, that our credence function at t should not rule out that $\mathrm{ch}_t(A) = x$, i. e. we should have $\mathrm{cr}_t(\mathrm{ch}_t(A) = x) \neq 0$, otherwise the credence in (3.1) will be ill-defined; second, and more subtly, that our credence function at t should not incorporate information that is 'inadmissible' at t. To make this intuitive, think of the chance at t of A as fixing the objectively fair odds for betting on A. Clearly, if in advance of making the bet either you or the bookie had some more direct information about whether or not A will come to pass (e. g. one of you is clairvoyant), that would be cheating. In order for the notion of 'objectively fair odds' at time t to make sense, we need to have an appropriate notion of what is admissible in shaping our credences at t. Not everything can be admissible, for then we would allow A, or not-A, and the only propositions about chance that could satisfy (3.1) would correspondingly be propositions implying $\mathrm{ch}_t(A) = 1$, or propositions implying $\mathrm{ch}_t(A) = 0$, and all chances would be trivial. On the other hand, some propositions must be admissible, since $\mathrm{ch}_t(A) = x$ clearly is. We can write our credence function $\mathrm{cr}_t(A)$ as $\mathrm{cr}(A|E_t \,\&\, T)$, where $\mathrm{cr}(A)$ is a 'reasonable initial credence function', T represents any background assumptions that we make (which might perhaps better go in a subscript), and E_t is our epistemic context at time t (by which I mean all propositions about matters of fact known at time t). If we leave fixed the background assumptions T, then $\mathrm{cr}_t(A)$ just evolves by Bayesian conditionalisation upon the new facts that we learn between any two times t and t'.[12]

[11]Two proposals of note are the one by Deutsch and by Wallace in the context of Everettian quantum mechanics, who derive the quantum mechanical Born rule from what they argue are rationality constraints on decision making in the context of Everett (see e.g. Wallace 2010, or Brown and Ben Porath (2020)), and the notion of 'Humean objective chances' along the lines of the neo-Humean 'best systems' approach (Loewer 2004; Frigg and Hoefer 2010). Note that I am explicitly distinguishing between Lewisian chances and Humean objective chances. The latter are an attempt to find something that will satisfy the definition of Lewisian chances, but there is no consensus as to whether it is a successful attempt. (See Sect. 3.10 below for some further comments on neo-Humeanism.) The Deutsch-Wallace approach is often claimed to provide the only known example of Lewisian chances, but again this claim is controversial, as it appears that some of the constraints that are used cannot be thought of as rationality assumptions but merely as naturalness assumptions. On this issue see Brown and Ben Porath (2020) as well as the further discussion in Saunders et al. (2010). The difficulty in finding anything that satisfies the definition of Lewisian chances is (of course) an argument in favour of subjectivism.

[12]I shall assume throughout that *degrees of belief* attach to propositions about matters of fact (in the actual world), which incidentally need not be restricted to Lewis's 'particular' matters of fact (e.g. I shall count as matters of fact also holistic facts about quantum states – if such facts there be). I shall call these *material* or *empirical* propositions.

Of course, some of the background assumptions that we make in choosing our degrees of belief may not just refer to matters of fact, but will be *theoretical* propositions (say, referring to possible worlds). In that case, I would prefer not to talk about degrees of belief attaching to such theoretical propositions, but about *degrees of acceptance*.

Substituting into (3.1), we have

$$\mathrm{cr}\Big(A \,\Big|\, (\mathrm{ch}_t(A) = x) \;\&\; E_t \;\&\; T\Big) = x. \tag{3.2}$$

And the requirement that (3.1) should hold whatever the form of our credence function at t (subject to the provisos) now translates into the requirement that (3.2) should hold for any 'reasonable initial credence function' (which we shall keep fixed) and for any admissible propositions E_t and T, which for Lewis are, indeed, ones that are compatible with the chance of A being x and in addition do not provide information about the occurrence of A over and above what is provided by knowing the chance of A (if chance is to guide credence, we must not allow information that trumps the chances). This is Lewis's 'Principal Principle' (PP) relating credence and chance. It can be taken, is often taken, and I shall take it as providing the implicit definition of objective probabilities: iff there is a proposition X such that, for all choices of reasonable initial credence function and admissible propositions,

$$\mathrm{cr}\Big(A \,\Big|\, X \;\&\; E_t \;\&\; T\Big) = x, \tag{3.3}$$

The matter is further complicated by the fact that even full acceptance of a theory need not imply belief in all propositions about matters of fact following from the theory. For instance, an empiricist may believe only a subset of the propositions about actual matters of fact following from a theory they accept (say, only propositions about observable events), thus drawing the empirical-theoretical distinction differently and making a distinction between material propositions and empirical propositions, which for the purposes of this paper I shall ignore. (For my own very real empiricist leanings, see Bacciagaluppi 2019.)

For simplicity I shall also ignore the distinction between material propositions and actual propositions that one makes in a many-worlds theory, where the material world consists of several 'possible' branches, even though I shall occasionally refer to Everett for illustration of some point.

One might reflect the distinction between material (or empirical) and theoretical propositions by not conditionalising on the latter, but letting them parameterise our credence functions. Another way out of these complications may be to assign subjective probabilities to all propositions alike, but interpret the probabilities differently depending on what propositions they attach to: in general they will be degrees of acceptance, but in some cases (all material propositions, or a more restricted class of empirical propositions), one may think of them as straightforward degrees of belief. Such an additional layer of interpretation of subjective probabilities will, however, not affect the role they play in guiding our actions (degrees of acceptance will generally be degrees of 'as-if-belief'). In any case, the stroke notation makes clear that assuming a theory – even hypothetically – is meant to affect our credences as if we were conditionalising on a proposition stating the theory (I believe this is needed to derive some of Lewis's results).

Note that for Lewis, T will typically include theoretical propositions about chances themselves in the form of history-to-chance conditionals, and because in Lewis's doctrine of Humean supervenience chances in fact supervene on matters of fact in the actual world, such background assumptions may well be propositions that degrees of belief can attach to (at least depending on how one interprets the conditionals), even if one wishes to restrict the latter (as I have just sketched) to propositions about matters of fact in the actual world.

We shall return to the material-theoretical distinction in Sect. 3.7.

then there is such a thing as the objective chance at t of A and its value is x. The weakest such proposition X is then $\mathrm{ch}_t(A) = x$.[13]

More precisely, Lewis restricts admissible propositions about matters of fact to historical propositions up to time t (at least 'as a rule': no clairvoyance, time travel and such, and we shall follow suit for now – although he appears to leave it open that one may rethink the analysis if one should wish to consider such less-than-standard situations (Lewis 1980, p. 274)). Admissible propositions about matters of fact can thus be taken to be what can be reasonably thought to be knowable in principle at t, and Lewis allows them to include the complete history H_t of the world up to that time. He also allows hypothetical propositions about the theory of chance, i. e. about how chances may depend on historical facts.

Even more propositions might be admissible at time t, but this partial characterisation of admissibility already makes the PP a useful and precise enough characterisation of objective chances. Indeed, taking the conjunction $H_t \,\&\, T_C$ of the complete history of the world up to t and a complete and correct theory T_C of how chances depend on history, the following will be an instance of the PP:

$$\mathrm{cr}\Big(A \,\Big|\, (\mathrm{ch}_t(A) = x) \,\&\, H_t \,\&\, T_C\Big) = x. \tag{3.4}$$

Since T_C is a complete and correct theory of chance, $H_t \,\&\, T_C$ already implies that $\mathrm{ch}_t(A) = x$, and we can rewrite (3.4) as

$$\mathrm{ch}_t(A) = \mathrm{cr}(A | H_t \,\&\, T_C). \tag{3.5}$$

This instance of the PP is used by Lewis to derive various properties of chances using known properties of credences, e. g. that they satisfy the probability calculus, and that they evolve by conditionalisation on intervening events. In particular, according to Lewis, chances of past events are all 0 or 1. Taking $\mathrm{ch}(A)$ as the chance function at some arbitrary time $t = 0$, we can thus for all later times write $\mathrm{ch}_t(A) = \mathrm{ch}(A|H_t)$.

3.3 Epistemic vs Ontic Probabilties

A systematic treatment of the epistemic-ontic distinction comparable to Lewis's treatement of subjective and objective probabilities is not available. But we can get a reasonably good idea of such a distinction, which will be enough for the time being, by looking at common ways of using the term 'epistemic probabilities'. (We shall revisit this distinction in Sect. 3.7.)

Epistemic probabilities are probabilities that attach to propositions of which we are ignorant. An obvious example is: I flip a coin and do not look at the result

[13]Note that X is a proposition about (actual) matters of fact. Thus, chances in the sense of Lewis indeed satisfy the intuition that objective probabilities are properties of objects or more generally determined by actual facts. Thanks to Jossi Berkovitz for discussion.

yet – what is the probability of Heads? In this situation, there is a matter of fact about whether the coin has landed Heads or Tails, and if I knew it, I would assign to Heads probability 0 or 1 (depending on what that matter of fact was).

Another common example is modelling coin flipping as a deterministic Newtonian process: even before the coin has landed, the details of the flip (initial height, velocity in the vertical direction, and angular momentum[14]) determine what the outcome will be, but we generally do not know these details. In this case, we say that the probabilities of Heads and Tails are epistemic, even if the event of the coin landing Heads or Tails is in the future, because we can understand such events as labels for sets of unknown initial conditions.

If we are looking for examples of ontic probabilities, we thus presumably need to look at cases of indeterminism. The most plausible example is that of quantum probabilities, say, the probability for a given atom to decay within the next minute. (Of course, whether this is a genuine case of indeterminism will depend on one's views about quantum mechanics, in particular about the existence of 'hidden variables'. For the sake of the example we shall take quantum mechanics to be genuinely indeterministic.)

Still, if all actual matters of fact past, present and future are taken to be knowable in principle, then all probabilities become purely epistemic, and (at least for now) it appears we cannot draw an interesting distinction.

One way of ensuring that the future is unknowable, of course, is to postulate that the future is *open*, i. e. that at the present moment there is no (single) future yet. This is a very thick metaphysical reading of indeterminism, however, which I wish to avoid for the purpose of drawing the epistemic-ontic distinction.[15] Indeterminism can be defined in a non-metaphysical way in terms of the allowable models of a theory (or nomologically possible worlds, if worlds are also understood in a metaphysically thin way): a theory is (future-)deterministic iff any two models agreeing up to (or at) time t agree also for all future times (and with suitable modifications for relativistic theories).

For the purpose of this first pass at drawing the epistemic-ontic distinction we can then simply stipulate that the notion of ignorance only pertains to facts in the actual world that are knowable in principle, and that at time t only facts up to t are knowable in principle (which is in fact the stipulation we already made in the context of Lewis's analysis of chances – we may want again to make a mental note to adapt the notion of 'knowable in principle' if we allow for clairvoyance, time

[14]For the classic Newtonian analysis of coin flipping see Diaconis et al. 2007.

[15]Or for any other purpose – as a matter of fact, I shall assume throughout a block universe picture. Specifically, in the case of indeterminism one should think of *each possible world* as a block universe. The actual world may be more real than the merely possible ones, or all possible worlds may be equally real (whether they are or not is inessential to the following, although a number of remarks below may make it clear that I am no modal realist, nor in fact believe in objective modality). But *within* each world, all events will be equally real. (Note that at a fundamental level the Everett theory is also deterministic and Everettians emphasise that the Everettian multiverse is indeed a block universe.)

travel and such like). If one so wishes, one can of course talk about 'ignorance' of which possible world is the actual world, or of which way events in our world will be actualised. But if probabilities in cases of indeterminism are to count as 'ontic', then it simply is the case that in practice we make a more restrictive use of the term 'ignorance' when we draw the epistemic-ontic distinction.

The general intuition is now that epistemic probabilities lie in the *gap* between probability assignments given what is knowable in principle (call it the *ontic context*), and probability assignments given what is actually known (the *epistemic context*). By the same token, ontic probabilities are any probabilities that remain non-trivial if we assume that we know everything that is knowable in principle.

In general, probabilities will be a *mix* of epistemic and ontic probabilities. For instance, we ask for the probability that any atom in a given radioactive probe may decay in the next minute, but the composition of the probe is unknown: the probability of decay will be given by the ontic probabilities corresponding to each possible radioactive component, *weighted* by the epistemic probabilities corresponding to each component.

The EPR setup in quantum mechanics (within a standard collapse picture) provides another good example: before Bob performs a spin measurement, assuming the global state is the singlet state, Alice has ontic probability 0.5 of getting spin up in any direction. After Bob performs his measurement, Alice has ontic probabilities $\cos^2(\alpha/2)$ and $\sin^2(\alpha/2)$ of getting up or down, where α is the angle between the directions measured on the two sides. However, she is ignorant of Bob's outcome, and her epistemic probabilities for his having obtained up or down are both 0.5. Thus, given her epistemic context, her (now mixed) probabilities for her own outcomes are unchanged.[16]

As a final example, assume we have some ontic probability that is also an objective chance in the sense of Lewis. According to the PP in the form (3.5), this ontic probability will be given by

$$\mathrm{ch}_t(A) = \mathrm{cr}(A|H_t \,\&\, T_C). \tag{3.6}$$

Now compare it with $\mathrm{cr}(A|E_t \,\&\, T_C)$, our credence in A at t given the same theoretical assumptions but our actual epistemic context E_t. This now is a mixed (partially epistemic) probability. Indeed, let H_t^i be the various epistemic possibilities (compatible with our epistemic context E_t) for the actual history up to t (i. e. for the actual ontic context). We then have

$$\mathrm{cr}(A|E_t \,\&\, T_C) = \sum_i \mathrm{cr}(A|H_t^i \,\&\, E_t \,\&\, T_C)\mathrm{cr}(H_t^i|E_t \,\&\, T_C), \tag{3.7}$$

[16]This is of course the quantum mechanical no-signalling theorem, as seen in the standard collapse picture. (It is instructive to think through what the theorem means in pilot-wave theory or in Everett. Thanks to Harvey Brown for enjoyable discussion of this point.) And of course, mixed quantum states are the classic example of the distinction in quantum mechanics between probabilities that are ignorance-interpretable ('proper mixtures') and probabilities that are not ('improper mixtures').

which simplifies to

$$\mathrm{cr}(A|E_t \,\&\, T_C) = \sum_i \mathrm{cr}(A|H_t^i \,\&\, T_C)\mathrm{cr}(H_t^i|E_t \,\&\, T_C), \tag{3.8}$$

because the H_t^i are fuller specifications of the past history at t than the one provided by E_t. In (3.8), now, the credences $\mathrm{cr}(H_t^i|E_t \,\&\, T_C)$ are purely epistemic probabilities, and the $\mathrm{cr}(A|H_t^i \,\&\, T_C)$ are purely ontic probabilities corresponding to each of the epistemically possible ontic contexts. The various H_t^i could e. g. fix the composition of the radioactive probe in our informal example above.[17]

Judging by the examples so far, the subjective-objective distinction and the epistemic-ontic distinction could well be co-extensive. Indeed, both subjective and epistemic probabilities seem to have to do with us, while both objective and ontic probabilities seem to have to do with the world. Epistemic probabilities depend on epistemic contexts, as do subjective ones; and chances depend on past history, which we have also taken as the context in which ontic probabilities are defined. But appearances may deceive.

In order to argue that the subjective-objective distinction and the epistemic-ontic distinction are to be drawn truly independently, I need to convince you that there are *bona fide* examples of probabilities that are both subjective (in fact purely subjective) and ontic, and of probabilities that are both epistemic and objective. These will be the respective objectives of the next two sections.

3.4 Subjective Ontic Probabilities

Let us start with the case of subjective ontic probabilities. I have discussed this case before, in a paper (Bacciagaluppi 2014) in which I comment on Chris Fuchs and co-workers' radical subjectivist approach to quantum mechanics, known as qBism (see e.g. Fuchs 2002, 2014),[18] and in which I spell out how I believe radical subjectivism *à la* de Finetti (1970) can be applied to quantum mechanics not only along qBist lines but also within any of the traditional foundational approaches (Bohm, GRW and Everett). The idea that subjectivism is applicable to quantum mechanics had

[17]In this example, it is of course possible to think of mixed epistemic and ontic probabilities as arising not only through ignorance of matters of fact in the ontic context H_t, but also through ignorance of the correct theory of chance T_C. We might indeed know the exact composition of the radioactive probe, but take ourselves to be ignorant of the correct decay laws for the various isotopes, with different credences about which laws are correct. I shall neglect this possibility in the following, because as noted already in Footnote 12, I prefer not to think of ignorance when talking about theoretical assumptions. In any case, I shall not need this further possible case for the arguments below.

[18]I am deeply grateful to Chris Fuchs for many lovely discussions about qBism and subjectivism over a number of years.

been already worked out by Jossi Berkovitz in the years 1989–1991 (prompted by Itamar), with specific reference to de Finetti's idea that probabilities are only definable for verifiable events.

For the purposes of the present paper, I need to make the case that *purely* subjective probabilities can be ontic, thus radical subjectivism about quantum probabilities (in a context in which quantum mechanics is taken to be irreducibly indeterministic) provides the ideal example, and in the following I shall rely heavily on my own previous discussion. (Alternatively, I could arguably have relied on Berkovitz 2012.) Note that what I need to convince you of is not that such a position is correct, but only that it is coherent, i. e. that there is an interesting sense in which one can combine the notion of purely subjective probabilities with that of ontic probabilities.

Itamar himself (Pitowsky 2003, 2007), Bub (2007, 2010) and Fuchs and co-workers have all defended approaches to quantum mechanics in which probabilities are subjective in the sense of being degrees of belief, and some modifications or additions to Bayesian rationality are introduced in order to deal with situations, respectively, in which one deals with 'quantum bets' (incompatible bets possibly with some outcomes in common), in which one assumes no-cloning as a fundamental principle, or in which gathering evidence means irreducibly intervening into the world. I take it that all of these authors (as well as de Finetti himself) agree on the assumption that in general there are no present or past facts that will determine the outcomes of quantum bets or any quantum measurements, and thus that the probabilities in their approaches are ontic in the sense used in the previous section.

It is less clear that the probabilities involved are purely subjective. Indeed, both Pitowsky and Bub see subjective probability assignments as severely constrained by the structure of quantum bets or quantum observables, in particular through Gleason's theorem and (relatedly) through the mathematical structure of collapse. These might be seen as rationality constraints, so that the probabilities in these approaches might in fact be seen as at least partly objective.[19] As Itamar puts it (Pitowsky 2003, p. 408):

---

[19]Whether or not they are will make little difference to my argument (although if they should be purely subjective I would have further concrete examples of subjective ontic probabilities in the sense I need). In Bub's case, he explicitly states that the no-cloning principle is an assumption and could turn out to be false. That means that the constraints hold only given some specific theoretical assumptions, and are not rationally compelling in the strong sense used by Lewis. In Pitowsky's case, the constraints seem stronger, in the sense that the 'quantum bets' are not *per se* theoretical assumptions about the world, but are just betting situations whose logical structure is non-classical. Whether or not there are rational constraints on our probability assignments will be a well-posed question only *after* you and the bookie have agreed on the logical structure of the bet. On the other hand, whenever we want to evaluate credences in a particular situation, we need to make a theoretical judgement as to what the logic of that situation is. That means, I believe, that whether we can have rationally compelling reasons to set our credences in any particular situation will still depend on theoretical assumptions, and fall short of Lewis's very stringent criterion. See Bacciagaluppi (2016) for further discussion of identifying events in the context of quantum probability, and Sect. 1.2 of Pitowsky (2003) for Itamar's own view of this issue. Cf. also Brown and Ben Porath (2020).

> For a given single physical system, Gleason's theorem dictates that all agents share a common prior probability distribution or, in the worst case, they start using the same probability distribution after a single (maximal) measurement.

Indeed, since pure quantum states always assign probability 1 to results of some particular measurement, the constraints may be so strong that even some dogmatic credences are forced upon us.

Fuchs instead clearly wishes to think of his probabilities as radically subjective (which I want to exploit). Precisely because of the strong constraints imposed by collapse, he feels compelled to not only take probability assignments as subjective, but also quantum states themselves and even Hamiltonians (Fuchs 2002, Sect. 7). To a certain extent, however, I think this is throwing out the baby without getting rid of the bath water. Indeed, I believe the (empirically motivated but normative) additions to Bayesian rationality to which Fuchs is committed imply already by themselves very strong constraints on qBist probability assignments (Bacciagaluppi 2014, Sect. 2.2). On the other hand, as I shall presently explain, I also believe that radical subjectivism can be maintained even within a context in which quantum states and Hamiltonians are taken as (theoretical) elements in our ontology. While in some ways less radical than Fuchs's own qBism, this is the position that I propose in my paper, and which I take to prove by example that a coherent case can be made for (purely) subjective ontic probabilities. It can be summarised as follows.

Subjectivism about probabilities (de Finetti 1970) holds that there are *no* objective probabilities, only subjective ones. Subjective probability assignments are strategies we use for navigating the world. Some of these strategies may turn out to be more useful than others, but that makes them no less subjective. As an example, take again (classical) coin flipping: we use subjective probabilities to guide our betting behaviour, and these probabilities get updated as we go along (i.e. based on past performance). But it is only a contingent matter that we should be successful at all in predicting frequencies of outcomes. Indeed, *any* given sequence of outcomes (with any given frequency of Heads and Tails) will result from appropriate initial conditions.

Under certain assumptions the judgements of different agents will converge with subjective probability 1 when conditionalised on sufficient common evidence, thus establishing an intersubjective connection between degrees of belief and frequencies. This is de Finetti's theorem (see e.g. Fuchs 2002, Sect. 9). One might be tempted to take it as a sign that subjective probabilities are tracking something 'out there'.[20] This, however, does not follow – because the assumption of 'exchangeability' under which the theorem can be proved is itself a subjective feature of our probability assignments. The only objective aspect relevant to the theorem is the common evidence, which is again just *past* performance.

For the subjectivist, there is no sense in which our probability judgements are right or wrong. Or as de Finetti (1970, Preface, quoted in Fuchs 2014) famously

[20] Such a reading is e.g. perhaps present in Greaves and Myrvold (2010), or even in Pitowsky (2003).

expressed it: 'PROBABILITIES DO NOT EXIST'. While this sounds radical, it is indeed perfectly unremarkable as long as we assume that states of affairs are always determinate and evolve deterministically. In that case, Lewis himself will say that objective chances are all 0 or 1, so that all non-trivial probabilities are indeed subjective and (non-trivial) chances do not exist.

What we need to consider is the case of indeterminism. Surely the situation is completely different in quantum mechanics? As Lewis puts it in the introduction to the 'Subjectivist's Guide', 'the practice and analysis of science require both [subjective and objective probabilities]'. He accordingly demands that also subjectivists should be able to make sense of the proposition that 'any tritium atom that now exists has a certain chance of decaying within a year' (Lewis 1980, p. 263). Thus, we are told, de Finetti's eliminative strategy towards chance cannot be applied to quantum mechanics.

But, actually, why not? In the deterministic case, de Finetti's analysis shows us how we can have a useful and even intersubjectively robust notion of probability that relates to actual matters of fact but is conceptually quite autonomous from them. And after all, it is especially when we take determinism to fail and the world around us to behave randomly that we may most need (probabilistic) strategies to navigate the world.

What goes unappreciated in the sweeping intuition that subjectivism must fail in the case of indeterminism is precisely the role and status of the *strategies* we apply in choosing our probability assignments. Subjectivists may well use certain strategies or shortcuts to decide what subjective probability assignments to make. These are just what we called *theoretical assumptions* or *models* in Sect. 3.2. They are not Lewisian rationality constraints on probability assumptions, but systematic ways in which we subjectively choose our credences. Data may well 'confirm' such models, but again that just means that our subjective probability assignments have performed well in the past.

The form theoretical models may typically take is that they will systematically connect certain non-probabilistic features of a given situation with certain probability assignments. For instance, we may choose to assign certain subjective probabilities to rolls of dice given their internal weight distribution. For the subjectivist, these probabilities are and remain purely subjective. Indeed, there is no necessary connection between the weight distribution in a die and the outcomes of our rolls, not just in the Humean sense that these are independent matters of fact, but even in the sense that, given a Newtonian model of dice rolling, for any possible sequence of results there are initial conditions that will produce it. Thus, there is no Lewisian compelling rational justification for letting our credences depend on the weight distribution in the die, only pragmatic criteria such as simplicity, again past performance, etc.

From this perspective, *quantum mechanics* is just such a theoretical model for choosing our subjective probabilities, and indeed a very successful one. One can take the analogy with probabilistic models of dice rolling quite far: unlike Fuchs, who *identifies* quantum states with catalogues of subjective probabilities, I believe a subjectivist may take the quantum state itself as a non-probabilistic, ontic feature of the world and use it (like weight distribution) as the basis for choosing our subjective probabilities.[21] There is no need for a qBist like Fuchs to reject the putative onticity of quantum states,[22] precisely because that by itself does not make our probability assignments objective (forgive me, Itamar!).

As further described in Bacciagaluppi (2014), this general perspective can be applied to all the main approaches to quantum mechanics that take the quantum state as ontic, so that one can be a de Finettian subjectivist about probabilities in the context of Bohm, GRW or Everett. In the case of Bohm, this is unremarkable of course, because the Bohm theory is deterministic and probabilities are epistemic in the sense used above. In the case of GRW, it is an unfamiliar position. The default intuition in GRW is presumably to take probabilities as some kind of propensities, but subjectivism may prove more attractive.[23] Finally, the Deutsch-Wallace decision-theoretic framework for the Everett theory already analyses probabilities as subjective. Deutsch and Wallace themselves (see e.g. Wallace 2010) argue that the weights in the universal wavefunction in fact *satisfy* the Principal Principle (and are in fact the only concrete example of something that qualifies as objective chances in Lewis's sense!). Many authors, however, dispute that the Deutsch-Wallace theorem rationally compels one to adopt the Born rule. I concur, but see that as no reason to reject the decision-theoretic framework itself. One can simply use the Deutsch-Wallace theorem as an ingredient in our subjective strategy for assigning probabilities on the basis of quantum states. I take it that this is in fact the subjectivist (or 'pragmatist') position on Everett proposed by Price (2010).

Thus, in all these cases (with the exception of Bohm of course), we have a view in which the probabilities of quantum mechanics are coherently understood as both purely subjective and ontic, which is what we set out to establish in this section.

3.5 Epistemic Objective Probabilities

We now need to discuss the case of epistemic objective probabilities. We shall do so in the context of Lewis's treatment of objective probabilities, which however immediately raises an obstacle of principle. Recall that in Sect. 3.3 we identified

[21]The quantum state is thus taken to be a matter of fact, but presumably not a 'particular matter of fact' in the sense of Lewis, because of the holistic aspects of the quantum state. (Cf. Footnote 12 above.) Note also that historically the notion of quantum state *preceded* its 'statistical interpretation', famously introduced by Max Born (for discussion see Bacciagaluppi 2008).

[22]For other, even more forceful arguments to this conclusion, see Brown (2019) and Myrvold (2020).

[23]Frigg and Hoefer (2007) have argued that the propensity concept is inadequate to the setting of GRW. Their own preference is to use 'Humean objective chances' instead (cf. Frigg and Hoefer 2010). Note that if the alternative subjectivist strategy includes taking the quantum state as a non-probabilistic physical object, there are severe constraints on doing this for the case of *relativistic* spontaneous collapse theories (see also below, Sects. 3.6 and 3.9).

epistemic probabilities as situated in the gap between what is knowable in principle and what is actually known, and that we settled (at least provisionally) on taking what is knowable in principle at time t to be the history of the world H_t up to that time. Now take our credence at t in some proposition A (about some events at time later than t). As mentioned, $\mathrm{cr}_t(A) = \mathrm{cr}(A|E_t \,\&\, T)$, where E_t are the facts at times up to t that we actually know at that time, and T are our theoretical assumptions. As in the case of (3.8) above, $\mathrm{cr}_t(A)$ has thus the general form

$$\mathrm{cr}(A|E_t \,\&\, T) = \sum_i \mathrm{cr}(A|H_t^i \,\&\, T)\mathrm{cr}(H_t^i|E_t \,\&\, T). \tag{3.9}$$

Here, unless the theoretical assumptions T describe the situation as deterministic, the probabilities $\mathrm{cr}(A|H_t^i \,\&\, T)$ will be non-trivial *ontic* probabilities. (In fact, it is only now that this statement is unproblematic, after we have argued in the last section that ontic probabilities can be equally subjective or objective.[24]) The epistemic component of our credence is given by the various $\mathrm{cr}(H_t^i|E_t \,\&\, T)$. But if these probabilities are non-trivial, then according to Lewis they cannot be objective (i.e. Lewisian chances). Indeed, $\mathrm{cr}(H_t^i|E_t \,\&\, T)$ is just our credence $\mathrm{cr}_t(H_t^i)$ at t in the proposition H_t^i. But that is a proposition about past history, and according to Lewis chances of past events are always 0 or 1. Thus, non-trivial epistemic probabilities $\mathrm{cr}(H_t^i|E_t \,\&\, T)$ cannot be objective.

We see that, given the PP, epistemic objective probabilities would seem to be an oxymoron: if we are ignorant of certain matters of fact, and if propositions about these facts are admissible in the sense of Lewis, then our non-trivial credences about these facts are not objective, because chances of past events are 0 or 1. It appears that Lewis is ruling out by definition that epistemic probabilities could ever be objective. As seemed to be the case with Popper, also for Lewis epistemic probabilities must be subjective.

For the sake of arguing that epistemic objective probabilities do make sense after all, we shall of course grant the existence of objective probabilities in the first place (*pace* de Finetti). Thus, assume that e.g. a quantum measurement of electron spin displays objective probabilities, say 50-50 chances of 'up' or 'down'. We shall take these as objective, in the Lewisian sense that if I ask you to bet on the outcome of the next measurement, then conditionally on the chances it is rationally compelling for you to bet at equal odds.

Now consider the case in which I perform such a measurement but *do not show you the outcome yet*, and I ask you to bet. I make the following two claims:

- Your (subjective) probabilities about the outcome are epistemic: there is an outcome out there which is in principle knowable, and you employ probabilities only because you do not in fact know it.

[24]Indeed, we defined these probabilities as ontic, so unless we accept subjective ontic probabilities, we are committed to regarding these probabilities either as objective or (more generally!) as objectively wrong.

- You are still going to bet at equal odds, and in fact this is *just as rationally compelling* as betting at those odds before the measurement.

The first claim is uncontroversial. The second one – your epistemic probabilities after the measurement are rationally compelling, granted the probabilities before the measurement are – I take to be incontrovertible. From a Bayesian perspective, for instance, you have no new evidence. So it would be irrational to change your subjective probabilities after I have performed the experiment. But what is more, I shall now argue that your non-trivial probabilities are objective *for the same reason* as in Lewis's discussion. That is, the rational justification for your probability assignment is still completely in the spirit of the PP, even though it does not respect its letter.

Indeed, recall what the PP (3.3) says, as an implicit definition of objective probabilities: iff there is a proposition X such that, for all choices of reasonable initial credence function and admissible propositions,

$$\mathrm{cr}\Big(A \,\Big|\, X \,\&\, E_t \,\&\, T\Big) = x, \tag{3.10}$$

then A has an objective probability and its value is x. Crucially, Lewis sees chances as functions of time. Thus he requires that our credence function at t should not incorporate information that is inadmissible *at the time* t. This corresponds to a betting situation in which you are allowed to know any matters of fact up to time t, and the chances fix the objectively fair odds for betting in this situation.

But in the case we are considering now, *the betting situation is different*. Indeed, you are not allowed to know what has happened after I have performed the measurement (even if the measurement was at a time $t' < t$). If in this betting situation you knew the outcome of the measurement, *that* now would be cheating. In this situation, the proposition that the chance at t is equal to x (in particular 0 or 1) is *itself* inadmissible information, and it is so for the same reason that Lewis holds a proposition about the future chance of an event to be inadmissible.

Indeed, in his version of the PP, Lewis is considering a betting context restricted to what we can in principle know at t (i.e. the past history H_t), and he asks for the best we can do in setting our rational credence in a future event A. The answer is that the best we can do is to set our credence to the chance of A at that time t. Information about how the chances evolve after t is inadmissible because it illicitly imports knowledge exceeding what we are allowed to know at t. If we consider restricting our betting context further so that what we are allowed to know is what we could in principle have known at an earlier time t', and now ask for the best we can do in setting our rational credence in an event at t, then information about the chance at t of that event has become inadmissible, again precisely because it exceeds the assumed restriction on what we are allowed to know. The best we can do is to set our credence to the chance of A at the earlier time t'.

The point is perhaps even clearer if we take Lewis's reformulated version (3.5) of the PP:

$$\mathrm{ch}(A|H_t) = \mathrm{cr}(A|H_t \,\&\, T_C). \tag{3.11}$$

This tells us that if at t we include all admissible information, namely H_t and T_C, this information already implies what the chances are at t, so that the best we can do in terms of setting our credence at t, is to take any reasonable initial credence function and conditionalise on H_t and T_C.

But now if we assume that our epistemic context E_t is restricted to $H_{t'}$, clearly the best we can do is to take the information $H_{t'}$ we actually have and feed it into the complete theory of chance T_C, yielding as credence $\mathrm{cr}(A|E_t \,\&\, T_C) = \mathrm{ch}(A|H_{t'})$. *This* credence will now be rationally compelling, even if it is not equal to $\mathrm{ch}(A|H_t)$.

The intuition is that in order to judge whether a subjective probability conditional on the epistemic context $E_t = H_{t'}$ is in fact objective, we need to compare it not to the objective probability conditional on the history H_t at the time of evaluation of our credence, but to the objective probability conditional on the history $H_{t'}$ that *matches* our epistemic context. And this now indeed provides the room for non-trivial objective probabilities even for events that could in principle be known.

3.6 Redrawing the Subjective-Objective Distinction

We have argued above that one can make good sense both of subjective (indeed, purely subjective) ontic probabilities and of epistemic objective probabilities, thus arguing in favour of two unusual conceptual possibilities. But in other respects we had to be quite conservative. Indeed, the arguments above would have been weakened had we not tried to keep as much as possible to the accepted use of the terms 'epistemic', 'ontic', 'subjective' and 'objective'. Once we have seen the usefulness of admitting subjective ontic probabilities and epistemic objective probabilities, however, we might be interested in trying to redraw the subjective-objective distinction and the epistemic-ontic distinction in ways that are both more systematic and cleaner in the light of our new insights.

Let us start with the subjective-objective distinction. What we have just argued for in Sect. 3.5 is that there are more cases of rationally compelling credences than Lewis in fact allows for.[25]

Lewis's intuition is that chances are functions of *time*, and in the formulation of the PP this effectively forces him to privilege the maximal epistemic context H_t comprising all that is knowable in principle at t, or equivalently to *quantify* over all epistemic contexts E_t comprising what may be actually known at t. Thus, according to Lewis, we need to ask whether propositions exist that would fix our credences across all epistemic contexts admissible at time t.

But we have just argued that we can also ask whether propositions exist that would fix our credences when we are in a *particular* one of these epistemic contexts. Specifically, we considered the situation in which at t we only know all matters of

[25]Many thanks to Ronnie Hermens for noting that this point was not spelled out clearly enough in a previous draft.

fact up to $t' < t$. In that case, if we assume Lewisian chances exist, we know in fact that there are such propositions, namely the chances at t'.

Now we can ask that same question *in general*: do propositions exist that would fix our credences when we are in an *arbitrary* epistemic context E (e. g. any ever so partial specification of the history up to a time t)? If so, we shall talk of the chances not *at the time* t, but the chances *given the context* E. Lewisian chances at t are then just the special case when E has the form H_t.

Writing it down in formulas, we have a natural generalisation of the PP and of objective probabilities: given a context E, iff there is a proposition X such that, for all choices of reasonable initial credence function and any further theoretical assumptions T (compatible with X),

$$\mathrm{cr}\Big(A \Big| X \,\&\, E \,\&\, T\Big) = x, \tag{3.12}$$

then A has an objective probability given the context E and its value is x. The weakest such proposition will be denoted $\mathrm{ch}_E(A) = x$.

Now, similarly as with Lewis's own reformulation (3.5) of the PP, note that a complete theory of chance will now include also history-to-chance conditionals for any antecedent E for which chances given E exist. Therefore, not only will the following be an instance of the generalised PP (3.12),

$$\mathrm{cr}\Big(A \Big| (\mathrm{ch}_E(A) = x) \,\&\, E \,\&\, T_C\Big) = x, \tag{3.13}$$

but we can further rewrite it as

$$\mathrm{ch}_E(A) = \mathrm{cr}(A|E \,\&\, T_C). \tag{3.14}$$

Thus, also our generalised chances will have the kind of properties that Lewis derives for his own chances from his reformulation (3.5) of the original PP (in particular, they will obey the probability calculus).

Further, (3.14) makes vivid the intuition that if we have a complete and correct theory of (generalised) chance, the best we can rationally do if we are in the epistemic context E, is just to plug the information we *actually* have into the complete and correct theory of chance, and let that guide our expectations.

Finally, and very significantly, we see from (3.14) that chances in the more general sense are not functions of propositions and times, but functions of *pairs of propositions*, and they have the form of *conditional probabilities*. That is, in this picture *chances are fundamentally conditional*. I believe this has many advantages.[26]

[26]Note that also epistemic and ontic probabilities as discussed here are naturally seen as conditional (on epistemic and ontic contexts, respectively). That all probabilities should be fundamentally conditional has been championed by a number of authors (see e.g. Hájek 2011 and references therein). The idea in fact fits nicely with Lewis's own emphasis that already the notion of possibility

From the point of view of the philosophy of physics, such a liberalisation of the PP means that all kinds of other probabilities used in physics can now be included in what could potentially be objective chances.

For instance, it allows one to consider the possibility of *high-level probabilities* being objective. By high-level probabilities I mean probabilites featuring in theories such as Boltzmannian statistical mechanics, which is a theory about *macrostates*, identified as coarse-grained descriptions of physical systems. Our generalised PP allows one among other things to apply a Lewisian-style analysis to probabilities conditional on such a macrostate of a physical system, thus introducing the possibility of objective chances even in case we are ignorant of the microstate of a system such as a gas (*pace* Popper).[27]

Or take again the EPR setup: Bob performs a measurement, usually taken to collapse the state also of Alice's photon, so that the probabilities for her own measurement results are now $\cos^2(\alpha/2)$ and $\sin^2(\alpha/2)$. Let us grant that these probabilities are objective. In Sect. 3.3 we considered Alice's probability assignments in the epistemic context in which she does not know yet the result of the measurement on Bob's side, and treated them as partially epistemic. Now, however, we can argue that in Alice's epistemic context what is rationally compelling for her is to keep her probabilities unchanged, because we compare them with the quantum probabilities conditional on events only on *her* side.

Indeed, the situation is very much analogous to the simple example in which I perform a measurement but do not tell you the result. If anything, in the EPR example the case is even *more* compelling, because there is no way even in principle that Alice could know Bob's result if the two measurements are performed at spacelike separation. Thus, if we take probabilities to be encoded or determined by the quantum state, we have now made space for the idea that collapse does not act instantaneously at a distance, but along the *future light cone* (of Bob's measurement in this case).

Quite generally, it is clear that Lewis's treatment of chances taken literally is incompatible with relativity, for Lewis's chances as functions of time presuppose a notion of absolute simultaneity. In fact, the idea that chance could be relativised to space-like hypersurfaces or more generally to circumstances which are independent of absolute time, e.g. properties of space-time regions, was proposed in Berkovitz (1998), who discusses this proposal with respect to causality and non-locality in the quantum realm.

---

is indeed to be naturally understood as a notion of *compossibility*, of what is possible keeping fixed a certain context. (Recall his lovely example in Lewis (1976) of whether he can speak Finnish as opposed to an ape, or cannot as opposed to a native speaker.)

[27] And *pace* Lewis, who explicitly ruled out the possibility of such deterministic chances (Lewis (1986), p. 118). Note that high-level deterministic chances relate to all kinds of interesting questions (e.g. emergence, the trade-off between predictive accuracy and explanatory power, the demarcation between laws and initial conditions, etc.). For recent literature on the debate about deterministic chances, see e.g. Glynn (2009), Lyon (2011), Emery (2013), and references therein.

A relativistic generalisation of Lewisian chances will automatically require more liberal contexts E – as we have proposed here – because there is no longer an absolute notion of simultaneity, or of 'past', or of 'becoming', determining which events we need to conditionalise upon to temporally evolve our chances. As is well known, in special relativity these notions can be modified in two ways: one either takes simultaneity to be relative to inertial frames or more generally to spacelike foliations, or one localises the notion of absolute past to the past light cone of a spacetime event (or spacelike segment or bounded spacetime region). Both indeed require generalising chances from just defining them relative to 'past histories' H_t to contexts that are the past of arbitrary spacelike hyperplanes or hypersurfaces, or, respectively, to contexts that are the past light cones of arbitrary events (or spacelike segments or bounded regions).

Conditionalising on the past of arbitrary hypersurfaces is what Myrvold (2000) has proposed as a way of conceptualising relativistic collapse in the case of quantum mechanical probabilities.[28] Conditionalising on past light cones instead is what Ismael (2011) has proposed as a way of conceptualising chances in the case of relativity. Applied to quantum mechanics, this corresponds to the EPR example as we have just revisited it, with a conception of collapse along the future light cone.[29]

For yet another example, in recent work Adlam (2018a,b) has argued that 'temporal locality' and 'objective chances' risk becoming two mutually supporting dogmas of modern physics. Maybe slightly oversimplifying, the point is that the idea of objective chances defined at a time t and that of a physical state at time t being all that is needed to predict the physical state at times $t + \varepsilon$ in fact seem to naturally relate to and support each other. Adlam argues that this creates an obstacle to the development of e. g. non-Markovian, retrocausal, and 'all-at-once' theories.[30] But in fact, it is only the standard Lewisian form of objective chances *at time t* that may be inimical to such developments, and the present generalisation of the PP removes this particular obstacle.

For instance, suppose we have some physical theory involving non-Markovian probabilities. The Lewisian definition of chances actually does allow for non-Markovian chances, in the sense that $\mathrm{ch}_t(A) = \mathrm{ch}(A|H_t)$ now in general is a proposition that depends on the *whole* of the history H_t. However, if at t we only

[28] I actually prefer the 2000 archived version to the published paper (Myrvold 2002). More recently, Myrvold (2017a) has shown that any such theories of foliation-dependent collapse must lead to infinite energy production. I believe, however, that Myrvold's idea that probabilities in special relativity (and in quantum field theory) can be defined conditional on events to the past of an arbitrary hypersurface can still be implemented (see Sect. 3.9 below).

[29] I am indebted to both of these proposals and their authors in more ways than I can say. See also my Bacciagaluppi (2010a).

[30] Retrocausal theories (in particular in the context of quantum mechanics) have been vigorously championed by Huw Price. All-at-once theories are ones in which probabilities for events are specified given general boundary conditions, rather than just initial (or final) conditions, and have been championed in particular by Ken Wharton. See e.g. Price (1997), Wharton (2014) and the other references in Adlam (2018a). For further discussion see also Rijken (2018) (whom I thank for discussion of Adlam's work).

have information about some previous times $t_1, \ldots, t_n$, the original PP has nothing to recommend us except to try and figure out the missing information, so as to be able to determine the chances $\mathrm{ch}(A|H_t)$ at t. Our physical theory, instead, does contain also probabilities of the form $p(A|E_{t_1}\&\ldots\&E_{t_n})$, where E_{t_i} describes the state of our physical system at time t_i. In order to be able to say that also these probabilities featuring in our non-Markovian theory are objective, we have to generalise chances to allow for contexts of the 'gappy' form $E = E_{t_1}\&\ldots\&E_{t_n}$ in (3.12)–(3.14).

Note that $p(A|E_{t_1}\&\ldots\&E_{t_n})$ is just a coarse-graining of probabilities of the form $p(A|H_t^j)$, where H_t^j is a possible history of the world up to t compatible with $E_{t_1}\&\ldots\&E_{t_n}$, i.e. a coarse-graining of probabilities which in turn do allow a reading as Lewisian chances. In this sense, our generalisation (3.12) appears to be required by sheer consistency, if we want objective probabilities to be closed under coarse-grainings.[31]

Finally, even leaving aside motivations from philosophy of physics, what may perhaps be decisive from the point of view of the philosopny of probability is that such a generalisation of the PP completely eliminates the need to restrict epistemic contexts to 'admissible propositions'. The reason for worrying about admissibility was that chances were tacitly or explicitly thought to always refer to the context H_t, and the actual epistemic context had to be restricted to admissible propositions in order not to bring in any information not contained in H_t. By requiring that the (ontic) context of the chances exactly match the (epistemic) context of the credences, no such possibility of mismatch can arise, and no additional conditions of admissibility are needed.

I suspect Lewis himself might have welcomed such a simplification in the formulation of the PP, especially given his own qualified formulations when treating of admissible propositions in connection with the possibility of non-standard epistemic contexts in the presence of time travel, clairvoyance, and other unusual cases (as will in fact be the retrocausal and all-at-once theories considered by Adlam).

3.7 Redrawing the Epistemic-Ontic Distinction

Generalising chances to arbitrary contexts E in our discussion of subjective and objective probabilities suggests we might want to do something similar in our discussion of epistemic and ontic probabilities. While sufficient for the purpose of our examples so far, taking epistemic probabilities to be probabilities of past events (or any events determined by such events) is now too restrictive. As we did in the

[31]Of course also in the case of $E = H_{t'}$, the chance given E is the coarse-graining over all possible intervening histories between t' and t. But this particular coarse-graining does yield again a Lewisian chance, namely the chance at t'.

case of chances, we should not only want to be able to generalise to a relativistic setting, but also to be able to get rid of the restriction to past events altogether, allowing in fact for all kinds of non-standard epistemic contexts (clairvoyance, backwards-directed agents with memories of the future, omniscience, and what not), and allowing conversely for taking an ontic perspective also on probabilities that are about past events (e.g. in the case of coarse-graining).

This, however, also seems to raise a problem, namely that the distinction between epistemic and ontic probabilities becomes a purely pragmatic one at best. If we can extend or restrict at will which propositions we assume to be knowable in principle, then any probability can be alternatively seen as epistemic or as ontic, and there is no substantive difference between the two.

To a large extent, we can bite the bullet, because the distinction between epistemic and ontic probabilities clearly does have such a pragmatic aspect. When in Sect. 3.5 we considered the example of the spin measurement with unknown outcome, in the given epistemic context those probabilities are of course epistemic. But if we are asked how we choose our probability assignments in that situation, we will say that we are simply applying the quantum probabilities that obtained at the time of the measurement, i. e. we are mentally *switching* to a context in which the indeterministic outcomes are still before us, and in which the probabilities predicted by quantum mechanics (which there we assumed are objective) are thus ontic, so that our theoretical assumptions about that ontic context can inform our probability assignments in our current epistemic context.

In so doing, however, we also seem to be performing a second switch, one that was in fact crucial to our arguing in Sect. 3.4 that ontic probabilities could also be subjective (e.g. when adopting a subjectivist position about probabilities of atomic decay). Indeed, while epistemic probabilities are about unknown matters of fact in the actual world, ontic probabilities as we used them in Sect. 3.4 are *theoretical* probabilities about alternative possibilities in our models.[32]

We mentioned already in Footnote 12 above a distinction between *material* propositions (propositions about matters of fact in the actual world) and *theoretical* propositions (propositions within our theoretical models about the world).[33] We should thus distinguish accordingly between different *dimensions of possibility*: material possibilities if we are wondering about what *is* the case in the material world, and theoretical possibilities if we are wondering about what *could be* or *might have been* the case.

In this sense, epistemic probabilities are clearly probabilities about material possibilities: this is the dimension of possibility in which beliefs live, and epistemic probabilities are measures over unknown matters of fact in the actual world

---

[32] And, indeed, already in Sect. 3.3 we suggested that ontic probabilities tend to be associated with indeterministic contexts, which we defined in terms of multiple possible futures in our theoretical models compatible with the present (or present and past) state of the world.

[33] As mentioned above, one could of course distinguish further between the actual world and the material world (if one is a many-world theorist) or take some material propositions to be theoretical (if one is an empiricist), but for simplicity I neglect these possibilities.

(conditional on keeping the epistemic context fixed). But now we recognise an ambiguity in the standard use of 'ontic probabilities' as construed so far: we have been sliding between on the one hand taking also ontic probabilities to be *material* probabilities, namely measures over unknowable facts in the actual world (conditional on keeping the ontic context fixed, i.e. what is knowable in principle), and on the other hand taking ontic probabilities to be *theoretical* probabilities, namely measures over alternative possible worlds (now thinking of the fixed ontic context as what we are conditioning on *in the model*). In the radioactive decay example this means sliding between taking ontic probabilities as about what the actual facts in the future are, and taking them as measures over alternative possible worlds. Our examples of subjective ontic probabilities can be read in either way, and remain subjective under either reading, whether we identify ontic probabilities directly with the theoretical probabilities of our models (which in Sect. 3.4 we took to be purely subjective), or whether we take the ontic probabilities to be material, and *use* our subjective theoretical probabilities in fixing them.

It is not essential to choose between the two readings of ontic probabilities, as long as one can distinguish if required. Insofar as ontic probabilities feature in our expectations, one might prefer to think of them as material probabilities (since our expectations are about the actual world). On the other hand, our probabilistic statements about, say, future quantum events will tend to have modal force (in the sense of supporting causal and/or law-like statements), indicating that more often than not we intend such probabilities to be theoretical.[34]

In this context, while in some of my examples I use the terminology of possible worlds, or refer to ontic probabilities as law-like, the important thing is indeed that our theoretical models and the probabilities that feature in them have modal force, however understood. Our theoretical models may feature probabilities in terms of probabilistic laws, or in terms of probabilistic causes; if you object to possible worlds as a formalisation of modality (even though I take them in a very thin sense) or to thinking of probabilities as law-like, any alternative account of modal force will do.[35]

Another point to note is that, while we have said that insofar as epistemic probabilities are material probabilities they do not have modal force, not all theoretical probabilities need in fact have modal force. Some probabilities in our models may be understood as contingent elements of a model. Such is typically the case for initial distributions, or more generally single-time distributions, even in theoretical models as, say, classical statistical mechanics. And, indeed, when discussing Popper's

[34]Material probabilities by definition support only indicative conditionals: 'If Beethoven was not born on 16th December, he was born on the 15th' expresses our next best guess. Instead, when talking about atomic decay we more commonly use subjunctive or counterfactual conditionals, such as 'Should this atom not decay in the next hour, it would have probability 0.5 of decaying during the following hour'.

[35]My thanks to Niels van Miltenburg and to Albert Visser for pressing me on this point.

worries in Sect. 3.1 we claimed as self-evident that the probability distributions in classical statistical mechanics are epistemic. But again we need to distinguish: as long as we are talking of our credences about what is in fact the case inside a bottle, these are indeed material and epistemic probabilities; as long, instead, as these are theoretical probability distributions over different alternative theoretical possibilities in our model, they do not range over epistemic possibilities. Rather, they appear to represent the contingent aspects of our theoretical model.[36] Thus we see that there is a certain ambiguity between material and theoretical probabilities also when we are talking about epistemic probabilities. Again, this ambiguity need not be worrying, as long as one is clearly aware of the material-theoretical distinction.[37]

Indeed, to revisit one more time our standard example of mixed probabilities (unknown composition of a radioactive probe, known decay laws for the various isotopes), we can now give multiple readings of these probabilities. We can see them as expressing our expectations about a material proposition in our future, which we evaluate by mixing our epistemic probabilities for the composition of the probe and our ontic probabilities (in the material sense) about what will happen in the various cases. In this case, we fix the values of the latter probabilities using the law-like probabilities of our theoretical models. Or we see them throughout as probabilities in a theoretical model, but with a contingent distribution over the isotopes in the radioactive probe. In that case, we can fix that distribution using our epistemic probabilities for the composition of the probe. Or yet again, we can think of such mixed probabilities as a *sui generis* mix of material and theoretical probabilities, which we however commonly resort to in practice.

How to use theoretical models (i.e. which probabilities to use) will depend pragmatically on our epistemic situation. Any given context can be seen as the epistemic context of some putative agent wondering about the actual world, or as an ontic context from which to judge the accessibility of other possible worlds. But in any *actual* epistemic context in which we wonder about what we do not know in the material world, we will happily and often resort to theoretical models to guide our expectations.

[36]We shall refine this point in our discussion of time-symmetric theoretical models in Sect. 3.8.

[37]It may, however, contribute to muddle the waters in the Popper and Jaynes examples, or indeed in modern discussions about 'typicality' in statistical mechanics: are we talking about the actual world (of which there is only one), or about possible worlds (of which there are many, at least in our models)? It is also not always clear which aspects of a model are indeed law-like and which are contingent. We shall see a class of examples in Sect. 3.8, but a much-discussed one would be whether or not the 'Past Hypothesis' can be thought of as a law – which it is in the neo-Humean 'best systems' approach (cf. Loewer 2007); thanks to Sean Gryb for raising this issue. For further discussion of typicality and/or the past hypothesis, see e.g. Goldstein (2001), Frigg and Werndl (2012), Wallace (2011), Pitowsky (2012), and Lazarovici and Reichert (2015).

3.8 Remarks on Time Symmetry

Many more things could be said about applying the newly-drawn distinctions in practice. I shall limit myself in this section to a few remarks on the issue of time symmetry, and in the next one to some further remarks on the quantum state.

I mentioned in Sect. 3.3 that for the purpose of distinguishing epistemic and ontic probabilities I did not want to commit myself to a metaphysically thick notion of indeterminism, with a 'fixed' past and an 'open' future. Now we can see that such a notion is in fact incompatible with the way I have suggested drawing the epistemic-ontic distinction.

Indeed, as drawn in Sect. 3.7, the epistemic-ontic distinction allows for ontic probabilities also in the case we theoretically assume *backwards* indeterminism (where models that coincide at a time t or for times $s \geq t$ need not coincide at times earlier than t). We can take the entire history to the *future* of t as an ontic context (call it $\tilde{H}_t$), and thus probabilities $p(A|\tilde{H}_t)$ about events *earlier* than t as ontic. This is plainly incompatible with tying ontic probabilities (whether in the material or the theoretical sense) to the idea of a fixed past and open future.

There are other reasons for rejecting the idea of fixed past and open future.[38] But my suggestion above seems to fly in the face of a number of arguments *for* a different metaphysical status of past and future themselves based on probabilistic considerations. It is these arguments I wish to address here, specifically by briefly recalling and reframing what I have already written in my Bacciagaluppi (2010b) against claims that past-to-future conditional probabilities will have a different status (namely ontic) from that of future-to-past conditional probabilities (allegedly epistemic). I refer to that paper for details and further references.

Of course we are all familiar with the notion that probabilistic systems can display strongly time-asymmetric behaviour. To fix the ideas, take atomic decay, modelled simplistically as a two-level classical Markov process (we shall discuss quantum mechanics proper in the next section), with an excited state and a de-excited state, a (comparatively large) probability α for decay from the excited state to the de-excited state in unit time, and a (comparatively small) probability ε for spontaneous re-excitation in unit time. We shall take the process to be homogeneous, i.e. the transition probabilities to be time-translation invariant.

Any initial probability distribution over the two states will converge in time to an equilibrium distribution in which most of the atoms are de-excited (a fraction $\frac{\alpha}{\alpha+\varepsilon}$), and only a few are excited (a fraction $\frac{\varepsilon}{\alpha+\varepsilon}$), and the transition rate from the

[38] In particular, I agree with Saunders (2002) that standard moves to save 'relativistic becoming' in fact undermine such metaphysically thick notions. Rather, I believe that the correct way of understanding the 'openness' of the future, the 'flow' of time and similar intuitions is via the logic of temporal perspectives as developed by Ismael (2017), which straightforwardly generalises to relativistic contexts, since it exactly captures the perspective of an IGUS ('Information Gathering and Utilising System') along a time-like worldline. For introducing me to the pleasures of time symmetry I would like to thank in particular Huw Price.

excited to the de-excited state will asymptotically match the transition rate from the de-excited to the excited state (so-called 'detailed balance' condition).

This convergence to equilibrium is familiar time-directed behaviour. Of course, given the joint distributions for two times $s > t$, transition probabilities can be defined both from t to s and from s to t (by conditionalising on the state at t and at s, respectively), but the example can be used in a number of ways to argue that these transition probabilities forwards and backwards in time must have a different status. In the language of this paper:

(A) Were backwards probabilities also ontic and equal to the forwards probabilities, then the system would converge to equilibrium also towards the past, and the system would have to be stationary (Sober 1993).
(B) While by assumption the forwards transition probabilities are time-translation invariant, the backwards transition probabilities in general are not, suggesting that they have a different status (Arntzenius 1995).
(C) If in addition to the forwards transition probabilities also the backwards transition probabilities were law-like, the initial distribution would also inherit this law-like status, while it is clearly contingent – indeed can be freely chosen (Watanabe 1965).

The problem with all these arguments is that they take the observed backwards *frequencies* as estimates for the putative ontic backwards *probabilities*. But frequencies can be used as estimates for probabilities only when we have reasons to think our samples are unbiased.[39] For instance, we would never use post-selected ensembles to estimate forwards transition probabilities. However, we routinely use *pre*-selected ensembles, especially when we set up experiments ourselves and in fact freely choose the initial conditions. Thus, while we normally use such experiments to estimate forwards transition probabilities, we plainly cannot rely on them to estimate the backwards ones.

It turns out that this time-asymmetric behaviour (and a large class of more general examples) can indeed be modelled using stationary stochastic processes in which the transition probabilities are time-symmetric, and in which the time asymmetry is introduced by the choice of special initial (rather than final) *frequencies*. Thus, in modelling such time-asymmetric behaviour there is no need to use time-asymmetric ontic probabilities or to deny ontic status altogether to the backwards transition probabilities of the process. The apparent asymmetry comes in because we generally have asymmetric *epistemic access* to a system: we tend to know aspects of its past history, and thus are able to pre-select, while its future history will generally be unknowable.[40]

---

[39] That this is not always licensed is clear also from discussion of causal loops (Berkovitz 1998, 2001).

[40] Of course, this strategy is analogous to that of modelling time-directed behaviour in the deterministic case using time-symmetric laws and special initial conditions. The same point was made independently and at the same time by (Uffink 2007, Section 7), specifically in the context of probabilistic modelling of classical statistical mechanics. A further explicit example of time-

If one assumes that the transition probabilities are time-symmetric (or time-symmetric except for some overall drift[41]), this places severe constraints on the overall (multi-time) distributions that define the process. Not only is the process stationary if it is homogeneous, but the process may well be uniquely determined by its forwards and backwards transition probabilities.[42] Thus, in particular, the entire process inherits the ontic status of the transiton probabilities.

This will seem puzzling, if we are used to thinking of transition probabilities as ontic, and of initial distributions as epistemic, so that the probabilities of the process should have a mixed character (as in the examples we discussed beginning in Sect. 3.3). But as we saw in Sect. 3.7, we standardly align the distinction between epistemic and ontic probabilities with the distinction between contingent and law-like aspects of our theoretical probabilistic models. The *theoretical probabilities* of the process may very well be even uniquely determined by the forwards and backwards transition probabilities, but our theoretical models will still provide scope for contingent elements through the notion of *samples* of the process. We then feed our epistemic probabilities into the theoretical model via constraints on the frequencies in the samples, which will typically be initial constraints because of our time-asymmetric access to the empirical world.[43]

3.9 Further Remarks on the Quantum State

We have just seen that, classically, a stochastic process may be nothing over and above an encoding of its forwards and backwards transition probabilities. I now

symmetric ontic probabilities is provided by Nelson's (1966,1985) stochastic mechanics, where the quantum mechanical wavefunction and the Schrödinger equation are derived from an underlying time-symmetric diffusion process. See also Bacciagaluppi (2005) and references therein. The notorious gap in Nelson's derivation pointed out by Wallstrom (1989) has now been convincingly filled through the modification of the theory recently proposed by Derakhshani (2017).

[41]In the case of Nelson's mechanics, for instance, one has a systematic drift velocity that equals the familiar de Broglie-Bohm velocity (which in fact also changes sign under time reversal). If one subtracts the systematic component, transition probabilities are then exactly time-symmetric.

[42]More precisely, it will be uniquely determined within each 'ergodic class'. (Note that ergodicity results are far easier to prove for stochastic processes than for deterministic dynamical systems. See again the references in Bacciagaluppi (2010b) for details.)

[43]Without going into details, these distinctions will be relevant to debates about the 'neo-Boltzmannian' notion of typicality (see the references in Footnote 37), and about the notion of 'sub-quantum disequilibrium' in theories like pilot-wave theory or Nelsonian mechanics (for the former see e.g. Valentini 1996, for the latter see Bacciagaluppi 2012).

wish to briefly hint at the possibility that the same may be true also in quantum mechanics.[44]

In quantum mechanics, the role of determining transition probabilities belongs to the quantum state (either the fixed Heisenberg-picture state, or the totality of the time-evolving Schrödinger-picture or interaction-picture states[45]). The quantum state, however, is traditionally seen as more than simply encoding transition probabilities, rather as a *physical object* in itself, which in particular determines these probabilities. One obvious exception is qBism, where however I have argued in Sect. 3.4 above that the rejection of the quantum state as a physical object is not sufficiently motivated.[46]

In this section I want to take a look at a different approach to quantum theory, where I believe there are strong constraints to taking the quantum state as an independently existing object, and reasons to take it indeed as just encoding (ontic, law-like) transition probabilities, largely independently of the issue of whether these probabilities should be subjective or objective.[47]

This approach is the foliation-dependent approach to relativistic collapse. As is well-known, non-relativistic formulations of collapse will postulate that the quantum state collapses instantaneously at a distance (whether the collapse is spontaneous or the traditional collapse upon measurement). So, for instance, in an EPR scenario the collapse will depend on whether Alice's or Bob's measurement takes place first. However, *predictions* for the outcomes of the measurements are independent of their order because the measurements commute: we obtain the same probabilities for the same pairs of measurement results. This suggests that one might be able to obtain a relativistically invariant formulation of collapse after all. One of the main strategies for attempting this takes quantum states to be defined relative to spacelike hyperplanes or more generally relative to arbitrary spacelike hypersurfaces, and collapse accordingly to be defined relative to spacelike foliations. This option has been pioneered for a long time by Fleming (see e.g.

---

[44]This section can be skipped without affecting the rest of the paper. The discussion is substantially influenced by Myrvold (2000), but should be taken as an elaboration rather than exposition of his views. See also Bacciagaluppi (2010a). For Myrvold's own recent views on the subject, see Myrvold (2017b, 2019).

[45]The interaction picture is obtained from the Schrödinger picture by absorbing the free evolution into the operators. In other words, one encodes the free terms of the evolution in the evolution of the operators, and encodes the interaction in the evolution of the state.

[46]Another obvious exception is Nelson's stochastic mechanics (mentioned in Footnote 40), but there it is not the quantum mechanical transition probabilities that are fundamental but the transition probabilities of the underlying diffusion process. While also a number of Bohmians (cf. Goldstein and Zanghì 2013) think of the wavefunction not as a material entity but as nomological, as a codification of the law-like motion of the particles, there is no mathematical derivation in pilot-wave theory of the wavefunction and Schrödinger equation from more fundamental entities comparable to that in stochastic mechanics, and the wavefunction can be thought of as derivative only if one takes a quite radical relationist view, like the one proposed by Esfeld (forthcoming).

[47]For a sketch of how I think this relates to the PBR theorem (Pusey et al. 2012), see below.

1986,1989,1996), and analysed conceptually in great detail by Myrvold (2000) (see also Bacciagaluppi 2010a).[48]

A convenient formal framework for foliation-dependent collapse is provided by the Tomonaga-Schwinger formalism of quantum field theory, in which the (interaction-picture) quantum state is explicitly hypersurface-dependent, and unitary evolution in the presence of interactions is represented as a local evolution of the quantum state from one hypersurface to the next (and is independent of the choice of foliation between two hypersurfaces S and S'). A non-unitary collapse evolution can be included in the same way, as proposed explicitly e.g. by Nicrosini and Rimini (2003), provided one can ensure that the evolution is also independent of the foliation chosen.

The substantial difference between the unitary and non-unitary case is that in the unitary case, the local state on a segment of a spacelike hypersurface (say, the reduced state of Alice's electron) is independent of whether the segment, say, $T \cap T'$ is considered to be a segment of a hypersurface T or of a different hypersurface T'. In the non-unitary case, in general the restriction $\psi_{T|T\cap T'}$ on $T \cap T'$ of a state ψ_T on T is different from the restriction $\psi_{T'|T\cap T'}$ on $T \cap T'$ of the state $\psi_{T'}$ on T'. Indeed, in the EPR case ψ_T could be the singlet state, so that $\psi_{T|T\cap T'}$ is maximally mixed, while T' could be to the future of Bob's measurement, so that $\psi_{T'|T\cap T'}$ is, say, the state $|+_y\rangle$. In particular, whether Alice's electron is entangled with Bob's will depend on the hypersurface one considers (or the frame of reference – Myrvold (2000) aptly calls this the 'relativity of entanglement').

This now raises issues of how to interpret such a collapse theory, specifically in a way that could give rise to the manifest image of localised objects in space and time. The standard options for doing so go under the names of 'wavefunction ontology', 'mass-density ontology', and 'flash ontology', depending on whether one takes the physical objects of the theory to be the quantum states, local mass densities defined via the states, or the collapse events themselves. Hypersurface-dependence of quantum states appears to rule out the wavefunction ontology, precisely because if quantum states are hypersurface-dependent, they do not have unique local restrictions, so an ontology based on them fails to underpin any local reality. Similarly, mass-density ontology is ruled out, because hypersurface-dependent states do not define uniquely any local mass densities. The only option that could give rise to a localised manifest image of reality appears to be the flash ontology. This means that hypersurface-dependent states indeed merely encode *probabilities for physical events to the future of a spacelike hypersurface, conditional on physical events to the past of that hypersurface*. If so, as Myrvold (2000, Footnote 7) remarks, it is unsatisfactory that we should not yet have a formulation of the theory directly in terms of transition probabilities, but require an additional theoretical object, the

[48]The other main strategy is to consider collapse to be occurring along light cones (either along the future light cone or along the past light cone, as in the proposal by Hellwig and Kraus (1970)), and it arguably allows one to retain the picture of quantum states as physical objects (Bacciagaluppi 2010a). No such theory has been developed in detail, however.

quantum state, to determine these probabilities. (As we saw in Sect. 3.6, also Adlam (2018a,b) is critical of thinking of a separate physical object as determining the quantum probabilities.)

What I wish to suggest is that, as in the classical case of Sect. 3.8, the use of a separate quantum state to calculate transition probabilities may be an artefact of our own limited epistemic perspective, and that one needs no more than forwards and backwards transition probabilities corresponding to a 'stationary process'. The form of this process is easy to guess, reasoning by time symmetry from the fact that the future boundary condition we standardly use is the maximally mixed state. If we take that to be the case because we happen to lack information about the future, and therefore do not post-select, that suggests that if we want to remove the pre-selection that we normally do, and thus get to the correct theoretical probabilities, we need to take the maximally mixed state also as the past boundary condition. The appearance of a contingent quantum state would then follow from the fact that we have epistemic access to past collapses, and conditionalise upon them when making predictions, in general mixing our epistemic probabilities about such past collapses ('results of measurements') and ontic transition probabilities provided by the theory.

Of course in general the maximally mixed state is unnormalisable, but *conditional* probabilities will be well-defined. In such a 'no-state' quantum theory not only are conditional probabilities fundamental, but unconditional probabilities in general do not even exist.[49]

No-state foliation-dependent flash-ontology quantum field theory would thus be a way of directly implementing Myrvold's idea that conditional probabilities should be determined by the theory directly, and not by way of the quantum state. Better predictions will be obtained the more events one conditionalises upon stretching back into the past. The theory is thus essentially non-Markovian (thus temporally non-local, in line with Adlam's views). As mentioned in Footnote 28, Myrvold (2017a), developing remarks by Pearle (1993), has shown that foliation-dependent collapse theories always run into a problem of infinite energy production, but I believe a no-state theory in fact escapes this conclusion.[50]

[49]Rather related ideas are common in the decoherent histories or consistent histories literature. Griffiths (1984) does not use an initial and a final state, but an initial and a final projection of the same kind as the other projections in a history, so his consistent histories formalism is in fact a no-state quantum theory. Hartle (1998) suggests that the fundamental formula for the decoherence functional is the two-state one, but that we are well-shielded from the future boundary condition so that two-state quantum theory is predictively equivalent to standard one-state quantum theory (i.e. we can assume the maximally mixed state as future boundary condition). If in one-state quantum theory we assume that we are similarly shielded from the initial boundary condition, then by analogy one-state quantum theory is in fact predictively equivalent to no-state quantum theory (i.e. we can assume the maximally mixed state also as past boundary condition). Thus, it could just as well be that no-state quantum theory is fundamental. (Of course we need to include pre-selection on the known preparation history.)

[50]The idea behind Myrvold's proof is extremely simple. Leaving technicalities aside, because of the Lorentz-invariance of the vacuum state, infinite energy production would arise if localised collapse operators did not leave the vacuum invariant on average. Thus, one should require that

Such a proposal has in fact been put forward in the non-relativistic case by Bedingham and Maroney (2017), in a paper in which they discuss in general the possibility of time-symmetric collapse theories. As with the requirement of Lorentz invariance, the requirement of time symmetry makes it problematic to think of the collapsing quantum state as a physical object, because the sequence of collapsing states defined by the collapse equations applied in the past-to-future direction is not the time reversal of the sequence of states defined by the collapse equations applied in the future-to-past direction. That is, the quantum state is time-direction-dependent. However, under very mild conditions, from the application of the equations in the two directions of time one does obtain the same probabilities for the same collapse events. This suggests, again as in the relativistic case, that a time-symmetric theory of collapse requires a flash ontology. And, as I am suggesting in the case of hypersurface-dependence, Bedingham and Maroney suggest one might want to get rid altogether of the quantum state (since the collapse mechanism suggests it is asymptotically maximally mixed).[51]

One can find an analogy for the approach discussed in this section by looking at Heisenberg's views on 'quantum mechanics' (i.e. matrix mechanics) as expressed in the 1920s in opposition to Schrödinger's views on wave mechanics.[52] The evidence suggests that Heisenberg did not believe in the existence of the quantum state as a physical object, or of collapse as a physical process. Rather, the original physical picture behind matrix mechanics was one in which quantum systems performed

local collapse operators leave the vacuum invariant. But since a local operator commutes with local operators at spacelike separation, and since (by the Reeh-Schlieder theorem) applications of the latter can approximate any global state, it follows that in order to avoid infinite energy production collapse operators must leave every state invariant. Thus there is no collapse.

One way out that is already present in the literature (and which motivates Myrvold's theorem) is to tie the collapse to some 'non-standard' fields (which commute with themselves at spacelike and timelike separation), as done by both Bedingham (2011) and Pearle (2015) himself. Another possibility, favoured by Myrvold (2018a), is to interpret the result as supporting the idea that collapse is tied to gravity: in the curvature-free case of Minkowski spacetime, the theorem shows there is no collapse. As mentioned, however, I believe that a no-state proposal will also circumvent the problem of infinite energy production, because collapse operators will automatically leave the (Lorentz-invariant) maximally mixed state invariant on average.

[51]I should briefly remark on how I think these suggestions relate to the work of Pusey et al. (2012), who famously draw an epistemic-ontic distinction for the quantum state, and who conclude from their PBR theorem that the quantum state is ontic. An ontic state for PBR is an independently existing object that determines the ontic probabilities (in my sense) for results of measurements. And an epistemic reading of the quantum state for PBR is a reading of it as an epistemic probability distribution (in my sense) over the ontic states of a system. Thus, the PBR theorem seems to push for a wavefunction ontology. In a no-state proposal (as here or in Bedingham and Maroney 2017), however, the probabilities for future (or past) collapse events are fully determined by the set of past (or future) collapse events. The contingent quantum state (in fact whether defined along hypersurfaces or along lightcones) can thus be *identified* with such a set of events, which is indeed ontic, thus removing the apparent contradiction. Whether this is the correct way of thinking about this issue of course deserves further scrutiny.

[52]For more detailed accounts, see Bacciagaluppi and Valentini (2009), Bacciagaluppi (2008), Bacciagaluppi and Crull (2009), and Bacciagaluppi et al. (2017).

quantum jumps between stationary states. This then evolved into a picture of quantum systems performing stochastic transitions between values of measured quantities. Transition probabilities for Heisenberg were 'out there' in the world (i.e. presumably objective as well as ontic). If a quantity had been measured, then it would perform transitions according to the quantum probabilities. The observed statistics would conform to the usual rules of the probability calculus, in particular the law of total probability. But if a quantity had not been measured, there would be nothing to perform the transitions, giving rise to the phenomenon of 'interference of probabilities'.

Schrödinger's quantum states were for Heisenberg nothing but a convenient tool for calculating (forwards) transition probabilities, and any state that led to the correct probabilities could be used. (Note that, classically, forwards transition probabilities are independent of the chosen initial conditions.) Hence in particular, the movability of the 'Heisenberg cut' and Heisenberg's cavalier attitude to collapse,[53] and the peculiar mix of 'subjective' and 'objective' elements in talking about quantum probabilities. And now, finally, despite the shrouding in 'Copenhagen mist', we can perhaps start to make sense of some of Heisenberg's intuitions as expressed in 'The Copenhagen interpretation of quantum theory' (Heisenberg 1958, Chap. 3).

3.10 Hume and de Finetti

This paper has suggested a framework for thinking of probabilities in a way that distinguishes between subjective and epistemic probabilities. Two strategies have been used: liberalising the contexts with respect to which one can define various kinds of probabilities, and making a liberal use of probabilities over theoretical possibilities. I wish to conclude by putting the latter into a wider perspective and very briefly sketching the kind of subjectivist and anti-metaphysical position that I wish to endorse.[54]

In forming theoretical models, whether they involve causal, probabilistic or any other form of reasoning, we aim at describing the behaviour of our target systems in a way that guides our expectations. To do so, our models need to have modal force, to describe law-like rather than accidental behaviour. It is up for grabs what this modal force is. We can have an anti-Humean view of causes, laws or probabilities, as some kind of necessary connections, but Hume (setting aside the finer scruples of Hume scholarship) famously denied the existence of such necessary connections. This traditionally gives rise to the problem of induction. Hume himself proposed

[53]For an example of (Born and) Heisenberg using a very unusual choice of 'collapsed states' to calculate quantum probabilities, see (Bacciagaluppi and Valentini 2009, Sect. 6.1.2).

[54]To this position, I would wish to add a measure of empiricism (Bacciagaluppi 2019). For a similar view, see Brown and Ben Porath (2020). On these matters (and several others dealt with in this paper) I am earnestly indebted to many discussions with Jenann Ismael over the years. Any confusions are entirely my own, however.

a 'sceptical solution': we are predisposed to come to consider certain connections between events as causal through a process of habituation, which then forms the basis of our very sophisticated reasonings about matters of fact ('Elasticity, gravity, cohesion of parts, communication of motion by impulse; these are probably the ultimate causes and principles which we shall ever discover in nature; and we may esteem ourselves sufficiently happy, if, by accurate enquiry and reasoning, we can trace up the particular phenomena to, or near to, these general principles', Hume 1748, IV.I.26). There is no guarantee that our expectations will come out true, but these forms of reasoning have served us well.

There is no need, however, for Humeans to restrict themselves to thinking of theoretical models in terms of Hume's own analysis of causation. Any theoretical models are indeed patterns we project onto the world of matters of fact in order to reason about it. Modern-day pragmatists about causation are essentially following Hume, even though they substitute manipulation for habituation (Menzies and Price 1993). And the same is true of subjectivists about probability. Subjective probabilities are patterns that we form in our head, and impose on matters of fact to try and order them for our pragmatic purposes.

Self-styled latter-day Humeans often attempt to give objective underpinnings to the laws and probabilities in our models. The best systems approach takes Humean laws to be the axioms in the best possible systematisation of the totality of events, and Lewis's 'Subjectivist Guide' is a blueprint for the objective underpinning of subjective probabilities.

But the Humean may well have no need of such objective underpinnings. Radical subjectivists about probability have been saying so for a long time. There is no sense in which probabilistic judgements are right or wrong, just as there is no necessary connection between causes and effects. By and large through a process similar to habituation (Bayesian conditioning!) we form the idea of probabilistic connections, which then forms the basis of our further reasonings about matters of fact. There is no guarantee that our expectations will come out true, but these forms of reasoning have served us well.

For the subjectivist, or pragmatist, such are all our theoretical models of science, including quantum mechanics. And we judge them in terms of the standard criteria of science: first and foremost the empirical criterion of past performance, as well as other criteria such as simplicity, expectation of fruitfulness, and so on. In this sense, there are no objective chances in the sense of strict Lewisian rationality, no objective laws nor objective modality in either an anti-Humean or even neo-Humean sense, but merely in the sense of the pragmatic rationality of science.

Indeed, the best systems analysis provides a plausible if schematic picture of our inductive practices, but there is no point to applying it at the 'end of time', when we no longer need to project any predicates. Any laws that we formulate along the way, any notions of modality that we construe are only the result of analysing the epistemically available data, and as such are only projections of what has been successful so far. From a Humean point of view, this seems to me as it should be, just as Hume himself was content with his sceptical solution to the problem of induction.

And, as a matter of fact, in the relatively neglected Section VI of the *Enquiry*, Hume explicitly discusses probabilities in terms of forming expectations, again through habit, about what will happen in certain proportions of cases. Thus Hume was perhaps the first de Finettian...

Acknowledgements I am deeply grateful to Meir Hemmo and Orly Shenker for the invitation to contribute to this wonderful volume, as well as for their patience. I have given thanks along the way, both for matters of detail and for inspiration and stimuli stretching back many years, but I must add my gratitude to students at Aberdeen, Utrecht and the 5th Tübingen Summer School in the History and Philosophy of Science, where I used bits of this material for teaching, and to the audience at two Philosophy of Science seminars at Utrecht, where I presented earlier versions of this paper, in particular to Sean Gryb, Niels van Miltenburg, Wim Mol, Albert Visser and Nick Wiggershaus. Special thanks go to Ronnie Hermens for precious comments on the first written draft, further thanks to Harvey Brown, Alan Hájek and Jenann Ismael for comments and encouraging remarks on the final draft, and very special thanks to Jossi Berkovitz for a close reading of the final draft and many detailed suggestions and comments.

References

Adlam, E. (2018a). Spooky action at a temporal distance. *Entropy, 20*(1), 41–60.

Adlam, E. (2018b). A tale of two anachronisms. Plenary talk given at Foundations 2018, Utrecht University, 13 July 2018. https://foundations2018.sites.uu.nl/.

Arntzenius, F. (1995). Indeterminism and the direction of time. *Topoi, 14*, 67–81.

Bacciagaluppi, G. (2005). A conceptual introduction to Nelson's mechanics. In R. Buccheri, M. Saniga, & E. Avshalom (Eds.), *Endophysics, time, quantum and the subjective* (pp. 367–388). Singapore: World Scientific. Revised version at http://philsci-archive.pitt.edu/8853/.

Bacciagaluppi, G. (2008). The statistical interpretation according to Born and Heisenberg. In C. Joas, C. Lehner, & J. Renn (Eds.), *HQ-1: Conference on the history of quantum physics* (MPIWG preprint series, Vol. 350, pp. 269–288). Berlin: MPIWG. http://www.mpiwg-berlin.mpg.de/en/resources/preprints.html.

Bacciagaluppi, G. (2010a). Collapse theories as beable theories. *Manuscrito, 33*(1), 19–54. http://philsci-archive.pitt.edu/8876/.

Bacciagaluppi, G. (2010b). Probability and time symmetry in classical Markov processes. In M. Suárez (Ed.), *Probabilities, causes and propensities in physics* (Synthese library, Vol. 347, pp. 41–60). Dordrecht: Springer. http://philsci-archive.pitt.edu/archive/00003534/.

Bacciagaluppi, G. (2012). Non-equilibrium in stochastic mechanics. *Journal of Physics: Conference Series, 361*(1), article 012017. http://philsci-archive.pitt.edu/9120/.

Bacciagaluppi, G. (2014). A critic looks at qBism. In M. C. Galavotti, D. Dieks, W. Gonzalez, S. Hartmann, T. Uebel, & M. Weber (Eds.), *New directions in the philosophy of science* (pp. 403–416). Cham: Springer. http://philsci-archive.pitt.edu/9803/.

Bacciagaluppi, G. (2016). Quantum probability – An introduction. In A. Hájek & C. Hitchcock (Eds.), *The Oxford handbook of probability and philosophy* (pp. 545–572). Oxford: Oxford University Press. http://philsci-archive.pitt.edu/10614/.

Bacciagaluppi, G. (2019). Adaptive empiricism. In G. M. D'Ariano, A. Robbiati Bianchi, & S. Veca (Eds.), *Lost in physics and metaphysics – Questioni di realismo scientifico* (pp. 99–113). Milano: Istituto Lombardo Accademia di Scienze e Lettere. https://doi.org/10.4081/incontri.2019.465.

Bacciagaluppi, G., & Crull, E. (2009). Heisenberg (and Schrödinger, and Pauli) on hidden variables. *Studies in History and Philosophy of Modern Physics, 40*, 374–382. http://philsci-archive.pitt.edu/archive/00004759/.

Bacciagaluppi, G., & Valentini, A. (2009). *Quantum theory at the crossroads – Reconsidering the 1927 Solvay conference*. Cambridge: Cambridge University Press.
Bacciagaluppi, G., Crull, E., & Maroney, O. (2017). Jordan's derivation of blackbody fluctuations. *Studies in History and Philosophy of Modern Physics, 60*, 23–34. http://philsci-archive.pitt.edu/13021/.
Bedingham, D. (2011). Relativistic state reduction dynamics. *Foundations of Physics, 41*, 686–704.
Bedingham, D., & Maroney, O. (2017). Time reversal symmetry and collapse models. *Foundations of Physics, 47*, 670–696.
Bell, J. S. (1984). Beables for quantum field theory. Preprint CERN-TH-4035. Reprinted in: (1987). *Speakable and unspeakable in quantum mechanics* (pp. 173–180). Cambridge: Cambridge University Press.
Berkovitz, J. (1998). Aspects of quantum non-locality II – Superluminal causation and relativity. *Studies in History and Philosophy of Modern Physics, 29*(4), 509–545.
Berkovitz, J. (2001). On chance in causal loops. *Mind, 110*(437), 1–23.
Berkovitz, J. (2012). The world according to de Finetti – On de Finetti's theory of probability and its application to quantum mechanics. In Y. Ben-Menahem & M. Hemmo (Eds.), *Probability in physics* (pp. 249–280). Heidelberg: Springer.
Berkovitz, J. (2019). On de Finetti's instrumentalist philosophy of probability. *European Journal for Philosophy of Science, 9*, article 25.
Brown, H. (2019). The reality of the wavefunction – Old arguments and new. In A. Cordero (Ed.), *Philosophers look at quantum mechanics* (Synthese library, Vol. 406, pp. 63–86). Cham: Springer.
Brown, H., & Ben Porath, G. (2020). Everettian probabilities, the Deutsch-Wallace theorem and the principal principle. This volume, pp. 165–198.
Bub, J. (2007). Quantum probabilities as degrees of belief. *Studies in History and Philosophy of Modern Physics, 38*, 232–254.
Bub, J. (2010). Quantum probabilities – An information-theoretic interpretation. In S. Hartmann & C. Beisbart (Eds.), *Probabilities in physics* (pp. 231–262). Oxford: Oxford University Press.
Bub, J., & Pitowsky, I. (1985). Critical notice – Sir Karl R. Popper, Postscript to the logic of scientific discovery. *Canadian Journal of Philosophy, 15*(3), 539–552.
de Finetti, B. (1970). *Theory of probability*. New York: Wiley.
Derakhshani, M. (2017). Stochastic mechanics without ad hoc quantization – Theory and applications to semiclassical gravity. Doctoral dissertation, Utrecht University. https://arxiv.org/pdf/1804.01394.pdf.
Diaconis, P., Holmes, S., & Montgomery, R. (2007). Dynamical bias in the coin toss. *SIAM Review, 49*(2), 211–235.
Emery, N. (2013). Chance, possibility, and explanation. *The British Journal for the Philosophy of Science, 66*(1), 95–120.
Esfeld, M. (forthcoming). A proposal for a minimalist ontology. Forthcoming in *Synthese*, https://doi.org/10.1007/s11229-017-14268.
Fleming, G. (1986). On a Lorentz invariant quantum theory of measurement. In D. M. Greenberger (Ed.), *New techniques and ideas in quantum measurement theory* (Annals of the New York Academy of Sciences, Vol. 480, pp. 574–575). New York: New York Academy of Sciences.
Fleming, G. (1989). Lorentz invariant state reduction, and localization. In A. Fine & M. Forbes (Eds.), *PSA 1988* (Vol. 2, pp. 112–126). East Lansing: Philosophy of Science Association.
Fleming, G. (1996). Just how radical is hyperplane dependence? In R. Clifton (Ed.), *Perspectives on quantum reality – Non-relativistic, relativistic, and field-theoretic* (pp. 11–28). Dordrecht: Kluwer Academic.
Frigg, R., & Hoefer, C. (2007). Probability in GRW theory. *Studies in History and Philosophy of Modern Physics, 38*(2), 371–389.
Frigg, R., & Hoefer, C. (2010). Determinism and chance from a Humean perspective. In D. Dieks, W. Gonzalez, S. Hartmann, M. Weber, F. Stadler, & T. Uebel (Eds.), *The present situation in the philosophy of science* (pp. 351–371). Berlin: Springer.
Frigg, R., & Werndl, C. (2012). Demystifying typicality. *Philosophy of Science, 79*(5), 917–929.

Fuchs, C. A. (2002). Quantum mechanics as quantum information (and only a little more). arXiv preprint www.quant-ph/0205039.
Fuchs, C. A. (2014). Introducing qBism. In M. C. Galavotti, D. Dieks, W. Gonzalez, S. Hartmann, T. Uebel, & M. Weber (Eds.), *New directions in the philosophy of science* (pp. 385–402). Cham: Springer.
Glynn, L. (2009). Deterministic chance. *The British Journal for the Philosophy of Science, 61*(1), 51–80.
Goldstein, S. (2001). Boltzmann's approach to statistical mechanics. In J. Bricmont, D. Dürr, M. C. Galavotti, G.C. Ghirardi, F. Petruccione, & N. Zanghì (Eds.), *Chance in physics* (Lecture notes in physics, Vol. 574, pp. 39–54). Berlin: Springer.
Goldstein, S., & Zanghì, N. (2013). Reality and the role of the wave function in quantum theory. In A. Ney, & D. Z. Albert (Eds.), *The wave function – Essays on the metaphysics of quantum mechanics* (pp. 91–109). Oxford: Oxford University Press.
Greaves, H., & Myrvold, W. (2010). Everett and evidence. In S. Saunders, J. Barrett, A. Kent, & D. Wallace (Eds.), *Many worlds? Everett, quantum theory, and reality* (pp. 264–304). Oxford: Oxford University Press.
Griffiths, R. B. (1984). Consistent histories and the interpretation of quantum mechanics. *Journal of Statistical Physics, 36*(1–2), 219–272.
Hájek, A. (2011). Conditional probability. In D. M. Gabbay, P. Thagard, J. Woods, P. S. Bandyopadhyay, & M. R. Forster (Eds.), *Philosophy of statistics*, Handbook of the philosophy of science (Vol. 7, pp. 99–135). Dordrecht: North-Holland.
Hartle, J. B. (1998). Quantum pasts and the utility of history. *Physica Scripta T, 76*, 67–77.
Heisenberg, W. (1958). *Physics and philosophy – The revolution in modern science*. New York: Harper.
Hellwig, K.-E., & Kraus, K. (1970). Formal description of measurements in local quantum field theory. *Physical Review D, 1*, 566–571.
Howard, D. (2004). Who invented the Copenhagen interpretation? A study in mythology. *Philosophy of Science, 71*(5), 669–682.
Hume, D. (1748). *An enquiry concerning human understanding*. London: A. Millar.
Ismael, J. (2011). A modest proposal about chance. *The Journal of Philosophy, 108*(8), 416–442.
Ismael, J. (2017). Passage, flow, and the logic of temporal perspectives. In C. Bouton & P. Huneman (Eds.), *Time of nature and the nature of time* (Boston studies in the philosophy and history of science, Vol. 326, pp. 23–38). Cham: Springer.
Jaynes, E. T. (1985). Some random observations. *Synthese, 63*(1), 115–138.
Lazarovici, D., & Reichert, P. (2015). Typicality, irreversibility and the status of macroscopic laws. *Erkenntnis, 80*(4), 689–716.
Lewis, D. (1976). The paradoxes of time travel. *American Philosophical Quarterly, 13*(2), 145–152.
Lewis D. (1980). A subjectivist's guide to objective chance. In R. C. Jeffrey (Ed.), *Studies in inductive logic and probability* (Vol. 2, pp. 263–293). Reprinted in: (1986). *Philosophical papers* (Vol. 2, pp. 83–113). Oxford: Oxford University Press.
Lewis, D. (1986). Postscripts to 'A subjectivist's guide to objective chance'. In *Philosophical papers* (Vol. 2, pp. 114–132). Oxford: Oxford University Press.
Loewer, B. (2004). David Lewis's Humean theory of objective chance. *Philosophy of Science, 71*, 1115–1125.
Loewer, B. (2007). Counterfactuals and the second law. In H. Price & R. Corry (Eds.), *Causation, physics, and the constitution of reality – Russell's republic revisited* (pp. 293–326). Oxford: Oxford University Press.
Lyon, A. (2011). Deterministic probability – Neither chance nor credence. *Synthese, 182*(3), 413–432.
Menzies, P., & Price, H. (1993). Causation as a secondary quality. *British Journal for the Philosophy of Science, 44*, 187–203.
Myrvold, W. (2000). Einstein's untimely burial. PhilSci preprint http://philsci-archive.pitt.edu/222/.

Myrvold, W. (2002). On peaceful coexistence – Is the collapse postulate incompatible with relativity? *Studies in History and Philosophy of Modern Physics, 33*(3), 435–466.
Myrvold, W. (2017a). Relativistic Markovian dynamical collapse theories must employ nonstandard degrees of freedom. *Physical Review A, 96*(6), 062116.
Myrvold, W. (2017b). Ontology for collapse theories. In Shan Gao (Ed.), *Collapse of the wave function – Models, ontology, origin, and implications* (pp. 99–126). Cambridge: Cambridge University Press.
Myrvold, W. (2018a). Private communication, New Directions in Philosophy of Physics, Viterbo, June 2018. http://carnap.umd.edu/philphysics/newdirections18.html.
Myrvold, W. (2018b). Philosophical issues in quantum theory. In E. N. Zalta (Ed.), *The Stanford encyclopedia of philosophy* (Fall 2018 Edition). https://plato.stanford.edu/archives/fall2018/entries/qt-issues/.
Myrvold, W. (2019). Ontology for relativistic collapse theories. In O. Lombardi, S. Fortin, C. López, & F. Holik (Eds.), *Quantum worlds* (pp. 9–31). Cambridge: Cambridge University Press. http://philsci-archive.pitt.edu/14718/.
Myrvold, W. (2020). Subjectivists about quantum probabilities should be realists about quantum states. This volume, pp. 449–465.
Nelson, E. (1966). Derivation of the Schrödinger equation from Newtonian mechanics. *Physical Review, 150*, 1079–1085.
Nelson, E. (1985). *Quantum fluctuations*. Princeton: Princeton University Press.
Nicrosini, O., & Rimini, A. (2003). Relativistic spontaneous localization – A proposal. *Foundations of Physics, 33*(7), 1061–1084.
Pearle, P. (1993). Ways to describe dynamical state-vector reduction. *Physical Review A, 48*(2), 913–923.
Pearle, P. (2015). Relativistic dynamical collapse model. *Physical Review D, 91*, article 105012.
Pitowsky, I. (1989). *Quantum probability, quantum logic* (Lecture notes in physics, Vol. 321). Berlin: Springer.
Pitowsky, I. (1994). George Boole's 'conditions of possible experience' and the quantum puzzle. *British Journal for the Philosophy of Science, 45*, 95–125.
Pitowsky, I. (2003). Betting on the outcomes of measurements – A Bayesian theory of quantum probability. *Studies in History and Philosophy of Modern Physics, 34*(3), 395–414.
Pitowsky, I. (2007). Quantum mechanics as a theory of probability. In W. Demopoulos & I. Pitowsky (Eds.), *Physical theory and its interpretation – Festschrift in honor of Jeffrey Bub* (Western Ontario series in philosophy of science, Vol. 72, pp. 213–240). New York: Springer.
Pitowsky, I. (2012). Typicality and the role of the Lebesgue measure in statistical mechanics. In Y. Ben-Menahem & M. Hemmo (Eds.), *Probability in physics* (pp. 41–58). Heidelberg: Springer.
Popper, K. R. (1982). *Quantum theory and the schism in physics: Postscript to the logic of scientific discovery, Vol. III*. W. W. Bartley III (Ed.). London: Hutchinson.
Price, H. (1997). *Time's arrow and archimedes' point – New directions for the physics of time*. Oxford: Oxford University Press.
Price, H. (2010). Decisions, decisions, decisions – Can Savage salvage Everettian probability? In S. Saunders, J. Barrett, A. Kent, & D. Wallace (Eds.), *Many worlds? Everett, quantum theory, and reality* (pp. 369–391). Oxford: Oxford University Press.
Pusey, M. F., Barrett, J., & Rudolph, T. (2012). On the reality of the quantum state. *Nature Physics, 8*(6), 475–478.
Rijken, S. (2018). Spacelike and timelike non-locality. Masters Thesis, History and Philosophy of Science, Utrecht University. https://dspace.library.uu.nl/handle/1874/376363.
Saunders, S. (2002). How relativity contradicts presentism. *Royal Institute of Philosophy Supplements, 50*, 277–292.
Saunders, S., Barrett, J., Kent, A., & Wallace, D. (Eds.). (2010). *Many worlds? Everett, quantum theory, and reality*. Oxford: Oxford University Press.
Sober, E. (1993). Temporally oriented laws. *Synthese, 94*, 171–189.

Uffink, J. (2007). Compendium of the foundations of classical statistical physics. In J. Butterfield & J. Earman (Eds.), *Handbook of the philosophy of physics: Part B* (pp. 923–1074). Amsterdam: North-Holland.

Valentini, A. (1996). Pilot-wave theory of fields, gravitation and cosmology. In J. T. Cushing, A. Fine, & S. Goldstein (Eds.), *Bohmian mechanics and quantum theory – An appraisal* (pp. 45–66). Dordrecht: Springer.

Wallstrom, T. C. (1989). On the derivation of the Schrödinger equation from stochastic mechanics. *Foundations of Physics Letters, 2*(2), 113–126.

Watanabe, S. (1965). Conditional probability in physics. *Progress of Theoretical Physics Supplement, E65*, 135–167.

Wharton, K. (2014). Quantum states as ordinary information. *Information, 5*, 190–208.

Wallace, D. (2010). How to prove the Born rule. In S. Saunders, J. Barrett, A. Kent, & D. Wallace (Eds.), *Many worlds? Everett, quantum theory, and reality* (pp. 227–263). Oxford: Oxford University Press.

Wallace, D. (2011). The Logic of the past hypothesis. PhilSci preprint http://philsci-archive.pitt.edu/8894/.

Chapter 4
Wigner's Friend as a Rational Agent

Veronika Baumann and Časlav Brukner

Abstract In a joint paper Jeff Bub and Itamar Pitowsky argued that the quantum state represents *"the credence function of a rational agent [...] who is updating probabilities on the basis of events that occur"*. In the famous thought experiment designed by Wigner, Wigner's friend performs a measurement in an isolated laboratory which in turn is measured by Wigner. Here we consider Wigner's friend as a rational agent and ask what her "credence function" is. We find experimental situations in which the friend can convince herself that updating the probabilities on the basis of events that happen solely inside her laboratory is not rational and that conditioning needs to be extended to the information that is available outside of her laboratory. Since the latter can be transmitted into her laboratory, we conclude that the friend is entitled to employ Wigner's perspective on quantum theory when making predictions about the measurements performed on the entire laboratory, in addition to her own perspective, when making predictions about the measurements performed inside the laboratory.

Keywords Wigner's friend · (Observer) paradox · Measurement problem · Interpretations of quantum theory · Realism · Subjectivism

---

V. Baumann (✉)
Vienna Center for Quantum Science and Technology (VCQ), Faculty of Physics, University of Vienna, Vienna, Austria

Faculty of Informatics, Università della Svizzera italiana, Lugano, Switzerland
e-mail: baumav@usi.ch

Č. Brukner
Vienna Center for Quantum Science and Technology (VCQ), Faculty of Physics, University of Vienna, Vienna, Austria

Institute of Quantum Optics and Quantum Information (IQOQI), Austrian Academy of Sciences, Vienna, Austria
e-mail: caslav.brukner@univie.ac.at

M. Hemmo, O. Shenker (eds.), *Quantum, Probability, Logic*,
Jerusalem Studies in Philosophy and History of Science,
https://doi.org/10.1007/978-3-030-34316-3_4

4.1 Introduction

Wigner's-friend-type experiments comprise observations of observers who themselves have performed measurements on other systems. In the standard version of the thought experiment (Wigner 1963) there is an observer called Wigner's friend measuring a quantum system in a laboratory, and a so-called superobserver Wigner, to whom the laboratory including the friend constitutes one joint quantum system. From the point of view of the friend, the measurement produces a definite outcome, while for Wigner the friend and the system undergo a unitary evolution. The discussion surrounding the Wigner's-friend gedankenexperiment revolves around the questions: Has the state of the physical system collapsed in the experiment? Has it collapsed only for the friend or also for Wigner?

Recently, extended versions of Wigner's-friend scenarios have been devised in order to show the mutual incompatibility of a certain set of (seemingly) plausible assumptions. In Brukner (2015, 2018) a combination of two spatially distant Wigner's-friend setups is used as an argument against the idea of "observer-independent facts" via the violation of a Bell-inequality. In Frauchiger and Renner (2018) two standard Wigner's-friend setups are combined such that one arrives at a contradiction when combining the supposed knowledge and reasoning of all the agents involved. Both arguments are closely related to the quantum measurement problem and attempted resolutions involve formal as well as interpretational aspects (Brukner 2015; Baumann and Wolf 2018). At the core of the formal aspects is the standard measurement-update rule. Without further ontological commitment regarding a "collapse of the wave function" the latter can be considered to be the rational-update-belief rule for agents acting in laboratories. The updated state is the one that gives the correct probabilities for subsequent measurements and would, hence, be used by a rational agent to make her predictions. That a subjective (observer dependent) application of the measurement-update rule would lead to the friend and Wigner making different predictions was already discussed in Hagar and Hemmo (2006). In general, however, this would be admissible as long as these predictions cannot directly be tested against each other, i.e. have no factual relevance.

Here we find a situation in which the friend can *verify* that her prior prediction about future measurement outcomes – which is based on applying the standard state-update rule on the basis of the events that happen inside her laboratory – is incorrect. In this way, the friend can convince herself, that her prediction cannot be conditioned solely on the outcomes registered inside her laboratory, but needs to take into account the information that is available outside of the laboratory as well. The latter can be transmitted to her by Wigner or an automatic machine from outside her laboratory. Equipped with this additional information the friend is entitled to adopt both Wigner's and her own perspective on the use of quantum theory.

4.2 Wigner, Friend and Contradictions

We first consider a simple version of our Wigner's-friend experiment, see also Finkelstein (2005), moving later to a more general case. Let a source emit a qubit S (spin-1/2 particle) in state $|\phi\rangle_S = 1/\sqrt{2}(|\uparrow\rangle_S + |\downarrow\rangle_S)$, where $|\uparrow\rangle_S$ is identified with "spin up" and $|\downarrow\rangle_S$ with "spin down" along the z-direction. The friend F measures the particle in the σ_z-basis (Pauli z-spin operator) with the projectors M_F: $\{|\uparrow\rangle\langle\uparrow|_S, |\downarrow\rangle\langle\downarrow|_S\}$. To Wigner W, who is outside and describes the friend's laboratory as a quantum system, the measurement of the qubit by the friend constitutes a unitary evolution entangling certain degrees of freedom of the friend – her memory that is initially in state $|0\rangle$ (i.e. the register in state "no measurement") – with the qubit. Suppose that this results in the overall state

$$|\phi\rangle_S|0\rangle_F \mapsto \frac{1}{\sqrt{2}}(|\uparrow\rangle_S|U\rangle_F + |\downarrow\rangle_S|D\rangle_F) =: |\Phi^+\rangle_{SF}, \tag{4.1}$$

where $|Z\rangle_F$ with $Z \in \{U, D\}$ is the state of the friend having registered outcome $z \in \{$"up", "down"$\}$ respectively. Wigner can perform a highly degenerate projective measurement M_W: $\{|\Phi^+\rangle\langle\Phi^+|_{FS}, \mathbb{1} - |\Phi^+\rangle\langle\Phi^+|_{FS}\}$ to verify his state assignment. The respective results of the measurement will occur with probabilities $p(+) = 1$ and $p(-) = 0$. We note that the specific relative phase between the two amplitudes in Eq. (4.1) (there chosen to be zero) is determined by the interaction Hamiltonian between the friend and the system, which models the measurement and is assumed to be known to and in control of Wigner.

We next introduce a protocol through which Wigner's friend will realise that her predictions will be incorrect, if she bases them on the state-update rule conditioned only on the outcomes observed inside her laboratory. The protocol consists of three steps (see Fig. 4.1): (a) The friend performs measurement M_F and registers an outcome; (b) The friend makes a prediction about the outcome of measurement

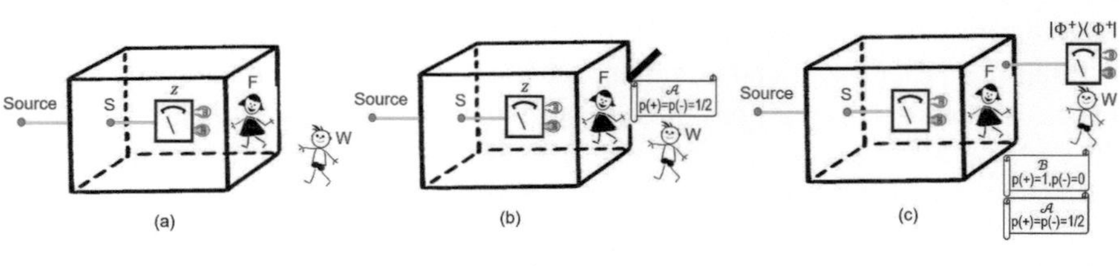

Fig. 4.1 Three steps of the protocol: (**a**) Wigner's friend (F) performs a measurement M_F on the system S in the laboratory and registers an outcome. (**b**) The friend makes a prediction about the outcome of a future measurement M_W on the laboratory (including the system and her memory), for example, by writing it down on a piece of paper $\mathcal{A}$. She communicates her prediction to Wigner (W). (**c**) Wigner performs measurement M_W. The three steps (**a–c**) are repeated until sufficient statistics for measurement M_W is collected. The statistics is noted, for example, on another piece of paper $\mathcal{B}$. Afterwards, the "lab is opened" and the friend can compare the two lists $\mathcal{A}$ and $\mathcal{B}$. An alternative protocol is given in the main text

M_W and communicates it to Wigner; (c) Wigner performs measurement M_W Subsequently, Wigner "opens the lab" and the friend can exchange information with him. Note that apart from the outcome of measurement M_F no further information is given to the friend.

Step (a): Following the laboratory practice and textbook quantum mechanics, when the friend observers outcome z, she applies the measurement-update rule, and consequently predicts the probabilities: $p(+|z) = |\langle z|\langle Z|\Phi^+\rangle_{SF}|^2 = \frac{1}{2} = p(-|z)$, where $|z\rangle_S|Z\rangle_F$ can be either $|\uparrow\rangle_S|U\rangle_F$ or $|\downarrow\rangle_S|D\rangle_F$ depending on which outcome the friend observes. It is important to note that the friend's prediction of the result of M_W does not reveal the result of M_F, because for either outcome z the predicted conditional probability is the same, i.e. $p(+) = p(-) = \frac{1}{2}$. This means that the friend's prediction could be sent out of the laboratory and communicated to Wigner without changing the superposition state $|\Phi^+\rangle_{SF}$.

Note that the predictions made by the friend in Wigner's-friend-type experiments require her to assign a state to herself, which is sometimes regarded as problematic. One might reject this possibility altogether and, therefore, claim quantum theory does not allow the friend to make predictions about Wigner's measurement results. Nonetheless, we will see that there is an operationally well-defined procedure, which if followed by the friend, would give a correct prediction for the measurement. Moreover, since the measurement bases of both the friend and Wigner are fixed, it is always possible to have a classical joint probability distribution for their outcomes, whose marginals reproduce the observed distributions by the friend and Wigner. However, such a classical description of the experiment is no longer possible if one extends the protocol into one with more measurement choices and Bell-type set-ups (Brukner 2015, 2018).

Step (b): The friend opens the laboratory in a manner that allows a record of her prediction (e.g., a specific message $\mathcal{A}$ written on a piece of paper) to be passed outside to Wigner, keeping all other degrees of freedom fully isolated and in this way preserving the coherence of the state. The piece of paper waits outside of the laboratory to be evaluated later in the protocol. The state that Wigner assigns to the friend's laboratory at the present stage is $|\Phi^+\rangle_{SF}|p(+)=p(-)=1/2\rangle_{\mathcal{A}}$, where the second ket refers to the message.

Step (c): After the friend communicates her prediction to the outside, Wigner (or an automatic machine) performs measurement M_W and records the result, for example on a separate piece of paper $\mathcal{B}$. At this stage of the experiment, Wigner assigns to the friend's laboratory and the two messages the state $|\Phi^+\rangle_{SF}|p(+) = p(-) = 1/2\rangle_{\mathcal{A}}|p(+) = 1, p(-) = 0\rangle_{\mathcal{B}}$ with the two last kets representing the two lists.

We note that the measured state $|\Phi^+\rangle_{SF}$ is an eigenstate of the measured operator. This implies that outcome "+" will occur with unit probability, and furthermore that the measurement does not change the state of the laboratory. Hence the three steps (a–c) of the protocol can be repeated again and again without changing the initial superposition state until a sufficient statistics in measurement M_W is collected. As a result, the two pieces of paper, one with the friend's prediction

$\mathcal{A} = \{p(+) = \frac{1}{2}, p(-) = \frac{1}{2}\}$, and one with the actually observed relative number of counts for the two outcomes $\mathcal{B} = \{p(+) = 1, p(-) = 0\}$, display a statistically significant difference. In the very last step of the protocol "the laboratory is opened" (which is equivalent to Wigner performing a measurement in the basis $\{|z\rangle\langle z|_S \otimes |Z'\rangle\langle Z'|_F\}$), such that the joint states for the systems and the friend's memory are reduced to the form $|z\rangle_S|Z\rangle_F$. Irrespective of the specific state to which the friend is reduced, she can now compare the two messages and convince herself that her prior prediction deviates from the actually observed statistics.

In an alternative protocol the friend resides in the laboratory for the duration of the whole protocol. Steps (a) and (c) are retained, but instead of step (b), the friend either receives the result of measurement M_W from Wigner after each measurement run or the entire list $\mathcal{B}$ at the end of the protocol. In either case the friend can compare the two lists inside the laboratory and arrive at the same conclusion as in the previous protocol. One may object that in the present case the friend's conclusion relies on Wigner being trustful and reliable when providing her with the measurement result of M_W. As mentioned above, this is irrelevant as Wigner can be replaced in the protocol by an automatic machine that the friend could have pre-programmed herself.

The discrepancy between the friend's prediction and the actual statistics can be made as high as possible in a limit. Consider that the source emits a higher dimensional quantum system in state $|\phi_d\rangle_S = \frac{1}{\sqrt{d}}\sum_{j=1}^{d}|j\rangle_S$ and the friend measures in the respective basis, $M_F : \{|j\rangle\langle j|_S\}$ with $j = 1 \dots d$. Wigner describes the measurement of the friend as a unitary process that results in state

$$|\phi_d\rangle_S|0\rangle_F \mapsto \frac{1}{\sqrt{d}}\sum_{j=1}^{d}|j\rangle_S|\alpha_j\rangle_F =: |\Phi_d^+\rangle_{SF}, \tag{4.2}$$

where $|\alpha_j\rangle_F$ is the state of the friend's memory after having observed outcome j. For M_W we choose: $\{|\Phi_d^+\rangle\langle\Phi_d^+|_{SF}, \mathbb{1} - |\Phi_d^+\rangle\langle\Phi_d^+|_{SF}\}$. Wigner records the "+" result with unit probability, i.e. $p(+) = 1$ and $p(-) = 0$. Using the measurement-update rule the friend, however, predicts

$$\begin{aligned} p(+|j) &= |\langle j|\langle\alpha_j|\Phi_d^+\rangle_{SF}|^2 = \frac{1}{d} \underset{d\to\infty}{\longrightarrow} 0 \\ p(-|j) &= 1 - \frac{1}{d} \underset{d\to\infty}{\longrightarrow} 1, \end{aligned} \tag{4.3}$$

independently of the actual outcome she registers in measurement M_F.

Either the three steps (a–c) of the protocol in Fig. 4.1 can be repeatedly applied or Winger and the friend follow the alternative protocol. In both cases, the friend can convince herself, without relying on the knowledge other agents may have, that she made an incorrect prediction of future statistics using the state-update rule conditioned solely on the observations made within her laboratory. As the dimension of the measured system and the number of outcomes increases, the discrepancy between her prediction and the actual statistics becomes maximal, giving rise to an "all-versus-nothing" argument.

4.3 Discussion

We now discuss possible responses of different interpretations and modifications of quantum theory to the experimental situation described above.

(1) *Theories that deny universal validity of quantum mechanics.* The theories postulate a modification of standard quantum mechanics, such as spontaneous (Ghirardi et al. 1986) and gravity-induced (Penrose 2000; Diósi 2014) collapse models, which become significant at the macroscopic scale. These models could exclude superpositions of observers as required in Wigner's-friend-type experiments and hence the protocol cannot be run. In our view this is the most radical position as it would give rise to new physics.
(2) *Many-worlds interpretation* (Wheeler 1957). These incorrect prediction of the friend is due to her lack of knowledge of the total state of the laboratory including the friend's memory (i.e. "the wave function of the many-worlds"). If the friend would know this state, she could make a prediction that is in accordance with actually observed statistics.
(3) *Quantum Bayesian* (Fuchs 2010), *neo-Copenhagen* (Brukner 2015), *and relational interpretation* (Rovelli 1996). The quantum state is always given relative to an observer and represents her/his knowledge or degree of belief. It is natural for an agent to update his or her degree of belief in face of new information. As a consequence of newly acquired knowledge in the protocol the friend would update her degree of belief and assign a new state to the laboratory that is in agreement with the observations.

Note that in both interpretations (2) and (3) the friend *can learn* the state of the laboratory as described by Wigner. Wigner can simply communicate either the initial state $|\phi\rangle_S|0\rangle_F$ of the system together with the Hamiltonian of the laboratory, or the final state $|\Phi_d^+\rangle_{SF}$ of the laboratory to the friend before step (b) begins. The important feature is that this can be done without destroying the coherence of the quantum state of the laboratory. In this way, the friend can learn the overall state of her laboratory from the perspective of an outside observer and include it in making her prediction. As a consequence, the friend may operate with a pair of states $\{|z\rangle_S, |\Phi_d^+\rangle_{SF}\}$. The first component is used for predictions of measurements made on the system alone and is conditional on the registered outcome in the laboratory; the second component is used for predictions of measurements performed on the entire laboratory (= "system + friend"). When an outcome of a measurement on the system is registered in the laboratory, the friend applies the state-update rule on the first component, without affecting the second component of the pair of states.

Alternatively, the friend can use only the quantum state Wigner assigns to the laboratory, but apply a modification of the Born rule as introduced in Baumann and Wolf (2018). The new rule enables to evaluate conditional probabilities in a sequence of measurements. In the case when the sequence of measurements is performed on the same system, it restores the predictions of the standard state-update rule. However, the rule enables making predictions in Wigner's-friend scenarios as well, where the first measurement is performed on the system by Wigner's friend, and the subsequent measurement is performed on the "system + friend" by Wigner. In this case, the rule recovers the prediction as given by Eq. (4.3). This modified Born rule gains an operational meaning in the present setup where Wigner's friend has access to outcomes of both measurements and can evaluate the computed conditional probabilities (Note that Wigner does not have this status, since he has no access to the outcome of the friend's measurement.). While Bub and Pitowsky claim in Bub and Pitowsky (2010) that *"Conditionalizing on a measurement outcome leads to a nonclassical updating of the credence function represented by the quantum state via the von Neumann-Lüders rule..."* we argue that in case of Wigner's friend experiments both the "outside" and "inside" perspective should contribute to "the credence function" for making predictions.

Modified Born Rule (Baumann and Wolf 2018)

We apply the modified Born rule of Baumann and Wolf (2018) to the present situation. The rule defines the conditional probabilities in a sequence of measurements. According to it, every measurement is described as a unitary evolution analogous to Eq. (4.1). Consider a sequence of two measurements, the first one described by unitary U and the subsequent one by unitary V. The sequence of unitaries correlate the results of the two measurements with two memory registers 1 and 2. The overall state of the system and the two registers evolve as

$$|\phi\rangle_S|0\rangle_1|0\rangle_2 \xrightarrow{U} \sum_j \langle j|\phi\rangle|j\rangle_S|\alpha_j\rangle_1|0\rangle_2 \tag{4.4}$$

$$\xrightarrow{V} \sum_{jk} \langle j|\phi\rangle\langle\tilde{\beta}_{jk}|j,\alpha_j\rangle|\tilde{\beta}_{jk}\rangle_{S1}|\beta_k\rangle_2 = |\Phi_{tot}\rangle.$$

The conditional probability for observing outcome $\bar{k}$ given that outcome $\bar{j}$ has been observed in the previous measurement is given by

$$p(\bar{k}|\bar{j}) = \frac{p(\bar{k},\bar{j})}{\sum_{\bar{k}} p(\bar{k},\bar{j})}. \tag{4.5}$$

Here $p(\bar{k}, \bar{j})$ is the joint probability for observing the two outcomes and is given by the projection of the overall state $|\Phi_{tot}\rangle$ on the states $|\alpha_{\bar{j}}\rangle_1$ and $|\beta_{\bar{k}}\rangle_2$ of the two registers:

$$p(\bar{j}, \bar{k}) = \mathrm{Tr}[(\mathbb{1}_S \otimes |\alpha_{\bar{j}}\rangle\langle\alpha_{\bar{j}}|_1 \otimes |\beta_{\bar{k}}\rangle\langle\beta_{\bar{k}}|_2)|\Phi_{tot}\rangle\langle\Phi_{tot}|] \tag{4.6}$$

In the case when the sequence of the measurements are performed on a single system, the rule restores the prediction of the standard state-update rule, i.e. $p(\bar{k}, \bar{j}) = |\langle\bar{k}|\bar{j}\rangle|^2$. This corresponds to the state

$$|\Phi_{tot}\rangle = \sum_{jk} \langle j|\phi\rangle\langle k|j\rangle|k\rangle_S|\alpha_j\rangle_1|\beta_k\rangle_2. \tag{4.7}$$

For Wigner's-friend scenarios, however, the friend would predict conditional probabilities for Wigner's result $+$ given her result j that are in accordance with Wigner's observations Eq. (4.3). In this case the overall state is

$$|\Phi_{tot}\rangle = \sum_{j=1}^{d} \langle j|\phi\rangle|j\rangle_S|\alpha_j\rangle_1|+\rangle_2 \tag{4.8}$$

where $|+\rangle_2$ is the state of the memory register corresponding to observing the result of $|\Phi_d^+\rangle\langle\Phi_d^+|_{S1}$, and the joint probability is given by

$$p(+, j) = \mathrm{Tr}[(\mathbb{1}_S \otimes |\alpha_j\rangle\langle\alpha_j|_1 \otimes |+\rangle\langle+|_2)|\Phi_{tot}\rangle\langle\Phi_{tot}|]. \tag{4.9}$$

This means that full unitary quantum theory together with the adapted Born rule allows for consistent predictions in Wigner's-friend-type scenarios, while giving the same predictions as the standard Born and measurement-update rules for standard observations.

Acknowledgements We thank Borivoje Dakic, Fabio Costa, Flavio del Santo, Renato Rennner and Ruediger Schack for useful discussion. We also thank Jeff Bub for the discussions and for the idea of the alternative protocol. We acknowledge the support of the Austrian Science Fund (FWF) through the Doctoral Programme CoQuS (Project no. W1210-N25) and the projects I-2526-N27 and I-2906. This work was funded by a grant from the Foundational Questions Institute (FQXi) Fund. The publication was made possible through the support of a grant from the John Templeton Foundation. The opinions expressed in this publication are those of the authors and do not necessarily reflect the views of the John Templeton Foundation.

References

Baumann, V., & Wolf, S. (2018). On formalisms and interpretations. *Quantum, 2*, 99.

Brukner, Č. (2015). On the quantum measurement problem. arXiv:150705255. https://arxiv.org/abs/1507.05255.

Brukner, Č. (2018). A no-go theorem for observer-independent facts. *Entropy, 20*(5), 350.

Bub, J., & Pitowsky, I. (2010). Two dogmas about quantum mechanics. In *Many worlds?: Everett, quantum theory, & reality* (pp. 433–459). Oxford: Oxford University Press.

Diósi, L. (2014). Gravity-related wave function collapse. *Foundations of Physics, 44*(5), 483–491.

Finkelstein, J. (2005). "quantum-states-as-information" meets Wigner's friend: a comment on hagar and hemmo. arXiv preprint quant-ph/0512203.

Frauchiger, D., & Renner, R. (2018). Quantum theory cannot consistently describe the use of itself. *Nature Communications, 9*(1), 3711.

Fuchs, C. A. (2010). Qbism, the perimeter of quantum bayesianism. arXiv:10035209. https://arxiv.org/abs/1003.5209.

Ghirardi, G. C., Rimini, A., & Weber, T. (1986). Unified dynamics for microscopic and macroscopic systems. *Physical Review D, 34*(2):470. https://doi.org/10.1103/PhysRevD.34.470.

Hagar, A., & Hemmo, M. (2006). Explaining the unobserved—Why quantum mechanics ain't only about information. *Foundations of Physics, 36*(9), 1295–1324.

Penrose, R. (2000). Wavefunction collapse as a real gravitational effect. In *Mathematical physics 2000* (pp. 266–282). Singapore: World Scientific.

Rovelli, C. (1996). Relational quantum mechanics. *International Journal of Theoretical Physics, 35*(8), 1637–1678.

Wheeler, J. A. (1957). Assessment of everett's "relative state" formulation of quantum theory. *Reviews of Modern Physics, 29*(3), 463.

Wigner, E. P. (1963). The problem of measurement. *American Journal of Physics, 31*(1), 6–15. https://doi.org/10.1119/1.1969254.

Chapter 5
Pitowsky's Epistemic Interpretation of Quantum Mechanics and the PBR Theorem

Yemima Ben-Menahem

Abstract In his work on the foundations of Quantum Mechanics, Itamar Pitowsky set forth an epistemic interpretation of quantum state functions. Shortly after his untimely death, Pusey, Barrett and Rudolph published a theorem—now known as the PBR theorem—that purports to rule out epistemic interpretations of quantum states and render their realist interpretations mandatory. If one accepts this conclusion, the theorem is fatal to Pitowsky's interpretation. The aim of this paper is to cast doubt on this verdict and show that Pitowsky's position is actually unaffected by the PBR theorem. Had it been a form of instrumentalism, the defense of Pitowsky's view would have been trivial, for according to its authors, the PBR theorem does not apply to instrumentalist readings of quantum mechanics. I argue, however, that Pitowsky's position differs significantly from instrumentalism, and yet, remains viable in the face of the PBR theorem.

Keywords Realism · Instrumentalism · Probability · Subjective probability · Quantum mechanics · PBR theorem · Epistemic interpretation

5.1 Introduction

The probabilistic interpretation of Schrödinger's equation, offered by Max Born in 1926, constitutes as essential component of standard interpretations of quantum mechanics (QM).[1] In comparison with the use of probability in physics up to that

This paper draws on an earlier paper on the PBR theorem (Ben-Menahem 2017).

[1]Standard interpretations are mostly descendants of the Copenhagen interpretation to which Born contributed and subscribed. Standard interpretations need not embrace such Copenhagen tenets

Y. Ben-Menahem (✉)
Department of Philosophy, The Hebrew University of Jerusalem, Jerusalem, Israel
e-mail: yemima.ben-menahem@mail.huji.ac.il

M. Hemmo, O. Shenker (eds.), *Quantum, Probability, Logic*, Jerusalem Studies in Philosophy and History of Science,
https://doi.org/10.1007/978-3-030-34316-3_5

point, the novelty of Born's interpretation was that it construed quantum probabilities as fundamental in the sense that they were believed to be *irreducible*, that is, they were not alleged to arise from averaging over an underlying deterministic theory. In other words, quantum probabilities were said to be different from those of other probabilistic theories, statistical mechanics in particular, where the existence of an underlying deterministic theory—classical mechanics—is taken for granted. The proclaimed irreducibility gave rise to serious conceptual puzzles, (more on these below), but, from the pragmatic/empirical point of view, the notion that quantum states (as represented and calculated by QM) stand for probability amplitudes of measurement results was sufficient to make quantum mechanics a workable (and highly successful) theory.

Itamar Pitowsky endorsed the irreducibly probabilistic nature of QM, adding to it a major twist. On the basis of developments such as Bell's inequalities, Gleason's theorem and other no-go theorems, he argued that QM is not merely a probabilistic theory, but a *non-classical* probabilistic theory.[2] Studying the structure of the new theory in both algebraic and geometrical terms, he was able to demonstrate that quantum probabilities deviate from classical probabilities in allowing *more correlations* between events than classical probability theory would allow. He saw his interpretation as a response to the threat of non-locality and as a key to the solution of other puzzles such as the notorious measurement problem. In addition, his understanding of QM as a theory of non-classical probability also constitutes a new take on the status of the theory of probability, turning it from an *a priori* extension of logic into an empirical theory, which may or may not apply to particular empirical domains.[3]

The meaning of the concept of probability, which has been debated among scholars for a long time, is of particular concern in the context of QM. An option that has attracted a great deal of attention in the last few decades is an epistemic interpretation of probability (also known as information-theoretic, or Bayesian), according to which probabilities in general, and quantum probabilities in particular, represent states of knowledge or belief or information. How are we to complete this sentence—knowledge (belief, information) about *what*? It turns out that the

---

as complementarity (which had actually been debated by members of the Copenhagen school), but from the perspective of this paper and the problems about probability it discusses, the term 'standard interpretation' will do. I will not discuss the notion of probability in non-standard approaches such as Bohm's theory or the Many Worlds interpretation.

[2]The non-classical character of quantum probability had been recognized, for example, by Richard Feynman: "But far more fundamental was the discovery that in nature the laws of combining probabilities were not those of the classical probability theory of Laplace" (Feynman 1951, p. 533). In actually developing and axiomatizing the non-classical calculus of quantum probability, Pitowsky went far beyond that general recognition. I will not undertake a comparison of Pitowsky's work with that of his predecessors or contemporaries.

[3]As a theory of rational belief, the theory of probability can be seen as an extension of logic and the idea that the theory of probability is empirical—as extension of the view that logic is empirical. Pitowsky's interpretation of QM is thus related to the earlier suggestion that QM obeys a non-classical logic—quantum logic. The connection is explicit in the title of Pitowsky's (1989) *Quanyum Probability—Quantum Logic*.

question is far from trivial; theorists who share an epistemic interpretation of quantum mechanics may still differ significantly regarding its answer.[4] As Pitowsky subscribed to an epistemic interpretation of quantum probabilities, his answer is central to what follows and will be examined in detail.

In 2011, a year after Pitowsky's premature death, Pusey, Barrett and Rudolph published the theorem now known as the PBR theorem (Pusey et al. 2012), which purports to rule out epistemic interpretations of QM and vindicate realist ones.[5] Pitowsky's epistemic interpretation thus appeared to be undermined. I will argue, however, that it is not. It should be stressed right away that Pitowsky's interpretation is not a form of instrumentalism. If it were, the argument would be trivial, for—as its authors concede—the PBR theorem does not apply to instrumentalist understandings of quantum mechanics. Exploring the difference between Pitowsky's position and instrumentalism in terms of their empirical implications, I will show that despite its departure from instrumentalism, Pitowsky's interpretation, remains viable in the face of the PBR theorem. The structure of the paper is the following: In Sect. 5.2, I describe some of the above-mentioned conceptual problems that, from early on, plagued the probabilistic interpretation of QM. Section 5.3 provides a summary of Pitowsky's epistemic interpretation and Sect. 5.4—a summary of the PBR theorem. In Sect. 5.5, I explain why Pitowsky's interpretation is not vulnerable to the PBR theorem. Section 5.6 addresses the thorny issue of instrumentalism.

5.2 Born's Probabilistic Interpretation and Its Problems

In 1926, upon realizing that Schrödinger's wavefunction cannot represent an ordinary three-dimensional wave, Born offered his probabilistic interpretation according to which quantum states, described by Schrödinger's Ψ function, represent probability amplitudes. Born was explicit about the curious character of Ψ under this interpretation: Although particles follow probabilistic laws, he said, the probability itself transforms "in accordance with the causal principle," i.e., deterministically. Furthermore, QM only answers well-posed statistical questions, and remains silent about the individual process. Born therefore characterized QM as a "peculiar blend of mechanics and statistics,"—"*eine eigenartige Verschmelzung von Mechanik und Statistik*".[6] This formulation is almost identical to that given by Jaynes and quoted in the PBR paper

[4]Prior to the publication of the PBR theorem, the difference between possible answers to this question was rather obscure. One of the contributions of the theorem is that it sheds light on this difference.

[5]The 2011 version of the paper was published in the **arXiv quant-ph11113328.**

[6]Born [1926] (1963), p. 234. "Die Bewegung der partikeln folgt Wharscheinlichkeitsgesetzen, die wharscheinlichkeit selbst aber breitet sich im Einklang mit dem Kausalgesetz aus."... Die Quantenmechanik allenfals nur Antwort gibt auf richtig gestellte statistische Fragen, aber im allgemeinen die Frage nach dem Ablauf sines Einzelprozesses unbeantwortet laesst."

> Our present [quantum mechanical] formalism ... is a peculiar mixture describing in part realities of Nature, in part incomplete human information about Nature, all scrambled up by Heisenberg and Bohr into an omelette that nobody has seen how to unscramble. (Pusey et al. 2012, p. 475).[7]

The majority of quantum physicists were willing to tolerate the paradoxical situation diagnosed by Born. They treated QM both as a probabilistic theory and as a fundamental physical theory in which the wave equation plays the role of the equations of motion in classical mechanics. The question of whether the concept of probability could sustain this *Verschmelzung* was left hanging in midair. A dissident minority, including Einstein, preferred to bite the bullet and understand QM as a full-blown probabilistic theory, that is, a statistical description of an *ensemble* of similar systems that makes no definite claims about individual members of the ensemble.[8] Hence, the immediate analogy with statistical mechanics.

Statistical mechanics does indeed hold promise for peaceful coexistence between the probabilistic description of an ensemble and the precise physical description of each individual system comprising the ensemble.[9] Even if this precise description is inaccessible in practice, it is taken to be conceivable in principle. The analogy between statistical and quantum mechanics suggested that quantum states correspond to macrostates in statistical mechanics, implying that in QM too, individual systems could occupy determinate physical states. These determinate states, though, like microstates in statistical mechanics, are not directly represented by quantum states; the latter only represent the statistical behavior of the ensemble. The picture of an underlying fundamental level, the description of which bears to QM the relation that classical mechanics bears to statistical mechanics, seemed to diminish the novelty and uniqueness of QM. It was only natural, then, that advocates of the Copenhagen interpretation opposed this picture and the analogy with statistical mechanics on which it was based. For them, QM was a novel theory—complete, fundamental and irreducible—that *replaces* classical mechanics rather than supplementing it with a convenient way of treating multi-particle systems. They therefore disagreed with ensemble theorists about the prospects of completing QM by means of 'hidden variable' theories.

---

[7] Jaynes presupposes an epistemic interpretation of probability, but the scrambling he refers to does not depend on it. We could replace the epistemic construal of probability with an objective ensemble interpretation and the question would remain the same: What, exactly, does the state function represent? Nonetheless, there are significant differences between the ensemble and epistemic interpretations, which the PBR theorem helps clarify, and which will be addressed later.

[8] The ensemble interpretation was recommended, for example by Blokhintsev (1968), Ballentine (1970), Popper (1967).

[9] It is well-known that there are serious problems regarding the compatibility of statistical mechanics and classical mechanics, but these pertain mainly to the issue of irreversibility and will not concern us here. See Albert (2000), Hemmo and Shenker (2012), and the literature there cited. Note, further, that despite the fact that the central laws of statistical mechanics are probabilistic, there is no binding reason to construe these probabilities as subjective.

Neither of these competing interpretations of quantum probabilities was epistemic. In this respect, there is a significant difference between the alternatives debated in the first decades of QM and the contemporary alternatives juxtaposed in the PBR argument, where the divide is between objective (physical, ontic) and epistemic interpretations. Given that epistemic interpretations of probability in general (outside of the quantum context) came of age later than objective interpretations, this difference should not surprise us.[10] It is significant, though, that when, in the wake of work by de Finetti and Savage, epistemic interpretations became popular, they were understood as diverging from traditional interpretations only in their conceptual basis, not in the probabilities calculated by them. Indeed, it was considered to be the great achievement of these epistemic interpretations that despite their subjective starting point, they recovered the rules of objective theories of probability and proved to be equivalent to them.

The equivalence between subjective and objective interpretations of the concept of probability holds for most applications of probability theory. In ordinary contexts, such as gambling or insurance, one could think of probabilities as reflecting our ignorance about the actual result rather than in terms of frequencies, ensembles or propensities, and it wouldn't matter. The equivalence also holds in statistical mechanics. In the above quotation from Jaynes, for instance, he adopts an epistemic interpretation of statistical mechanical probabilities, but this subjective interpretation does not yield different predictions than those of standard statistical mechanics. The epistemic interpretation of QM, however, presumes to make an *empirical* difference—this, as we will see, is the crux of the PBR theorem. Bearing this difference in mind (and returning to it later), let us probe a bit deeper into the controversy between the majority view that swallowed Born's *Verschmelzung* and the dissidents promoting the ensemble interpretation. The focal question on which the parties were divided was whether quantum states tell us anything about the *individual* system. Here are some of the pros and cons of the conflicting positions on this question.

The great merit of the ensemble interpretation was its readymade response to the problem regarding the 'collapse' of the wavefunction. If the wavefunction represents the physical state of an individual system, its instantaneous 'collapse' (to a sharp value) upon measurement is worrisome. As a physical process, it should be restricted by the known constraints on physical processes, in particular, the spatiotemporal constraints of the Special Theory of Relativity (STR). But once quantum probabilities are assumed to refer to an ensemble of similar systems, the fact that measurements on individual systems yield definite values is exactly what one would expect! The quantum collapse, under this interpretation, is no more troubling than the 'collapse' of a flipped coin from a state of probability 0.5, which

[10]The epistemic interpretation already existed in 1930 but took some time to make an impact outside of the community of probability theorists. Leading advocates of the subjective/epistemic interpretation of probability as degree of belief are Ramsey (1931), de Finetti and Savage (1954). The delay in impact is clearly reflected in the fact that de Finetti's works of the 1930s had to wait several decades to be collected and translated into English (de Finetti 1974).

characterizes an ensemble, to either heads or tails when a single coin is flipped. In other words, since, on the ensemble interpretation, the Ψ function does not stand for a physical state of an individual system in the first place, worries about the physical properties of the collapse, its Lorentz invariance etc., are totally misplaced.

Despite this advantage, the ensemble interpretation also faces a number of serious questions.

(a) Interference and other periodic effects. What do quantum phenomena such as superposition, interference and entanglement mean under an ensemble interpretation? Superposition and interference had an intuitive meaning in Schrödinger's original picture of wave mechanics, but he too became disillusioned when realizing that the wave function was a multi-dimensional wave in configuration space, not an ordinary wave in three-dimensional space. Born's interpretation of Ψ as representing a probability amplitude had the advantage that Ψ no longer needed to be situated in three dimensional space, and the disadvantage that the meaning of superposition and interference became mysterious. Getting a pattern on the screen from the interference of abstract probability waves has been compared to getting soaked from the probability of rain.[11] One could also formulate this question in terms of the meaning of the phase of the wave function.[12] Periodicity appears in all versions of QM, but the expression representing this periodicity–the phase–disappears in the calculation of probabilities. If we are only interested in probabilities and deny the reality of the wave, there is no good explanation for the significance of the phase. An intriguing example that suggests a physical meaning of the phase is the Aharonov-Bohm effect where an interference pattern resulting from a phase-shift in a field-free region is produced. It is difficult to explain this interference from the perspective of an ensemble interpretation.[13]

(b) The uncertainty relations. What is the status of Heisenberg's uncertainty principle from the perspective of the ensemble interpretation? Presumably, the principle should be construed as a statistical law, which constrains the dispersion of values in an ensemble of systems, not the definiteness of values representing properties of individual systems. It should thus be compatible with QM for an individual system to have sharp values of all variables including canonically conjugate ones. In other words, on the ensemble interpretation, the uncertainty relations could be violated! Proponents of the ensemble interpretation in fact entertained such violation. In a 1935 letter to Einstein,

[11] I am not sure about the origin of this analogy; I may have heard it from David Finkelstein.

[12] The meaning of the phase is being examined by Guy Hetzroni in his Ph.D dissertation *The Quantum Phase and Quantum Reality* submitted to the Hebrew University in July 2019 and approved in October 2019. See also Hetzroni 2019a.

[13] The Aharonov Bohm paper was published in 1959 but the problem about the phases could have been raised in 1926. There are different interpretations of the Aharonov Bohm effect (see, for example Aharonov and Rohrlich 2005, Chaps. 4, 5 and 6 and Healey 2007, Chap. 2), but, none of them is intuitive from the perspective of an ensemble interpretation.

Popper proposed a thought experiment designed to illustrate a system whose position and momentum are both well-defined. In his response (written a short while after the EPR paper; Einstein et al. 1935), Einstein rejected Popper's attempt to violate the uncertainty relations directly, but argued that one could, nonetheless, circumvent them in the split system described in the EPR paper.[14] Evidently, Einstein did not see the uncertainty relations as applying to individual systems; in accordance with the ensemble interpretation, he took the values of conjugate observables to be well-determined, though not directly measurable. This statistical understanding of the uncertainly relations was unacceptable to the Copenhagen school and was later challenged on both theoretical and empirical grounds.

(c) Disturbance by measurement. Thought experiments by Heisenberg and others motivated the uncertainty principle by showing how measurement of one quantum observable, say position, destroys the value of another observable (measured on the same particle or system), in this case, momentum.[15] On the ensemble interpretation, it appears, this 'disturbance' could be avoided, for this interpretation takes QM to pertain to a multitude of measurements made on many different particles or many different systems. In that case, however, the lesson of the convincing thought experiments becomes dubious. Is it possible to get around disturbance? The Copenhagen school declined this possibility and treated disturbance as fundamental. Later experiments failed to eliminate disturbance, thus constituting another difficulty for the ensemble interpretation.[16]

With these questions in mind, we can return to the contentious analogy between QM and statistical mechanics. The ensemble theorist who embraces this analogy, we now realize, is not only committed to a more fundamental level than the quantum level (i.e., underlying 'hidden variables'), but is also at variance with the standard interpretation about the meaning of central quantum laws such as the uncertainty relations! It is therefore slightly misleading to use the term 'interpretation' in this

[14]The letter is reproduced in Popper (1968) pp. 457–464. Incidentally, Einstein's letter repeats the main argument of the EPR paper in terms quite similar to those used in the published version and makes it clear that the meaning of the uncertainty relations is at the center of the argument. The letter thus disproves the widespread allegations that Einstein was unhappy with the published argument and that it was not his aim to question the uncertainty relations.

[15]It was only in the first few years of QM that Heisenberg and Bohr understood the uncertainty relations as reflecting concrete disturbance by measurement, that is, as reflecting a *causal* process. In his response to the EPR paper, Bohr realized that no concrete disturbance takes place in the EPR situation. He construed the uncertainty relations, instead, as integral to the formalism, and as such, holding for entangled states even when separated to spacelike distances so that they cannot exert any causal influence on one another.

[16]In contrast to the older ensemble interpretation, current epistemic interpretations of QM do not seek to evade disturbance. Rather, disturbance is understood in these interpretations as loss of information, a quantum characteristic resulting from the underlying assumption of maximal, and yet less-than-complete, information.

context; the two positions actually represent different *theories* diverging in their empirical implications.[17]

Thus far, I have concentrated on the distinction between the ensemble interpretation and standard QM rather than the distinction between epistemic and ontic models that is the focus of the PBR theorem. The reason for using the former as backdrop for the latter is that in general, ensemble and epistemic interpretations have not been properly distinguished by their proponents and do indeed have much in common; neither of them construes quantum states as a unique representation of physical states of individual systems and both can dismiss the worry about the collapse of the wavefunction. In principle, however, epistemic and ensemble interpretations could diverge: The epistemic theorist could go *radical* and decline the assumption of an underlying level of well-determined physical states that was the *raison d'être* of the ensemble interpretation. This was in fact the route taken by Pitowsky.

5.3 Pitowsky's Interpretation of QM

Pitowsky's approach to QM is based on three very general insights. First, the intriguing nature of QM—its departure from classical mechanics—already manifests itself at the surface level of quantum events and their correlations. Thus, before undertaking a theoretical explanation (in terms superposition, interference, entanglement, collapse, nonlocality, duality, complementarity, and so on), one must identify and systematize the *observed* phenomena. Secondly, classical probability theory seems to be at odds with various observable quantum phenomena.[18] Pitowsky therefore conjectured that a *non-classical probability theory* that deviates from its classical counterpart in its empirical implications provides a better framework for QM. Thirdly, once we follow this route and construe QM as a non-classical theory of probability, no further causal or dynamical explanation of quantum peculiarities is required. Characteristic quantum phenomena such as nonlocality fall into place in the new framework in the same way that rod contraction (in the direction of motion) and time dilation fall into place in the spacetime structure of STR. In both cases, the new framework renders the causal/dynamical explanation redundant.

The non-classical nature of quantum probability manifests itself in the violation of basic classical constraints on the probabilities of interrelated events. For example, in the classical theory of probability, it is obvious that for two events E_1 and E_2 with probabilities p_1 and p_2, and an intersection whose probability is $p_1.p_2$, the

---

[17]Construing the diverging interpretations as different theories does not mean that the existing empirical tests are conclusive, but that in principle, such tests are feasible. Even the evidence accumulated so far, however, makes violation of the uncertainty relations in the individual case highly unlikely.

[18]There are of course other approaches to QM, some of which retain classical probability theory. The ensemble interpretation discussed above and Bohm's theory, which I do not discuss here are examples of such approaches.

probability of the union $(E_1 U E_2)$ is $p_1 + p_2 - p_{12}$, and cannot exceed the sum of the probabilities $(p_1 + p_2)$.

$$0 \leq p_1 + p_2 - p_{12} \leq p_1 + p_2 \leq 1$$

The violation of this constraint is already apparent in simple paradigm cases such as the two-slit experiment, for there are areas on the screen that get *more* hits when the two slits are both open for a certain time interval Δt, than when each slit is open separately for the same interval Δt. In other words, contrary to the classical principle, we get a *higher* probability for the union than the sum of the probabilities of the individual events. (Since we get this violation in different experiments—different samples—it does not constitute an outright logical contradiction.) This quantum phenomenon is usually described in terms of interference, superposition, the wave particle duality, the nonlocal influence of one open slit on a particle passing through the other, and so on. Pitowsky's point is that regardless of the explanation of the pattern predicted by QM (and confirmed by experiment), we must acknowledge the bizarre phenomenon that it displays—nothing less than a violation of a highly intuitive principle of the classical theory of probability.

QM predicts analogous violations of other classical conditions. The most famous of these violations is that of the Bell inequalities which are violated by QM (and experiment). Pitowsky showed that from the analogue of the above rule for three events, it is just a short step to Bell's inequalities. He therefore linked the inequalities directly to Boole's "conditions of possible experience" (the basic conditions of classical probability) and their violation to the need for an alternative probability calculus. To appreciate the significance of this move for the debate over the epistemic interpretation, it is crucial to understand that what makes the violated conditions *classical* is their underlying assumption of determinate states occupied by the entities in question (or properties they obtain) independently of measurement: Just as balls in an urn have definite properties such as being red or wooden, to derive Bell's inequalities it is assumed that particles have a definite polarization or a definite spin in a certain direction, and so on. The violation of the classical principles of probability compels us to discard this picture, replacing it with a new understanding of quantum states and quantum properties. Rather than an analogue of the classical state, which represents physical entities and their properties prior to measurement, Pitowsky saw the quantum state function as keeping track of the probabilities of measurement results, "a device for the bookkeeping of probabilities" (2006, p. 214).[19]

This understanding of the state function, in turn, led Pitowsky to two further observations: First, he interpreted the book-keeping picture subjectively, i.e., quan-

[19]Pitowsky's answer is similar to that of Schrödinger, who characterizes quantum states as "the momentarily-attained sum of theoretically based future expectations, somewhat laid down as in a *catalog*. It is the determinacy bridge between measurements and measurements" (1935, p. 158). Pitowsky only became aware of this similarity around 1990. For more on Schrödinger's views and their similarity to those of Pitowsky see note number 25 below and Ben-Menahem (2012).

tum probabilities are understood as degrees of partial belief of a rational observer of quantum phenomena. Second (and here he joins other epistemic interpretations), the notorious collapse problem is not as formidable as when the state function is construed realistically, for if what collapses is not a real entity in physical space, there is no reason for the collapse to abide by the constraints of locality and Lorentz invariance.[20] Consequently, he renounced what he and Bub, in their joint paper (2010), dub "two dogmas" of the received view, namely the reality of the state function and the need for a dynamic account of the measurement process.

In his (2006), Pitowsky proposed an axiomatic system for QM that elaborates the earlier axiomatization by Birkhoff and von Neumann. In particular, he reworked the relation between the Hilbert space structure of quantum events and projective geometry. Furthermore, he sought to incorporate later developments such as Gleason's theorem and identify their connection with the axiom system. The ramifications of the non-classical structure of the quantum probability space, he argued, include indeterminism, the loss of information upon measurement, entanglement and Bell-type inequalities. As Pitowsky and Bub argued in their joint 2010, the implications of the axiom system are also closely related to the information-theoretic principles of no-cloning(for pure states) or no broadcasting (for mixed states) according to which an arbitrary quantum state cannot be copied (Park 1970; Wootters and Zurek 1982).

Pitowsky took the uncertainty relations to be "the centerpiece that demarcates between the classical and quantum domain" (2006, p. 214). The only non-classical axiom in the Birkoff von Neumann axiomatization, and thus the logical anchor of the uncertainty relations, is the axiom of *irreducibility*.[21] While a classical probability space is a Boolean algebra where for all events x and z (and where the event $z^{\perp}$ is the complement to z):

$$x = (x \cap z) \cup \left(x \cap z^{\perp} \right) \text{ (reducibility)}$$

in QM, we get irreducibility, i.e., (with *0* as the null event and *1* the certain event):

$$\text{If for some } z \text{ and for all } x, x = (x \cap z) \cup \left(x \cap z^{\perp} \right) \text{ then } z = 0 \text{ or } z = 1$$

Irreducibility signifies the non-Boolean nature of the algebra of possible events, for the only irreducible Boolean algebra is the trivial one $\{0,1\}$. According to Birkhoff and von Neumann, irreducibility means that there are no 'neutral' elements

[20] The idea that the collapse can always be understood epistemically has been challenged by Meir Hemmo; see Hagar and Hemmo (2006).

[21] Pitowsky's formulation is slightly different from that of Birkhoff and von Neumann, but the difference is immaterial. Pitowsky makes significant progress, however, in his treatment of the representation theorem for the axiom system, in particular in his discussion of Solér's theorem. The theorem, and the representation problem in general, is crucial for the application of Gleason's theorem (Gleason 1957), but will not concern us here.

$z, z \neq 0\ z \neq 1$ such that for all x, $x = (x \cap z) \cup (x \cap z^{\perp})$. If there would be such 'neutral' events, we would have non-trivial projection operators commuting with all other projection operators. Intuitively, irreducibility embodies the uncertainty relations for when x cannot be presented as the union of its intersection with z and its intersection with $z^{\perp}$, then x and z cannot be assigned definite values at the same time. Thus, whenever $x \neq (x \cap z) \cup (x \cap z^{\perp})$—$x$ and z are incompatible and, consequently, a measurement of one yields no information about the other. The axiom further implies genuine uncertainty, or indeterminism—probabilities strictly between (unequal to) 0 and 1. This result follows from a theorem Pitowsky calls the *logical indeterminacy principle,* and which proves that for incompatible events x and y

$$p(x) + p(y) < 2$$

The loss of information upon measurement—the phenomenon called 'disturbance' by the founders of QM—also emerges as a formal consequence of the probabilistic picture. The interpretation of the uncertainty relations that emerges from this axiomatization is obviously at variance with the ensemble interpretation described in the previous section. Whereas the latter tolerates the ascription of well-defined values to non-commuting operators/observables and construes the uncertainty relations as constraining only the distribution of values over the ensemble, the constraint derived from the Birkhoff- von Neumann-Pitowsky axiomatization applies to the individual quantum system.

Having shown that his axiom system entails genuine uncertainty, Pitowsky moved on to demonstrate the violation of the Bell-inequalities, i.e. the phenomena of entanglement and nonlocality. Violations already appear in finite-dimensional cases and follow from the calculation of the probabilities of the intersection of the subspaces of the Hilbert space representing the (compatible) measurement results at the two ends of the entangled system. Pitowsky showed, in both logical and geometrical terms, that the quantum range of possibilities is indeed larger than the classical range so that we get more correlation than is allowed by the classical conditions. Whereas the standard response to this phenomenon consists in attempts to discover the dynamic that makes it possible, Pitowsky emphasizes that this is a logical-conceptual argument, independent of specific physical considerations over and above those that follow from the non-Boolean character of the event structure. He says

> Altogether, in our approach there is no problem with locality and the analysis remains intact no matter what the kinematic or the dynamic situation is; the violation of the inequality is a purely probabilistic effect. The derivation of Clauser-Horne inequalities… is blocked since it is based on the Boolean view of probabilities as weighted averages of truth values. This, in turn, involves the metaphysical assumption that there is, simultaneously, a matter of fact concerning the truth-values of incompatible propositions…. [F]rom our perspective the commotion about locality can only come from one who sincerely believes that Boole's conditions are really conditions of possible experience. ….But if one accepts that one is

> simply dealing with a different notion of probability, then all space-time considerations become irrelevant (2006, pp. 231–232).

Recall that in order to countenance nonlocality without breaking with STR, the no signaling constraint must be observed. As nonlocality is construed by Pitowsky in formal terms—a manifestation of the quantum mechanical probability calculus, uncommitted to a particular dynamic—it stands to reason that no signaling will likewise be derived from probabilistic considerations. Indeed, it turns out that the no signaling constraint can be understood as an instance of the more general principle known as the non-contextuality of measurement (Barnum et al. 2000). In the spirit of the probabilistic approach to QM, Pitowsky and Bub therefore maintain that

> No signaling is not specifically a relativistic constraint on superluminal signaling. It is simply a condition imposed on the marginal probabilities of events for separated systems requiring that the marginal probability of a B-event is independent of the particular set of mutually exclusive and collectively exhaustive events selected at A, and conversely" (Bub and Pitowsky 2010, p. 443).[22]

To sum up: Pitowsky proposed a radical epistemic interpretation of QM. It is epistemic in the sense that the probabilities represent partial knowledge (information, rational belief) of the observer and it is radical in the sense that it is not assumed to be knowledge (information, rational belief) about the physical state of the system, but rather about observable measurement outcomes. It therefore deviates from epistemic interpretations of probability in other areas such as gambling or statistical mechanics, where the reality of underlying well-defined states is taken for granted. Pitowsky justified the denial of the reality of the physical state by urging that the assumption of determinate physical states leads to classical probabilities, whereas an adequate systematization of quantum measurement results requires a non-classical theory of probability. In the framework of this non-classical theory, he claimed, the troubling features of QM—nonlocality in particular—are grounded in the quantum formalism. No deeper explanation is needed. Just as the structure of Minkowski's spacetime accounts for relativistic effects that in earlier theories received dynamical explanations, the structure of QM as a non-classical theory of probability accounts for quantum effects that according to other interpretations call for dynamical explanations.

[22]In the literature, following in particular Jarrett 1984, it is customary to distinguish outcome independence, violated in QM, from parameter independence, which is observed, a combination that makes possible the peaceful coexistence with STR. The non-contextuality of measurement amounts to parameter independence. See, however Redhead (1987) and Maudlin (2011), among others, for a detailed exposition and critical discussion of the distinction between outcome and parameter independence and its implications for the compatibility with STR. More recently, Brandenburger and Yanofsky (2008) also refine Jarrett's distinction. I thank Meir Hemmo for drawing my attention to this paper. See also the paper by Levy and Hemmo in this volume.

5.4 The PBR Theorem

Pusey, Barrett and Rudolph summarize the no-go argument that has come to be known as the PBR theorem in the following sentence:

> We show that any model in which a quantum state represents mere information about an underlying physical state of the system, and in which systems that are prepared independently have independent physical states, must make predictions that contradict those of quantum theory. (Pusey et al. 2012, p. 475)

Clearly, the theorem purports to target the epistemic interpretation of the wave function, thereby supporting its realist interpretation. This is indeed the received reading of the theorem and the source of its appeal. Under the title "Get Real" Scott Aaronson (2012) announced the new result as follows:

> Do quantum states offer a faithful representation of reality or merely encode the partial knowledge of the experimenter? A new theorem illustrates how the latter can lead to a contradiction with quantum mechanics. (2012, p. 443)

Aaronson distinguishes between the more common question of whether a quantum state may correspond to different physical states, and the PBR question—whether a physical state can correspond to different quantum states. It is the latter that the PBR theorem purports to answer in the negative (See Fig. 5.1).

> It is crucial to understand that we're not discussing whether the same wavefunction can be compatible with multiple states of reality, but a different and less familiar question: whether the same state of reality can be compatible with multiple wavefunctions. Intuitively, the reason we're interested in this question is that the wavefunction seems more 'real' if the answer is no, and more 'statistical' if the answer is yes (2012, p. 443).

Since the PBR theorem does indeed answer the said question in the negative, the conclusion is that the wave function has gained (more) reality.

Why is the negative answer to this question a refutation of the epistemic interpretation? Here Pusey et al. build on a distinction between Ψ -ontic models and Ψ-epistemic models introduced by Harrigan and Spekkens (2010): In Ψ -

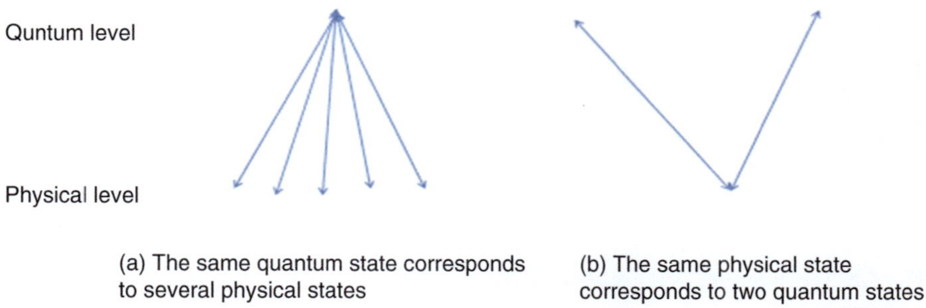

Fig. 5.1 Two types of relation between the physical level and the quantum level. (**a**) The same quantum state corresponds to several physical states. (**b**) The same physical state corresponds to two quantum states

ontic models the Ψ function corresponds to the physical state of the system; in Ψ-epistemic models, Ψ represents knowledge about the physical state of the system. Consequently, there are also two varieties of incompleteness: Ψ could give us a partial description of the physical state or a partial representation of our knowledge about that state. If a Ψ-ontic model is incomplete, it is conceivable that Ψ could be supplemented with further parameters—'hidden variables'. In this case, the same Ψ function could correspond to various physical states of the system, distinguishable by means of the values of the additional hidden variables. Presumably, Ψ-epistemic models can also be complete or incomplete but completing them cannot be accomplished by hidden variables of the former kind. Note that this analysis presupposes a definite answer to the question posed at the end of the introduction to this paper–what, according to the epistemic view, is the knowledge (belief, information) represented by the quantum state *about*? The assumption underlying the theorem is that it is knowledge about the *physical state* of the system.

So far, Harrigan and Spekkens' distinction is merely terminological, but they proceed to offer a *criterion* that distinguishes Ψ-*e*pistemic from Ψ-ontic models: If the Ψ function is understood epistemically, they claim, it can stand in a non-functional relation to the physical state of the system, that is, the same physical state may correspond to different (non-identical but also not orthogonal) Ψ functions. Or, when probabilities rather than sharp values are considered, the supports of the probability distributions corresponding to different Ψ functions can overlap for some physical states (See Fig. 5.2 for an illustration of this criterion in the case of two Ψ functions).[23]

Harrigan and Spekkens maintain that the possibility of overlap only makes sense under the epistemic interpretation of Ψ, for, they argue, knowledge about the physical state could be updated without any change to the physical state itself.

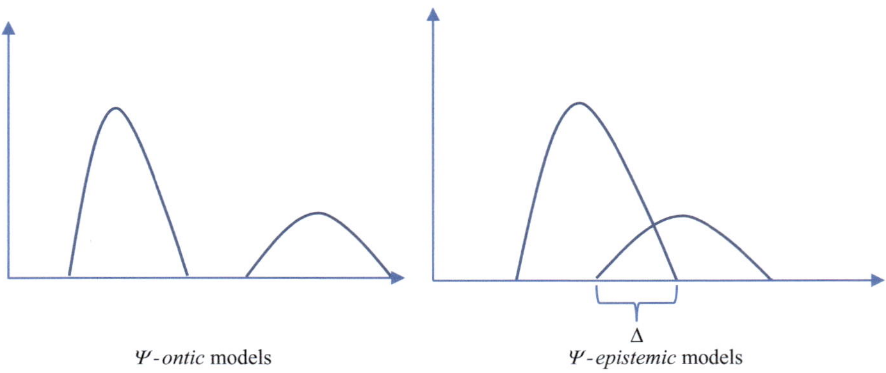

Fig. 5.2 The difference between Ψ-*ontic* and Ψ-*epistemic* models: (**a**) Ψ-*ontic* models. (**b**) Ψ-*epistemic* models

[23]Figure 5.2 follows the schematic illustration in the PBR paper.

In short, a criterion for a model being Ψ-epistemic is precisely the possibility of such overlap of the supports of different probability functions. Getting ahead of my argument, I should stress that allowing such a non-functional relation between the physical state of the system and the quantum state, namely, allowing that the quantum state is not uniquely determined by the physical state, is a non-trivial assumption. The question of whether an epistemic theorist is actually committed to it is crucial for assessing the verdict of the PBR theorem. In philosophical terminology, the ontic option illustrated in figure 1(a), where every physical state corresponds to a single quantum state is referred to as a relation of supervenience. (There is no assumption of supervenience in the reverse direction). In figure 1(b), however, illustrating what Harrigan and Spekkens call an epistemic model, the quantum state does not even supervene on the physical state. This failure of supervenience is not a necessary feature of probabilistic theories and is indeed being debated in other physical theories involving probabilities, such as statistical mechanics. I will return to this point below.

The authors of the PBR theorem make two assumptions: 1.The system has a definite physical state—not necessarily identical with the quantum state. 2. Systems that are prepared independently have independent physical states. PBR prove that for distinct quantum states, if epistemic models are allowed, that is (using the Harrigan Spekkens criterion), if an overlap of the probability distributions is allowed so that the intersection of the supports has a non-zero measure—then a contradiction with the predictions of QM could be derived. They derive the contradiction first for the special case in which there are two systems such that $| < \Psi_0 | \Psi_1 > | = 1/\sqrt{2}$ and then generalize to multiple (non-orthogonal) states. In the special case the proof is based on the following steps.

1. The authors consider two independent preparations of a system with states $|\Psi_0>$ and $|\Psi_1>$.
2. They choose a basis for the Hilbert space for which $|\Psi_0> = |0>$ and $|\Psi_1> = (|0>+|1>)/\sqrt{2}$. (Such a basis can always be found).
3. They consider an epistemic model, namely, the probability distributions for the two states overlap in a non-zero region Δ. Thus, there is a positive probability $q > 0$ to get a physical state from that region in each one of the systems and, because of independence, there is a positive probability $q^2 > 0$ to get such a value from the region Δ for both systems together.
4. They assume that the two systems are brought together and their *entangled* state measured.
5. They show that QM assigns zero probability to all four orthogonal states onto which such a measurements projects, whereas by the assumption made in step 3, the probability is non-zero.
   Contradiction!

The conclusion PBR draw is that epistemic models are incompatible with the predictions of QM, and thus ruled out. We are therefore left with ontic models, which are "more real," to use Aaronson's formulation.

5.5 Possible Responses by the Epistemic Theorist

Naturally, the strength of a theorem depends on the strength of its assumptions. In the case of the PBR theorem, its second assumption—independence—has in fact been called into question (Schlosshauer and Fine 2012, 2014), and will not be discussed here. One could, however, also question the *first* assumption, namely, the assumption that the system has a definite physical state and that in epistemic models the quantum state represents the experimenter's knowledge (belief, information) *about that physical state.* Clearly, the assumption regarding the reality of a physical state of the system is a realist assumption. The theorem therefore undermines a particular combination of a realist assumption about physical states and a non-realist interpretation of quantum states. The contradiction it derives may indicate that this peculiar combination of realist and non-realist ingredients does not work. From the logical point of view, one could also take the realist assumption itself to be responsible for the contradiction, in which case the theorem could be taken to challenge, rather than support a realist interpretation of quantum states. As we have seen, the common reading of the theorem is that it demonstrates the failure of an epistemic interpretation that allows a non-functional relation between the physical state and the quantum state (i.e., it allows several quantum states corresponding to a single physical state). It follows that an epistemic theorist who declines this particular understanding of the epistemic approach, thereby rejecting the Harrigan Spekkens criterion for epistemic models, will not be threatened by the theorem. Let us consider these options in more detail.

Recall that the existence of definite physical states underlying quantum states has been a matter of controversy ever since the early days of QM and yet, Pusey et al. make no effort to motivate their assumption.[24] They note briefly that instrumentalists would deny it, implying that *only* instrumentalists would deny it (and perhaps implying, further, that instrumentalism is too ridiculous a position to be argued against). In his "Extended Review" of the literature on the reality of the quantum state, Leifer (2014) elucidates the situation by distinguishing *realist* epistemic interpretations (namely, those assigning well-defined physical states to quantum systems) from "anti-realist, instrumentalist, or positivist" interpretations, which he dubs "neo Copenhagen" (2014, p. 72). He acknowledges that neo-Copenhagen interpretations are not affected by the PBR theorem, but, like the authors of the theorem, he is somewhat dismissive about them and more importantly, he does not even consider the possibility of epistemic interpretations at variance with instrumentalism. We will see, however (Sect. 5.6), that Pitowsky's epistemic interpretation, although denying the definite-state assumption, still differs significantly from instrumentalism. Moreover, Leifer construes epistemic states as subjective in the sense that an epistemic state is "something that exists in the mind of the

[24]Interpretations such as Bohm's theory and the many world interpretation show that the assumption is not ruled out by QM, but it is, nonetheless, sufficiently contentious to require some justification.

observer rather than in the external physical world," (2014, p. 69) and can therefore vary from one observer to another. Pitowsky, I suggest, did not understand the epistemic interpretation in that way. QM, he thought, is an algorithm that determines degrees of *rational* belief, or rational betting procedures in the quantum context. Even if the *belief* is in an individual's mind, the algorithm is objective. According to an epistemic interpretation of this kind, an observer using classical probability theory rather than QM, is bound to get results that deviate from QM, or, in the betting jargon often adopted by epistemic theorists, is bound to lose. The epistemic interpretation construes quantum probabilities as *constraints* on what can be known, or rationally believed, about particular measurement outcomes. By itself, it does not imply more liberty (about the ascription of probabilities to certain outcomes) than realist interpretations.

To what extent is an epistemic theorist committed to the Harrigan-Spekkens condition of non-supervenience as characterizing Ψ-epistemic models? Let's assume that the condition holds and there are well-determined physical states that 'correspond' to different quantum states. Obviously, this correspondence is not meant to be an identity for a physical state cannot be identical with different quantum states. A possible way of unpacking this correspondence is as follows: a well-defined physical state exists (prior to measurement) and, upon measurement 'generates' or 'gives rise to' different results. Such as interpretation is at odds with the received understanding of Bell's inequalities and, more specifically, with Pitowsky's interpretation of QM as a non-classical theory of probability. We have seen that Pitowsky explicitly denied the classical picture of well-defined physical states generating quantum states, a picture that leads to Bool's classical conditions, which, in turn, are violated by QM. Indeed, the language of a physical state 'giving rise' to, or 'generating,' the result of a quantum measurement would be rejected by Pitowsky even if it referred to a physical state corresponding to *single* quantum state, not merely to different ones as is the assumption in the Harrigan-Spekkens analysis. If, on the other hand, by the said 'correspondence' we mean the physical state of the system at the very instant it is being measured (or thereafter, but before another measurement is made), we are no longer running into conflict with Bell's theorem, but then, there is no reason to accept that this physical state corresponds to *different* quantum states. In other words, there is no reason to deny supervenience. The Epistemic theorist can take quantum states to determine the probabilities of measurement results, and understand probabilities as degrees of rational belief, and still deny the possibility that the same physical state corresponds to different quantum states. It would actually be very strange if the epistemic theorist would endorse such a possibility. Clearly, the quantum state—a probability amplitude—is compatible with different measurement outcomes, but the probability itself is fixed by QM.

Moreover, if quantum probabilities are conceived as the *maximal* information QM makes available, then surely we cannot expect different probability amplitudes to correspond the same physical state.[25] The idea of maximal information plays an important role in Spekkens (2005), where, on the basis of purely information-theoretic constraints, a toy model of QM is constructed. The basic principle of this toy model is that in a state of maximal knowledge, the amount of knowledge is equal to the amount of uncertainty, that is, the number of questions about the physical state of a system that can be answered is equal to the number of questions that receive no answer. (This was also Schrödinger's intuition cited in the previous footnote). Spekkens succeeds in deriving from this principle many of the characteristic features of QM, for example, interference, disturbance, the noncommutative nature of measurement and no cloning.[26] Either way, then—whether the real physical state allegedly corresponding to different quantum states is taken to exist prior to measurement or thereafter—the epistemic theorist I have in mind—Pitowsky in particular—need not accept the Harrigan-Spekkens condition, let alone endorse it as a criterion for epistemic models.

A comparison with statistical mechanics is again useful. The goal of statistical mechanics is to derive thermodynamic behavior, described in terms of macro-properties such as temperature, pressure and entropy, from the description of the underlying level of micro-states in terms of classical mechanics. For some macroscopic parameters, the connection with micro-properties is relatively clear—it is quite intuitive, for instance, to correlate the pressure exerted by a gas on its container with the average impact (per unit area) of micro-particles on the container. But this is not the case of other macro-properties; entropy, in particular. Recovering the notion of entropy is essential for the recovery of the second law of thermodynamics and constitutes a major challenge of statistical mechanics. The fundamental insight underlying the connection between entropy and the micro-

[25]The idea of maximal information or completeness is explicit in Schrödinger, who took the Ψ function to represent a maximal catalog of possible measurements. The completeness of the catalog, he maintained, entails that we cannot have a more complete catalog, that is, we cannot have two Ψ functions of the same system one of which is included in the other. For Schrödinger this is the basis for 'disturbance': Any additional information, arrived at by measurement, must change the previous catalog by *deleting* information from it, meaning that at least some of the previous values have been destroyed. Entanglement is also explained by Schrödinger on the basis of the maximality or completeness of Ψ. He argues as follows: A complete catalog for two separate systems is, *ipso facto*, also a complete catalog of the combined system, but the reverse does not follow. *"Maximal knowledge of a total system does not necessarily include total knowledge of all its parts, not even when these are fully separated from each other and at the moment are not influencing each other at all"* (1935, p. 160, italics in original). The reason we cannot infer such total information is that the maximal catalog of the combined system may contain conditional statements of the form: *if* a measurement on the first system yields the value *x*, a measurement on the second will yield the value *y*, and so on. Schrödinger sums up: "Best possible knowledge of a whole does not necessarily include the same for its parts . . . The whole is in a definite state, the parts taken individually are not" (1935, p. 161). In other words, separated systems can be correlated or entangled via the Ψ function of the combined system, but this does not mean that their individual states are already determined.

[26]Spekkens also notes that the model fails to reproduce the Kochen-Specker theorem and the violation of Bell-inequalities. In that sense it is still a toy model.

level was that macrostates are *multiply realizable* by microstates. The implication is that in general, since the same macrostate could be realized by numerous different microstates, the detailed description of the microstate of the system is neither sufficient nor necessary for understanding the properties of its macrostate. What is crucial for the characterization of a macrostate in terms of the micro-structure of the system realizing it, however, is the *number of ways* (or its measure-theoretic analogue for continuous variables) in which a macrostate can be realized by the microstates comprising it. Each macrostate is thus realizable by all the microstates corresponding to points that belong to the phase-space volume representing this macrostate—a volume that can vary enormously from one macrostate to another. Note the nature of the correspondence between microstates and macrostates in this case: When considering a particular system, at each moment it is both in a specific microstate and in a specific macrostate. In other words, microstates neither give rise to macrostates nor generate them! The system is in single well-defined state, but it can be *described* either in terms of its micro-properties or in terms of its macro-properties. For an individual state, the distinction between macro and micro levels is not a distinction between different states but rather between different descriptions of the *same* state. It is only the *type,* characterized in terms of macro-properties, that has numerous different microstates belonging to it.[27]

The insight that macrostates differ in the number of their possible realizations (or it measure-theoretic analogue) led to the identification of the volume representing a macrostate in phase space with the probability of this macrostate and to the definition of entropy in terms of this probability. On this conception, the maximal entropy of the equilibrium state in thermodynamic terms is a manifestation of its high probability in statistical-mechanical terms.[28] The link between entropy and probability constituted a crucial step in transforming thermodynamics into statistical mechanics, but it was not the end of the story. For one thing, the meaning of probability in this context still needed refinement, for another, the connection between the defined probability and the dynamic of the system had to be established.[29]

[27]It seems that this is not the way Harrigan and Spekkens or Puesy et al. conceive of the relation between the physical state and the quantum state. This obscurity might be the source of the problems raised here regarding their characterization of the epistemic interpretation.

[28]The probability of a macrostate can be understood objectively in terms of the 'size' of that macrostate (in phase space) or subjectively, as representing our ignorance about the actual microstate (as in the above quotation from Jaynes). This difference in interpretation does not affect the probabilities of macrostates. Note that from the perspective of a single microstate (and certainly from the perspective of a single atom or molecule), it should not matter whether it belongs to a macrostate of high entropy, that is, a highly probable macrostate to which numerous *other* microstates also belong, or to a low-entropy state—an improbable one, to which only a tiny fraction of microstates belongs. In so far as only the microlevel is concerned, entropy does not constitute an observable physical category.

[29]A natural account of probability would involve an ensemble of identical systems (possibly an infinite ensemble), where probabilities of properties are correlated with fractions of the ensemble. In such an ensemble of systems we should expect to find more systems in probable macrostates

Different approaches to statistical mechanics differ in their response to these problems. Whereas Boltzmann's approach is more focused on the dynamical evolution of the individual system, in Gibbs' statistical mechanics it is the mathematical precision of the notion of probability, and thus the ensemble of identical systems, that takes priority. One of the differences between these approaches is that in Boltzmann's approach macrostates supervene on microstates, while in Gibbs' ensemble approach they do not. The epistemic interpretation of quantum states as construed in the PBR theorem is in this respect closer to Gibbs' statistical mechanics than to Boltzmann's. But since the PBR theorem rules out this kind of epistemic interpretation, it ultimately points to a *disanalogy* between OM and Gibbs' statistical mechanics. At the same time, the epistemic interpretation (as understood by the PBR theorem), also differs from Boltzmann's statistical mechanics in allowing the failure of supervenience in the first place, and exposes once again a disanalogy between QM and statistical mechanics. By contrast, what Harrigan and Spekkens call an ontic model does supervene on the physical state of the system and is in that sense closer to Boltzmann's statistical mechanics than to Gibbs'. But we are then left with the above mentioned difficulties concerning the legitimacy of postulating a level of well-defined physical states underlying quantum states in the same way that microstates in statistical mechanics underlie macrostates. Whichever approach to statistical mechanics we choose, the analogy between statistical mechanics and QM does not seem to work.[30]

These considerations are sufficient to indicate that the PBR account of the epistemic interpretation as taking quantum states to represent knowledge (belief, information) about the physical state of the system—especially when combined with the Harrigan -Spekkens condition of non supervenience—is not as innocent as it appears. Pitowsky's interpretation exemplifies an epistemic view that does not share the PBR assumption of non-supervenience and is therefore not refuted by the theorem. It could now be objected that there was no need to go to such length to defend Pitowsky's view: the restriction to the surface level of measurement outcomes amounts to instrumentalism, a position that according to Pusey et al. is immune to their theorem. Let me now turn to this objection.

---

than in improbable ones. And yet, how does this consideration reflect on the *individual* system we are looking at? Assuming an actual system that happens to be in an *im*probable macrostate—it would certainly not follow from the above probabilistic considerations that it should, in the course of time, move from its improbable state to a more probable one. If one's goal is to account for the evolution of an individual system (in particular, its evolution to equilibrium), there is a gap in the argument. In addition, there is the formidable problem of deriving thermodynamic irreversibility from the statistical-mechanical description, which is based on the time-reversal-symmetric laws of mechanics. Since this problem has no analogue in the present discussion of the interpretation of quantum states, we can set it aside.

[30]As mentioned in Sect. 5.1, the analogy was rejected by the Copenhagen school from its earliest days.

5.6 Instrumentalism

Admittedly, the instrumentalist will not assume the existence of physical states that cannot be observed and will thus avoid the first assumption of the PBR theorem. In this respect instrumentalism and Pitowsky's radical epistemic interpretation are alike. But here they part ways. To see why, it is useful to characterize instrumentalism more precisely as committed to a verificationist (non-realist) theory of meaning.[31] On this theory, the meaning of a sentence depends on the existence of a verification procedure—where none exists, the sentence is meaningless—and its truth (falsity) on the existence of a proof (refutation). Unlike the realist, who accepts bivalence (every sentence is either true or false regardless of whether or not we can prove it), the verificationist admits a third category of *indeterminate* sentences—sentences that are neither provable, nor disprovable. Consider the difference between realism and verificationism when entertaining the claim that Julius Cesar, say, could not stand raw fish or that he had a scar on his left toe. Assuming that there is no evidence to settle the matter, the verificationist will deem these claims indeterminate, whereas the realist will insist that although we may never know, they must be either true or false. What if we *stipulate* that Julius Cesar detested raw fish, or that he did in fact have a scar on his toe? On the assumption that there is no evidence, such stipulation cannot get us into conflict with the facts. We are therefore free to stipulate the truth values of indeterminate sentences. This freedom, I claim, is *constitutive* of indeterminacy. The continuum hypothesis provides a nice example from mathematics: Since it is independent of the axioms of set theory, it might be true in some models of set theory and false in others. If by stipulating the truth value of the hypothesis we could generate a contradiction with set theory, we could no longer regard it as undecidable. The axiom of parallels provides another example of the legitimacy of stipulation. Since it is independent of the other axioms its negation(s) can be added to the other axioms without generating a contradiction. In QM, questions regarding the meaning of quantum states were for a long time undecided. If they were indeed undecid*able*, that is undecidable in principle, the verificationist (instrumentalist) could stipulate the answer by fiat. Schrödinger provided the first clue that this is not the case—the very stipulation of determinate values would lead to conflict with QM. Later no-go theorems and experiments confirmed this conclusion. As we saw, Pitowsky argued that the reason for the violation of Bell's inequalities is that in order to derive the inequalities, one makes the classical assumption of determinate states underlying quantum states. The violation of the inequalities, he thought, indicates that this classical assumption is not merely indeterminate but actually false—leading to a classical rather than quantum probability space. The constitutive characteristic of

---

[31]In this section I repeat an argument made in Ben-Menahem (1997). The difference between instrumentalism and the radical epistemic interpretation stands even for an instrumentalist who does not explicitly endorse verificationism, for even such an instrumentalist would not envisage the empirical implications that the radical epistemist calls attention to.

freedom is therefore missing: If the stipulation of definite states leads to conflict with QM, it is no longer up to the instrumentalist to approve of their existence or deny it. The radical epistemic interpretation proposed by Pitowsky is not neutral with regard to the assumption of definite physical states. Unlike the instrumentalist, who assumes there is no fact of the matter, and thus withholds judgement, radical epistemic theorists take a definite position, albeit a negative one. They deny the assumption and expect this denial to have empirical implications. To identify the negative with the neutral is a category mistake: There is a fundamental difference between provable negative assertions—no largest prime number—and no-fact-of-the-matter assertions.

I have reviewed four accounts of quantum probabilities: The Copenhagen conception—a paradoxical '*Verschmelzung*' of probabilistic and physical elements; the ensemble interpretation, which is probabilistic throughout; the epistemic conception assumed by PBR; the radical epistemic interpretation proposed by Pitowsky. These accounts correspond to the following responses to the question of what quantum probabilities refer to (respectively): observable states of individual systems, distribution of physical states in an ensemble of similar systems, the observer's knowledge about the physical state of an individual system, objective constraints on the information that can be gleaned from QM on possible measurement outcomes. The 'knowledge about what?' question clearly distinguishes the epistemic interpretation discussed by PBR from the radical interpretation proposed by Pitowsky. Several differences between the role of probabilities in QM and statistical mechanics have been noted, primarily differences regarding the supervenience of higher-level states on an underlying level of well-defined physical states. We have seen that, due to their impact on the meaning of fundamental quantum laws (the uncertainty relations in particular), questions regarding the meaning of probability are vital in the context of QM, but not in statistical mechanics, where different interpretations of probability do not change the contents of the theory. Finally, the difference between instrumentalism and Pitowsky's radical epistemic interpretation has been clarified.

The PBR theorem targets an epistemic interpretation that combines a realist assumption—the existence of definite physical states underlying quantum states—with a non-realist construal of quantum probabilities. Inspired by Pitowsky, I have argued that this is not the only available epistemic interpretation, and not the one most commonly upheld by epistemic theorists. An interpretation polar to that of PBR is available: One can let go of the assumption of definite physical states, but construe quantum probabilities as objective constraints on the information made available by measurement. From the perspective of such a radical epistemic interpretation, the Harrigan-Spekkens criterion of non-supervenience does not make sense. If I am right, the widespread reading of the PBR theorem as dealing a fatal blow to the epistemic interpretation should be reconsidered.

References

Aaronson, S. (2012). Get real. *Nature Physics, 8*, 443–444.

Aharonov, Y., & Bohm, D. (1959). Significance of electromagnetic potentials in the quantum theory. *Physics Review, 115*, 485–489.

Aharonov, Y., & Rohrlich, D. (2005). *Quantum paradoxes: Quantum theory for the perplexed.* Winheim: Wiley-VCH.

Albert, D. (2000). *Time and chance*. Cambridge, MA: Harvard University Press.

Ballentine, L. E. (1970). The statistical interpretation of quantum mechanics. *Review of Modern Physics, 42*, 358–381.

Barnum, H., Caves, C. M., Finkelstein, J., Fuchs, C. A., & Schack, R. (2000). Quantum probability from decision theory? *Proceedings of the Royal Society of London A, 456*, 1175–1190.

Ben-Menahem, Y. (1997). Dummett vs Bell on quantum mechanics. *Studies in History and Philosophy of Modern Physics, 28*, 277–290.

Ben-Menahem, Y. (2012). Locality and determinism: The odd couple. In Y. Ben-Menahem & M. Hemmo (Eds.), *Probability in physics* (Frontiers in Science Series) (pp. 149–166). Berlin: Springer.

Ben-Menahem, Y. (2017). The PBR theorem: Whose side is it on? *Studies in History Philosophy of Modern Physics, 57*, 80–88.

Birkhoff, G., & von Neumann, J. (1936). The logic of quantum mechanics. *Annals of Mathematics, 37*, 823–845.

Blokhintsev, D. I. (1968). *The philosophy of quantum mechanics*. Amsterdam: Reidel.

Born, M. [1926] (1963). Quantenmechanik der Stossvorgange. *Zeitschrift f. Physik, 38*, 803–827. In: Ausgewahlte Abhandlungen II (233–252). Gottingen: Vandenhoek Ruprecht.

Brandenburger, A., & Yanofsky, N. (2008). A classification of hidden variable properties. *Journal of Physics A. Mathematical and Theoretical, 41*(42)5302, 1–21.

Bub, J., & Pitowsky, I. (2010). Two dogmas about quantum mechanics. In S. Saunders, J. Barrett, A. Kent, & D. Wallace (Eds.), *Many worlds? Everett, quantum theory and reality* (pp. 433–459). Oxford: Oxford University Press.

De Finetti, B. (1974). *Theory of probability*. New York: Wiley.

Einstein, A., Podolsky, B., & Rosen, N. (1935). Can quantum-mechanical description of physical reality be considered complete? *Physics Review, 47*, 777–780.

Feynman, R. P. (1951). *The concept of probability in quantum mechanics*. In Second Berkeley symposium on mathematical statistics and probability, pp. 533–541.

Gleason, A. M. (1957). Measurement on closed subspaces of a Hilbert space. *Journal of Mathematics and Mechanics, 6*, 885–893.

Hagar, A., & Hemmo, M. (2006). Explaining the unobserved—Why quantum mechanics ain't only about information. *Foundations of Physics, 36*, 1295–1324.

Harrigan, N., & Spekkens, R. W. (2010). Einstein, incompleteness, and the epistemic view of quantum states. *Foundations of Physics, 40*, 125–157.

Healey, R. (2007). *Gauging what's real*. Oxford: Oxford University Press.

Hemmo, M., & Shenker, O. (2012). *The road to Maxwell's demon*. Cambridge: Cambridge University Press.

Hetzroni, G. (2019a). Gauge and ghosts (forthcoming). *The British Journal for the Philosophy of Science*. https://doi.org/10.1093/bjps/axz021.

Hetzroni, G. (2019b). *The quantum phase and quantum reality*. Ph.D. dissertation (submitted to the Hebrew University of Jerusalem.

Kochen, S., & Specker, E. P. (1967). The problem of hidden variables in quantum mechanics. *Journal of Mathematics and Mechanics, 17*, 59–89.

Leifer, M. S. (2014). Is the quantum state real? An extended review of Ψ–ontology theorems. *Quanta, 3*, 67–155.

Maudlin, T. (2011). *Quantum non-locality and relativity*. Oxford: Wiley-Blackwell.

Park, J. (1970). The concept of transition in quantum mechanics. *Foundations of Physics, 1*, 23–33.

Pitowsky, I. (1989). *Quantum probability, quantum logic* (Lecture notes in physics) (Vol. 321). Heidelberg: Springer.
Pitowsky, I. (2006). Quantum mechanics as a theory of probability. In Demopoulos & Pitowsky (Eds.), *Physical theory and its interpretation* (pp. 213–239). Dordrecht: Springer.
Popper, K. R. (1967). Quantum mechanics without 'The observer'. In M. Bunge (Ed.), *Quantum theory and reality* (pp. 7–43). Berlin: Springer.
Popper, K. R. (1968). *The logic of scientific discovery* (Revised edn). London: Hutchinson.
Pusey, M. F., Barrett, J., & Rudolph, T. (2012). On the reality of the quantum state. *Nature Physics, 8*, 475–478.
Ramsey, F. P. (1931). Truth and probability. In R. B. Braithwait (Ed.), *The foundations of mathematics and other logical essays* (pp. 156–198). London: Kegan Paul.
Redhead, M. (1987). *Incompleteness, nonlocality and realism*. Oxford: Clarendon Press.
Savage, L. J. (1954). *The foundations of statistics*. New York: Wiley.
Schlosshauer, M., & Fine, A. (2012). Implications of the Pusey-Barrett-Rudolph quantum no-go theorem. *Physical Review Letters, 108*, 260404.
Schlosshauer, M., & Fine, A. (2014). No-go theorem for the composition of quantum systems. *Physical Review Letters, 112*, 070407.
Schrödinger, E. (1935). The present situation in quantum mechanics (trans. J. D. Trimmer). In: Wheeler & Zurek (Eds.), *Quantum theory and measurement* (pp. 152–167). Princeton: Princeton University Press.
Spekkens, R. W. (2005). *In defense of the epistemic view of quantum states: A toy theory*. arXive:quant-ph/0401052v2.
Wootters, W., & Zurek, W. H. (1982). A single quantum cannot be cloned. *Nature, 299*, 802–803.

Chapter 6
On the Mathematical Constitution and Explanation of Physical Facts

Joseph Berkovitz

Abstract The mathematical nature of modern physics suggests that mathematics is bound to play some role in explaining physical reality. Yet, there is an ongoing controversy about the prospects of mathematical explanations of physical facts and their nature. A common view has it that mathematics provides a rich and indispensable language for representing physical reality but that, ontologically, physical facts are not mathematical and, accordingly, mathematical facts cannot really explain physical facts. In what follows, I challenge this common view. I argue that, in modern natural science, mathematics is constitutive of the physical. Given the mathematical constitution of the physical, I propose an account of explanation in which mathematical frameworks, structures, and facts explain physical facts. In this account, mathematical explanations of physical facts are either species of physical explanations of physical facts in which the mathematical constitution of some physical facts in the *explanans* are highlighted, or simply explanations in which the mathematical constitution of physical facts are highlighted. In highlighting the mathematical constitution of physical facts, mathematical explanations of physical facts make the explained facts intelligible or deepen and expand the scope of our understanding of them. I argue that, unlike other accounts of mathematical explanations of physical facts, the proposed account is not subject to the objection that mathematics only represents the physical facts that actually do the explanation. I conclude by briefly considering the implications that the mathematical constitution of the physical has for the question of the unreasonable effectiveness of the use of mathematics in physics.

Keywords Mathematical explanations · Mathematical constitution of the physical · Relationships between mathematics and physics · Effectiveness of

---

J. Berkovitz (✉)
Institute for the History and Philosophy of Science and Technology, University of Toronto, Victoria College, Toronto, ON, Canada
e-mail: joseph.berkovitz@utoronto.ca

M. Hemmo, O. Shenker (eds.), *Quantum, Probability, Logic*,
Jerusalem Studies in Philosophy and History of Science,
https://doi.org/10.1007/978-3-030-34316-3_6

mathematics in physics · Mathematical representations · Pythagorean · neo-Kantian

6.1 The Orthodoxy

Modern physics is highly mathematical, and its enormous success seems to suggest that mathematics is very effective as a tool for representing the physical realm. Nobel Laureate Eugene Wigner famously expressed this view in "The unreasonable effectiveness of mathematics in the natural sciences":

> the mathematical formulation of the physicist's often crude experience leads in an uncanny number of cases to an amazingly accurate description of a large class of phenomena (Wigner 1960, p. 8).

The great success of mathematical physics has led many to wonder about the causes of and reasons for the effectiveness of mathematics in representing the physical realm. Wigner thought that this success is unreasonable and even miraculous.

> The miracle of the appropriateness of the language of mathematics for the formulation of the laws of physics is a wonderful gift which we neither understand nor deserve. (Ibid., 1960, p. 11)

Mathematics is commonly conceived as the study of purely abstract concepts and structures and the question is how such concepts and structures could be so successful in representing the physical realm.[1]

The ubiquity and great success of mathematics in physics does not only raise the puzzle of the "unreasonable" effectiveness of mathematics in physics. It also suggests that mathematics is bound to play a role in explaining physical reality. Yet,

---

[1]Steiner (1998, p. 9) argues that there are two problems with Wigner's formulation of question of the "unreasonable effectiveness of mathematics in the natural sciences." First, Wigner ignores the cases in which "scientists fail to find appropriate mathematical descriptions of natural phenomena" and the "mathematical concepts that never have found an application." Second, Wigner focuses on individual cases of successful applications of mathematical concepts, and these successes might have nothing to do with the fact that a mathematical concept was applied. Steiner seeks to formulate the question of the astonishing success of the applicability of mathematics in the natural sciences in a way that escapes these objections. He argues that in their discoveries of new theories, scientists relied on mathematical analogies. Often these analogies were 'Pythagorean', meaning that they were inexpressible in any other language but that of pure mathematics. That is, often the strategy that physicists pursued to guess the laws of nature was Pythagorean: "they used the relations between the structures and even the notations of mathematics to frame analogies and guess according to those analogies" (ibid., pp. 4–5). Steiner argues that although not every guess, or even a large percentage of the guesses, was correct, this global strategy was astonishingly successful.

Steiner's reasoning and examples are intriguing and deserve an in-depth study, which, for want of space, I need to postpone for another opportunity. I believe, though, that such a study will not change the main thrust of my analysis of the question of the mathematical constitution of the physical and its implications for the questions of mathematical explanations of physical facts and the "unreasonable effectiveness of mathematics in the natural sciences".

there is an ongoing controversy about the prospects and nature of mathematical explanations of physical facts. (Henceforth, the term 'physical fact' is meant to subsume all aspects of the physical, such as laws, principles, properties, relations, etc. Further, for the sake of simplicity and continuity with the literature, I will use the term 'physical facts' to be more general so as to include natural facts.) A common view has it that mathematics provides a rich and indispensable language for representing physical reality but could not play a role in explaining physical facts. A related prevalent view is that, ontologically, the physical is to be sharply distinguished from the mathematical. Thus, it is common to think that sharing mathematical properties does not entail sharing physical properties. The idea is that, fundamentally, physical facts are not mathematical, and that mathematics only provides a language for representing the physical realm, even if this language is indispensable in the sense of being by far the most effective. These common views about the nature of physical facts and the role that mathematics plays in physics seem to be dogmas of contemporary mainstream schools of philosophy of science. Accordingly, the idea that mathematical facts could explain physical facts seems perplexing: How could facts about abstract, non-physical entities, which obtain in all 'possible worlds' (including 'non-physical' ones) and are ontologically separated from the physical, explain physical facts?

It should not be surprising then that until recently the subject of mathematical explanations of physical facts drew little attention in the philosophical literature. Yet, a recent interest in philosophical questions concerning the application of mathematics in the natural sciences has led philosophers to study the prospects and nature of mathematical explanations of physical facts, and various accounts of such explanations have been proposed (see, for example, Steiner 1978; Clifton 1998; Colyvan 2001, 2002; Batterman 2002, 2010, 2018; Pincock 2004, 2011a, b, 2012, 2015; Baker 2005, 2009, 2017; Bokulich 2008a, b, 2011; Dorato and Felline 2011; Lyon 2012; Batterman and Rice 2014; Lange 2016, 2018; Baron et al. 2017; Bueno and French 2018; Felline 2018; and Mancosu 2018). These accounts are clearly intended to offer non-causal explanations of physical facts. Yet, the exact nature of the explanations on offer is unclear. For example, some of these accounts attempt to show that mathematical facts or mathematical structures 'could make a difference' to physical facts (Bokulich 2008a, b, 2011; Pincock 2011b, 2015; Baron et al. 2017). But granted the common view that, ontologically, there is a sharp distinction between the physical and the mathematical, it is not clear how mathematical structures or mathematical facts could make such a difference and what the nature of the difference making is.

Further, the term 'mathematical explanation of physical facts' is ambiguous. In what follows, by this term I mean explanations in which mathematical frameworks, structures, or facts explain physical facts, and by 'mathematical facts' I mean mathematical truths, such as $2 + 2 = 4$. This kind of explanation is to be distinguished from explanations of physical facts that merely appeal to mathematics in order to represent physical facts. In the literature, there are various putative examples of mathematical explanations of physical facts and attempts to subsume them under a general account. In Sect. 9, I consider four such accounts. I argue that the nature of explanation in these accounts is unclear and that they are subject to the objection that the mathematical frameworks, structures, or facts they appeal to play a representational rather than explanatory role.

6.2 An Alternative Perspective

Although those who advocate the existence of mathematical explanations of physical facts endorse the idea of the indispensability of mathematics in physics, they seem to presuppose that mathematics does not play a constitutive role in defining what the physical is. I shall propose below that a way to circumvent the conundrum that current accounts of mathematical explanations of physical facts encounter is to acknowledge that, in modern natural science, the mathematical is constitutive of the physical. I believe that there are good reasons to think that contemporary natural science is in effect committed to such constitution. In particular, I will argue that if physical facts are fundamentally non-mathematical, the common wisdom about how mathematical models represent physical systems – which is based on the so-called 'mapping account' – becomes a mystery. Accordingly, the idea that there is a mathematical constitution of the physical should appeal to anybody who accepts the mapping account as a necessary part of mathematical representation of physical facts, independently of their view about whether mathematical explanations of physical facts exist. In fact, reflecting on the nature of the physical in contemporary natural science, I will argue that this idea should also be compelling to those who reject the mapping account as a necessary part of mathematical representations.

Given the mathematical constitution of the physical, I shall propose a new account of mathematical explanation of physical facts. The main idea of this account is that a mathematical explanation of physical facts highlights the mathematical constitution of some physical facts, thus making the explained facts intelligible or deepening and expanding the scope of our understanding of them. While this account could be applied to many of the putative examples of mathematical explanations of physical facts in the literature, the question whether it could account for all kinds of putative mathematical explanations of physical facts is beyond the scope of this paper for various reasons. First, as mentioned above, the term 'mathematical explanation of physical facts' is ambiguous and my understanding of it need not agree with all of its applications in the literature. Second, for lack of space and to keep things as simple as possible, I will only sketch the proposed account and consider it mainly in the context of simple examples of mathematical explanations of physical facts. Third, the range of putative examples I have encountered is not sufficiently broad to support such a grand claim. Yet, I think that there are good reasons to believe that the proposed account could be applied to all the examples of mathematical explanations of physical facts in which mathematical frameworks, structures, or facts are supposed to explain physical facts.

The current study was inspired by an exchange that I had with participants of the workshop The Role of Mathematics in Science at the Institute for the History and Philosophy of Science and Technology at the University of Toronto in October 2010. To my surprise, the common view was that while mathematics is very important for representing physical systems/states/properties/laws, it does not play any constitutive role in physics. Even more surprising, this view was also shared by supporters of ontic structural realism, for whom such a constitution seems to be a natural premise. The workshop Mathematical and Geometrical Explanations at the Universitat Autònoma de Barcelona in March 2012 and subsequent conferences and colloquia provided me with the opportunity to present the proposed account of mathematical explanation of physical facts, and the invitation to contribute to the second volume in honor of Itamar Pitowsky prompted me to write up this paper.

Pitowsky had a strong interest in mathematics and its use in physics. His research covers various topics in these fields: the philosophical foundations of probability; the nature of quantum probability and its relation to classical probability and quantum logic; the interpretation of quantum mechanics, in general, and as a theory of probability, in particular; quantum non-locality; computation complexity and quantum computation; and probability in statistical mechanics. This research has inspired a new perspective on the non-classical nature of quantum probability, quantum logic, and quantum non-locality, and influenced our understanding of these issues.[2]

Pitowsky's view of the relationship between mathematics and physics is less known, however. Based on his commentary on Lévy-Leblond's reflections on the use and effectiveness of mathematics in physics in the Bar-Hillel colloquium in the mid 1980s, in the next section I present his thoughts on this topic. Lévy-Leblond's and Pitowsky's reflections will provide a background for the main focus of the paper: the mathematical constitution and mathematical explanations of physical facts. Both Lévy-Leblond and Pitowsky embrace the idea of the mathematical constitution of the physical. Yet, it is not clear from their discussion what the exact nature of this constitution is. In Sect. 4, I review two traditional conceptions of mathematical constitution of physical facts – the Pythagorean and the neo-Kantian – and briefly give an example of mathematical constitution of physical facts. In Sect. 5, I review the common view of how mathematical models/theories represent physical facts. The main idea of this view is that a mathematical model/theory represents the properties of its target physical entities on the basis of a similarity between the relevant mathematical structure in the model/theory and the structure of the properties of these entities. More precisely, the idea is that there is an appropriate structure-preserving mapping between the mathematical system in the model/theory and the properties and relations of the represented physical entities. I argue that the mapping account undermines the idea that physical facts are non-mathematical. Further, in Sect. 6, I argue that a reflection on the concept of the physical in modern natural science supports the idea that the physical is constituted by the mathematical. In Sect. 7, I suggest that the philosophical interpretative frameworks of scientific

[2]For a list of Pitowsky's publications, see https://www.researchgate.net/scientific-contributions/72394418_Itamar_Pitowsky

theories/models in natural science determine the scope rather than the existence of the mathematical constitution of the physical. In Sect. 8, I propose a new account of mathematical explanation of physical facts. This account is sufficiently general so as to be integrated into the frameworks of various existing accounts of explanation. Accordingly, such accounts of explanation can be revised along the lines of the proposed account to embody the idea that mathematical explanations of physical facts highlight the mathematical constitution of physical facts and thus make the explained facts intelligible or deepen and expand the scope of our understanding of them. In Sect. 9, I consider four existing accounts of mathematical explanation of physical facts. I argue that the exact nature of explanation in these accounts is unclear and, moreover, that they are subject to the objection that mathematics only plays a representational role, representing the physical facts that actually have the explanatory power. I then show that these accounts can be revised along the lines of the proposed account so as to circumvent the above objections. I conclude in Sect. 10 by briefly commenting on the implications of the mathematical constitution of the physical for the question of the unreasonable effectiveness of mathematics in physics.

6.3 On the Relationship Between Mathematics and Physics

There is a broad consensus among physicists and philosophers that mathematics constitutes the language of modern physics. Poincaré expressed this view succinctly.

> [O]rdinary language is too poor and too vague to express relationships so delicate, so rich, and so precise. This, then, is a primary reason why the physicist cannot do without mathematics; it furnishes him with the only language he can speak (Poincaré 1913, p. 141).

Lévy-Leblond (1992, p. 147) notes that the picture of mathematics as the language of physics is ambiguous and admits two different interpretations: mathematics as the intrinsic language of Nature, which has to be mastered in order to understand her workings; and mathematics as a special language into which facts of Nature need to be translated in order to be comprehended. He refers to Galileo and Einstein as advocates of the first view, and Heisenberg as an exponent of the second view. He does not think of these views as contradictory but rather as extreme positions on a continuous spectrum of opinions. Lévy-Leblond explores the question of the peculiar relation between mathematics and physics, and he contends that the view of mathematics as a special language of nature does not provide an answer to this question. Instead of explaining why mathematics applies so well to physics, this view is concerned with the supposed adequacy of mathematics as a language for acquiring knowledge of nature in general. In his view, in other natural sciences, such as chemistry, biology, and the earth sciences, we could talk about the relation between mathematics and the particular science as one of application, with the implication that "we are concerned with an instrumental relation, in which mathematics appears as a purely technical tool, external to its domain of implementation and independent of the concepts proper to that domain." By contrast, he argues that mathematics plays a much deeper role in physics and that the relation between mathematics and physics is not instrumental (ibid., p. 148). He maintains that mathematics is internalized by physics and, like Bachelard (1965), he characterizes the relation of mathematics to physics as that of "constitutivity" (ibid., p. 149).

Lévy-Leblond does not specify what he means by this relation. He notes, though, that by mathematics being constitutive of physics, he does not mean that a physical concept is to be identified with or reduced to its underlying mathematical concept(s). He also does not think of this relation as implying that a physical concept is a mathematical concept plus a physical interpretation: "the mathematical concept is not a skeleton fleshed out by physics, or an empty abstract form to be filled with concrete content by physics." And, again, following Bachelard, he does not conceive of the "distinction between a physical concept and its mathematical characterization" as static:

> mathematics... is a dynamical 'way of thinking', leading from the experimental facts to the constitution of the physical concept (ibid., p. 149).

Further, Lévy-Leblond rejects the Platonic interpretation of the connection between mathematics and physics, "according to which the role of the physicist is to decipher Nature so as to reveal 'the hidden harmony of things' (Poincaré), as expressed by mathematical formulas, underlying the untidy complexity of physical facts." For him, the mathematical constitution of the physical does not imply "that every physical concept is the contingent materialization of an absolute mathematical being which is supposed to express its deepest truth, its ultimate essence" (ibid., p. 150). He takes the "mathematical polymorphism" of physics – namely, the possibility that physical laws and concepts may be endowed with several mathematizations – as evidence against the Platonic viewpoint and for the dynamical nature of the mathematical constitution of the physical (ibid., pp. 150–152).

Lévy-Leblond also rejects the view that there is a preestablished harmony between physical and mathematical concepts. First, he draws attention to "the physical multivalence of mathematical structures (which is the converse of the mathematical polymorphism of physical laws), i.e., to the possibility that the same mathematical theory may be adequate to several completely different physical domains" (ibid., p. 152). And, following Feynman et al. (1963, vol. 2, Section 12-7), he maintains that equations from different phenomena are so similar not because of the physical identity of the represented systems, properties, or laws. Rather, it is due to "a common process of approximation/abstraction" that these phenomena are "encompassed by analogous mathematizations" (ibid., pp. 152–153). Second, he argues that, "contrary to some rash assertions", it is not true that the abstract constructions elaborated by mathematics are all useful for the physicist. He contends that, as mathematics develops on its own, the fraction of mathematical theories and ideas that are utilized by physics is decreasing rather rapidly: while the use of mathematics in physics keeps on growing, it does so much less rapidly than the development of mathematics that is not utilized by physics (ibid., p. 154).

Lévy-Leblond's reflections above are intended to probe the nature of the mathematical constitution of the physical and the special relationship between physics and mathematics. His main conclusions are that there is no pre-established harmony between physics and mathematics and that the singularity of physics in its relationship to mathematics is very difficult to account for on the view that mathematics is merely a language for representing the physical reality. Lévy-Leblond holds that if mathematics is conceived as merely a language, it "would necessarily have to be a universal one, applicable in principle to each and every natural science", and the specificity of physics "must then be treated as only a matter of degree, and criteria must be found to distinguish it from other sciences" (ibid., p. 156). He considers two possible ways to try to explain the specificity of physics as a matter of degree. One is the common view that physics is more advanced. The idea here is that "more precise experimental methods, a firmer control of experimental conditions, permit the quantitative measurement of physical notions, which are thus transformed into numerical magnitudes" (ibid., p. 157). The other explanation localizes

> the specificity of physics in its objects of inquiry ... It is commonly stated that physics is "more fundamental" than the other natural sciences. Dealing with the deepest structures of the world, it aims to bring to light the most general laws of Nature, which are implicitly considered as the "simplest" and thus, in an Aristotelian way, as the "most mathematical" (ibid.).

Lévy-Leblond rejects both explanations. He contends that the first explanation *per se* fails to explain the privileged status of physics, and, moreover, it relies on a very naive view of mathematics. And he claims that the second explanation implies that "mathematicity assumes a normative character and becomes a criterion of scientificity", but the development of other sciences, such as chemistry, geology, and molecular biology seems to contradict this norm. He argues that these sciences are characterized by a clear-cut separation between their conceptual equipment and the mathematical techniques used – a separation that allows them to maintain a degree of autonomy even in the face of new developments in relevant branches of physics (ibid., pp. 148, 157–158).

In the end, Lévy-Leblond fails to find an explanation for the singular relationship between mathematics and physics. He thus opts for a radical solution according to which physics is defined as any area of the natural sciences where mathematics is constitutive (ibid., p. 158).

In his commentary on Lévy-Leblond's reflections, Pitowsky (1992, p. 166) finds Lévy-Leblond's characterization of physics very attractive. He points out, though, that it is incomplete, as the question arises as to what makes a given use of mathematics constitutive rather than a mere (non-constitutive) application. Indeed, as one could see from the above review, the questions of the nature of the mathematical constitution of the physical and the distinction between constitutive and non-constitutive uses of mathematics remain open. While Pitowsky shares

Lévy-Leblond's intuition that the use of mathematics in physics is different and deeper, he finds it difficult to provide an analytical framework for this observation.

Pitowsky also addresses the question whether Lévy-Leblond's view of the relationship between mathematics and physics fits in with the reductionist ethos of science. He remarks that while "[o]ne of the central ideals in the scientific enterprise is to try and provide a unified explanation of all natural phenomena", the sharp lines of demarcation that Lévy-Leblond draws "between the sciences appears to obstruct this idea, which has had some remarkable manifestations" (ibid., pp. 166–167).

Further, Pitowsky considers Lévy-Leblond's observation that the ratio of mathematics to mathematical physics increases with time. While he finds this observation correct, he does not think that it settles the question of whether there is a harmony between mathematical structures and physical phenomena for two reasons (ibid., p. 163):

(a) It is impossible to foretell which portion of mathematics is relevant to physics and which is not. There is no clear a priori distinction between pure mathematics and mathematical physics.
(b) There is far more abstract mathematics involved in physics than one usually assumes.

To motivate these claims, Pitowsky discusses Bolzano's unintuitive example of a continuous function which is nowhere differentiable (ibid., pp. 163–165). The example was introduced at the beginning of the nineteenth century, but was ignored for a few decades only to be resurrected and expanded by Weierstrass around the middle of the nineteenth century.[3] About 50 years later, Wiener (1923) showed, for the probability space of the set of all possible continuous trajectories of Brownian particles, that (with probability one) the path taken by a Brownian particle is nowhere differentiable. As Pitowsky (ibid., p. 165) notes, "the Brownian particle simply *does not have a velocity*", and the consequent 'pathological' creatures of 'continuous nowhere differentiable curves' were instrumental in explaining why a very large error had occurred in previous calculations. Continuous nowhere differentiable functions and similar creatures also play a role in more recent theories as well, such as the theory of fractals (Mandelbrot 1977). This and other examples suggest that abstruse mathematics plays a subtle role in physical theories, which is not sufficiently appreciated, and that it is impossible to tell a priori whether any such piece of mathematics will figure in future theories (Pitowsky 1992, p. 165).

Pitowsky believes that such examples and the "entanglement of mathematical structure with physical fact" suggest that "there exists no a priori principle which distinguishes pure mathematics from applied mathematics, mathematical physics in particular", and that mathematical physics is much broader than it is commonly

[3]Bolzano's example is believed to have been produced in the 1830s but the manuscript was only discovered in 1922 (Kowalewski 1923) and published in 1930 (Bolzano 1930). Neuenschwander (1978) notes that Weierstrass presented a continuous nowhere differentiable function before the Royal Academy of Sciences in Berlin on 18 July 1872 and that it was published first, with his assent, in Bois-Reymond (1875) and later in Weierstrass (1895).

conceived. Thus, the question of the "correspondence" between mathematics and physics is unavoidable, and "one can understand it in a naive, realistic sense, or defend a thesis with a Kantian flavor" (ibid., pp. 165–166).

I find Lévy-Leblond's and Pitowsky's reflections on the relationship between mathematics and physics and the effectiveness of the use of mathematics in physics intriguing. I share with them the view that, in modern physics, mathematics is constitutive of the physical and that the use of mathematics in physics is more successful than in other natural sciences. I disagree, though, with Lévy-Leblond's opinion that mathematics is only constitutive in physics. I believe that mathematics is also constitutive in other natural sciences. Indeed, in both of the traditional frameworks of portraying the physical as constituted by the mathematical – the Pythagorean and the neo-Kantian – the constitution pertains to the physical broadly conceived across the natural sciences. Yet, having a unified framework does not entail that the natural sciences are reducible to physics.

I also share with Lévy-Leblond and Pitowsky the view that the question of the particular success of the use of mathematics in physics is very interesting, important, and open. Recall that Lévy-Leblond rejects two possible explanations for the specificity of physics among the natural sciences: one is that physics is more advanced, and the other is that the objects of physics are the simplest and most fundamental. I don't find his reasoning here compelling, and I believe that both of these lines of reasoning are worth pursuing. For example, Lévy-Leblond argues against the second explanation that it implies a false hierarchy of sciences where physics is the most scientific and the idea that all the natural sciences are reducible to physics. It is not clear to me, however, why the view that the objects of physics are the simplest and most fundamental has such implications. Indeed, one may maintain this view yet reject the suggested implications.

Finally, I share with Pitowsky the view that in modern physics there is a "correspondence" or "harmony" between the physical and the mathematical, which could be understood along Pythagorean or Kantian lines. I believe that such a correspondence/harmony can be accounted for by the mathematical constitution of the physical, and in the next section I will briefly review the nature of this constitution in the context of Pythagorean and neo-Kantian frameworks. I also agree with Pitowsky that Lévy-Leblond's observation that the ratio of mathematics to mathematical physics increases with time is not evidence against the correspondence/harmony between the physical and the mathematical, and, moreover, that there is no a priori distinction between pure and applied mathematics. And, similarly to him, I believe that a reflection on the history of mathematical natural science could support the absence of such a distinction.

The question of the specificity of physics among the natural sciences deserves an in-depth consideration. Yet, for want of space, I will have to postpone its discussion to another opportunity and focus on the constitutive role that mathematics plays in physics and its implications for mathematical explanations of physical facts and the question of the unreasonable effectiveness of the use of mathematics in physics. In Sects. 5–7, I motivate the mathematical constitution of the physical and comment on its scope. Given the mathematical constitution of the physical, in Sect. 8 I propose a new account of mathematical explanation of physical facts, in Sect. 9 I compare it to exiting accounts, and in Sect. 10 I briefly discuss the question of the unreasonable effectiveness of the use of mathematics in physics. But, first, I turn to briefly review two ways of thinking about the mathematical constitution of physical facts.

6.4 On Conceptions of Mathematical Constitution of the Physical

There are two traditional ways of conceiving the mathematical constitution of the physical. One is along what I shall call the 'Pythagorean picture'. There is no canonical depiction of this picture. In what follows, I will consider three versions of it, due to Aristotle, Galileo, and Einstein.

Aristotle describes the Pythagorean picture in the *Metaphysics,* Book 1, Part 5.

> [T]he so-called Pythagoreans, who were the first to take up mathematics, not only advanced this study, but also having been brought up in it they thought its principles were the principles of all things. Since of these principles numbers are by nature the first, and in numbers they seemed to see many resemblances to the things that exist and come into being – more than in fire and earth and water (such and such a modification of numbers being justice, another being soul and reason, another being opportunity – and similarly almost all other things being numerically expressible); since, again, they saw that the modifications and the ratios of the musical scales were expressible in numbers; since, then, all other things seemed in their whole nature to be modelled on numbers, and numbers seemed to be the first things in the whole of nature, they supposed the elements of numbers to be the elements of all things, and the whole heaven to be a musical scale and a number. (Aristotle 1924)

In *The Assayer* Galileo endorses a different version of the Pythagorean picture. In his characterization, the universe is written in the language of mathematics, the characters of which are geometrical figures.

> Philosophy is written in this grand book — I mean the universe — which stands continually open to our gaze, but it cannot be understood unless one first learns to comprehend the language and interpret the characters in which it is written. It is written in the language of mathematics and its characters are triangles, circles, and other geometrical figures without which it is humanly impossible to understand a single word of it; without these, one is wandering a dark labyrinth. (Galileo 1623/1960)

In his lecture "On the method of theoretical physics", Einstein also seems to endorse a version of the Pythagorean picture. He depicts nature as the realization of mathematical ideas.

> Our experience hitherto justifies us in believing that nature is the realization of the simplest conceivable mathematical ideas. I am convinced that we can discover by means of purely mathematical constructions the concepts and the laws connecting them with each other, which furnish the key to the understanding of natural phenomena. (Einstein 1933/1954)

The Pythagorean picture has also been embraced by a number of other notable natural philosophers and physicists. While the exact nature of this picture varies from one version to another, all share the idea that mathematics constitutes all the elements of physical reality. Physical objects, properties, relations, facts, principles, and laws are mathematical by their very nature, so that ontologically one cannot separate the physical from the mathematical. This does not mean that physical things are numbers, as it may be tempting to interpret Aristotle's characterization in the *Metaphysics* and as it is frequently suggested. Rather, the idea is that mathematics defines the fundamental nature of physical quantities, relations, and structures. Without their mathematical properties and relations, physical things would have been essentially different from what they are. The Pythagoreans see the mathematical features of the physical as intrinsic to it, so that mathematics is not regarded merely as a 'language' or conceptual framework within which the physical is represented and studied.

The main challenge for the Pythagoreans is to identify the mathematical frameworks that define the nature of the physical, especially given that both mathematics and natural science continue to develop. The Pythagoreans may argue, though, that it is the role of mathematics and natural science to discover these frameworks.

The second traditional way of conceiving the mathematical constitution of the physical is along neo-Kantian lines, especially those of Herman Cohen and Ernest Cassirer of the Marburg school. Cassirer (1912/2005, p. 97) comments that Cohen held that

> the most general, fundamental meaning of the concept of object itself, which even physiology presupposes, cannot be determined rigorously and securely except in the language of mathematical physics.

For example, in mechanics,

> [w]hat motion "is" cannot be expressed except in concepts of quantity; understanding these presupposes a fundamental system of a pure doctrine of Quantity. Consequently, the principles and axioms of mathematics become the specific foundation that must be taken as fixed in order to give content and sense to any statement of natural science about actuality.

Cassirer held that mathematics is crucial for furnishing the fundamental scaffolding of physics, the intellectual work of understanding, and the construction of physical reality.

> "Pure" experience, in the sense of a mere inductive collection of isolated observations, can never furnish the fundamental scaffolding of physics; for it is denied the power of giving mathematical form. The intellectual work of understanding, which connects the bare fact systematically with the totality of phenomena, only begins when the fact is represented and replaced by a mathematical symbol. (Cassirer 1910/1923, p. 147)

The idea is that "the chaos of impressions becomes a system of numbers" and the "objective value" is obtained in "the transformation of impression into the mathematical 'symbol'." The physical analysis of an object "into the totality of its numerically constants" is a judgment in which

> the concrete impression first changes into the physically determinate object. The sensuous quality of a thing becomes a physical object, when it is transformed into a serial determination. The "thing" now changes from a sum of properties into a mathematical system of values, which are established with reference to some scale of comparison. Each

> of the different physical concepts defines such a scale, and thereby renders possible an increasingly intimate connection and arrangement of the elements of the given. (Ibid., p. 149)

More generally, in Cassirer's view, the concepts of natural science are "products of constructive mathematical procedure". It is only when the values of physical constants are inserted in the formulae of general laws, "that the manifold of experiences gains that fixed and definite structure, that makes it 'nature'." The reality that natural science studies is a construction, and the construction is in mathematical terms.

> The scientific construction of reality is only completed when there are found, along with the general causal equations, definite, empirically established, quantitative values for particular groups of processes: as, for example, when the general principle of the conservation of energy is supplemented by giving the fixed equivalence-numbers, in accordance with which the exchange of energy takes place between two different fields. (Ibid., p. 230)

In the neo-Kantian school, the mathematical constitution of natural science is the outcome of a historical process. The reality that mathematical natural science studies is a continuous serial process, and along this process the exact nature of the mathematics that constitutes the physical evolves.

A notable example of the mathematical constitution of the physical is the role that the calculus plays in shaping the concepts of modern physics. For example, the concept of the mathematical limit constitutes the concept of instantaneous velocity and more generally the concept of instantaneous change. This constitution marks a very significant change from the concepts of instantaneous velocity and instantaneous change before the calculus revolution. Consider, for instance, Zeno's arrow paradox.[4] The paradox starts from the premise that time is composed of moments, which are indivisible, and that during any such moment the arrow has no time to move. Thus, since the arrow does not have time to move in any particular moment, it is concluded that the arrow can never be in motion. Zeno's paradox suggests that in the pre-calculus era instantaneous velocity does not make sense. By contrast, when, based on the calculus, instantaneous velocity is defined as the limit of a series of average velocities as the time interval approaches zero, Zeno's paradox can be circumvented.[5]

---

[4]For the arrow paradox and its analysis, see for example Huggett (1999, 2019) and references therein.

[5]Two comments: 1. The introduction of the calculus is often presented as a solution to Zeno's arrow paradox. I believe that this presentation is somewhat misleading. The calculus does not really solve Zeno's paradox. It just evades it by *fiat*, i. e. by redefining the concept of instantaneous velocity. 2. For a recent example of the role that the calculus plays in constituting physical facts, see Stemeroff's (2018, Chap. 3) analysis of Norton's Dome – the thought experiment that is purported to show that non-deterministic behaviour could exist within the scope of Newtonian mechanics. Stemeroff argues that: (i) this thought experiment overlooks the constraints that the calculus imposes on Newton's theory; and that (ii) when these constraints are taken into account, Norton's Dome is ruled out as impossible in Newtonian universes.

6.5 On the Common View of How Mathematical Models Represent Physical Reality

In modern physics, the physical is characterized in mathematical terms. Yet, the common view is that, while mathematics provides a rich and indispensable language for representing the physical, the physical is fundamentally non-mathematical. If the physical is fundamentally non-mathematical, how could a mathematical model represent adequately physical systems?

The replies to this question are often based on a widespread view that a mathematical model represents in terms of the similarity or identity between the mathematical structures in the model and the structures of the physical things that the model represents. More precisely, the idea is that a mathematical model represents in terms of an appropriate structure-preserving mapping from the mathematical structures that the model posits to the structures of the physical things it represents. Chris Pincock (2004, 2011a, b, 2012, Chap. 2) calls this conception of representation the 'mapping account'.

For example, a model based on Newton's three laws of motion (together with the relevant 'boundary conditions') may represent the solar system. Newton's laws of motion are expressed in terms of mathematical relations.

First law: The net force F on an object is zero just in case its velocity v is constant:

$$\sum F = 0 \Longleftrightarrow dv/dt = 0; \tag{6.1}$$

where t denotes time and dv/dt denotes the derivative of velocity with respect to time.

Second law: The net force F on an object is equal to the rate of change of its linear momentum p, $p = mv$, where m denotes the object's mass:

$$F = \frac{dp}{dt} = \frac{d(mv)}{dt} = m\frac{dv}{dt}. \tag{6.2}$$

Third law: For every action there is an equal and opposite re-action. That is, for every force exerted by object A, F_A, on object B, B exerts a reaction force, F_B, that is equal in size, but opposite in direction:

$$F_A = -F_B. \tag{6.3}$$

These mathematical relations are supposed to correspond to the physical relations that obtain between the forces acting on the planets and the planets' masses, linear momenta, velocities, and accelerations, and the correspondence is achieved by a mathematical mapping from the model to the solar system.

In its simplest form, the mapping account posits that a mathematical model/theory represents physical systems because the mathematical structures it posits are isomorphic to the structures of properties of, and relations between these systems. Intuitively, the idea is that a mathematical model and properties and relations of the physical systems it represents share the same structures. More precisely, there is a one-to-one mapping between the mathematical structures in the model and properties of, and relations between the represented systems. Mathematical models/theories could also represent by means of weaker notions of structural identity, such as partial isomorphism (French and Ladyman 1998, Bueno et al. 2002, Bueno and French 2011, 2018), embedding (Redhead 2001), or homomorphism (Mundy 1986). Some other accounts of how mathematical models/theories represent take the mapping account to be incomplete. In particular, Bueno and Colyvan (2011, p. 347) argue that, for the mapping account to get started, we need "something like a pre-theoretic structure of the world (or at least a pre-modeling structure of the world)." While it is common to think of mathematical structure as a set of objects and a set of relations on them (Resnik 1997, Shapiro 1997), Bueno and Colyvan remark that "the world does not come equipped with a set of objects ... and sets of relations on those. These are either constructs of our theories of the world or identified by our theories of the world." Thus, the mapping account requires having what they call "an assumed structure" (ibid.). Yet, Bueno and Colyvan accept the mapping account as a necessary part of mathematical representation of the physical world.

While the mapping account is supposed to support the common view that mathematical models could represent physical systems even if the physical is fundamentally non-mathematical, it actually undermines it. If the notion of representation is to be adequate, the notion of identity between a mathematical structure and the physical structure it represents has to be sufficiently precise. But it is difficult to see how the notion of identity between mathematical and physical structures could be sufficiently precise for the purposes of modern physics if physical structures did not have a mathematical structure. Thus, it seems that any adequate account of mathematical representation of the physical that is based on the mapping account (or some similar cousin of it) will have to presuppose that the represented physical things have mathematical structures.

6.6 On the Notion of the Physical

The above argument for the mathematical constitution of the physical is focused on accounts of representation which are based on the mapping account of mathematical representations. There have been various objections to the mapping account (see, for example, Frigg and Nguyen (2018) for criticism of structuralist accounts of representation and references therein). I believe that the main idea of the above argument can be extended to any adequate account of mathematical representation of the physical, as any such account would have to include some aspects of the mapping account as a necessary component. Indeed, it is difficult to see how mathematics could have the potency it has in modern natural science if mathematical representations did not involve mapping. In any case, there is another weighty reason for rejecting the idea that the physical is non-mathematical. It is related to the question of the nature of the physical, and it is applicable to all accounts of scientific representation, independently of whether they embody the mapping account.

A key motivation for thinking about the physical as non-mathematical is an equivocation between two notions of the physical:

(i) the physical as determined by the theoretical and experimental frameworks of modern physics and, more generally, modern natural science; and
(ii) the physical as some kind of stuff out there, the nature of which is not defined by these frameworks.

The second notion of the physical is too vague and rudimentary to play any significant role in most contemporary scientific applications. In particular, theories and models in physics actually refer to the first notion of the physical, where the physical is constituted by the mathematical in the sense that fundamental features of the physical are in effect mathematical. Indeed, physical objects, properties, relations, and laws in modern physics are by their very essence characterized in mathematical terms. Yet, although the gap between the above notions of the physical seems unbridgeable, it is common to assume without justification that our theories and models, which embody the first notion of the physical, are about the second notion of the physical.

The considerations above suggest that in modern natural science:

(a) The common view – that the physical is ontologically separated from the mathematical – fails to explain how mathematical models/theories could represent the physical.
(b) The mapping account is an essential component of mathematical representations of the physical.
(c) The mapping account requires that physical facts have mathematical structure.
(d) The mapping account supports the idea of the mathematical constitution of the physical, and this constitution makes sense of the mapping account as an essential component of mathematical representations of the physical.

6.7 On the Scope of the Mathematical Constitution of the Physical

So far, I have not considered whether the interpretation of scientific theories/models along realist or instrumentalist lines has relevance for the question of the mathematical constitution of the physical. The reason for overlooking this issue earlier on is that the philosophical interpretative framework of theories/models in current natural science pertains to the scope rather than the existence of mathematical constitution of the physical. Under both realist and instrumentalist interpretations, the physical phenomena that current theories/models account for are mathematically constituted. The raw data that scientists collect are typically quantitative. Further, these data often lack the systematicity, stability, and unity required for constructing these phenomena. The phenomena that our theories/models have to answer for are constructed from raw data in terms of various postulates and statistical inferences which are mathematically constituted. The scope of the mathematical constitution of physical facts beyond the phenomena depends on the interpretative framework. If all the current theories/models in natural science were interpreted as purely instrumental, so that they are not supposed to represent anything beyond the phenomena, the mathematical constitution of the physical would only concern the phenomena; for under such interpretative framework, natural science is not supposed to represent anything beyond physical phenomena. While there are reasons to consider parts of modern natural science as purely instrumental, it is doubtful that there is any adequate understanding of all of it as purely instrumental. Accordingly, it is plausible to conceive the mathematical constitution of the physical as pertaining also to various aspects of the physical reality beyond the phenomena. In any case, in the account of mathematical explanation of physical facts that I now turn to propose the extent of the mathematical constitution of the physical has implications for the scope rather than nature of mathematical explanations of physical facts.

6.8 A Sketch of a New Account of Mathematical Explanation of Physical Facts

In thinking about mathematical explanations of physical facts, it is important to distinguish between:

(I) explanations in which mathematical frameworks, structures, or facts explain physical facts; and
(II) explanations of physical facts that merely appeal to mathematical frameworks, structures, or facts in order to represent physical facts.

The account suggested here aims at the first type of explanations. In these explanations, mathematics plays an essential role, which surpasses its representational role. While the requirement to surpass the representational role may seem trivial, as we shall see in the next section current accounts of mathematical explanation of physical facts struggle to meet it.

In the proposed account, mathematical explanations of physical facts are of two related kinds. One kind of explanation is a subspecies of physical explanations of physical facts. In a physical explanation of physical facts, physical facts are explained by physical facts. In the proposed account, a mathematical explanation of physical facts is a physical explanation of physical facts that *focuses* on the mathematical *constitution* of some of the physical facts in the *explanans*. That is, it is a physical explanation of physical facts that highlights the mathematical constitution of some of the physical facts in the *explanans*, and by highlighting this constitution it makes the *explanandum* intelligible or deepens and expands the scope of our understanding of it. The second kind of mathematical explanation of physical facts simply highlights the mathematical constitution of physical facts and, similarly to the first kind of explanation, by highlighting this constitution it makes the *explanandum* intelligible or deepens and expands the scope of our understanding of it.

By physical explanation of physical facts, I shall mean explanations of: how physical facts (could) come about; how physical facts (could) influence or make a difference to other physical facts; how physical patterns or regularities (could) come about; how physical facts, patterns or regularities follow from, or are related to other physical facts, patterns, or regularities; how physical principles/laws and physical facts entail other physical facts; how physical principles/laws entail or are related to other physical principles/laws; etc.

The proposed account is sufficiently open-ended to be based on existing accounts of explanation, such as the D-N, causal, and unification accounts of explanation. However, it cannot be identified with any of these accounts since it requires that the mathematical constitution of some physical facts in the *explanans* be highlighted. Given the account's openness, it will be easier to demonstrate how it works by analyzing the way it circumvents the difficulties that current accounts of mathematical explanations of physical facts encounter. In the next section, I will consider four such accounts.

Finally, recall that, in this study, the term 'physical facts' is used broadly to include natural facts. The proposed account is supposed to cover mathematical explanations of natural facts. I believe that this account could also be extended to social facts. For, arguably, in current social science there is also a mathematical constitution of various social facts, economic facts being a prime example. But, for lack of space, I leave the consideration of this issue for another opportunity.

6.9 On Mathematical Explanations of Physical Facts

Recently, the literature on the mathematical explanation of physical facts has grown steadily and various accounts have been proposed. Here I can only consider some of them, and the choice reflects the way the current study has developed so far. Analyzing putative examples of mathematical explanations of physical facts, I will argue that the structure of the explanation in these accounts is unclear and that they are susceptible to the objection that the mathematical frameworks, structures, or facts that they appeal to play a representational rather than explanatory role. I will then show how these accounts could be revised along the lines of the proposed account to circumvent these challenges.

6.9.1 On a D-N Mathematical Explanation of the Life Cycle of 'Periodical' Cicadas

Baker (2005, 2009, 2017) proposes a D-N-like account of mathematical explanation of physical facts. In presenting the account, he focuses on a natural phenomenon that is drawn from evolutionary biology: the life-cycle of the so-called 'periodical' cicada. Certain species of the North-American cicada share the same kind of unusual life-cycle, where the nymph remains in the soil for a lengthy period, and then the adult cicada emerges after either 13 or 17 years, depending on the geographical area. "[S]trikingly, this emergence is synchronized among all the members of a cicada species in any given area. The adults all emerge within the same few days, they mate, die a few weeks later and then the cycle repeats itself" (Baker 2005, p. 229). Baker's explanation concentrates on the prime-numbered-year cicada life-cycle and it proceeds as follows (Baker 2005, pp. 230–233, 2009, p. 614, 2017, p. 195).

Explanation of the Periodical Cicada Life-Cycle

(1) Having a life-cycle period that minimizes intersection with other (nearby/lower) periods is evolutionarily advantageous. [biological law]
(2) Prime periods minimize intersection (compared to non-prime periods). [number theoretic theorem]

---

(3) Hence organisms with periodic life-cycles are likely to evolve periods that are prime. ['mixed' biological/mathematical law] (from (1) and (2))
(4) Cicadas in ecosystem-type E are limited by biological constraints to periods from 14 to 18 years. [ecological constraint]

---

(5) Hence cicadas in ecosystem-type E are likely to evolve 17-year periods. (from (3) and (4))

Baker's core thesis is that "the cicada case study is an example of an indispensable, mathematical explanation of a purely natural phenomenon" (Baker 2009, p. 614).

Before turning to discuss this explanation, a few remarks are in place. First, the explanation is actually incomplete. It is based on the implicit assumption that if something is evolutionary advantageous, it is likely to occur. This assumption is required for (1) and (2) to imply (3). Second, in appealing to the D-N account of explanation, Baker (2005, p. 235) broadens "the category of laws of nature to include mathematical theorems and principles, which share commonly cited features such as universality and necessity." Third, whether the explanation is in fact a good explanation of the periodical cicada life-cycle will not matter for the analysis of the nature of Baker's proposed D-N-like account of mathematical explanation of physical facts.

There have been various objections to the above explanation, such as that the choice of time units (e.g. years rather than seasons) is arbitrary, that the explanation begs the question against nominalism, and that the mathematical facts in the explanation play a representational rather than explanatory role (Melia 2000, 2002; Leng 2005; Bangu 2008; Baker 2009; Daly and Langford 2009; Saatsi 2011, 2016, 2018; Koo 2015). Since my main interest here is to identify the nature of Baker's proposed explanation, I will focus on the last objection.

Daly and Langford (2009, p. 657) argue that

> [i]t is this property of the cicadas' life-cycle duration – this periodic intersection with the life-cycles of certain predatory kinds of organism – that plays an explanatory role in why their life-cycle has the duration it has. Primes have similar properties, and so successfully index the duration. But then so also do various non-primes: measuring the life-cycle in seasons produces analogous patterns.

That is, the idea is that the durations of the life-cycles of the cicada and of the other relevant organisms in its environment (and the forces of evolution) explain the cicada life-cycle duration, and the number of units these durations amount to (relative to a given measuring system) only index the durations measured. Saatsi (2011, 2016) expresses a similar objection, proposing that Baker's explanation could proceed with an alternative premise to (2) and (3):

> (2/3) For periods in the range of 14–18 years the intersection-minimizing period is 17. [fact about time]

While Daly, Langford and Saatsi have nominalist sympathies, the objection that the mathematical fact in Baker's explanation plays only a representational role has also been submitted by Brown (2012, p. 11) who advocates Platonism.

Baker (2017, p. 198) replies that the above nominalist perspective lacks a *scope generality*: it is inapplicable to other situations in which the ecological conditions are different. As he points out, this is not only a hypothetical point, as there are subspecies of the periodical cicada with 13-year life cycles. Indeed, (2/3) could be generalized into a schema (Saatsi 2011, p. 152):

> (2/3)* There is a unique intersection minimizing period T_x for periods in the range [T_1, ..., T_2] years.

Yet, Baker (ibid., p. 199) notes that while (2/3)* has a more general scope, the explanation that this schema provides is less unified and has less depth than the original one. Further, Baker argues that the nominalist perspective also lacks *topic generality* (ibid., pp. 200–208). A mathematical explanation of physical facts has a higher topic generality if the mathematical facts that are supposed to do the explanation could apply to explanations of other topics. For example, Baker shows that a revised version of the explanation in (1)–(5), which meets scope generality and topic generality, could share the same core as an explanation of why in brakeless fixed-gear bicycles the most popular gear ratios are 46/17, 48/17, 50/17, and 46/15.

Baker argues that: (a) for an explanation to have a high level of scope and topic generality, mathematics is indispensable; (b) the interpretation of the mathematical facts in a mathematical explanation of physical facts as representations of physical facts limits the topic generality of the explanation; and (c) the optimal version of the mathematical explanation of the cicada life-cycle "has an explanatory core that is topic general, and is not 'about' any designated class of physical facts, such as facts about time, or facts about durations" (ibid., p. 201). To establish (c), he proposes a generalized explanation of the cicada life-cycle, which I discuss below.

I agree with Baker's observation that the nominalist perspective restricts the level of scope generality and topic generality of explanations. But I don't think that his response meets the objection that, in his account, mathematics only represents the physical facts that actually do the explanation. In what follows, I will present the challenge for Baker's account, first considering the original explanation of the periodical cicada life-cycle and then the generalized version.

Baker's explanation in (1)–(5) is ambiguous because premise (2) is ambiguous. It could be interpreted as:

(2i) a fact about time, i.e. a physical fact; or
(2ii) a theorem about number theory, i.e. a purely mathematical fact.

For the above D-N-like explanation to be valid, it requires (2i) rather than (2ii): since (2ii) is a pure mathematical fact, (1) and (2ii) *per se* do not imply (3). Thus, it may be argued that it is the physical fact about time in (2i) that explains why the cicadas' life-cycle is likely to be prime, and the mathematical fact in (2ii) only represents this physical fact; and the impression that (1) and (2ii) entail (3) is due to an equivocation between (2i) and (2ii). The upshot is that Baker fails to demonstrate that the explanation of the periodical cicada life-cycle above is a mathematical explanation of a physical fact. If we interpret premise (2) as a physical fact about the nature of time, i.e. as (2i), the D-N-like explanation in (1)–(5) is a physical explanation of a physical fact, and it is not clear what explanatory role the mathematical fact about prime numbers plays. If, on the other hand, we interpret (2) as a purely mathematical fact, i.e. as (2ii), the explanation becomes either invalid or unclear: if the explanation is supposed to be a D-N-like explanation, where the *explanandum* follows deductively from the *explanans*, it is invalid; and if the explanation is not intended as a D-N-like explanation, it is not clear what kind of explanation it is.

It may be tempting to reply that the above objection does not apply to Baker's generalized explanation of the cicada life-cycle, where the explanatory core is not 'about' any designated class of physical facts because of its topic generality. Yet, as we shall see below, the generalized explanation is still subject to the same challenge.

Generalized explanation of the periodical cicada life-cycle

(M1) The lowest common multiple (LCM) of two numbers, m, n, is maximal if and only if m and n are coprime. [pure mathematical fact]

(UC1) The gap between successive co-occurrences of the same pair of cycle elements of two unit cycles is equal to the LCM of their respective lengths.

(UC2) Hence any pair of unit cycles with periods m and n maximizes the gap between successive co-occurrences of the same pair of cycle elements if and only if m and n are coprime. (from M1 and UC1)

(M2) All and only prime numbers are coprime with all smaller numbers. [pure mathematical fact]

(UC3) Hence, given a unit cycle, p_m, of length m and a range of unit cycles, q_i, of lengths shorter than m, p_m maximizes the gap between successive intersections with each q_i if and only if m is prime. (from UC2, M2)

(1G) For periodical organisms, having a life-cycle period that maximizes the gap between successive co-occurrences with periodical predators is evolutionarily advantageous. [biological law]

(2G) Periodical organisms with periodical predators whose life cycles are restricted to multiples of a common base unit can be modeled as pairs of unit cycles.

(3G) Hence organisms with periodic life-cycles that are exposed to periodic predators with shorter life-cycles, and whose life cycles are restricted to multiples of a common base unit, are likely to evolve periods that are prime. ['mixed' biological/mathematical law] (from UC3, 1G, 2G)

(4G) North American periodical cicadas fit the application conditions stated in premise (3G).

(5G) Hence periodical cicadas are likely to evolve periods that are prime. (from 3G, 4G)

(6G) Cicadas in ecosystem-type E are limited by biological constraints to periods from 14 to 18 years. [ecological constraint]

(7G) 17 is the only prime number between 14 and 18. [pure mathematical fact]

(8G) Hence cicadas in ecosystem-type E are likely to evolve 17-year periods. (from 5G, 6G, 7G)

As it is not difficult to see, in the above explanation the explanatory core – (M1), (M2), and (UC1) – (UC3) – is topic general. Yet, similarly to the original explanation, one could argue that for the inference from (UC3), (1G) and (2G) to (3G) to be valid, (UC3) has to be interpreted as a proposition about physical systems – all the physical systems that satisfy the requirement of topic generality under consideration – and that the mathematical fact in (UC3) is only a representation of this universal physical fact. Since (UC3) follows from (UC1) and (UC2) (and (M1) and (M2)), for the explanation to be valid, (UC1) and (UC2) also have to be interpreted as propositions about physical systems – again, all the physical systems that satisfy the requirement of topic generality under consideration. Thus, like in the original explanation, the generalized explanation of the periodical cicada life-cycle is subject to the challenge that it is either a physical explanation of a physical fact, invalid, or unclear.

The objection above can be divided into two related objections. (α) For the conclusion (8G) to follow from the premises in the generalized explanation of the cicada life-cycle, (UC1) – (UC3) have to be interpreted as propositions about physical facts. Thus, the explanation is a physical explanation of a physical fact, and it is not clear in what sense it is a mathematical explanation of a physical fact. (β) Like in the original explanation of the cicada life-cycle, it may be argued that mathematics plays a representational rather than explanatory role: it represents the physical facts that actually explains the cicada life-cycle.

The second objection is particularly compelling against those who assume that the physical can ontologically be separated from the mathematical. It presupposes that the facts about physical co-occurrences can ontologically be separated from the corresponding mathematical facts. Similarly, the objection to the original explanation of the cicada life-cycle presupposes that the physical fact that prime time periods minimize intersection with other (nearby/lower) periods can ontologically be separated from the corresponding mathematical fact. But these presuppositions are unwarranted for those who take the mathematical to constitute the physical. The physical fact that prime time periods minimize intersection with other (nearby/lower) periods is constituted by number theory, in general, and its theorems about prime numbers, in particular. Accordingly, the physical fact that prime time periods minimize intersection with other (nearby/lower) periods cannot be separated from the corresponding mathematical fact. Likewise, facts about physical co-occurrences are constituted by number theory, in general, and its theorems about prime numbers, in particular. Thus, such facts cannot be separated from the corresponding mathematical facts.

Given the mathematical constitution of the physical, Baker's D-N-like account of mathematical explanation of physical facts can be revised as follows to circumvent the objections in (α) and (β). A mathematical explanation of physical facts is a physical explanation of physical facts along the D-N model in which the mathematical constitution of some physical facts in the *explanans* is highlighted. The idea here is that by highlighting the mathematical constitution of certain physical facts in the *explanans*, the explanation deepens and expands the scope of the understanding of the physical fact in the *explanandum*. Focusing on the original explanation of the periodical cicada life-cycle, the revised account has two parts. First, it derives deductively the likelihood of the prime-numbered-year life-cycle of the periodical cicada from general biological facts, particular facts about the periodical cicada and its eco-system, and facts about physical time. Second, it highlights the mathematical constitution of physical time, in general, and the mathematical constitution of the fact that prime periods of physical time minimize intersection with other time periods, in particular. The theorem of number theory in (2ii) is a reflection of this constitution. Time is modelled as a line of real numbers and accordingly time intervals with a length equal to a natural number are subject to the theorems of number theory. In highlighting the mathematical constitution of these physical facts, the explanation deepens our understanding of the curious life-cycle of the periodical cicada in eco-system type E and expands its potential scope to the life-cycle of other subspecies of the periodical cicada. Turning to the generalized explanation of the periodical cicada life-cycle, in addition to showing how the likelihood of the prime-numbered-year life-cycle of the periodical cicada follows deductively from general biological facts, particular facts about the periodical cicada and its eco-system, and universal physical facts about successive co-occurrences of the same pair of cycle elements of two unit cycles, the explanation highlights the mathematical constitution of these latter physical facts. In highlighting these facts, the explanation deepens the understanding of the periodical cicada life-cycle in ecosystem-type E and expands its potential scope to explanations of other subspecies of the periodical cicada – namely, periodical cicada in other types of ecosystems – as well as to explanations of various other natural facts.

As it is not difficult to see, the proposed revision of Baker's account of mathematical explanation of physical facts is not subject to the objections in (α) and (β). Premise (2) in the original explanation of the periodical cicada life-cycle and premises (UC1) – (UC3) in the generalized explanation of this life-cycle are interpreted as statements about intervals of time and physical unit cycles, respectively. Thus, the objection that the *explanans* in these explanations do not imply their *explanandum* does not apply. Further, the logical structure of the explanation is clear, and, by construction, mathematics plays an explanatory role. That is, the second part of the original/generalized explanation of the cicada life-cycle highlights the mathematical constitution of the physical facts in (2)/(UC1)–(UC3) and thus deepens and expands the scope of the understanding of this life-cycle.

6.9.2 On Structural Explanation of the Uncertainty Relations

Dorato and Felline (2011, p. 161) comment that the ongoing controversy concerning the interpretation of the formalism of quantum mechanics may explain why philosophers have often contrasted the poor explanatory power of quantum theory to its unparalleled predictive capacity. Yet, they claim, "quantum theory provides a kind of *mathematical* explanation of the *physical* phenomena it is about", which they refer to as *structural explanation*. To demonstrate their claim, they present two case studies: one involves the quantum uncertainty relations between position and momentum, and the other focuses on quantum nonlocality.

Following Clifton (1998, p. 7),[6] Dorato and Felline (2011, p. 163) hold that

> we explain some feature B of the physical world by displaying a mathematical model of part of the world and demonstrating that there is a feature A of the model that corresponds to B, and is not explicit in the definition of the model.

The idea here is that the *explanandum* B is made intelligible via its structure similarities with its formal representative, the *explanans* A. How do these structural similarities render the *explanandum* intelligible? Dorato and Felline propose that "in order for such a representational relation to also be sufficient for a structural explanation, … we have to accept the idea that we understand the physical phenomenon in terms of its formal representative, by locating the latter in the appropriate model" (ibid., p. 165).

Consider, for example, a structural explanation of why position and momentum cannot assume simultaneously definite values. In standard quantum mechanics, systems' position and momentum are related through Fourier transforms, and Dorato and Felline locate the *explanans* as the mathematical properties of these Fourier transforms. The formal representation of the momentum (position) of a particle is a Fourier transform of the formal representation of its position (momentum), and Dorato and Felline propose that these Fourier transforms are required to make intelligible the uncertainty relations between position and momentum. These transformations are supposed to constitute the answer to the question "*why do position and momentum not assume simultaneously sharp magnitudes?*": "*because their formal representatives in the mathematical model have a property that makes this impossible*" (ibid., p 166).

Since 'because' here is not supposed to have a causal interpretation, the question arises as to its exact meaning. Dorato and Felline do not address this question explicitly. Their structural account of explanation seems to be based on the following key ideas: (i) A structural explanation makes the *explanandum* intelligible; (ii) the assumption that the properties of a physical system exemplify the relevant parts of the mathematical model that represents it allows one to use the properties of the latter to make intelligible the properties of the former; (iii) there is a structure-preserving morphism from the representing mathematical model to the represented physical fact, and this relation ensures that the represented fact can be made intelligible by locating its representative in the mathematical model. Yet, these ideas leave the logical structure of the explanation unclear. In particular, the extent to which the structural explanation is different from mere representation is unclear. Indeed, as Dorato and Felline acknowledge, one may raise the objection that the proposed explanation is a mere translation or redescription of the physical explanation to be given (ibid., p. 166). They concede that, in a sense, the physical properties carry the explanatory burden. They consider, for example, a balance with eight identical apples, five on one pan and three on the other.

[6]Clifton (1998) appropriates, with minor modifications, a definition of explanation given by Hughes (1993).

> If someone explained the dropping of the pan with five apples (or the rising of the side with three) by simply saying "5 > 3", he/she would not have provided a genuine explanation. The side with five apples drops because it is heavier and because of the role that its gravitational mass has vis-à-vis the earth, not because 5 > 3!

In reply, Dorato and Felline claim that "structural explanations are not so easily translatable into non-mathematical terms without loss of explicative power" (ibid., p. 167). The problem with this reply is that it does not really address the above objection. The question under consideration is not whether to replace mathematical explanations with non-mathematical ones. Rather, it is a question about the nature of the mathematical explanation on offer. In particular, it is not clear in what sense the proposed structural explanation is a mathematical explanation of physical facts rather than a physical explanation of physical facts in which mathematics only plays a representational role. It is not clear, for example, that, by pointing to Fourier transforms between the functions that represent position and momentum, we are providing a mathematical explanation of a physical fact, and not merely a physical explanation of a physical fact in which mathematics plays a representational role.

Another aspect that is unclear in Dorato and Felline's structural explanation is related to the question whether the proposed pattern of explanation requires an interpretation of the representing mathematical model. Dorato and Felline conceive the structural explanation as providing "a common ground for understanding [the uncertainty relations between position and momentum], independently of the various different ontologies underlying the different interpretations of quantum theory" (ibid., p. 165). The question arises then: How does this conception square with the fact that their structural explanation relies on the idea that for a mathematical model to explain properties of a physical system the physical system has to exemplify the relevant parts of the model? The problem here is that the quantum-mechanical formalism represents different things under different interpretations. Consider, for example, Bohmian mechanics (Goldstein 2017 and references therein). The functions that represent systems' positions and momentum in the standard interpretation of the quantum-mechanical formalism do not represent their position and momentum in Bohmian mechanics. Under this alternative interpretation, quantum systems always have definite position and momentum, and systems' position and momentum do not exemplify the Fourier transform relations. Thus, the explanation that the quantum-mechanical formal representatives – the Fourier transforms between position and momentum – have a property that makes it impossible for systems to have simultaneously definite position and momentum is not correct in this case. The functions that represent systems' position and momentum in the standard interpretation represent in Bohmian mechanics the range of possible outcomes of measurements of positions and momentum and their probabilities; and the Fourier transform relations between these functions reflect the epistemic limitation on the knowledge of systems' position and momentum. The upshot is that Dorato & Felline's structural explanation cannot circumvent the question of the interpretation of the quantum-mechanical formalism.

Given the mathematical constitution of the physical, Dorato & Felline's structural account of explanation can be revised along the lines proposed in Sect. 8 so as to meet the above challenges. The main idea of the revised account is that a reference to the mathematical representatives of physical facts is explanatory if it highlights the mathematical constitution of these facts and thus makes them intelligible. For example, the structural explanation of why, in standard quantum mechanics, it is impossible for position and momentum to assume simultaneously definite values highlights the mathematical constitution of the relations between position and momentum and thus makes the *explanandum* intelligible. In Bohmian mechanics, a reference to the mathematical representatives of position and momentum in the quantum-mechanical formalism (i.e. to the position and momentum 'observables') also highlights the mathematical constitution of physical facts, but the physical facts are not the same as in standard quantum mechanics. In Bohmian mechanics, the highlighting is of the mathematical constitution of the limitation on simultaneous measurements, and accordingly knowledge, of a system's position and momentum at any given time.[7]

6.9.3 On Abstract Mathematical Explanation of the Impossibility of a Minimal Tour Across the Bridges of Königsberg

Pincock (2007, 2011a, b, 2012, 2015) proposes an account of mathematical explanation which he calls *abstract explanation*. Like most causal accounts of explanation, abstract explanation is based on the idea of an objective dependence relation between the *explanandum* and the *explanans*, but the notion of dependence in play is different from that of causal dependence (Pincock 2015, p. 877). In the case of abstract explanation, the dependence is on an abstract entity which is more abstract than the state of affairs being explained (ibid., p. 879). Pincock has applied this account to various examples, one of which is the explanation of the impossibility of a 'minimal' tour across the bridges of Königsberg.

The citizens of eighteenth century Königsberg wished to make a *minimal* tour across the city's bridges (see Fig. 6.1), crossing each bridge exactly once and returning to the starting point. But they failed. Euler saw that the network of bridges and islands yields a certain abstract structure. Represented in graph theory, this structure has edges which correspond to bridges and vertices which correspond to islands or banks (see Fig. 6.2). Let us call a graph 'Eulerian' just in case it has a continuous path from a vertex that crosses each edge exactly once and return to that vertex (for an example of such a graph, see Fig. 6.3). Then, Euler's (1736/1956)

[7]In Bohmian mechanics, a system's momentum is not an intrinsic property. Rather, it is a relational property that is generally different from the system's momentum 'observable'. Thus, the momentum observable does not generally reflect the system's momentum (Dürr et al. 2013, Chap. 7; Lazarovici et al. 2018).

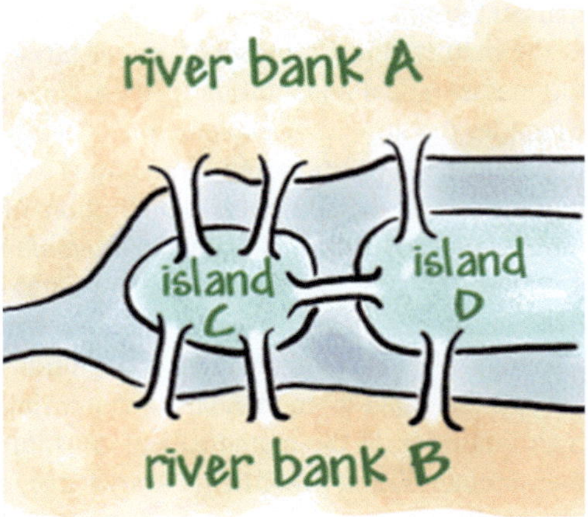

Fig. 6.1 The bridges of Königsberg

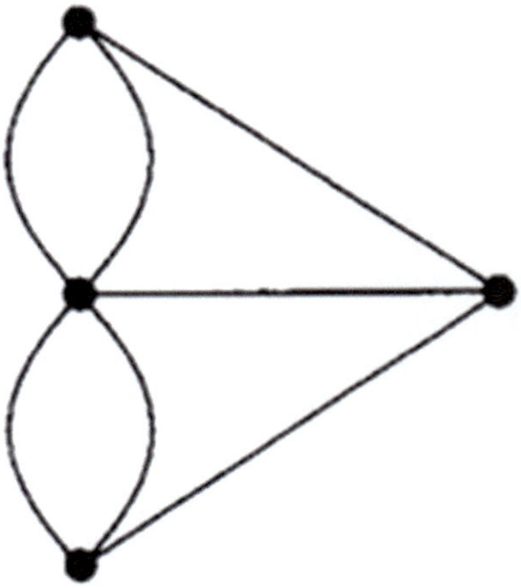

Fig. 6.2 The graph of the bridges of Königsberg: Edges correspond to bridges and vertices correspond to islands or banks

conclusions about minimal tours across bridge systems can be reformulated as the following theorem in graph theory.[8]

> *Euler's theorem*: A graph is Eulerian if and only if it contains no odd vertices; where a vertex is called 'odd' just in case it has an odd number of edges leading to it.

Pincock's proposed abstract explanation for the impossibility of a minimal tour across the bridges of Königsberg is that such a tour is impossible *because* the corresponding graph is not Eulerian. Put another way, the impossibility of a minimal tour across the bridges of Königsberg is *abstractly dependent* on the corresponding

[8]Since graph theory did not exist at the time, this is obviously an anachronistic account of Euler's reasoning. That is not a problem for the current discussion, as the aim is not to reconstruct Euler's own analysis but rather to consider Pincock's account of mathematical explanation of physical facts. For Euler's reasoning, see Euler (1736/1956), and for a reconstruction of it, see, for example, Hopkins and Wilson (2004).

Fig. 6.3 An example of Eulerian graph

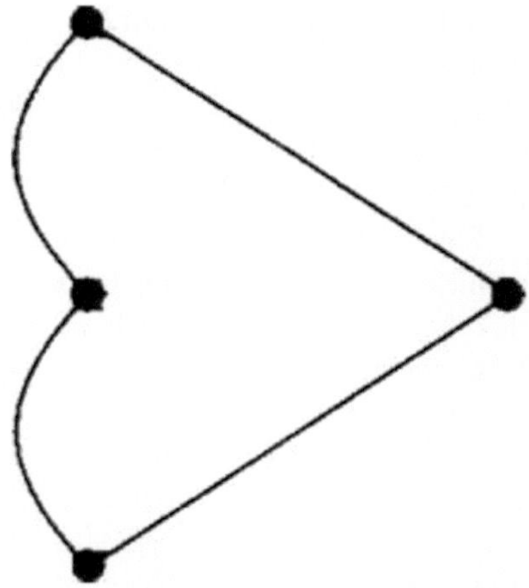

graph not being Eulerian. Similarly, the impossibility of a minimal tour across this bridge system is *abstractly dependent* on the corresponding graph having odd vertices.

Pincock's motivation for an abstract explanation of the impossibility of a minimal walk across the Königsberg bridge system is that it seems superior because it gets to the root cause of why such a walk is impossible by focusing on the abstract structure of this system. More generally, abstracting away from various physical facts, such as the bridge materials and dimensions, size of the islands and river banks, etc., scientists can often give better explanations of features of physical systems (Pincock 2007, p. 260).

Since abstract dependence is supposed to be objective but not causal, the question arises as to what kind of dependence it is. Pincock considers a few candidates. First, he (2015) discusses Woodward's (2003, p. 221) proposal that

> the common element in many forms of explanation, both causal and noncausal, is that they must answer what-if-things-had-been-different questions. . . . When a theory or derivation answers a what-if-things-had-been-different question but we cannot interpret this as an answer to a question about what would happen under an intervention, we may have a noncausal explanation of some sort.

In the case of abstract explanation, the question that is answered concerns the "the systematic relationship between the more abstract objects and their properties [the *explanans*], and the more concrete objects and their properties [the *explanandum*]." Pincock comments that Woodward's proposal could be applicable to abstract explanations "if the sort of explanatory question here could be suitably clarified" (ibid., p. 869).

Next, Pincock (ibid., p. 869–871) discusses Strevens' (2008) Kairetic account of explanation. He notes that the Kairetic account "allows important explanatory contributions from more abstract entities like mathematical objects" (ibid., p. 869). He considers whether this account – which, like Woodward's account, is based on the idea of difference making, and moreover aims to reduce all explanations to causal explanations – could be a candidate for explicating the notion of abstract dependence. Pincock rejects this route on the ground that abstract dependence violates Strevens' 'cohesion' condition; where a causal model is cohesive "if its realizers constitute a contiguous set in causal similarity space" (Strevens 2008, p. 104).

Pincock (2011a) also considers another notion of difference making which is explicated in terms of comparisons across relevant range of possibilities. Applying this notion to the explanation of why a minimal tour across the bridges of Königsberg is impossible, his reasoning is basically as follows. Let *V* be a binary variable that denotes whether all vertices of a graph are even (where a vertex is called 'even' just in case it has an even number of edges leading to it), *E* be a binary variable that denotes whether the graph is Eulerian, and *M* be a binary variable that denotes whether there is a minimal tour across the bridges of Königsberg. The graph of the bridges of Königsberg has odd vertices and it fails to be Eulerian, but many similar graphs have all even vertices and each of these graphs is Eulerian. Thus, *V* makes a difference for *E*. Further, the bridges of Königsberg fail to have a corresponding graph that is Eulerian and they fail to have a minimal tour, but many similar bridge systems have a corresponding graph that is Eulerian and each of these other bridge systems has a minimal tour. Thus, *E* makes a difference for *M*.

In the above proposal, the exact nature of abstract dependence is still unclear. In particular, Pincock does not explicate the type of similarity he has in mind, so it is difficult to evaluate the nature of modality involved in his proposed notion of difference making. Relatedly, it may be argued that Pincock's reasoning does not establish that *E* (*V*) makes a difference for *M* but rather that there is a correspondence between *E* (*V*) and *M*, and that such a correspondence *per se* does not qualify as a difference making. It is also unclear why one needs to appeal to the notion of difference making in order to establish the modal relation between *V* and *E* given that one could simply provide a mathematical proof of Euler's theorem.

Finally, Pincock (2015) considers Koslicki's (2012) notion of ontological dependence as a possible route to explicating the nature of abstract dependence. One variety of such a dependence is the following.

> *Constituent Dependence*: An entity, Φ, is constituent dependent on an entity (or entities), Ψ, just in case Ψ is an essential constituent (or are essential constituents) of Φ. (Koslicki, ibid., p. 205)

An example of such ontological dependence is lightning: "for lightning to occur is just for energy to be discharged by some electrons in a certain way, and when lightning occurs, these electrons are constituents of the lightning" (Pincock 2015, p. 878). Pincock remarks that "it might be tempting to try to reduce abstract dependence to constituent dependence" and "[it] is hard to argue that this cannot be done." But

> there is one *prima facie* barrier that seems difficult to overcome. ... [a] distinguishing feature of abstract explanation and abstract dependence is that we appeal to a more abstract entity that has a more concrete entity as an instance. ... By contrast, in the constituent

cases, the entities that we appeal to in the explanation are constituents of the fact to be explained. (Ibid., p. 879)

As the brief review of Pincock's attempts to explicate abstract dependence demonstrates, the nature of this dependence remains unclear. Further, the abstract account of explanation is open to the objection that mathematics just plays a representational role. The idea here is that the graph that corresponds to the bridges of Königsberg only represents the relevant physical structure of this bridge system which explains the impossibility of a minimal tour (Jansson and Saatsi 2019).[9] This objection has more traction in view of the difficulties in explicating abstract dependence, and it is even more pressing for those who deny the mathematical constitution of the physical; for it is difficult to see how the impossibility of a minimal tour – a physical fact – could depend on a mathematical object – a graph – if the physical is fundamentally non-mathematical.

While the prospects of explicating abstract dependence in terms of Constituent Dependence seem dim, there are other notions of ontological dependence that escape this fate. In particular, the notions of mathematical constitution of physical facts we reviewed in Sect. 4 are not subject to Pincock's concern about Constituent Dependence. When abstract dependence involves a dependence of physical facts on mathematical structures or facts, it could be explicated in terms of these notions of mathematical constitution. Consider, again, the bridges of Königsberg. The impossibility of a minimal tour across the Königsberg bridge system is due to its physical structure. This physical structure has a mathematical constitution. Some aspects of this constitution, which concern the topology of the Königsberg bridge system, are highlighted by graph theory, in general, and Euler's theorem, in particular. Thus, abstract explanation of the impossibility of a minimal tour across the Königsberg bridge system could be conceived as highlighting the aspects of the mathematical constitution of the physical structure of this bridge system that render such a tour impossible. Envisaged in this way, the explanation highlights the ontological dependence of the physical structure of the Königsberg bridge system on a certain mathematical structure, which is expressed in terms of graph theory. Highlighting the relevant aspects of the mathematical constitution of this physical structure makes the impossibility of a minimal tour across the Königsberg bridge system intelligible. It also deepens and expands the scope of our understanding by relating this impossibility to various other cases of possible and impossible minimal tours.

[9]Vineberg (2018) suggests that the objection above does not apply to the structuralist understanding of mathematics. It is not clear, however, how an appeal to this conception of mathematics could help here. One may accept the structuralist rejection of mathematical objects yet argue that mathematical structures only represent the physical structures which actually do the explanation.

6.9.4 *On Explanations by Constraints that Are More Necessary than Laws of Nature*

In *Because without cause: Non-causal explanations in science and mathematics*, Lange (2016) considers examples of "distinctively mathematical explanations". For him, distinctively mathematical explanations are subspecies of a more general type of non-causal explanation: "explanations by constraint".

> Explanations by constraint work not by describing the world's causal relations, but rather by describing how the explanandum arises from certain facts ("constraints") possessing some variety of necessity stronger than ordinary laws of nature possess. The mathematical facts figuring in distinctively mathematical explanations possess one such stronger variety of necessity: mathematical necessity. (Ibid., p. 10)

Lange suggests that a distinctively mathematical explanation appeals "only to facts (including but not always limited to mathematical facts) that are modally stronger than ordinary laws of nature, together with contingent conditions that are contextually understood to be constitutive of the arrangement or task at issue in the why question" (ibid., pp. 9–10). That is, in such explanations the *explanans* consist of mathematically necessary facts and possibly other facts which possess some variety of necessity stronger than that of ordinary laws of nature (the "constraints"), as well as contingent conditions which are presupposed by the why question under consideration. The explanation shows that "the fact to be explained could not have been otherwise – indeed, was inevitable to a stronger degree than could result from the action of causal powers" (ibid., pp. 5–6). Under the presupposed contingent conditions, the *explanadum* arises from the "constraints", and it is necessary in a stronger sense than the necessity that ordinary laws of nature mandate. Thus, the explanation works

> not by describing the world's actual causal structure, but rather by showing how the explanandum arises from the framework that any *possible* physical system (whether or not it figures in causal relations) must inhabit, where the "possible" systems extend well beyond those that are logically consistent with all of the actual natural laws. (Ibid., pp. 30–31)

For example, the fact that a mother fails repeatedly to distribute evenly 23 strawberries among her 3 children without cutting any strawberry is explained by the mathematical fact that 23 cannot be divided evenly by 3 (ibid., p. 6). The explanation shows that mother's success is impossible in a stronger sense than causal considerations underwrite (ibid., p. 30). The *explanans* in this case consist of the mathematical fact that 23 cannot be divided evenly by 3 and the contingent facts presupposed by the why-question under consideration: namely, that there are 23 strawberries and 3 children, and the distribution is of uncut strawberries. And the explanation shows that, under such contingent conditions, an even distribution of the 23 strawberries among the 3 children is impossible, where the impossibility is stronger than natural impossibility.

Saatsi (2018) poses two challenges for Lange's account of explanation. One challenge is that "information about the strong degree of necessity involved [in explanation by constraint] risks being *too cheap*: the exalted modal aspect of the explanandum can be communicated without doing much explaining, and it can be grasped without having much understanding" (ibid., p. 5). Another challenge is to pin down the difference due to which explanations-by-constraint work so differently from causal explanations. Saatsi proposes, as an alternative to Lange's approach, that causal and non-causal explanations alike "can explain by virtue of providing what-if-things-had-been-different information that captures a dependence relation between the explanandum and the explanans." He believes that this counterfactual-dependence perspective suggests that, in various distinctively mathematical explanations, at the very least "we should not hang the analysis of explanatoriness entirely on the hook of modal 'constraint'." For example, "[i]f we squeeze out, as it were, all the modal information regarding how Mother's predicament would differ as a function of the number of strawberries/kids, it looks that we are left with a very shallow explanation at best, *even if we fully retain* the information concerning the exalted modal status of the explanandum" (ibid., p. 6).

Lange (2016, 2018, p. 32) resists the proposal to explicate distinctively mathematical explanations in terms of counterfactual dependence. He maintains that some explanations by constraint, and more particularly some distinctively mathematical explanations, are associated with no pattern of counterfactual dependence. Further, he (2018, p. 33) also argues that "[s]ometimes the explanans and explanandum of explanations by constraint do figure in patterns of counterfactual dependence, but these patterns fail to track explanatory relations". Yet, while Saatsi's view that causal and non-causal explanations could be accounted in terms of counterfactual dependence is controversial, his objections highlight the fact that the exact nature of Lange's distinctively mathematical explanations is unclear. Consider, for example, Saatsi's question (ibid., p. 8):

> Why is it that given that mass is additive, if A has the mass of 1 kg, and B has the mass of 1 kg, then the union A+B has the mass of 2 kg?

Lange (2018, p. 35) admits that "perhaps Saatsi is correct that the answer 'Because 1+1=2' is 'utterly shallow'." But he suggests that

> that impression may arise from everyone's knowing that 1+1=2 and that this fact suffices to necessitate the explanandum, so it is difficult to see what information someone asking Saatsi's question might want. Furthermore, "Because 1+1=2" may not explain A+B's mass at all, if A and B can chemically interact when "united". (Ibid.)

Indeed, the mathematical fact $1 + 1 = 2$ and its intuitive relation to the *explanandum* are very familiar. But this familiarity conceals the fact that the exact structure of Lange's account of distinctively mathematical explanation is unclear. It is not clear from Lange's account how the "because" of explanation is supposed to work here: how a mathematical fact *per se* could explain a fact about the nature of physical objects? One may argue that the fact that $1 + 1 = 2$ seems to necessitate the *explanandum* is due to equivocation between: (i) the mathematical fact that $1 + 1 = 2$; and (ii) the universal physical fact that, for any two possible physical objects of 1 kg, which are 'united' without being changed, the total mass is 2 kg. That is, one may argue that it is the physical fact in (ii), rather than the mathematical fact in (i), that actually does the explanation, and that the mathematical fact in

(i) just represents the physical fact in (ii). Put another way, one may argue that the intuitive appeal of Lange's distinctively mathematical explanation is due to the above equivocation, taking the explanation to be a physical explanation of a physical fact – it is an explanation of a particular physical fact in terms of a corresponding universal physical fact – in which the mathematical fact $1 + 1 = 2$ only plays a representational role.

The above objection is particularly compelling for those who maintain that the physical is ontologically separated from the mathematical. It can be circumvented by acknowledging the mathematical constitution of the physical. Distinctively mathematical explanations of physical facts could then be understood as explanations of physical facts in which the mathematical constitution of some physical facts is highlighted by stating the constraints that this constitution mandates. For example, in explaining "why the union A+B has the mass of 2 kg" by pointing out that "$1 + 1 = 2$", one appeals to the universal physical fact that for any two possible physical objects of 1 kg, A and B, which are 'united' without being changed, the total mass of A + B is 2 kg. Yet, one observes that this universal physical fact is constituted by number theory, in general, and a mathematical fact that follows from it, $1 + 1 = 2$, in particular. This constitution is highlighted by stating the constraint $1+1=2$ that it mandates. In this kind of explanation, there is no need to appeal to counterfactual dependence. Indeed, it is not clear how counterfactual dependence could be of any help here.

6.10 Is the Effectiveness of Mathematics in Physics Unreasonable?

It seems to be a dogma of contemporary mainstream philosophy of science that, fundamentally, physical facts are not mathematical, and that mathematics only provides a language for representing the physical realm, even if this language is indispensable. Thus, the idea of mathematical explanations of physical facts may naturally appear puzzling or even paradoxical. I argued above that the view that the physical is ontologically separated from the mathematical fails to make sense of the common conception of how mathematical models and theories represent physical phenomena and reality in modern natural science. I suggested that this conundrum could be avoided if we accept the idea that the physical is constituted by the mathematical. I then reviewed two traditional ways of conceiving such a constitution: the Pythagorean and the neo-Kantian. Granted the mathematical constitution of the physical, I proposed a new account of mathematical explanation of physical facts. In this account, there are two related kinds of explanations. One kind of mathematical explanation of physical facts consists in: explaining physical facts by physical facts along the lines of the D-N, causal, unification, or any other acceptable account of explanation; and highlighting the mathematical constitution of some of the physical facts in the *explanans* and thus deepening and expanding the scope of the understanding of the explained physical facts. The second kind of mathematical explanation of physical facts consists in highlighting the mathematical constitution of physical facts and thus making the explained facts intelligible or deepening and expanding the scope of our understanding of them. I also considered four other accounts of mathematical explanation of physical facts. I argued that, unlike the proposed account, they are open to the objections that the nature of their explanation is unclear and that mathematics plays only a representational role, representing the physical facts that actually do the explanation. I then suggested that these accounts could be revised along the lines of the proposed account so as to circumvent both challenges.

The proposed account is neutral with respect to the controversy in the philosophy of mathematics about the ontological status of 'mathematical objects' (numbers, sets, relations, functions, etc.). Further, it applies to both realist and instrumentalist interpretations of theories/models. The interpretation of the ontological status of theoretical terms only determines the scope of the mathematical constitution of the physical beyond the phenomena.

In conclusion, I turn briefly to comment on how the idea of the mathematical constitution of the physical reflects on the question whether the effectiveness of the use of mathematics in physics is unreasonable. It is reasonable to assume that one's conception of mathematics and its relation to physics is important for adjudicating this question. Let us consider then Wigner's conception. Wigner (1960, pp. 2–3) holds that while

> the concepts of elementary mathematics and particularly elementary geometry were formulated to describe entities which are directly suggested by the actual world, the same does not seem to be true of the more advanced concepts, in particular the concepts which play such an important role in physics. ... Most more advanced mathematical concepts, such as complex numbers, algebras, linear operators, Borel sets – and this list could be continued almost indefinitely – were so devised that they are apt subjects on which the mathematician can demonstrate his ingenuity and sense of formal beauty.

He (ibid.) conceives mathematics as

> the science of skillful operations with concepts and rules invented just for this purpose. The principal emphasis is on the invention of concepts. ... The depth of thought which goes into the formulation of the mathematical concepts is later justified by the skill with which these concepts are used. The great mathematician fully, almost ruthlessly, exploits the domain of permissible reasoning and skirts the impermissible. That his recklessness does not lead him into a morass of contradictions is a miracle in itself ... The principal point which will have to be recalled later is that the mathematician could formulate only a handful of interesting theorems without defining concepts beyond those contained in the axioms and that the concepts outside those contained in the axioms are defined with a view of permitting ingenious logical operations which appeal to our aesthetic sense both as operations and also in their results of great generality and simplicity.

Based on this conception and consideration of the application of various "advanced mathematical concepts" in physics, Wigner concludes that the appropriateness of the language of mathematics for the formulation of the laws of physics is a miracle, and he hopes that this miracle will continue in future research.

On Wigner's conception of mathematics, the relation between the physical and mathematical is not intrinsic. Advanced mathematics largely develops on its own and is then picked up by physicists to yield great success. It may thus be natural to see the success of the application of the language of mathematics in physics as surprising, mysterious, and even unreasonable. However, Wigner's conception is controversial (see, for example, Lützen 2011, Ferreirós 2017, Islami 2017, and references therein). Some have argued that while Wigner's view of mathematics seems to have been inspired by the formalist philosophy of mathematics, this philosophy has lost its credibility by the second half of the twentieth century (Lützen 2011, Ferreirós 2017). Further, historically, the development of mathematics has been entangled with that of the natural sciences, in general, and physics, in particular, and Wigner's conception of mathematics fails to reflect this fact. Consider, for example, complex numbers. Wigner singles them out as one of the prime examples of most advanced mathematical concepts that mathematicians invented for the sole purpose of demonstrating their ingenuity and sense of formal beauty without any regard to possible applications. Indeed, it seems that complex numbers were introduced by Scipione del Ferro, Niccola Fontana (also known as Tartaglia, the stammerer), Gerolamo Cardano, Ludovico Ferrari, and Rafael Bombelli in the sixteenth century with no intent in mind to apply them. Yet, the context of the introduction was the attempt to solve the general forms of quadratic and cubic equations, which by that time had a long history of applications (Katz 2009). Thus, taking into account this broader context, the claim that the introduction of complex numbers was solely for the purpose of demonstrating ingenuity and sense of formal beauty is misleading.

That is not to argue that all the developments in mathematics were connected to physics, neither to deny that in many cases future applications of mathematical concepts could not be anticipated at the time of their introduction. Yet, attention to the historical entanglement between the developments of mathematics and physics casts doubt on the validity and scope of Wigner's argument for the unreasonable success of the application of mathematics in physics.

In any case, while Wigner's view of mathematics may provide some support to the view that the success of the application of mathematics in physics is unreasonable, things are very different if we conceive the mathematical as constitutive of the physical. In the context of such a conception, the relationship between mathematics and physics is intrinsic. The physical is characterized in mathematical terms. The physicist's crude experience is formulated in precise mathematical terms, often in statistical models of the phenomena, and accordingly the gap between the phenomena and the theoretical models that account for them diminishes. There is no essential gap between the concepts of elementary mathematics and geometry and more advanced mathematical concepts. The division between applied and theoretical mathematics is neither *a priori* nor fundamental. And the appropriateness of the use of the language of mathematics in future physics is not in doubt. Thus, it is reasonable to expect the language of mathematics to be appropriate for representing physical reality.

Of course, such a conception of the role of mathematics in physics does not obliterate the sense of wonder that one has with respect to the fact that "in spite of the baffling complexity of the world, certain regularities in the events could be discovered" (Wigner, ibid., p. 4). Yet, the reasons to conceive this wonder in miraculous terms are not as compelling as for those who deny the mathematical constitution of physical facts.

Acknowledgments I owe a great debt to Itamar Pitowsky. Itamar's graduate course in the philosophy of probability stimulated my interest in the philosophical foundations of probability and quantum mechanics. Itamar supervised my course essay and MA thesis on the application of de Finetti's theory of probability to the interpretation of quantum probabilities. During my work on this research project, I learned from Itamar a great deal about the curious nature of probabilities in quantum mechanics. Itamar was very generous with his time and the discussions with him were always enlightening. Our conversations continued for many years to come, and likewise they were always helpful in clarifying and developing my thoughts and ideas. Itamar's untimely death has been a great loss. Whenever I have an idea I would like to test, I think of him and wish we could talk about it. I sorely miss the meetings with him and I wish I could discuss with him the questions and ideas considered above.

I am very grateful to the volume editors, Meir Hemmo and Orly Shenker, for inviting me to contribute, and to Meir for drawing my attention to Itamar's commentary on the relationship between mathematics and physics. The main ideas of the proposed account of mathematical explanations of physical facts were first presented in the workshop on Mathematical and Geometrical Explanations at the Universitat Autònoma de Barcelona (March 2012), and I thank Laura Felline for inviting me to participate in the workshop. Earlier versions of this paper were also presented in the 39th, 40th, 43rd, and 46th Dubrovnik Philosophy of Science conferences, IHPST, Paris, IHPST, University of Toronto, Philosophy, Università degli Studi Roma Tre, Philosophy, Università degli Studi di Firenze, CSHPS, Victoria, CPNSS, LSE, Philosophy, Leibniz Universität Hannover, ISHPS, Jerusalem, Munich Center for Mathematical Philosophy, LMU, Faculty of Sciences, Universidade de Lisboa. I would like to thank the audiences in these venues for their helpful comments. For discussions and comments on earlier drafts of the paper, I am very grateful to Jim Brown, Donald Gillies, Laura Felline, Craig Fraser, Aaron Kenna, Flavia Padovani, Noah Stemeroff, and an anonymous referee. The research for this paper was supported by SSHRC Insight and SIG grants as well as Victoria College travel grants.

References

Aristotle. (1924). *Metaphysics*. A revised text with introduction and commentary by W. D. Ross (2 vols). Oxford: Clarendon Press.

Bachelard, G. (1965). *L' activité rationaliste de la physique contemporaine*. Paris: Presses Universitaires de France.

Baker, A. (2005). Are there genuine mathematical explanations of physical phenomena? *Mind, 114*(454), 223–238.

Baker, A. (2009). Mathematical explanations in science. *British Journal for the Philosophy of Science, 60*(3), 611–633.

Baker, A. (2017). Mathematics and explanatory generality. *Philosophia Mathematica, 25*(2), 194–209.

Bangu, S. (2008). Inference to the best explanation and mathematical explanation. *Synthese, 160*(1), 13–20.

Baron, S., Colyvan, M., & Ripley, D. (2017). How mathematics can make a difference. *Philosophers' Imprint, 17*(3), 1–19.

Batterman, R. (2002). Asymptotics and the role of minimal models. *British Journal for the Philosophy of Science, 53*(1), 21–38.

Batterman, R. (2010). On the explanatory role of mathematics in empirical science. *British Journal for the Philosophy of Science, 61*(1), 1–25.

Batterman, R. (2018). Autonomy of theories: An explanatory problem. *Noûs, 52*(4), 858–873.
Batterman, R., & Rice, C. (2014). Minimal model explanations. *Philosophy of Science, 81*(3), 349–376.
Bokulich, A. (2008a). Can classical structures explain quantum phenomena? *British Journal for the Philosophy of Science, 59*(2), 217–235.
Bokulich, A. (2008b). *Reexamining the quantum-classical relation: Beyond reductionism and pluralism*. Cambridge: Cambridge University Press.
Bokulich, A. (2011). How scientific models can explain. *Synthese, 180*(1), 33–45.
Bolzano, B. (1930). Functionenlehre. In K. Rychlik (Ed.) *Spisy Bernarda Bolzana* Vol. 1. Prague: Royal Bohemian Academy of Sciences.
Brown, J. R. (2012). *Plantonism, naturalism, and mathematical knowledge*. London: Routledge.
Bueno, O., & Colyvan, M. (2011). An inferential conception of the application of mathematics. *Nous, 45*(2), 345–374.
Bueno, O., & French, S. (2011). How theories represent. *British Journal for the Philosophy of Science, 62*(4), 857–894.
Bueno, O., & French, S. (2018). *Applying mathematics: Immersion, inference, interpretation*. Oxford: Oxford University Press.
Bueno, O., French, S., & Ladyman, J. (2002). On representing the relationship between the mathematical and the empirical. *Philosophy of Science, 69*(3), 497–518.
Cassirer, E. (1910/1923). *Substanzbegriff und funktionsbegriff. Untersuchungen über die Grundfragen der Erkenntniskritik. Berlin: Bruno Cassirer. Translated as Substance and function.* Chicago: Open Court.
Cassirer, E. (1912/2005). Herman Cohen and the renewal of Kantian philosophy (trans. By Lydia Patton). *Angelaki, 10*(1), 95–104.
Clifton, R. (1998). Scientific explanation in quantum theory. *PhilSci Archive*. http://philsci-archive.pitt.edu/91/
Colyvan, M. (2001). *The indispensability of mathematics*. Oxford: Oxford University Press.
Colyvan, M. (2002). Mathematics and aesthetic considerations in science. *Mind, 111*(441), 69–74.
Daly, C., & Langford, S. (2009). Mathematical explanation and indispensability arguments. *Philosophical Quarterly, 59*(237), 641–658.
Dorato, M., & Felline, L. (2011). Scientific explanation and scientific structuralism. In A. Bokulich & P. Bokulich (Eds.), *Scientific structuralism* (Boston studies in the philosophy of science) (pp. 161–177). Dordrecht: Springer.
du Bois-Reymond, P. (1875). Versuch einer classification der willkürlichen functionen reeller argumente nach ihren aenderungen in den kleinsten intervallen. *Journal für die reine und angewandte Mathematik, 79*, 21–37.
Dürr, D., Goldstein, S., & Zanghì, N. (2013). *Quantum physics without quantum philosophy*. Berlin: Springer.
Einstein, A. (1933/1954). On the method of theoretical physics. In A. Einstein, *Ideas and opinions* (new translations and revisions by S. Bargmann), New York: Bonanza Books (1954), pp. 270–276.
Euler, L. (1736/1956). Solutio problematis ad geometriam situs pertinentis. *Commentarii Academiae Scientiarum Imperialis Petropolitanae, 8*, 128–140.
Felline, L. (2018). Mechanisms meet structural explanations. *Synthese, 195*(1), 99–114.
Ferreirós, J. (2017). Wigner's "unreasonable effectiveness" in context. *Mathematical Intelligencer, 39*(2), 64–71.
Feynman, R., et al. (1963). *The Feynman lectures in physics* (Vol. 2). Reading: Addison-Wesley.
French, S., & Ladyman, J. (1998). A semantic perspective on idealization in quantum mechanics. In N. Shanks (Ed.), *Idealization IX: Idealization in contemporary physics* (Pozna'n Studies in the Philosophy of the Sciences and the Humanities) (pp. 51–73). Amsterdam: Rodopi.
Frigg, R., & Nguyen, J. (2018). Scientific representation. *The Stanford Encyclopedia of philosophy* (Winter 2018 Edition). In E. N. Zalta (Ed.). URL=https://plato.stanford.edu/archives/win2018/entries/scientific-representations/

Galileo, G. (1623/1960). *The Assayer. In the controversy of the comets of 1618: Galileo Galilei, Horatio Grassi, Mario Guiducci, Johann Kepler* (S. Drake, & C. D. O'Malley, Trans.) Philadelphia: The University of Pennsylvania Press.

Goldstein, S. (2017). Bohmian mechanics. *The Stanford Encyclopedia of Philosophy* (Summer 2017 Edition), E. N. Zalta (Ed.). URL=https://plato.stanford.edu/archives/sum2017/entries/qm-Bohm/

Hopkins, B., & Wilson, R. J. (2004). The truth about Königsberg. *The College Mathematics Journal, 35*(3), 198–207.

Huggett, N. (Ed.). (1999). *Space from Zeno to Einstein: Classic readings with a contemporary commentary*. Cambridge, MA: MIT Press.

Huggett, N. (2019). Zeno paradoxes. *The Stanford Encylopedia of Philosophy* (Spring 2019 Edition), E. N. Zalta (ed.). URL= https://plato.stanford.edu/archives/win2019/entries/paradox-zeno/

Hughes, R. I. G. (1993). Theoretical explanation. *Midwest Studies in Philosophy, XVIII*, 132–153.

Islami, A. (2017). A match not made in heaven: On the applicability of mathematics in physics. *Synthese, 194*(12), 4839–4861.

Jansson, L., & Saatsi, J. (2019). Explanatory abstraction. *British Journal for the Philosophy of Science, 70*(3), 817–844.

Katz, V. J. (2009). *A history of mathematics: An introduction* (3rd ed.). Reading: Addision-Wesley.

Koo, A. (2015). *Mathematical explanation in science*. PhD thesis, IHPST, University of Toronto.

Koslicki, K. (2012). Varieties of ontological dependence. In F. Correia & B. Schnieder (Eds.), *Metaphysical grounding: Understanding the structure of reality* (pp. 186–213). Cambridge: Cambridge University Press.

Kowalewski, G. (1923). Über Bolzanos nichtdiffrenzierbare stetige funktion. *Acta Mathematica, 44*, 315–319.

Lange, M. (2016). *Because without cause: Non-causal explanations in science and mathematics*. Oxford: Oxford University Press.

Lange, M. (2018). Reply to my critics: On explanations by constraint. *Metascience, 27*(1), 27–36.

Lazarovici, D., Oldofredi, A., & Esfeld, M. (2018). Observables and unobservables in quantum mechanics: How the no-hidden-variables theorems support the Bohmian particle ontology. *Entropy, 20*(5), 116–132.

Leng, M. (2005). Mathematical explanation. In C. Cellucci & D. Gillies (Eds.), *Mathematical reasoning and heuristics* (pp. 167–189). London: King's College Publishing.

Lévy-Leblond, J.-M. (1992). Why does physics need mathematics? In E. Ullmann-Margalit (Ed.), *The scientific enterprise* (Boston Studies in the Philosophy of Science) (Vol. 146, pp. 145–161). Dordrecht: Springer.

Lützen, J. (2011). The physical origin of physically useful mathematics. *Interdisciplinary Science Reviews, 36*(3), 229–243.

Lyon, A. (2012). Mathematical explanations of empirical facts, and mathematical realism. *Australasian Journal of Philosophy, 90*(3), 559–578.

Mancosu, P. (2018). Explanation in Mathematics. *The Stanford Encyclopedia of Philosophy* (Summer 2018 Edition), E. N. Zalta (Ed.). URL=https://plato.stanford.edu/archives/sum2018/entries/mathematics-explanation/

Mandelbrot, B. (1977). *Fractals, form, chance and dimension*. San Francisco: W.H. Freeman.

Melia, J. (2000). Weaseling away the indispensability argument. *Mind, 109*(435), 455–479.

Melia, J. (2002). Response to Colyvan. *Mind, 111*(441), 75–79.

Mundy, B. (1986). On the general theory of meaningful representation. *Synthese, 67*(3), 391–437.

Neuenschwander, E. (1978). Riemann's example of a continuous "nondifferentiable" function. *Mathematical Intelligencer, 1*, 40–44.

Pincock, C. (2004). A new perspective on the problem of applying mathematics. *Philosophia Mathematica, 12*(2), 135–161.

Pincock, C. (2007). A role for mathematics in the physical sciences. *Noûs, 41*(2), 253–275.

Pincock, C. (2011a). Abstract explanation and difference making. A colloquium presentation in the Munich Center for Mathematical Philosophy, LMU, December 12, 2011.

Pincock, C. (2011b). Discussion note: Batterman's "On the explanatory role of mathematics in empirical science". *British Journal for the Philosophy of Science, 62*(1), 211–217.

Pincock, C. (2012). *Mathematics and scientific representation*. Oxford: Oxford University Press.

Pincock, C. (2015). Abstract explanations in science. *British Journal for the Philosophy of Science, 66*(4), 857–882.

Pitowsky, I. (1992). Why does physics need mathematics? A comment. In E. Ullmann-Margalit (Ed.), *The scientific enterprise* (Boston Studies in the Philosophy of Science) (Vol. 146, pp. 163–167). Dordrecht: Springer.

Poincaré, H. (1913). *La valeur de la science*. Geneva: Editions du Cheval Ailé.

Redhead, M. (2001). Quests of a realist: Review of Stathis Psillos's *Scientific Realism: How science tracks truth. Metascience, 10*(3), 341–347.

Resnik, M. D. (1997). *Mathematics as a science of patterns*. Oxford: Clarendon Press.

Saatsi, J. (2011). The enhanced indispensability argument: Representational versus explanatory role of mathematics in science. *British Journal for the Philosophy of Science, 62*(1), 143–154.

Saatsi, J. (2016). On the "indispensable explanatory role" of mathematics. *Mind, 125*(500), 1045–1070.

Saatsi, J. (2018). A pluralist account of non-causal explanations in science and mathematics: Review of M. Lange's *Because without cause: Non-causal explanation in science and mathematics. Metascience, 27*(1), 3–9.

Shapiro, S. (1997). *Philosophy of mathematics: Structure and ontology*. Oxford: Oxford University Press.

Steiner, M. (1978). Mathematical explanations. *Philosophical Studies, 34*(2), 135–151.

Steiner, M. (1998). *The applicability of mathematics as a philosophical problem*. Cambridge, MA: Harvard University Press.

Stemeroff, N. (2018). *Mathematics, structuralism, and the promise of realism: A study of the ontological and epistemological implications of mathematical representation in the physical sciences*. A PhD thesis submitted to the University of Toronto.

Strevens, M. (2008). *Depth: An account of scientific explanation*. Cambridge, MA: Harvard University Press.

Vineberg, S. (2018). Mathematical explanation and indispensability. *Theoria, 33*(2), 233–247.

Weierstrass, K. (1895). Über continuirliche functionen eines reellen arguments, die für keinen werth des letzteren einen bestimmten differentialquotienten besitzen (read 1872). In *Mathematische werke von Karl Weierstrass* (Vol. 2, pp. 71–76). Berlin.

Wiener, N. (1923). Differential space. *Journal of Mathematical Physics, 2*, 131–174.

Wigner, E. (1960). The unreasonable effectiveness of mathematics in the natural sciences. *Communications in Pure and Applied Mathematics, 13*(1), 1–14.

Woodward, J. (2003). *Making things happen: A theory of causal explanation*. Oxford: Oxford University Press.

Chapter 7
Everettian Probabilities, The Deutsch-Wallace Theorem and the Principal Principle

Harvey R. Brown and Gal Ben Porath

> *Chance, when strictly examined, is a mere negative word, and means not any real power which has anywhere a being in nature.* David Hume (Hume 2008)
>
> *[The Deutsch-Wallace theorem] permits what philosophy would hitherto have regarded as a formal impossibility, akin to deriving an ought from an is, namely deriving a probability statement from a factual statement. This could be called deriving a tends to from a does.* David Deutsch (Deutsch 1999)
>
> *[The Deutsch-Wallace theorem] is a landmark in decision theory. Nothing comparable has been achieved in any chance theory. ... [It] is little short of a philosophical sensation ... it shows why credences should conform to [quantum chances].* Simon Saunders (Saunders 2020)

Abstract This paper is concerned with the nature of probability in physics, and in quantum mechanics in particular. It starts with a brief discussion of the evolution of Itamar Pitowsky's thinking about probability in quantum theory from 1994 to 2008, and the role of Gleason's 1957 theorem in his derivation of the Born Rule. Pitowsky's defence of probability therein as a logic of partial belief leads us into a broader discussion of probability in physics, in which the existence of objective "chances" is questioned, and the status of David Lewis influential

---

The original version of this chapter was revised: spelling error in the chapter title was corrected now. The correction to this chapter is available at https://doi.org/10.1007/978-3-030-34316-3_28

H. R. Brown (✉)
Faculty of Philosophy, University of Oxford, Oxford, UK
e-mail: harvey.brown@philosophy.ox.ac.uk

G. Ben Porath
Department of History and Philosophy of Science, University of Pittsburgh, Pittsburgh, PA, USA

M. Hemmo, O. Shenker (eds.), *Quantum, Probability, Logic*,
Jerusalem Studies in Philosophy and History of Science,
https://doi.org/10.1007/978-3-030-34316-3_7

Principal Principle is critically examined. This is followed by a sketch of the work by David Deutsch and David Wallace which resulted in the Deutsch-Wallace (DW) theorem in Everettian quantum mechanics. It is noteworthy that the authors of this important decision-theoretic derivation of the Born Rule have different views concerning the meaning of probability. The theorem, which was the subject of a 2007 critique by Meir Hemmo and Pitowsky, is critically examined, along with recent related work by John Earman. Here our main argument is that the DW theorem does not provide a justification of the Principal Principle, contrary to the claims by Wallace and Simon Saunders. A final section analyses recent claims to the effect that the DW theorem is redundant, a conclusion that seems to be reinforced by consideration of probabilities in "deviant" branches of the Everettian multiverse.

Keywords Probability · Quantum mechanics · Everett interpretation · Deutsch-Wallace theorem · Principal Principle · Gleason's theorem

7.1 Introduction

Itamar Pitowsky was convinced that the heart of the quantum revolution concerned the role of probability in physics, and in 2008 defended the notion that quantum mechanics is essentially a new theory of probability in which the Hilbert space formalism is a logic of partial belief (Sect. 7.2 below). In the present paper we likewise defend a subjectivist interpretation of probability in physics, but not restricted to quantum theory (Sect. 7.3). We then turn (Sect. 7.4) to the important work by David Deutsch and David Wallace over the past two decades that has resulted in the so-called Deutsch-Wallace (DW) theorem, providing a decision-theoretic derivation of the Born Rule in the context of Everettian quantum mechanics. This derivation is arguably stronger than that proposed by Pitowsky in 2008 based on the 1957 Gleason theorem and further developed by John Earman, but it is noteworthy that Deutsch and Wallace have differing views on the meaning of probability. In Sect. 7.5 we attempt to rebut the claim by Wallace and Simon Saunders – not Deutsch – that the DW theorem provides a justification of Lewis' influential Principal Principle. Finally in Sect. 7.6 we analyse the 2007 critique of the DW theorem by Meir Hemmo and Pitowsky as well as Wallaces reply. We also consider recent claims to the effect that the theorem is superfluous, a conclusion that seems to be reinforced by consideration of probabilities in "deviant" branches within the Everettian multiverse.

7.2 Quantum Probability and Gleason's Theorem

Itamar Pitowsky, in attempting to explain why the development of quantum mechanics was a revolution in physics, argued in 1994 that at the deepest level, the reason has to do with probability.

> . . . the difference between classical and quantum phenomena is that relative frequencies of microscopic events, which are measured on distinct samples, often systematically violate some of Boole's conditions of possible experience.[1]

Pitowsky's insightful exploration of the connections between's George Boole's 1862 work on probability, polytopes in geometry, and the Bell inequality in quantum mechanics is well known. In 1994, he regarded the experimental violation of the Bell inequality as "the edge of a logical contradiction". After all, weren't Boole's conditions of possible experience what any rational thinker would come up with who was concerned about the undeniable practical connection between probability and finite frequencies?

Not quite. Pitowsky was clear that a strict logical inconsistency only comes about if frequencies violating the Boolean conditions are taken from a single sample. Frequencies taken from a batch of samples (as is the case with Bell-type experiments) need not, for a number of reasons. The trouble for Pitowsky in 1994 was that none of the explanations for the Bell violations that had been advanced until then in quantum theory (such as Fine's prism model or non-local hidden variables) seemed attractive to him.

This strikes us as more a problem, in so far as it is one, in physics than logic. At any rate, it is noteworthy that in a 2006 paper, Pitowsky came to view the violation of Bell-type inequalities with much more equanimity, and in particular less concern about lurking logical contradictions. Now it is a "purely probabilistic effect" that is intelligible once the meaning of probability is correctly spelt out in the correct axiomatisation of quantum mechanics. What is this notion of probability?

In his 1994 paper, Pitowsky did not define it. He stated that Boole's analysis did not require adherence to any statement about what probability means; it is enough to accept that in the case of repeatable (exchangeable or independent) events, probability is *manifested* in, but not necessarily defined by, frequency. However in 2006, he was more committal:

> . . . a theory of probability is a theory of inference, and as such, a guide to the formulation of rational expectations.[2]

In the same paper, Pitowsky argued that quantum mechanics is essentially a new theory of probability, in which the Hilbert space formalism is a "logic of partial belief" in the sense of Frank Ramsey.[3] Pitowsky now questioned Boole's view of probabilities as weighted averages of truth values, which leads to the erroneous "metaphysical assumption" that incompatible (in the quantum sense) propositions have simultaneous truth values. Whereas the 1994 paper expressed puzzlement over the quantum violations of Boole's conditions of possible experience by way

[1] Pitowsky (1994, p. 108).

[2] Pitowsky (2006, p. 4).

[3] *Ibid.*

of experimental violations of Bell-type inequalities, the 2006 paper proffered a resolution of this and other "paradoxes" of quantum mechanics.[4]

We happen to have sympathy for Pitowsky's view on the meaning of probability, and will return to it below. For the moment we are interested rather in Pitowsky's 2006 derivation of the Born Rule in quantum mechanics by way of the celebrated 1957 theorem of Gleason (Gleason 1957).

Gleason showed that the a measure function that is defined over the closed subspaces of a separable Hilbert space $\mathcal{H}$ with dimension $D \geq 3$ takes the form of the trace of the product of two operators, one being the orthogonal projection on the subspace in question, the other being a semidefinite trace class operator. In Pitowsky's 2006 axiomatisation of quantum mechanics, the closed subspaces of the Hilbert space representing a system correspond to "events, or possible events, or possible outcomes of experiments". If the trace class operator in Gleason's theorem is read as the statistical (density) operator representing the state of the system, then probabilities of events are precisely those given by the familiar Born Rule. But there is a key condition in reaching this conclusion: such probabilities are a priori "non-contextual". That, of course, is the rub. As Pitowsky himself admitted, it is natural to ask why the probability assigned to the outcome of a measurement B should be the same whether the measurement is simultaneous with A or C, when A and C are incompatible but each is compatible with B.

Pitowsky's argument for probabilistic non-contextualism hinges on the commitment to "a Ramsey type logic of partial belief" while representing the event structure in quantum mechanics in terms of the lattice of closed subspaces of Hilbert space. A certain identity involving closed subspaces is invoked, which when interpreted in terms of "measurement events", implies that the event $\{B = b_j\}$ (b_j being the outcome of a B measurement) is defined independently of how the measurement process is set up (specifically, of which observables compatible with B are being measured simultaneously).[5] Pitowsky then invokes the rule that *identical events always have the same probability*, and the non-contextualism result follows. (In terms of projection operators, the probability of a projector is independent of which Boolean sublattice it belongs to.)

[4]We will not discuss here Pitowsky's 2006 resolution of the familiar measurement problem, other than to say that it relies heavily on his view that the quantum state is nothing more than a device for the bookkeeping of probabilities, and that it implies that "we cannot consistently maintain that the proposition 'the [Schrödinger] cat is alive' has a truth value. *Op. cit.* p. 28. For a recent defence of the view that the quantum state is ontic, and not just a bookkeeping device, see Brown (2019).

[5]The identity is

$$\bigcup_{i=1}^{k} \left(\{B = b_j\} \cap \{A = a_i\}\right) = \{B = b_j\} = \bigcup_{i=1}^{l} \left(\{B = b_j\} \cap \{C = c_i\}\right) \tag{7.1}$$

where A, B and C measurements have possible outcomes $a_1, a_2, \ldots, a_k$; $b_1, b_2, \ldots, b_r$ and $c_1, c_2, \ldots, c_l$, respectively.

The argument is, to us, unconvincing. Pitowsky accepted that the non-contextualism issue is an empirical one, but seems to have resolved it by fiat. (It is noteworthy that in his opinion it is by *not committing to the lattice of subspaces as the event structure* that advocates of the (Everett) many worlds interpretation require a separate justification of probabilistic non-contextualism.[6]) Is making such a commitment any more natural in this context than was, say, Kochen and Specker's ill-fated presumption (Kochen and Specker 1967) that truth values (probabilities 0 or 1) associated with propositions regarding the values of observables in a deterministic hidden variable theory must be non-contextual?[7] Arguably, the fact that Pitowsky's interpretation of quantum mechanics makes no appeal to such theories – which must assign contextual values under pain of contradiction with the Hilbert space structure of quantum states – may weaken the force of this question. We shall return to this point in Sect. 7.4.3 below when discussing the extent to which probabilistic non-contextualism avoids axiomatic status in the Deutsch-Wallace theorem in the Everett picture.

John Earman (Earman 2018) is one among other commentators who also regard Gleason's theorem as the essential link in quantum mechanics between subjective probabilities and the Born Rule. But he questions Pitowksy's claim that in the light of Gleason's theorem the event structure dictates the quantum probability rule when he reminds us that Gleason showed only that in the case of $D > 2$, and where $\mathcal{H}$ is separable when $D = \infty$, the probability measure on the lattice of subspaces is represented uniquely by a density operator on $\mathcal{H}$ *iff* it is *countably* additive (Earman 2018). And countable additivity does not hold in all subjective interpretations of probability. (de Finetti, for instance, famously restricted probability to finite additivity.)

In our view, the two main limitations of such Gleason-based derivations of the Born Rule are the assumption of non-contextualism and the awkward fact that the Gleason theorem fails in the case of $D < 3$, and thus can have no probabilistic implications for qubits.[8] One might also question whether probabilities in quantum mechanics need be subjective. Indeed, Earman holds a dualist view, involving objective probabilities as well (and which, as we shall see later, have a separate underpinning in quantum mechanics).

The possibility of contextual probabilities admittedly raises the spectre of a violation of the no-signalling principle, analogously to the way the (compulsory) contextuality in hidden variable theories leads to nonlocality.[9] It is noteworthy that *assuming* no superluminal signalling and the projection postulate (which Pitowsky

[6]*Op. cit.* Footnote 2.

[7]For reasons to be skeptical ab initio about non-contextualism in hidden variable theories, see Bell (1966).

[8]Note that a Gleason-type theorem for systems with $D \geq 2$ was provided by Busch (2003) but for POMs (positive operator-valued measures) rather than PVMs (projection-valued measures). More recent, stronger Gleason-type results are discussed in Wright and Weigert (2019).

[9]See for example Brown and Svetlichny (1990).

adopts) in the case of measurements on entangled systems, Svetlichny showed in 1998 that the probabilities must be non-contextual and used Gleason's theorem to infer the Born Rule.[10]

Prominent Everettians have in recent years defended a derivation of the Born Rule based on decision theoretic principles, one which has gained considerable attention. Critics often see probability as the Achilles' Heel in the many worlds picture, but for some Everettians (including the authors of the proof) the treatment of probability is a philosophical triumph. It is curious, then, that the two authors, David Deutsch and David Wallace, appear to disagree as to what the result means. The disagreement is rooted in the question of what probability means – to which we now turn.

7.3 The Riddle of Probability

7.3.1 Chances

The meaning of probability in the physical sciences has been a long-standing subject of debate amongst philosophers (and to a lesser extent physicists), and it remains controversial. Let us start with something that we think all discussants of the notion of probability can agree on. That is the existence in Nature of "chance processes or set-ups" that lead to more-or-less stable relative frequencies of outcomes over the long term. Such an existence claim is non-trivial, and of a straightforwardly empirical, and hence objective, nature. "Chances", the term widely used by philosophers for objective probabilities, are distilled in some way from these frequencies, or at least connected somehow with them (see below). The situation is somewhat reminiscent of Poincaré's remark that time in physics is the great simplifier. It is a non-trivial feature of the non-gravitational interactions that choosing the "right" temporal parameter in the fundamental equations in the (quantum) theory of each interaction – equations which contain derivatives with respect to this parameter – results in the greatest simplification of such equations.[11] There is a fairly clear sense in which the universal standard metric of time in non-gravitational physics corresponds to the workings of Nature, even if there is nowhere in Nature a system, apart from the Universe itself, that acts as a perfect clock, and even if choosing a non-standard metric does not make physics impossible, just more complicated. But whether time itself is to be reified is a moot point; our own inclination is to deny it. Is objective probability, or chance, any different? Is probability in physics any more of a candidate for reification than time is?

[10]See Svetlichny (1998). For more recent derivations of the Born Rule based on no-signalling, see Barnum (2003), and McQueen and Vaidman (2019). A useful critical review of derivations of the Born Rule – including those of Deutsch and Wallace (see below) – is found in Vaidman (2020).

[11]The fact that it is the same parameter in each case makes the phenomenon even more remarkable.

When philosophers write about objective probabilities, or "chances", an example that comes up repeatedly is the probability associated with the decay of a radioactive atomic nucleus. Perhaps the most famous example of such an isotope is that of Carbon-14, or ^{14}C, given its widespread use in anthropology and geology in the dating of objects containing organic material. Its half-life is far better established experimentally than the probability of heads for any given coin (another widely cited example) and the decay process, being quantum, is often regarded as intrinsically stochastic.[12] Donald Gillies has argued that the spontaneity of radioactive decay makes such probabilities "objective" rather than "artifactual", the latter sort being instantiated by repetitive experiments in physics, which necessarily involve human intervention.[13] Furthermore:

> Any standard textbook of atomic physics which discusses radioactive elements will give values for their atomic weights, atomic numbers, etc., and also for the probability of their disintegrating in a given time period. These probabilities appear to be objective physical constants like atomic weights, etc., and, like these last quantities, their values are determined by a combination of theory and experiment. *In determining the values of such probabilities no bets at all are made or even considered.* Yet all competent physicists interested in the matter agree on certain standard values–just as they agree on the values of atomic weights, etc. [our emphasis][14]

In a similar vein, Tim Maudlin writes:

> The half-life of tritium ... is about 4499 days. Further experimentation could refine the number ... Scientific practice proceeds as if there is a real, objective, physical probability density here, not just ... degrees of belief. *The value of the half-life has nothing to do with the existence or otherwise of cognizers. ...* [our emphasis][15]

David Wallace similarly states that

> If probabilities are personal things, reflecting an agent's own preferences and judgements, then it is hard to see how we could be right or wrong about those probabilities, or how they can be measured in the physicist's laboratory. But in scientific contexts at least, both of these seem commonplace. One can erroneously believe a loaded die to be fair (and thus erroneously believe that the probability of it showing '6' is one sixth); *one can measure the cross-section of a reaction or the half-life of an isotope. ...*
>
> ... as well as the personal probabilities (sometimes called subjective probabilities or credences) there also appear to be objective probabilities (sometimes called chances), which do not vary from agent to agent, and which are the things scientists are talking about when they make statements about the probabilities of reactions and the like. [our emphasis][16]

---

[12]There are good reasons, however, for considering the source of the unpredictability in some if not all classical chance processes as having a quantum origin too; see Albrecht and Phillips (2014). But whether strict indeterminism is actually in play in the quantum realm is far from clear, as we see below.

[13]Gillies (2000), pp. 177, 178.

[14]Gillies (1972), pp. 150, 151.

[15]Maudlin (2007).

[16]Wallace (2012), p. 137.

In fact, Wallace goes so far as to say that denial of the existence of objective probabilities is incompatible with a realist stance in the philosophy of science, unless one entertains a "radical revision of our extant science".[17] At any rate, Wallace is far from unique in denying that such probabilities can be *defined* in terms of frequencies (finite or otherwise)[18] so it seems to follow that the quantitative element of reality he invokes is distinct from the presumably uncontentious, qualitative one related to stable frequencies, referred to at the beginning of this section.

Many philosophers have tried to elucidate what this element of reality is; the range of ideas, which we will not attempt to itemize here, runs from the more-or-less intuitive notion of "propensity" to the abstract "best system" approach to extracting probabilities from frequencies in the Humean mosaic (or universal landscape of events in physics, past, present and future).[19] The plethora of interpretations on the part of philosophers has a curious feature: the existence of chance is widely assumed before clarification is achieved as to what it is! And the presumption is based on physics, or what is taken to be the lesson of physics.[20]

7.3.2 *Carbon-14 and the Neutron*

Until relatively recently, the half-life (or 0.693 of the mean lifetime) of ^{14}C (which by definition is a probabilistic notion[21]) was anomalous from a theoretical point

[17]*Op. cit.*, p. 138. It should be noted, however, that although Wallace claims that the certain probabilities in statistical mechanics must be objective, he accepts that such a claim is hard to make sense of without bringing in quantum mechanics (see Wallace 2014). For further discussion of Wallace's views on probability in physics, see (Brown, 2017).

[18]For discussion of the problems associated with such a definition, see, e.g., Saunders (2005), Greaves and Myrvold (2010), Wallace (2012), p. 247, and Myrvold, W. C., *Beyond Chance and Credence* [unpublished manuscript], Sect. 3.2.

[19]This last approach is due to David Lewis (1980); an interesting analysis within the Everettian picture of a major difficulty in the best system approach, namely "undermining", is found in Saunders (2020). It should also be recognised that the best system approach involves a judicious almagam of criteria *we* impose on our description of the world, such as simplicity and strength; the physical laws and probabilities in them are not strict representations of the world but rather our systematization of it. Be that as it may, Lewis stressed that whatever physical chance is, it must satisfy his Principal Principle linking it to subjective probability; see below.

[20]Relatively few philosophers follow Hume and doubt the existence of chances in the physical world; one is Toby Handfield, whose lucid 2012 book on the subject (Handfield 2012) started out as a defence and turned into a rejection of chances. Another skeptic is Ismael (1996); her recent work (Ismael 2019) puts emphasis on the kind of weak objectivity mentioned at the start of this section. See also in this connection recent work of Bacciagaluppi, whose view on the meaning of probability is essentially the same as that defended in this paper: (Bacciagaluppi, 2020), Sect. 10.

[21]The half-life is not, *pace* Google, the amount of time needed for a given sample to decay by one half! (What if the number of nuclei is odd?) It is the amount of time needed for any of the nuclei in the sample to have decayed with a probability of 0.5. Of course this latter definition will

of view. It is many orders of magnitude larger than both the half-lives of isotopes of other light elements undergoing the same decay process, and what standard calculations for nucleon-nucleon interactions in Gamow-Teller beta decay would indicate.[22]

Measurements of the half-life of ^{14}C have been going on since 1946, involving a number of laboratories worldwide and two distinct methods. Results up until 1961 are now discarded from the mix; they varied between 4.7 and 7.2 K years. The supposedly more accurate technique involving mass-spectrometry has led to the latest estimate of 5700 $\pm$ 30 years (Bé and Chechev 2012). This has again involved compiling frequencies from different laboratories using a weighted average (the weights depending on estimated systematic errors in the procedures within the various laboratories).

In the case of the humble neutron, experimental determination of its half-life is, as we write, in a curious state. Again, there are two standard techniques, the so-called 'beam' and 'bottle' measurements. The latest and most accurate measurement to date using the bottle approach estimates the lifetime as 877.7 $\pm$ 0.7 s (Pattie 2018), but there is a 4 standard deviations disagreement with the half-life for neutron beta decay determined by using the beam technique which is currently not well understood.

It should be clear that measurements involve frequencies of decays in all these cases. The procedures might be compared to the measurement of, say, the mass of the neutron. As Maudlin says, further experimentation could refine the numbers. But for those who (correctly) deny that probabilities can be defined in terms of finite frequencies, the question obviously arises: in what precise sense does taking measurements of frequencies constitute a measurement of probability (in the present case, half-life)? Is Gillies right that nothing like a bet is being made?

The difference between measuring mass or atomic weights, say, and half-life is not that one case involves complicated, on-going interactions between theory and experiment and the other doesn't. The essential difference is that for the latter, *if it is thought to be objective*, no matter how accurate the decay frequency measurements are and how many runs are taken, the resulting frequencies could in principle all be a fluke, and thereby significantly misleading. That is the nature of chance procedures, whether involving isotopes or coins. Here is de Finetti, taken out of context:

> It is often thought that these objections [to the claim that probability theory and other exact sciences] may be escaped by making the relations between probabilities and frequencies precise is analogous to the practical impossibility that is encountered in all the experimental sciences of relating exactly the abstract notions of the theory and the empirical realities. The analogy is, in my view illusory ... in the calculus of probability it is the theory itself which obliges us to admit the possibility of all frequencies. In the other sciences the uncertainty

---

be expected to closely approximate the former when the sample is large, given the law of large numbers (see below).

[22]In 2011, (Maris et al. 2011) showed for the first time, with the aid of a supercomputer, that allowing for nucleon-nucleon-nucleon interactions in a dynamical no-core shell model of beta decay explains the long lifetime of ^{14}C.

> flows indeed from the imperfect connection between the theory and the facts; in our case, on the contrary, it does not have its origin in this link, but in the body of the theory itself ....[23]

In practice, of course, given "enough" runs, the weighted averages of the frequencies are taken as reliable guides to the "objective" half-life, given the (weak) law of large numbers. It would be highly improbable were such averaged frequencies to be significantly different from the "real" half-life (assuming we trust our experimental methods). Note first that this meta-probability is in the nature of a belief, or "credence" on the part of the physicist; it is not based on frequencies on pain of an endless regress. So this is a useful reminder that in physics whether or not we need to appeal to the existence of objective chances, we certainly need to appeal in an important way to a subjective, or personal notion of probability, even in quantum mechanics. And objectivists, at least, *are* effectively betting – that the observed frequencies are "typical", so as to arrive at something close to the "objective" half-life.[24]

[23]de Finetti (1964), p. 117. de Finetti's argument, as Gillies (2000) stresses on pages 103 and 159, was made in the context of the falsifiability of probabilistic predictions, even in the case of subjective probabilities. But the argument also holds in the case of inferring "objective" probabilities by way of frequencies.

[24]Saunders puts the point succinctly:

> Chance is measured by statistics, and perhaps, among observable quantities, only by statistics, but only with high chance. (Saunders 2010)

This point is repeated in Saunders (2020, Sect. 7.2). To repeat, this "high chance" translates into subjective confidence; Saunders, like many others, believes that chances and subjective probabilities are linked by way of the "Principal Principle" (see below). Wallace's 2012 version of the law of large numbers explicitly involves both "personal" and objective probabilities. His personal prior probabilities $\Pr(\cdot|C)$ are conditional on the proposition expressing all of his current beliefs, denoted by C. X_p denotes the hypothesis that the objective probability of E_n is p, where as usual it is assumed that p is an *iid*. Then, Wallace claims, the Principal Principle ensures that his personal probability of the proposition Y_i, that heads occurs in the ith repetition of the random process, is p: symbolically $\Pr(Y_i|X_p\&C) = p$. Given the *iid* assumption for p and the usual updating rule for personal probabilities, it follows from combinatorics that the personal probability $\Pr(K_M|X_p\&C)$ is very small unless $p = M/N$, where K_M is the proposition that the experiment results in M heads out of N. Wallace concludes that

> ...as more and more experiments are carried out, any agent conforming to the Principal Principle will become more and more confident that the objective probability is close to the observed relativity frequency. (Wallace 2012, p. 141.)

Note that there are versions of the "objective" position that avoid the typicality assumption mentioned above, such as that defended in Gillies (2000), Chap. 7, and (Greaves and Myrvold 2010). For more discussion of the latter, see Sect. 2.3 (ii) below.

7.3.3 *The Law of Large Numbers*

(i) Whether we are objectivists, subjectivists or dualists about probability, for such processes as the tossing of a bent coin, or the decay of Carbon-14, we will need to be guided by experience in order to arrive at the relevant probability. For the objectivist, this is letting relative frequencies uncover an initially unknown objective probability or chance. For the subjectivist, it is using the standard rule of updating a (suitably restricted) subjective prior probability in the light of new statistical evidence, subject to a certain constraint on the priors.

Suppose we have a chance set-up, and we are interested in the probability of a particular outcome (say heads in the case of coin tossing, or decay for the nucleus of an isotope within a specified time) in the $n + 1$th repetition of the experiment, when that outcome has occurred k times in the preceding n repetitions. For the objectivist, the chance p at each repetition ("Bernoulli trial") is typically assumed to be an identically distributed and independent distribution (*iid*), which means that the objective probability in question is p, whatever the value of M is. But when p is unknown, we expect frequencies to guide us. It is their version of the (weak) law of large numbers that allows the objectivists to learn from experience.[25]

The celebrated physicist Richard Feynman defined the probability of an event as an essentially time-asymmetric notion, *viz*. our estimate of the most likely relative frequency of the event in N future trials.[26] This reliance on the role of the estimator already introduces an irreducibly subjective element into the discussion.[27] And note how Feynman describes the experimental determination of probabilities in the case of tossing a coin or a similar "chancy" process.

> We have defined $P(H) = \langle N_H \rangle / N$ [where $P(H)$ is the probability of heads, and $\langle N_H \rangle$ is the expected number of heads in N tosses]. How shall we know what to *expect* for N_H? In some cases, the best we can do is observe the number of heads obtained in large numbers of tosses. For want of anything better, we must set $\langle N_H \rangle = N_H$(observed). (How could we expect anything else?) We must understand, however, that in such a case a different experiment, or a different observer, might conclude that $P(H)$ was different. We would *expect*, however, that the various answers should agree within the deviation $1/2\sqrt{N}$ [if $P(H)$ is near one-half]. An experimental physicist usually says that an "experimentally determined" probability has an "error", and writes
>
> $$P(H) = \frac{N_H}{N} \pm \frac{1}{2\sqrt{N}} \tag{7.2}$$
>
> There is an implication in such an expression that there is a "true" or "correct" probability which *could* be computed if we knew enough, and that the observation may be in "error"

---

[25]This is not to say that the law of large numbers allows for probabilities to be *defined* in terms of frequencies, a claim justly criticised in Gillies (1973), pp. 112–116.

[26]Feynman et al. (1965), Sect. 6.1.

[27]See Brown (2011). In so far as agents are brought into the picture, who remember the past and not the future, probability requires the existence of an entropic arrow of time in the cosmos; see in this connection (Handfield 2012, chapter 11), Myrvold (2016) and (Brown, 2017), Sect. 7.

> due to a fluctuation. There is, however, no way to make such thinking logically consistent. It is probably better to realize that the probability concept is in a sense subjective, that it is always based on uncertain knowledge, and that its quantitative evaluation is subject to change as we obtain more information.[28]

Now how far this position takes Feynman away from the objectivist stance is perhaps debatable; note that Feynman does not refer here to quantum indeterminacies. But other physicists have openly rejected an objective interpretation of probability; we have in mind examples such as E. T. Jaynes (1963), Frank Tipler (2014), Don Page (1995) and Jean Bricmont (2001). In 1995, Page memorably compared interpreting the unconscious quantum world probabilistically to the myth of animism, i.e. ascribing living properties to inanimate objects.[29]

(ii) It is indeed worth reminding ourselves of the fact that the brazen subjectivists can make a good – though far from uncontroversial – case for the claim that their practice is consistent with standard experimental procedure in physics.

The subjectivist specifies a prior subjective probability for such a probability (presumably conditional on background knowledge of some kind) and appeals to Bayes' rule for updating probability when new evidence (frequencies) accrues. Following the work principally of de Finetti, the crucial assumption that replaces the *iid* condition for objectivists is the constraint of "exchangeability" on the prior probabilities, which opens the door to learning from the past. de Finetti's 1937 representation theorem based on exchangeability leads to an analogue of the Bernoulli law of large numbers.[30] In the words of Brian Skyrms,

> de Finetti ... looks at the *role* that chance plays in standard statistical reasoning, and argues that role can be fulfilled perfectly well without the metaphysical assumption that chances exist. (Skyrms 1984)

A detailed discussion of the role of the de Finetti representation theorem in physics was provided in 2010 by Greaves and Myrvold, which appears to question de Finetti's own purely subjectivist interpretation of the theorem. Consider an

[28]Feynman et al. (1965), Sect. 6.3. We are grateful to Jeremy Steeger for bringing this passage to our attention. For his careful defence of a more objective variant of Feynman's notion of probability, see (Steeger, 'The Sufficient Coherence of Quantum States' [unpublished manuscript]).

[29]See Page (1995). Page has also defended the view that truly testable probabilities are limited to conditional probabilities associated with events defined at the same time (and perhaps the same place); see Page (1994). This view is motivated by the fact that strictly speaking we have no direct access to the past or future, but only present records of the past, including of course memories. The point is well-taken, but radical skepticism about the veracity of our records/memories would make science impossible, and as for probabilities concerning future events conditional on the past (or rather our present memories/records of the past) *they will be testable when the time comes*, given the survival of the relevant records/memories. So it is unclear whether recognition of Page's concern should make much difference to standard practice.

[30]de Finetti (1964); for a review of de Finetti's work see Galavotti (1989). A helpful introduction to the representation theorem is given by Gillies (2000, chapter 4); in this book Gillies provides a sustained criticism of the subjectivist account of probabilities.

agent looking for the optimal betting strategy for sequences of a relevant class of experimental setups. Then

> ... when he conditionalizes on the results of elements of the sequence, he learns about what the optimal strategy is, and he is certain that any agent with non-dogmatic priors on which the sequence of experiments is exchangeable will converge to the same optimal strategy. If this is not the same as believing that there are objective chances, then it is something that serves the same purpose. Rather than eliminate the notion of objective chance, we have uncovered, in [the agent's] belief state, implicit beliefs about chances, – or, at least, about something that plays the same role in his epistemic life. ...
>
> ... like it or not, an agent with suitable preferences acts as if she believes that there are objective chances associated with outcomes of the experiments, about which she can learn, provided she is non-dogmatic. ... There may be more to be said about the nature and ontological status of such chances, but, whatever more is said, it should not affect the basic picture of confirmation we have sketched.[31]

Note that in explicitly leaving the question of the "nature and ontological status" of chances open, Greaves and Myrvold's analysis appeals only to the intersubjectivity of the probabilistic conclusions drawn by (non-dogmatic) rational betting agents who learn from experience according to Bayesian updating. Whatever one calls the resulting probabilities, it is questionable whether they are out there in the world independent of the existence of rational agents, and playing the role of elements of reality seemingly demanded by Gillies, Maudlin and Wallace as spelt out in Sect. 2.1 above.[32,33]

(iii) In anticipation of the discussion of the *refutability* of probabilistic assertions in science below (Sect. 7.6), it is worth pausing briefly to consider an important feature of subjectivism in this sense. A subjective probability for some event, either construed as a prior, or conditional on background information (say relevant past frequencies), is not open to revision. Such a probability is immutable; it is not to be "repudiated or corrected".[34] The same holds for the view that conditional probabilities are rational inferences arising from incomplete information. What the introduction of relevant new empirical information does is not to correct the initial probability but replace it (by Bayesian updating) with a new one conditional on the new (and past, if any) evidence, which in turn is immutable.[35]

---

[31] Greaves and Myrvold (2010), Sect. 3.3.

[32] It is noteworthy that in Footnote 4 (*op. cit.*), Greaves and Myrvold entertain the view that "it is via [the Principal Principle] that we ascribe beliefs about chances to the agent". This Principle is the topic of the next section of our paper, in which we report Myrvold's more recent claim that it is not about chances at all.

[33] Greaves and Wallace also take issue (p. 22) with the claim made in the final sentence of Sect. 2.2 above that objectivists need to introduce the substantive additional assumption that the statistical data is "typical". As far as we can see, this is again because the notion of chance they are considering is ontologically neutral.

[34] See de Finetti (1964, pp. 146, 147).

[35] Gillies (2000, pp. 74, 83, 84) and Wallace (2012, p. 138), point out that de Finetti's scheme of updating by Bayesian conditionalisation yields reasonable results only if the prior probability function is appropriate. If it isn't, no amount of future evidence can yield adequate predictions based on Bayesian updating. This a serious concern, which will not be dealt with here (see in this

(iv) We note finally that, as Myrvold has recently stressed, there is a tradition dating back to the nineteenth century of showing another way initial credences can "converge", which does not involve the accumulation of new information. The "method of arbitrary functions", propounded in particular by Poincaré, involves dynamical systems in which somewhat arbitrary prior probability distributions are "swamped" by the physics, which is to say that owing to the dynamics of the system a wide range of prior distributions converge over time to a final distribution. It is noteworthy that Myrvold accepts that the priors can be epistemic in nature, but regards the final probabilities as "hybrid", neither epistemic nor objective chances. He calls them *epistemic chances*.[36]

7.3.4 The Principal Principle

If one issue has dominated the philosophical literature on chance in recent decades – and which will be relevant to our discussion of the Deutsch-Wallace theorem below – it is the *Principal Principle* (henceforth PP), a catchy term coined and made prominent by David Lewis in 1980 (Lewis 1980), but referring to a longstanding notion in the literature.[37]

This has to do with the second side of what Ian Hacking called in 1975 the Janus-faced nature of probability,[38] or rather its construal by objectivists. The first side concerns the issue touched on above, i.e. the estimation of probabilities on the basis of empirical frequencies. Papineau (1996) called this the *inferential link*between frequencies and objective probabilities (chances). The second side concerns what we do with these probabilities, or why we find them useful. Recall the famous dictum of Bishop Butler that probabilities are a guide to life. Objective probabilities determine, or should determine, our beliefs, or "credences", about the future (and less frequently the past) which if we are rational we will act on. Following Ramsey, it is widely accepted in the literature that this commitment can be operationalised, without too much collateral damage, by way of the act of betting.[39]

connection Greaves and Myrvold 2010). Note that it seems to be quite different from the reason raised by Wallace (see Sect. 2.1 above) as to why the half-life of a radioactive isotope, for example, must be an objective probability. See also Sect. 7.5.1 below.

[36]Myrvold, W. C., *Beyond Chance and Credence* [unpublished manuscript], Chap. 5.

[37]See Strevens (1999). In our view, a refreshing exception to the widespread philosophical interest in the PP is Gillies' excellent 2000 book (Gillies 2000), which makes no reference to it.

[38]Hacking (1984), though see Gillies (2000, pp. 18, 19) for references to earlier recognition of the dual nature of probability. Note that Wallace has argued that the PP grounds both sides; see Footnote 23 above.

[39]See Gillies (2000, chapter 4).

Papineau (1996) called the connection between putative *objective* probabilities (chances) and credences the *decision-theoretic link*. More famously, it is expressed by way of the Principal Principle; here is a simple version of it:

> For any number x, a rational agent's personal probability, or credence, of an event E conditional on the objective probability, or chance, of E being x, and on any other accessible background information, is also x.[40]

Now even modest familiarity with the literature reveals that the status of the PP is open to dispute. According to Ned Hall (2004), it is an analytic truth (i.e. true by virtue of the meaning of the terms "chance" and "credence") though this view is hardly consensual. The weaker position that chance is implicitly defined by the PP has been criticised by Carl Hoefer.[41] Amongst other philosophers who think the PP requires justification, Simon Saunders considers that the weakness in the principle has to do with the notoriously obscure nature of chances:

> Failing an account of what objective probabilities are, it is hard to see how the [PP] could be justified, for it seems that it ought to be facts about the physical world that dictate our subjective future expectations. (Saunders 2005)

It is precisely the Deutsch-Wallace theorem that Saunders regards as providing the missing account of what chances are, as we shall see. If there is a difference between his view and that of David Wallace, it is that the latter sees the *prima facie* difficulty with the PP as more having to do with the its inferential form than the clarification of one of its components (chances). Wallace argues that justification of the rational inference from chances to credences is generically hard to come by, unless by way of the principles of Everettian physics and decision theory (of which more below).[42] In the absence of such principles, Wallace's position seems close to that of Michael Strevens, who in 1999 wrote:

> ...in order to justify [the PP], it is not enough to simply define objective probability as whatever makes [the PP] rational. In addition, it must be shown that there is something in this world with which it is rational to coordinate subjective probabilities. ...
>
> As with Humean induction, so with probability coordination: we cannot conjure a connection between the past and the future, or between the probabilistic and the non-probabilistic, from mathematics and deductive logic alone.[43]

Whether the Deutsch-Wallace theorem is the ultimate conjuring trick will be examined shortly. We finish this brief and selective survey of interpretations of the PP with mention of the striking view of Wayne Myrvold, who, like Strevens,

---

[40]This version is essentially the same as that in Wallace p. 140. Note that Wallace here (i) does not strictly equate the PP with the decision-theoretic link and (ii) sees PP as underpinning not just the decision-theoretic link but also the inferential link (see Footnote 24 above). We return to point (i) below; we are also overlooking here the usual subtleties involved with characterising the background information as "accessible". For a more careful discussion of the PP, see (Bacciagaluppi 2020).

[41]Hoefer (2019), Sect. 1.3.4.

[42]Wallace (2012), Sects. 4.10, 4.12.

[43]Strevens (1999). See also (Hoefer, 2019, Sects. 1.2.3 and 1.3.4).

wonders how chances could possibly induce normative constraints on our credences. Myrvold cuts the Gordian Knot by claiming that the PP is concerned rather with our *beliefs about chances* – even if there are no non-trivial chances in the world.

> PP captures all we know about credences about chances, since it is not really about chances.[44]

Can there be any doubt that the meaning and justification of the PP are issues that are far from straightforward?

Subjectivists can of course watch the whole tangled debate with detachment, if not amusement: if there are no chances, nor credences about chances, the PP is empty. And Papineau's two links collapse into one: from frequencies to credences.[45] But now an obvious and old question needs to be addressed: can credences sometimes be determined without any reference to frequencies?

7.4 Quantum Probability Again

7.4.1 The Principle of Indifference

It has been recognised from the earliest systematic writings on probability that symmetry considerations can also play an important role in determining credences in chance processes. In fact, in what is often called the "classical" theory of probability, which, following Laplace, was dominant amongst mathematicians for nearly a century,[46] the application of the *principle of indifference* (or insufficient reason) based on symmetry considerations is the grounding of all credences in physics and games of chance. The closely related, so-called "objective" Bayesian approach, associated in particular with the name of E. T. Jaynes, likewise makes prominent use of the principle of indifference.

David Wallace voices the opinion of many commentators when he describes this principle as problematic.[47] It is certainly stymied in the case of chance processes lacking symmetry (such as involving a biased coin) and it is widely accepted that serious ambiguities arise when the sample space is *continuous*.[48] But what can be wrong with use of the principle when assigning the probability $P(H) = 1/2$ (see Sect. 7.3.3) in the case of a seemingly unbiased coin? And are we not obtaining in

[44] Myrvold, 'Beyond Chance and Credence' [unpublished manuscript], Sect. 2.5.

[45] See Brown (2011). In this paper it is briefly argued that the related philosophical "problem of induction" should be seen as a pseudo-problem.

[46] See Gillies (2000, chapter 2) for a useful introduction to the classical theory.

[47] See Wallace (2012, section 4.11).

[48] A lucid discussion is found in Gillies (2000, pp. 37–48), which contains an insightful critique of Jaynes' defence of the principle of indifference in physics – and conceding that the principle has been successfully applied in a number of cases in physics.

this case a probabilistic credence purely on the basis of a principle of rationality, deriving a 'tends to' from a 'does', in the words of David Deutsch?

Not according to Wallace. He points to the fact that in the case of a classical (deterministic) chance process, or even a genuinely stochastic process, the symmetry (if any) must ultimately be broken when any one of the possible outcomes is observed. The obvious culprit? Either suitably randomised initial conditions, or stochasticity itself.

> ...since only one outcome occurs, something must break the symmetry – be it actual microconditions of the system, or the actual process that occurs in a stochastic process. Either way, we have to build probabilistic assumptions into the symmetry-breaking process, and in doing so we effectively abandon the goal of explicating probability.[49]

But if, in a branching Everettian universe where, loosely speaking, everything that can happen does happen, a derivation of credences can be given on the basis of rationality principles that feature the symmetry properties of the wavefunction, no such symmetry need be broken, and a reductive analysis of probability is possible. It is precisely this feat that the Deutsch-Wallace theorem achieves.

7.4.2 Deutsch

In 1999, David Deutsch (Deutsch 1999) attempted to demonstrate two things within the context of Everettian quantum mechanics: (i) the meaning of probabilistic claims in quantum mechanics, which otherwise are ill-defined, and (ii) the fact that the Born Rule need not be considered an extra postulate in the theory. In a more recent paper, Deutsch stresses that (iii) his 1999 derivation of the Born Rule also overcomes what is commonly called the "incoherence problem" in the Everettian account of probabilities.[50]

As for (i), Deutsch interpreted probability as that concept which appears in a Savage-style representation theorem within the application of rational decision theory. In the case of quantum mechanics, Deutsch exploited the fragment of decision theory expunged of any probabilistic notions, applying it to quantum games in an Everettian universe – or preferences over the future of quantum mechanical measurement outcomes associated with certain payoffs. The emergent notion of probability is agent-dependent, in the sense that, as with Feynman's notion of probability (recall Sect. 7.3.3 above), probabilities arise from the actions of ideally rational agents and have no independent existence in the universe – a branching universe whose fundamental dynamics is deterministic.

> ...the usual probabilistic terminology of quantum theory is justifiable in light of result of this paper, provided one understands it all as referring ultimately to the behaviour of

[49] Wallace (2012), pp. 147, 148. See also Wallace (2010).

[50] This is the problem of understanding how probabilities can come about when everything that can happen does happen. See Wallace (2012, pp. 40, 41).

> rational decision makers. ... [probabilistic predictions] become implications of a purely factual theory, rather than axioms whose physical meanings are undefined.[51]

What distinguishes the Deutsch argument from the usual decision-theoretic representation theorem in analogous single-world scenarios is that rational agents are constrained to behave not just in conformity with probability theory, but the values of the probabilities are uniquely defined by the state of the system. It is essential in the 1999 derivation that preferences are based on states which are represented solely by the standard pure and mixed states in quantum mechanics; hidden variables are ruled out.

Deutsch stressed that the pivotal result concerns the special case where the pure state is a symmetric superposition of two eigenstates of the observable being measured (i.e. the coefficients, or amplitudes, in the superposition are equal). Based on several rules of rationality, Deutsch showed that in this case a rational decision maker behaves as if she believed that each of the two possible outcomes has equal probability, and that she was maximising the probabilistic expectation value of the payoff (expected utility). The equal probability conclusion in this case might be considered a simple consequence of the principle of indifference, but Deutsch is intent on showing by way of decision theory that it makes sense to assign preferences even in the case of "indeterminate" outcomes, i.e., Everettian branching (see point (iii) above).

7.4.3 Wallace

Wallace's version of the Deutsch-Wallace (DW) theorem evolved through a series of iterations, starting in 2003 and culminating in his magisterial book on the Everett interpretation of 2012. Wallace attempted to turn Deutsch's "minimalist" proof of the Born Rule into a highly rigorous decision-theoretic derivation based on weaker, but more numerous assumptions. Unlike Deutsch, Wallace adheres to a dualist interpretation of probability, involving subjective credences and objective chances, and sees both playing a role in the DW-theorem. We spell this out in more detail in Sect. 7.5.1 below.

What concerns us at this point is the question raised in Sect. 7.2 whether non-contextualism need be axiomatic in the Everettian picture. In the case of Deutsch's 1999 proof, it is a consequence of (but not equivalent to) an implicit assumption which Wallace was to identify and call *measurement neutrality*; Wallace made it an explicit assumption in his 2003 (Wallace 2003) reworking of the Deutsch argument. It would be generous to say that a *proof* of non-contextualism obtains

---

[51] Deutsch (1999), p. 14. As Hemmo and Pitowsky noted, Deutsch's proof, if successful, would give "strong support to the subjective approaches to probability in general." (Hemmo and Pitowsky 2007, p. 340).

in either account.[52] The situation is more complicated in Wallace's 2012 proof of the theorem.

In his book, Wallace is at pains to show the role of what he calls non-contextualism. His *non-contextual inference theorem* is the decision theoretic analogue of Gleason's theorem:

> This theorem states that any solution to a quantum decision problem, provided that the problem is richly structured and satisfies the assumptions of Chapter 5 and that the solution satisfies certain rationality constraints similar to those discussed in Chapter 5, is represented by a density operator iff it is noncontextual. (Wallace 2012, p. 214)

This is followed by proof of the claim that any solution to a quantum decision problem which is compatible with a state dependent solution must be non-contextual.[53]

But it is important to note how Wallace defines this condition. Informally, an agent's preferences conform to a probability rule that is non-contextualist in Wallace's terms if it assigns the same probabilities to the outcomes of a measurement of operator X whether or not a compatible operator Y is measured at the same time.[54] After giving a more formal decision-theoretic definition, Wallace explicitly admits that this is not exactly the principle used in Gleason's theorem, "but it embodies essentially the same idea".[55] We disagree. Wallace's principle, which for purposes of subsequent discussion we call *weak non-contextualism*, involves a single measurement procedure; its violation means that a rational agent prefers "a given act to the same (knowably the same act, in fact) under a different description, which violates state supervenience (and, I hope, is obviously irrational)."[56] But the Gleason-related principle, or *strong non-contextualism*, involves mutually incompatible procedures. Now Wallace appears to be referring to this strong non-contextualism when he writes

> It is fair to note, though, that just as a non-primitive approach to measurement allows one and the same physical process to count as multiple abstractly construed measurements, it also allows one and the same abstractly construed measurement to be performed by multiple physical processes. It is then a nontrivial fact, and in a sense a physical analogue of noncontextuality, that rational agents are indifferent to which particular process realizes a given measurement.[57]

But why should rational agents be so indifferent? Because according to Wallace, it is a consequence of the condition of measurement neutrality, which, while having axiomatic status in both Deutsch's 1999 version and Wallace's 2003 version of the DW-theorem, is a trivial corollary of the 2012 Born Rule theorem, which, to repeat, is based on weaker assumptions. It is therefore rationally required. And

---

[52] Similar qualms were voiced by Hemmo and Pitowsky (Hemmo and Pitowsky 2007).

[53] Wallace (2012), p. 196.

[54] *Op. cit.* p. 196.

[55] *Op. cit.* p. 214.

[56] *Op. cit.* p. 197.

[57] *Ibid.*

> The short answer as to why is that two acts which correspond to the same abstractly construed measurement can be transformed into the same act via processes to which rational agents are indifferent.[58]

Now it seems to us, as it did to Pitowsky (see Sect. 7.2), that to contemplate a *contextual* assignment of probabilities in quantum mechanics is *prima facie* far from irrational, given the non-commutative property of operators associated with measurements.[59] In our view, the most plausible justification of non-contextualism in the context of the DW theorem was given by Timpson (Timpson 2011; Sect. 5.1). It is based on consideration of the details of the dynamics of the measurement process in unitary quantum mechanics, and shows that nothing in the standard account of the process supports the possibility of contextualism. However, this argument presupposes that the standard account is independent of the Born Rule, a supposition which deserves attention. At any rate, it should not be overlooked that the DW-theorem applies to systems with Hilbert spaces of arbitrary dimensions, which is a significant advantage over the proof of the Born Rule for credences using Gleason's theorem.

7.5 A Quantum Justification of the Principal Principle?

7.5.1 Wallace and Saunders

(i) For David Wallace and Simon Saunders, it is of great significance that Everettian quantum mechanics (EQM) provides an underpinning, unprecedented in single-world physics, for the Principal Principle.[60] For Wallace,

> ... if an Everett-specific derivation of the Principal Principle can be given, then the Everett interpretation solves an important philosophical problem which could not be solved under the assumption that we do not live in a branching universe.[61]

For Saunders, "... nothing comparable has been achieved for any other physical theory of chance."[62]

---

[58] *Ibid.* Note that a different, recent approach to proving that credences should be non-contextual in quantum mechanics is urged by Steeger (Steeger, J., 'The Sufficient Coherence of Quantum States' [unpublished manuscript]).

[59] It was mentioned in Sect. 7.2 above that contextual probabilities may lead to the possibility of superluminal signalling. But this does not imply that contextualism is irrational. Indeed, violation of no-signalling is bound to happen in some "deviant" branches in the Everett multiverse; see Sect. 7.6 below.

[60] Perhaps this should be understood as an 'emancipated' PP, given that Lewis himself did not believe in the existence of chances in a deterministic universe.

[61] Wallace (2012), Footnote 26, pp. 150–151.

[62] Saunders (2010), p. 184.

And according to Wallace, his latest approach (see below) succeeds where Deutsch's 1999 fails in providing a vindication of the PP.

> Deutsch's theorem ... amounts simply to a proof of the decision theoretic link between objective probabilities and action.[63]

Clearly, Wallace has a very different reading of the Deutsch theorem to ours; we see no reference to objective probabilities therein. But Wallace is attempting to make a substantive point: that the PP in the context of classical decision theory is not equivalent to Papineau's decision-theoretic link between chances and credences. Here, briefly, is the reason.

The utility function $\mathcal{U}$ in the Deutsch theorem is, Wallace argues, not obviously the same as the utility function $\mathcal{V}$ in classical decision theory, where the agent's degrees of belief may refer to unknown information. $\mathcal{U}$ is a feature of what Wallace calls the *Minimal decision-theoretic link* between objective probability and action, which supposedly encompasses the Deutsch theorem and Wallace's Born Rule theorem (though see below). In contrast, the standard classical decision theory in which $\mathcal{V}$ is defined makes no mention of objective probabilities, and allows for bets when the physical state of the system is unknown, and even when there is uncertainty concerning the truth of the relevant physical theory.

For Wallace, the PP will be upheld in Everettian quantum mechanics only when $\mathcal{U}$ and $\mathcal{V}$ can be shown to be equivalent (up to a harmless positive affine transformation). If the quantum credences in the Minimal decision-theoretic link are just subjective, or personal probabilities, then the matter is immediately resolved. But in his 2012 book, Wallace feels the need to provide a "more direct" proof involving a thought experiment and some further mathematics, resulting in what he calls the *utility equivalence lemma*.[64] Such considerations are a testament to Wallace's exceptional rigour and attention to detail, but for the skeptic about chances and hence the PP (ourselves included) they seem like a considerable amount of work for no real gain.

Let's take another look at the Minimal decision theoretic link. The argument is, again, that if preferences over quantum games satisfy certain plausible constraints, the credences defined in the corresponding representation theorem are in turn constrained to agree with branch weights. The arrow of inference goes from suitably constrained credences to (real) numbers extracted from the (complex) state of the system. This is somewhat out of kilt with Papineau's decision-theoretic link, which involves an inference from knowledge of objective probabilities to credences. And this discrepancy is entirely innocuous, we claim, in Deutsch's own understanding of his 1999 theorem, where branch weights are not interpreted as objective, agent-independent chances.

(ii) As far as we can tell, the grip that the PP has on Wallace's thinking can be traced back to his conviction that probabilities in physics, both in classical statistical

[63] Wallace (2012), p. 237.

[64] Wallace (2012), pp. 208–210 and Appendix D.

mechanics and Everettian quantum mechanics (EQM), represent objective elements of reality, despite the underlying dynamics being fundamentally deterministic. For a philosophical realist to deny the existence of such elements of reality would thus be tantamount to self-refutation (see Sect. 7.3.1 above). More specifically, a subjective notion of probability, wrote Wallace in 2002,

> ...seems incompatible with the highly objective status played by probability in science in general, and physics in particular. Whilst it is coherent to advocate the abandonment of objective probabilities, it seems implausible: it commits one to believing, for instance, that the predicted decay rate of radioisotopes is purely a matter of belief.[65]

But subjectivists do *not* claim that the half-life of ^{14}C, for example, is purely a matter of belief. It is belief highly constrained by empirical facts. The half-life is arrived at through Bayesian updating based on the results of ever more accurate/plentiful statistical measurements of decay, as we saw in Sect. 7.3.3.[66]

For Wallace, in the case of quantum mechanics (and hence of classical statistical mechanics, which he correctly sees as the classical limit of quantum mechanics) the probabilistic elements of reality – chances – are (relative) branch weights, or mod-squared amplitudes. Now no one who is a realist about the quantum state would question whether amplitudes are agent-independent and supervenient on the ontic universal wavefunction. But are they intrinsically "chances" of the kind that defenders of the PP would recognise?

This is a hard question to answer, in part because the notion of chance in the literature is so elusive. Wallace and Saunders adopt the approach of "cautious functionalism". Essentially, this means that branch weights act as if they were chances, according to the PP. Here is Wallace:

> ...the Principal Principle can be used to provide what philosophers call a *functional* definition of objective probability: it defines objective probability to be *whatever thing* fits the 'objective probability' slot in the Principal Principle.[67]

In Saunders' words:

> [Wallace] shows that branching structures and the squared moduli of the amplitudes, in so far as they are known, *ought* to play the same decision theory role [as in Papineau's decision theoretic link] that chances play, in so far as they are known, in one-world theories.[68]

And again:

> The [DW] theorem demonstrates that the particular role ordinarily but mysteriously played by physical probabilities, whatever they are, in our rational lives, is played in a *wholly perspicuous and entirely unmysterious way* by branch weights and branching. It is

[65]Wallace (2002), Sect. 2.7.

[66]In his 2012 book, Wallace does acknowledge this point; see Wallace (2012, p. 138).

[67]Wallace (2012), p. 141.

[68]Saunders (2010), p. 184.

> establishing that this role is played by the branch weights, and establishing that they play all the other chance roles, that qualifies these quantities as probabilities.[69]

Saunders calls this reasoning a "derivation" of the PP,[70] but in our view *it amounts to presupposing the PP and showing that the elusive notion of chance can be cashed out in EQM* in terms of branch weights. It hardly seems consistent with Wallace's express hope

> …to find some alternative characterization of objective probability, independent of the Principal Principle, and then prove that the Principal Principle is true for that alternatively characterized notion.[71]

Note, however, that Wallace himself stresses that his Born Rule theorem is insufficient for deriving the PP, since it is "silent on what to do when the quantum state is not known, or indeed when the agent is uncertain about whether quantum mechanics is true." This leads Wallace to think that the theorem, like Deutsch's, merely establishes the *Minimal decision-theoretic link* between objective probability and action, as we have seen. As a consequence Wallace developed a decision-theoretic "unified approach" to probabilities in EQM,[72] which avoids the awkwardness of having the concepts of probability and utility derived twice, from his Born Rule theorem, and from appeal to something like the 2010 Greaves-Myrvold de Finetti-inspired solution to what they call the "evidential problem" in EQM, although a defence of the PP can still be gained by appeal (see above) to the utility equivalence lemma.[73]

In our view, any such defence relies on the claim that somewhere in the decision-theoretic reasoning an "agent-independent notion of objective probability"[74] emerges. We turn to this issue now.

(iii) In his 2010 treatment of EQM, Saunders gives an account of "what probabilities *actually are* (branching structures)".[75] It is important to recognise the role of the distinction between branching structures and amplitudes in this analysis. The former are 'emergent' and non-fundamental; the latter are fundamental and provide (in terms of their modulus squared) the numerical values of the chances:

> Just like other examples of reduction, …[probability] can no longer be viewed as fundamental. It can only have the status of the branching structure itself; it is 'emergent' …. Chance, like quasiclassicality, is then an 'effective' concept, its meaning at the microscopic level entirely derivative on the establishment of correlations, natural or man-made, with macroscopic branching. That doesn't mean that amplitudes in general …have no place in

---

[69] See Saunders (2020); this paper also contains a rebuttal of the claim that the process of decoherence, so essential to the meaning of branches in EQM, itself depends on probability assumptions.

[70] Saunders (2010).

[71] Wallace (2012), p. 144.

[72] *Op. cit.* Chap. 6.

[73] *Op. cit.* p. 234.

[74] *Op. cit.* p. 229.

[75] *Ibid.*

> the foundations of EQM – on the contrary, they are part of the fundamental ontology – but their link to probability is indirect. It is simply a mistake, if this reduction is successful, to see quantum theory as at bottom a theory of probability.[76]

Now Everettian subjectivists (about probabilities) concur completely with this conclusion, given that rational agents themselves have the emergent character described in this passage. At any rate, in Saunder's picture, at the microscopic level, amplitudes have nothing to do with chances. Amplitudes only become connected with chances when the appropriate experimental device correlates them with branching structure,[77] and only then, in our view, by way of the PP.

This notion of chance in EQM is weaker than that often adopted by advocates of objective probability; indeed Wallace himself states that the functional definition of chance has a different character from that of charge, mass or length.[78] The decision-theoretic approach to the Born Rule starts with rational agents and their preferences in relation to quantum games; without the credences emerging from the rationality and structure axioms, application of the PP to infer chances (branch weights) would be impossible. In what sense then are such chances "out there", independent of the rational cogitations of agents?

Wallace's answer to this question appears in the Second Interlude of his 2012 book, and in which the Author replies to objections or queries raised by the Sceptic. Here is the relevant passage:

> SCEPTIC: Do you really find it acceptable to regard quantum probabilities as defined via decision theory? Shouldn't things like the decay rate of tritium be objective, rather than defined via how much we're prepared to bet on them?
> AUTHOR: The quantum probabilities are objective. In fact, it's clearer in the Everettian context what those objective things are: they're relative branch weights. They'd be unchanged even if there were no bets being made.
> SCEPTIC: In that case, what's the point of the decision theory?
> AUTHOR: Although branch weights are objective, what makes it true that branch weight=probability is the way in which branch weights figure in the actions of (ideally) rational agents. ...[79]

It is easy to see the tension between the Author's replies if one thinks of the objective probabilities emerging from the DW-theorem as having an existence independent of betting agents.[80] But such is not what the application of the PP yields here. And note again that the branch weights yielded by the wavefunction are not objective probabilities (according to the argument) until the branches are effectively

[76]Saunders (2010), p. 182.

[77]Simon Saunders, private communication.

[78]Wallace (2012), p. 144. Compare this with the views expressed in Sect. 2.1 above.

[79]Wallace 2012, p. 249.

[80]Wallace, *ibid*, compares branch weights with physical money, in our view a very apt analogy. A dollar note, e.g., has no more intrinsic status as money than branch weights have the status of probability, until humans confer this status on it. This point is nicely brought out by Harari, who writes that the development of money "involved the creation of a new inter-subjective reality that exists solely in people's shared imagination." See Harari 2014, chapter 10.

defined by the process of decoherence. In an important sense, adoption of something like this derivative notion of chance in the Saunders-Wallace approach is inevitable for chance "realists" who defend EQM. If agent-independent chances existed, then there would be more to Everettian ontology than just the universal wavefunction, contrary to the aims of the approach.[81]

The subjectivist, on the other hand, who views the quantum probabilities as credences, may take (subject to doubts raised in the next section) the DW-theorem to show, remarkably, that there are objective features of the Everettian "multiverse" that together with rationality principles constrain credences (which are not fundamental elements in EQM) to align with the Born Rule. There is arguably no strict analogue in classical, one-world physics, as Deutsch emphasised. But to call relative branch weights "objective probabilities" is a mere *façon de parler*, and the temptation to functionally define them as such only reflects an a priori commitment of questionable merit to the validity of the PP.

It is noteworthy that in his 2010 treatment of the DW-theorem, Wallace says the following:

> The decision-theoretic language in which this paper is written is no doubt necessary to make a properly rigorous case and to respond to those who doubt the very coherence of Everettian probability, but in a way the central core of the argument is not decision-theoretic at all. What is really going on is that the quantum state has certain symmetries and the probabilities are being constrained by those symmetries.[82]

Note that the probabilities in the argument are credences; the exercise is at heart the application of a sophisticated variant of the principle of indifference based on symmetry. This makes the nature of quantum probabilities essentially something Laplace would recognise, but with the striking new feature mentioned at the beginning of this section. To reify the resulting quantitative values of the credences (branch weights) *as chances* seems to us both unnecessary and ill-advised; it would be like telling Laplace that his credence in the outcome 'heads' is a result (in part) of an objective probability inherent in the individual (unbiased) coin.[83]

---

[81] Wallace describes the Everettian program as interpreting the "bare quantum formalism" – which itself makes no reference to probability (Wallace 2012, p. 16) – in a "straightforwardly realist way" without modifying quantum mechanics (*op. cit.*, p. 36).

[82] Wallace (2010), pp. 259, 260. Saunders has recently also stressed the role of symmetry in the DW theorem; see Saunders (2020, Sect. 7.6).

[83] Our approach to probability in EQM is very similar to the "pragmatic" view of Bacciagaluppi and Ismael, in their thoughtful 2015 review of Wallace's 2012 book (of which more in Sect. 7.6 below):

> According to such a view, the ontological content of the theory makes no use of probabilities. There is a story that relates the ontology to the evolution of observables along a family of decoherent histories, and probability is something that plays a role in the cognitive life of an agent whose experience is confined to sampling observables along such a history. In so doing, one would still be doing EQM ... (Bacciagaluppi and Ismael 2015, Sect. 3.2).

7.5.2 Earman

Recently, John Earman (2018) has also argued that something like the PP is a "theorem" of non-relativistic quantum mechanics, but now in a single-world context. Briefly, the argument goes like this.

Earman starts the argument within the abstract algebraic formulation of the theory. A "normal" quantum state ω is defined as a normed positive linear functional on the von Neumann algebra of bounded observables $\mathcal{B}(\mathcal{H})$ operating on a separable Hilbert space $\mathcal{H}$, with a density operator representation. This means that there is a trace class operator ρ on $\mathcal{H}$ with $Tr(\rho) = 1$, such that $\omega(A) = Tr(\rho A)$ for all $A \in \mathcal{B}(\mathcal{H})$. Normal quantum states induce quantum probability functions Pr^{ω} on the lattice of projections $\mathcal{P}(\mathcal{B}(\mathcal{H}))$: $Pr^{\omega}(E) = \omega(E) = Tr(\rho E)$ for all $E \in \mathcal{P}(\mathcal{B}(\mathcal{H}))$. *Earman takes the quantum state to be an objective feature of the physical system, and infers that the probabilities induced by them are objective.*[84] This is spelt out by showing how pure state preparation can be understood non-probabilistically using the von Neumann projection postulate in the case of yes-no experiments, and using the law of large numbers to relate the subsequent probabilities induced by the prepared state to frequencies in repeated measurements. All this Earman calls the "top-down" approach to objective quantum probabilities.

Credences are now introduced into the argument by considering a "bottom-up" approach' in which probability functions on $\mathcal{P}(\mathcal{B}(\mathcal{H}))$ are construed as the credence functions of actual or potential Bayesian agents. States are construed as bookkeeping devices used to keep track of credence functions. By considering again a (non-probabilistic) state preparation procedure, and the familiar Lüders updating rule, Earman argues that a rational agent will in this case, in the light of Gleason's theorem, adopt a credence function on $\mathcal{P}(\mathcal{B}(\mathcal{H}))$ which is precisely the objective probability induced by the prepared pure state (as defined in the previous paragraph).[85] Thus, Earman is intent on showing

> that there is a straightforward sense in which no new principle of rationality is needed to bring rational credences over quantum events into line with the events' objective chances – the alignment is guaranteed by as a theorem of quantum probability, assuming the credences satisfy a suitable form of additivity. (Earman 2018)

Here are a few remarks on Earman's argument.

1. Unless the PP in its original guise is interpreted as analytically true, or providing an implicit definition of chance, or, following Myrvold, it is not about real

However, we do not follow these authors in their attempt to define chances consonant with such probabilities and the PP. (We are grateful to David Wallace for drawing to our attention this review paper.)

[84] *Op. Cit.* p. 16.

[85] Recall that in the typical case of the infinite dimensional, separable Hilbert space $\mathcal{H}$, a condition of Gleason's theorem is that the probability function must be countably additive; see Sect. 7.2 above.

chances at all, the principle is virtually by definition a principle or rule of rationality.[86] Were it valid, the appeal to Gleason's theorem in Earman's account would be redundant: credences would track the chances supposedly embedded in the algebraic formalism. Earman purports to show that chances and credences in quantum mechanics have separate underpinnings, and that their numerical values coincide when referring to the same measurement outcomes. Rather than providing a quantum theoretical derivation of the PP, Earman is essentially rejecting the PP and attempting to prove a connection between chances and credences whose nature is not that of an extra principle of rationality.

2. The argument in the top-down view is, however, restricted to the algebraic formulation of quantum mechanics that Earman advocates. Other approaches, such as the early de Broglie-Bohm theory and the Everett picture, also take the quantum state – or that part of it related to the Hilbert space – to represent a feature of reality. But it is essentially defined therein as a solution to the time (in-)dependent Schödinger equation(s), with no a priori connection between the state and probabilities of measurement outcomes, unless in the case of the Everett picture the decisions of agents are taken into account.[87] In the algebraic approach, this connection is baked in, at a very heavy cost: the terms "measurement" and "objective probability" are primitive.
3. Earman's argument is subject to the same criticism that was raised above in relation to Pitowsky's derivation of the Born Rule: to the extent that they both rely on Gleason's theorem, credence functions are assumed to be not just countably additive but non-contextual. And the derivation breaks down for Hilbert spaces with dimensions less than 3, as Earman recognises.

7.6 The Refutability Issue for Everettian Probability

There is an important aspect of probability as a *scientific* notion that has been overlooked so far in this paper. It is the falsifiable status of probabilistic claims, as stressed by Karl Popper and more recently Donald Gillies.[88] Deutsch himself gives it prominent status in his 2016 paper:

> If a theory attaches numbers p_i to possible results a_i of an experiment, and calls those numbers 'probabilities', and if, in one or more instances of the experiment, the observed *frequencies* of the a_i differ significantly, according to some [preordained] statistical test, from the p_i, then a scientific problem should be deemed to exist. (Deutsch 2016)

In relation to dynamical collapse theories and de Broglie-Bohm pilot wave theory, Deutsch argues that without this methodological rule the probabilities therein are merely decorative, and with it the theories are problematic. The rule, says Deutsch,

[86] See, for instance, (Wallace 2012, section 4.10).

[87] But recall the complication referred to in footnote 64 above.

[88] See Gillies (2000, chapter 7).

"is not a supposed law of nature, nor is it any factual claim about what happens in nature (the explicanda), nor is it derived from one. ... And *one cannot make an explanation problematic merely by declaring it so.*" p. 29.

For Deutsch, it is a triumph of the DW-theorem that in Everettian quantum mechanics this problem – that the methodological falsifiability rule is mere whim – is avoided. The theorem shows that

> ... rational gamblers who knew Everettian quantum theory, ... and have ... made no probabilistic assumptions, when playing games in which randomisers were replaced by quantum measurements, would place their bets as if those were randomisers, i.e. using the [Born] probabilistic rule ... according to the methodological [falsifiability] rule [above]. p. 31.

Furthermore, it is argued that "the experimenter – who is now aware of the same evidence and theories as they are – must agree with them" when one of two gamblers regards his theory as having been refuted. The argument is subtle, if not convoluted, and requires appeal on the part of gamblers to a non-probabilistic notion of expectation that Deutsch introduces early in the 2016 paper.

Perhaps a simpler approach works. Subjectivists, as much as objectivists, aim to learn from experience; despite the immutability of their conditional probabilities (see Sect. 7.3.3) they of course update their probabilities conditional on new information. As we saw, this is the basis of their version of the law of large numbers. In fact, it might be considered part of the scientific method for subjectivists, or of the very meaning of probabilities in science, that updating is deemed necessary in cases where something like the Popper-Gillies-Deutsch falsifiability rule renders the theory in question problematic. In practice, such behaviour is indistinguishable from that of objectivists about probability in cases of falsification (even in the tentative sense advocated by Deutsch and others). But this line of reasoning does not make special appeal to a branching universe, and thus is not in the spirit of Deutsch's argument.

The reader can decide which, if either, of these approaches is right. But now an interesting issue arises regarding the DW-theorem, as a number of commentators have noticed.

Given the branching created in (inter alia) repeated measurement processes, it is inevitable that in some branches statistics of measurement outcomes will be obtained that fail to adhere to the Born Rule, whatever reasonable preordained statistical test is chosen. These are sometimes called *deviant* branches. What will Everettian-favourable observers in such branches – conclude?[89] If they adhere to

[89] In his 1999 paper, Deutsch assumed that agents involved in his argument were initially adherents of the non-probabilistic part of Everettian theory. Wallace was influenced by the work of Greaves and Myrvold (Greaves and Myrvold 2010) who developed a confirmation theory suitable for branching as well as non-branching universes, but he went on to develop his own confirmation theory as part of what he called a "unified approach" to probabilities in EQM (see Sect. 7.4.1(ii) above). However, Deutsch, in his 2016 paper, follows Popperian philosophy and rejects any notion of theory confirmation, thereby explicitly sidelining the work of Greaves and Myrvold, despite the fact that it contains a falsifiability element as well as confirmation.

the Popper-Gillies-Deutsch falsifiability rule, they *must* conclude that their theory is in trouble. Wallace and Saunders correctly point out that from the point of view of the DW-theorem, they happen just to be unlucky.[90] But such observers will have no option but to question Everettian theory. To say that statistics *trump* theory, including principles of rationality ultimately based on symmetries,[91] is just to say that the falsifiability rule is embraced, and not to embrace it for subjectivists like Deutsch is to confine the probabilistic component of quantum mechanics to human psychology.

But what part of the theory is to be questioned? Read has recently argued that it is the non-probabilistic fragment of EQM – the theory an Everettian accepts before adopting the Born Rule. In this case, observers in deviant branches could question the physical assumptions involved in the DW-theorem (for example, that the agent's probabilities supervene solely on the wavefunction), and thus consider it inapplicable to their circumstances.

Hemmo and Pitowsky, on the other hand, argued in 2007 (Hemmo and Pitowsky 2007) that such an observer could reasonably question the rationality axioms that lead to non-contextualism in (either version of) the DW-theorem. Here is Wallace's 2012 reply.

> ...it should be clear that the justification used in this chapter is not available to operationalists, for (stripping away the technical detail) the Everettian defence of non-contextuality is that two processes are decision-theoretically equivalent if they have the same effect on the physical state of the system, whatever it is. Since operationalists deny that there *is* a physical state of the system, this route is closed to them.[92]

Now we saw in Sect. 7.2 (footnote 4) that Pitowsky, in particular, denies the physical reality of the quantum state. But it is unclear to us whether this is relevant to the matter at hand. Wallace seems to be referring to what we called in Sect. 7.4.3 weak non-contextualism, whereas what Hemmo and Pitowsky have in mind is strong non-contextualism. Wallace proceeds to give a detailed classical decision problem (analogous to the quantum decision problem) which contains a condition of "classical noncontextuality" that again is justified if it assumed "that the state space really is a space of physical states". But in a classical theory, no analogue exists of non-commutativity and the incompatibility of measurement procedures involved in strong non-contextualism. It seems to us that the concern raised by Hemmo and Pitowsky is immune to Wallace's criticism; it does not depend on adopting a non-"operationalist" stand in relation to the quantum state, and applies just as much to non-deviant as to deviant branches.[93] But we mentioned in Sect. 7.3.3 above that a strong plausibility argument for non-contextualism in Wallace's decision-theoretic approach has been provided by Timpson.

---

[90] See Wallace (2012, p. 196) and Saunders (2020). Saunders again correctly points out that in an infinite non-branching universe an analogous situation holds.

[91] Read (2018), p. 138.

[92] Wallace (2012), p. 226.

[93] However, Hemmo and Pitowsky also argue that

In the Copenhagen interpretation of quantum mechanics, quantum bets *are* in effect inferred from observed frequencies: the Born Rule has as its justification nothing more than past statistics. This option is open to the Everettian (though adherents of the DW-theorem hope for more). Indeed, whether they are objectivists or subjectivists concerning probability, their reasoning here would be no different from that described in Sect. 7.3.3 in relation to the determination of the half-life of ^{14}C. And this threatens now to make the DW-theorem redundant in non-deviant branches.[94] After all, if statistics trump theory (see above) this should be the case as much for agents in non-deviant as in deviant branches.

Read, following Richard Dawid and Karim Thébault (Dawid and Thébault 2014), has essentially endorsed the redundancy – or what they call the irrelevancy – conclusion, with one caveat.

> ... the central case in which DW could have any relevance is the ... scenario in which EQM is believed, but the agent in question has no statistical evidence for or against the theory. DW might deliver a tighter link between subjective probabilities and branch weights, and therefore put the justification of PP, and the status of quantum mechanical branch weights as objective probabilities, on firmer footing. However, DW is not necessary to establish the rationality of betting in accordance with Born rule probabilities *tout court*.[95]

While we have doubts above as to whether the DW-theorem does justify the PP, Read is obviously right that the past-statistics-driven route to the Born Rule is not available to an agent ignorant of the statistics! But such an epistemologically-limited agent would be very hard to find in practice.

> ... in the many worlds theory, not only Born's rule but any probability rule is meaningless. The only way to solve the problem ... is by adding to the many worlds theory some stochastic element. (Hemmo and Pitowsky 2007, p. 334)

This follows from the claim that the shared reality of all of the multiple branches in EQM entails "one cannot claim that the quantum probabilities might be inferred (say, as an empirical conjecture) from the observed frequencies." (p. 337). This is a variant of the "incoherence" objection to probabilities in EQM (see footnote 45 above), but in our opinion it is *not* obviously incoherent to bet on quantum games in a branching scenario (see Wallace 2010).

[94]Although in their above-mentioned review, Bacciagaluppi and Ismael do not go quite this far, they do not regard the DW-theorem as necessary for "establishing the intelligibility of probabilities" in EQM. By appealing to standard inductive practice based on past frequencies, they argue:

> The inputs [frequencies] are solidly objective, but the outputs [probabilities] need not be reified.... In either the classical or the Everettian setting, probability emerges uniformly as a secondary or derived notion, and we can tell an intelligible story about the role it plays in setting credences. (Bacciagaluppi and Ismael 2015, section 3.2).

[95]Read (2018), p. 140. Note that Read, like Dawid and Thébault, rests the rationality of the statistics-driven route on the Bayesian confirmation analysis given by Greaves and Myrvold *op. cit.*.

7.7 Conclusions

The principal motivation for this paper was the elucidation of the difference between the views of David Deutsch and David Wallace as regards the significance of the DW-theorem. Deutsch interprets the probabilities therein solely within the Ramsay-de Finetti-Savage tradition of rational choice, now in the context of quantum games. Wallace defends a dualistic interpretation involving objective probabilities (chances) and credences, and, like Simon Saunders, argues that his more rigorous version of the theorem provides, at long last, a "derivation" or "proof" of David Lewis' Principal Principle.

We have questioned whether this derivation of the PP is sound. In our view, the Wallace-Saunders argument presupposes the PP and provides within it a number to fill the notorious slot representing chances. For proponents of reified chances in the world, this is, nonetheless, a remarkable result. But it raises awkward questions about the scope of the ontology in EQM, and for subjectivists about probability like Deutsch the concerns with the PP should seem misguided in the first place. Our own inclinations are closer to those of Deutsch, and in key respects mirror those of Bacciagaluppi and Ismael in their discussion of the DW-theorem.

We have also questioned the doubts expressed by Hemmo and Pitowsky as to whether all the rationality assumptions in the DW-theorem are compelling, with particular emphasis on the role of non-contextualism. Finally, by considering the falsifiable nature of probabilities in science, we regard the complaint by Dawid and Thébault, and largely endorsed by Read, that the DW-theorem is, for all practical purposes, redundant, a serious challenge to those who endorse the arguments of the authors of the theorem.

Acknowledgements The authors are grateful to the editors for the invitation to contribute to this volume. HRB would like to acknowledge the support of the Notre Dame Institute for Advanced Study; this project was started while he was a Residential Fellow during the spring semester of 2018. Stimulating discussions over many years with Simon Saunders and David Wallace are acknowledged. We also thank David Deutsch, Chiara Marletto, James Read and Christopher Timpson for discussions, and Guido Bacciagaluppi, Donald Gillies and Simon Saunders for helpful critical comments on the first draft of this paper. We dedicate this article to the memory of Itamar Pitowsky, who over many years was known to and admired by HRB, and to Donald Gillies, teacher to HRB and guide to the philosophy of probability.

References

Albrecht, A., & Phillips D. (2014). Origin of probabilities and their application to the multiverse. *Physical Review D, 90*, 123514.

Bacciagaluppi, G., & Ismael, J. (2015). Essay review: The emergent multiverse. *Philosophy of Science, 82*, 1–20.

Bacciagaluppi, G. (2020). Unscrambling subjective and epistemic probabilities. This volume, http://philsci-archive.pitt.edu/16393/1/probability%20paper%20v4.pdf

Barnum, H. (2003). No-signalling-based version of Zurek's derivation of quantum probabilities: A note on "environment-assisted invariance, entanglement, and probabilities in quantum physics". quant-ph/0312150.

Bé, M. M., & Chechev, V. P. (2012). ^{14}C – Comments on evaluation of decay data. www.nucleide.org/DDEP_WG/Nucleides/C-14_com.pdf. Accessed 01 June 2019.

Bell, J. S. (1966). On the problem of hidden variables in quantum mechanics. *Reviews of Modern Physics, 38*, 447–452.

Bricmont, J. (2001). Bayes, Boltzmann and Bohm: Probabilities in physics. In J. Bricmont, D. Dürr, M. C. Galavotti, G. Ghirardi, F. Petruccione, & N. Zangi (Eds.), *Chance in physics. Foundations and perspectives*. Berlin/Heidelberg/New York: Springer.

Brown, H. R. (2011). Curious and sublime: The connection between uncertainty and probability in physics. *Philosophical Transactions of the Royal Society, 369*, 1–15. https://doi.org/10.1098/rsta.2011.0075

Brown, H. R. (2017). Once and for all; The curious role of probability in the Past Hypothesis. http://philsci-archive.pitt.edu/id/eprint/13008. Forthcoming In D. Bedingham, O. Maroney, & C. Timpson (Eds.). (2020). *The quantum foundations of statistical mechanics*. Oxford: Oxford University Press.

Brown, H. R. (2019). The reality of the wavefunction: Old arguments and new. In A. Cordero (Ed.), *Philosophers look at quantum mechanics* (Synthese Library 406, pp. 63–86). Springer. http://philsci-archive.pitt.edu/id/eprint/12978

Brown, H. R., & Svetlichny, G. (1990). Nonlocality and Gleason's Lemma. Part I. Deterministic theories. *Foundations of Physics, 20*, 1379–1387.

Busch, P. (2003). Quantum states and generalized observables: A simple proof of Gleasons theorem. *Physical Review Letters, 91*, 120403.

Dawid, R., & Thébault, K. (2014). Against the empirical viability of the Deutsch-Wallace-Everett approach to quantum mechanics. *Studies in History and Philosophy of Modern Physics, 47*, 55–61.

de Finetti, B. (1964). Foresight: Its logical laws, its subjective sources. English translation. In: J. E. Kyburg & H. E. Smokler (Eds.), *Studies in subjective probability* (pp. 93–158). New York: Wiley.

Deutsch, D. (1999). Quantum theory of probability and decisions. *Proceedings of the Royal Society of London, A455*, 3129–3137.

Deutsch, D. (2016). The logic of experimental tests, particularly of Everettian quantum theory. *Studies in History and Philosophy of Modern Physics, 55*, 24–33.

Earman, J. (2018) The relation between credence and chance: Lewis' "Principal Principle" is a theorem of quantum probability theory. http://philsci-archive.pitt.edu/14822/

Feynman, R. P., Leighton, R. B., & Matthew Sands, M. (1965). *The Feynman lectures on physics* (Vol. 1). Reading: Addison-Wesley.

Gillies, D. (1972). The subjective theory of probability. *British Journal for the Philosophy of Science, 23*(2), 138–157.

Gillies, D. (1973). *An objective theory of probability*. London: Methuen & Co.

Gillies, D. (2000). *Philosophical theories of probability*. London: Routledge.

Galavotti, M. C. (1989). Anti-realism in the philosophy of probability: Bruno de Finetti's Subjectivism. *Erkenntnis, 31*, 239–261.

Gleason, A. (1957). Measures on the closed subspaces of a Hilbert space. *Journal of Mathematics and Mechanics, 6*, 885–894.

Greaves, H., & Myrvold, W. (2010). Everett and evidence. In S. Saunders, J. Barrett, A. Kent, & D. Wallace (Eds.), *Many worlds? Everett, quantum theory, and reality* (pp. 264–304). Oxford: Oxford University Press. http://philsci-archive.pitt.edu/archive/0004222

Hacking, I. (1984). *The emergence of probability*. Cambridge: Cambridge University Press.

Hall, N. (2004). Two mistakes about credence and chance. *Australasian Journal of Philosophy, 82*, 93–111.

Handfield, T. (2012). *A philosophical guide to chance*. Cambridge: Cambridge University Press.

Harari, Y. N. (2014). *Sapiens. A brief history of humankind*. London: Harvill-Secker.

Hemmo, M., & Pitowsky, I. (2007). Quantum probability and many worlds. *Studies in History and Philosophy of Modern Physics, 38*, 333–350.
Hoefer, C. (2019). *Chance in the world: A Humean guide to objective chance*. (Oxford Studies in Philosophy of Science). Oxford, Oxford University Press, 2019.
Hume, D. (2008). *An enquiry concerning human understanding*. Oxford: Oxford University Press.
Ismael, J. (1996). What chances could not be. *British Journal for the Philosophy of Science, 47*(1), 79–91.
Ismael, J. (2019, forthcoming). On Chance (or, Why I am only Half-Humean). In S. Dasgupta (Ed.), *Current controversies in philosophy of science*. London: Routledge.
Jaynes, E. T. (1963). Information theory and statistical mechanics. In: G. E. Uhlenbeck et al. (Eds.), Statistical physics. (1962 Brandeis lectures in theoretical physics, Vol. 3, pp. 181–218). New York: W.A. Benjamin.
Kochen, S., & Specker, E. P. (1967). The problem of hidden variables in quantum mechanics. *Journal of Mathematics and Mechanics, 17*, 59–87.
Lewis, D. (1980). A subjectivist's guide to objective chance. In: R. C. Jeffrey (Ed.), *Studies in inductive logic and probability* (Vol. 2, pp. 263–293). University of California Press (1980). Reprinted in: Lewis, D. *Philosophical papers* (Vol. 2, pp. 83–132). Oxford: Oxford University Press.
Maris, J. P, Navrátil, P., Ormand, W. E., Nam, H., & Dean, D. J. (2011). Origin of the anomalous long lifetime of ^{14}C. *Physical Review Letters, 106*, 202502.
Maudlin, T. (2007). What could be objective about probabilities? *Studies in History and Philosophy of Modern Physics, 38*, 275–291.
McQueen, K. J., & Vaidman, L. (2019). In defence of the self-location uncertainty account of probability in the many-worlds interpretation. *Studies in History and Philosophy of Modern Physics, 66*, 14–23.
Myrvold, W. C. (2016). Probabilities in statistical mechanics, In C. Hitchcock & A. Hájek (Eds.), *Oxford handbook of probability and philosophy*. Oxford: Oxford University Press. Available at http://philsci-archive.pitt.edu/9957/
Page, D. N. (1994). Clock time and entropy. In J. J. Halliwell, J. Pérez-Mercader, & W. H. Zurek (Eds.), *The physical origins of time asymmetry*, (pp. 287–298). Cambridge: Cambridge University Press.
Page, D. N. (1995). Sensible quantum mechanics: Are probabilities only in the mind? arXiv:gr-qc/950702v1.
Papineau, D. (1996). Many minds are no worse than one. *British Journal for the Philosophy of Science, 47*(2), 233–241.
Pattie, R. W. Jr., et al. (2018). Measurement of the neutron lifetime using a magneto-gravitational trap and in situ detection. *Science, 360*, 627–632.
Pitowsky, I. (1994). George Boole's 'conditions of possible experience' and the quantum puzzle. *British Journal for the Philosophy of Science, 45*(1), 95–125.
Pitowsky, I. (2006). Quantum mechanics as a theory of probability. In W. Demopoulos & I. Pitowsky (Eds.), *Physical theory and its interpretation* (pp. 213–240). Springer. arXiv:quant-phys/0510095v1.
Read, J. (2018). In defence of Everettian decision theory. *Studies in History and Philosophy of Modern Physics, 63*, 136–140.
Saunders, S. (2005). What is probability? In A. Elitzur, S. Dolev, & N. Kolenda (Eds.), *Quo vadis quantum mechanics?* (pp. 209–238). Berlin: Springer.
Saunders, S. (2010). Chance in the Everett interpretation. In S. Saunders, J. Barrett, A. Kent, & D. Wallace (Eds.), *Many worlds? Everett, quantum theory, and reality* (pp. 181–205). Oxford: Oxford University Press.
Saunders, S. (2020), The Everett interpretation: Probability. To appear in E. Knox & A. Wilson (Eds.), *Routledge Companion to the Philosophy of Physics*. London: Routledge.
Skyrms, B. (1984). *Pragmatism and empiricism*. New Haven: Yale University Press.
Strevens, M. (1999). Objective probability as a guide to the world, philosophical studies. *An International Journal for Philosophy in the Analytic Tradition, 95*(3), 243–275.

Svetlichny, G. (1998). Quantum formalism with state-collapse and superluminal communication. *Foundations of Physics, 28*(2), 131–155.

Timpson, C. (2011). Probabilities in realist views of quantum mechanics, chapter 6. In C. Beisbart & S. Hartmann (Eds.), *Probabilities in physics*. Oxford: Oxford University Press.

Tipler, F. J. (2014). Quantum nonlocality does not exist. *Proceedings of the National Academy of Sciences, 111*(31), 11281–11286.

Vaidman, L. (2020). Derivations of the Born Rule. This volume, http://philsci-archive.pitt.edu/15943/1/BornRule24-4-19.pdf

Wallace, D. (2002). Quantum probability and decision theory, Revisited, arXiv:quant-ph/0211104v1.

Wallace, D. (2003). Everettian rationality: Defending Deutsch's approach to probability in the Everett interpretation. *Studies in History and Philosophy of Modern Physics, 34*, 415–439.

Wallace, D. (2010). How to prove the born rule. In: S. Saunders, J. Barrett, A. Kent, & D. Wallace (Eds.), *Many worlds? Everett, quantum theory, and reality* (pp. 227–263). Oxford: Oxford University Press.

Wallace, D. (2012). *The emergent universe. Quantum theory according to the Everett interpretation*. Oxford: Oxford University Press.

Wallace, D. (2014). Probability in physics: Statistical, stochastic, quantum. In A. Wilson (Ed.), *Chance and temporal asymmetry* (pp. 194–220). Oxford: Oxford University Press. http://philsci-archive.pitt.edu/9815/1/wilson.pdf

Wright, V. J., & Weigert, S. (2019). A Gleason-type theorem for qubits based on mixtures of projective measurements. *Journal of Physics A: Mathematical and Theoretical, 52*(5), 055301.

Chapter 8
'Two Dogmas' Redux

Jeffrey Bub

Abstract I revisit the paper 'Two dogmas about quantum mechanics,' co-authored with Itamar Pitowsky, in which we outlined an information-theoretic interpretation of quantum mechanics as an alternative to the Everett interpretation. Following the analysis by Frauchiger and Renner of 'encapsulated' measurements (where a super-observer, with unrestricted ability to measure any arbitrary observable of a complex quantum system, measures the memory of an observer system after that system measures the spin of a qubit), I show that the Everett interpretation leads to modal contradictions. In this sense, the Everett interpretation is inconsistent.

Keywords Measurement problem · Information-theoretic interpretation · Everett interpretation · Wigner's friend · Encapsulated measurements · Frauchiger-Renner argument

8.1 Introduction

About ten years ago, Itamar Pitowsky and I wrote a paper, 'Two dogmas about quantum mechanics' (Bub and Pitowsky 2010), in which we outlined an information-theoretic interpretation of quantum mechanics as an alternative to the Everett interpretation. Here I revisit the paper and, following Frauchiger and Renner (2018), I show that the Everett interpretation leads to modal contradictions in 'Wigner's-Friend'-type scenarios that involve 'encapsulated' measurements, where a super-observer (which could be a quantum automaton), with unrestricted ability to measure any arbitrary observable of a complex quantum system, measures the

J. Bub (✉)
Department of Philosophy, Institute for Physical Science and Technology, Joint Center for Quantum Information and Computer Science, University of Maryland, College Park, MD, USA
e-mail: jbub@umd.edu

M. Hemmo, O. Shenker (eds.), *Quantum, Probability, Logic*,
Jerusalem Studies in Philosophy and History of Science,
https://doi.org/10.1007/978-3-030-34316-3_8

memory of an observer system (also possibly a quantum automaton) after that system measures the spin of a qubit. In this sense, the Everett interpretation is inconsistent.

8.2 The Information-Theoretic Interpretation

The salient difference between classical and quantum mechanics is the noncommutativity of the algebra of observables of a quantum system, equivalently the non-Booleanity of the algebra of two-valued observables representing properties (for example, the property that the energy of the system lies in a certain range of values, with the two eigenvalues representing 'yes' or 'no'), or propositions (the proposition asserting that the value of the energy lies in this range, with the two eigenvalues representing 'true' or 'false'). The two-valued observables of a classical system form a Boolean algebra, isomorphic to the Borel subsets of the phase space of the system. The transition from classical to quantum mechanics replaces this Boolean algebra by a family of 'intertwined' Boolean algebras, to use Gleason's term (Gleason 1957), one for each set of commuting two-valued observables, represented by projection operators in a Hilbert space. The intertwinement precludes the possibility of embedding the whole collection into one inclusive Boolean algebra, so you can't assign truth values consistently to the propositions about observable values in all these Boolean algebras. Putting it differently, there are Boolean algebras in the family of Boolean algebras of a quantum system, for example the Boolean algebras for position and momentum, or for spin components in different directions, that don't fit together into a single Boolean algebra, unlike the corresponding family for a classical system. In this non-Boolean theory, probabilities are, as von Neumann put it, 'sui generis' (von Neumann 1937), and 'uniquely given from the start' (von Neumann 1954, p. 245) via Gleason's theorem, or the Born rule, as a feature of the geometry of Hilbert space, related to the angle between rays in Hilbert space representing 'pure' quantum states.[1]

The central interpretative question for quantum mechanics as a non-Boolean theory is how we should understand these 'sui generis' probabilities, since they are not probabilities of spontaneous transitions between quantum states, nor can they be interpreted as measures of ignorance about quantum properties associated with the actual values of observables prior to measurement.

The information-theoretic interpretation is the proposal to take Hilbert space as the kinematic framework for the physics of an indeterministic universe, just as Minkowski space provides the kinematic framework for the physics of a non-

[1]This conceptual picture applies to quantum mechanics on a finite-dimensional Hilbert space. A restriction to 'normal' quantum states is required for quantum mechanics formulated with respect to a general von Neumann algebra, where a generalized Gleason's theorem holds even for quantum probability functions that are not countably additive. Thanks to a reviewer for pointing this out.

Newtonian, relativistic universe.[2] In special relativity, the geometry of Minkowski space imposes spatio-temporal constraints on events to which the relativistic dynamics is required to conform. In quantum mechanics, the non-Boolean projective geometry of Hilbert space imposes objective kinematic (i.e., pre-dynamic) probabilistic constraints on correlations between events to which a quantum dynamics of matter and fields is required to conform. In this non-Boolean theory, new sorts of nonlocal probabilistic correlations are possible for 'entangled' quantum states of separated systems, where the correlated events are intrinsically random, not merely apparently random like coin tosses (see Bub 2016, Chapter 4).

In (Pitowsky 2007), Pitowsky distinguished between a 'big' measurement problem and a 'small' measurement problem. The 'big' measurement problem is the problem of explaining how measurements can have definite outcomes, given the unitary dynamics of the theory. The 'small' measurement problem is the problem of accounting for our familiar experience of a classical or Boolean macroworld, given the non-Boolean character of the underlying quantum event space. The 'big' problem is the problem of explaining *how individual measurement outcomes come about dynamically*, i.e., how something indefinite in the quantum state of a system before measurement can *become definite* in a measurement. There is nothing analogous to this sort of transition in classical physics, where transitions are always between an initial state of affairs specified by what is and is not the case (equivalent to a 2-valued homomorphism on a Boolean algebra) to a final state of affairs with a different specification of what is and is not the case. The 'small' problem is the problem of explaining the *emergence of an effectively classical probability space for the macro-events we observe in a measurement.*

The 'big' problem is deflated as a pseudo-problem if we reject two 'dogmas' about quantum mechanics. The first dogma is the view that measurement in a fundamental mechanical theory should be treated as a dynamical process, so it should be possible, in principle, to give a complete dynamical analysis of how individual measurement outcomes come about. The second dogma is the interpretation of a quantum state as a representation of physical reality, i.e., as the 'truthmaker' for propositions about the occurrence and non-occurrence of events, analogous to the ontological significance of a classical state.

The first dogma about measurement is an entirely reasonable demand for a fundamental theory of motion like classical mechanics. But noncomutativity or non-Booleanity makes quantum mechanics quite unlike any theory we have dealt with before in the history of physics, and there is no reason, apart from tradition, to assume that the theory can provide the sort of *representational* explanation we are familiar with in a theory that is commutative or Boolean at the fundamental level. The 'sui generis' quantum probabilities can't be understood as quantifying ignorance about the pre-measurement value of an observable, as in a Boolean theory, but cash out in terms of 'what you'll find if you measure,' which involves

[2]See (Janssen 2009) for a defense of this view of special relativity contra Harvey Brown (2006).

considering the outcome, at the Boolean macrolevel, of manipulating a quantum system in a certain way.

A quantum 'measurement' is not really the same sort of thing as a measurement of a physical quantity of a classical system. It involves putting a microsystem, like a photon, in a macroscopic environment, say a beamsplitter or an analyzing filter, where the photon is forced to make an intrinsically random transition recorded as one of a number of macroscopically distinct alternatives in a macroscopic device like a photon detector. The registration of the measurement outcome at the Boolean macrolevel is crucial, because it is only with respect to a suitable structure of alternative possibilities that it makes sense to talk about an event as definitely occurring or not occurring, and this structure—characterized by Boole as capturing the 'conditions of possible experience' (Pitowsky 1994)—is a Boolean algebra.

In special relativity, Lorentz contraction is a kinematic effect of motion in a non-Newtonian space-time. The contraction is *consistent* with a dynamical account, but such an account takes the forces involved to be Lorentz covariant, which is to say that the dynamics is assumed to have symmetries that respect Lorentz contraction as a kinematic effect of relative motion. (By contrast, in Lorentz's theory, the contraction is explained as a dynamical effect in Newtonian space-time.) Analogously, the loss of information in a quantum measurement—Bohr's 'irreducible and uncontrollable disturbance'—is a *kinematic* effect of *any* process of gaining information of the relevant sort, irrespective of the dynamical processes involved in the measurement process. A solution to the 'small' measurement problem—a dynamical explanation for the effective emergence of classicality, i.e., Booleanity, at the macrolevel—would amount to a proof that the unitary quantum dynamics is consistent with the kinematics, analogous to a proof that relativistic dynamics is consistent with the kinematics.

We argued in 'Two dogmas' that the 'small' measurement problem can be resolved as a consistency problem by considering the dynamics of the measurement process and the role of decoherence in the emergence of an effectively classical probability space of macro-events to which the Born probabilities refer. (The proposal was that decoherence solves the 'small' measurement problem, not the 'big' measurement problem—decoherence does not provide a dynamical explanation of how an indefinite outcome in a quantum superposition becomes definite in the unitary evolution of a measurement process.) An alternative solution is suggested by Pitowsky's combinatorial treatment of macroscopic objects in quantum mechanics (Pitowsky 2004). Entanglement witnesses are observables that distinguish between separable and entangled states. An entanglement witness for an entangled state is an observable whose expectation value lies in a bounded interval for a separable state, but is outside this interval for the entangled state. Pitowsky showed, for a large class of entanglement witnesses, that for composite systems where measurement of an entanglement witness requires many manipulations of individual particles, entangled states that can be distinguished from separable states become rarer and rarer as the number of particles increases, and he conjectured that this is true in general. If Pitowsky's conjecture is correct, a macrosystem in quantum mechanics can be characterized as a many-particle system for which the measure of the set of entangled states that can be distinguished from separable states tends to zero. In this sense, a macrosystem is effectively a commutative or Boolean system.

On the information-theoretic interpretation, quantum mechanics is a new sort of *non-representational* theory for an indeterministic universe, in which the quantum state is a bookkeeping device for keeping track of probabilities and probabilistic correlations between intrinsically random events. Probabilities are defined with respect to a single Boolean perspective, the Boolean algebra generated by the 'pointer-readings' of what Bohr referred to as the 'ultimate measuring instruments,' which are 'kept outside the system subject to quantum mechanical treatment' (Bohr 1939, pp. 23–24):

> In the system to which the quantum mechanical formalism is applied, it is of course possible to include any intermediate auxiliary agency employed in the measuring processes. ... The only significant point is that in each case some ultimate measuring instruments, like the scales and clocks which determine the frame of space-time coordination—on which, in the last resort, even the definition of momentum and energy quantities rest—must always be described entirely on classical lines, and consequently be kept outside the system subject to quantum mechanical treatment.

Bohr did not, of course, refer to Boolean algebras, but the concept is simply a precise way of codifying a significant aspect of what Bohr meant by a description 'on classical lines' or 'in classical terms' in his constant insistence that (his emphasis) (Bohr 1949, p. 209)

> however far the phenomena transcend the scope of classical physical explanation, the account of all evidence must be expressed in classical terms.

by which he meant 'unambiguous language with suitable application of the terminology of classical physics'—for the simple reason, as he put it, that we need to be able 'to tell others what we have done and what we have learned.' Formally speaking, the significance of 'classical' here as being able 'to tell others what we have done and what we have learned' is that the events in question should fit together as a Boolean algebra, so conforming to Boole's 'conditions of possible experience.'

The solution to the 'small' measurement problem as a consistency problem does not show that unitarity is suppressed at a certain level of size or complexity, so that the non-Boolean possibility structure becomes Boolean and quantum becomes classical at the macrolevel. Rather, the claim is that the unitary dynamics of quantum mechanics is consistent with the kinematics, in the sense that treating the measurement process dynamically and ignoring certain information that is in practice inaccessible makes it extremely difficult to detect the phenomena of interference and entanglement associated with non-Booleanity. In this sense, an effectively classical or Boolean probability space of Born probabilities can be associated with our observations at the macrolevel.

Any system, of any complexity, is fundamentally a non-Boolean quantum system and can be treated as such, in principle, which is to say that a unitary dynamical analysis can be applied to whatever level of precision you like. But we see actual events happen at the Boolean macrolevel. At the end of a chain of instruments and recording devices, some particular system, M, functions as the 'ultimate measuring instrument' with respect to which an event corresponding to a definite measurement outcome occurs in an associated Boolean algebra, whose selection is not the outcome of a dynamical evolution described by the theory. The system M, or any part of M, can be treated quantum mechanically, but then some other system, M', treated as classical or commutative or Boolean, plays the role of the ultimate measuring instrument in any application of the theory. The outcome of a measurement is an intrinsically random event at the macrolevel, *something that actually happens*, not described by the deterministic unitary dynamics, so outside the theory. Putting it differently, the 'collapse,' as a conditionalization of the quantum state, is something you put in by hand after recording the actual outcome. The physics doesn't give it to you.

As Pauli put it (Pauli 1954):

> Observation thereby takes on the character of *irrational, unique actuality* with unpredictable outcome. ... Contrasted with this *irrational* aspect of concrete phenomena which are determined in their *actuality*, there stands the *rational* aspect of an abstract ordering of the possibilities of statements by means of the mathematical concept of probability and the ψ-function.

A representational theory proposes a primitive ontology, perhaps of particles or fields of a certain sort, and a dynamics that describes how things change over time. The non-Boolean physics of quantum mechanics does not provide a representational explanation of phenomena. Rather, a quantum mechanical explanation involves showing how a Boolean output (a measurement outcome) is obtained from a Boolean input (a state preparation) via a non-Boolean link. If a 'world' in which truth and falsity, 'this' rather than 'that,' makes sense is Boolean, then there is no quantum 'world,' as Bohr is reported to have said,[3] and it would be misleading to attempt to picture it. In this sense, a quantum mechanical explanation is 'operational,' but this is not simply a matter of convenience or philosophical persuasion. We adopt quantum mechanics—the theoretical formalism of the non-Boolean link between Boolean input and output—for empirical reasons, the failure of classical physics to explain certain phenomena, and because there is no satisfactory representational theory of such phenomena.

Quantum mechanics on the Everett interpretation is regarded, certainly by its proponents, as a perfectly good representational theory: it explains phenomena in terms of an underlying ontology and a dynamics that accounts for change over time. As I will show in the following section, the Everett interpretation leads to modal contradictions in scenarios that involve 'encapsulated' measurements.

---

[3] Aage Petersen (Petersen 1963, p. 12): 'When asked whether the algorithm of quantum mechanics could be considered as somehow mirroring an underlying quantum world, Bohr would answer, "There is no quantum world. There is only an abstract quantum physical description. It is wrong to think that the task of physics is to find out how nature *is*. Physics concerns what we can *say* about nature."

8.3 Encapsulated Measurements and the Everett Interpretation

According to the Born rule, the probability of finding the outcome a in a measurement of an observable A on a system S in the state $|\psi\rangle = \sum_a \langle a|\psi\rangle |a\rangle$ is:

$$p_\psi(a) = \mathrm{Tr}(P_a P_\psi) = |\langle a|\psi\rangle|^2 \tag{8.1}$$

where P_a is the projection operator onto the eigenstate $|a\rangle$ and P_ψ is the projection operator onto the quantum state. After the measurement, the state is updated to

$$|\psi\rangle \longrightarrow |a\rangle \tag{8.2}$$

On the information-theoretic interpretation, the state update is understood as conditionalization (with necessary loss of information) on the measurement outcome.

For a sequence of measurements, perhaps by two different observers, of A followed by B on the same system S initially in the state $|\psi\rangle$, where A and B need not commute, the conditional probability of the outcome b of B given the outcome a of A is

$$p_\psi(b|a) = \frac{p_\psi(a,b)}{\sum_b p_\psi(a,b)} = \frac{p_\psi(a,b)}{p_\psi(a)} = \mathrm{Tr}(P_b P_a) = |\langle b|a\rangle|^2 \tag{8.3}$$

with $p_\psi(a,b) = |\langle a|\psi\rangle|^2 \cdot |\langle b|a\rangle|^2 = p_\psi(a) \cdot |\langle b|a\rangle|^2$.

According to Everett's relative state interpretation, a measurement is a unitary transformation, or equivalently an isometry, V, that correlates the pointer-reading state of the measuring instrument and the associated memory state of an observer with the state of the observed system. (In the following, I'll simply refer to the memory state, with the understanding that this includes the measuring instrument and recording device and any other systems involved in the measurement.) In the case of a projective measurement by an observer O, the transformation $\mathcal{H}_S \xrightarrow{V} \mathcal{H}_S \otimes \mathcal{H}_O$ maps $|a\rangle_S$ onto $|a\rangle_S \otimes |A_a\rangle_O$, where $|A_a\rangle_O$ is the memory state correlated with the eigenstate $|a\rangle_S$, so

$$|\psi\rangle \xrightarrow{V} |\Psi\rangle = \sum_a \langle a|\psi\rangle |a\rangle_S \otimes |A_a\rangle_O \tag{8.4}$$

Baumann and Wolf (2018) formulate the Born rule for the memory state $|A_a\rangle_O$ of the observer O having observed a in the relative state interpretation, where there is no commitment to the existence of a physical record of the measurement outcome as classical information. The probability of the observer O observing a in this sense is

$$q_\psi(a) = \mathrm{Tr}(\mathbb{1} \otimes P_{A_a} \cdot V P_\psi V^\dagger) \tag{8.5}$$

where P_{A_a} is the projection operator onto the memory state $|A_a\rangle_O$. This is equivalent to the probability $p_\psi(a)$ in the standard theory.

This equivalence can also be interpreted as showing the movability of the notorious Heisenberg 'cut' between the observed system and the composite system doing the observing, if the observing system is treated as a classical or Boolean system. The cut can be placed between the system and the measuring instrument plus observer, so that only the system is treated quantum mechanically, or between the measuring instrument and observer, so that the system plus instrument is treated as a composite quantum system. Similar shifts are possible if the measuring instrument is subdivided into several component systems, e.g., if a recording device is regarded as a separate component system, or if the observer is subdivided into component systems involved in the registration of a measurement outcome.

On the relative state interpretation, the probability that S is in the state $|a'\rangle$ but the observer sees a is

$$q_\psi(|a'\rangle, a) = \mathrm{Tr}(P_{a'} \otimes P_{A_a} \cdot V P_\psi V^\dagger) = \delta_{a',a} \tag{8.6}$$

So the probability that S is in the state $|a\rangle$ after O observes a is 1, as in the standard theory according to the state update rule.

For a sequence of measurements of observables A and B by two observers, O_1 and O_2, on the same system S, the conditional probability for the outcome b given the outcome a is

$$q_\psi(b|a) == \frac{q_\psi(a,b)}{\sum_b q_\psi(a,b)} = \frac{\mathrm{Tr}(\mathbb{1} \otimes P_{A_a} \otimes P_{B_b} \cdot V_{O_2} V_{O_1} P_\psi V_{O_1}^\dagger V_{O_2}^\dagger)}{\sum_b \mathrm{Tr}(\mathbb{1} \otimes P_{A_a} \otimes P_{B_b} \cdot V_{O_2} V_{O_1} P_\psi V_{O_1}^\dagger V_{O_2}^\dagger)} \tag{8.7}$$

which turns out to be the same as the conditional probability given by the state update rule: $q_\psi(b|a) = p_\psi(b|a)$.

So Everett's relative state interpretation and the information-theoretic interpretation make the same predictions, both for the probability of an outcome of an A-measurement on a system S, and for the conditional probability of an outcome of a B-measurement, given an outcome of a prior A-measurement on the same system S by two observers, O_1 and O_2.

As Baumann and Wolf show, this is no longer the case for conditional probabilities involving encapsulated measurements at different levels of observation, where an observer O measures an observable of a system S and a super-observer measures an arbitrary observable of the joint system $S + O$. In other words, the movability of the cut is restricted to measurements that are not encapsulated. Note that both the observer and the super-observer could be quantum automata, so the analysis is not restricted to observers as conscious agents.

Suppose an observer O measures an observable with eigenstates $|a\rangle_S$ on a system S in a state $|\psi\rangle_S = \sum_a \langle a|\psi\rangle |a\rangle_S$, and a super-observer SO then measures an observable with eigenstates $|b\rangle_{S+O}$ in the Hilbert space of the composite system $S + O$.

On the information-theoretic interpretation of the standard theory, the state of $S+O$ after the measurement by O is $|a \otimes A_a\rangle_{S+O} = |a\rangle_S \otimes |A_a\rangle_O$, where $|A_a\rangle_O$ is the memory state of O correlated with the outcome a. The probability of the super-observer finding the outcome b given that the observer O obtained the outcome a is then

$$p_\psi(b|a) = |\langle b|a \otimes A_a\rangle_{S+O}|^2 \tag{8.8}$$

On the relative state interpretation, after the unitary evolutions associated with the observer O's measurement of the S-observable with eigenstates $|a\rangle$, and the super-observer SO's measurement of the $(S+O)$-observable with eigenstates $|b\rangle$, the state of the composite system $S+O+SO$ is

$$|\Psi\rangle = \sum_{a,b} \langle a|\psi\rangle\langle b|a \otimes A_a\rangle|b\rangle_{S+O}|B_b\rangle_{SO} \tag{8.9}$$

where $|B_b\rangle_O$ is the memory state of SO correlated with the outcome b. The probability of the super-observer SO finding the outcome b given that the observer O obtained the outcome a is then $q_\psi(b|a)$ as given by (8.7), with $|\Psi\rangle\langle\Psi|$ for $V_{O_2}V_{O_1}P_\psi V_{O_1}^\dagger V_{O_2}^\dagger$. The numerator of this expression, after taking the trace over the $(S+O)$-space followed by the trace over the O-space of the projection onto P_{B_b} followed by the projection onto P_{A_a}, becomes

$$p_\psi(b|a) \sum_{a',a''} \langle a'|\psi\rangle\langle\psi|a''\rangle\langle b|a' \otimes A_{a'}\rangle\langle a'' \otimes A_{a''}|b\rangle \tag{8.10}$$

so

$$q_\psi(b|a) = \frac{p_\psi(b|a)\sum_{a',a''}\langle a'|\psi\rangle\langle\psi|a''\rangle\langle b|a' \otimes A_{a'}\rangle\langle a'' \otimes A_{a''}|b\rangle}{\sum_b \mathrm{Tr}(\mathbb{1} \otimes P_{A_a} \otimes P_{B_b} \cdot |\Psi\rangle\langle\Psi|)} \tag{8.11}$$

Baumann and Wolf point out that this is equal to $p_\psi(b|a)$ in (8.8) only if $|b\rangle_{S+O} = |a\rangle_S \otimes |A_a\rangle_O$ for all b, i.e., if $|b\rangle_{S+O}$ is a product state of an S-state $|a\rangle_S$ and an O-state $|A_a\rangle_O$, which is not the case for general encapsulated measurements.

Here O's measurement outcome states $|A_a\rangle_O$ are understood as relative to states of SO, and there needn't be a record of O's measurement outcome as classical or Boolean information. If there is a classical record of the outcome then, as Baumann and Wolf show, SO's conditional probability is the same as the conditional probability in the standard theory.

To see that the relative state interpretation leads to a modal contradiction, consider the Frauchiger-Renner modification of the 'Wigner's Friend' thought experiment (Frauchiger and Renner 2018). Frauchiger and Renner add a second Friend ($\overline{F}$) and a second Wigner ($\overline{W}$). Friend $\overline{F}$ measures an observable with eigenvalues 'heads' or 'tails' and eigenstates $|h\rangle, |t\rangle$ on a system R in the state

$\frac{1}{\sqrt{3}}|h\rangle + \sqrt{\frac{2}{3}}|t\rangle$. One could say that $\overline{F}$ 'tosses a biased quantum coin' with probabilities 1/3 for heads and 2/3 for tails. She prepares a qubit S in the state $|\downarrow\rangle$ if the outcome is h, or in the state $|\rightarrow\rangle = \frac{1}{\sqrt{2}}(|\downarrow\rangle + |\uparrow\rangle)$ if the outcome is t, and sends S to F. When F receives S, he measures an S-observable with eigenstates $|\downarrow\rangle, |\uparrow\rangle$ and corresponding eigenvalues $-\frac{1}{2}, \frac{1}{2}$. The Friends $\overline{F}$ and F are in two laboratories, $\overline{L}$ and L, which are assumed to be completely isolated from each other, except briefly when $\overline{F}$ sends the qubit S to F.

Now suppose that $\overline{W}$ is able to measure an $\overline{L}$-observable with eigenstates

$$|\text{fail}\rangle_{\overline{L}} = \frac{1}{\sqrt{2}}(|h\rangle_{\overline{L}} + |t\rangle_{\overline{L}})$$

$$|\text{ok}\rangle_{\overline{L}} = \frac{1}{\sqrt{2}}(|h\rangle_{\overline{L}} - |t\rangle_{\overline{L}})$$

and W is able to measure an L-observable with eigenstates

$$|\text{fail}\rangle_L = \frac{1}{\sqrt{2}}(|-\frac{1}{2}\rangle_L + |\frac{1}{2}\rangle_L)$$

$$|\text{ok}\rangle_L = \frac{1}{\sqrt{2}}(|-\frac{1}{2}\rangle_L - |\frac{1}{2}\rangle_L)$$

Suppose W measures first and we stop the experiment at that point. On the relative state interpretation, $\overline{F}$'s memory and all the systems in $\overline{F}$'s laboratory involved in the measurement of R become entangled by the unitary transformation, and similarly F's memory and all the systems in F's laboratory involved in the measurement of S become entangled. If R is in the state $|h\rangle_R$ and $\overline{F}$ measures R, the entire laboratory $\overline{L}$ evolves to the state $|h\rangle_{\overline{L}} = |h\rangle_R|h\rangle_{\overline{F}}$, where $|h\rangle_{\overline{F}}$ represents the state of $\overline{F}$'s memory plus measuring and recording device plus all the systems in $\overline{F}$'s lab connected to the measuring and recording device after the registration of outcome h. Similarly for the state $|t\rangle_{\overline{L}}$, and for the states $|-\frac{1}{2}\rangle_L = |-\frac{1}{2}\rangle_S|-\frac{1}{2}\rangle_F$ and $|\frac{1}{2}\rangle_L = |\frac{1}{2}\rangle_S|\frac{1}{2}\rangle_F$ of F's lab L with respect to the registration of the outcomes $\pm\frac{1}{2}$ of the spin measurement. So after the evolutions associated with the measurements by $\overline{F}$ and F, the state of the two laboratories is

$$|\Psi\rangle = \frac{1}{\sqrt{3}}|h\rangle_{\overline{L}}|-\frac{1}{2}\rangle_L + \sqrt{\frac{2}{3}}|t\rangle_{\overline{L}}\frac{|-\frac{1}{2}\rangle_L + |\frac{1}{2}\rangle_L}{\sqrt{2}} \tag{8.12}$$

$$= \frac{1}{\sqrt{3}}|h\rangle_{\overline{L}}|-\frac{1}{2}\rangle_L + \sqrt{\frac{2}{3}}|t\rangle_{\overline{L}}|\text{fail}\rangle_L \tag{8.13}$$

According to the state (8.13), the probability is zero that W gets the outcome 'ok' for his measurement, given that $\overline{F}$ obtained the outcome t for her measurement.

Now suppose that $\overline{W}$ and W both measure, and $\overline{W}$ measures before W, as in the Frauchiger-Renner scenario. What is the probability that W will get 'ok' at the later time given that $\overline{F}$ obtained 'tails' at the earlier time? The global entangled state at any time simply expresses a correlation between measurements—the time order of the measurements is irrelevant. The two measurements by W and $\overline{W}$ could be spacelike connected, and in that case the order in which the measurements are carried out clearly can't make a different to the probability.[4]

Since observers in differently moving reference frames agree about which events occur, even if they disagree about the order of events, an event that has zero probability in some reference frame cannot occur for any observer in any reference frame.[5] In a reference frame in which W measures before $\overline{W}$ the probability is zero that W gets 'ok' if $\overline{F}$ gets 'tails,' because this must be the same as the probability if we take $\overline{W}$ out of the picture. It follows that if $\overline{F}$ gets 'tails,' the 'ok' measurement outcome event cannot occur in a reference frame in which $\overline{W}$ measures before W.

Think of it from $\overline{F}$'s perspective. If she is certain that she found 'tails,' she can predict with certainty that W's later measurement (timelike connected to her 'tails' event) will not result in the outcome 'ok.' She wouldn't (shouldn't!) change her prediction because of the possible occurrence of an event spacelike connected to W's measurement—not if she wants to be consistent with special relativity. W finding 'ok' is a zero probability event in the absolute future of $\overline{F}$'s prediction that cannot occur for $\overline{F}$ given the 'tails' outcome of her R-measurement, and hence cannot occur in any reference frame. The fact that $\overline{W}$'s measurement, an event spacelike connected to W's measurement, might occur after her prediction doesn't support altering the prediction, even though it would make $\overline{F}$'s memory of the earlier event (and any trace of the earlier event in $\overline{F}$'s laboratory L) indefinite, so no observer could be in a position to check whether or not $\overline{F}$ obtained the earlier outcome t when W gets 'ok.'

The state (8.13) can be expressed as

$$|\Psi\rangle = \frac{1}{\sqrt{12}}|\text{ok}\rangle_{\overline{L}}|\text{ok}\rangle_L - \frac{1}{\sqrt{12}}|\text{ok}\rangle_{\overline{L}}|\text{fail}\rangle_L + \frac{1}{\sqrt{12}}|\text{fail}\rangle_{\overline{L}}|\text{ok}\rangle_L + \frac{\sqrt{3}}{2}|\text{fail}\rangle_{\overline{L}}|\text{fail}\rangle_L \tag{8.14}$$

It follows that the probability that both Wigners obtain the outcome 'ok' is 1/12. Since the two Wigners measure commuting observables on separate systems, $\overline{W}$ can communicate the outcome 'ok' of her measurement to W, and her prediction that she is certain, given the outcome 'ok,' that W will obtain 'fail,' without 'collapsing' the global entangled state. Then in a round in which W obtains the outcome 'ok' for his measurement and so is certain that the outcome is 'ok,' he is also certain that the outcome of his measurement is not 'ok.'

---

[4]Thanks to Renato Renner for pointing this out.

[5]In Bub and Stairs (2014), Allen Stairs and I proposed this as a consistency condition to avoid potential contradictions in quantum interactions with closed timelike curves.

Frauchiger and Renner derive this modal contradiction on the basis of three assumptions, Q, S, and C. Assumption Q (for 'quantum') says that an agent can 'be certain' of the value of an observable at time t if the quantum state assigns probability 1 to this value after a measurement that is completed at time t, and the agent has established that the system is in this state. An agent can also 'be certain' that the outcome does not occur if the probability is zero. One should also include under Q the assumption that measurements are quantum interactions that transform the global quantum state unitarily. Assumption S (for 'single value') says, with respect to 'being certain,' that an agent can't 'be certain' that the value of an observable is equal to v at time t and also 'be certain' that the value of this observable is not equal to v at time t.[6] Finally, assumption C (for 'consistency') is a transitivity condition for 'being certain': if an agent 'is certain' that another agent, reasoning with the same theory, 'is certain' of the value of an observable at time t, then the first agent can also 'be certain' of this value at time t.

As used in the Frauchiger-Renner argument, 'being certain' is a technical term entirely specified by the three assumptions. In other words, 'being certain' can mean whatever you like, provided that the only inferences involving the term are those sanctioned by the three assumptions. In particular, it is not part of the Frauchiger-Renner argument that if an agent 'is certain' that the value of an observable is v at time t, then the value of the observable is indeed v at time t—the observable might have multiple values, as in the Everett interpretation. So it does not follow that the proposition 'the value of the observable is v at time t' is true, or that the system has the property corresponding to the value v of the observable at time t. Also, it is not part of the Frauchiger-Renner argument that if an agent 'is certain' of the value of an observable, then the agent knows the value, in any sense of 'know' that entails the truth of what the agent knows.

The argument involves measurements, and inferences by the agents via the notion of 'being certain,' that can all be modeled as unitary transformations to the agents' memories. So picture the agents as quantum automata, evolving unitarily, where these unitary evolutions result in changes to the global quantum state of the two friends and the two Wigners that reflect changes in the agents' memories. What Frauchiger and Renner show is that these assumptions lead to memory changes that end up in a modal contradiction.

Obviously, the relative state interpretation is inconsistent with the assumption S in the sense that different branches of the quantum state can be associated with different values of an observable after a measurement. So, with respect to a particular branch, an agent can be certain that an observable has a certain value and, with respect to a different branch, an agent associated with that branch can be certain that the observable has a different value. This is not a problem. Rather, the

[6]The standard axiom system for the modal logic of the operator 'I am certain that' includes the axiom 'if I am certain that A, then it is not the case that I am certain that not-A' (assuming the accessibility relation is serial and does not have any dead-ends). Thanks to Eric Pacuit for pointing this out.

problem for the relative state interpretation is that there is *a branch* of the global quantum state after W's measurement, which has a non-zero probability of 1/12, with a memory entry that is inconsistent with S: W is certain that the outcome of his measurement is 'fail' and also certain that the outcome is 'ok.'

Renner gives a full description of the state of the global system as it evolves unitarily with the measurements and inferential steps at every stage of the experiment, which is repeated over many rounds until $\overline{W}$ and W both obtain the outcome 'ok' (Renner 2019). The measurements are assumed to be performed by the agents $\overline{F}, F, \overline{W}, W$ in sequence beginning at times 00, 10, 20, and 30 (and ending at times 01, 11, 21, and 31).

So, for example, in round n after F's measurement the global entangled state is $|\Psi\rangle^{n:11}$ (cf. (8.12)):

$$\frac{1}{\sqrt{3}}|h\rangle_{\overline{R}}|h, \text{no conclusion}\rangle_{\overline{F}}|\downarrow\rangle_S|z=-\tfrac{1}{2}\rangle_F$$
$$+\sqrt{\frac{1}{3}}|t\rangle_{\overline{R}}|t, \text{so I am certain that } w = \text{fail at n:31}\rangle_{\overline{F}}|\downarrow\rangle_S|z=-\tfrac{1}{2}\rangle_F$$
$$+\sqrt{\frac{1}{3}}|t\rangle_{\overline{R}}|t, \text{so I am certain that } w = \text{fail at n:31}\rangle_{\overline{F}}|\uparrow\rangle_S|z=\tfrac{1}{2}\rangle_F \tag{8.15}$$

After the measurements by $\overline{W}$ and W with outcomes $\bar{w}$ and w, the global entangled state is $|\Psi\rangle^{n:31}$:

$$\sqrt{\frac{1}{12}}|\text{ok}\rangle_{\overline{L}}|\text{ok, so I am certain that } W = \text{fail}\rangle_{\overline{W}}|\text{ok}\rangle_L|\text{I am certain that } W = \text{fail; I observe } W = \text{ok!}\rangle_W$$
$$-\sqrt{\frac{1}{12}}|\text{ok}\rangle_{\overline{L}}|\text{ok, so I am certain that } W = \text{fail}\rangle_{\overline{W}}|\text{fail}\rangle_L|\text{I am certain that } W = \text{fail; I observe } W = \text{fail}\rangle_W$$
$$+\sqrt{\frac{1}{12}}|\text{fail}\rangle_{\overline{L}}|\text{fail, no conclusion}\rangle_{\overline{W}}|\text{ok}\rangle_L|\text{no conclusion previously; I observe } W = \text{ok}\rangle_W$$
$$+\frac{\sqrt{3}}{2}|\text{fail}\rangle_{\overline{L}}|\text{fail, no conclusion}\rangle_{\overline{W}}|\text{fail}\rangle_L|\text{no conclusion previously; I observe } W = \text{fail}\rangle_W \tag{8.16}$$

The inconsistency with S is apparent on the first branch.

8.4 A Clarification

Baumann and Wolf argue that standard quantum mechanics with the 'collapse' rule for measurements, and quantum mechanics on Everett's relative state interpretation, are really two different *theories*, not different *interpretations* of the same theory. For the Frauchiger-Renner scenario, they derive the conditional probability

$$p_{\Psi}(\text{ok}_W | t_{\overline{F}}) = 0 \tag{8.17}$$

for the standard theory by supposing that $\overline{F}$ updates the quantum state on the basis of the outcome of her measurement of R and corresponding state preparation of S, but F's measurement is described as a unitary transformation from $\overline{F}$'s perspective, so the state of the two labs after the measurements by $\overline{F}$ (assuming $\overline{F}$ found 'tails') and F is $|t\rangle_{\overline{L}}(|-\frac{1}{2}\rangle_L + |+\frac{1}{2}\rangle_L) = |t\rangle_{\overline{L}}|\text{fail}\rangle_L$. For the relative state theory, they derive the conditional probability

$$q_{\Psi}(\text{ok}_W | t_{\overline{F}}) = 1/6 \tag{8.18}$$

This would seem to be in conflict with the claim in the previous section that the probability is zero that W gets the outcome 'ok' for his measurement, given that $\overline{F}$ obtained the outcome 'tails' for her measurement, whether or not $\overline{W}$ measures before W.

To understand the meaning of the probability $q_{\Psi}(\text{ok}_W | t_{\overline{F}})$ as Baumann and Wolf define it, note that after the unitary evolution associated with $\overline{W}$'s measurement on the lab $\overline{L}$, the state of the two laboratories and $\overline{W}$ is (from (8.14))

$$\begin{aligned}\Phi = &\frac{1}{\sqrt{12}}|\text{ok}\rangle_{\overline{L}}|\text{ok}\rangle_{\overline{W}}|\text{ok}\rangle_L - \frac{1}{\sqrt{12}}|\text{ok}\rangle_{\overline{L}}|\text{ok}\rangle_{\overline{W}}|\text{fail}\rangle_L \\ &+\frac{1}{\sqrt{12}}|\text{fail}\rangle_{\overline{L}}|\text{fail}\rangle_{\overline{W}}|\text{ok}\rangle_L + \frac{\sqrt{3}}{2}|\text{fail}\rangle_{\overline{L}}|\text{fail}\rangle_{\overline{W}}|\text{fail}\rangle_L\end{aligned} \tag{8.19}$$

which can be expressed as

$$\begin{aligned}\Phi = &\frac{1}{\sqrt{2}}|h\rangle_{\overline{L}}\Big[\sqrt{\frac{5}{6}}\Big(\frac{3}{\sqrt{10}}|\text{fail}\rangle_{\overline{W}} - \frac{1}{\sqrt{10}}|\text{ok}\rangle_{\overline{W}}\Big)|\text{fail}\rangle_L \\ &+\frac{1}{\sqrt{6}}\Big(\frac{1}{\sqrt{2}}|\text{fail}\rangle_{\overline{W}} + \frac{1}{\sqrt{2}}|\text{ok}\rangle_{\overline{W}}\Big)|\text{ok}\rangle_L\Big] \\ &\frac{1}{\sqrt{2}}|t\rangle_{\overline{L}}\Big[\sqrt{\frac{5}{6}}\Big(\frac{3}{\sqrt{10}}|\text{fail}\rangle_{\overline{W}} + \frac{1}{\sqrt{10}}|\text{ok}\rangle_{\overline{W}}\Big)|\text{fail}\rangle_L \\ &+\frac{1}{\sqrt{6}}\Big(\frac{1}{\sqrt{2}}|\text{fail}\rangle_{\overline{W}} - \frac{1}{\sqrt{2}}|\text{ok}\rangle_{\overline{W}}\Big)|\text{ok}\rangle_L\Big]\end{aligned} \tag{8.20}$$

Equation (8.20) is equivalent to (19) in (Healey 2018).

It follows from (8.20) that the joint probability $q_{\Phi}(\text{ok}_W, t_{\overline{L}})$ is defined after $\overline{W}$'s measurement and equals 1/12. Since $q_{\Phi}(\text{ok}_W, t_{\overline{L}}) + q_{\Phi}(\text{fail}_W, t_{\overline{L}}) = 1/12 + 5/12 = 1/2$,

$$q_{\Phi}(\text{ok}_W | t_{\overline{L}}) = \frac{q_{\Phi}(\text{ok}_W, t_{\overline{L}})}{q_{\Phi}(\text{ok}_W, t_{\overline{L}}) + q_{\Phi}(\text{fail}_W, t_{\overline{L}})} = 1/6 \tag{8.21}$$

The conditional probability $q_\Phi(\text{ok}_W|t_{\overline{L}})$ is derived from the *jointly observable statistics* at the latest time in the unitary evolution, so at the time immediately after W's measurement, following $\overline{W}$'s prior measurement.[7] After $\overline{W}$'s measurement of the observable with eigenvalues 'ok,' 'fail,', the $\overline{L}$-observable with eigenvalues 'heads,' 'tails' is indefinite. The probability $q_\Phi(\text{ok}_W|t_{\overline{L}})$ refers to a situation in which a super-observer $\overline{\overline{W}}$ measures an observable with eigenvalues 'heads' or 'tails' on $\overline{L}$ and notes the fraction of cases where W gets 'ok' to those in which this measurement results in the outcome 'tails.' The 'tails' value in this measurement is randomly related to the 'tails' outcome of F's previous measurement before $\overline{W}$'s intervention, so the probability $q_\Phi(\text{ok}_W|t_{\overline{L}})$, which Baumann and Wolf identify with $q_\Phi(\text{ok}_W|t_{\overline{F}})$, is not relevant to $\overline{F}$'s prediction that W will find 'fail' in the cases where she finds 'tails.' Certainly, given the disruptive nature of $\overline{W}$'s measurement, it is not the case that $\overline{\overline{W}}$ will find 'tails' after $\overline{W}$'s measurement if and only if $\overline{F}$'s measurement resulted in the outcome 'tails' at the earlier time before $\overline{W}$'s measurement.[8]

This is not a critique of Baumann and Wolf, who make no such claim. The purpose of this section is simply to clarify the difference between the conditional probability $q_\Phi(\text{ok}_W|t_{\overline{F}})$, as Baumann and Wolf define it, and $\overline{F}$'s conditional prediction that W will find 'fail' in the cases where she finds 'tails.'

8.5 Concluding Remarks

Frauchiger and Renner present their argument as demonstrating that any interpretation of quantum mechanics that accepts assumptions Q, S, and C is inconsistent, and they point out which assumptions are rejected by specific interpretations. For the relative state interpretation and the many-worlds interpretation, they say that these interpretations conflict with assumption S, because

> any quantum measurement results in a branching into different 'worlds,' in each of which one of the possible measurement outcomes occurs.

But this is not the issue. Rather, Frauchiger and Renner show something much more devastating: that for a particular scenario with encapsulated measurements involving multiple agents, there is a branch of the global quantum state with a contradictory memory entry.

An interpretation of quantum mechanics is a proposal to reformulate quantum mechanics as a Boolean theory in the representational sense, either by introducing hidden variables, or by proposing that every possible outcome occurs in a measurement, or in some other way. An implicit assumption of the Frauchiger-Renner argument is that quantum mechanics is understood as a representational theory,

[7]Thanks to Veronika Baumann for clarifying this.

[8]Thanks to Renato Renner for this observation.

in the minimal sense that observers can be represented as physical systems, with the possibility that observers can observe other observers. What the Frauchiger-Renner argument really shows is that quantum mechanics can't be interpreted as a representational theory at all.

Acknowledgements Thanks to Veronika Baumann, Michael Dascal, Allen Stairs, and Tony Sudbery for critical comments on earlier drafts.

References

Baumann, V., & Wolf, S. (2018). On formalisms and interpretations. *Quantum, 2*, 99–111.

Bell, J. (1987). Quantum mechanics for cosmologists. In J. S. Bell (Ed.), *Speakable and unspeakable in quantum mechanics*. Cambridge: Cambridge University Press.

Bohr, N. (1939). The causality problem in atomic physics. In *New theories in physics* (pp. 11–30). Warsaw: International Institute of Intellectual Cooperation.

Bohr, N. (1949). Discussions with Einstein on epistemological problems in modern physics. In P. A. Schilpp (Ed.), *Albert Einstein: Philosopher-scientist* (The library of living philosophers, vol. 7, pp. 201–241). Evanston: Open Court.

Brown, H. R. (2006). *Physical relativity: Space-time structure from a dynamical perspective*. Oxford: Clarendon Press.

Bub, J. (2016). *Bananaworld: Quantum mechanics for primates*. Oxford: Oxford University Press.

Bub, J., & Pitowsky, I. (2010). Two dogmas about quantum mechanics. In S. Saunders, J. Barrett, A. Kent, & D. Wallace (Eds.), *Many worlds? Everett, quantum theory, and reality* (pp. 431–456). Oxford: Oxford University Press.

Bub, J., & Stairs, A. (2014). Quantum interactions with closed timelike curves and superluminal signaling. *Physical Review A, 89*, 022311.

Frauchiger, D., & Renner, R. (2018). Quantum theory cannot consistently describe the use of itself. *Nature Communications, 9*, article number 3711.

Gleason, A. N. (1957). Measures on the closed subspaces of Hilbert space. *Journal of Mathematics and Mechanics, 6*, 885–893.

Healey, R. (2018). Quantum theory and the limits of objectivity. *Foundations of Physics, 49*, 1568–1589.

Janssen, M. (2009). Drawing the line between kinematics and dynamics in special relativity. *Studies in History and Philosophy of Modern Physics, 40*, 26–52.

Pauli, W. (1954). 'Probability and physics,' In C. P. Enz & K. von Meyenn (Eds.), *Wolfgang Pauli: Writings on physics and philosophy* (p. 46). Berlin: Springer. The article was first published in (1954) *Dialectica, 8*, 112–124.

Petersen, A. (1963). The philosophy of Niels Bohr. *Bulletin of the Atomic Scientists, 19*, 8–14.

Pitowsky, I. (1994). George Boole's 'conditions of possible experience' and the quantum puzzle. *British Journal for the Philosophy of Science, 45*, 95–125. 'These ... may be termed the conditions of possible experience. When satisfied they indicate that the data *may* have, when not satisfied they indicate that the data *cannot* have, resulted from actual observation.' Quoted by Pitowsky from George Boole, 'On the theory of probabilities,' *Philosophical Transactions of the Royal Society of London, 152*, 225–252 (1862). The quotation is from p. 229.

Pitowsky, I. (2004). Macroscopic objects in quantum mechanics: A combinatorial approach. *Physical Review A, 70*, 022103–1–6.

Pitowsky, I. (2007). Quantum mechanics as a theory of probability. In W. Demopoulos & I. Pitowsky (Eds.), *Festschrift in honor of Jeffrey Bub* (Western Ontario Series in Philosophy of Science). New York: Springer.

Renner, R. (2019). Notes on the discussion of 'Quantum theory cannot consistently describe the use of itself.' Unpublished.

von Neumann, J. (1937). Quantum logics: Strict- and probability-logics. A 1937 unfinished manuscript published in A. H. Taub (Ed.), *Collected works of John von Neumann* (vol. 4, pp. 195–197). Oxford/New York: Pergamon Press.

von Neumann, J. (1954). Unsolved problems in mathematics. An address to the international mathematical congress, Amsterdam, 2 Sept 1954. In M. Rédei & M. Stöltzner (Eds.), *John von Neumann and the foundations of quantum physics* (pp. 231–245). Dordrecht: Kluwer Academic Publishers, 2001.

Chapter 9
Physical Computability Theses

B. Jack Copeland and Oron Shagrir

Abstract The Church-Turing thesis asserts that every effectively computable function is Turing computable. On the other hand, the *physical* Church-Turing Thesis (PCTT) concerns the computational power of physical systems, regardless of whether these perform effective computations. We distinguish three variants of PCTT – modest, bold and super-bold – and examine some objections to each. We highlight Itamar Pitowsky's contributions to the formulation of these three variants of PCTT, and discuss his insightful remarks regarding their validity. The distinction between the modest and bold variants was originally advanced by Piccinini (Br J Philos Sci 62:733–769, 2011). The modest variant concerns the behavior of physical computing systems, while the bold variant is about the behavior of physical systems more generally. Both say that this behavior, when formulated in terms of some mathematical function, is Turing computable. We distinguish these two variants from a third – the super-bold variant – concerning decidability questions about the behavior of physical systems. This says, roughly, that every physical aspect of the behavior of physical systems – e.g., stability, periodicity – is decidable (i.e. Turing computable). We then examine some potential challenges to these three variants, drawn from relativity theory, quantum mechanics, and elsewhere. We conclude that all three variants are best viewed as open empirical hypotheses.

Keywords Physical computation · Church-Turing thesis · Decidability in physics · Computation over the reals · Relativistic computation · Computing the halting function · Quantum undecidability

---

B. J. Copeland
Department of Philosophy, University of Canterbury, Christchurch, New Zealand
e-mail: jack.copeland@canterbury.ac.nz

O. Shagrir (✉)
Department of Philosophy, The Hebrew University of Jerusalem, Jerusalem, Israel

M. Hemmo, O. Shenker (eds.), *Quantum, Probability, Logic*,
Jerusalem Studies in Philosophy and History of Science,
https://doi.org/10.1007/978-3-030-34316-3_9

9.1 Introduction

The physical Church-Turing Thesis (PCTT) limits the behavior of physical systems to Turing computability. We will distinguish several versions of PCTT, and will discuss some possible empirical considerations against these. We give special emphasis to Itamar Pitowsky's contributions to the formulation of the physical Church-Turing Thesis, and to his remarks concerning its validity.

The important distinction between 'modest' and 'bold' variants of PCTT was noted by Gualtiero Piccinini (2011). Modest variants concern only computing systems, while bold variants concern the behavior of physical systems without restriction. The literature contains numerous examples of both modest and bold formulations of PCTT; e.g., bold formulations appear in Deutsch (1985) and Wolfram (1985), and modest formulations in Gandy (1980) and Copeland (2000).

We will distinguish the modest and bold variants from a third variant of PCTT, which we term "super-bold". This variant goes beyond the other two in including decidability questions within its scope, saying, roughly, that every physical aspect of the behavior of any physical system – e.g., stability, periodicity – is Turing computable.

Once the distinction between the modest, bold, and super-bold variants is drawn, we will give three different formulations of PCTT: a modest version PCTT-M, a bold version PCTT-B, and a super-bold version PCTT-S. We will then review some potential challenges to these three versions, drawn from relativity theory, quantum mechanics, and elsewhere. We will conclude that all three are to be viewed as open empirical hypotheses.

9.2 Three Physicality Theses: Modest, Bold and Super-Bold

The issue of whether every aspect of the physical world is Turing computable was raised by several authors in the 1960s and 1970s, and the topic rose to prominence in the mid-1980s. In 1985, Stephan Wolfram formulated a thesis that he described as "a physical form of the Church-Turing hypothesis": this says that the universal Turing machine can simulate any physical system (1985: 735, 738). Wolfram put it as follows: "[U]niversal computers are as powerful in their computational capacities as any physically realizable system can be, so that they can simulate any physical system" (Wolfram 1985: 735). In the same year David Deutsch (who laid the foundations of quantum computation) formulated a principle that he also called "the physical version of the Church-Turing principle" (Deutsch 1985: 99). Other formulations were advanced by Earman (1986), Pour-El and Richards (1989) and others.

Pitowsky also formulated a version of CTTP, in his paper "The Physical Church Thesis and Physical Computational Complexity" (Pitowsky 1990), based on his

1987 lecture in the Eighth Jerusalem Philosophical Encounter Workshop.[1] He said: "Wolfram has recently proposed a thesis—'a physical form of the Church-Turing thesis'—which maintains, among other things, that no non-recursive function is physically computable" (1990: 86). The "other things" pertain to computational complexity: Pitowsky interpreted Wolfram as also claiming that the universal Turing machine *efficiently* simulates physical processes, and Pitowsky challenged this further contention (see also Pitowsky 1996, 2002). We will not discuss issues of computational complexity here (but see Copeland and Shagrir 2019 for some relevant discussion of the so-called "Extended Church-Turing Thesis").

Many have confused PCTT with the original Church-Turing Thesis, formulated by Alonzo Church (1936) and Alan Turing (1936); see Copeland (2017) for discussion of misunderstandings of the original thesis. It is now becoming better understood that, by 'computation', both Church and Turing meant a certain *human* activity, numerical computation; in their day, computation was done by rote-workers called "computers", or, more rarely, "computors" (see e.g. Turing 1947: 387, 391). Pitowsky correctly emphasized that PCTT and the original form of the thesis are very different:

> It should be noted that Wolfram's contention has nothing to do with the original Church's thesis. By 'every computable function is recursive,' Church meant that the best analysis of our pre-analytic notion of 'computation' is provided by the precise notion of recursiveness. Indeed one sometimes refers to Church's thesis as 'empirical,' but the meaning of that statement, too, has nothing to do with physics. (Pitowsky 1990: 86)

We will use the term *physical* to refer to systems whose operations are in accord with the actual laws of nature. These include not only actually existing systems, but also idealized physical systems (systems that operate in some idealized conditions), and physically possible systems that do not actually exist, but that *could* exist, or will exist, or did exist, e.g., in the universe's first moments. (Of course, there is no consensus about exactly what counts as an idealized or possible physical system, but this is not our concern here.)

We start by formulating a modest version of PCTT:

> **Modest Physical Church-Turing Thesis (PCTT-M)** *Every function computed by any physical computing system is Turing computable.*

The functions referred to need not necessarily be defined over discrete values (e.g., integers). Many physical systems presumably operate on real-valued magnitudes; and the same is true of physical computers, e.g., analog computers. The nervous system too might have analog computing components. All this requires consideration of computability over the reals. The extension of Turing computability to real-valued domains (or non-denumerable domains more generally) was initiated by Turing, who talked about real computable numbers in his (1936). Definitions of

[1] Some papers from the Workshop were published in 1990, in a special volume of *Iyyun*. The volume also contains papers by Avishai Margalit, Charles Parsons, Warren Goldfarb, William Tait, and Mark Steiner.

real-valued computable functions have been provided by Grzegorczyk (1955, 1957), Lacombe (1955), Mazur (1963), Pour-El (1974), Pour-El and Richards (1989), and others. The definitions are related to one another but are not equivalent. The central idea behind the definitions is that a universal Turing machine can *approximate* (or simulate) the values of a function over the reals, to any degree of accuracy. We describe one of the definitions in Sect. 9.4.

Bold theses, on the other hand, omit the restriction to computing systems: they concern all (finite) physical systems, whether computing systems or not. Piccinini emphasized, correctly, that the bold versions proposed by different writers are often "logically independent of one another", and exhibit "lack of confluence" (2011: 747–748). The following bold thesis is based on the theses put forward independently by Wolfram and Deutsch (Wolfram 1985; Deutsch 1985):

> **Bold Physical Church-Turing Thesis (PCTT-B)** *Every finite physical system can be simulated to any specified degree of accuracy by a universal Turing machine.*

Pitowsky in fact interpreted Wolfram as advancing a modest version of the thesis, namely "that no non-recursive function is physically computable" (1990: 86). However, this is because Pitowsky was treating every physical process as computation; he said, for example: "According to this rather simple picture, the planets in their orbits may be conceived as 'performing computations'" (1990: 84). Under this assumption, according to which all physical processes are computing processes, there is no difference at all between modest and bold versions. Piccinini, on the other hand, sensibly distinguished between computational and non-computational physical processes; he took it that the planets in their orbits do not perform computations. Against the backdrop of this distinction, both Wolfram's formulation and Deutsch's formulation are bold: they concern physical systems in general and not just computing systems.

In a recent paper, we introduced a new, stronger, form of PCTT, the "super-bold" form, here named PCTT-S (Copeland et al. 2018). (The entailments between PCTT-S and PCTT-B and PCTT-M are: PCTT-S entails PCTT-B, and, since PCTT-B entails PCTT-M, PCTT-S also entails PCTT-M.) Unlike bold versions, the super-bold form concerns not only the ability of the universal Turing machine to simulate the behavior of physical systems (to any required degree of precision), but additionally concerns decidability questions about this behavior, questions that go beyond the simulation (or prediction) of behavior. Pitowsky (1996) provided some instructive examples of yes/no questions that reach beyond the simulation or prediction of behavior:

> There are, however, questions about the future that do not involve any specific time but refer to all the future. For example: 'Is the solar system *stable?*', 'Is the motion of a given system, in a known initial state, *periodic?*' These are typical questions asked by physicists and involve (unbounded) quantification over time. Thus, the question of periodicity is: 'Does there exist some T such that for all times t, $x_i(T+t) = x_i(t)$ for $i = 1, 2, \ldots n$? Similarly the question concerning stability is: 'Does there exist some D such that for all times t the maximal distance between the particles does not exceed D?'. (1996: 163)

The physical processes involved in these scenarios – the motion and stability of physical systems – may (so far as we know at present) be Turing computable, in the sense that the motions of the planets may admit of simulation by a Turing machine, to any required degree of accuracy. (Another way to put this is that possibly a Turing machine could be used to predict the locations of the planets at every specific moment.) Yet the answers to certain physical questions about physical systems – e.g., whether (under ideal conditions) the system's motion eventually terminates – may nevertheless be uncomputable. The situation is similar in the case of the universal Turing machine itself: the machine's behavior (consisting of the physical actions of the read/write head) is always Turing computable in the sense under discussion, since the behavior is produced by the Turing machine's program; yet the answers to some yes/no questions about the behavior, such as whether or not the machine halts given certain inputs, are not Turing computable. Undecidable questions also arise concerning the dynamics of cellular automata and many other idealized physical systems.

We express this form of the physical thesis as follows:

Super-Bold Physical Church-Turing Thesis (PCTT-S): *Every physical aspect of the behavior of any physical system is Turing computable (decidable).*

Are these three physical versions of the Church-Turing Thesis true, or even well-evidenced? We discuss the modest version first.

9.3 Challenging the Modest Thesis: Relativistic Computation

There have been several attempts to cook up idealized physical machines able to compute functions that no Turing machine can compute. Perhaps the most interesting of these are "supertask" machines—machines that complete infinitely many computational steps in a finite span of time. Among such machines are accelerating machines (Copeland 1998, Copeland and Shagrir 2011), shrinking machines (Davies 2001), and relativistic machines (Pitowsky 1990; Hogarth 1994; Andréka et al. 2009). Pitowsky proposed a relativistic machine in the 1987 lecture mentioned earlier. At the same time, Istvan Németi also proposed a relativistic machine, which we outline below.

The fundamental idea behind relativistic machines is intriguing: these machines operate in spacetime structures with the property that the entire endless lifetime of one participant is included in the finite chronological past of a second participant—sometimes called "the observer". Thus the first participant could carry out an endless computation, such as calculating each digit of π, in what is a finite timespan from the observer's point of view, say 1 hour. Pitowsky described a setup with extreme acceleration that nevertheless functions in accordance with Special Relativity. His example is of a mathematician "dying to know whether Fermat's conjecture is true or false" (the conjecture was unproved back then). The mathematician takes a trip in a satellite orbiting the earth, while his students (and then their students, and then their students . . .) "examine Fermat's conjecture one case after another, that is, they take quadruples of natural numbers (*x*,*y*,*z*,*n*), with $n \geq 3$, and check on a conventional computer whether $x^n + y^n = z^n$" (Pitowsky 1990: 83).

Pitowsky suggested that similar set-ups could be replicated by spacetime structures in General Relativity (now sometimes called Malament-Hogarth spacetimes). Mark Hogarth (1994) pointed out the non-recursive computational powers of devices operating in these spacetimes. More recently, Etesi and Németi (2002), Hogarth (2004), Welch (2008), Button (2009), and Barrett and Aitken (2010) have further explored the computational powers of such devices, within and beyond the arithmetical hierarchy. In what follows, we describe a relativistic machine *RM* that arguably computes the halting function (we follow Shagrir and Pitowsky (2003)).

RM consists of a pair of communicating Turing machines T_A and T_B: T_A, the observer, is in motion relative to T_B, a universal machine. When the input (*m*,*n*)—asking whether the *m*th Turing machine (in some enumeration of the Turing machines) halts or not, when started on input *n*—enters T_A, T_A first prints 0 (meaning "never halts") in its designated output cell and then transmits (*m*,*n*) to T_B. T_B simulates the computation performed by the *m*th Turing machine when started on input *n* and sends a signal back to T_A if and only if the simulation terminates. If T_A receives a signal from T_B, it deletes the 0 it previously wrote in its output cell and writes 1 there instead (meaning "halts"). After 1 hour, T_A's output cell shows 1 if the *m*th Turing machine halts on input *n* and shows 0 if the *m*th machine does not halt on *n*. Since *RM* is able to do this for any input pair (*m*,*n*), *RM* will deliver any desired value of the halting function. There is further discussion of *RM* in Copeland and Shagrir (2007).

Here we turn to the question of whether *RM* is a counterexample to PCTT-M. This depends on whether *RM* is *physical* and on whether it really *computes* the halting function. First, is *RM* physical? Németi and his colleagues provide the most physically realistic construction, locating machines like *RM* in setups that include huge slowly rotating Kerr-type black holes (Andréka et al. 2009). They emphasize that the computation is physical in the sense that "the principles of quantum mechanics are not violated" and *RM* is "not in conflict with presently accepted scientific principles"; and they suggest that humans might "even build" a relativistic computer "sometime in the future" (Andréka et al. 2009: 501). Naturally, all this is controversial. John Earman and John Norton pointed out that communication between the participants is not trivially achieved, due to extreme blue-shift effects, including the possibility of the signal destroying the receiving participant (Earman and Norton 1993). Subsequently, several potential solutions to this signaling problem have been proposed; see Etesi and Németi (2002), Németi and Dávid (2006) and Andréka et al. (2009: 508–9). An additional potential objection is that infinitary computation requires infinite memory, and so requires infinite computation space (Pitowsky 1990: 84). Another way of putting the objection is that the infinitary computation requires an unbounded amount of matter-energy, which seems to violate the basic principles of quantum gravity (Aaronson 2005)—although Németi and Dávid (2006) offer a proposed solution to this problem. We return to the infinite memory problem in a later section.

Second, does *RM compute* the halting function? The answer depends on what is included under the heading "physical computation". We cannot even summarize here the diverse array of competing accounts of physical computation found in the current literature. But we can say that *RM* computes in the senses of "compute" staked out by several of these accounts: the semantic account (Shagrir 2006; Sprevak 2010), the mechanistic account (Copeland 2000; Miłkowski 2013; Fresco 2014; Piccinini 2015), the causal account (Chalmers 2011), and the BCC (broad conception of computation) account (Copeland 1997). According to all these accounts, *RM* is a counterexample to the modest thesis if *RM* is physical; at the very least, *RM* seems to show that non-Turing physical computation is logically possible. However, if computation is construed as the execution of an algorithm in the classical sense, then *RM* does not compute.

9.4 Challenging the Bold Thesis

PCTT-B says that the behavior of every physical system can be *simulated* (to any required degree of precision) by a Turing machine. Speculation that there may be physical processes whose behavior cannot be calculated by the universal Turing machine stretches back over several decades (see Copeland 2002 for a survey). The focus has been not so much on idealized constructions such as *RM* (which, if physical, is a counterexample to PCTT-B, as well as PCTT-M, since PCTT-B entails PCTT-M); rather, the focus has been on whether the mathematical equations governing the dynamics of physical systems are or are not Turing computable. Early papers by Scarpellini (1963), Komar (1964), and Kreisel (1965, 1967) raised this question. Georg Kreisel stated "There is no evidence that even present day quantum theory is a mechanistic, i.e. recursive theory in the sense that a recursively described system has recursive behavior" (1967: 270). Roger Penrose (1989, 1994) conjectured that some mathematical insights are non-recursive. Assuming that this mathematical thinking is carried out by some physical processes in the brain, the bold thesis must then be false. But Penrose's conjecture is highly controversial.

Another challenge to the bold thesis derives from the postulate of genuine physical randomness (as opposed to quasi-randomness). Church showed in 1940 that any infinite, genuinely random sequence is uncomputable (Church 1940: 134–135). Some have argued that, under certain conditions relating to unboundedness, PCTT-B is false in a universe containing a random element (to use Turing's term from his 1950: 445; see also Turing 1948: 416). A random element is a system that generates random sequences of bits. It is argued that if physical systems include systems capable of producing unboundedly many digits of an infinite random binary sequence, then PCTT-B is false (Copeland 2000, 2004; Calude and Svozil 2008; Calude et al. 2010; Piccinini 2011). One of us, Copeland, also argues that (again under unboundedness conditions) a digital computer using a random element forms a counterexample to PCTT-M (Copeland 2002). However, the latter claim, unlike the corresponding claim concerning PCTT-B, depends crucially upon one's account of computation—Shagrir denies that a digital computer with a random element *computes* the generated uncomputable sequences. In any case, though, it is an open question whether genuine random elements, able to generate unboundedly many digits of random binary sequences, exist in the physical universe, or physically could exist.

A further challenge to the bold thesis was formulated by Piccinini (2011). One of his argument's premises is: "if our physical theories are correct, most transformations of the relevant physical properties are transformations of Turing-uncomputable quantities into one another" (2011: 748). Another premise is: "a transformation of one Turing-uncomputable value into another Turing-uncomputable value is certainly a Turing-uncomputable operation" (2011: 748–749). The observation that in a continuous physical world, not all arithmetical operations are Turing computable is certainly correct (Copeland 1997). Where x and y are uncomputable real numbers, $x + y$ is in general not Turing computable, since the inputs x and y cannot be inscribed on a Turing machine's tape (except in the special case where x and y have been given proper names, e.g., where the halting number is named "τ"—but since there are only countably many proper names, most Turing uncomputable real numbers must remain nameless). If, therefore, the bold thesis is simply that "Any physical process is Turing computable" (Piccinini 2011: 746), then the thesis is indeed false in a continuous universe; as Piccinini argued, a Turing machine can receive at most a denumerable number of different inputs, and so the falsity of the bold thesis results from the cardinality gap between the physical functions, defined over non-denumerable domains, and the Turing computable functions, defined over denumerable domains.

However, this simple argument shows merely that Piccinini's version of the bold thesis is of little interest if the physical world is assumed to be continuous. Our own version of the thesis is sensitive to these considerations and requires only that physical processes be simulable to any specified degree of accuracy. Our version of the thesis is responsive to an account of (Turing machine) computation over the reals according to which—contra Piccinini's second premiss—the transformation of one Turing uncomputable value into another Turing uncomputable value can be a (Turing machine) computable operation. Relative to this account, the real-valued functions of **plus**, **identity**, and **inverse** are computable by Turing machine, even though these functions sometimes map Turing uncomputable inputs to Turing uncomputable outputs: the definitions of real computable functions impose a continuity constraint, enabling the approximation (simulation) of uncomputable arguments and values. There are several (non-equivalent) characterizations of this continuity constraint; for the purpose of illustration, we select the characterization given by Andrzej Grzegorczyk (1955, 1957), and we adapt the exposition of Earman (1986).

We start with numbers:

Definition 1: A sequence of rational numbers $\{x_n\}$ is said to be effectively computable if there exist three Turing computable functions (over N) a,b,c such that $x_n = (-1)^{c(n)} a(n) \div b(n)$.

Definition 2: A real number r is said to be effectively computable if there is an effectively computable sequence of rational numbers that converges effectively to r. ('Converges effectively' means that there is an effectively computable function d over N such that $|r - x_n| < 1 \div 2^m$ whenever $n \geq d(m)$.)

Now to functions:

Definition 3: A function f is an effectively computable function over the reals if:

(i) f is *sequentially computable*, i.e., for each effectively computable sequence $\{r_n\}$ of reals $\{f(r_n)\}$ is also effectively computable;
(ii) f is *effectively uniformly continuous* on rational intervals, i.e., if $\{x_n\}$ is an effective enumeration of the rationals without repetitions then there is a three-place Turing computable function g such that $|f(r) - f(r')| < 1 \div 2^k$ whenever $x_m < r$, $r' < x_n$ and $|r - r'| < 1 \div g(m,n,k)$ for all $r,r' \in R$ and all $m,n,k \in N$.

(If we confine f to a closed and bounded interval with computable end points then the above definition simplifies: no enumeration is necessary and g is only a function of k.)

Importantly, given that **plus** is *effectively uniformly continuous* on rational intervals, **plus** is a computable function, even though **plus** maps Turing uncomputable reals to Turing uncomputable reals.

It is interesting that, by and large, known physical laws give rise to functions over the reals that are computable in the sense just defined. A well-known exception was discovered by Marian Pour-El and Ian Richards (1981), who showed that the wave equation produces non-computable sequences for some computable initial conditions (i.e., for some computable input sequences). In that respect, the wave equation violates condition (i) in the definition of an effectively computable function over the reals (Definition 3). Clearly, this exception forms a potential counterexample to (our version of) the bold thesis.

However, this result of Pour-El and Richards is at the purely mathematical level, not the physical level. In his discussion of Pour-El and Richards, Pitowsky (1990) argued that their result does not refute the bold thesis, for two reasons:

> Firstly, the function f in the initial condition, though a recursive real function, is an extremely complex function. One can hardly expect such an initial condition to arise 'naturally' in any real physical situation. Secondly, we deal with recursive real functions, and in physics we never get beyond a few decimal digits of accuracy anyway. (Pitowsky 1990: 86–87)

Nevertheless, the result does demonstrate that being recursive

> is not a natural physical property. Physical processes do not necessarily preserve it. (Pitowsky 1990: 87)

Even if, in our world, the initial conditions envisaged in the Pour-El & Richards example do not occur, nevertheless these conditions could occur in some other physically possible world in which the wave equation holds, so showing, as Pitowsky said, that physical processes do not necessarily preserve recursiveness.

9.5 Challenging the Super-Bold Thesis

We will turn next to PCTT-S. It might be objected that PCTT-S is immediately false, as may be shown by considering a universal Turing machine implemented as a physical system: many interesting questions about such a system are undecidable. Pitowsky (1996) describes one such construction, due to Moore (1990), in which a universal machine is realized in a moving particle bouncing between parabolic and linear mirrors in a unit square. This system can certainly be simulated by a Turing machine (since it is a Turing machine). But there are nevertheless undecidable questions about its behavior:

> At this stage we can apply Turing's theorem on the undecidability of the halting problem. It says: *There is no algorithm to decide whether a universal Turing machine halts on a given input.* Translating this assertion into physical language, it means that there is no algorithm to predict whether the particle ever enters the subset of the square corresponding to the halting state. This assertion is valid when we know the exact initial conditions with unbounded accuracy, or even with actually infinite accuracy. Therefore, to answer the question: 'Is the particle ever going to reach this region of space?', Laplace's Demon needs computational powers exceeding any algorithm. In other words, he needs to consult an oracle. (Pitowsky 1996: 171)

Pitowsky further noted that many other yes/no questions about the system are computationally undecidable. In fact, it follows from Rice's theorem that "almost every conceivable question about the unbounded future of a universal Turing machine turns out to be computationally undecidable" (Pitowsky 1996: 171; see also Harel and Feldman 1992). Rice's theorem says: any nontrivial property about the language recognized by a Turing machine is undecidable, where a property is nontrivial if at least one Turing machine has the property and at least one does not.

However, the objection that PCTT-S is straight-out false can hardly be sustained. It would by no means be facile to implement a Turing machine, with its infinite tape, in a finite physical object. The assumption that the physical universe is able to supply an infinite memory tape is controversial. If the number of particles in the cosmos is finite, and each cell of the tape requires at least one particle for its implementation, then clearly the assumption of an infinite physical tape must be false. Against this, it might be pointed out that there are a number of well-known constructions for shoehorning an infinite amount of memory into a finite object. For example, a single strip of tape 1 metre long can be used: the first cell occupies the first half-metre, the second cell the next quarter-metre, the third cell the next eighth of a metre, and so on. However, this observation does not help matters. Obviously, this construction can be realized only in a universe whose physics allows for the infinite divisibility of the tape—again by no means an evidently true assumption.

Another way to implement the Turing machine with its infinite tape is in the continuous (or rational) values of a physical magnitude, as described in Moore's system. Assume that each (potential) configuration of the Turing machine is encoded in a (potentially infinite) sequence of 0s and 1s. The goal now is to efficiently realize each sequence in a unique location of the particle in the (finite) unit square. Much like the example of the 1 metre strip of tape above, this realization can be achieved if we use potentially infinitely many different locations (x,y) within a unit square, where x and y are continuous (or rational) values between 0 and 1. This realization, however, requires that the (idealized) mirrors have the ability to bounce the particle, accurately, into potentially infinitely many different positions within the unit square. Moreover, if Laplace's Demon wants to predict the future position of the particle after k rounds, the Demon will have to be able to measure the differences between arbitrarily close positions. As Pitowsky notes: "[T]o predict the particle position with accuracy of 1/2 the demon has to measure initial conditions with accuracy $2^{-(k+1)}$. The ratio of final error to initial error is 2^k, and growing exponentially with time k (measured by the number of rounds)" (1996, 167). This means that implementing a Turing machine in Moore's system requires determining a particle's position with practically infinite precision (an accuracy of $2^{-(k+1)}$ for unbounded k), and it is questionable whether this implementation is physically feasible. At any rate, the claim that this is physically feasible is, again, far from obvious.

To summarize this discussion, PCTT-S hypothesizes in part that the universe is physically unable to supply an infinite amount of memory, since if PCTT-S is true, the resources for constructing a universal computing machine must be unavailable (the other necessary resources, aside from the infinite memory, being physically undemanding). This point helps illuminate the relationships between PCTT-S and PCTT-M. Returning to the discussion of the relativistic machine RM and the infinite memory problem raised in Sect. 9.3, it is clearly the case that, since RM requires infinite memory, PCTT-S rules out RM (and this is to be expected, since PCTT-M rules out RM, and PCTT-S entails PCTT-M). Nevertheless, the falsity of PCTT-S, and the availability of infinite memory, would be insufficient to falsify PCTT-M—a universe that consists of nothing but a universal Turing machine with its infinite tape does not falsify PCTT-M. Thus, counterexamples to PCTT-M must postulate not only the availability of infinite memory but also additional physical principles of some sort, such as gravitational time dilation or unbounded acceleration or unbounded shrinking of components. Relativistic, accelerating and shrinking machines arguably invoke these principles successfully, and, hence, provide counterexamples to PCTT-M.

Moving on to challenges to PCTT-S at the quantum level, there are undecidable questions concerning the behavior of quantum systems. In 1986, Robert Geroch and James Hartle argued that *un*decidable physical theories "should be no more unsettling to physics than has the existence of well-posed problems unsolvable by any algorithm have been to mathematics"; and they suggested such theories may be "forced upon us" in the quantum domain (Geroch and Hartle 1986: 534, 549). Arthur Komar raised "the issue of the macroscopic distinguishability of quantum states" in 1964, claiming there is no effective procedure "for determining whether two arbitrarily given physical states can be superposed to show interference effects" (Komar 1964: 543–544). More recently, Jens Eisert, Markus Müller and Christian Gogolin showed that "the very natural physical problem of determining whether certain outcome sequences cannot occur in repeated quantum measurements is undecidable, even though the same problem for classical measurements is readily decidable" (Eisert et al. 2012: 260501-1). (This is an example of a problem that refers unboundedly to the future, but not to any specific time, as in Pitowsky's examples mentioned earlier.) Eisert, Müller and Gogolin went on to suggest that "a plethora of problems" in quantum many-body physics and quantum computing may be undecidable (2012: 260501-1 – 260501-4).

Dramatically, a 2015 *Nature* article by Toby Cubitt, David Perez-Garcia, and Michael Wolf outlined a proof that "the spectral gap problem is algorithmically undecidable: there cannot exist any algorithm that, given a description of the local interactions, determines whether the resultant model is gapped or gapless" (Cubitt et al. 2015: 207). Cubitt describes this as the "first undecidability result for a major physics problem that people would really try to solve" (in Castelvecchi 2015).

The spectral gap, an important determinant of a material's properties, refers to the energy spectrum immediately above the ground energy level of a quantum many-body system (assuming that a well-defined least energy level of the system exists); the system is said to be gapless if this spectrum is continuous and gapped if there is a well-defined next least energy level. The spectral gap problem for a quantum many-body system is the problem of determining whether the system is gapped or gapless, given the finite matrices describing the local interactions of the system.

In their proof, Cubitt et al. encode the halting problem in the spectral gap problem, so showing that the latter is at least as hard as the former. The proof involves an infinite family of 2-dimensional lattices of atoms; but they point out that their result also applies to finite systems whose size increases: "Not only can the lattice size at which the system switches from gapless to gapped be arbitrarily large, the threshold at which this transition occurs is uncomputable" (Cubitt et al. 2015: 210–211). Their proof offers an interesting countermodel to the super-bold thesis. The countermodel involves a physically relevant example of a finite system, of increasing size, that lacks a Turing computable procedure for extrapolating future behavior from (complete descriptions of) its current and past states.

It is debatable whether any of these quantum models matches the real quantum world. Cubitt et al. admit that the model used in their proof is highly artificial, saying "Whether the results can be extended to more natural models is yet to be determined" (Cubitt et al. 2015: 211). There is also the question of whether the spectral gap problem could become computable when only local Hilbert spaces of realistically low dimensionality are considered. Nevertheless, these results are certainly suggestive. The super-bold thesis cannot be taken for granted—even in a finite quantum universe.

9.6 Conclusion

We have distinguished three theses about physical computability, and have discussed some empirical evidence that might challenge these. Concerning the boldest versions, PCTT-B and PCTT-S, both are false if the physical universe permits infinite memory and genuine randomness. Even assuming that the physical universe is deterministic, the most that can be said for PCTT-B and PCTT-S is that, to date, there seems to be no decisive empirical evidence against them. PCTT-B and PCTT-S are both thoroughly empirical theses; but matters are more complex in the case of the modest thesis PCTT-M, since a conceptual issue also bears on the truth or falsity of this thesis (even in a universe containing genuine randomness)—namely, the difficult issue of what counts as physical computation. Our conclusion is that, at the present stage of physical enquiry, it is unknown whether any of the theses is true.

Acknowledgements An early version of this chapter was presented at the *Workshop on Physics and Computation* (UCNC2018) in Fontainebleau, at the *International Association for Computing and Philosophy meeting (*IACAP2018*)* in Warsaw, and at the research logic seminar in Tel Aviv University. Thanks to the audiences for helpful discussion. We are also grateful to Piccinini for his comments on a draft of this paper. Shagrir's research was supported by the Israel Science Foundation grant 830/18.

References

Aaronson, S. (2005). Guest column: NP-complete problems and physical reality. *ACM SIGACT News, 36*(1), 30–52.

Andréka, H., Németi, I., & Németi, P. (2009). General relativistic hypercomputing and foundation of mathematics. *Natural Computing, 8*, 499–516.

Barrett, J. A., & Aitken, W. (2010). A note on the physical possibility of ordinal computation. *The British Journal for the Philosophy of Science, 61*, 867–874.

Button, T. (2009). SAD computers and two versions of the Church-Turing thesis. *The British Journal for the Philosophy of Science, 60*, 765–792.

Calude, C. S., Dinneen, M. J., Dumitrescu, M., & Svozil, K. (2010). Experimental evidence of quantum randomness incomputability. *Physical Review A, 82*: 022102–1-022102–8.

Calude, C. S., & Svozil, K. (2008). Quantum randomness and value indefiniteness. *Advanced Science Letters, 1*, 165–168.

Castelvecchi, D. (2015). Paradox at the heart of mathematics makes physics problem unanswerable. *Nature, 528*, 207.

Chalmers, D. J. (2011). A computational foundation for the study of cognition. *Journal of Cognitive Science, 12*, 323–357.

Church, A. (1936). An unsolvable problem of elementary number theory. *American Journal of Mathematics, 58*, 345–363.

Church, A. (1940). On the concept of a random sequence. *American Mathematical Society Bulletin, 46*, 130–135.

Copeland, B. J. (1997). The broad conception of computation. *American Behavioral Scientist, 40*, 690–716.

Copeland, B. J. (1998). Even Turing machines can compute uncomputable functions. In C. Calude, J. Casti, & M. Dinneen (Eds.), *Unconventional models of computation*. New York: Springer.

Copeland, B. J. (2000). Narrow versus wide mechanism. *Journal of Philosophy 96*, 5–32. Reprinted in M. Scheutz (Ed.) (2002). *Computationalism*. Cambridge, MA: MIT Press.

Copeland, B. J. (2002). Hypercomputation. *Minds and Machines, 12*, 461–502.

Copeland, B. J. (2004). Hypercomputation: Philosophical issues. *Theoretical Computer Science, 317*, 251–267.

Copeland, B. J. (2017). The Church-Turing thesis. In E. N. Zalta (Ed.), *The Stanford encyclopedia of philosophy*. https://plato.stanford.edu/archives/win2017/entries/church-turing/

Copeland, B. J., & Shagrir, O. (2007). Physical computation: How general are Gandy's principles for mechanisms? *Minds and Machines, 17*, 217–231.

Copeland, B. J., & Shagrir, O. (2011). Do accelerating Turing machines compute the uncomputable? *Minds and Machines, 21*, 221–239.

Copeland, B. J., & Shagrir, O. (2019). The Church–Turing thesis—Logical limit, or breachable barrier? *Communications of the ACM, 62*, 66–74.

Copeland, B. J., Shagrir, O., & Sprevak, M. (2018). Zuse's thesis, Gandy's thesis, and Penrose's thesis. In M. Cuffaro & S. Fletcher (Eds.), *Physical perspectives on computation, computational perspectives on physics* (pp. 39–59). Cambridge: Cambridge University Press.

Cubitt, T. S., Perez-Garcia, D., & Wolf, M. M. (2015). Undecidability of the spectral gap. *Nature, 528*(7581), 207–211.

Davies, B. E. (2001). Building infinite machines. *The British Journal for the Philosophy of Science, 52*, 671–682.

Deutsch, D. (1985). Quantum theory, the Church-Turing principle and the universal quantum computer. *Proceedings of the Royal Society of London A: Mathematical, Physical and Engineering Sciences, 400*(1818), 97–117.

Earman, J. (1986). *A primer on determinism*. Dordrecht: Reidel.

Earman, J., & Norton, J. D. (1993). Forever is a day: Supertasks in Pitowsky and Malament-Hogarth spacetimes. *Philosophy of Science, 60*, 22–42.

Eisert, J., Müller, M. P., & Gogolin, C. (2012). Quantum measurement occurrence is undecidable. *Physical Review Letters, 108*(26), 260501.

Etesi, G., & Németi, I. (2002). Non-Turing computations via Malament-Hogarth space-times. *International Journal of Theoretical Physics, 41*, 341–370.

Fresco, N. (2014). *Physical computation and cognitive science*. Berlin: Springer.

Gandy, R. O. (1980). Church's thesis and principles of mechanisms. In S. C. Kleene, J. Barwise, H. J. Keisler, & K. Kunen (Eds.), *The Kleene symposium*. Amsterdam: North-Holland.

Geroch, R., & Hartle, J. B. (1986). Computability and physical theories. *Foundations of Physics, 16*, 533–550.

Grzegorczyk, A. (1955). Computable functionals. *Fundamenta Mathematicae, 42*, 168–203.

Grzegorczyk, A. (1957). On the definitions of computable real continuous functions. *Fundamenta Mathematicae, 44*, 61–71.

Harel, D., & Feldman, Y. A. (1992). *Algorithmics: The spirit of computing* (3rd ed.). Harlow: Addison-Wesley.

Hogarth, M. (1994). Non-Turing computers and non-Turing computability. *Proceedings of the Biennial Meeting of the Philosophy of Science Association, 1*, 126–138.

Hogarth, M. (2004). Deciding arithmetic using SAD computers. *The British Journal for the Philosophy of Science, 55*, 681–691.

Komar, A. (1964). Undecidability of macroscopically distinguishable states in quantum field theory. *Physical Review*, second series, 133B, 542–544.

Kreisel, G. (1965). Mathematical logic. In T. L. Saaty (Ed.), *Lectures on modern mathematics* (Vol. 3). New York: Wiley.

Kreisel, G. (1967). Mathematical logic: What has it done for the philosophy of mathematics? In R. Schoenman (Ed.), *Bertrand Russell: Philosopher of the century*. London: Allen and Unwin.

Lacombe, D. (1955). Extension de la notion de fonction récursive aux fonctions d'une ou plusieurs variables réelles III. *Comptes Rendus Académie des Sciences Paris, 241*, 151–153.

Mazur, S. (1963). *Computational analysis*. Warsaw: Razprawy Matematyczne.

Miłkowski, M. (2013). *Explaining the computational mind*. Cambridge: MIT Press.

Moore, C. (1990). Unpredictability and undecidability in dynamical systems. *Physical Review Letters, 64*, 2354–2357.

Németi, I., & Dávid, G. (2006). Relativistic computers and the Turing barrier. *Journal of Applied Mathematics and Computation, 178*, 118–142.

Penrose, R. (1989). *The emperor's new mind: Concerning computers, minds and the laws of physics*. Oxford: Oxford University Press.
Penrose, R. (1994). *Shadows of the mind*. Oxford: Oxford University Press.
Piccinini, G. (2011). The physical Church-Turing thesis: Modest or bold? *The British Journal for the Philosophy of Science, 62*, 733–769.
Piccinini, G. (2015). *Physical computation: A mechanistic account*. Oxford: Oxford University Press.
Pitowsky, I. (1990). The physical Church thesis and physical computational complexity. *Iyyun, 39*, 81–99.
Pitowsky, I. (1996). Laplace's demon consults an oracle: The computational complexity of prediction. *Studies in History and Philosophy of Science Part B, 27*(2), 161–180.
Pitowsky, I. (2002). Quantum speed-up of computations. *Philosophy of Science, 69*, S168–S177.
Pour-El, M. B. (1974). Abstract computability and its relation to the general purpose analog computer (some connections between logic, differential equations and analog computers). *Transactions of the American Mathematical Society, 199*, 1–28.
Pour-El, M. B., & Richards, I. J. (1981). The wave equation with computable initial data such that its unique solution is not computable. *Advances in Mathematics, 39*, 215–239.
Pour-El, M. B., & Richards, I. J. (1989). *Computability in analysis and physics*. Berlin: Springer.
Scarpellini, B. (1963). Zwei unentscheidbare Probleme der Analysis. *Zeitschrift für mathematische Logik und Grundlagen der Mathematik, 9*, 265–289. English translation in *Minds and Machines* 12 (2002): 461–502.
Shagrir, O. (2006). Why we view the brain as a computer. *Synthese, 153*, 393–416.
Shagrir, O., & Pitowsky, I. (2003). Physical hypercomputation and the Church–Turing thesis. *Minds and Machines, 13*, 87–101.
Sprevak, M. (2010). Computation, individuation, and the received view on representation. *Studies in History and Philosophy of Science, 41*, 260–270.
Turing, A. M. (1936). On computable numbers, with an application to the Entscheidungsproblem. *Proceedings of the London Mathematical Society, Series, 2*(42), 230–265.
Turing, A. M. (1947). Lecture on the Automatic Computing Engine. In Copeland, B. J. (2004). *The essential Turing: Seminal writings in computing, logic, philosophy, artificial intelligence, and artificial life, plus the secrets of Enigma*. Oxford University Press.
Turing, A. M. (1948). Intelligent machinery. In *The essential Turing*. Oxford University Press.
Turing, A.M. (1950). Computing machinery and intelligence. In *The essential Turing*. Oxford University Press.
Welch, P. D. (2008). The extent of computation in Malament-Hogarth spacetimes. *The British Journal for the Philosophy of Science, 59*, 659–674.
Wolfram, S. (1985). Undecidability and intractability in theoretical physics. *Physical Review Letters, 54*, 735–738.

Chapter 10
Agents in Healey's Pragmatist Quantum Theory: A Comparison with Pitowsky's Approach to Quantum Mechanics

Mauro Dorato

Abstract In a series of related papers and in a book (Healey R. The quantum revolution in philosophy. Oxford University Press, Oxford, 2017b), Richard Healey has proposed and articulated a new pragmatist approach to quantum theory that I will refer to as PQT. After briefly reviewing PQT by putting it into a more general philosophical context, I will discuss Healey's self-proclaimed quantum realism by stressing that his agent-centered approach to quantum mechanics makes his position rather close to a sophisticated form of instrumentalist, despite his claims to the contrary. My discussion of the possible sources of incompatibility between PQT and the view that agents can be regarded as physical systems (agent physicalism) will allow me to compare Healey's view of quantum theory with Pitowsky's. In particular, by focusing on the measurement problem, I will discuss the role of observers and the notion of information in their respective philosophical approach.

Keywords Agents · Healey's pragmatist quantum theory · Physicalism · Pitowski

10.1 Introduction

In a series of related papers (2012a, b, 2015, Healey 2017c, 2018a, 2018b) and in a book (2017b), Richard Healey has proposed and articulated a new pragmatist approach to quantum theory (PQT henceforth) that so far has not received the attention it deserves. And yet, Healey's approach – besides trying to clarify and solve (or dissolve) well-known problems in the foundations of quantum mechanics from a technical point of view – has relevant consequences also for philosophy in general, in particular in areas like scientific realism, metaphysics, the philosophy of

M. Dorato (✉)
Department of Philosophy Communication and Performing Arts, Università degli Studi Roma 3, Rome, Italy
e-mail: mauro.dorato@uniroma3.it

M. Hemmo, O. Shenker (eds.), *Quantum, Probability, Logic*, Jerusalem Studies in Philosophy and History of Science,
https://doi.org/10.1007/978-3-030-34316-3_10

probability, and the philosophy of language. Given that quantum theory (even in its non-relativistic form) is one of two pillars of contemporary physics,[1] any attempt to offer a coherent and general outlook on the physical world that, like his, is not superficial or merely suggestive, is certainly welcome.

The main aim of this paper is to show that, as a consequence of this ambition, PQT raises important questions related to its compatibility with a narrowly defined form of physicalism that enables us to clearly specify in what sense *agents can be regarded as quantum systems*. It is important to note from the start that this problem is related to the question whether quantum theory can be applied to all physical systems – *and therefore also to agents who use it* (see Frauchiger and Renner 2018; Healey 2018b) – to the extent that this applicability is a necessary condition to consider them as physical systems. In what follows, I will not discuss no-go arguments of this kind, since in the relationist context that Healey at least initially advocated and that will be the subject of my discussion, he himself grants that agents *can* be regarded as quantum systems by other agents.[2]

In this perspective, let me refer to this particular form of physicalism as "agent physicalism", in order to distinguish it from other approaches to physicalism that Healey does not explicitly reject – among these, *supervenience physicalism* regarded as the aim of physics[3]- and that therefore will *not* be the target of my paper. In fact, Healey grants that a particular form of physicalism – according to which our most fundamental theory provides a complete ontology on which everything else must supervene – *is* incompatible with PQT. From the outset, he claims that quantum theory "lacking beables of its own, has no physical ontology and states no fact about physical objects and events", i.e. is not a fundamental *description* of the physical world, but it rather amounts to a series of effective *prescriptions* yielding correct attributions of quantum states to physical systems (Healey 2017b, pp. 11–12). PQT's normative approach to quantum theory is evident from the terminology that in his papers abound ("prescribes", "provides good advice", etc.) and that presupposes rules or norms that rational agents should follow). PQT postulates in fact non-representational relations between physically situated agents' epistemic states (inter-subjectively shared credences) on the one hand, and mind-independent probabilistic correlations instantiated by physical systems on the other. As we will

---

[1]The other being, of course, general relativity. In this paper, I will neglect any considerations emerging from quantum field theory to which, in the intention of his proposer, PQT certainly applies. This fact is also clear from the examples that he discusses.

[2]There are various forms of physicalism in the literature and here I am not interested in discussing them all.

[3]"Quine (1981, p. 98) offered this argument for a nonreductive physicalism: "Nothing happens in the world, not the flutter of an eyelid, not the flicker of a thought, without some redistribution of microphysical states. . . If the physicist suspected there was any event that did not consist in a redistribution of the elementary states allowed for by his physical theory, he would seek a way of supplementing his theory" (Healey 2017a, 235). Even though Healey never rejected supervenience physicalism, he adds that: "If anything, the quantum revolution set back progress toward Quine's goal" (Healey 2017a, b, p. 236).

see, one of the main reasons of tension between agent physicalism and Healey's pragmatist approach to quantum mechanics is that, lacking a precise characterization of agents that Healey does not provide, we don't know what PQT is about. To the extent that agent physicalism requires an exact notion of agent, PQT is incompatible with it and quantum theory cannot give a quantum description of an agent. In this sense, this particular form of physicalism is therefore incompatible with PQT.[4]

I will begin by briefly reviewing PQT by putting it into a more general philosophical frame. In the second section I will discuss Healey's self-proclaimed quantum realism by arguing that, aside from an implicit endorsement of a contextual entity realism,[5] his position is really instrumentalist despite his claims to the contrary Healey 2018a. This conclusion will be important to clarify in more details the ontological status of Healey's agents and the related issue of agent physicalism. In the third section, I will discuss the problem of providing some criterion to distinguish between agents that are quantum systems from agents that are not and cannot be (like electrons or quarks). As we will argue later, the fact that Healey's characterization of agents plausibly relies on the irreducibly epistemic notion of information is a first argument in favor of the incompatibility between agent physicalism and PQT.[6] This argument will allow me to compare Healey view of quantum theory with Pitowsky's approach to the measurement problem, by discussing in particular the role of observers and the notion of information in their philosophy.

Before beginning, let me put forth two points to be kept in mind in what follows. First, in order to be as faithful as possible to Healey's claims and avoid possible misunderstandings on my part, it will be necessary to resort to a few quotations from his texts. The second point is terminological but important nevertheless: "realism about the wave function Ψ" does not entail a commitment to the existence of a complex-valued function defined on an abstract mathematical space but entails the view that Ψ denotes a real physical state that, following many authors and Healey himself, I will call "*the quantum state*".

10.2 A Brief Review of PQT[7]

It is difficult to summarize an approach to quantum theory as complex as Healey's in a few words, also because it has been presented in various papers from different angles. Here I will just touch on three points that are closely connected to the main topic of the paper and that are common to all the published versions of PQT, namely

[4]The question is whether the failure of agent physicalism also entails the failure of supervenience physicalism will not be an object of this paper.

[5]In PQT, also the reality of quantum particles is a contextual matter.

[6]Of course, agents can be characterized in terms of supervenience physicalism.

[7]See also Glick (2018).

(P1) Healey's endorsement of Brandom's inferentialist theory of meaning,
(P2) His view of the task of the philosopher of quantum theory, and its problematic reliance on non-quantum magnitude claims (Healey 2012a, p. 740, 2015, p. 11) or "canonical magnitude claims" as they are later defined (Healey 2017a, b, 80) and finally,
(P3), the relational nature of quantum states ascriptions, due to the different perspectives of "physically situated agents" using the Born rule, an expression often repeated in his texts but that needs some clarification.

P1. In all his presentations, Healey explicitly refers to Brandom's inferentialist theory of concepts (Brandom 2000) as a general philosophical background against which PQT is constructed. There are two aspects that in our context are worth stressing.

P1.1 The first is that in Brandom's view, the meaning of any statement (including those appearing in the formulation of scientific theories) is given by its *use*, and in particular by the inferences that it entitles to draw, and not by its representational capacity. As we will see however – and unlike Suárez more general inferentialist, non-representationalist approach (2004) – Healey applies his anti-representationalism explicitly only to mathematical models employed by quantum theory, without raising the question whether this claim can be extended also to classical physics. He seems to be committed to the view that *all* physical models are non-representational: otherwise he would have to explain how to demarcate classical mathematical models from quantum mathematical models and what it is about the former that make it different from the latter.

P1.2 The second aspect is that, beyond Brandom's philosophical influence,[8] the reasons to embrace anti-representationalism are internal to the philosophy of quantum mechanics. Therefore, if successful, PQT gives an independent and important support to the general pragmatist agenda proposed by Brandom.[9]

It is for this reason that PQT might have decisive consequences not just for the philosophy of quantum mechanics regarded as a fundamental theory of the physical world, but also for general epistemology and metaphysics. As far as the former is concerned, PQT invites us to abandon a spectator theory of knowledge, in which knowing essentially means representing and not intervening in the physical world (Hacking 1983). Considering that our brain evolved as a means to interact successfully with the natural world, "agents" and their "knowing how" are to be regarded as essential components of any kind of naturalistic metaphysics aiming at understanding the place of human beings in the vast scheme of things.

P2. The second point follows from the first: all physicists know how to *use* the theory, but there is no agreement among the different philosophical schools working on foundations of quantum theory. Philosophers of quantum mechanics

[8]Healey's former student Price (2011) has certainly played a role too.

[9]A recent, rather original approach to quantum theory that for various aspects is similar to Healey's but more inspired by Wittgenstein is Friederich's (2015).

have typically assumed that their main task is to answer the question: "what can the world be like if quantum theory is true?". Thirty years ago, Healey himself adopted this viewpoint, since the words in scare quotes are taken from his older book on quantum mechanics (Healey 1989, p. 6). As he sees it now, however, the problem with the above question is that it is responsible for a long-standing dissent in the philosophy of quantum mechanics, which should be put to an end. The question of interpreting the formalism of the theory in a realistic framework typically presupposes the so-called "beables", whose postulation is needed as an attempt to answer what I take to be an inescapable question: "what is quantum theory *about*?" For scientific realist, a refusal to answer this question, is equivalent to betraying the aim of physics.[10]

However, according to Healey, in the philosophy of quantum mechanics this is the wrong question to ask, since it leads to the formulation of theories (Bohm's, GRW's, and many worlds') underdetermined by their key parameters and by the evidence.[11] For instance, a instrumentalistic view of the quantum state and Bohm's realism about the latter state in principle cannot make any predictive difference. This point has been repeatedly insisted upon also by Pitowsky: "I think that Weinberg's point and also Bohm's theory are justified only to the extent that they yield *new discoveries* in physics (as Weinberg certainly hoped). So far they haven't" (Pitowsky 2005, p. 4, my emphasis).[12]

The new task that Healey assigns to PQT is not to explain the success of the Born rules by appealing to some constructive theory in the sense of Einstein (1919), but rather to explain how quantum theory can be so empirically successful even refusing to grant both representational power to its models and "reality" to the quantum state and the Born's rule, in a sense of "reality" to be carefully specified below (Healey 2018a).[13] Somewhat surprisingly, he claims that his pragmatist approach, somehow strongly suggested by the theory itself, constitutes the *new revolutionary* aspect of quantum mechanics (2012a, b, 2017b, p. 121). This claim should strike the reader as particularly controversial for at least two reasons. Firstly, as this section illustrates, a pragmatistic philosophy is at least in part *presupposed* by Healey and, like any philosophical approach to QT, cannot be read just from the data and the theory. Secondly, despite some differences that basically involve the idea of complementarity, Healey's PQT rather than being so new, is rather *close* to Bohr's philosophy of quantum mechanics, especially if the latter is purified from myths surrounding the so-called "Copenhagen interpretation" (Howard 2004) and

---

[10]"What is the theory is about" presupposes simpleminded answers like: classical mechanics is about particles acted upon by forces, and classical electromagnetism is about fields generated by charged particles.

[11]For this position, see also Bub (2005).

[12]Pitowsky is referring to Weinberg's attempt to reconstruct gravity on a flat spacetime (ibid.).

[13]In particular when discussing a crucial but obscure passage in which he claims that "a quantum states is not an element of reality (it is not a beable") since "the wave function does not represent a physical system or its properties. But is real nonetheless" Healey (2017c, p. 172).

read more charitably in light of more careful historical reconstructions (Faye and Folse 2017).[14]

In order to convince ourselves of the proximity of PQT and Bohr's philosophy, consider the first, following point. It is true that Healey's empiricist understanding of quantum theory, unlike Bohr's, presupposes decoherence, but Howard (1994) and Camilleri and Schlosshauer (2008, 2015) have put forward the claim that this notion was in some sense already implicit in Bohr's philosophy of quantum mechanics, despite the lack at his time of the relevant physical knowledge. Secondly, even if this somewhat ahistorical claim were rejected, one must consider PQT's treatment of the classical-quantum separation. Without a clear criterion that separates the quantum from the classical regime, the set of magnitudes that Healey refers to as "non-quantum" can be typically regarded as those described by classical physics. Healey, however, explicitly *denies* that "non-quantum" refers to "classical" since, he claims, it also refers to the property of any ordinary physical object". However, the term '*ordinary object'* is rather vague, unless it is intended to refer to objects with which we interact in our ordinary life and in our ordinary environment. Consequently, the reports of experimental results *via* what he calls NQMC (non-quantum magnitudes claims)[15] involve objects that can be described only by classical physics. This claim resembles Bohr's appeal to classical physics as the condition of possibility to describe the quantum world: "indeed, if quantum theory itself provides no resources for describing or representing physical reality, then all present and future ways of describing or representing it will be non-quantum" (Healey 2012a, p. 740).

In addition, there are two close analogies between PQT and Bohr's empiricist philosophy of quantum mechanics. The first involves Bohr's *pragmatism*, whose importance for his philosophy has been supported by highly plausible historical evidence (see among others Murdoch 1987 and Folse 1985). Furthermore, in force of his holistic view of quantum mechanics, Bohr – analogously to Healey, – would *not* deny that there is (a weak) sense in which a quantum state refers to or describes genuinely physical, mind-independent probabilistic correlations between measurement apparata and quantum systems. If Healey denied *any* representation force to the wave function, he would be in the difficult position to justify the claim that *only* classical models, unlike quantum models, represent or describe physical entities. Otherwise, whence this epistemic separation if one cannot make recourse to an ontological distinction (remember that in PQT there are no beables)?

These remarks, which are put forward to show the persisting though often unrecognized influence of Bohr's philosophy of quantum mechanics,[16] are not

---

[14]For a neo Copenhagen interpretation that here I cannot discuss, see Brukner (2017).

[15]NQMCs (or canonical magnitude claims) are of the form "the value of dynamical variable M on physical system S lies in a certain interval Δ".

[16]After the 1935 confrontation with Einstein, physicists attributed to Bohr a definite victory over Einstein's criticism. But since the late 1960s surge of interest in the foundations of physics caused by Bell's theorem and his sympathy for alternative formulations of quantum mechanics, Bohr has become to be regarded as responsible – and not just by philosophers – for having "brainwashed a whole generation of physicists into thinking that the job was done 50 years ago" (Gell-Mann

intended to deny that PQT diverges from the latter and that therefore significantly enriches the contemporary debate in the philosophy of quantum mechanics.

P3. The last notable aspect, the relationality of quantum systems, touches more on the metaphysical aspects of PQT, and if even if it was more stressed in the earlier version of the formulation of PQT, is the most relevant for my purposes and in any case interesting for its own sake. In order to summarize its main thrust, it will be helpful to compare it with Rovelli's relational quantum mechanics (Rovelli 1996; Rovelli and Smerlak 2007; Laudisa and Rovelli 2008; Dorato 2016). Formally, PQT does not accept the eigenstate-eigenvalue link. Philosophically, the difference involves Healey's stress on a pragmatist aspect. According to Rovelli's view, any interaction of a system with another system (not necessarily an agent) yields definite results (events). In PQT ascriptions of quantum states are not relative "to any other system, but only to physically instantiated agents, *whether conscious or not*" (Healey 2012a, p. 750, emphasis added). Healey is aware that his relationism may conflict with agent physicalism but claims to have a way out: "Since agents are themselves in the physical world, in principle one agent may apply a model to another agent regarded as part of the physical world. This second agent then becomes the target of the first, who wields the model. I say "in principle" since human agents are so complex and so "mixed up" with their physical environment that to effectively model a human (or a cat!) in quantum theory is totally impracticable" (Healey 2017b, p. 10). In short, his essential proposal, to be discussed in what follows, is that "all agents are quantum systems with respect to other users, but not all quantum systems are agents" (Healey 2012a, b, p. 750).[17] Except for the relational aspect, this statement is analogous to the claim that every mental event is a physical event, but not all physical events are mental. Independently or the relational aspect of the claim, the quotation raises another crucial question to be discussed below: how do we distinguish between quantum systems that are agents and those that are non-agents? Notice that this question is essential to any philosophical view that stresses the role of agents in *using* the quantum formalism.

10.3 Is PQT Really Realist?

I conclude my brief sketch of PQT by assessing its implications for the issue of scientific realism which is importanto also for (Healey 2018a). It might be objected that this is a desperate or useless enterprise, for at least two reasons. Firstly,

1976, p. 29). It is desirable to have a more balanced view of his contribution to the philosophy of quantum mechanics.

[17]While here I will focus on this relational version of PQT as expressed in the 2017's quotation, but I will show why the same questions would be relevant also to later non-relational versions of PQT.

each philosopher has her brand of realism and following Fine (1984) and others, we might legitimately have the temptation to drop the whole issue in favor of a natural ontological attitude. Secondly, it is not easy to locate Healey's position in the continuous spectrum that goes from full-blown scientific realism to radical instrumentalism, since he is very careful to stir away from both positions. But for my purpose however, attempts to solve this problem is important to find out whether some kind of scientific realism about quantum mechanics is necessary for agent physicalism.

Here we should distinguish between entity realism in Hacking's "interventionist sense" (1983) and any other sort of scientific realism. As far as the former is concerned, there is no doubt about the fact that Healey *is* a scientific realist: PQT describes the unobservable because entities (like quarks or muons) that agents apply quantum theory to exist independently of them. However, the following six points ought to convince the reader that, despite Healey's claim to the contrary (2017a, p. 253 (Healey 2018a)), PQT is a typical instrumentalist position in any other possible sense of "scientific realism".

1. Healey claims that his PQT is not outright instrumentalist because of the *normative* role of the quantum state, which is justified by the fact that its use yields *correct*, empirically confirmed, mind-independent probabilistic predictions. His "realism" about quantum mechanics could at most imply that the quantum state can be taken to "refer" or "represent" *mind-independent, statistical correlations*. However, statistical correlations about what? Since Healey does not explicitly raise this crucial question, we may want to consider two possible interpretations. The first is to read these correlations as holding between *observations* relating what he calls "backing conditions" (inputs) via "advice conditions" to outputs or predictions that must be formulated in the language of *classical physics* via NQMC (see note 16). The idea that a theory connects mind-independent empirical correlations between observations and predictions is one of the features of instrumentalism, so that this first reading leads to antirealism. In referring to those conditions, the second reading omits any explicit reference to observations. However, a distinctive feature of PQT is that these empirical correlations are *not* to be explained bottom-up by postulating some beables. On the contrary, it is plausible to attribute Healey the view that they ought to be regarded as universally valid, *empirical regularities* like those that one encounters in thermodynamics or, more generally, in the so-called "theories of principle" (Einstein 1919). The idea that quantum mechanics ought to be regarded as a theory of principle in this sense expresses an *empiricist* philosophical attitude that is shared also by Bub and Pitowsky (2010). This attitude, however, is not a mark of a realist approach to physical theories, at least to the extent that such an approach tries to build *constructive* theories to explain those empirical regularities.

2. As in all pragmatist theories of truth, there is no attempt to explain the effectiveness of the Born rule by the claim that quantum theory is approximately true, since in the framework of a pragmatistic theory there is nothing to explain! Antirealist philosophers are suspicious of explanations of effectiveness, since if the quantum algorithm were not effective, it would not be empirically adequate (a typical refrain of van Fraassen). To the extent that this effectiveness is a brute

fact, truth as correspondence is either a superfluous honorific title or a purely metaphorical reference to a "mirroring" of models. In this respect, Healey follows Price in defending minimalist theories of truth (see Price 2011). However, the legitimate refusal to endorse *an inference to the best explanation* (van Fraassen 1985, p. 252) based on a more robust representational power of quantum models is a mark of empiricist/antirealist philosophies of science, whether quantum theory is involved or not (see Psillos 1999). If one grants that PQT explains in the sense of Healey (2015), as it does, "more realistic" approaches to quantum theory become useless or redundant.

3. The third argument in favor of the claim that PQT is deep down instrumentalistic and involves Healey's partially spelled out treatment of *quantum probabilities*. In order to make it more explicit, it is first useful to distinguish between the *inter-subjective validity* (objectivity$_1$) and *mind independence* (objectivity$_2$) of probability claims, and then note that Healey explicitly *rejects* Qbism's subjective views of probability. In this sense he is distant from Pitowsky's subjectivist account of probability (Bub and Pitowsky 2010). Independently of what probability and information correspond to in the physical world, Healey holds that by using the *information* (*nota bene*, an epistemic term) contained in the quantum state, any physically situated agents would assign the *same* probability to any measurement outcome. Given the reliance on information, it necessarily follows that in PQT probabilities are to be regarded as inter-subjectively valid (i.e. objective$_1$) but still *epistemic*. The fact that "quantum states are objective *informational* bridges" (Healey 2017a) reinforces my point: I know of no approach to 'information' that accounts for it in mind-independent terms:

However, there is a sense in which according to Healey these probabilities are *also* mind-independent (objective$_2$): "there are *real patterns* of statistical correlation in the physical world. Correctly assigned quantum states *reliably track* these through legitimate applications of the Born rule" (2017c, p. 172, my emphasis) and the Born rules "offer a correct or "good advice" to agents whose aim is predicting, explaining and controlling natural phenomena" (Healey 2017b, p. 102). In Healey's philosophy of quantum probability, however, there is an oscillation between these two senses of "objective", which are never clearly distinguished. We just read that, according to Healey, the Born rules do capture objective$_2$, mind-independent statistical correlations, a claim that partially *explains* why they offer "good advice" and are effective for predictions. On the other hand, 'to track these correlations" for him *cannot* be equivalent to 'represent', because the Born rules do not describe. One reason why they don't is that "the frequencies they display are not in exact conformity to that rule—the unexpected does happen..." (2017c, p. 172). It follows that the expression 'to track correctly' can only refer to an unexplained link between the predictive success ensured by the rules and mind-independent, probabilistic patterns. The link is unexplained because the Born rules are *not* derived from some deeper posit of the theory (as in Bohmian mechanics' initial distributions of particles, for example). Healey's is a legitimate empiricist position but, lacking a more exact account of the notion of probability while denying representation force to Born rule, it amounts to a form of probabilistic instrumentalism. Healey could rebut that postulating objective physical probabilities is *not* incompatible *per se* with the claim that these are *not* reducible or supervenient upon propensities, frequencies or Lewisian chances. However, these are the most often discussed *objective* accounts of probabilities in the market and postulating objective but "*theoretical probabilities*" (a theoretical entity that, according to Healey's anti-representationalism, is *not "represented" but tracked* by the Born rules) is certainly *against* a pragmatist spirit. Therefore, until we are told what these objective probabilities are and what "track" exactly means, we be entitled to conclude that in PQT the Born rules are just an effective bookkeeping device, given that the assignment of the quantum state depends entirely on the "algorithm provided by the Born Rule for generating quantum probabilities" (2012a, b, p. 731). In a word, given the probabilistic nature of quantum theory, it is safe to assume that Healey's anti-representationalism about the Born rule (and the quantum state) essentially depends on his philosophy of probability, and therefore on the fact that PQT's probabilities are objective$_1$ (or intersubjectively valid for all agents) but they are not objective$_2$. The question is whether this epistemic view can be regarded as consistent with agent physicalism and treating agents as physical systems.

4. Another piece of evidence of Healey's antirealistic approach is his view of relations between quantum system: "By assigning a quantum state one does not represent any physical relation among its subsystems: entanglement is *not* a physical relation" (Healey 2017b, p. 248, my emphasis). Now, even if Schrödinger was wrong in claiming that entanglement is "the characteristic trait of quantum mechanics" (Schrödinger 1935, p. 555), there is no question that, in more realistically-inclined approaches to quantum theory, entangled states are considered to be real physical relations between physical systems that have no separate identity. *Ontic* structural realism is an example of a philosophical approach that takes relations between quantum entities as ontically prior (French 2016).

This approach to entanglement creates tension with PQT's reliance on decoherence as a way to solve *practically* the measurement problem, to explain the emergence of the classical world and to justify its reliance on the epistemic rock given by NQMC. In particular, decoherence is needed because quantum theory does not imply and cannot by itself explain the definiteness of our experience (2017a, p. 99). To the extent that invoking decoherence does not exclude more realistic attempts at solving the measurement problem – dynamical collapse models, Bohmian mechanics, Everett-type interpretations – PQT *is* more antirealistic than these positions that grant reality to the quantum state and yet also rely on decoherence. Moreover, decoherence presupposes a spreading of entanglement, which is typically regarded as a genuine, irreversible physical process. But if "entanglement is not a physical relation" but a mathematical relation among state vectors (2017a, p. 248), it is problematic to rely on this relation to claim that our ordinary classical has an approximately classical appearance.

5. Despite what the author claims about the fact that PQT does not explicitly mention "observables" or "measurement" (2012b, p. 44) – the typical vocabulary of instrumentalist positions – the centrality of NQMC in Healey's philosophical project is incompatible with the view that his approach to quantum theory is

realist. To begin with, non-quantum physical magnitudes *obtained in a lab* are measurements outcomes describable with classical physics. And even if many of these claims refer to systems that are outside the lab, these magnitudes are "...not part of the ontology of quantum mechanics itself. To provide authoritative advice on credences in magnitude claims, quantum mechanics must be supplied with relevant magnitudes—it does not introduce them as novel ontological posits" (Healey 2017b, p. 233). To this fifth objection, Healey could reply that we cannot and should not ask more in the case of quantum theory, since its function is to produce observational results that, as already stressed by Bohr, must necessarily be expressed in classical language. In this case, however, he owes us an explanation as to why classical models do not represent just empirical correlations, but can be used to yield deeper explanations of the latter in terms of entities or properties that help us to "construct" them, in the same sense in which statistical mechanics grounds thermodynamical regularities (Einstein 1919).

His response to this point is contained in the following paragraph, hinting to the contextual and relational nature of the quantum ontology, an aspect that also characterizes quantum field theory: "the unknown particles, strings, and fields would still not be in the ontology of the quantum field theory itself: in that sense they would still be "classical" rather than quantum, even though no prior theory of classical physics posited their existence. A contextual ontology could not provide the ultimate building blocks of the physical world" (2017b, pp. 233–234). In a word, quantum fields "are not a novel ontological posit of quantum field theories: they are assumables, not beables (ibid.)". Two counter-objections to this passage can be raised. First, "assumable" is closely related to "useful posit", or even fictional entity, as atoms were for Mach and Poincaré: as a consequence, Healey's brand of entity realism is even weaker than that defended by Hacking (1983). Secondly, it is not clear why contextuality *per se* should be incompatible with an ontological commitment to entities whose *properties* are not intrinsic or monadic, but fundamentally relational or dispositional, as they are in Rovelli's relational quantum mechanics.[18] But PQT's contextualism concerns entities and not the metaphysical nature of their properties and in this sense it is weaker than Rovelli's.

6. In general, pragmatist approaches to knowledge, and therefore also to physical theories (PQT included), seem to be particular fit to justify a wide-ranging, naturalistic worldview. Within naturalism as it usually intended, beliefs in general (scientific ones included) are typically regarded as more or less effective *guides to action*. Implicitly, this applies also to PQT: the reliability of the agents' beliefs about the effectiveness of attributing a certain quantum state to physical systems depends on the past predictive success of this attribution. As the previous section has made abundantly clear, PQT has a definite empiricist and instrumentalist flavor, witnessed by the fact that typical empiricist epistemologies rely on conditions of assertibility for certain hypotheses, and therefore stress the epistemic limitations of knowers –

[18] Dorato (2016).

or, as Healey has, of "physically situated agents or users". Some of these conditions are obviously dictated by the theory itself.

It seems to me that the many senses in which one can be a realist about quantum mechanics discussed above are sufficient to refute the claim that PQT is a non-instrumentalist approach to quantum theory. Insisting that PQT is "realist" about the quantum state while denying the status of a beable is realism only in name, since PQT's "realism" about the wave function amounts to the claim that assigning a quantum state is an effective instrument to predict the outcomes of an experiment. As hinted above, this conclusion is not just the product of an exegetic effort, but will be important in the next section, when I will deal with the question of agent physicalism.[19]

10.4 The Incompatibility Between PQT and Agent Physicalism

Interpretations of quantum theory are doubtlessly controversial, and physicalism is also controversial and vaguely defined Stoljar (2010), but the current scientific outlook has it that conscious agents are complex "physical systems". It is not plausible to suppose that advances in the neurocognitive sciences may require a change in the way we interpret quantum theory. However, even without defending agent physicalism, I take it that it would still be important to establish whether PQT is compatible with it or not. If agent physicalism were more credible than any approach to quantum theory that violates it (something that here I will not discuss), a defense of PQT would become more difficult.

Given that agents are physically situated users, Healey's empiricist denial of realism about the wave function and the existence of beables *prima facie* cannot be regarded as *incompatible* with some generic of weak forms of reducibility (or supervenience) of the mental states of agents to (over) their physical states and therefore with agent physicalism or the possibility of applying quantum mechanics to agents. Unfortunately, even granting to Healey that quantum theory does not provide a fundamental and complete ontology of the physical world on which

[19]In fairness to Healey, it is very important to qualify these six points: PQT is not incompatible with the idea that realism ought to be the *goals* of scientific theorizing. I think Healey would agree with the idea that realism, like physicalism, are *stances* or *attitudes* (van Fraassen 1985; Ladyman and Ross 2007), that is, they are theses about the aim of science. One way to express this desideratum is as follows: "a physical theory *ought* to be about something and therefore describe physical reality". If he were right to argue that quantum theory has not beables of its own, he would be entitled to be against the view that quantum theory, in its present form at least, has the resources or instruments to achieve this aim, but he could nevertheless argue that realism about the wave function as a beable may have an important heuristic role, similar to an idea of reason in the sense of Kant's transcendental dialectic. Another possibility is that by endorsing some kind of "piecemeal realism" (Fine 1991), this conclusion may just apply to posits within quantum theory and not to quantum theory in general or to other physical theories.

everything else supervenes, I will now show that the attempt to rescue PQT from the charge of incompatibility with agent physicalism is doomed to fail.

Let us begin by noting that in his original 2012a's formulation of PQT, Healey states the following principle:

RPQT *any physical system whatsoever (agents included) can be a quantum system and therefore be assigned a quantum state only relative to an agent applying the algorithm of quantum theory*. Healey 2012a, p. 750)

RPQT is a necessary consequence of any agent-centered approach to quantum theory. The essentially relational aspect of PQT depends in fact on the meaning of "agent" and cannot be avoided by the kind of pragmatist approach to quantum theory defended by Healey. Notice in addition that Healey obviously admits that not all quantum systems can be regarded as agents: "all agents are quantum systems with respect to other agents, but *not all quantum systems are agents*" (Healey 2012a, p. 750, my emphasis),

Consider the following argument, where "physical" means describable in principle by current physical theories

1. All quantum systems are physical systems;
2. According to RPQT above, all physical systems (electrons, quarks, atom, agents etc.) possess a quantum state and are quantum systems not intrinsically, but only relationally, that is, only with respect to real (or hypothetical) agents who use information provided by these systems.
3. Some quantum systems are not agents, but all agents are quantum systems only relative to other agents.
4. PQT does not provide any *intrinsic and exact* criterion to distinguish between quantum systems that are agents from those that are not agents.
5. agent physicalism requires an *intrinsic and exact* physical characterization of the notion of agent or a physically instantiated user of key notions of the theory.

Conclusion

PQT is incompatible with physicalism

Let us discuss these premises in turn. Premise (1) must be granted by any non-dualistic approach to quantum theory and is accepted by PQT[20]: if it were false, some quantum objects (agents are the most plausible candidates) would be non-physical and my claim would be proved. Premise 2 expresses the *essential*

[20]Also, the converse of (1) holds, since it is reasonable to claim that *all* physical objects are quantum objects. Notoriously size, which looks the most plausible candidate to be an intrinsic criterion to separate the two realms, is a non-starter. Microscopic systems like electrons, neutrons or photons are quantum systems par excellence, but also mesoscopic quantum systems need quantum mechanics for their description: C60 fullerene molecules, to which also Healey refers, are visible with electron microscopes. Very recently, scientists in California managed to put an object visible to the naked eye into a mixed quantum state of moving and not moving (O'Connelly et al. 2010). It is possible to claim that even stars are quantum systems, given that in their nuclei there are nuclear reactions.

commitment of PQT in any of its form[21] and therefore must be accepted by any agent-centered view of quantum theory: the recourse to agents or users is necessary to PQT, otherwise Healey's approach would not be pragmatist. Premise 3 is a quotation from Healey's text (see above) and must be granted: not all physical systems (say electrons) can apply the algorithm provided by the Born rule, but only agents can, whatever an agent is.

As far as I can tell, the truth of premise 4 also relies on textual evidence and I will now show why, as a consequence of the vagueness of the notion of 'a physically instantiated agent' – and despite the fact that needs physical agent for its formulation (see premise 2) – PQT does not tell us *what quantum theory is about.*

Relying on a passage from Healey (2017c, p.137), he could rebut this claim by noting that quantum theory is *about the mathematical structures* that figure in its models.[22] However, as a proposal to understanding the theory, PQT must of course *refer* to the role that agents have in making sense of the functions of the key terms of the theory (Born rule, quantum state, etc.) by applying them to quantum systems. Note furthermore that the passage mentioned above justifies the claim that PQT must refer to agents, not in the simpleminded sense that any physical theory needs agents for its formulation and implementation. Even in model-theoretical accounts of scientific theories, a physical theory must be able to interpret its formulas, lest it is devoid of any *physical* meaning: classical mechanics is not just about the family of its mathematical models. And given the refusal to give a representationalist account of the models of quantum theory, PQT relationalism as expressed by 2 implies that PQT is also about agents relationally regarded as special quantum systems.

Now the problem is that in order to distinguish in an exact way quantum systems that are agents from those that are not we need an *intrinsic* criterion, which Healey, unfortunately, does not provide: "*Agents are generic, physically situated systems in interaction with other physical systems*: ... the term 'agent' is being used very broadly so as to apply to any physically instantiated user of quantum theory, whether human, merely conscious, or neither" (2012a, p. 752, my emphasis). However, we do have an intuitive idea of what agents are: they may be either conscious or unconscious robots acting like conscious agents (as Healey himself has it in the same passage, 2012a, p. 752), but atoms, electrons, photons and quarks are certainly *not* agents in any sense of the word.

Once again, if we want to know what the theory is about, the distinction between quantum systems that are agent and quantum systems that are not must be given in intrinsic terms and must be exact in Bell's sense (see note 19). The point of the argument is that the possibility of formulating agent physicalism needs an exact formulation of the notion a physically instantiated user of the mathematical apparatus of the theory that, due to the irremediable vagueness of "agent", PQT is not able to give. It follows that, in its current form at least, PQT is incompatible with agent physicalism and this conclusion holds independently of the claim that

[21] As a response to a criticism, this must hold also in later, less relational versions of 2.

[22] This quotation has been reminded to me by one of the referees.

quantum theory provides a fundamental ontology of the physical world (realism about the wave function), a view of physicalism which Healey rejects and that would be unfair to attribute to him.

10.5 What Is an Agent in PQT?

The conclusion of the previous section raises immediate questions: why should PQT be saddled with the task to describe physically instantiated agents in quantum terms? Shouldn't this problem be left to cognitive scientists? "Quantum theory, like all scientific theories, was developed by (human) agents for the use of agents (not necessarily human: while insisting that any agent be physically situated, I direct further inquiry on the constitution of agents to cognitive scientists") (Healey 2017a, p. 137). In view of this, don't alternative formulations or approaches to quantum theory also suffer from the same problem? Of course, if supervenience physicalism holds in these latter approaches, it also holds in PQT, since in PQT agents are quantum systems, even though relatively to other agents.

However, the problem is that, unlike what happens in Everettian, Bohmian and dynamical-collapse quantum mechanics – PQT cannot account for agents in quantum terms (and therefore in physical terms) *even in principle*. In Bohmian mechanics, agents are physical systems that *can* in principle be described in quantum terms: there is no difference between quantum systems that are agents and quantum systems that are not. An analogous conclusion holds in Everettian quantum mechanics, whose bizarre ontological consequences (many worlds or many minds) were put forth to avoid considering agents as different in principle from any other quantum systems. Dynamical collapse models, in the flash version for instance, regard agents as constituted by galaxies of "flashes" in the same sense in which a table is. Analogous considerations hold for the matter field version of GRW. The fact that in PQT agent physicalism fails explains why it is different from these approaches: agents must be treated in a different way from non-agents but we are not told in exact physically terms why this should be so. How can we make sense in a clearer way of the status of agents qua physically instantiated user of a mathematical model on the one hand and the physical electrons and quarks on the other?

In order to avoid explaining why agents are distinguished from non-agents, Healey could regard the notion of 'agent' as "primitive". In one sense, "primitive" as referred to agents would entail that deeper, constructive explanations (possibly involving beables) of how the physical interactions or information exchange between agents and quantum systems can occur can be altogether avoided. This move would also eschew the charge that PQT, which, as we have argued must essentially be about agents, is silent about the demarcation line separating them from system that are clearly non-agents: the interaction between agents and non-agents must be accepted as one that does not require any explanation, since it is presupposed by an application of PQT by agents.

In this perspective, PQT's postulated existence of a division line between agents and non-agents becomes strikingly similar to the well-known demarcation between the quantum realm and the classical realm advocated by Bohr. As a matter of fact, PQT failure of agent physicalism suffers from the same problems. The remarkable similarities between Bohr's philosophy of quantum theory and PQT, already sketched in Sect. 2 above, here can be made more explicit.

First of all, Bohr's approach to quantum theory is as anti-subjectivist as Healey's (for this claim about Bohr, see Howard 1994; Dorato 2017). Secondly, and despite frequent misunderstandings of his position – suggested by the fact that according to Bohr any experimental report about quantum systems had to be given in the language of classical physics – in Bohr's opinion the distinction between the quantum and the classical realms is to be regarded as *contextual and relational* (Zinkernagel 2016). According to Bohr in fact, in certain experimental situations it is legitimate to treat a classical object like a quantum object: the choice depends on the whole experimental context.[23] Healey refers to the whole experimental context with the expression "physical situatedness", but I submit that, despite the verbal differences, linked to the fact that Bohr's interpretation does not mention agents, PQT's is very similar to Bohr's. In fact, the relationality of PQT leads us to claim that an agent A can *both* apply quantum theory to quantum systems that are *not* agents (an electron) and, in other physical contexts, be the target of such an application by *another* agent B. These remarks about Healey's philosophy of quantum mechanics are not meant to detract from its originality but rather to put the pragmatist approach in the correct historical light.

While here I cannot argue in more details for this claim, in Bohr's thought classical physics has been plausibly likened to the Kantian transcendental condition of possibility for describing the otherwise unattainable, "noumenal" quantum world (see MacKinnon 1982; Murdoch 1987): *exactly as in PQT, according to Bohr there cannot be a quantum description of quantum mechanics*. Now, while in Healey's texts there is no explicit evidence for this Kantian interpretation, agents could be regarded as those quantum systems that – in virtue of some distinguishing physical property P that PQT cannot in principle specify in quantum terms – *must be presupposed to regard any physical system as a quantum system*. That such a P is needed depends on the fact that an electron's properties do not suffice to apply quantum theory to other quantum systems.

This property P can either physical or epistemic. We have seen that, in virtue of the failure of agent primitivism, we cannot describe agents in exact quantum *physical* terms (we don't know what distinguishes physically instantiated agents from non-agents). Consequently, one might be tempted to suppose – against Healey's explicit position – that there exist some transcendental *epistemic* capacity P in virtue of which agents can play their role as users, a role that electrons cannot

---

[23]See his response to Einstein's objection, involving the moving screen with the double slit as reported in Bacciagaluppi and Valentini (2009).

have. By invoking this transcendental reading of PQT, we could give a possible explanation as of why agents cannot be described in quantum mechanical terms.

However, given the failure of agent physicalism, 'transcendental condition of possibility' would imply 'epistemic condition of possibility'. In this case, agents' epistemic conditions of possibility to apply the quantum formalism in PQT would become primitive. In fact, unlike what happens in other philosophical approaches to quantum mechanics – any quantum theory that might be invoked to reduce or explain these states, for its application would *presuppose* them![24] In a word, to the extent that agents' epistemic states play the role of the distinguishing property P conditions of possibility for applying PQT to the physical world, agents and their epistemic states would acquire an epistemically primitive status.

If this Kantian reading of the primitive role of agents in quantum theory were unconvincing because not grounded in the text, one could try to clarify the implications of PQT by characterizing the property in virtue of which electrons and quantum agents differ in terms of the notion of *information*. Agents are primitive because they have the capacity to manipulate information. In fact, there is an asymmetry between the acquisition of information on the part of an agent *A* on a non-agential quantum system *S*, and the flow of information going in the opposite direction: if *S* is an electron, *it* cannot acquire information on an agent *A* except in a Pickwickian way, at least if "information" is (controversially) regarded as an irreducibly epistemic term. Once again, the distinction between agents and non-agents must be regarded as intrinsic, since only agents can be characterized as *users* of the information provided by a quantum system, where this information can be used with the *purpose* of predicting and explaining its behavior.

The problem is not given by the teleological characters of terms like user's 'aims' or 'goals' relative to predictions and explanation of the physical world: Deep Blue can be attributed the intention of defeating a human chess player. The rub rather lies is the vagueness of the meaning of "information" (see Timpson 2013). Healey, however, takes it in "the ordinary sense of the word" (2017c, p. 179), and therefore in an essentially epistemic sense: it is only an agent that can have information about something.[25] Without a more detailed analysis, information has no clear physical meaning, despite the fact that it can be measured by Shannon's formula. This justifies why it can be regarded as physically primitive: if we cannot be told in what precise sense and in which way 'information' can be considered to be a relation between two physical systems, we can regard agents manipulating it as primitive posits of quantum theory. In a word, and similarly to what was argued before, if information qua epistemic notion is needed to apply quantum theory to a quantum system, any reducing quantum theory would have to presuppose

[24]This conclusion is somewhat similar to Frauricher and Renner's claim that "Quantum theory cannot consistently describe the use of itself" (2018), an interesting paper which here I cannot discuss.

[25]Compare Bell's famous questions: "whose information, information about what?"

information, and therefore, in Healey's understanding of the term, agents' epistemic states *irreducibly*. If a single epistemic state is irreducible, agent physicalism is false.

Since I leave to the next, final section the task to provide additional arguments in favor of the view that contemporary information-theoretic approach to quantum theory are epistemic, I conclude the current section by discussing the following example, which I take as providing additional arguments to the conclusion that PQT and agent physicalism are incompatible, and which is a very simplified elaboration of Wigner's- type thought experiments (Wigner 1967; Frauchiger and Renner 2018; Healey 2018a).

Consider a quantum system *S* that, in Healey's term, is *not* a physically situated agent. Let *A* be a physically situated agent that extracts (and then uses) information from *S* by physically interacting or correlating with it. Suppose that after this interaction the experiment reveals "spin up". Given the relational character of PQT, *S* has spin up relatively to agent *A*, who is certain of this result. At the same time, the joint system $A + S$ can be regarded as a quantum system by another physically situated agent *B*. Relative to *B*, *before* her information-gathering physical interaction with the quantum system $A + S$, *A* and *S* are still in a superposition of states. Consequently, given the relativity of probability assignments in PQT, it may happen with a certain probability that *after* the interaction between the coupled quantum systems $A + S$ and *B*, the physically situated agent *B* reports: "relative to me, *A* observes spin *down* and *S* has spin *down*" even if from *A*'s perspective, *A* has observed spin *up*. Suppose this is exactly what happens after B's measurement of $A + S$.

It is very important to notice that, as in Rovelli's interpretation of quantum mechanics (1996), this fact does not contradict *A*'s previous report, since it is simply a consequence of the relativity of the ascription of quantum states by quantum agents. Since *B* can be regarded as a *quantum* system by *A,* when *B* and *A* correlate, *A* will claim "from agent's *B* perspective, I have observed spin down even if from my perspective I have observed spin up" and *B* will agree with this report.

However, it is one thing to claim that there is no fact of the matter as to what the spin of the system is independently of the experimental setup and the physical situatedness of the agents. It is quite another to have to deny the claim that "agent *A* has either observed that the spin is up or she has observed that it is down." Agents' observations cannot be relative to different interacting systems. There seem to be only two ways out of this problem. (i) The first is to claim that PQT is incompatible with the view that measurements have definite results, so that it requires some kind of Everettian relative-state mending, which however, PQT's instrumentalism about the quantum state cannot advocate: remember that in Everettian quantum mechanics the wave function is all there is. Furthermore, without some many worlds formulation, we would have *two* different epistemic states of the same agents *A* relative to the same measurement result, which might be regarded as a failure of supervenience: different conscious perceptions correlate to the same physical outcomes, so that the former don't fix the latter. (ii) Quantum theory is not universally applicable, and in particular not applicable to agents, which must be regarded *intrinsically* as classical physical systems having informational access, via decoherence, to non-quantum magnitude claims. But this way out introduces once again an unclear distinction between quantum systems and classical systems whose vagueness the relationality of PQT wanted to avoid.

10.6 Why Pitowsky's Information-Theoretic Approach to Quantum Theory Also Needs Agents

In this concluding section, I want to very briefly comment on Healey's and Pitowsky's approaches to the related questions of realism, probability and information.

10.6.1 Realism

As to realism, that there has been an evolution in Pitowsky view of quantum theory. In chapter 6 of *Quantum Probability and Quantum Logic* (Pitowsky 1989), he seems to defend a form of underdetermination of theories by data, and his preference for Bohmian mechanics, were one to choose a hidden variable theory, is clearly stated: "... Can we produce a hidden variable theory which goes beyond quantum theory and predicts new phenomena? Psychologically this question seems to be connected to the problem of realism. It seems that realists are more inclined to consider hidden variable theories. Logically however, the physical merits of hidden variable theories are independent of the metaphysical debate. One can take an instrumentalist position regarding hidden variables and still appreciate their potential application in predicting new phenomena. Alternatively, one can believe in the existence of hidden causes, while still holding on to the view that these causes lie beyond our powers of manipulation. Hidden variable theories may very well lead to new predictions, this is a matter of empirical judgment. My own bias tends to side with non-local theories such as Bohm (1952). In this and similarly detailed dynamical theories we may be able to find a way to manipulate and control the hidden variables and arrive at a new phenomenological level. In any case, I can conceive of no a-priori argument why this could not occur" (Pitowsky 1989, p. 181).

While this manipulation is today regarded as out of question, in his earlies production Pitowsky's open-mindedness about quantum theory distinguishes his position from Healey's more recent work. It is significant, however, that in order to break the underdetermination, he does not regard the greater explanatory virtue of Bohm's theory as sufficient to rule out instrumentalist approaches, as long as the latter are regarded as having the same epistemic force. The decisive element would be the production of new experimental evidence, an aspect that today is unreasonably regarded with skepticism.

In later works, however, Pitowsky has argued that quantum theory can be characterized structurally as a theory about the logic and probability of quantum events, that the measurement problem is a pseudo-problem, that the quantum state does not represent reality but just provides an algorithm to assign probabilities. Unlike the position defended by Healey, according to Pitowsky probabilities are *credences* that reflect objective correlations and that are assigned to certain outcomes. As PQT, also according to Pitowsky decoherence explains the emergence of a classical world (see, for instance, Bub and Pitowsky 2010, p. 1). These are the typical traits of instrumentalistic approaches to quantum theory, at least insofar as the quantum state, denoted by the wavefunction, is regarded as the central notion used by quantum theory and there is no attempt at giving a dynamic explanation of the definiteness of the measurement outcomes that does not rely on decoherence. This particular form of antirealism about the wave function is certainly shared also by Healey, even he does not explicitly argue that the claim that measurement should not be appear in the primitive of the theory is the result of "a dogma".

10.6.2 Probability

The theory of probability defended by the two authors prima facie seems different. Healey rejects subjectivist view of probability, since, modulo the remarks above, PQT is described as capable of tracking mind-independent probabilistic correlations. According to Bub and Pitowsky, the quantum state is credence function (2010), and the objective nature of probability is given by the kinematic constraints represented by the projective geometry of Hilbert space. Yet, also Bub and Pitowsky argue that angles in such a geometry are associated with "structural probabilistic constraints on correlations between events", where these events are to be regarded as physical, in analogy with events in Minkowski spacetime, also structured by spatiotemporal relations. Despite the fact that Healey does not explicitly mention intersubjectively shared credences, also the probabilistic approach of the two philosophers is strikingly similar.

The real difference between Healey and Pitowsky's (and Bub's) philosophies of quantum theory involves the relational character of PQT, which is called for by the agent-centered view of Healey's approach. We have seen why it is the role of agents that makes PQT into a relational theory and therefore creates troubles *vis à vis* agent physicalism. I will now show that there is an important pragmatic, even if only implicit component also in Pitowsky's approach, which *is a consequence of the pivotal role that information has in his late production* and that pushes toward a relational approach, involving conscious agents manipulating and transferring information.

10.6.3 Information

In order to argue in favor of this claim it is important to figure out in what sense information can be regarded as physical, given that according to Pitowsky and

Bub, quantum theory can be reconstructed as a theory about the transmission and manipulation of information, "constrained by the possibilities and impossibilities of information-transfer in our world" (Bub 2005).[26] Recall that according to Healey information has a merely epistemic sense and plays a minor role in his theory. However, the fact that agents are necessarily presupposed also in information theoretic approaches to quantum theory is evident from the fact that only *users can manipulate information regarded as a physical property of objects*. Think of questions like: "what is an information source?", what does it mean "to broadcast the outcome of a source?" Is the "no-signaling principle" understandable without the presence of an agent, and therefore in purely physical terms?

Let us agree to begin with that the no cloning principle applies to machines, since it claims that "there is no cloning machine that can copy the outcome of an arbitrary information source" (2010, p. 7). The main question therefore involves the notion of a *source of information*. A fair coin, that either land head or tail, is a source of information (i.e., a classical bit) but from the physical viewpoint it is merely a short cylinder with two different faces. Only agents, whether conscious or unconscious, can *use* these physical properties with some goal or intention (toss it decide who must go shopping for example), but without a use of this property by someone or something, there is no *source* of information, even if information has no semantic sense, as pointed out by Shannon in his original paper.

Signaling is a word that has a univocal, *intentional* sense: only agents can signal to other agents by using physical instruments with the purpose of communicating something, but the existence of a probabilistic correlation between two experimental outcomes by itself can carry no information. Consider now a photon, which is a qubit with respect to its polarization. In this sense, it's purely physical (possibly dispositional) properties are analogous to that of a coin, despite the fact that a qubit is associated to "an infinite set of noncommuting two-values observables that cannot have all their value simultaneously" (Bub 2016, p. 30). The physical, agent-independent, property of the photon is its elliptical or linear polarization, the information it carries and the possibility of using it as a source thereof presupposes the existence of a user, whether conscious or not, with some intention to do something. In this sense, information theoretic approaches *à* la Pitowsky and Bub are much more similar to Healey's PQT than it might be thought at first sight. This should be of no surprise, given the shared claim that the wave function has no representative power and has a pure algorithmic role. There is a sense according to which instrumentalist philosophies are more anthropocentric than their more realistically inclined rivals.

[26]"quantum mechanics [ought to be regarded] as a theory about the representation and manipulation of information constrained by the possibilities and impossibilities of information-transfer in our world (a fundamental change in the aim of physics), rather than a theory about the behavior of non-classical waves and particles" (Bub 2005, p. 542).

Acknowledgments I thank two anonymous referees for their very precious help on previous version of this manuscript. In particular, I owe many remarks that I added to the second version of the paper to the second referee.

References

Bacciagaluppi, G., & Valentini, A. (2009). *Quantum theory at the crossroad*. Cambrigde: Cambrigde University Press.

Bell, J. (1989). Towards an exact quantum mechanics. In S. Deser & R. J. Finkelstein (Eds.), *Themes in contemporary physics II. Essays in honor of Julian Schwinger's 70th birthday*. Singapore: World Scientific.

Bohm, D. (1952). A suggested Interpretation of the quantum theory in terms of 'Hidden' variables, I and II. *Physical Review, 85*(2), 166–193. https://doi.org/10.1103/PhysRev.85.166.

Brandom, R. (2000). *Articulating reasons: An introduction to Inferentialism*. Cambridge, MA: Harvard University Press.

Breuer, T. (1993). The impossibility of accurate state self-measurements. *Philosophy of Science, 62*, 197–214.

Brukner, Č. (2017). On the quantum measurement problem. In R. Bertlmann & A. Zeilinger (Eds.), *Quantum (un)speakables II* (pp. 95–117). Cham: Springer.

Bub, J. (2005). Quantum theory is about quantum information. *Foundations of Physics, 35*(4), 541–560.

Bub, J. (2016). *Banana world. Quantum mechanics for primates*. Oxford: Oxford University Press.

Bub, J., & Pitowsky, I. (2010). Two dogmas about quantum mechanics. In S. Saunders, J. Barrett, A. Kent, & D. Wallace (Eds.), *Many worlds?: Everett, quantum theory & reality* (pp. 433–459). Oxford: Oxford University Press.

Camilleri, K., & Schlosshauer, M. (2008). The quantum-to-classical transition: Bohr's doctrine of classical concepts, emergent classicality, and decoherence. arXiv:0804.1609v1.

Camilleri, K., & Schlosshauer, M. (2015). Niels Bohr as philosopher of experiment: Does decoherence theory challenge Bohr's doctrine of classical concepts? *Studies in History and Philosophy of Modern Physics, 49*, 73–83.

Churchland, P., & Hooker, C. (Eds.). (1985). *Images of science: Essays on realism and empiricism (with a reply from bas C. van Fraassen)*. Chicago: University of Chicago Press.

Dalla Chiara, M. L. (1977). Logical self-reference, set theoretical paradoxes and the measurement problem in quantum mechanics. *Journal of Philosophical Logic, 6*, 331–347.

Dorato, M. (2016). Rovelli's relation quantum mechanics, antimonism and quantum becoming. In A. Marmodoro & D. Yates (Eds.), *The metaphysics of relations* (pp. 235–261). Oxford: Oxford University Press.

Dorato, M. (2017). Bohr's relational holism and the classical-quantum interaction. In H. Folse & J. Faye (Eds.), *Niels Bohr and philosophy of physics: Twenty first century perspectives* (pp. 133–154). London: Bloomsbury Publishing.

Dürr, D., Goldstein, S., Tumulka, R., & Zanghì, N. Bohmian mechanics. https://arxiv.org/abs/0903.2601v1

Einstein, A. (1919, November 28). Time, space, and gravitation. *Times (London)*, pp. 13–14.

Faye, J., & Folse, H. (Eds.). (2017). *Niels Bohr and the philosophy of physics, twenty-first-century perspectives*. London: Bloomsbury Academic.

Fine, A. (1984). The natural ontological attitude. In J. Leplin (Ed.), *Scientific realism*. Berkeley: University of California Press.

Fine, A. (1991). Piecemeal realism. *Philosophical Studies, 61*, 79–96.

Folse, H. (1985). *The philosophy of Niels Bohr. The framework of complementarity*. Amsterdam: North Holland.

Folse, H. (2017). Complementarity and pragmatic epistemology. In J. Faye & H. Folse (Eds.), *Niels Bohr and the philosophy of physics. A-twenty-first-century perspective* (pp. 91–114). London: Bloomsbury.

French, S. (2016). *The structure of the world*. Oxford: Oxford University Press.

Frauchiger, D., & Renner, R. (2018). Quantum theory cannot consistently describe the use of itself. *Nature Communications, 9*, Article number: 3711. https://doi.org/10.1038/s41467-018-05739-8.

Friederich, S. (2015). *Interpreting quantum theory. A therapeutic approach*. New York: Palgrave Macmillan.

Gell-Mann, M. (1976). What are the building blocks of matter? In H. Douglas & O. Prewitt (Eds.), *The nature of the physical universe: Nobel conference* (pp. 27–45). New York: Wiley.

Glick, D. (2018). *Review of the quantum revolution in philosophy by Richard Healey*. philsci-archive.pitt.edu/14272/2/healey_review.pdf

Godfrey-Smith, P. (2003). *Theory and reality*. Chicago: University of Chicago Press.

Hacking, I. (1983). *Representing and intervening*. Cambridge: Cambridge University Press.

Healey, R. (1989). *The philosophy of quantum mechanics*. Cambridge: Cambridge University Press.

Healey, R. (2012a). Quantum theory: A pragmatist approach. *British Journal for the Philosophy of Science, 63*, 729–771.

Healey R. (2012b). Quantum decoherence in a pragmatist view. Resolving the measurement problem. arXiv.org > quant-ph > arXiv.

Healey, R. (2015). How quantum theory helps us explain. *British Journal for Philosophy of Science, 66*, 1–43.

Healey R. (2017a). Quantum-Bayesian and pragmatist views of quantum theory. In E. N. Zalta (Ed.), *The Stanford encyclopedia of philosophy* (Spring 201a edition). URL: https://plato.stanford.edu/archives/spr2017/entries/quantum-bayesian/

Healey, R. (2017b). *The quantum revolution in philosophy*. Oxford: Oxford University Press.

Healey, R. (2017c). Quantum states as objective informational bridges. *Foundations of Physics, 47*, 161–173.

Healey, R. (2018a, unpublished manuscript). Pragmatist quantum realism. *British Journal for Philosophy of Science*.

Healey, R. (2018b). Quantum theory and the limits of objectivity. *Foundations of Physics, 48*(11), 1568–1589.

Howard, D. (1994). What makes a classical concept classical? Toward a reconstruction of Niels Bohr's philosophy of physics. In *Niels Bohr and contemporary philosophy* (Vol. 158 of Boston studies in the philosophy of science) (pp. 201–229). Dordrecht: Kluwer.

Howard, D. (2004). Who invented the 'Copenhagen interpretation?' A study in mythology. *Philosophy of Science, 71*, 669–682.

Ladyman, J., & Ross, S. (2007). *Everything must go. Metaphysics naturalized*. Oxford: Oxford University Press.

Laudisa, F., & Rovelli, C. (2008). Relational quantum mechanics. In E. N. Zalta (Ed.), *The Stanford encyclopedia of philosophy* (Fall 2008 edition). URL: http://plato.stanford.edu/archives/fall2008/entries/qm-relational/

MacKinnon, E. (1982). *Scientific explanation and atomic physics*. Chicago: Chicago University Press.

Maudlin, T. (2007). *The metaphysics within physics*. Oxford: Oxford University Press.

McGinn, C. (1993). *Problems in philosophy: The limits of inquiry*. Cambridge: Basil Blackwell.

McLaughlin, B., & Bennett, K. (2014). Supervenience. In E. N. Zalta (Ed.), *The Stanford encyclopedia of philosophy* (Spring 2014 edition). URL: https://plato.stanfrd.edu/archives/spr2014/entries/supervenience/

Minkowski, H. (1952). Space and time. In H. A. Lorentz, A. Einstein, H. Minkowski, & H. Weyl (Eds.), *The principle of relativity: A collection of original memoirs on the special and general theory of relativity* (pp. 75–91). New York: Dover.

Murdoch, R. (1987). *Niels Bohr's philosophy of physics*. Cambridge: Cambridge University Press.

O'Connell, A. D., Hofheinz, M., Ansmann, M., Bialczak, R. C., Lucero, E., Neeley, M., Sank, D., Wang, H., Weides, M., Wenner, J., Martinis, J. M., & Cleland, A. N. (2010). Quantum ground state and single-phonon control of a mechanical resonator. *Nature, 464*, 697–703.

Pitowsky, I. (1989). *Quantum probability, quantum logic*. Berlin: Springer.

Pitowsky, I. (2005). Quantum mechanics as a theory of probability. arXiv:quant-ph/0510095v1.

Price, H. (2011). *Naturalism without mirrors*. Oxford: Oxford University Press.

Psillos, S. (1999). *How science tracks truth*. London/New York: Routledge.

Quine, W. V. (1981). *Theories and things*. New York: The Belknap Press.

Rovelli, C. (1996). Relational quantum mechanics. *International Journal of Theoretical Physics, 35*, 1637–1678.

Rovelli, C., & Smerlak, M. (2007). Relational EPR. *Foundations of Physics, 37*, 427–445.

Schrödinger, E. (1935). Discussion of probability relations between separated systems. *Proceedings of the Cambridge Philosophical Society, 31*, 555–563; 32, 446–451.

Stoljar, D. (2010). *Physicalism*. New York: Routledge.

Stoljar, D. (2017). Physicalism. *The Stanford Encyclopedia of Philosophy* (Winter 2017 Edition) (E. N. Zalta (Ed.)). https://plato.stanford.edu/archives/win2017/entries/physicalism/

Suárez, M. (2004). An inferential conception of scientific representation. *Philosophy of Science, 71*(Supplement), S767–S779.

Timpson, C. (2013). *Quantum information theory*. Oxford: Clarendon Press.

Van Fraassen, B. (1985). Empiricism in the philosophy of science. In P. Churchland & C. Hooker (Eds.), Images of science: Essays on realism and empiricism (pp. 245–308). Chicago: University of Chicago Press.

Van Fraassen, B. (2002). *The empirical stance*. Princeton: Princeton University Press.

Wigner, E. P. (1967). Remarks on the mind–body question. In *Symmetries and reflections* (pp. 171–184). Bloomington: Indiana University Press.

Zinkernagel, H. (2016). Niels Bohr on the wave function and the classical/quantum divide. *Studies in History and Philosophy of Modern Physics, 53*, 9–19.

Chapter 11
Quantum Mechanics As a Theory of Observables and States (And, Thereby, As a Theory of Probability)

John Earman and Laura Ruetsche

Abstract Itamar Pitowsky contends that quantum states are derived entities, bookkeeping devices for quantum probabilities, which he understands to reflect the odds rational agents would accept on the outcomes of quantum gambles. On his view, quantum probability is subjective, and so are quantum states. We disagree. We take quantum states, and the probabilities they encode, to be objective matters of physics. Our disagreement has both technical and conceptual aspects. We advocate an interpretation of Gleason's theorem and its generalizations more nuanced—and less directly supportive of subjectivism—than Itamar's. And we contend that taking quantum states to be physical makes available explanatory resources unavailable to subjectivists, explanatory resources that help make sense of quantum state preparation.

Keywords Probability · Quantum theory · Subjectivism · State preparation · Pitowsky · Quantum field theory

11.1 Introduction

Our title pays homage to Itamar Pitowsky's brilliant "Quantum Mechanics As a Theory of Probability" (Pitowsky 2006), and at the same time it indicates our (partial) disagreement with the view of the foundations of quantum mechanics

J. Earman
Department of History and Philosophy of Science, University of Pittsburgh, Pittsburgh, PA, USA
e-mail: jearman@pitt.edu

L. Ruetsche (✉)
Department of Philosophy, University of Michigan, Ann Arbor, MI, USA
e-mail: ruetsche@umich.edu

M. Hemmo, O. Shenker (eds.), *Quantum, Probability, Logic*,
Jerusalem Studies in Philosophy and History of Science,
https://doi.org/10.1007/978-3-030-34316-3_11

promoted in this paper.[1] We cannot begin to do justice here to the richness of Itamar's paper, and we confine our discussion principally to two of Itamar's central theses:

T1: The lattice of closed subspaces of Hilbert space is "the structure that represents the 'elements of reality' in quantum theory" (p. 214)
T2: "The quantum state is a derived entity, it is a device for bookkeeping quantum probabilities" (ibid.), where these are understood as the credences ideally rational agents would assign events whose structure is described by T1.

We agree very much with the spirit of T1, but we emphasize that T1 needs to be appropriately generalized to cover the branches of quantum theory that go beyond ordinary non-relativistic QM, which was Itamar's focus. And we would underscore, even more strongly than Itamar did, how the structure that quantum theory gives to the 'elements of reality'—or as we would prefer to say, to quantum events or propositions—shapes quantum probabilities.

Our disagreement with Itamar concerns T2, and has technical and conceptual dimensions. Our fundamental unhappiness is conceptual. We contend that T2 mischaracterizes the nature of quantum probabilities, and we propose inverting T2 to the counter thesis

T2∗: Not only is the quantum state not a derived entity, and so not merely a device for bookkeeping quantum probabilities, rather the quantum state has ontological primacy in that quantum probabilities are properties of quantum systems that are derived from the state.

Although T2∗ gainsays T2's claim that quantum states are derived entities, the theses differ on a deeper matter. Taking quantum states to codify objective features of quantum systems T2∗ implies that quantum probabilities are objective, observer-independent probabilities—objective chances if you care to use that label. The territory, between T2∗ and a view asserting quantum states to be derived from *objective* probabilities, may not be worth quibbling over. However, Itamar denies that quantum probabilities are objective. His T2 claims not only *that* quantum states are derived entities but also explains *why* they are derived entities. On Itamar's view, the event structure T1 come first, and then come the probabilities, in the form of credences rational agents would assign events so structured. The states are bookkeeping devices for these credences. Itamar's interpretation of quantum probability is personalist, and an element of a striking and original interpretation of QM. Itamar's personalism widens the distance between T2 and T2∗ to include territory that is worth fighting over.

Let's briefly explore that territory. Naively viewed, quantum theory describes an *event structure* and a set of *probability distributions over that event structure*. The sophisticated might question the division of labor this naive picture presupposes. Which of the elements italicized above does *physics* supply and which (if any) do *we*, qua rational agents, supply? For Itamar, physics supplies only the event structure

[1] See also Pitowsky (2003) and Bub and Pitowsky (2010).

T1; the rest is the work of rational agents. Quantum probability distributions reflect betting odds ideal rational agents would accept on "quantum gambles"; quantum states are just bookkeeping devices for those probability distributions.[2] By contrast, while we also accept T1, we expect physics to do more work. And so we assert T2∗: We take it that physics supplies quantum probabilities, which probabilities are objective, independent of us, and determined by physical quantum states.

On Itamar's view, rationality and only rationality constrains what quantum probabilities, and so what quantum states, are available. Our view countenances additional constraints. If probabilities are physical, both physical law and physical contingency can govern what quantum probabilities are available for physical purposes. It's clear that our explanatory resources are richer than Itamar's. Rationality on its own can't explain Kepler's laws, but Newton's Law of Universal Gravitation can. And LUG on its own can't explain why Halley's comet appears every 75 years, but celestial contingencies (in collaboration with physical law) can. What's not so obvious: whether there are any explanations *essential to making sense of quantum mechanics* to which our resources are adequate but Itamar's are not.

Enough has been said to introduce and situate our technical disagreement with T2. The main result Itamar cites in support of T2 is known as Gleason's theorem. By itself the original Gleason theorem does *not* yield T2; by itself the theorem says nothing about the ontological priority of probability measures over states, nor does it rule on whether quantum probabilities are personalist or objective. Finally, Itamar's claim that the Born Rule follows from Gleason's theorem rests on a tacit restriction on admissible probability distributions/physically realizable states.[3] As we will discuss, how one justifies these restrictions depends on one's interpretation of quantum probabilities and/or the status assigned to states. Arguably proponents of T2∗ have better grounds than proponents of T2 for adopting the restrictions, which do critical work in attempts to make sense of quantum state preparation.

§§2–4 will explore these technical matters in more depth. §5 and beyond elaborate and adjudicate our conceptual disagreement with Itamar. The standard by which both views are judged is: do they explain why QM works as well as it does? That is, do they make sense of our remarkably successful laboratory practice of preparing systems, attributing them quantum states, extracting statistical predictions from those states, and performing measurements whose outcomes confirm those predictions? With respect to this standard, Itamar's version of T2 holds both promise and peril. The promise is that it leads to a dissolution of the notorious measurement problem.[4] The peril is that T2 looks shaky in the face of two explanatory demands: first, the demand to explain the fact that (some) quantum states can be prepared; and,

---

[2]For details of Pitowsky's account of quantum gambles, see his (2003, Sec. 1).

[3]A further technical quibble, readily addressed, is that the version Itamar discusses applies only to the special setting of non-relativistic QM; §4.2 extends his discussion.

[4]Pitowsky (2003) and Bub and Pitowsky (2010) call it the "'big' measurement problem," which they distinguish from the "small" measurement problem. More on this below.

second, the demand to explain the fact that the probabilities induced by prepared states yield the correct statistics for outcomes of experiments.

The first demand is the topic of §§5–8, which describe how each view fields the challenge of state preparation, both in ordinary QM and in more general settings (QFT, thermodynamic limit). An attractive account of state preparation appeals to physical states—which seems to favor T2∗. In fairness, however, we indicate how the formal account of state preparation offered by quantum theory can be turned into an alternative account of state preparation as belief state fixation—a repurposing congenial to advocates of T2. However, the repurposing requires a restriction on the space of admissible credences that may not be motivated by demands of rationality alone. We also confess that we are in danger of being hoisted on our own petard. Because the quantum account of state preparation seems to falter in QFT, our insistence that T1 be generalized to cover branches of quantum theory beyond ordinary QM appears to weaken the support consideration of state preparation lends T2∗. We review some results that promise to circumvent this roadblock.

Turning to the second demand, to make sense of our capacity to complete measurements confirming quantum statistical predictions, §9 confesses that T2∗ runs smack into the measurement problem. It contends that Itamar's view also founders on this explanatory demand. We argue that the measurement problem is a real problem that needs to be solved rather than dissolved. A concluding section recaps the main points of our discussion, and relates them to another foundations question dear to Itamar's heart: just how much realism about QM is it possible to have?

We leave it to the reader to render final judgment on the relative merits of T2 vs. T2∗. All we insist upon is that knocking heads over T2 vs. T2∗ is a good way of sharpening some of the most fundamental issues in the foundations of quantum theory.

We begin our discussion with a reformulation of thesis T1 designed to cover all branches of quantum theory.

11.2 Restatement of Thesis 1

We take a quantum system to be characterized by an algebra of observables and a set of states on the algebra. For the moment we focus on the algebra of observables, which we take to be a von Neumann algebra $\mathfrak{N}$ acting on a Hilbert space $\mathcal{H}$.[5] The elements of quantum reality—or, as we would prefer to say, the quantum events or propositions—are represented by the elements of the projection lattice $\mathcal{P}(\mathfrak{N})$ of $\mathfrak{N}$ (see Appendix 1), and quantum probability is then the study of quantum probability measures on $\mathcal{P}(\mathfrak{N})$. This, which we denote by T1∗, is the promised generalization

[5]For the relevant operator algebra theory the reader may consult Bratelli and Robinson (1987) and Kadison and Ringrose (1987).

of Itamar's T1. We feel confident that he would endorse it. It is hardly a novel or controversial thesis; indeed, T1$*$ is the orthodoxy among mathematical physicists who use the algebraic approach to quantum theory (see, for example, Hamhalter 2003).[6]

A quantum probability measure on $\mathcal{P}(\mathfrak{N})$ is a map $\Pr : \mathcal{P}(\mathfrak{N}) \to [0, 1]$ satisfying the quantum analogs of the classical probability axioms:

(i) $\Pr(I) = 1$ (I the identity projection)

(ii) $\Pr(E_1 \vee E_2) = \Pr(E_1) + \Pr(E_2)$ whenever projections $E_1, E_2 \in \mathcal{P}(\mathfrak{N})$ are mutually orthogonal.

The requirement (ii) of finite additivity can be strengthened to complete additivity

(ii*) For any family $\{E_a\} \in \mathcal{P}(\mathfrak{N})$ of mutually orthogonal projections, $\Pr(\vee_a E_a) = \Pr(\sum\limits_a E_a) = \sum\limits_a \Pr(E_a)$.[7]

When $\mathfrak{N}$ acts on a separable $\mathcal{H}$ any family of mutually orthogonal projections is countable and, thus, complete additivity reduces to countable additivity. When $\mathfrak{N}$ acts on a finite dimensional $\mathcal{H}$ any family of mutually orthogonal projections contains only a finite number of members and so only finite additivity is in play. The different additivity requirements will play an important role in what follows.

Different branches of quantum theory posit different von Neumann algebras of observables and, thus, according to T1$*$, different quantum event structures. In the case of ordinary QM (sans superselection rules) the algebra of observables is taken to be the Type I factor algebra $\mathfrak{B}(\mathcal{H})$, the algebra of all bounded operators acting on $\mathcal{H}$. Here the elements of $\mathcal{P}(\mathfrak{N})$ are in one-one correspondence with the closed subspaces of $\mathcal{H}$, in accord with Itamar's T1. In relativistic QFT the algebras of observables associated with open bounded regions of spacetime are typically Type III von Neumann algebras. This requires a significant modification of T1 since these algebras contain only infinite dimensional projections. In contrast to Type III algebras, Type II algebras contain finite dimensional projections; and in contrast to Type I algebras they contain no minimal projections. Although ergodic measure theory played a role in the construction of the first examples of Type II algebras (see Petz and Rédei 1996), these algebras haven't found wide physical application. Superselection rules provide a more subtle way in which T1 has to be modified. In the case of non-relativistic QM a superselection rule is modeled by a decomposition of the Hilbert space into a direct sum $\mathcal{H} = \oplus_j \mathcal{H}_j$ and a decomposition of the algebra of observables into the direct sum algebra $\oplus_j \mathfrak{B}(\mathcal{H}_j)$. A projection onto a ray of $\mathcal{H}$ that cuts across the superselection sectors of Hilbert space is not in the algebra of observables.

[6]This is not to say that T1$*$ is uncontroversial among philosophers of physics. Bohmians would deny that T1$*$ captures the fundamental event structure.

[7]The sum of probabilities for a transfinite collection is understood as the sup of sums of finite subcollections. The sup exists because the sequence of sums over ever larger finite subcollections gives a non-decreasing bounded sequence of real numbers, and every such sequence has a least upper bound.

The different event structures T1$*$ extracts from different von Neumann algebras naturally bring with them differences in the structure of quantum probabilities. In general once the algebra of observables $\mathfrak{N}$ is fixed so are the non-measurable events: the latter consist of the complement of $\mathcal{P}(\mathfrak{N})$ in the set of all projections on $\mathcal{H}$. In ordinary QM there is no analog of the phenomenon of non-measurable events in classical probability. All projections on $\mathcal{H}$ are in the ordinary QM event algebra; therefore all lie in the range of a probability measure defined on the algebra. In relativistic QFT, by contrast, there are non-measurable events galore. The Type III local algebras endemic there contain no finite dimensional projections and, thus, every finite dimensional projection on a Hilbert space $\mathcal{H}$ on which a Type III algebra acts corresponds to a non-measurable event. Furthermore, the theory offers a prima facie explanation of why some events are non-measurable: they are not assigned probabilities because they do not correspond to observables.

Different types of algebras of observables have different implications for the status of additivity requirements for quantum probabilities. In particular, in ordinary QM where the algebra of observables is a Type I factor, countable additivity entails complete additivity unless the dimension of the Hilbert space is as great as the least measurable cardinal (see Drish 1979 and Eilers and Horst 1975). This entailment does not hold for the Type III algebras encountered in QFT.

In the following sections we will see that differences in the algebras of observables also make for differences in the possible states, which in turn imply differences in the possible probability measures.

11.3 The Relation Between Quantum States and Quantum Probabilities

11.3.1 Quantum States

Quantum states are expectation value functionals on the observable algebra $\mathfrak{N}$—technically, complex valued, normalized positive linear functionals. A state ω is said to be mixed (or impure) iff it can be written as a convex linear combination of distinct states, i.e. there are distinct states ϕ_1, ϕ_2 such that $\omega = \lambda_1\phi_1 + \lambda_2\phi_2$ with $0 < \phi_1, \phi_2 < 1$ and $\phi_1 + \phi_2 = 1$. A vector state ω is a state such that there is a $|\psi\rangle \in \mathcal{H}$ where $\omega(A) = \langle\psi|A|\psi\rangle$ for all $A \in \mathfrak{N}$. Vector states are among the normal states, which can be given a number of equivalent characterizations. Here we quote one relevant result:

Theorem 1. The following conditions are equivalent for a state ω on a von Neumann algebra $\mathfrak{N}$ acting on $\mathcal{H}$:

(a) ω is completely additive, i.e. $\omega(\sum_a E_a) = \sum_a \omega(E_a)$ for any family $\{E_a\}$ of mutually orthogonal projections in $\mathfrak{N}$,

(b) there is a density operator ϱ, i.e., a trace class operator on $\mathcal{H}$ with $Tr(\varrho) = 1$, such that $\omega(A) = Tr(\varrho A)$ for all $A \in \mathfrak{N}$.[8]

Condition (b) is often referred to as the "Born rule."

11.3.2 From States to Probabilities, and from Probabilities to States

A quantum state, whether pure or impure, normal or non-normal, induces a probability measure on the projection lattice $\mathcal{P}(\mathfrak{N})$: it is easily checked that if ω is a state on $\mathfrak{N}$ then $\Pr^{\omega}(E) := \omega(E)$, $E \in \mathcal{P}(\mathfrak{N})$, satisfies the conditions listed in §2 for a quantum probability measure. From Theorem 1, the probability $\Pr^{\omega}$ induced by the state ω is completely additive iff ω is normal. Again, complete additivity reduces to countable additivity when $\mathcal{H}$ is separable and to finite additivity when $\mathcal{H}$ is finite dimensional.

Moving in the other direction, from probability measures on $\mathcal{P}(\mathfrak{N})$ to states on $\mathfrak{N}$, requires the use of Gleason's theorem, which is the main technical result Itamar cites in support of his T2. Below we will discuss the relevance of Gleason's theorem for T2, but in this section we confine ourselves to an exposition of the theorem. The version of the theorem Itamar quotes applies only to ordinary non-relativistic QM and, thus, in concert with the needed generalization of T1 the theorem needs to be generalized to cover the branches of quantum theory that go beyond ordinary QM. Thanks to twenty years of labor by a number of mathematicians the needed generalization is available.

The original version of Gleason's theorem for ordinary QM can be given the following formulation:

Theorem 2 (Gleason). Let Pr be a quantum probability measure on $\mathcal{P}(\mathfrak{B}(\mathcal{H}))$ where $\mathcal{H}$ is separable and $\dim(\mathcal{H}) > 2$. If Pr is countably additive then it extends uniquely to a normal state on $\mathfrak{B}(\mathcal{H})$.

This result has been generalized to cover non-separable Hilbert spaces and, more significantly, quite general von Neumann algebras, so that it is applicable to all branches of quantum theory:

Theorem 3 (Gleason generalized). Let $\mathfrak{N}$ be a von Neumann algebra acting on a Hilbert space $\mathcal{H}$, separable or non-separable. Suppose that $\mathfrak{N}$ does not contain any direct summands of Type I_2. Then for any quantum probability measure Pr on $\mathcal{P}(\mathfrak{N})$ there is a unique extension of Pr to a quantum state $\omega^{\Pr}$ on $\mathfrak{N}$. Further, the state $\omega^{\Pr}$ is normal iff Pr is completely additive.[9]

---

[8] See Kadison and Ringrose (1987, Vol. 2, Theorem 7.1.12) and Bratelli and Robinson (1987, Theorem 2.4.21).

[9] This Theorem requires proof techniques that are different from the one used for the original Gleason theorem; see Hamhalter (2003) and Maeda (1990) detailed treatments of this crucial theorem. A Type I_2 summand is isomorphic to the algebra of bounded operators on a two-dimensional Hilbert space.

11.3.3 The Born Rule and a Challenge for T2 and T2∗

The generalized Gleason's theorem opens the door to those who want to endorse T2 and relegate quantum states to bookkeeping devices across the board in quantum theory. But why go through this door? One motivation comes from those who want to apply Bruno de Finetti's personalism to quantum theory and construe quantum probabilities as the degrees of belief rational agents have about quantum events $\mathcal{P}(\mathfrak{N})$. Gleason's theorem allows the personalists to claim that quantum states have no independent ontological standing but function only to represent and track the credences of Bayesian agents. This answer leads directly to Itamar's T2: quantum states are bookkeeping devices for quantum probabilities, which reflect rational responsiveness to quantum event structures limned by T1.

Itamar writes: "The remarkable feature exposed by Gleason's theorem is that the event structure dictates the quantum probability rule. It is one of the strongest pieces of evidence in support of the claim that the Hilbert space formalism is just a new kind of probability theory" (Pitowsky 2003, p. 222). By "the quantum probability rule" he means the Born rule. The Gleason theorem by itself does *not* yield the Born rule for calculating probabilities and expectation values. The Born rule uses a density operator representation of expectation values, and from the point of view of states the validity of this representation comes from the restriction of possible states to normal states; and from the point of view of probabilities the validity derives from the restriction to completely additive measures.[10] And it is exactly here that a challenge arises for both of the competing theses T2 and T2∗.

The use of the Born Rule is, we claim, baked into the practice of applying quantum theory. In the following section we offer evidence for this claim. The challenge for the proponents of the opposing T2 and T2∗ is to explain and justify this practice, or explain when and why the practice should be violated.

11.4 Justifying Normality/Complete Additivity

11.4.1 Normality and the Practice of QM

In this subsection we use 'state' in way that is neutral with respect to T2 vs T2∗; so if you are a fan of T2 feel free to understand 'state' as qua representative of a

[10] An awkwardness for Itamar's account is the existence, in the case where $\dim(\mathcal{H}) = 2$, of probability measures on $\mathcal{P}(\mathfrak{B}(\mathcal{H}))$ that don't extend to quantum states on $\mathfrak{B}(\mathcal{H})$. In order to maintain that quantum states are just bookkeeping devices for the credences of rational agents contemplating bets on quantum event structures, it seems that Itamar must either deny that two dimensional Hilbert spaces afford quantum event structures, or articulate and motivate rationality constraints (in addition to the constraint of conformity to the axioms of the probability calculus!) that prevent agents from adopting "stateless" credences.

quantum probability measure. Our purpose is to indicate how the practice of QM has baked into it the restriction to normal states.

Begin by noting that most standard texts on quantum theory use 'state' in a way that makes it synonymous with normal state, e.g. it is assumed that expectation values of observables follow the Born rule. A substantive example of the role of normal states comes from the generally accepted analysis of quantum entanglement, which is based on normal states. Let $\mathfrak{N}_1$ and $\mathfrak{N}_2$ be mutually commuting von Neumann algebras acting on a common Hilbert space. Think of $\mathfrak{N}_1$ and $\mathfrak{N}_2$ as characterizing the observables of two subsystems of a system whose algebra has the tensor product structure $\mathfrak{N}_1\overline{\otimes}\mathfrak{N}_2$. A state ω on the composite system algebra $\mathfrak{N}_1\overline{\otimes}\mathfrak{N}_2$ is a product state iff $\omega(AB) = \omega(BA) = \omega(A)\omega(B)$ for all $A \in \mathfrak{N}_1$ and $B \in \mathfrak{N}_2$. A normal state ω on $\mathfrak{N}_1\overline{\otimes}\mathfrak{N}_2$ is said to be quantum entangled over $\mathfrak{N}_1$ and $\mathfrak{N}_2$ iff it cannot be approximated by convex linear combinations of normal product states with respect to $\mathfrak{N}_1$ and $\mathfrak{N}_2$, i.e. ω fails to lie in the norm closure of the hull of convex linear combinations of normal product states. The following result links entanglement and the non-abelian character of quantum observables:

Theorem 4 (Raggio 1988). Let $\mathfrak{N}_1$ and $\mathfrak{N}_2$ be mutually commuting von Neumann algebras acting on a common Hilbert space. Then the following two conditions are equivalent:

(R1) At least one of $\mathfrak{N}_1$ and $\mathfrak{N}_2$ is abelian
(R2) No normal state on $\mathfrak{N}_1\overline{\otimes}\mathfrak{N}_2$ is entangled over $\mathfrak{N}_1$ and $\mathfrak{N}_2$.

Conversely, the possibility of quantum entanglement requires that both subsystem algebras have a quantum (= non-abelian) character. Thus, one can hold that the key feature that separates classical from quantum physics is the difference between abelian vs. non-abelian observables and at the same time agree with Schrödinger that entanglement is the key feature separating quantum from the classical. This result does *not* hold for non-normal states.

Such examples can be multiplied: check the theorems presented in textbooks on quantum information and quantum computing and find that often the theorems presuppose that 'state' means normal state. For future reference note how the restriction to normal states affects the allowable pure states. When $\mathfrak{N}=\mathfrak{B}(\mathcal{H})$, as in ordinary QM sans superselection rules, the class of normal pure states is identical with the class of vector states. In ordinary QM with superselection rules some vector states are not pure—in particular, states that superpose across superselection sectors. This, rather than the incorrect notion that superselection rules represent a limitation on the superposition principle, is the way to bring home one of the implications of superselection. When $\mathfrak{N}$ is Type II or Type III there are no normal pure states; and for Type III all normal states are vector states.

To say that the exclusive use of normal states is baked into the practice of quantum theory is not to justify the practice, except insofar as 'It works, so it must be right' is a justification. What one takes as a satisfying justification of the practice depends upon whether one takes the side of Itamar's T2 or our alternative T2∗. We examine the contrasting justifications in turn.

11.4.2 Justifying Normality: T2$*$

Our T2$*$ takes quantum states to codify objective, observer-independent features of quantum system. From this perspective a justification of the exclusive reliance on normal states would have to take the form of an argument showing that only normal states are physically realizable. Here we give an example, drawn from the context of the algebraic formulation of QFT, of such an argument.

AQFT postulates an association $\mathcal{O} \mapsto \mathfrak{N}(\mathcal{O})$ of von Neumann algebras $\mathfrak{N}(\mathcal{O})$ with open bounded regions $\mathcal{O}$ of Minkowski spacetime $\mathcal{M}$ satisfying the net property, viz. if $\mathcal{O} \subseteq \mathcal{O}'$ then $\mathfrak{N}(\mathcal{O}) \subseteq \mathfrak{N}(\mathcal{O}')$. The quasi-local global algebra $\mathfrak{N}(\mathcal{M})$ is taken to be the smallest von Neumann algebra generated by the $\mathfrak{N}(\mathcal{O})$ as $\mathcal{O}$ ranges over all of the open bounded regions of $\mathcal{M}$. Since Minkowski spacetime is based on a Hausdorff manifold, for any spacetime point $p \in \mathcal{M}$ there is a sequence $\{\mathcal{O}_n\}_{n=1}^{\infty}$ of open regions such that $\mathcal{O}_{n+1} \subset \mathcal{O}_n$ for all $n \in \mathbb{N}$ and $\cap_{n=1}^{\infty}\mathcal{O}_n = \{p\}$. The property of local definiteness is designed to capture the notion that there are no observables "at a point" so that physically realizable states become indistinguishable in sufficiently small regions: technically, the requirement is that for any physically realizable states ω and φ on $\mathfrak{N}(\mathcal{M})$, $||(\omega - \varphi)|_{\mathfrak{N}(\mathcal{O}_n)}|| \to 0$ as $n \to \infty$ (see Haag 1992, p. 131). The only other assumption needed is the unobjectionable posit that the physically realizable states on $\mathfrak{N}(\mathcal{M})$ contain at least one normal state φ. It follows that all physically realizable states ω are locally normal, i.e. for any $p \in \mathcal{M}$ there is a neighborhood $\widetilde{\mathcal{O}}$ of p such that $\omega|_{\mathfrak{N}(\widetilde{\mathcal{O}})}$ is normal (see Appendix 2 for a proof).

This result is modest since it gives no assurance that the neighborhood on which the state is normal is large enough to accommodate an experiment one might hope to conduct; and even such as it is its effectiveness turns on the plausibility of the property of local definiteness. Nevertheless, it serves as a useful example of how normality can be grounded in physical properties. Notice that, unless it can be established a priori that there are no observables at a point, considerations of rationality alone are inadequate to secure such a grounding.

There are much more ambitious arguments for normality that revolve around the state preparation. We will postpone an examination of these arguments until §5. But before reviewing these arguments we should also mention that on the other side of the ledger there are occasional calls for the use of non-normal states. For example, Srinvas (1980) argues that non-normal states are needed to define the joint probabilities associated with successive measurements of quantum observables with continuous spectra. Relatedly, Halvorson (2001) considers the use of non-normal states as a means of assigning sharp values to the position of a particle and, thus, as providing a basis for interpreting the modulus square $|\psi(q)|^2$ of the wave function (in the usual $L^2_{\mathbb{C}}$ realization) of a spinless particle as the probability density that the particle has exact position q.[11] Non-normal states also provide a means to overcome

[11]For a contrary viewpoint, see Feintzeig and Weatherall (2019).

no-go results on the existence of conditional expectations for normal states such as that of Takesaki (1972). Such considerations do not undermine our support for T2$*$; indeed they strengthen it by indicating that the sorts of considerations physicists use to decide the admissibility of quantum states are quite different from the considerations of rationality of credences the proponents of T2 would bring to bear.

11.4.3 Justifying Normality: T2

From the point of view of T2 the justification for normality can only be that the credence functions rational agents put on the projection lattice $\mathcal{P}(\mathfrak{N})$ must be completely additive and, thus, by the generalized Gleason theorem the states that bookkeep these credences are normal.

One way to make good on the claim that the credence functions of rational agents are completely additive is to mount a Dutch book argument. An agent who uses her credence function to assign betting quotients to events is subject to Dutch book just in case there is a family of bets, each of which she accepts but collectively have the property that she loses money in every possible outcome. The goal is to show that conforming credences to the axioms of probability with complete additivity is both necessary and sufficient for being immune to Dutch book. The ‘only if’ part of the goal can be attained if the agent is required to stand ready to accept any bet she regards as merely fair, but not if she is only required to accept bets that she regards as strictly favorable (see Skyrms 1992). This may be regarded as a minor glitch in that for most applications of quantum theory a separable Hilbert space suffices; in these cases complete additivity reduces to countable additivity, and the Dutch book argument goes through even if the agent only accepts what she regards as strictly favorable bets.

Nevertheless, personalists are on thin ice here since other constructions—such as scoring rule arguments—personalists have used to try to prove that rational credences should conform to the axioms of probability do not reach countable, much less complete, additivity. Additionally, Bruno de Finetti, the patron saint of the personalist interpretation of probability, adamantly denied that credences need be countably additive much less completely additive (de Finetti 1972, 1974). The only restrictions on admissible probability functions to which personalists may appeal are restrictions justified by considerations of rationality alone. If the patron saint of personalism is rational, it seems that the status of countable and complete additivity is something about which rational parties can disagree. And if countable/complete additivity can’t be rationally compelled then the restriction of quantum probability measures to countably and completely additive ones—and ergo the restriction of quantum states to normal ones—eludes the reach of personalist explanatory resources.[12]

Apart from the additivity issue the main challenge to the personalist interpretation of quantum probabilities comes from the implication that there are as many quantum states as there are rational agents. But, we'll suggest, the desideratum that state preparation be explicable supports T2$*$ against T2, and supports the existence of a unique agent-independent state against the existence of as many agent-dependent states as there are rationally admissible credence assignments.

11.4.4 Finitizing

We have been pressing the line that advocates of T2$*$ have better resources for justifying normality, as well as for explicating departures from normality, than advocates of T2 have. But if normality requires no justification, this line evaporates, taking with it our best grounds for preferring our own objectivist interpretation of quantum probabilities to Itamar's personalist interpretation of those probabilities. And Itamar has at his disposal a maneuver that would eradicate the need to justify normality. That maneuver is to finitize. In this subsection, we'll sketch it briefly, then explain why we think even Itamar should resist it.

Throughout his exposition of quantum mechanics as a theory of probability, Itamar assumes the von Neumann algebras $\mathfrak{B}(\mathcal{H})$—the algebras framing quantum bets and affording quantum event structures—to be *finite dimensional*. We can find motivation for the assumption in his 2006: "A proposition describing a possible event in a probability space is of a rather special kind. It is constrained by the requirement that there should be a viable procedure to determine whether the event occurs" (p. 217). This is because "only a recordable event can be the object of a gamble" (p. 218). Itamar's point is that the outcomes subject to quantum bets have to be the sorts of thing we can settle. Otherwise we wouldn't be able to resolve the bets we've made. Given certain assumptions about our perceptual limitations and our attention (and life) spans, this constrains quantum bets to concern events with only finitely many possible outcomes—and thereby constrains quantum event structures to be given by the projection lattices of *finite dimensional* von Neumann algebras. And, of course, with quantum events structures thus constrained, complete additivity collapses to finite additivity, and the rock-bottom requirement that quantum probability functions be finitely additive implies that quantum states are automatically normal. Normality requires no further justification; it makes no sense to assess accounts of quantum probability on the basis of how well they justify normality.

---

[12]Bayesian statisticians are divided on the issue of additivity requirements. Some versions of decision theory, such as that of Savage (1972), operate with mere finite additivity. Kadane et al. (1986) argue that various anomalies and paradoxes of Bayesian statistics can be resolved by the use of so-called improper priors and that such priors can be interpreted in terms of finitely additive probabilities. For more considerations in favor of mere finite additivity see Seidenfeld (2001).

While we concede the formal merits of this maneuver, we contend that even Itamar should resist it, because it fundamentally undermines the spirit and interest of his interpretation of quantum mechanics. Baldly put, we think the finitizing move is hubris. It amounts to demanding that the quantum event structure—the topic of T1/T1∗, and the element all hands agree to be supplied by the physics—conform to packages of gambles mortal human agents can entertain and settle. That is, the finitizing maneuver sculpts the physical event structure by appeal to the limitations of agents. We acknowledge and admire the resourcefulness and integrity of personalist attempts to derive constraints on *credences* from features of the agents entertaining them. But we take the present finitizing move to inject agents where they don't belong: into the *physical event structures* themselves. Itamar's conjunction of T1 and T2 is brilliant and provocative in no small part due to the delicate division of labor it posits: physics supplies the event structure; we supply everything else. Upsetting that division of labor, the finitizing move diffuses the distinctive force of Itamar's position. Without finitizing, anyone who agrees that the quantum event structure is physical must take Itamar's position seriously. With finitizing, opponents might well complain that Itamar's pulled a bait and switch, and offered subjectified quantum event structures in lieu of physical ones.

11.5 Ontological Priority of States: State Preparation and Objective Probabilities

To repeat, the view we are promoting is that quantum states codify objective features of quantum systems, and the probabilities they induce are thereby objective. In support of this view we offer the facts that (some) quantum states can be prepared and that the probabilities they induce are borne out in the statistics of the outcomes of experiments. Quantum theory itself contains the ingredients of a formal account of state preparation.

The account contains two main ingredients, the first of which is the von Neumann/Lüders projection postulate:

vN/L Postulate: If the pre-measurement state of a system with observable algebra $\mathfrak{N}$ is ω and a measurement of $F \in \mathcal{P}(\mathfrak{N})$ returns a Yes answer, then the immediate post-measurement state is $\omega_F(A) := \dfrac{\omega(FAF)}{\omega(F)}$, $A \in \mathfrak{N}$, provided that $\omega(F) \neq 0$.[13]

Note that if ω is a normal state then so is ω_F. The second ingredient is the concept of a filter for states:

Def. A projection $F_\varphi \in \mathcal{P}(\mathfrak{N})$ is a filter for a normal state φ on $\mathfrak{N}$ iff for any normal state ω on $\mathfrak{N}$ such that $\omega(F_\varphi) \neq 0$,

[13] von Neumann's preferred version of the projection postulate differs from the version stated here in the case of observables with degenerate spectra. Preliminary experimental evidence seems to favor the Lüders version stated here; see Hegerfeldt and Mayato (2012) and Kumar et al. (2016).

$$\omega_{F_\varphi}(A) := \frac{\omega(F_\varphi A F_\varphi)}{\omega(F_\varphi)} = \varphi(A) \text{ for all } A \in \mathfrak{N}.$$

In ordinary QM with $\mathfrak{N} = \mathfrak{B}(\mathcal{H})$ the normal pure states are the vector states, and as the reader can easily check any such state ψ has a filter $F_\psi \in \mathcal{P}(\mathfrak{B}(\mathcal{H}))$, which is the projection onto the ray spanned by the vector $|\psi\rangle$ representing ψ. So suppose that F_ψ is measured and that the measurement returns a Yes answer. Then by the vN/L postulate, if the pre-measurement state is ω then the post-measurement state is ω_{F_ψ}. And since F_ψ is a filter for ψ the post-measurement state is ψ regardless of what the pre-measurement state ω is, provided only that ω is a normal state and that $\omega(F_\psi) \neq 0$.

To repeat, this formal account of state preparation meshes with the statistics of experiments. E.g. prepare a system in the normal pure state ψ. Immediately after preparation make a measurement of some $E \in \mathcal{P}(\mathfrak{B}(\mathcal{H}))$ and record the result. Repeat the preparation of ψ on the system or a similar system; then repeat the measurement of E and record the result. Repeat .... Since the trials are independent and identically distributed the strong law of large numbers implies that as the number of trials is increased the frequency of Yes answers for E almost surely approaches its theoretical value $\psi(E)$. This is exactly what happens in actual cases—at least there is always apparent convergence towards $\psi(E)$.

The ansatz that only normal states are physically realizable has played a silent but crucial role in the above account of state preparation. We have already suggested that this ansatz fits more comfortably with our T2∗—which countenances *physical* constraints on the space of admissible states—than it does with Itamar's T2—which countenances only rationality constraints. The restriction to normal states has played a crucial rule because if a system were initially in a non-normal state it would be inexplicable how it could be prepared in a normal state since neither unitary evolution nor Lüders/von Neumann projection can cause a tiger to change its stripes: if ω is non-normal then so is $\omega'(\bullet) = \omega(U \bullet U^{-1})$ for any unitary $U : \mathcal{H} \to \mathcal{H}$; and if ω is non-normal then so is $\omega'(\bullet) = \dfrac{\omega(F \bullet F)}{\omega(F)}$ for any $F \in \mathcal{P}(\mathfrak{B}(\mathcal{H}))$ such that $\omega(F) \neq 0$ (see Ruetsche 2011). Turning this around, the success in preparing systems in normal states provides a transcendental argument for the ansatz. Transcendental arguments are a priori, if they are arguments at all. So perhaps an unholy marriage of Kant and Itamar will avail advocates of T2 of the resources necessary, on this story, to make sense of state preparation.

Another attitude toward the ansatz that only normal states are physically realizable might be available to advocates of T2∗ embarrassed by transcendental arguments but not embarrassed by certain dialectical moves made on behalf of the Past Hypothesis (see e.g. Loewer 2012). This is the hypothesis that the initial macrostate of the universe was exceedingly low entropy. Some fans of the Past Hypothesis claim that (i) it helps explain prevalent time asymmetries integral to our epistemic practices; that (ii) it thereby figures as an axiom in the simplest, strongest systematization of physical fact; and that (iii) this qualifies the Past Hypothesis as a physical law by the lights of the Best Systems Account of laws of nature. Despite recoiling at this line of thought, we observe that less discriminating advocates of T2$*$ might adapt its basic architecture to argue for the lawlike status of a "Normality Hypothesis," that all physically realizable quantum states are normal. After all, the Normality Hypothesis, by making sense of our central epistemic practice of state preparation, earns its chops as an element of the simplest, strongest systematization of physical fact, and so a physical law. And we also observe that this maneuver is not available to advocates of Itamar's T2. The maneuver explains the state restrictions required to make sense of state preparation by appeal to physical law (the Normality Hypothesis). Itamar would translate the restriction on physically realizable states to a restriction on rationally admissible credence functions. But the explanation of the restriction defies translation: physical laws and physical contingency can explain physical restrictions, but only rationality constraints can explain restrictions on credences.

In §7 below we will consider a difficulty in extending the account of state preparation developed above to more general von Neumann algebras of observables. But in the next section we note the proponents of T2 and a personalist interpretation of quantum probabilities can piggyback on the above formalism to give an alternative account of state preparation. And in addition, they claim, this alternative account dissolves the notorious measurement problem—a consideration that, if correct, would give Itamar's T2 a huge lead over our T2$*$ in the race to make sense of quantum theory's undeniable empirical success.

11.6 Fighting Back: An Alternative Account of 'State Preparation'

11.6.1 Updating Quantum Probabilities

The proponent of T2 owes an alternative account of what happens in 'state preparation' if states are just bookkeeping devices. Such an account starts with a rule for updating probabilities on measurement outcomes. The generally accepted updating rule for quantum probabilities uses Lüders conditionalization, which is motivated by the following result:

Theorem 5 Let Pr be a completely additive quantum probability measure on the projection lattice $\mathcal{P}(\mathfrak{N})$ of a von Neumann algebra $\mathfrak{N}$ which does not contain any direct summands of Type I_2, and let $F \in \mathcal{P}(\mathfrak{N})$ be such that $\Pr(F) \neq 0$. Then there is a unique functional $\Pr(\bullet//F)$ on $\mathcal{P}(\mathfrak{N})$ such that (a) $\Pr(\bullet//F)$ is a quantum probability, and (b) for all $E \in \mathcal{P}(\mathfrak{N})$ such that $E \leq F$, $\Pr(E//F) = \Pr(E)/\Pr(F)$.[14]

Granted that conditions (a) and (b) capture features we want updating to have, the uniquely determined updating rule specifies that if the pre-measurement probability

[14]This is a slight generalization of the result of Cassinelli and Zanghi (1983).

measure is Pr and if a measurement of $F \in \mathcal{P}(\mathfrak{N})$ returns a Yes answer then the updated post-measurement probability measure is $\Pr(\bullet//F)$.

To see what the updated measure comes to, note that the conditions of the Theorem 5 make the generalized Gleason theorem applicable so that Pr extends uniquely to a normal state ω on $\mathfrak{N}$. Then

$$\Pr(E//F) = \frac{\omega(FEF)}{\omega(F)} = \frac{\omega(FEF)}{\Pr(F)} \quad \text{for all } E, F \in \mathcal{P}(\mathfrak{N}). \tag{L}$$

When E and F commute $EF = FE = E \wedge F$ and (L) reduces to

$$\Pr(E//F) = \frac{\Pr(EF)}{\Pr(F)} = \frac{\Pr(E \wedge F)}{\Pr(F)} \tag{B}$$

which is the rule of Bayes updating used in classical probability. However, when E and F do not commute $\omega(FEF)$ cannot be written as $\Pr(FEF)$ since $FEF \notin \mathcal{P}(\mathfrak{N})$ and is not assigned a probability. It should be embarrassing to the proponent of T2 that states have to be utilized in the updating rule, but let this pass.

11.6.2 'State Preparation' According to T2

Call $X_\varphi \in \mathcal{P}(\mathfrak{N})$ a belief filter for a normal state φ on $\mathfrak{N}$ iff for any completely additive Pr on $\mathcal{P}(\mathfrak{N})$ such that $\Pr(X_\varphi) \neq 0$, $\Pr(E//X_\varphi) = \varphi(E)$ for all $E \in \mathcal{P}(\mathfrak{N})$. When the generalized Gleason theorem applies an application of (L) shows that a state filter for φ is a belief filter for φ. The converse is also true: when the generalized Gleason theorem applies a belief filter for φ is also a state filter for φ (see Appendix 4 for a proof). The equivalence of belief filters and state filters allows the proponents of T2 and a personalist interpretation of probability to claim various virtues.

They can say that they can account for 'state preparation' without having to carry the ontological baggage of T2$*$. What the proponents T2$*$ call a state filter for φ, they call a belief filter for φ. And in virtue of X_φ being a belief filter, all agents who assign a non-zero prior probability to the belief filter X_φ and who operate with completely additive credence functions will, upon updating on the belief filter X_φ, agree that posterior probabilities are equal to those assigned by φ. The state φ has been 'prepared', but this just means that the bookkeeping device φ represents the mutually shared updated opinion.

But please note that the personalist appropriation of §5's formal account of state preparation requires a restriction to completely additive probabilities—a restriction, we've suggested, that fails to follow from considerations of rationality alone, unless those considerations encompass transcendental arguments of a novel and surprising nature. And, to review the discussion of §4.4, while the maneuver of finitizing quantum event structures accomplishes the "restriction" in one fell swoop, it also revises the division of labor so compellingly effected by Itamar's T1 and T2. To our minds, proponents of T2$*$, who can invoke the dumb luck of physical contingency, are better positioned to make sense of state preparation. Insofar as we're assessing proposals with respect to their capacity to deliver the goods of making sense of quantum theory's empirical success, this tips the balance in favor of our T2$*$.

Upsetting the balance is the possibility that our strategy for making sense of state preparation might not generalize beyond Type I von Neumann algebras. Given our insistence on such generalization, such a possibility would hoist us on our own petard. The next section gives life to the possibility, but also explains how we can avoid being hoisted.

11.7 State Preparation in QFT

11.7.1 The Conundrum

The account of state preparation given in §5 focused on normal pure states. We have already explained the restriction to normal states. The restriction to pure states is explained by the fact that mixed states do not have filters and, thus, are outside the ambit of the formal account of state preparation offered above (see Appendix 3 for a proof). This fact leads a puzzle about how and whether state preparation is possible in QFT.

It is widely assumed that local observers can only make measurements in bounded regions of spacetime. If this is so it seems that such observers cannot prepare global states, i.e. states on the quasi-local algebra $\mathfrak{N}(\mathcal{M})$ for the entirely of Minkowski spacetime $\mathcal{M}$. If $\mathfrak{N}(\mathcal{M})$ is Type I, as is typically the case, it admits normal pure states, and such states have filters. But a state filter is a minimal projection, and the local Type III algebras $\mathfrak{N}(\mathcal{O})$ associated with open bounded regions $\mathcal{O} \subset \mathcal{M}$ do not contain any such projections. This is just as well, you may say, for if a global state could be prepared by making measurements in a bounded $\mathcal{O} \subset \mathcal{M}$ some horrible form of non-locality would be in the offing. But things seem just as bad at the local level. Since a local algebra $\mathfrak{N}(\mathcal{O})$ is Type III it does not admit any normal pure states, and filters for mixed states on $\mathfrak{N}(\mathcal{O})$ do not exist in $\mathfrak{N}(\mathcal{O})$.

11.7.2 The Buchholz, Doplicher, and Longo Theorem

A way out of this seeming conundrum was sketched by Buchholz et al. (1986) by showing that, under certain conditions on the net $\mathcal{O} \mapsto \mathfrak{N}(\mathcal{O})$ of local algebras, local state preparation is possible. The exposition begins with a definition:

Def. Let φ be a normal state on $\mathfrak{N}(\mathcal{M})$. A projection $E_\varphi \in \mathfrak{N}(\mathcal{M})$ is a local filter for the restriction $\varphi|_{\mathfrak{N}(\mathcal{O})}$ of φ to the local algebra $\mathfrak{N}(\mathcal{O})$ iff for any normal state ω on $\mathfrak{N}(\mathcal{M})$

$$\omega(E_\varphi A E_\varphi) = \omega(E_\varphi)\varphi(A) \text{ for all } A \in \mathfrak{N}(\mathcal{O}).$$

We know that E_φ cannot belong to $\mathfrak{N}(\mathcal{O})$ itself. But ideally it can be located in some $\mathfrak{N}(\widehat{\mathcal{O}})$ where $\widehat{\mathcal{O}}$ only slightly larger than $\mathcal{O}$. So the hope is that for any normal state φ on $\mathfrak{N}(\mathcal{M})$ and any $\mathcal{O} \subset \mathcal{M}$ there is a local filter $E_\varphi \in \mathfrak{N}(\widehat{\mathcal{O}})$ for $\varphi|_{\mathfrak{N}(\mathcal{O})}$ with $\widehat{\mathcal{O}}$ only slightly larger than $\mathcal{O}$.

The hope for local filters can be fulfilled if the local algebras exhibit a property called split inclusion:

Def. Let $\mathfrak{N}_1$ and $\mathfrak{N}_2$ be von Neumann algebras with $\mathfrak{N}_1 \subseteq \mathfrak{N}_2$. The pair $(\mathfrak{N}_1, \mathfrak{N}_2)$ is a split inclusion iff there is a Type I factor $\mathfrak{F}$ such $\mathfrak{N}_1 \subseteq \mathfrak{F} \subseteq \mathfrak{N}_2$.

The net $\mathcal{O} \mapsto \mathfrak{N}(\mathcal{O})$ has the split inclusion property iff for any $\mathcal{O}$ and $\widehat{\mathcal{O}}$ with $\widehat{\mathcal{O}} \supset \mathcal{O}$, the pair $(\mathfrak{N}(\mathcal{O}), \mathfrak{N}(\widehat{\mathcal{O}}))$ is a split inclusion.

Theorem 6 (Buchholz et al. 1986). The split inclusion property of the local algebras is both necessary and sufficient for the existence of local filters.

The proof of the sufficiency is given in Appendix 5.

Despite the importance of this theorem it is rarely mentioned in textbooks. An exception is Araki's *Mathematical Theory of Quantum Fields*. His summary of the significance of the theorem:

> The existence of the local filter for a state in a certain domain means that the state can be prepared by using some apparatus which is a little larger than the original one, and this assumption is equivalent with the assumption of split inclusion. (Araki 2009, 191).

In terms of our discussion the moral is that the competing stories the proponents of T2 and T2$*$ told in the preceding section about state preparation in ordinary QM can be retold in QFT for states on local algebras just in case the split inclusion property holds.[15]

So what is the status of the split inclusion property? It is known to hold in some models of AQFT and fail in others. That the dividing line is being crossed is signaled by the number of degrees of freedom in the model. It is often assumed that the vacuum state for $\mathfrak{N}(\mathcal{M})$ is a vector state whose generating vector is cyclic and separating. Halvorson (2007) shows

Theorem 7. Let $\mathcal{O} \mapsto \mathfrak{N}(\mathcal{O})$ be a net of von Neumann algebras acting on $\mathcal{H}$, and suppose that there is a vector $|\Omega\rangle \in \mathcal{H}$ that is cyclic and separating for all the local algebras in the net. If the split inclusion property holds then $\mathcal{H}$ is separable.[16]

Physical models of AQFT feature vacuum states that are cyclic and separating. Further, the split inclusion property is known to be entailed by the nuclearity condition, which expresses a precise form of the idea that as the energy of a system increases the energy level density should not increase too rapidly (see Buchholz and Wichmann 1986). Perhaps a case can be made that the split inclusion property picks out "physically reasonable" models of AQFT, but it is clearly a contingent property.

---

[15]One of us (LR) has reservations about the interpretational morals being drawn from the BDL theorem; see Appendix 6.

[16]Let $\mathfrak{N}$ be a von Neumann algebra acting on $\mathcal{H}$. A vector $|\Omega\rangle \in \mathcal{H}$ is cyclic iff $\mathfrak{N}|\Omega\rangle$ is dense in $\mathcal{H}$. It is separating vector iff $A|\Omega\rangle = 0$ implies $A = 0$ for any $A \in \mathfrak{N}$ such that $[A, \mathfrak{N}] = 0$.

Of course, its contingency isn't a challenge for Itamar's T2. The property restricts which algebras afford QFT's quantum event structure. The topic of Itamar's T1 and our friendly amendment thereof to T1∗, quantum event structures are something Itamar regards as a matter of physics.

11.8 Taking Stock

The proponents of T2 might claim to have achieved a standoff with the proponents of T2∗. In ordinary QM where filters exist for all normal pure states the argument for T2∗ from state preparation can be rewired to give an alternative account of 'state preparation' in terms of belief fixation that is in accord with T2. In QFT the situation is a bit more complex. At the global level filters for states on the global algebra are unavailable to local observers so that the state preparation argument for T2∗ falters. At the local level the story divides. In models where the split inclusion property holds there are local filters and, thus, there is a standoff over local state preparation parallel to the situation in ordinary QM; and in models where the split inclusion property fails there are no local filters so that the state preparation argument for T2∗ falters.

The proponents of T2∗ may wish to put a different spin on matters. At the global level in AQFT there is no preparation story for states on $\mathfrak{N}(\mathcal{M})$ for the personalist to piggyback on. The proponents of T2∗ can not only admit this but explain why measurements by local observers cannot be used to prepare global states. And they will also note that insofar as the practice of physics presupposes that there is a unique global state, the proponents of T2 are at a loss to explain the practice. Similarly, at the local level there is no preparation story for local states when the split inclusion property fails. Again proponents of T2∗ can explain why measurements by local observers cannot be used to prepare local states; and they will also note that insofar as the practice of physics presupposes that there is on any local $\mathfrak{N}(\mathcal{O})$ a unique local state, the proponents of T2 are at a loss to explain the practice.

Setting aside QFT impediments, we take the proponents of T2∗ to have an advantage over the proponent of T2. Both the formal T2∗ and the appropriated T2 accounts of state preparation by filtration need to restrict admissible quantum probability functions to completely additive ones. The expanded explanatory resources afforded proponents of T2∗ give them the means to do so. Proponents of T2 need to anchor such restrictions in considerations of rationality alone. It is not at all clear that such considerations suffice.

Of course, if proponents of T2 could solve the measurement problem, there would be good reason to set such quibbles aside. The next section contends that they cannot.

11.9 The Measurement Problem

The von Neumann/Lüders projection postulate gives the empirically correct prescription for calculating post-measurement probabilities. If, following T2$*$, quantum states codify objective features of quantum systems and if, following John Bell's injunction, 'measurement' is not to be taken as primitive term of quantum theory then there is a problem in accounting for the success of the vN/L postulate. For numerous no-go results show that treating a quantum measurement as an interaction between an apparatus and an object system and describing the composite apparatus-object system interaction as Schrödinger/Heisenberg evolution does not yield the desired result.

It is tempting to think that the problem can be dissolved by rejecting Bell's injunction and also rejecting T2$*$ in favor of T2. For these twin rejections open the way to say: The change of state postulated in vN/L does take place but this change is innocuous and calls for no further explanation; for the change is simply a change in the bookkeeping entries used to track the change in credence functions when updating on measurement outcomes. In more detail, when the information that a measurement of $F \in \mathcal{P}(\mathfrak{N})$ gives a Yes result is acquired a completely additive credence function Pr on $\mathcal{P}(\mathfrak{N})$ is updated by Lüders conditionalization on F to $\Pr'(\bullet) = \Pr(\bullet//F)$. Assuming that the generalized Gleason theorem applies, it follows that there are unique normal states ω and ω' bookkeeping Pr and Pr′ respectively, and it follows from the definition of Lüders conditionalization that these bookkeepers are related by $\omega'(\bullet) = \dfrac{\omega(F \bullet F)}{\omega(F)}$, just as vN/L requires. That's all there is to it.[17]

We demur: there is more, and the more steers us back towards the original measurement problem. On T2$*$ there is a unique but, perhaps, unknown pre-measurement state, and the vN/L postulate refers to the change in this state. By contrast, on T2 there are as many different bookkeeping states as there are agents with different pre-measurement credence functions, and unless the F in question is a belief filter there are many different post-measurement bookkeeping states. Only one of these post-measurement bookkeeping states gives correct predictions about observed frequencies, something easily understood on T2$*$ since only one of these states induces the objective chances of post-measurement results. Thus, on T2 there is the 0th measurement problem of how to explain uniqueness of states: how, in the parlance of T2, to explain why exactly one of the myriad admissible probability functions available accords with subsequent observation. The problem becomes particularly pressing in cases where there are no belief filters, as in models of QFT where the split property fails.

---

[17] Or rather that is all there is supposed to be to what Pitowsky (2003) and Bub and Pitowsky (2010) call the 'big' measurement problem. They think that after the big problem is dissolved there still remains the 'small' measurement problem, viz. to explain the emergence of the classical world we observe.

The 0th measurement problem threatens to spill over into the BIG measurement problem. It is our experience that after measurements, the credences available to observers narrow to those corresponding to full belief in one outcome or another. That is, upon updating on a measurement outcome, I obtain a credence assigning probability 1 to an element of the set $\{E_i\}$ of spectral projections of the pointer observable. I never obtain a credence assigning probability 1 to a member of a set of projections $\{E_i'\}$ incompatible with the pointer observable. It is not clear how Itamar can explain why our options are invariably narrowed in these characteristic ways. And even if he can, he faces a further mystery: why does a measurement deliver me to full belief in one rather than another of the $\{E_i\}$? "Because that outcome—E_n (say) was observed!" he might reply, rehearsing §6's account of belief fixation via filtration. But §6's account posits an outcome on which all believers update—and it is not clear that Itamar has the wherewithal to deliver the posit. Concerning quantum event structures, T1 says nothing about which quantum events are realized in any given physical situation. Thus it's unclear that Itamar can say what the physical facts are, much less that they include measurement outcomes. Compare our view, which can marshal the physical quantum state to characterize the facts. The physical state, along with some account of what's true of a system in that state (e.g. the unjustly maligned eigenvector-eigenvalue link) is the right kind of resource for saying what the physical facts are. Alas, to date, no one's articulated a rule that manages, without legerdemain, to number measurement outcomes among the physical facts.

Additionally, even if 'measurement' is taken as a primitive it makes sense to ask how physically embodied agents acquire information about measurement results. Are we to take this as an unanalyzed process? Or do we search for an account using quantum theory to analyze the interaction of agents with the rest of the world? The former strengthens the suspicion that real problems are being swept under the rug. The latter points to rerun of the original problem.

11.10 Conclusion

The key question separating our position from Itamar's is how to understand quantum probabilities and quantum states. Are they a matter of physics (our T2$*$) or a matter of rationality (Itamar's T2)? But there is another question in the vicinity, illuminating further points of contention. The question is: just how much realism about QM is possible?

One way to approach the question is see how well candidate realisms accommodate the key explanandum that QM works. A stability requirement for a would-be abductive realist is that what she believes when she believes QM must be consistent with the evidence she cites to support her belief in QM. That evidence takes the form of a history of preparing physical systems, attributing them quantum states, extracting probabilistic predictions from those states, and verifying those predictions by means of subsequent measurements. Interpretations that render preparation infeasible, statistical prediction unreachable, or measurement impossible fail this stability requirement. They are self-undermining. If an interpretation of QM is self-undermining, belief in QM under that interpretation is more realism about QM than is possible. (Although we concede that the foregoing applies only to abductively-supported realisms, we consider realisms motivated by other means to be tasteless.)

Itamar takes his subjective interpretation to afford as much realism as you can have about QM. He calls it "realism about quantum gambles." We can explicate it as realism about QM understood in terms of T1 and T2: quantum event structures are physically real; quantum probabilities are all in our heads. One could be even *less* of a realist about QM than Itamar is. One might, with Bohmians, deny T1 by asserting physical events to have some structure different from that given by the lattice of closed subspaces of Hilbert space.[18] But you can't be more of a realist, Itamar cautions. One way to be more of a realist is to believe QM understood in terms of T1 (and its generalization T1∗) and T2∗. On this doubly realist view, both quantum event structures and quantum probabilities/states are physically real. For Itamar, the reason you can't have this much realism is the measurement problem. Understanding quantum states as physical renders understanding measurement impossible. A realism guided by T1∗ and T2∗ is self-undermining.

We have suggested here that both our line and Itamar's fail stability, but for interestingly different reasons. Itamar's fails to accommodate state preparation, and fails because rationality considerations are the only grounds a personalist can give for restricting admissible credences. Restricting admissible states to normal states (and bracketing momentarily complications encountered in QFT), we can invoke vN/L to make sense of state preparation (§5). Taking states to be physical, we have the resources to justify the restriction we need: the restriction might obtain as a matter of physical contingency or as matter of metaphysically contingent physical law. Itamar's personalist can co-opt the account of state preparation developed in §5. But the co-option requires restricting admissible credences to completely additive ones. And this restriction is heterodox when gauged against the personalist gospel according to de Finetti. This leaves it contentious at best whether it's a constraint dictated by the demands of rationality alone.

We confess that QFT poses impediments to the account of state preparation available to advocates of T2∗. §7 details how those advocates might deal with them. Advocates of T2 have an additional maneuver by which they might negotiate these same impediments. That maneuver is the one discussed in §4.4., of coarse-graining and/or finitizing in a way that bleaches physical significance from the infinitary and otherwise exotic algebras encountered in the setting of QFT. The finitizing move also helps T2 advocates cope with preparation in more homely settings. Collapsing countable to finite additivity, it effects the restriction on credences needed, only by something like fiat. The fiat is the price here, and we don't buy it.

[18]This is hardly to deny that Bohmians are realists. They're just realists about something else—the Bohm theory. Standard QM stands to the Bohm theory as an effective theory stands to a more fundamental theory whose coarse-grained implications it mimics. Thus realism about the Bohm theory induces something like "effective realism" about QM. The point is that this represents less realism about the theory than Itamar's position. For more on effective realism, see Williams (2017).

What about the measurement problem? Both views encounter it in some form. Both are self-undermining. But at least our view does *not* undermine physics. For advocates of T2∗, the measurement problem signals that something physical is eluding our grasp—and that signal is a call to develop new physics. Advocates of T2 deny that anything is missing from the physics. Indeed, they contend that the measurement problem results from putting something in the physics—quantum states that reflect objective probabilities—that doesn't belong there. Taking the measurement problem to be thereby dissolved, they predict that no fruitful new physics is to be had by confronting the problem.

In the end, one's attitude towards the measurement problem comes down to a bet on what is the most fruitful way to advance physics. Our bet is that path starts from taking the measurement problem as a problem to be solved rather than dissolved; and further, our bet is that advancement down this path will require new physics, perhaps something along the lines of a stochastic mechanism to produce the changes postulated by vN/L or perhaps a radically new theory.

Appendix 1 The Lattice of Projections

A projection $E \in \mathfrak{N}$ is an "observable," i.e. a self-adjoint operator, and it is idempotent, i.e., $E^2 = E$. The collection of projections $\mathcal{P}(\mathfrak{N})$ is equipped with a natural partial order, viz. for $E, F \in \mathcal{P}(\mathfrak{N})$, $F \leq F$ iff $range(E) \subseteq range(F)$. $\mathcal{P}(\mathfrak{N})$ is a lattice because it is closed under meet $\wedge$ (the least upper bound) and join $\vee$ (greatest lower bound) in this partial order. E_1,$E_2 \in \mathcal{P}(\mathfrak{N})$ are mutually orthogonal iff $E_1E_2 = E_2E_1 = 0$ (the null projection). When E_1 and E_2 are mutually orthogonal $E_1 \vee E_2 = E_1 + E_2$. Complementation in $\mathcal{P}(\mathfrak{N})$ is orthocomplementation, i.e. $E^c = E^{\perp} = I - E$.

Appendix 2 Locally Normal States in Algebraic QFT

Assume that among the physically realizable states on the quasi-local global algebra $\mathfrak{N}(\mathcal{M})$ there is at least one normal state φ. Assume also the property of local definiteness. It follows that any physically realizable state ω is locally normal, i.e. for any $p \in \mathcal{M}$ there is a neighborhood $\widetilde{\mathcal{O}}$ of p such that $\omega|_{\mathfrak{N}(\widetilde{\mathcal{O}})}$ is normal. To see this, consider a sequence $\{\mathcal{O}_n\}_{n=1}^{\infty}$ of open neighborhoods in $\mathcal{M}$ descending to p. By local definiteness there must be a sufficiently large $n_0 \in \mathbb{N}$ such that $||(\omega - \varphi)|_{\mathfrak{N}(\mathcal{O}_{n_0})}|| < 2$. This implies that $\omega|_{\mathfrak{N}(\mathcal{O}_{n_0})}$ and $\varphi|_{\mathfrak{N}(\mathcal{O}_{n_0})}$ belong to the same folium, and since $\varphi|_{\mathfrak{N}(\mathcal{O}_{n_0})}$ is normal so is $\omega|_{\mathfrak{N}(\mathcal{O}_{n_0})}$. To complete the proof set $\widetilde{\mathcal{O}} = \mathcal{O}_{n_0}$.

Appendix 3 No Filters for Mixed States

As a preliminary note that if ω is a state on $\mathfrak{N}$ such that $\omega(E) = 1$ for $E \in \mathcal{P}(\mathfrak{N})$ then it follows from the Cauchy-Schwartz inequality that $\omega(EFE) = \omega(F)$ for all $F \in \mathcal{P}(\mathfrak{N})$.

Now let $F_\varphi \in \mathcal{P}(\mathfrak{N})$ be a filter for a normal state φ on $\mathfrak{N}$. Since $F_\varphi \neq O$ there must be a normal state ω such that $\omega(F_\varphi) > 0$. Then from the filter property $\dfrac{\omega(F_\varphi F_\varphi F_\varphi)}{\omega(F_\varphi)} = 1 = \varphi(F_\varphi)$. Suppose that φ is a mixed that: $\varphi := \lambda_1\phi_1 + \lambda_2\phi_2$, where $0 < \lambda_1, \lambda_2 < 1$ and $\lambda_1 + \lambda_2 = 1$. Since $\varphi(F_\varphi) = 1$ it must be the case that $\phi_1(F_\varphi) = \phi_2(F_\varphi) = 1$. Now consider some other normal mixed state $\varphi' := \lambda_1'\phi_1 + \lambda_2'\phi_2$ with $\lambda_1' \neq \lambda_1$ and $\lambda_2' \neq \lambda_2$. Note that $\varphi'(F_\varphi) = 1$ and, therefore (as seen above), $\varphi'(F_\varphi E F_\varphi) = \varphi'(E)$ for all $E \in \mathcal{P}(\mathfrak{N})$. Using again the filter property, $\dfrac{\varphi'(F_\varphi E F_\varphi)}{\varphi'(F_\varphi)} = \varphi'(E) = \varphi(E)$ for all $E \in \mathcal{P}(\mathfrak{N})$. Since $\mathfrak{N}$ is the weak closure of $\mathcal{P}(\mathfrak{N})$ and since normal states are weakly continuous it follows that $\phi' = \phi$, a contradiction.

Appendix 4 Belief Filters Are State Filters

Suppose that $X_\varphi \in \mathcal{P}(\mathfrak{N})$ a belief filter for the normal state φ, and let Pr be a completely additive probability on $\mathcal{P}(\mathfrak{N})$ such that $\Pr(X_\varphi) \neq 0$. By the generalized Gleason theorem it follows that $\omega(X_\varphi) \neq 0$ and that

$$\frac{\omega(X_\varphi E X_\varphi)}{\omega(X_\varphi)} = \varphi(E) \text{ for all } E \in \mathcal{P}(\mathfrak{N}) \quad (*)$$

where ω is the normal state that extends Pr. Now use the facts that $\mathfrak{N}$ is the weak closure of $\mathcal{P}(\mathfrak{N})$ and that normal states are continuous in the weak topology to conclude that

$$\frac{\omega(X_\varphi A X_\varphi)}{\omega(X_\varphi)} = \varphi(A) \text{ for all } A \in \mathfrak{N} \quad (**)$$

Since Pr was arbitrary and since completely additive probabilities are in one-one correspondence with normal states, (**) holds for any normal state ω such that $\omega(X_\varphi) \neq 0$, which is to say that X_φ is a state filter for φ.

Appendix 5 The Buchholz-Doplicher-Longo Theorem

Here we sketch the proof that the split inclusion property is sufficient for the existence of local filters in AQFT; for a proof of the converse the reader is referred to Buchholz et al. (1986),

Consider any open $\mathcal{O} \subset \mathcal{M}$. By the split inclusion property of the net $\mathcal{O} \mapsto \mathfrak{N}(\mathcal{O})$ of von Neumann algebras, if $\widehat{\mathcal{O}}$ is even slightly larger than $\mathcal{O}$ there is a Type I factor $\mathfrak{F}$ such that $\mathfrak{N}(\mathcal{O}) \subseteq \mathfrak{F} \subseteq \mathfrak{N}(\widehat{\mathcal{O}})$. If $E \in \mathfrak{F}$ is a minimal projection for $\mathfrak{F}$ then

$$EAE = \mathbb{C}(A)E \text{ for all } A \in \mathfrak{F} \tag{11.1}$$

where $\mathbb{C}(A)$ is a complex number depending on A. The support projection E_ψ for a normal pure (= vector) state ψ on $\mathfrak{F} \simeq \mathfrak{B}(\mathcal{H})$ is the projection onto the ray spanned by the vector $|\psi\rangle \in \mathcal{H}$ representing ψ. Since E_ψ is a minimal projection and since $\psi(E_\psi) = 1$ and $\psi(E_\psi AE_\psi) = \psi(A)$, it follows from (11.1) that $\psi(E_\psi AE_\psi) = \psi(A) = \mathbb{C}(A)$ and, thus,

$$E_\psi AE_\psi = \psi(A)E_\psi \text{ for all } A \in \mathfrak{F}. \tag{11.2}$$

We can conclude that for any normal state ω on $\mathfrak{N}(\mathcal{M})$

$$\omega(E_\psi AE_\psi) = \omega(E_\psi)\psi(A) \text{ for all } A \in \mathfrak{F}. \tag{11.3}$$

And a fortiori (11.3) holds for all $A \in \mathfrak{N}(\mathcal{O}) \subset \mathfrak{F}$. The upshot is that if φ is a normal state on $\mathfrak{N}(\mathcal{M})$ whose restriction $\varphi|_{\mathfrak{N}(\mathcal{O})}$ to the local algebra $\mathfrak{N}(\mathcal{O})$ extends to a normal pure state ψ on $\mathfrak{F}$ then the support projection E_ψ for ψ serves as a local filter for φ and, as hoped, E_ψ belongs to $\mathfrak{N}(\widehat{\mathcal{O}})$ associated with a $\widehat{\mathcal{O}}$ only slightly larger than $\mathcal{O}$. It remains only to note that any normal state on $\mathfrak{N}(\mathcal{O})$ extends to a vector state on $\mathfrak{F}$. Since $\mathfrak{N}(\mathcal{O})$ is Type III, any normal state ξ on $\mathfrak{N}(\mathcal{O})$ is a vector state. So there is a $|\xi\rangle \in \mathcal{H}$ such that $\xi(A) = \langle\xi|A|\xi\rangle$ for all $A \in \mathfrak{N}(\mathcal{O})$. This state obviously extends to the vector state $\bar{\xi}$ on $\mathfrak{F} \simeq \mathfrak{B}(\mathcal{H})$ where $\bar{\xi}(B) = \langle\xi|B|\xi\rangle$ for all $B \in \mathfrak{F}$. The support projection for this state $E_{\bar{\xi}} \in \mathfrak{F} \subset \mathfrak{N}(\widehat{\mathcal{O}})$ is then the desired local filter.

Appendix 6 Interpreting the Buchholz-Doplicher-Longo Theorem

While the Buchholz-Doplicher-Longo theorem is airtight, its *physical interpretation* poses some puzzles. No filter for any state on $\mathfrak{N}(\mathcal{O})$ lies inside that algebra, which is assumed to contain all observables pertinent to the region $\mathcal{O}$. Even in the best case that a filter for a state on $\mathfrak{N}(\mathcal{O})$ belongs to an algebra associated with a slightly larger region that contains $\mathcal{O}$, this has the odd consequence that an experimenter who wants to prepare her laboratory, assumed to be coextensive with $\mathcal{O}$, in a state has to leave her laboratory in order to succeed. This predicament might strike some as in tension with the spirit of *local* quantum physics. A further oddity is the fact that, while $E_\varphi \in \mathfrak{N}(\widehat{\mathcal{O}})$ might be a filter for normal state φ on $\mathfrak{N}(\mathcal{O})$, it is not a filter for any normal state on its home algebra $\mathfrak{N}(\widehat{\mathcal{O}})$. A Type III algebra, $\mathfrak{N}(\widehat{\mathcal{O}})$ admits no filters for its normal states. And this short-circuits what might otherwise have been a comforting physical story about Buchholtz-Doplicher-Longo preparations. The comforting story is that we are preparing a state φ on $\mathfrak{N}(\mathcal{O})$ by preparing a state on a larger algebra containing $\mathfrak{N}(\mathcal{O})$; φ is just the restriction to $\mathfrak{N}(\mathcal{O})$ of the prepared superstate. (The comforting story incidentally explains why, contrary to ordinary expectations, the states we prepare in QFT are mixed. The explanation is that in the presence of entanglement, subsystem states are inevitably mixed, and subsystem states are what BDL preparations prepare.) The comforting story is short-circuited because in the BDL scenario, we aren't preparing a state on the larger algebra $\mathfrak{N}(\widehat{\mathcal{O}})$ /the larger region $\widehat{\mathcal{O}}$. We aren't because we can't: normal states on $\mathfrak{N}(\widehat{\mathcal{O}})$ lack filters in $\widehat{\mathcal{O}}$. While such puzzles are no reason to disbelieve the theorem, they leave some of us wondering, sometimes, what we're believing when we believe this piece of mathematics.

Acknowledgements This work reflects a collaboration with the late Aristides Arageorgis extending over many years. We are indebted to him. And we thank Gordon Belot, Jeff Bub, and the editors of this volume for valuable guidance and feedback on earlier drafts.

References

Araki, H. (2009). *Mathematical theory of quantum fields*. Oxford: Oxford University Press.

Bratelli, O., & Robinson, D. W. (1987). *Operator algebras and quantum statistical mechanics I* (2nd ed.). Berlin: Springer.

Bub, J., & Pitowsky, I. (2010). Two dogmas of quantum mechanics. In S. Saunders, J. Barrett, A. Kent, & D. Wallace (Eds.), *Many worlds? Everett, quantum theory, & reality* (pp. 433–459). Oxford: Oxford University Press.

Buchholz, D., & Wichmann, E. H. (1986). Causal independence and the energy-level density of states in local quantum field theory. *Communications in Mathematical Physics, 106*, 321–344.

Buchholz, D., Doplicher, S., & Longo, R. (1986). On Noether's theorem in quantum field theory. *Annals of Physics, 170*, 1–17.

Cassinelli, G., & Zanghi, N. (1983). Conditional probabilities in quantum mechanics. *Nuovo Cimento B, 73*, 237–245.

de Finetti, B. (1972). *Probability, induction and statistics*. London: Wiley.

de Finetti, B. (1974). *Theory of probability* (2 Vols). New York: Wiley.

Drish, T. (1979). Generalizations of Gleason's theorem. *International Journal of Theoretical Physics, 18*, 239–243.

Eilers, M., & Horst, E. (1975). Theorem of Gleason for nonseparable Hilbert spaces. *International Journal of Theoretical Physics, 13*, 419–424.

Feintzeig, B., & Weatherall, J. (2019). Why be regular? Part II. *Studies in History and Philosophy of Science Part B: Studies in History and Philosophy of Modern Physics, 65*, 133–144.

Haag, R. (1992). *Local quantum physics*. Berlin: Springer.

Halvorson, H. (2001). On the nature of continuous physical quantities in classical and quantum mechanics. *Journal of Philosophical Logic, 30*, 27–50.

Halvorson, H. (2007). Algebraic quantum field theory. In J. Earman & J. Butterfield (Eds.), *Handbook of the philosophy of science, philosophy of physics part A* (pp. 731–864). Amsterdam: Elsevier.

Hamhalter, J. (2003). *Quantum measure theory*. Dordrecht: Kluwer Academic.

Hegerfeldt, G. C., & Mayato, R. (2012). Discriminating between von Neumann and Lüders reduction rule. *Physical Review A, 85*, 032116.
Kadane, J. B, Schervish, M. J., & Seidenfeld, T. (1986). Statistical implications of finitely additive probability. In P. K. Goel & A. Zellner (Eds.), *Bayesian inference and decision techniques* (pp. 59–76). Amsterdam: Elsevier.
Kadison, R. V., & Ringrose, J. R. (1987). *Fundamentals of the theory of operator algebras* (2 Vols). Providence: Mathematical Society.
Kumar, C. S., Shukla, A., & Mahesh, T. S. (2016). Discriminating between Lüders and von Neumann measuring devices: An NMR investigation. *Physics Letters A, 380*, 3612–2616.
Loewer, B. (2012). Two accounts of laws and time. *Philosophical Studies, 160*, 115–137.
Maeda, S. (1990). Probability measures on projections in von Neumann algebras. *Reviews in Mathematical Physics, 1*, 235–290.
Petz, D., & Rédei, M. (1996). John von Neumann and the theory of operator algebras. *The Neumann Compendium*, 163–185.
Pitowsky, I. (2003). Betting on the outcomes of measurements: A Bayesian theory of quantum probability. *Studies in the History and Philosophy of Modern Physics, 34*, 395–414.
Pitowsky, I. (2006). Quantum mechanics as a theory of probability. In: W. Demopoulos & I. Pitowsky (Eds.), *Physical theory and its implications; essays in honor of Jeffrey Bub* (pp. 213–240). Berlin: Springer.
Raggio, G. A. (1988). A remark on Bell's inequality and decomposable normal states. *Letters in Mathematical Physics, 15*, 27–29.
Ruetsche, L. (2011). Why be normal? *Studies in History and Philosophy of Modern Physics, 42*, 107–115.
Savage, L. J. (1972). *Foundations of statistics* (2nd ed.). New York: Wiley.
Seidenfeld, T. (2001). Remarks on the theory of conditional probability. In V. F. Hendricks, et al. (Eds.), *Probability theory* (pp. 167–178). Dordrecht: Kluwer Academic.
Skyrms, B. (1992). Coherence, probability and induction. *Philosophical Issues, 2*, 215–226.
Srinvas, M. D. (1980). Collapse postulate for observables with continuous spectra. *Communications in Mathematical Physics, 71*, 131–158.
Takesaki, M. (1972). Conditional expectations in von Neumann algebras. *Journal of Functional Analysis, 9*, 306–321.
Williams, P. (2017). Scientific realism made effective. *The British Journal for the Philosophy of Science, 70*, 209–237.

Chapter 12
The Measurement Problem and Two Dogmas About Quantum Mechanics

Laura Felline

Abstract According to a nowadays widely discussed analysis by Itamar Pitowsky, the theoretical problems of QT are originated from two 'dogmas': the first, forbidding the use of the notion of measurement in the fundamental axioms of the theory; the second, imposing an interpretation of the quantum state as representing a system's objectively possessed properties and evolution. In this paper I argue that, contrarily to Pitowsky analysis, depriving the quantum state of its ontological commitment is not sufficient to solve the conceptual issues that affect the foundations of QT.

In order to test Pitowsky's analysis I make use of an argument elaborated by Amit Hagar and Meir Hemmo, showing how some probabilistic interpretations of QT fail at dictating coherent predictions in Wigner's Friend situations. More specifically, I evaluate three different probabilistic approaches: qBism, as a representative of the epistemic subjective interpretation of the quantum state; Jeff Bub's information-theoretic interpretation of QT, as an example of the ontic approach to the quantum state; Itamar Pitowsky's probabilistic interpretation, as an epistemic but objective interpretation. I argue that qBism maintains self-consistency in Wigner's Friend scenarios, although, in the resulting picture, the real subject matter of QT clashes alarmingly with scientific practice. The other two approaches, instead, strictly fail when confronted with Wigner's Friend scenarios.

Keywords Measurement · Quantum state · Wigner's Friend · Quantum information theory · Quantum Bayesianism

---

L. Felline (✉)
Department of Philosophy, University of Roma Tre, Rome, Italy

M. Hemmo, O. Shenker (eds.), *Quantum, Probability, Logic*,
Jerusalem Studies in Philosophy and History of Science,
https://doi.org/10.1007/978-3-030-34316-3_12

12.1 Introduction

There is a time-honoured approach to the foundations of Quantum Theory (QT), according to which the oddities of this theory originate from the improper interpretation of its subject study as concerning properties and behaviour of material objects, being them corpuscles or waves, represented by the quantum state. This interpretation is often held responsible, for instance, for the difficulties in explaining quantum correlations, or the measurement problem.

The argument behind this association is quite intuitive. Let's take the problem of (apparently) non-local quantum correlations, and the illustrative case of an EPR-Bohm entangled pair of particles A and B, which are sent apart. The EPR-Bohm state does not associate a determinate property to either particle, however, when a measurement on A is performed, the entangled state of A + B effectively and instantaneously collapses, associating now a specific value to the measured observable, not only in the system A, but in the system B as well. As a result, due to A's and B's perfect anticorrelation, knowing the result of a measurement on A automatically also informs us about the state of B.

Now, if the quantum state is interpreted realistically, as encoding the properties possessed by a system, and its evolution, then a straightforward interpretation takes the collapse as a physical change both in A's and in B's properties. But B was never disturbed by the observer's interaction with A, so an explanation is in order for B's behaviour. At the origin of the difficulty of the sought-after explanation is the tension between the natural assumption that the measurement over A has in some way caused the change in B's state, and the fact that causal processes can't be instantaneous.

This difficulty seems to disappear once we give up the assumption that quantum states, contrarily to classical states, represent properties and evolution of physical systems. If the quantum state is just an economical codification of a probability distribution over measurement results, then the collapse of the wave function does not represent an actual physical process, and B's state change does not have to be explained physically.

So, according to this argument, the problem of non-locality (and, as we are about to see, the measurement problem with it) is explained away once one rejects what I will call from now on a representational function of the quantum state.[1]

But this conclusion would be too quick. For instance, in the case of non-locality, a genuine solution to the problem must necessarily provide an explanation of why, given that there is no physical process connecting the occurrence of A's and B's measurement results, said results are always perfectly anti-correlated. But this is not an easy task. As a matter of fact, when it comes to provide an alternative explanation to quantum conundrums, non-representational approaches plod along as well.

[1]See later in this section ad especially note 2 for some clarifications about the terminology.

Because of these persisting difficulties, mainstream physics settles for the adoption of a pragmatic but inconsistent attitude towards the quantum state (see Wallace 2016), to be interpreted as representing a system or as a codification of probabilities of measurement results, depending on what works in that situation.

This, for instance, is physicist David Mermin, about the two-slit experiment:

> We know what goes through the two slits: the wave-function goes through, and then subsequently an electron condenses out of its nebulosity onto the screen behind. Nothing to it. Is the wave-function itself something real? Of course, because it's sensitive to the presence of both slits. Well then, what about radioactive decay: does every unstable nucleus have a bit of wave-function slowly oozing out of it? No, of course not, we're not supposed to take the wave-function that literally; it just encapsulates what we know about the alpha particle. So then it's what we know about the electron that goes through the two slits? Enough of this idle talk: back to serious things! (Mermin 1990, p. 187, cited in Maudlin 1997, p. 146)

This being said, in the last decades non-representational approaches to QT have experienced new life thanks to the evolutions of Information Theory. The hope is that the latter might provide a formal background powerful enough to allow a genuine solution to the theoretical problems of QT, without necessarily collapsing into instrumentalism. Accordingly, QT is not about particles or waves and their behaviour, but is rather about information.

A terminological clarification is in order at this point. The approaches denying that QT is about properties and behaviour of physical entities are often labelled anti-realist approaches to QT (see e.g. Myrvold 2018, §4.2); however, this characterization might be confusing, since not every such account is anti-realist. In fact, the majority of them openly rejects anti-realism (and this is definitely true of the three approaches examined in this paper): QT is not a theory about the elementary constituents of matter, but it is still realist with respect to its subject matter.

In this paper, I adopt Jeff Bub's terminology and label the two competing views probabilistic and representational approaches, where the former interprets QT as a theory of probability (and where we can further distinguish between an ontic and an epistemic interpretation of the quantum state), while the latter interpret QT as a theory about the elementary constituents of matter, their properties and their dynamical behaviour.[2]

As a more general point, I argue that what was already shown for more traditional probabilistic approaches, equally applies to the new information-based formulations: contrarily to the simplistic attitude often adopted even in recent illustrations of the subject, depriving the quantum state of its ontological commitment is not sufficient to solve the conceptual issues that affect the foundations of QT.

[2]A further clarification: Bub takes this terminology from Wallace (2016) where, however, probabilistic interpretations are characterized as those according to which observables always have determinate values, while the quantum state is merely "an economical way of coding a probability distribution over those observables" (p. 20). I think this characterization is incorrect of the three approaches here analysed, since they do not take observables as determinate all the time, therefore I don't refer to Wallace terminology.

In arguing for this general point, I focus on the measurement problem, typically taken to be explained away in non-representational accounts. As in the above illustrated example of non-locality, and contrarily to what often claimed by the literature, also in the case of the measurement problem, the majority of non-representational accounts fail to provide a genuine solution.

One of the most systematic expositions of the non-representational view and of its relationship with the measurement problem was put forward few years ago by Itamar Pitowsky, in a series of paper (Pitowsky 2003, 2004, 2006; Bub and Pitowsky 2010) where he puts forward his own Bayesian theory of quantum probability.

According to his analysis, the theoretical problems of QT are originated from two 'dogmas': the first forbidding the use of the notion of measurement in the fundamental axioms of the theory; the second imposing an interpretation of the quantum state as representing a system's objectively possessed properties and evolution.

In this paper I illustrate and criticise Pitowsky's analysis, taking the rejection of the two dogmas as an accurate articulation of the fundamental stances of the probabilistic approach to the measurement problem. In order to assess such stances, I will test them through the use of an argument elaborated by Amit Hagar and Meir Hemmo (2006), showing how some probabilistic interpretations of QT fail at dictating coherent predictions in Wigner's Friend situations.

More specifically, I evaluate three different probabilistic approaches: qBism (more specifically, the version articulated by Christopher Fuchs, e.g. Fuchs 2010; Fuchs et al. 2014) as a representative of the epistemic subjective interpretation of the quantum state; Bub's information-theoretic interpretation of QT (Bub 2016, 2018) as an example of the ontic approach to the quantum state; Pitowsky's probabilistic interpretation (Pitowsky 2003, 2004, 2006), as an epistemic but objective interpretation. I argue that qBism succeeds in providing a formal solution to the problem that does not lead to a self-contradictory picture, although the resulting interpretation would bring the real subject matter of QT to clash alarmingly with scientific practice. The other two approaches, instead, strictly fail when confronted to scenarios where the measurement problem is relevant, showing in such a way that they don't provide a genuine solution to the problem.

12.2 Two Dogmas and Two Problems

The idea that deflating the quantum state's ontological import is sufficient to explain away the measurement problem is a *leitmotif* in the literature about information-based accounts of QT, still shared by the vast majority of the advocates of the probabilistic view:

> The existence of two laws for the evolution of the state vector becomes problematical only if it is believed that the state vector is an objective property of the system. If, however, the state of a system is defined as a list of [experimental] propositions together with their [probabilities of occurrence], it is not surprising that after a measurement the state must be changed to be in accord with [any] new information. (Hartle 1968, as quoted in Fuchs 2010)

Itamar Pitowsky, also together with Jeff Bub (Pitowsky 2006; Bub and Pitowsky 2010), has identified two distinct issues behind the measurement problem. The first one, that he calls the small measurement problem, is the question "why is it hard to observe macroscopic entanglement, and what are the conditions in which it might be possible?" (2006, p. 28). The second issue, the big measurement problem, is the problem of accounting for the determinateness of our experiences:

> The 'big' measurement problem is the problem of explaining how measurements can have definite outcomes, given the unitary dynamics of the theory: it is the problem of explaining *how individual measurement outcomes come about dynamically*. (Bub and Pitowsky 2010, p. 5)

In this paper we are exclusively concerned with the big problem; therefore, from now on, I will refer to the latter more simply as the measurement problem. However, I reject Bub's and Pitowsky a priori request for a dynamical account; therefore I characterise the measurement problem as the problem of accounting for the determinateness of measurement results.

The second most notable contribution that Pitowsky has given to the analysis of the measurement problem concerns the identification of two assumptions, labelled as 'dogmas', in the interpretation of QT, that, according to the philosopher, are at the origin of the issues in the interpretation of QT.

> The first dogma is Bell's assertion (defended in [(Bell 1987)]) that measurement should never be introduced as a primitive process in a fundamental mechanical theory like classical or quantum mechanics, but should always be open to a complete analysis, in principle, of how the individual outcomes come about dynamically. The second dogma is the view that the quantum state has an ontological significance analogous to the ontological significance of the classical state as the 'truthmaker' for propositions about the occurrence and non-occurrence of events, i.e., that the quantum state is a representation of physical reality. (Pitowksy 2006, p. 5)

A couple of observations are in order here. First of all, the two claims: "the quantum state has an ontological significance analogous to the ontological significance of the classical state as the 'truthmaker' for propositions about the occurrence and non-occurrence of events" and "the quantum state is a representation of physical reality", are not equivalent, although they are here put forward as such. More specifically, one can deny that the quantum state represents 'traditionally' (Bub 2018) in the sense that it does not represent a system's properties, and therefore acting as a truthmaker for propositions about the occurrence and non-occurrence of events, without denying that the quantum state represents physical reality. In fact, that's the case for Bub's information-theoretic interpretation which, as I will argue below, denies that the quantum state represents a system's properties and dynamics, and yet interprets the quantum state ontically, as representing physical reality. In the following, therefore, I take the second dogma as the view that the quantum state does not represent traditionally, while the issue whether the quantum state represents physical reality or a mental state remains open.

Secondly, although probabilistic interpretations reject both dogmas, it is useful to keep in mind that, at least at a first sight, the two are logically distinct, and it still remains to be seen whether they need to come in package. The rejection of the first dogma leads to a black-box interpretation, whose traditional versions put forward an interpretation of the quantum state as representing the properties of physical systems (e.g. Bohr's interpretation). It might be argued that the more traditional black-box theories have proven inconsistent, and that once one rejects the first dogma, the explanation away of the measurement problem requires the rejection of the second. This is, for instance, the core of Pitowsky's and Bub's view, according to which not only the second dogma runs 'very quickly' into the measurement problem (Pitowsky 2006, p. 3), but the rejection of the latter is sufficient to solve the measurement problem in black-box theories.

In support of this last claim, over which the critical analysis of this paper is pivoting, Pitowsky outlines a straightforward argument. First of all, the measurement problem consists in the tension between the fact that measurements always yield determinate values, and the fact that they are also performed on systems whose states do not associate a determinate value to the measured observable. If this is so, for the achievement of a solution it would be sufficient to reject one of the horns of the dilemma:

> The BIG problem concerns those who believe that the quantum state is a real physical state which obeys Schrodinger's equation in all circumstances. In this picture a physical state in which my desk is in a superposition of being in Chicago and in Jerusalem is a real possibility; and similarly, a superposed alive-dead cat. [. . .]
>
> In our scheme quantum states are just assignments of probabilities to possible events, that is, possible measurement outcomes. This means that the updating of the probabilities during a measurement follows the Von Neumann-Luders projection postulate and not Schrodinger's dynamics. [. . .] So the BIG measurement problem does not arise. (Pitowsky 2006, pp. 26–27).

12.3 Incompatible Predictions in Black-Box Approaches

In this section I am going to introduce a somewhat neglected argument put forward in Hagar (2003) and in Hagar and Hemmo (2006), which introduces a scenario against which, according to HH, black-box interpretations of QT fail.

Let's take an experiment with two observers, Wigner and Friend, where the latter is inside a lab, isolated from Wigner and the rest of the environment. Before Friend enters the lab, she and Wigner agreed upon the following protocol: in the first part of the experiment Friend performs a z-spin measurement on a half-spin particle P in the state:

$$|\text{Psi } 0>_\text{P} = \left(1\sqrt{2}\,|+_\text{z}>_\text{P} + 1/\sqrt{2}\,|-_\text{z}>_\text{P}\right) \tag{12.1}$$

Therefore, the system composed of Friend and the particle (F + P) is in the state:

$$|\text{Psi } 0>_{\text{P+F}} = \left(1/\sqrt{2}\,|+_\text{z}>_\text{P} + 1/\sqrt{2}\,|-_\text{z}>_\text{P}\right)|\text{Psi } 0>_\text{F} \tag{12.2}$$

Since they agreed on this first part of the protocol, Wigner knows that Friend is going to perform the measurement; however, because the lab is isolated from Wigner and the rest of the environment, he does not know the result of her measurement.

Now, consider a second measurement, of an observable O of the system F + P, with eigenstate

$$| \mathrm{Psi}>_{\mathrm{P+F}} = 1/\sqrt{2}\,|+_{\mathrm{z}}>_{\mathrm{P}}\,|+_{\mathrm{z}}>_{\mathrm{F}} + 1/\sqrt{2}\,|-_{\mathrm{z}}>_{\mathrm{P}}|-_{\mathrm{z}}>_{\mathrm{F}} \quad (12.3)$$

whose eigenvalue is, say, YES.

HH's question is: according to a black-box approach, what predictions are dictated by QT for the results of this measurement?

On the one hand, F + P is an isolated system; therefore, its evolution should follow Schrodinger equation. If so, the state at the end of Friend's measurement should be

$$|\mathrm{Psi}\ 1>_{\mathrm{P+F}} = 1/\sqrt{2}\,|+_{\mathrm{z}}>_{\mathrm{P}}\,|+_{\mathrm{z}}>_{\mathrm{F}} + 1/\sqrt{2}\,|-_{\mathrm{z}}>_{\mathrm{P}}|-_{\mathrm{z}}>_{\mathrm{F}}, \quad (12.4)$$

which is an eigenstate of O, with eigenvalue YES. From this perspective, therefore, QT dictates that the measurement of O will yield YES with probability 1.

On the other hand, Friend's interaction with P is a measurement interaction, which, according to the black-box approach, requires the application of Luders' rule. It follows therefore that F + P's state should collapse into either one of two pure states

$$|\ \mathrm{Psi}\ 1> \ = \ |+_{\mathrm{z}}>_{\mathrm{P}}\,|+_{\mathrm{z}}>_{\mathrm{F}} \quad (12.5)$$

or

$$|\ \mathrm{Psi}\ 1> \ = \ |-_{\mathrm{z}}>_{\mathrm{P}}\,|-_{\mathrm{z}}>_{\mathrm{F}}, \quad (12.6)$$

both associating a probability ½ to the result YES in an O-measurement. According to this line of thought, therefore, QT dictates that the probabilities for a YES result over the measurement of O is ½.

To make things more complicated, each option adopts a specific stance about the evolution of the quantum state in a measurement process, ruling therefore out a black-box approach. If we take into consideration that the lab is an isolated system, and therefore apply Schrodinger equation to the first part of the experiment, this would correspond to a no-collapse approach. If, on the opposite, we apply Luders rule, then we are indeed applying a collapse view of quantum measurement.

According to Hagar and Hemmo (HH), this shows that any solution to this kind of Wigner's friend scenario (and therefore, a fortiori, of the measurement problem) implies an analysis of the processes behind measurement interaction. Since, they add, every black-box interpretation of QT is equally affected by this argument, they conclude that black-box approaches to quantum theory are unable to provide a coherent picture of the predictions of QT.

In the rest of the paper, the application of this argument to three different theoretical accounts will be used to explore the relationship between the two dogmas and the measurement problem.

12.4 QBism

I have already said that HH chose qBism as a case study to formulate their argument, while claiming that the same result applies to any black-box approach to QT. The succeeding debate around this specific approach, however, has shed some light about how different black-box approaches lead to different results when faced to Wigner's Friend scenarios. In order to formulate my criticism of qBism, in the following I will build on such a debate. I will first isolate the exact feature of qBism that allows it to formally escape HH's argument, and then show how this same feature leads to a philosophically problematic account of QT.

According to qBism, the quantum state represents the rational subjective belief of an observer, assembled from personal experience. The first direct consequence of this interpretation is that quantum probabilities are not a measure of anything in reality, but only of the degree of belief of an observer over future experience. Secondly, different observers might consistently attribute different quantum states to the same system. In order to solve the dilemma in the Wigner's Friend scenario, therefore, one needs to take into consideration each observer's individual perspective: given that she has gathered new data, QT prescribes *Friend* to update P's quantum state (and F + P's state, clearly). From her perspective, then, the prediction of the result in an O-measurement in the second stage of the experiment will be YES with probability 1/2. On the other hand, Wigner has registered no new data, therefore QT prescribes him to maintain the entangled state (12.4) for F + P. From his perspective, therefore, the O-measurement will yield YES with probability 1.

The surprising consequence is that, according to qBism, different observers can coherently hold incompatible predictions about the world. The subjectivism embraced by qBism turns QT into "a "single-user theory": probability assignments express the beliefs of the agent who makes them, and refer to that same agent's expectations for her subsequent experiences. The term "single-user" does not, however, mean that different users cannot each assign their own coherent probabilities" (Fuchs et al. 2014, p. 2).

According to Amit Hagar (2007) this strategy can't work: the (repeated) measurement of O will yield a series of results with a certain frequency that will (hopefully!) confirm one of the predictions and falsify the other. If this is so, at the end of the day one of the two alternatives (collapse or no-collapse) will be right, while the other will be wrong.

To Hagar, Timpson (2013) replies that this kind of objections are ineffective against Qbism:

> [n]o objection can be successful which takes the form: 'in such and such a situation, the quantum Bayesian position will give rise to, or will allow as a possibility, a state assignment which can be shown not to fit the facts', simply because the position denies that the requisite kind of relations between physical facts and probability assignments hold. (Timpson 2013, p. 209)

Here is my attempt to unpack Timpson's position: in the experiment here illustrated, the consequence of the subjective interpretation of probabilities is that Wigner's beliefs about the quantum state of F + P are partially determined by a previous subjective assignment, therefore the 'probability 1' of the result YES is also merely subjective, rather than an objective certainty. Let's say for instance that Wigner's experiment yields a different result from YES. According to the ontic view of the quantum state, probabilistic predictions are determined by an objective reality grasped by the theory. Wrong predictions, therefore, show that there is something wrong in Wigner's representation of reality: this is one of the pillars of empirical sciences reasoning. However, in qBism, the relation that holds between data and quantum state representation is not of this kind; probabilities do not represent anything in the world, which means that Wigner's quantum state representation can't be *wrong* about the world. As a consequence, dealing with 'surprising' experiences does not have to be cause of sweat for the qBist, who's reaction is simply to apply the same rules for updating expectations (s)he applies every time (s)he must associate (but not *discover*) previously unknown chances to an event.

However, one can still counter that, after a sufficiently large repetition of the experiment, the falsification (on either side) of one's predictions should lead a rational agent to abandon the unsuccessful rules she has been following for the formation of expectations (i.e. QT!).[3] But this means, as HH argued, that QT is, as it is, incomplete.

Another counterargument often used to block HH's conclusion is to hinge on the fact that there is no way for Wigner and Friend to compare their experience and predictions: "Results objectively available for anyone's inspection? This is the whole issue with "Wigner's friend" in the first place. If both agents could just "look" at the counter simultaneously with negligible effect in principle, we would not be having this discussion" (Fuchs and Stacey 2019, p. 9).

The argument here is that in the second measurement, P's z-spin and Friend's record of it will be erased, so that any ground for Friend's ½ predictions would fail. However, that this is not really the core of the issue has been already argued extensively,[4] and recently explicitly illustrated by more sophisticated versions of the Wigner's friend thought experiment showing that the same kind of contradicting predictions can be reproduced even in scenarios where the necessary information

[3] I am very grateful to Meir Hemmo and an anonymous referee for pushing me to provide a more thorough evaluation of Timpson's position.

[4] By HH themselves first, but see also, just to cite one, (Bacciagaluppi 2013) for an insightful take on the problematical character of the qBist solution.

is available to all the observers, while the lab is maintained in isolation (see for instance (Frauchiger and Renner 2018) and (Baumann and Brukner 2019)).

The core of the qBist solution to this problem lies instead in "some kind of solipsism or radical relativism, in which we care only about single individuals' credence's, and not about whether and how they ought to mesh" (Bacciagaluppi 2013, p. 6).

Now, it is difficult to formulate an argument with solid grounds in favour of an interpretation of QT, if your starting point is solipsism. However, it might be argued that the situation is not as bad as it might seem. Timpson (2013, §9.3, but see also Healey 2017, §2), for instance, denies that qBism implies solipsism, and claims instead that the former avoids the paradoxical or anyway pernicious consequences of the latter as a general epistemological thesis. In other words, although, according to qBism, QT fails to include the perspective of more than one observer, it does not follow that according to Qbism there is only one sentient being in the world, let alone that the rest of empirical science is affected by these limitations. This same defence strategy is used by Timpson against the charge of instrumentalism: it is true that, according to Qbism, QT is neither true or false of the world, and a mere instrument for bookkeeping of probabilities. However, in the case of Qbism, this conclusion follows, from specific considerations about the structure of QT and only applies to said theory. Because of these considerations, qBism does not imply the majority of the controversial features of instrumentalism as a general position about science.

The upshot of this rebuttal to criticisms is: QT might forbid inferences about the experiences of other observers as solipsism does, and it might be as void of knowledge about the world, as instrumentalism dictates – but this is OK, because qBism does not lead to the possibly paradoxical consequences of such thesis.

I am not going here to challenge Timpson's argument. Here, I take for granted that this defence provides qBism with a way-out from the conclusions of HH argument. This being said, it is legitimate to scrutinize the consequences of this unique status of QT with respect to the rest of empirical science – uniqueness that, we just concluded, lies at the core of the qBist success.

I am going to argue that this achievement is a pyrrhic victory, as it still fails to provide an acceptable picture of the scientific enterprise in the field of QT.

qBism succeeds in the context of Wigner's Friend scenarios, but also in explaining away other apparent oddities of the quantum world, because it characterizes QT not as a physical theory, but as a Bayesian theory for the formation of expectations. As such, it can coherently ignore the constraints imposed to an objective description of the world, when they are in the way of coherence. While a theory about the world must submit to the typical epistemic and methodological standards of empirical sciences, Bayesian epistemology is subordinate to very different standards, for the simple reason that different subject studies require different methodologies. The violation of the requirement of intersubjectivity in the qBist explanation away of non-locality (see Fuchs et al. 2014) and in the account of the Wigner's Friend thought experiment.

In other words, Qbism plays a different game, with different rules, with respect to physics and, more in general, with respect to natural sciences. And it escapes strict failure when confronted against HH's argument (as far as self-consistency is the only requirement to avoid strict failure, but see later) only as far as QT is not interpreted as natural science.

If this is so, a coherent and truthful endorsement of Qbism implies also the renunciation to constraints with a great value of guidance to scientific change like empirical adequacy, physical salience, or intersubjectivity because, in a quantum Bayesian theory about rational belief, such criteria fail to make sense. In fact, the renunciation to such criteria is the key for the Qbist explanation away of the oddities of QT.

And yet, it can't be ignored that giving up these criteria means throwing away a huge piece of scientific practice – a hard bullet to bite for working physicists using QT as an empirical science every day.

One might reply that this is a made-up problem: scientific practice does not, nor should, change because of philosophical debates, so there is nothing to be worried about. However, the unproblematic acknowledgment of such a chasm between scientific practice and the real content of scientific theories, is an even harder bullet to bite (if not straightforward failure) for philosophers.[5]

This is what one gets when submitting to a qBist view of QT. As we will see in the next sections, those that indeed think that this is too high price to pay for internal consistency, but still want to deny the representational role of QT, will have to do so while maintaining in some way the empirical import of the quantum state, denied by qBism.

---

[5]In his analysis of the virtues and problems of Qbism, Chris Timpson formulates a similar challenge to Qbism, but focusing on explanation:

> It seems that we do have very many extensive and detailed explanations deriving from quantum mechanics, yet if the quantum Bayesian view were correct, it is unclear how we would do so. (Timpson 2013, p. 226)

Timpson's criticism is that Qbism suffers of an explanatory deficit because it can't explain why QT explains; however, I think that the Qbist lack of a realist explanation for the explanatory power of QT is hardly an unsurmountable problem.

First of all, the Qbist can reply to Timpson's challenge that there is no reason why a theory that is neither true nor false can't do what false theories have done for centuries: being explanatory. If then one insists (wrongfully, I think) that only at least approximate truth have explanatory power, then the Qbist can simply deny a genuine explanatory power for QT: after all, the topic of explanation in QT is a time-honoured headache in philosophy of science, and the philosophers that deny the explanatory power of QT are not few. An objection hinging on the explanatory power of QT can go just as far.

12.5 Bub's Information-Theoretic Interpretation of QT

According to Jeff Bub, QT is a non-representational, probabilistic theory. Yet, while rejecting a traditional representational interpretation of the quantum state, his information-theoretic approach interprets the quantum state ontically, as a physical state, a complete description of a quantum system (Bub 2016, p. 222). More specifically, QT is a theory about information 'in the physical sense': a structure of correlations between intrinsically random events, which is different from the classical structure of correlations measured by Shannon information. Moreover, according to Bub, information is a new kind of physical primitive, whose structure "imposes objective pre-dynamic probabilistic constraints on correlations between events, analogous to the way in which Minkowski space-time imposes kinematic constraints on events" (Bub 2018, p. 5).

Bub adopts and contributes to Pitowsky's analysis of the two dogmas (Bub and Pitowsky 2010; Bub 2016, § 10) and in fact he also puts them at the origin of the issues gravitating around the measurement problem. However, the application of HH's argument to Bub's information-theoretic interpretation shows that the rejection of the two dogmas is not sufficient to solve the measurement problem.

In HH's thought experiment, in fact, QT provides contrasting instructions for the prediction of measurement results. The qBist way out to inconsistency is relativizing the quantum state to single observers. In Bub's interpretation, on the other hand, the quantum state represents an objective probabilistic structure that constrains the system's behaviour. This means that the probabilities codified by QT are objective, and therefore that different observers must associate the same measurement with the same predictions. In Bub's account, QT is not a single-user theory: through its use, Wigner and Friend must be able to come to agreement.

Let's wrap up the results achieved so far. Contrarily to HH's conclusions, their argument does not rule out black-box interpretations of QT; however, Qbism escapes failure by embracing an extremely subjective view of the quantum state that, although dragging QT out of the realm of empirical sciences, allows to consistently disattend constraints imposed to such sciences. Bub, on the other hand, puts forward an information-theoretic account of QT which has the merit of maintaining the empirical import not only of Hilbert space, but also of the quantum state. Regaining the empirical import of the theory, however, means losing the possibility to exploit the qBist solution to HH's test, because, as any other empirical theory, it must be empirically adequate and provide an intersubjective description of the non-perspectival features of our experiences.

In the next section I will discuss Pitowsky's information-theoretic interpretation, a kind of compromise in between the extreme subjectivism of Qbism, and Bub's ontic approach.

It has to be said that Bub has recently addressed the issue of Wigner's Friend scenarios in the information-theoretic interpretation. However, since this new contribution have several points of intersection with Pitowsky's account, it would be useful to discuss it also with his epistemic version of the information-theoretic approach in mind.

12.6 Pitowsky's Information-Theoretic Interpretation

Pitowsky puts forward an epistemic interpretation of the quantum state as a state of partial belief, in the sense of Bayesian probability. As we will see, differently from qBism, his view preserves at least part of the empirical character of the quantum state.

In Pitowsky's information-theoretic approach, closed subspaces of the Hilbert space correspond to possible states of the world, or events, while the quantum state represents an agent's uncertainty about it. In other words, here, as in qBism, the quantum state is a device for bookkeeping of probabilities (Pitowsky 2006, p. 4). Contrarily to Bub's information-theoretic approach, the quantum state is therefore a derived entity, and in fact a fundamental part of Pitowsky's interpretational work consists in the derivation of quantum probabilities from the structure of quantum events. The derivation starts from the axioms of QT as formulated by Birkoff and von Neumann (1936), of which the structure of quantum events is shown to be a model. According to the resulting representation theorem, the space of events L in quantum theory is the lattice of subspaces of a Hilbert space. The fact that, as a consequence of Gleason's theorem, quantum probabilities can be derivable from this sole structure, constitutes, according to Pitowsky, 'one of the strongest pieces of evidence in support of the claim that the Hilbert space formalism is just a new kind of probability theory' (2006, p. 14).

Finally, adopting Bayesian probability as a conceptual and formal background, means analysing probabilities through rational betting behaviour, i.e. they are measured 'by proposing a bet, and seeing what are the lowest odds he will accept' (Ramsey 1926, as cited in Pitowksy 2006, p. 15).

Few clarifications are here in order.

In Pitowsky's view, the notion of measurement appears in the axioms of QT: QT does not make predictions about generic occurrences and correlations between them – in this betting game, agents can only bet on measurements results, and, between them, only those that can be considered *facts*, where not every measurement result is a fact:

> by "fact" I mean here, and throughout, a recorded fact, an actual outcome of a measurement. Restricting the notion of "fact" in this way should not be understood, at this stage, as a metaphysical thesis about reality. It is simply the concept of "fact" that is analytically related to our notion of "event", in the sense that only a recordable event can potentially be the object of a gamble. (2006, p. 10)

If a measurement is performed of which no result is registered, then there is no fact of the matter about the result of that measurement. This does not mean that according to Pitowsky such results do not exist, but rather that they are not 'events' or 'facts' over which the theory is allowed to make predictions.

The second clarification concerns the second dogma, i.e. the reality of the quantum state. I have said that, in Pitowsky's view, the quantum state represents a bookkeeping device for keeping track of probabilities, which are different from classical probabilities. As we have seen, the risk for an epistemic views of the quantum state run the risk of as a bookkeeping device is that it turns out being void of physical content, or, even worse, instrumentalist; part of Pitowsky's conceptual work is therefore devoted to stressing how this approach avoids it.

One way to mark a distance from instrumentalism is to stress by stressing the explanatory power of quantum Bayesianism. Under an instrumentalist view (e.g. in the textbook approach to QT), in fact, the question 'why do the quantum events do not conform to classical probabilities?' has no answer. The information-theoretic view, on the other hand, have more tools to provide such an explanation, i.e. since it is realist towards the structure of quantum gambles, i.e. towards the structure of Hilbert space

> the Hilbert space, or more precisely, the lattice of its closed subspaces, [is] the structure that represents the "elements of reality" in quantum theory. (Pitowsky 2006, p. 4)

> Instrumentalists often take their "raw material" to be the set of space–time events: clicks in counters, traces in bubble chambers, dots on photographic plates, and so on. Quantum theory imposes on this set a definite structure. Certain blips in space–time are identified as instances of the same event. Some families of clicks in counters are assumed to have logical relations with other families, etc. What we call reality is not just the bare set of events, it is this set together with its structure, for all that is left without the structure is noise. [...]
>
> It is one thing to say that the only role of quantum theory is to "predict experimental outcome" and that different measurements are "complementary." It is quite another thing to provide an understanding of what it means for two experiments to be incompatible, and yet for their possible outcomes to be related; to show how these relations imply the uncertainty principle; and even, finally, to realize that the structure of events dictates the numerical values of the probabilities (Gleason's theorem). (Pitowksy 2003, p. 412)

The fact that such a structure has a physical reality allows the above-cited explanations to be genuine physical explanations rather than collapsing to the kind of explanations given by logic or formal epistemology. Said physical explanations are not dynamical, but structural, in the sense of (Felline 2018a, b): the explanandum phenomenon is explained as the instantiation of a physical structure, fundamental in the sense that it is not inferable from the properties and behaviour of underlying entities.

Pitowsky's realism goes further than this. I have said before that, according to Pitowsky's account, only registered measurement results are facts, about which the theory can make predictions. This claim, however, has an exception when the quantum state also describes an objective outside reality, and its predictions can be taken as genuine facts:

> In the Bayesian approach what constitutes a possible event is dictated by Nature, and the probability of the event represents the degree of belief we attach to its occurrence. This distinction, however, is not sharp; what is possible is also a matter of judgment in the sense that an event is judged impossible if it gets probability zero in all circumstances. In the present case we deal with physical events, and what is impossible is therefore dictated by the best available physical theory. Hence, probability considerations enter into the structure of the set of possible events. We represent by 0 the equivalence class of all events which our physical theory declares to be utterly impossible (never occur, and therefore always get probability zero) and by 1 what is certain (always occur, and therefore get probability one). (Pitowsky 2006, p. 5)

So, according to this stipulations, not only registered results are facts, but also results with probability 1. In this case, the quantum state represents something in the world.

This traces a crucial difference between Pitowsky's information-theoretic interpretation and Qbism. In the latter, given the strict Bayesian reading of probabilities, probability 1 only means that the person who associates this probability to an event E very strongly beliefs the occurrence of E (the same is valid, mutatis mutandis, for probability 0).

According to Pitowsky's analysis of the measurement problem, adopting an epistemic view is sufficient to explain away the measurement problem. In fact, the claim that the quantum state is a mental state blocks, at a first analysis, the consequences that we have seen in Bub's ontic approach.

This, however, is not a viable solution for Pitowsky's approach, due to the realism he acknowledges towards the quantum state.

In order to concretely see why, take again HH's argument and the predictions for the measurement of O. In a partially objective reading of the state like Pitowsky epistemic view, the fact that Wigner associates probability 1 to the result YES means that such result is an objective fact about reality. But if this is so, then there is a fact of the matter about what will be the result of the measurement of O and Friend's predictions are objectively wrong.

But this, we know, would mean adopting a non-collapse view of QT.

12.7 A Bohrian Escape

Recently (2018) Bub has formulated a new argument addressing the problem of incompatible predictions of QT in Wigner's Friend scenario. Bub does not directly cite HH's argument, and focuses instead on the more recent argument formulated in (Frauchiger and Renner 2018). In the following, for simplicity, I will keep using the argument as formulated by HH.

Bub stance about the notion of measurement is very much in line with Pitowsky's genuine facts. Accordingly, QT does not apply to, nor make predictions about, generic events, but only about measurement results. Here, as in Pitowsky, the notion of measurement is substantially revised:

> A quantum "measurement" is a bit of a misnomer and not really the same sort of thing as a measurement of a physical quantity of a classical system. It involves putting a microsystem, like a photon, in a situation, say a beamsplitter or an analyzing filter, where the photon is forced to make an intrinsically random transition recorded as one of two macroscopically distinct alternatives in a device like a photon detector. The registration of the measurement outcome at the Boolean macrolevel is crucial, because it is only with respect to a suitable structure of alternative possibilities that it makes sense to talk about an event as definitely occurring or not occurring, and this structure is a Boolean algebra. (Bub 2018, p. 6)

So, on the one hand Bub provides a physical characterization that partially explicates the notion of measurement (it involves an intrinsically random transition for the measurement system, whose final state must be registered macroscopically); on the other hand, the first dogma is still violated, since there is no criterion for whether or not a process counts as measurement (we don't know when a photon is forced to make an intrinsically random transition).

Bub explains the problems in Wigner's Friend scenario as originated by the structure of quantum events, composed by a "family of "interwined" Boolean algebras, one for each set of commuting observables [. . .]. The interwinement precludes the possibility of embedding the whole collection into one inclusive Boolean algebra, so you can't assign truth values consistently to the propositions about observable values in all these Boolean algebras" (Bub 2018, p. 5).

So, the algebra of observables is non-Boolean, but each observable (each measurement) picks up a Boolean algebra which associates different probabilities to same observables. In our thought experiment: Friend's and Wigner's measurements pick up different Boolean algebras that can't be consistently embedded into one single coherent framework.

According to Bub only one Boolean algebra provides the correct framework for the description of the state. In order to justify the selection of one single Boolean algebra, Bub starts from reminding how George Boole introduced Boolean constraints on probability as "conditions of possible experience". He therefore takes a 'Bohrian' perspective and suggests that the choice of a unique Boolean algebra is imposed by the necessity to "tell others what we have done and what we have learned" (Bohr, as cited in Bub 2018, p. 9).

The correct quantum state to associate to a system, therefore, is the one corresponding to what Bub calls the *ultimate measurement* (and relative *ultimate observer*). It is important to notice right from the start, and I will insist on this point later along the way, that in order for HH's objection to be met, all observers (not only the ultimate observer) must apply the same Boolean algebra, on pain of falling again into a scenario with inconsistent predictions.

But who is the ultimate observer in HH's thought experiment?

The only criterion for the identification of the ultimate measurement/observer is that the result of the measurement must be registered macroscopically.

"In a situation, as in the [Hagar and Hemmo] argument, where there are multiple candidate observers, there is a question as to whether [Friend is] "ultimate observer[..]," or whether only Wigner [is]". The difference has to do with whether [Friend] perform measurements of the observables [P] with definite outcomes at the Boolean macrolevel, or whether they are manipulated by Wigner [. . .] in unitary transformations that entangle [Friend] with systems in their laboratories, with no definite outcomes for the observables A and B. What actually happens to [Friend] is different in the two situations" (Bub 2018, p. 11).

It is important to clarify, about this passage, that the exclusive character of the disjunction 'Friend performs measurements registered at the macrolevel *or* Wigner manipulates observables' is misleading. The two disjuncts are actually not mutually excluding: both can be true. Let's assume, in fact, that both Wigner and Friend write down their results and are therefore ultimate observers. This is an unproblematic assumption, since, according to Bub's analysis,[6] both Friend and Wigner act as ultimate observers in different moments of the experiment: in the first part, when Friend performs her measurement, she is the ultimate observer, while Wigner becomes the ultimate observer in the second part of the experiment.

Now, Wigner knows (as we know) that Friend will write down her measurements' results, and so that she is the legitimate ultimate observer in the first part of the experiment. He also knows that he has to take this assignment very seriously, given that "What actually happens to [Friend] is different" whether or not Friend is the ultimate observer and "the difference between the two cases [...] is an objective fact at the macrolevel" (Bub 2018, p. 12).

This means that Wigner knows that in the first part of the experiment a legitimate ultimate measurement is being performed in the lab, and with it the uncontrollable disturbance that characterize, in measurement processes, the passage from non-Booleanity to Booleanity. If we were to use Pitowsky's Bayesian framework, we could say that Friend's measurement is a gamble over which Wigner can bet. He does not clearly know what is the result of the experiment, but he knows that she definitely gets a + or a− result, because this is what the 'ultimate Boolean algebra' predicts.

The point I am trying to make is that, if QT is not a 'single user theory' (as it is not according to Bub), and if the selection of a Boolean algebra does not apply uniquely to the ultimate observer (as it does not according to Bub), then after Friend's measurement, Both she and Wigner will have to predict that her measurement ends up either in the state (12.5) or in the state (12.6). The result of Friend's experiment is an objective fact about the world, which Wigner can't ignore.

The correct state that, according to this view, Wigner should associate to F + P after Friend's measurement is therefore a proper mixture of (12.5) and (12.6).

But if this is true, when Wigner makes his predictions about the measurement of O, he should use the probabilities dictated by a proper mixture of (12.5) and (12.6), rather than those dictated by the entangled state (12.4).

But this, again, "would seem to require a suspension of unitary evolution in favour of an unexplained "collapse" of the quantum state" (12.3), which the information-theoretic view clearly rejects.

[6] "If there are events at the macrolevel corresponding to definite measurement outcomes for Alice and Bob, then Alice and Bob represent "ultimate observers" and the final state of the combined quantum coin and qubit system is $|h\rangle_A|0\rangle_B$ or $|t\rangle_A|0\rangle_B$ or $|t\rangle_A|1\rangle_B$, depending on the outcomes. If Wigner and Friend subsequently measure the super-observables X, Y on the whole composite Alice-Bob system (so they are "ultimate observers" in the subsequent scenario), the probability of obtaining the pair of outcomes {ok, ok} is 1/4 for any of the product states $|h\rangle_A|0\rangle_B$ or $|t\rangle_A|0\rangle_B$ or $|t\rangle_A|1\rangle_B$" (Bub 2018).

12.8 Conclusions

In this paper I have tested the solutions to the big measurement problem provided by different interpretations of QT as a theory about information, by analysing how they behave in a thought experiment *à la* Wigner's Friend.

As analytical tools for the examination of such performances, I've used the two claims that, according to Itamar Pitowsky, lie at the basis of the conundrums of QT: the claim that the concept of measurement should not appear in the fundamental axioms of a physical theory and the claim that the quantum state represents a physical system, its properties and its dynamical evolution.

About the first dogma, HH argue that the conclusion of their argument is that black-box approaches are incoherent or not complete. If this is true, their argument can be seen as a justification of the first dogma, which, as a result, is not a dogma anymore: besides the already good reasons illustrated by John Bell, a reason to adopt Bell's dictum is that black-box theories can't provide consistent predictions. In the previous sections I have acknowledged that qBism might have a way-out of HH's conclusions. Should we conclude that HH's argument fails in vindicating the first dogma?

I think this conclusion would be wrong. Bell's criticism of the notion of measurement is a reflection over physical theories: measurement can't appear in a fundamental physical theory because it is not a physically fundamental notion. It should be already clear that, since the conclusion of my analysis is that qBism is not a physical theory, the first dogma does not apply to it.

When seen, as it was intended, as a criterion for fundamental physics, Bell's dictum seems quite reasonable. However, it is hardly surprising that the notion of measurement, and measurement results, appear in the axioms of a theory about beliefs and about how to update beliefs with new experience.[7]

The fact that Qbism is not affected by HH's argument, therefore, does not say much about the status of Bell's dictum as a dogma or as a justifiable request. As far as HH's argument is successful against physical black-box theories, it still provides a valid justification for the first dogma.

From the analysis put forward in this paper, therefore, Bell's dictum seems in great shape, and not a dogma at all.

Let's move to the analysis of the second dogma. The main target of this paper was the assumption, common to probabilistic and information-based approaches to QT, that the second dogma is responsible for the measurement problem, and that its rejection is sufficient to explain the problem away.

The failure of Bub's and Pitowsky's information-theoretic interpretations in the context of HH's thought experiment, on the one hand, and the discussed problems in the Qbist approach, show that this assumption is too simplistic: the claim that QT is about information does not solve the measurement problem, neither in the

[7]Although (Fuchs et al. 2014, p. 2) for a reflection over the notion of measurement.

physical interpretation of the notion of information (as in Bub), nor in its epistemic interpretation (as in Pitowsky's).

Acknowledgement The discussion of this paper at the Birmingham FraMEPhys Seminar series and at the MCMP-Western Ontario Workshop on Computation in Scientific Theory and Practice helped me to improve this paper. I am very thankful to the editors of this volume, and especially to Meir Hemmo for his valuable comments, and to Veronika Baumann for the same reason.

References

Bacciagaluppi, G. (2013). A critic looks at QBism. halshs-00996289.

Baumann, V., & Brukner, Č. (2019). Wigner's friend as a rational agent. arXiv preprint arXiv:1901.11274.

Bell, J. S. (1987). *Speakable and unspeakable in quantum mechanics*. Cambridge: Cambridge University Press.

Birkhoff, G., & von Neumann, J. (1936). The logic of quantum mechanics. *Annals of Mathematics, 37*, 823.

Bub, J. (2016). *Bananaworld: Quantum mechanics for primates*. Oxford: Oxford University Press.

Bub, J. (2018). In defense of a "single-world" interpretation of quantum mechanics. *Studies in History and Philosophy of Science Part B: Studies in History and Philosophy of Modern Physics*. arxiv.org/abs/1804.03267v1.

Bub, J., & Pitowsky, I. (2010). Two dogmas about quantum mechanics. In *Many worlds* (pp. 433–459). Oxford: Oxford University Press.

Felline, L. (2018a). Quantum theory is not only about information. *Studies in History and Philosophy of Science Part B: Studies in History and Philosophy of Modern Physics*. arXiv:1806.05323.

Felline, L. (2018b). Mechanisms meet structural explanation. *Synthese, 195*(1), 99–114.

Frauchiger, D., & Renner, R. (2018). Single-world interpretations of quantum mechanics cannot be self-consistent. arXiv eprint quant-ph/1604.0742.

Fuchs, C. A. (2010). QBism, the perimeter of quantum Bayesianism. arXiv preprint arXiv:1003.5209.

Fuchs, C. A., & Stacey, B. C. (2019, January). QBism: Quantum theory as a hero's handbook. *Proceedings of the International School of Physics "Enrico Fermi", 197*, 133–202. arXiv preprint arXiv:1612.07308.

Fuchs, C. A., Mermin, N. D., & Schack, R. (2014). An introduction to QBism with an application to the locality of quantum mechanics. *American Journal of Physics, 82*(8), 749–754.

Hagar, A. (2003). A philosopher looks at quantum information theory. *Philosophy of Science, 70*(4), 752–775.

Hagar, A. (2007). Experimental metaphysics2: The double standard in the quantum-information approach to the foundations of quantum theory. *Studies in History and Philosophy of Science Part B: Studies in History and Philosophy of Modern Physics, 38*(4), 906–919.

Hagar, A., & Hemmo, M. (2006). Explaining the unobserved—Why quantum mechanics ain't only about information. *Foundations of Physics, 36*(9), 1295–1324.

Healey, R. (2017). Quantum-Bayesian and pragmatist views of quantum theory. In E. N. Zalta (Ed.), *The Stanford encyclopedia of philosophy* (Spring 2017 Edition). https://plato.stanford.edu/archives/spr2017/entries/quantum-bayesian/

Maudlin, T. (1997). *Quantum non-locality and relativity: Metaphysical intimations of modern physics*. Malden: Wiley-Blackwell.

Myrvold, W. (2018). Philosophical issues in quantum theory. In E. N. Zalta (Ed.), *The Stanford encyclopedia of philosophy* (Fall 2018 edition). URL: https://plato.stanford.edu/archives/fall2018/entries/qt-issues/

Pitowsky, I. (2003). Betting on the outcomes of measurements: A Bayesian theory of quantum probability. *Studies in History and Philosophy of Science Part B: Studies in History and Philosophy of Modern Physics, 34*(3), 395–414.

Pitowsky, I. (2004). Macroscopic objects in quantum mechanics: A combinatorial approach. *Physical Review A, 70*(2), 022103.

Pitowsky, I. (2006). Quantum mechanics as a theory of probability. In *Physical theory and its interpretation* (pp. 213–240). Dordrecht: Springer. arXiv:quant-ph/0510095v1.

Ramsey, F. P. (1926). *Truth and probability*. Reprinted In D. H. Mellor (Ed.) *F. P. Ramsey: Philosophical Papers*. Cambridge: Cambridge University Press (1990).

Timpson, C. G. (2013). *Quantum information theory and the foundations of quantum mechanics*. Oxford: Oxford University Press.

Wallace, D. (2016). What is orthodox quantum mechanics? arXiv preprint arXiv:1604.05973.

Chapter 13
There Is More Than One Way to Skin a Cat: Quantum Information Principles in a Finite World

Amit Hagar

Abstract An analysis of two routes through which one may disentangle a quantum system from a measuring apparatus reveals how the no–cloning theorem can follow from an assumption on an infrared and ultraviolet cutoffs of energy in physical interactions.

Keywords Finite nature hypothesis · Quantum information · No cloning · No signaling · Quantum Zeno effect · Quantum adiabatic theorem

13.1 In Memory

Itamar Pitowsky was my teacher and my MA thesis advisor. He was the reason I became an academic. His classes were a sheer joy, and his deep voice and enthusiasm gave warmth to us all in those dreary Jerusalem winters. Itamar's genius became evident a decade earlier when he published his Ph.D. thesis in the *Physical Review* (Pitowsky 1983) and later in a Springer book (Pitowsky 1989) that has gone missing from so many university libraries because everybody steals it. In it he showed that one could still maintain a local realistic interpretation of quantum mechanical correlations if one would allow the existence of non-measurable sets. In a sense it was the beginning of his view of quantum mechanics as a theory of (non–Boolean) probability, a view which he has been advocating forcefully later in his career before his untimely death. I dedicate this paper to Itamar and his unique

I am thankful to G. Ortiz for discussion, to the audience in the IU Logic Colloquium and the UBC Physics Colloquium for their comments and questions on earlier versions of this paper, and to Itamar, the oracle from Givat Ram, whose advice has guided me since I met him for the first time more than 25 years ago.

A. Hagar (✉)
HPSC Department, Indiana University, Bloomington, IN, USA
e-mail: hagara@indiana.edu

M. Hemmo, O. Shenker (eds.), *Quantum, Probability, Logic*,
Jerusalem Studies in Philosophy and History of Science,
https://doi.org/10.1007/978-3-030-34316-3_13

empiricist view. The calm sunshine of his mind is among the things I will never forget.

13.2 Introduction

In the late 1920s von Neumann axiomatized non relativistic quantum mechanics (QM henceforth) with the Hilbert space, and ever since then generations of physicists have been accustomed to view this space as indispensable to the description of quantum phenomena. The Hilbert space with its inner product and the accompanying non commutative algebra of its orthomodular lattice of closed subspaces have also given rise to a manifold of "interpretations" of non relativistic QM, and to a subsequent plethora of puzzles and metaphysical speculations about the inhabitant of this space, the "state vector" (Nye and Albert 2013), its "branching" (Saunders et al. 2010), and its "collapse" (Ghirardi et al. 1986).

Such puzzles are not surprising. Faithful as they are to the Galilean imperative ("The book [of Nature] is written in mathematical language" Galilei 1623/1960), theoretical physicists often incline to interpret mathematical constructs that they use in the description of physical reality as designating pieces of that reality. In classical mechanics such a correspondence is relatively straightforward: the dynamical variables of the theory can be assigned values, and so can the attributes of the physical objects they represent. In QM, however, things are more complicated. To begin with, the mathematical structure can support two types for dynamical variables, the operators and the vectors they operate on (the historical reason for this duality can be traced back to the birth of quantum theory; see, e.g., Beller (1999, pp. 67–78)). Next, this mathematical structure doesn't yield itself to correspondence with physical reality as easily as in classical mechanics. One candidate for the dynamical variables, the operator, can be assigned (eigen)values only in specific circumstances (namely, when it operates on eigenvectors), and, furthermore, not all operators can be assigned (eigen)values simultaneously (Kochen and Specker 1967); the other candidate, the state vector, can be assigned values (amplitudes), but instead of actual attributes of a physical object, these (or rather their mod square) represent the *probabilities* of the object's having these attributes as an object represented by (or spanned by the eigenvectors of) the respective operator.

Those who insist on interpreting the state vector of a single quantum system as designating a real physical object can do so by (i) altering QM, namely by introducing a physical collapse of the state vector as a genuine physical process (Ghirardi et al. 1986), or by (ii) replacing QM with a deterministic and non–local alternative (Goldstein et al. 1992), or by (iii) keeping QM intact but committing to a metaphysical picture of multiple worlds (Saunders et al. 2010). Since neither of these routes is terribly attractive to physicists, a possible alternative is to regard the state vector as a mathematical object which does *not* represent pieces of reality, but rather our knowledge of that reality (Unruh 1994), knowledge which is inherently incomplete (Caves et al. 2002). On this view, QM is regarded a theory of probability

(Bub and Pitowsky 2010), with a non–Boolean structure that allows the physicist to make bets on the results of experiments.

The idea of QM as a theory of probability (and of the Hilbert space as a probability space) allows us to embark on all sorts of projects that, for example, aim to derive the structure of the Hilbert space from basic principles of (quantum) information (Clifton et al. 2003), or reconstruct QM as a theory of information (Fuchs 2003). In these projects interference and entanglement become manifestations of the non–Boolean structure of quantum probabilities, a structure that allows us to perform dense coding, teleportation, and cryptography and prevents us from signaling or cloning. One common feature of these projects is that they all take measurement results as primitives, and refrain from dynamically analyzing further how these results come about.

This strategy has gained a lot of support in the quantum foundations community (to the extent that quantum foundations itself has become by and large a branch of quantum information science). But no matter how attractive you find it, portraying QM as a theory of statistical inference still leaves us with an unresolved tension: given that on this view the probabilities that are employed in theory refer to *our* knowledge, then one cannot argue anymore that QM is a theory about the world; it is, rather, a theory about *our knowledge*. Contrast that with the advertised role of physics to describe the natural world (and not our state of mind while observing it).

Elsewhere I have argued for a way to ease this tension by interpreting "lack of knowledge" objectively as a natural limit on spatial resolution that can be traced to the finite nature of space (Hagar and Sergioli 2014; Hagar 2014, 2017). Here I would like to show how other features of the information–theoretic approach to QM can arise within such a finitist picture.

I will focus on the famous no–cloning theorem that says that an unknown quantum state cannot be cloned, and show how this information–theoretic principle can emerge from an examination of the limitations that exist on disentangling of a quantum system from its measurement apparatus in two generic scenarios, the quantum adiabatic theorem and the quantum Zeno effect.

After briefly introducing the no–cloning theorem and the finite nature hypothesis, in Sects. 13.3.1 and 13.3.2 I will show that in both scenarios there is no way to keep a quantum state undisturbed during a measurement process unless prior knowledge thereof exists. What prevents us from doing so in both cases, I shall demonstrate, is a finitist assumption on interaction energy. In this sense, I shall argue in Sect. 13.4, the no–cloning theorem can be seen as a result of the finite nature hypothesis.

13.3 Knowledge, or Lack Thereof

The famous eigenvector/eigenvalue rule in QM states that a quantum system possesses a definite value for a specific property when it is an eigenstate of the operator that represents that property. Now we certainly know how to prepare a system so that it will end up in a specific state, i.e., we measure the system in a certain orthonormal basis of an operator and "collapse" it to one of the eigenvectors of the operator. Note that the quantum state we prepared becomes "known" to us with certainty only *after* the measurement took place, and once it is, we can repeat the measurement many times without changing it.

That we need to accept this scenario and not analyze the measurement process any further is the essence of the information–theoretic view that regards measurement results as primitive and "collapse" as a subjective knowledge update and not as a physical process. But what if we asked the reverse question? how does one prepare an operator so that an unknown quantum state becomes its eigenstate?

The obstacle for doing so comes in the form of the no–cloning theorem, that states that a cloning device of the sort $|\psi\rangle \otimes |s\rangle$, where $|\psi\rangle$ is an unknown quantum state and $|s\rangle$ is the pure target state to which $|\psi\rangle$ should be cloned according to the unitary evolution

$$|\psi\rangle \otimes |s\rangle \rightarrow^{U} U(|\psi\rangle \otimes |s\rangle) = |\psi\rangle \otimes |\psi\rangle \tag{13.1}$$

can only work if $|s\rangle = |\psi\rangle$ (in which case $|\psi\rangle$ is known in advance) or if $|\psi\rangle$ and $|s\rangle$ are orthogonal. In other words, a general unknown quantum state cannot be cloned with unitary evolution (Dieks 1982; Wootters and Zurek 1982).

From an information–theoretic perspective, the theorem encapsulates the difference between quantum and classical information: in the classical case states such as $|0\rangle$ and $|1\rangle$ are always orthogonal but QM allows for additional non orthogonal states such as $\frac{1}{\sqrt{2}}(|0\rangle + |1\rangle)$. From an operational perspective, however, the no–cloning theorem is just another reminder that in QM the measurement process of an unknown state disturbs the state (Park 1970).[1] It is this feature that, or so I shall argue, the finite nature hypothesis can explain in the two generic measurement scenarios described below.

The finite nature hypothesis imposes an upper bound on energy in any physical interaction. It translates into physics a position in the philosophy of mathematics known as finitism, according to which the physical world is seen as "a large but finite system; finite in the amount of information in any volume of spacetime, and finite in the total volume of spacetime" (Fredkin 1990). On this view, space and time are only finitely extended and divisible, all physical quantities are discrete, and all physical processes are only finitely complex. What follows is an attempt to demonstrate how the operational principle behind the no–cloning theorem results from such a bound on energy in the interaction between the quantum system and its measuring device.

[1] If we have many identical copies of a qubit then it is possible to measure the mean value of non–commuting observables to completely determine the density matrix of the qubit Inherent in the conclusion that nonorthogonal states cannot be distinguished without disturbing them then is the implicit provision that it is not possible to make a perfect copy of a qubit.

13.3.1 How Slow Is Slow Enough

One way a construction of a measurement interaction that keeps the quantum state intact could take place is via the quantum adiabatic theorem. According to this theorem (Messiah 1961, pp. 739–746), an eigenstate (and in particular a ground state) of an initial time–dependent Hamiltonian of a quantum system can remain unperturbed in the process of deforming that Hamiltonian, if several conditions are met: (i) the initial and final Hamiltonians, as well as the ones the system is deformed through, are all non–degenerate, and their eigenvalues are piecewise differentiable in time (ii) there is no level–crossing (i.e., there always exists an energy gap between the ground state and the next excited state), and (iii) the deformation is adiabatic, i.e., infinitely slow. To take an everyday analogy, if your sound sleeper baby is sound asleep in her cradle at the living room, then moving the cradle as gently as possible from that room to the bedroom will not wake her up.

More precisely, consider a quantum system described in a Hilbert space $\mathcal{H}$ by a smoothly time–dependent Hamiltonian, that varies between H_{I} and H_{F}, for t ranging over $[t_0, t_1]$, with a total evolution time given by $T = t_1 - t_0$, and with the following convolution:

$$H = H(t) = \left(1 - \frac{t}{T}\right) H_{\text{I}} + \frac{t}{T} H_{\text{F}} \,. \tag{13.2}$$

When the above conditions (i–iii) of the adiabatic theorem are satisfied, then if the system is initially (in t_0) in an eigenstate of the initial Hamiltonian H_{I}, it will evolve at t_1 (up to a phase) to an eigenstate of H_{F}. In the special case where the eigenstate is the ground state, the adiabatic theorem ensures that in the limit $T \to \infty$ the system will remain in the ground state throughout its time–evolution.

Now, although in practice T is always finite, the more it satisfies the minimum energy gap condition (condition (ii) above), the smaller will be the probability that the system will deviate from the ground state. This means that the gap condition controls the evolution time of the process, in the exact following way:

$$g_{min} = \min_{0 \le t \le T} (E_1(t) - E_0(t)). \tag{13.3}$$

$$T \gg 1/g_{min}^2 . \tag{13.4}$$

where E_0 and E_1 are the ground state and the first excited state, respectively.

Prima facia the adiabatic theorem seems highly suitable for a measurement interaction that keeps the unknown quantum state intact, as it appears to allow one to remain agnostic about both the final state vector at t_1 and the final Hamiltonian H_{F}, while ensuring that the former is an eigenstate (and, in particular, a ground state, in the case one starts at t_0 from a state one knows to be the ground state of H_{I}) of the latter.

One may already point out that the adiabatic theorem is correct only at the limit $T \to \infty$. In actual physical scenarios, T is always finite, and so there is always a finite probability that the system will not remain in the ground state after all. In those cases, repeated measurements of the same state would send it to a state orthogonal thereto, and so one would not be able to fully characterize it.

However, here I shall argue that the issue isn't just with the impracticality of the above scenario, and the problem is actually much more serious: for even for a finite T, one can disentangle the system from the apparatus and keep the state undisturbed during the measurement interaction if and only if one *knows in advance* the Hamiltonian, which means that one also *knows in advance* its ground state.

As we shall see, that T is always finite and that prior knowledge of the Hamiltonian is required are two facets of the same fact, namely, that the question whether a state remains undisturbed during a process of the kind described above is an undecidable question. One could argue that there "exists" such a situation in which the state is undisturbed by the process, and that one can make the probability of failure of the adiabatic process as small as one wants by simply enlarging the finite evolution time T. The problem, however, is that even in the realistic case of a finite T, without prior knowledge of the Hamiltonian, the question *what T is* (i.e., *when to halt* the adiabatic process so that the state one ends up with is undisturbed), is undecidable; it is simply *impossible* to compute T in advance, unless one has prior knowledge of the final Hamiltonian (and thus of the ground state that can be calculated therefrom).

To see why, let us focus on the gap condition (Eq. 13.4). This condition is in the heart of the adiabatic theorem, as it ensures its validity, and, more important, as it governs the evolution time of the entire process. This spectral gap $g = E_1 - E_0$ may be calculated for H_{I} at t_0, since we ourselves construct this Hamiltonian and prepare the system to be in its ground state. But how does this gap behave along the convolution? Note that the question here is not whether the gap *exists* along the convolution. Saying that T is finite means that at some point a level–crossing would occur with a finite probability, and so the process would not keep the state undisturbed forever. But let's concede that a gap (hence an undisturbed state) does "exist". Conceding this, the question we ask now is how small is this gap in all instances other than t_0. For in the absence of a detailed spectral analysis, in general nobody knows what g is, how it behaves, or how to compute it!

This question of the gap behavior is of crucial importance. Recall that the gap controls the velocity of the process with which one would like to keep the state vector undisturbed throughout the measurement process. To achieve this, one needs to deform the initial *known* Hamiltonian H_{I} slow enough, until one arrives at the final *unknown* Hamiltonian H_{F}. But how slow is slow enough? How slow should one go, and, more important, when should one stop the process if one doesn't know *in advance* either (a) the spectral gap or (b) the final Hamiltonian H_{F}? The problem, therefore, is not that the evolution time T is finite, but that since g or H_{F} are unknown, T (and hence the "speed" T^{-1}), are finite alright, but also unboundedly large (slow).

Can we estimate T *in advance*? For each given "run" of the adiabatic process that ends with H_{F} we have to come up with a process whose rate of change is $\sim T_f^{-1}$. How do we know that we are implementing the correct rate of change while $H(t)$ is evolving? Apparently, by being able to measure differences of order T_f^{-1}, that is, having the sensitive "speedometer" at our disposal. But when the going gets tough, we approach very slow speeds of the order of $\sim T_f^{-1}$, which begs the question, since we can then compute T_f using our "speedometer", and hence compute H_{F} and determine its ground state without even running the adiabatic process.

Now if we don't have a "speedometer", then even if we decided to increase the running–time from the order of T_f to, say, the order of $T_f + 7$, we will have no clue that the machine is indeed doing $(T_f + 7)^{-1}$ and not T_f^{-1}, and that the state remains undisturbed. And so in the absence of knowledge *in advance* of g or H_{F}, we will never know how slow should we evolve our physical system. But then we will also fail to fulfill the adiabatic condition which ensures us that in any given moment the system remains in its ground state.

13.3.2 How Fast Is Fast Enough

Another route to realizing a scenario in which a measured state remains undisturbed is the well–known quantum Zeno effect (Degasperis et al. 1974; Peres 1980), also known as "the watched pot" effect. Here the idea is that the survival probability of an evolving quantum state is predicted to be altered by a process of repeated observations with a macroscopic apparatus. In the ideal case of continuous measurement process, the state remains "frozen" and doesn't change in time. In the more realistic case of a finite measurement process, the probability that the state changes vanishes as the frequency of the repeated measurements increases. In effect we have here a mirror image of the quantum adiabatic theorem: there, one could keep the (ground) state intact by interacting with it infinitely slow; here one can keep the (excited) state intact by interacting with it infinitely fast.

More precisely, consider a quantum system composed of an unknown state $|\psi\rangle$ which interacts with a measuring apparatus, and evolves according to the total time–independent Hamiltonian H (units are chosen so that $\hbar = 1$). The probability of the "survival" of the initial state is

$$|(\langle\psi|e^{-iHt}|\psi\rangle)|^2 \approx 1 - (\Delta H)^2 t^2 - \ldots, \tag{13.5}$$

where ΔH is the standard deviation defined as

$$(\Delta H)^2 = \langle H\psi|H\psi\rangle - (\langle\psi|H\psi\rangle)^2. \tag{13.6}$$

It follows that if the projection operator on the initial state ψ is measured after a short time t, the probability of finding the state unchanged is

$$p = 1 - (\Delta H)^2 t^2, \quad (13.7)$$

but if the measurement is performed n times, at intervals t/n, there is a probability

$$p = [1 - (\Delta H)^2 (t/n)^2]^n > 1 - (\Delta H)^2 t^2, \quad (13.8)$$

that in *all* measurements one will find the state unchanged. In fact, for $n \rightarrow \infty$, the left hand side in (13.8) tends to 1, and so in the limit when the frequency of measurements tends to infinity, the state will not change at all, and will remain "frozen".

From a mathematical perspective, the situation can be explained as a case in which, given a certain orthonormal basis for $H = H_0 + V$, all the diagonal terms reside in H_0, and V contains only off–diagonal elements. The eigenspace of H_0 is thus a "decoherence free subspace" (Lidar and Whaley 2003), a sector of the system's Hilbert space where the system is decoupled from the measuring apparatus (or more generally, from the environment) and thus its evolution is completely unitary. Physically this means that the additional term V in the Hamiltonian H rapidly alters the phases of the eigenvalues, and the decay products (responsible for the decaying of the state) acquire energies *different* from those of H, as to make the decay process more difficult, or, in the limit, impossible. Now if one could locate one's state vector in the eigenspace of H_0, then consecutive measurements would not change the state vector, and one could measure it without disturbing it, and violate the no cloning theorem.

The quantum Zeno effect was originally presented in terms of a paradox (Misra and Sudarshan 1977), the idea being that without any restrictions on the frequency of measurements (time being what it is, namely, a continuous parameter), QM fails to supply us predictions for the decay of a state which is monitored by a continuous measurement process, and so appears to be incomplete.

But as in the earlier case of the adiabatic theorem, the state can be guaranteed to undisturbed in the context of the quantum Zeno effect only at the limit $n \rightarrow \infty$ (i.e., $T \rightarrow 0$). In actual physical scenarios, $T \neq 0$ hence the frequency of the measurement is always finite, and so there is always a possibility that the state will be disturbed after all. As in the case of the adiabatic theorem, one could just replace actual infinity with potential unboundedness, and argue that a decoherence free subspace simply "exists", and, moreover, that one can make the probability of decay as small as one wants.

And yet, such a decoherence free subspace – a subspace in which one can measure a quantum state without disturbing it – can be constructed with certainty only if one knows *in advance* the state one wants to protect. Indeed, further analysis has revealed (Ghirardi et al. 1979) that a correct interpretation of the time–energy uncertainty relation (i) relates the measurement frequency to the uncertainty in energy of the measuring apparatus:

$$T \Delta E \geq 1 \quad (13.9)$$

where T is the time duration of the measurement and ΔE is the energy uncertainty, and (ii) controls, via this uncertainty, the rate of decay of the state:

$$|\langle\psi_0|\psi_t\rangle|^2 \geq \cos^2(\Delta E t), \quad \texttt{for } 0 \leq t \leq \frac{\pi}{2\Delta E}. \tag{13.10}$$

Before we go on to analyze these results, two remarks are in place. First, inequality (13.10) was first derived by Mandelstam and Tamm in 1945 for $\Delta E = \Delta H$, i.e., for the case where the uncertainty (spread) in energy is given in terms of the standard deviation, under the assumption that the latter is finite. Recently it was shown that since ΔH can diverge, a more suitable measure of uncertainty is required (Uffink 1993). To our purpose here this caveat is immaterial; even after a correct measure of uncertainty is chosen, the probability of decay of the state still depends on this measure of uncertainty. Second, while the duration of the measurement is a parameter and not a quantum operator, the application of the time–energy uncertainty relation is legitimate here because we are dealing with what may be called a static experiment (Aharonov and Bohm 1961): in this case T is not a preassigned or externally imposed parameter of the apparatus, but is on the contrary determined dynamically through the workings of the Schrödinger equation which describes the evolution of the wave packet through the measuring system. The wave packet may perhaps be said to interact with the apparatus during the time interval T and in this sense the time–energy uncertainty relation is valid, and constrains the precision of the energy of the interaction. Note that the time–energy uncertainty relation all by itself doesn't constrain T to be finite. It is the additional assumption that in realistic interaction scenarios energy is finite and bounded from above that does the trick. See, e.g., Hilgevoord (1998, p. 399).

With this preliminaries we can now state our negative result, which is similar to the one we have established in the case of the adiabatic theorem. For even in the case where the frequency T^{-1} is finite but unbounded (hence T is not exactly 0 but arbitrarily close to 0), one still needs to know the state *in advance* in order to know how fast one needs to repeat the measurements that project the state onto its protected subspace and thereby disentangle it from the measuring apparatus.

The reason is that also here, for each given "Zeno effect" process k, we have to come up with a repeated measurement process whose frequency is $\sim T_k^{-1}$. But how do we know that we are implementing the correct frequency? Apparently, by being able to measure differences of order T_k^{-1}, that is, having the sensitive "speedometer" at our disposal. But when the going gets tough we approach very fast speeds of the order of $\sim T_k^{-1}$, which begs the question, since we can then compute T_k using our "speedometer".

Now if we don't have a "speedometer", then even if we decided to increase the frequency of the repeated measurements from the order of T_k^{-1} to, say, the order of $(T_k - 7)^{-1}$, we will have no clue that the apparatus is indeed working at $(T_k - 7)^{-1}$ and not T_k^{-1}. And so also here, in the absence of knowledge *in advance* of the energy spread ΔE or the desired state (from which, along with the initial state, the energy spread can be calculated), we will never know how fast should we measure our physical system so that this state is protected. But then we will also fail to fulfill the conditions for the quantum Zeno effect which ensure us that the system remains undisturbed in its initial state.

13.4 Moral

In both scenarios presented above, the adiabatic evolution or the Zeno effect, the issue at stake is *not* the practical inability to realize actual infinity; in both cases, the upper and lower bounds on the duration of the measurement, while a practical necessity, signify a much deeper restriction.

What our analysis in section (3) has shown is that for the state to remain undisturbed, the decision *when* to halt the adiabatic process, or *how fast* to measure the state in the Zeno effect scenario requires *a–priori* knowledge of that state. But this means that the imposition of *some* upper or lower bound on measurement duration, ultraviolet or infrared cutoffs which signify the inability to resolve energy differences unboundedly small, or to apply unbounded energy with increasingly frequent measurements, respectively, is tantamount to *knowledge in advance* of the state to be measured. The actual duration or frequency in the two processes, respectively, yield the precision with which one can describe the desired known state.

In other words, if one had such unbounded resolution power or unbounded energy, one would be able to measure an unknown quantum state without disturbing it and discern between two non–orthogonal quantum states. Conversely, *at least in the two cases discussed here*, if one would like to measure an unknown quantum state without disturbing it, one must avail oneself to such unbounded resolution power or unbounded energy (this, of course, need not hold in general, as we only investigated two possible measurement scenarios, and there may be other scenarios in some domains which may allow one to circumvent the no–cloning theorem without availing oneself to unbounded energy. The sheer existence of those scenarios would entail the inapplicability of QM in those domains).[2] The upshot, in short, is that if the finite nature hypothesis holds, so does the no–cloning theorem.

---

[2]Note that even if one had bounded but large resolution one could not measure an unknown quantum state without disturbing it. Saying that such disturbance could be "very small" misses the point, because one would still need to know in advance what the result is supposed to be in order to say how small is "very small" disturbance. Regardless, even a small disturbance will be enough to validate no cloning. Indeed, Atia and Aharonov (2017) have recently shown – and I thank a referee for this reference – that *only when a Hamiltonian or its eigenvalues are known*, then there is a tradeoff between accuracy and efficiency of the kind envisioned in the "small disturbance" remark. Their result which connects quantum information to fundamental questions about precision measurements emphasizes a prior point made by Aharonov et al. (2002), and which is repeated here: once a-priori knowledge of the Hamiltonian or the state is allowed, the game has changed and one can bypass certain structural restrictions imposed by quantum mechanics. The whole point of this chapter is that these restrictions can arise from fundamental limitations on spatial resolution.

Can we give other quantum information–theoretic principles the same treatment? I believe we can, at least to the no–signaling principle which can be violated if one allows infinite precision in spatial resolution (a-la Bohm, see Hemmo and Shenker 2013). As for the no bit commitment principle, the answer is more involved. This third principle is different than the other two in that it actually constrains the quantum *dynamics* rather than the kinematics to certify that macroscopic superpositions are stable over time. Since we haven't yet observed such states, I see no reason to embark on a project to derive it from a more fundamental physical structure.

And the moral? Beginning with his Ph.D. thesis, Itamar has always traced the difference between the quantum and the classical back to the nature of the probability space that underlies the two descriptions. Here and elsewhere I have suggested that there may be a deeper reason for the non–Boolean nature of this probability space that underlies quantum non–commutativity, measurement disturbance, and, as argued here, no–cloning, namely, lower and upper bounds on energy in any physical interaction. Such a view also entails that the standard Hilbert space formalism is nothing but an ideal mathematical structure that describes an underlying physical reality, and that there is no point in attaching to this structure any physical significance.

We have arrived, for completely different reasons, to a similar conclusion as the proponents of the information–theoretic view of QM, namely, that the variety of discussions on "the reality of the wave function" or its "collapse", on "many worlds" or "branching", discussions which are all predicated on the applicability of the Hilbert space formalism to subatomic phenomena, and which have saturated the philosophy of QM since its inception, can be regarded as merely academic. Our moral, however, is that one should take a step further, and inquire about the underlying physics (and not only the information–theoretic principles) behind this formalism; physics which I have argued, is ultimately finite and discrete.

I would like to believe that such a moral would have resonated well with Itamar's empiricist mind.

References

Aharonov, Y., & Bohm, D. (1961). Time in the quantum theory and the uncertainty relation for time and energy. *Physical Review, 122*, 1649–1658.

Aharonov, Y., Massar, S., & Popescu, S. (2002). Measuring energy, estimating hamiltonians, and the time-energy uncertainty relation. *Physical Review A, 66*, 052107.

Atia, Y., & Aharonov, D. (2017). Fast-forwarding of hamiltonians and exponentially precise measurements. *Nature Communications, 8*(1), 1572.

Beller, M. (1999). *Quantum dialogues*. Chicago: University of Chicago Press.

Bub, J., & Pitowsky, I. (2010). Two dogmas about quantum mechanics. In S. Saunders et al. (Eds.), *Many worlds?* Oxford: Oxford University Press.

Caves, C., Fuchs, C., & Schack, R. (2002). Quantum probabilities as bayesian probabilities. *Physical Review A, 65*, 022305.

Clifton, R., Bub, J., & Halverson, H. (2003). Characterizing quantum theory in terms of information-theoretic constraints. *Foundations of Physics, 33*, 1561.

Degasperis, A., Fonda, L., & Ghirardi, G. C. (1974). Does the lifetime of an unstable system depend on the measuring apparatus? *Nuovo Cimento, 21A*, 471–484.

Dieks, D. (1982). Communication by EPR devices. *Physics Letters A, 92*, 271–272.

Fredkin, E. (1990). Digital mechanics: An informational process based on reversible universal cellular automata. *Physica D, 45*, 254–270.

Fuchs, C. (2003). Quantum mechanics as quantum information, mostly. *Journal of Modern Optics, 50*(6–7), 987–1023.

Galilei, G. (1623/1960). The assayer (Il Saggiatore). In *The controversy on the comets of 1618*. Philadelphia: University of Pennsylvania Press. English trans. S. Drake & C. D. O'Malley.

Ghirardi, G., Omero, C., Weber, T., & Rimini, A. (1979). Small-time behavior of quantum nondecay probability and zeno's paradox in quantum mechanics. *Nuovo Cimento, 52A*, 421–442.

Ghirardi, G., Rimini, A., & Weber, T. (1986). Unified dynamics for microscopic and macroscopic systems. *Physical Review D, 34*(2), 470–491.

Goldstein, S., Dürr, D., & Zanghi, N. (1992). Quantum equilibrium and the origin of absolute uncertainty. *The Journal of Statistical Physics, 67*(5/6). 843–907.

Hagar, A. (2014). *Discrete or continuous*? Cambridge: Cambridge University Press.

Hagar, A. (2017). On the tension between ontology and epistemology in quantum probabilities. In O. Lombardi et al. (Eds.), *What is quantum information?* Cambridge: Cambridge University Press.

Hagar, A., & Sergioli, G. (2014). Counting steps: a new approach to objective probability in physics. *Epistemologia, 37*(2), 262–275.

Hemmo, M., & Shenker, O. (2013). Probability zero in Bohm's theory. *Philosophy of Science, 80*(5), 1148–1158.

Hilgevoord, J. (1998). The uncertainty principle for energy and time. II. *American Journal of Physics, 66*(5), 396–402.

Kochen, S., & Specker, E. (1967). The problem of hidden variables in quantum mechanics. *Journal of Mathematics and Mechanics, 17*, 59–87.

Lidar, D., & Whaley, B. (2003). Decoherence–free subspaces and subsystems. In F. Benatti & R. Floreanini (Eds.), *Irreversible quantum dynamics* (Lecture notes in physics, Vol. 622, pp. 83–120). Berlin: Springer.

Messiah, A. (1961). *Quantum mechanics* (Vol. II). New York: Interscience Publishers.

Misra, B., & Sudarshan, E. C. G. (1977). The zeno paradox in quantum theory. *Journal of Mathematical Physics, 18*(4), 756–763.

Nye, A., & Albert, D., eds. (2013). *The wave function*. Oxford: Oxford University Press.

J. Park (1970). The concept of transition in quantum mechanics. *Foundations of Physics, 1*(1), 23–33.

Peres, A. (1980). Zeno paradox in quantum theory. *American Journal of Physics, 48*(11), 931–932.

Pitowsky, I. (1983). Deterministic model of spin and statistics. *Physical Review D, 27*, 2316.

Pitowsky, I. (1989). *Quantum probability, quantum logic* (Lectures note in physics, Vol. 321). Springer Berlin Heidelberg.

Saunders, S., Barrett, J., Kent, A., & Wallace, D. (Eds.). (2010). *Many worlds?* Oxford: Oxford University Press.

Uffink, J. (1993). The rate of evolution of a quantum state. *American Journal of Physics, 61*(10), 935–936.

Unruh, W. G. (1994). Reality and measurement of the wave function. *Physical Review A, 50*, 882–887.

Wootters, W. K., & Zurek, W. (1982). A single quantum cannot be cloned. *Nature, 299*, 802–803.

Chapter 14
Is Quantum Mechanics a New Theory of Probability?

Richard Healey

Abstract Some physicists but many philosophers believe that standard Hilbert space quantum mechanics faces a serious measurement problem, whose solution requires a new theory or at least a novel interpretation of standard quantum mechanics. Itamar Pitowsky did not. Instead, he argued in a major paper (Pitowsky, Quantum mechanics as a theory of probability. In: Demopoulos W, Pitowsky I (eds) Physical theory and its interpretation. Springer, Dordrecht, pp 213–240, 2006) that quantum mechanics offers a new theory of probability. In these and other respects his views paralleled those of QBists (quantum Bayesians): but their views on the objectivity of measurement outcomes diverged radically. Indeed, Itamar's view of quantum probability depended on his subtle treatment of the objectivity of outcomes as events whose collective structure underlay this new theory of probability. I've always been puzzled by the thesis that quantum mechanics requires a new theory of probability, as distinct from new ways of calculating probabilities that play the same role as other probabilities in physics and daily life. In this paper I will try to articulate the sources of my puzzlement. I'd like to be able to think of this paper as a dialog between Itamar and me on the nature and application of quantum probabilities. Sadly, that cannot be: by taking his part in the dialog I will inevitably impose my own distant, clouded perspective on his profound and carefully crafted thoughts.

Keywords Quantum probability · Dutch book argument · Measurement problem · Quantum gamble · Gleason's theorem · Observable outcomes · Pragmatism

---

R. Healey (✉)
Philosophy Department, University of Arizona, Tucson, AZ, USA
e-mail: rhealey@email.arizona.edu

M. Hemmo, O. Shenker (eds.), *Quantum, Probability, Logic*,
Jerusalem Studies in Philosophy and History of Science,
https://doi.org/10.1007/978-3-030-34316-3_14

14.1 Introduction

Itamar Pitowsky (2003, 2006) developed and defended the thesis that the Hilbert space formalism of quantum mechanics is just a new kind of probability theory. In developing this thesis he argued that all features of quantum probability, including the Born probability rule, can be derived from rational probability assignments to finite "quantum gambles". In defense of his view he argued that all experimental aspects of entanglement, including violations of the Bell inequalities, are explained as natural outcomes of the resulting probability structure. He further maintained that regarding the quantum state as a book-keeping device for quantum probabilities dissolves the BIG measurement problem that afflicts those who believe the quantum state is a real physical state whose evolution is always governed by a linear dynamical law; and that a residual small measurement problem may then be resolved by appropriate attention to the great practical difficulties attending measurement of any observable on a macroscopic system incompatible with readily executable measurements of macroscopic observables.

This is a bold view of quantum mechanics. It promises to transcend not only the frustratingly inconclusive debate among so-called realist interpretations (including Bohmian, Everettian, and "collapse" theories) but also the realist/instrumentalist dichotomy itself. In this way it resembles the pragmatist approach I have been developing myself (Healey 2012, 2017). Our views agree on more substantive issues, including the functional roles of quantum states and probabilities and the consequent dissolution of the measurement problem. But I remain puzzled by Itamar's central thesis. The best way to explain my puzzlement may be to show how the points on which we agree mesh with a contrasting conception of quantum probability. On this rival conception, quantum mechanics requires no new probability theory, but only new ways of calculating probabilities with the same formal features and the same role as other probabilities in physics and daily life.[1]

The rest of this contribution proceeds as follows. In the next section I say what Pitowsky meant by a probability theory and explain how he took quantum probability theory to differ formally from classical probability theory. The key difference arises from the different event structures over which these are defined. Whereas classical probabilities are defined over a σ-algebra of subsets of a set, quantum probabilities are defined over the lattice $\mathcal{L}$ of closed subspaces of a Hilbert space. Gleason's theorem plays a major role here: for a Hilbert space of dimension greater than 2, it excludes a classical truth-evaluation on $\mathcal{L}$ but at the same time completely characterizes the class of possible quantum probability measures on $\mathcal{L}$. Pitowsky sought to motivate the formal difference between quantum and classical probability theory by an extension of a Dutch book argument offered in support of a coherence requirement on rational degrees of belief (in the tradition of Ramsey

[1]That quantum mechanics involves a new way of calculating probabilities is a point made long ago by Feynman, as Pitowsky himself noted. But Feynman (1951) also says that the concept of probability is not thereby altered in quantum mechanics. Here I am in agreement with Feynman, though the understanding he offers of that concept is disappointingly shallow.

(1926) and De Finetti (1937)) to a class of what he called "quantum gambles" associated with possible measurements on a quantum system. Section 14.3 explains the extension but questions how well it justifies the proposed formal modification. A key issue here concerns the independence of the quantum probability of an event from the context in which a non-maximal observable containing it is measured.

A quantum gamble may be settled only if there is a viable procedure for determining which possible event assigned a quantum probability has occurred. So quantum probabilities are not assigned to past events of which we have only a partial record, or no record at all. This restricts "matters of fact" to include only observable records. Section 14.4 notes how this is in tension with applications of quantum theory in situations where we have no observable records. It goes on to consider hypothetical situations to which quantum theory may be applied in which records may be observable by some but not other observers before these records are permanently erased so they can no longer be read by any observers.

In Sect. 14.5 I defend an alternative pragmatist view of probability, explaining how it makes room for a notion of objective probability capable of rationally constraining the credences of a rational agent. On this view probabilities in quantum theory are defined over a family of Boolean algebras rather than a single non-Boolean lattice. This is because Born probabilities are defined over possible outcomes of a physical process represented in one, rather than some other, classical event space. Models of decoherence may provide a guide as to what that process is, even though they do not describe the dynamical evolution of a physical quantum state. Any agent who accepts quantum theory should adjust his or her credences to the Born probabilities given by the correct quantum state for one in that physical situation. Insofar as an agent's physical situation constrains what information is available, differently situated agents are correct to assign different quantum states and to adjust their credences accordingly. Each agent should update that quantum state (and associated Born probabilities) on accessing newly available information. This is one way a quantum state provides a book-keeping device for probabilities: the other is provided by its linear evolution while the system remains undisturbed.

14.2 Quantum Measure Theory

Pitowsky maintained that quantum probability theory differs from classical probability theory formally as well as conceptually. Following Kolmogorov, probability is usually characterized formally as a unit-normed, countably additive, measure Pr on a σ-algebra Σ of subsets E_i ($i \in \mathbb{N}$) of a set Ω: that is

$$\Pr : \Sigma \longrightarrow [0, 1] \text{ satisfies}$$

1. $\Pr(\Omega) = 1$ (Probability measure)
2. $\Pr \bigcup_i E_i = \sum_i \Pr E_i$ provided that $\forall i \neq j (E_i \cap E_j = \emptyset)$

Here Σ forms a lattice which is complemented and distributive, and hence a Boolean algebra. $\Pr(E_i)$ is the probability of the event E_i. Condition 2 is sometimes weakened to finite additivity.

The closed subspaces $\{S_i\}$ of a Hilbert space H also form a lattice $\mathcal{L}(H)$ whose meet $S_i \wedge S_j$ is $S_i \cap S_j$, and whose join $S_i \vee S_j$ is the smallest closed subspace containing all elements of S_i and S_j. H is the maximal element of $\mathcal{L}(H)$ (denoted by **1**) and the null subspace $\varnothing$ is the minimal element, denoted by **0**. Each element S has a complement $S^\perp$ consisting of every vector orthogonal to every vector in S. Indeed, $S^\perp$ is the unique orthocomplement of S. But this lattice is not distributive and so $\mathcal{L}(H)$ is not a Boolean algebra.

A (unit-normed) quantum measure μ on $\mathcal{L}(H)$ is defined as follows:

$$\mu : \mathcal{L}(H) \longrightarrow [0, 1] \text{ satisfies}$$

$$1.\ \ \mu(\mathbf{1}) = 1 \qquad \text{(Quantum measure)}$$

$$2. \mu(\bigvee_i S_i) = \sum_i S_i \text{ provided that } \forall i \neq j (S_i \perp S_j)$$

Condition 2 may be weakened to finite additivity.

The formal analogy between the conditions defining a probability measure and those defining a quantum measure motivated Pitowsky's proposal to interpret μ as a new kind of *quantum probability*. Accordingly, he referred to the closed subspaces of H as events, or possible events, or possible outcomes of experiments. He also took himself to be following a tradition that takes $\mathcal{L}(H)$ as the structure representing the"elements of reality" in quantum theory. Regardless of that tradition, others have also referred to quantum measures on non-Boolean lattices associated with quantum mechanics as quantum probability functions or measures (Earman 2018; Ruetsche and Earman 2012). The set of bounded self-adjoint operators on H forms a von Neumann algebra $\mathcal{B}(H)$: this includes as a subset the set $\mathcal{P}(\mathcal{B}(H))$ of projection operators $\hat{E}$ on H. By virtue of the one-one correspondence between the set of closed subspaces $\{S\}$ of H and the operators $\{\hat{E}_S\}$ that project onto them, $\mathcal{P}(\mathcal{B}(H))$ also forms a lattice isomorphic to $\mathcal{L}(H)$. This generalizes to von Neumann algebras $\mathcal{A}$ other than $\mathcal{P}(\mathcal{B}(H))$ that figure in what Ruetsche (2011) calls extraordinary quantum mechanics or QM_∞.

Gleason's theorem (Gleason 1957) completely characterizes the class of quantum measures on the lattice of closed subspaces of a Hilbert space of dimension greater than 2.

Gleason's Theorem: If H is a (separable) Hilbert space of dimension greater than 2 and μ is a (unit-normed) quantum measure on $\mathcal{L}(H)$, then there is a positive self-adjoint operator $\hat{W}$ of trace 1 such that, for every element $\hat{E}_S$ of $\mathcal{P}(\mathcal{B}(H))$, $\mu(S) = Tr(\hat{W}\hat{E}_S)$.

This theorem has two important consequences. The first is that there are no dispersion-free measures on $\mathcal{L}(H)$ if $\dim(H) \geqslant 3$, where a quantum measure μ is dispersion-free if and only if its range is the set $\{0, 1\}$. This is important since any truth-valuation on the set of all propositions stating that it is event S or instead event $S^\perp$ that represents an element of reality would have to give rise to a dispersion-free quantum measure on $\mathcal{L}(H)$. This, of course, is why Gleason's theorem and similar results (e.g. Kochen and Specker (1967), Bell (1966)) are considered important "no-hidden variable" results. While it may be understood as the outcome of an experiment, an event S may not be (uniformly) understood as simply the independently existing state of affairs that experiment serves to reveal.

The second important consequence of Gleason's theorem is that every (unit-normed) quantum measure on $\mathcal{L}(H)$ ($\dim(H) \geqslant 3$) uniquely extends to the Born probability measure associated with a quantum state, as represented by a vector in or density operator on H. Indeed, this is true even if the quantum measure is merely finitely additive. This consequence is not surprising, since the Born rule may be stated in the form

$$\Pr_{\hat{W}}(A \in \Delta) = Tr(\hat{W}\hat{E}(\Delta)) \qquad \text{(Born rule)}$$

where the quantum state is represented by density operator $\hat{W}$ and $\hat{E}(\Delta)$ is the relevant projection operator from the spectral family defined by the unique self-adjoint operator $\hat{A}$ corresponding to dynamical variable A.

Cognizant of the first consequence of Gleason's theorem, Pitowsky did not defend the radical thesis that quantum theory shows the world obeys a non-classical logic. But he took the second consequence as "one of the strongest pieces of evidence in support of the claim that the Hilbert space formalism is just a new kind of probability theory" (Pitowsky 2006, p.222).

14.3 Quantum Gambles

Pitowsky developed his conception of quantum probability within the Bayesian tradition pioneered by Ramsey (1926) and De Finetti (1937). This tradition locates probabilities in an agent's rationalized degrees of belief. A necessary, though possibly insufficient, condition for such degrees of belief to be rational is that they be (what has come to be called) coherent.

In this tradition, belief and its degrees are dispositions manifested not by self-avowal but in actions. The condition of coherence is understood in terms of the agent's dispositions to accept a set of wagers at various odds offered by a hypothetical bookie. A set of degrees of belief is coherent just in case it corresponds to a set of such dispositions that is not guaranteed to result in a loss if collectively manifested no matter what the outcome of all the bets in the set. Degrees of belief that are not coherent are said to allow a Dutch book to be made against the agent who has them. So-called synchronic Dutch book theorems are then taken to show that any coherent set of degrees of belief in a set of propositions may be represented by a (classical) probability measure over them.

Following Lewis (1980) an agent's coherent set of degrees of belief are called his or her *credences*. For a subjectivist like De Finetti, there is no further notion of *objective* probability at which a rational agent's credences should aim. Lewis (1980) disagreed. Instead he located a distinct concept of objective probability he called *chance*, of which all we know is that it provides a further rational constraint on an agent's credences through what he called the *Principal Principle*. I defer further consideration of this principle until Sect. 14.5, since it plays no role in Pitowsky's own view of quantum probability theory.

According to Pitowsky (2006), a quantum gamble consists of four steps:

1. A *single* physical system is prepared by a method known to everybody.
2. A finite set $\mathcal{M}$ of incompatible measurements, each with a finite number of possible outcomes, is announced by the bookie. The agent is asked to place bets on the possible outcomes of each one of them.
3. One of the measurements in the set $\mathcal{M}$ is chosen by the bookie and the money placed on all other measurements is promptly returned to the agent.
4. The chosen measurement is performed and the agent gains or loses in accordance with his bet on that measurement.

Here $\mathcal{M}$ is identified with a set of Boolean algebras $\{\mathcal{B}_1, \mathcal{B}_2, \ldots, \mathcal{B}_k\}$, each generated by the possible outcomes in $\mathcal{L}$ of the measurement to which it corresponds. The elements of $\mathcal{B} = \{S_1, S_2, \ldots S_m\} \in \mathcal{M}$ will not all be one-dimensional subspaces if $\mathcal{M}$ is not a maximal measurement. Pitowsky (2006, p.223) maintained that "by acting according to the standards of rationality the gambler will assign probabilities to the outcomes". He took the gambler in question to recognize identities in the logical structure consisting of the outcomes in $\mathcal{L}$, and in particular the cases in which the same outcome is shared by more than one experiment (i.e. type of measurement in $\mathcal{M}$.) But, crucially, this gambler was *not* assumed to know quantum mechanics.

Pitowsky then argued (2006, p.227: Corollary 7) that any such quantum gambler *not* meeting these standards of rationality must assign probability values to elements of a finite sublattice $\Gamma_0 \subset \mathcal{L}(H)$ ($\dim(H) \geq 3$) that cannot be extended to a quantum measure on a finite $\Gamma \supset \Gamma_0$. He took this conclusion to establish the claim that quantum theory is a new theory of probability.

Notice that this argument is not offered as a derivation of the Born rule insofar as it does not mention the quantum state $\hat{W}$ used in stating that rule. It concludes only that a rational agent's credences should be consistent with the probabilities specified by quantum mechanics through application of the Born rule to *some* quantum state. Instead, this conclusion is presented as justification for the claim that quantum probability is quantum measure theory: but is it warranted?

How do the standards of rationality constrain a quantum gambler's assignments of "probabilities" (i.e. degrees of belief) to outcomes in $\mathcal{L}$? Assume that for every $\mathcal{B} \in \mathcal{M}$ such an agent assigns degree of belief $cr(S|\mathcal{B})$ to outcome S of a measurement corresponding to $\mathcal{B}$. Pitowsky took the standards of rationality to require the agent to assign degrees of belief in accordance with two rules (in my notation):

RULE 1: For each measurement $\mathcal{B} \in \mathcal{M}$ the function $cr(\bullet|\mathcal{B})$ is a probability distribution on $\mathcal{B}$.

RULE 2: If $\mathcal{B}_1, \mathcal{B}_2 \in \mathcal{M}$, and $S \in \mathcal{B}_1 \cap \mathcal{B}_2$ then $cr(S|\mathcal{B}_1) = cr(S|\mathcal{B}_2)$.

Rule 1 is imposed to insure the coherence of the agent's degrees of belief concerning the possible outcomes of a single measurement: If the agent's degrees of belief do not confirm to this rule they will dispose him or her to accept bets on these outcomes guaranteed to lead to a sure loss. This is the standard Dutch book argument as to why a rational agent's degrees of belief must be representable as a (finitely additive) probability measure.

The force of such Dutch book arguments has been a topic of extended debate among philosophers of probability, and alternative arguments have been offered as to why a rational agent's degrees of belief should be representable as probabilities. Briefly stated, here are three standard objections seeking to undermine the Dutch book argument: A rational agent may simply refuse to bet: his betting behavior may fail to reveal his degrees of belief insofar as it is a function of the rest of his cognitive and affective state: agents have degrees of belief in propositions whose truth-value cannot be determined because there is no corresponding settleable outcome. I will not press either of the first two objections now. Pitowsky did anticipate this third objection by requiring that there be a viable procedure for determining which possible event assigned a quantum probability has occurred, and I will pursue this issue in Sect. 14.4.

Rule 2 requires a quantum gambler's credences to be non-contextual, in the sense that he or she have the same degree of belief in the truth of propositions stating the outcome of a measurement of a non-maximal observable no matter what type of measurement led to that outcome. It is a remarkable fact about the Born rule of quantum mechanics that the probabilities it yields are non-contextual in this sense. But Pitowsky's gambler cannot be assumed to know quantum mechanics. So why should the quantum gambler's credences be non-contextual?

Pitowsky (2006, p.216) first addresses this question in Sect. 2.1 as follows (emphases in the original):

> ...the *identity* of events which is encoded by the structure also involves judgments of probability in the sense that *identical events always have the same probability*. This is the meaning of accepting a structure as an algebra of events *in a probability space*.

Here he takes the structure of events to be the lattice $\mathcal{L}(H)$ of subspaces of some Hilbert space H whose Boolean sublattices correspond to measurements on a quantum system, some incompatible with others. Application of quantum theory's Born rule turns $\mathcal{L}(H)$ into a quantum measure space when quantum state $\hat{W}$ generates a subspace measure μ through $\mu(S) = Tr(\hat{W}\hat{P}_S)$, where $\hat{P}_S$ is the (unique) projection operator onto subspace $S \in \mathcal{L}(H)$. But so far we have been offered no reason to regard $\mathcal{L}(H)$ as a *probability* space.

The Dutch book argument Pitowsky took to require a rational quantum gambler's credences to conform to Rule 1 does not require that agent's credences to conform to Rule 2. To the extent that argument is successful, it establishes only that each Boolean subalgebra $\mathcal{B}$ of $\mathcal{L}(H)$ corresponding to a measurement in $\mathcal{M}$ may be taken to define a σ-algebra of subsets of $\mathbb{R}$ representing possible outcomes of that measurement, and in that sense $\mathcal{B}$ is a (subjective) probability space. Additional argument is needed to justify the claim that the full lattice $\mathcal{L}(H)$ is a probability space.

The argument Pitowsky (2006, p.216) offers in Sect. 2.1 proceeds by analogy to a classical application of probability theory to games of chance:

> Consider two measurements A, B, which can be performed together; and suppose that A has the possible outcomes $a_1, a_2, \ldots, a_k$, and B the possible outcomes $b_1, b_2, \ldots, b_r$. Denote by $\{A = a_i\}$ the event "the outcome of the measurement of A is a_i", and similarly for $\{B = b_j\}$. Now consider the identity:
>
> $$\{B = b_j\} = \bigcup_i^k (\{B = b_j\} \cap \{A = a_i\}) \tag{14.1}$$
>
> This is the distributivity rule which holds in this case as it also holds in all classical cases. This means, for instance, that if A represents the roll of a die with six possible outcomes and B the flip of a coin with two possible outcomes, then Eq. (14.1) is trivial. Consequently the probability of the left hand side of Eq. (14.1) equals the probability of the right hand side, for every probability measure.

In a classical joint probability space the event $\{B = b_j\}$ occurs if and only if some outcome $\{B = b_j\} \cap \{A = a_i\}$ occurs, and so the identity Eq. (14.1) holds trivially. If instead the event $\{B = b_j\}$ is the outcome of a trial with a set of possible outcomes $\{\{B = b_n\} : n = 1, \ldots, N\}$, then it is not trivial that the probability of the left hand side of Eq. (14.1) equals the probability of the right hand side, since these concern probabilities in different trials with different probability spaces. We would be astonished if a fair coin always came up heads whenever a die was rolled at the same time, but that is because we have empirical reasons to discount the influence of dice rolls on the outcomes of simultaneous coin tosses.

Pitowski draws an analogy between Eq. (14.1) (as applied to compatible quantum measurements or coin flips and dice rolls) and Eq. (14.2):

$$\bigcup_i^k (\{B = b_j\} \cap \{A = a_i\}) = \{B = b_j\} = \bigcup_i^l (\{B = b_j\} \cap \{C = c_i\}) \tag{14.2}$$

In this equation, A, B, C, are quantum observables such that $[A, B] = 0$, and $[B, C] = 0$ but $[A, C] \neq 0$, and $c_1, c_2, \ldots, c_l$ are the possible outcomes of C. Since A, C are incompatible observables they are not jointly measurable. Accordingly, the Born rule of quantum mechanics does not assign joint probabilities to events such as $\{A = a_m\}$, $\{C = c_n\}$. Unlike the events $\{A = a_m\}, \{B = b_n\}$ in Eq. (14.1), such events are not elements of any (classical) probability space acknowledged by quantum mechanics. Indeed, as Fine (1982) showed, the joint Born probabilities for some sets of pairwise compatible observables are not marginals of *any* (classical) joint probability distribution. Equation (14.2) expresses a trivial identity in a lattice $\mathcal{L}(H)$ if $\cup, \cap$ are read as join and meet operations in the lattice. But they cannot generally be read as set-theoretic union and intersection of elements of a σ-algebra of subsets of a set of outcomes of a (classical) probability space.

In a way, this was Pitowsky's point. He used the analogy only to motivate his proposal that quantum probability is *different* from classical probability in just this way. As he put it (Pitowsky 2006, pp.216–7)

> I assume that the **0** of the algebra of subspaces represents impossibility (zero probability in all circumstances) **1** represents certainty (probability one in all circumstances), and the identities such as Eqs. (14.1) and (14.2) represent identity of probability in all circumstances. This is the sense in which the lattice of closed subspaces of the Hilbert space is taken as an algebra of events. I take these judgments to be natural extensions of the classical case; a posteriori, they are all justified empirically.

However, his *argument* for this proposal was supposedly based on rationality requirements on the credences of a quantum gambler ignorant of quantum mechanics, and here empirical considerations are out of place. If probability *just is* rational credence, then a requirement of rationality cannot be based on empirical considerations of the kind that are taken to warrant acceptance of quantum mechanics and the non-contextuality of probabilities consequent on application of the Born rule. Immediately after stating Rule 2, Pitowsky (2006) says that this follows from an identity between events essentially equivalent to Eq. (14.2), and the principle that identical events in a probability space have equal probabilities. But, as we have seen, that principle applies automatically only when these events are part of a single probability space, corresponding to a single trial or measurement context. A lattice $\mathcal{L}(H)$ is naturally understood as a quantum measure space, but only its Boolean subspaces are naturally understood as *probability* spaces. Pitowsky's appeal to rationality requirements on a quantum gambler does not justify his thesis that quantum theory is a new kind of probability theory.

14.4 Objective Knowledge of Quantum Events

In the language of probability theory, the term 'event' may be used to refer either to a mathematical object (such as a set) or to a physical occurrence. When arguing that the Hilbert space formalism of quantum mechanics is a new theory of probability, Pitowsky took that theory to consist of an algebra of events, and the probability measure defined on it. Events in this sense are *mathematical* objects—subspaces of a Hilbert space. But to each such object he also associated a class of actual or possible *physical* occurrences—outcomes of an experiment in which a quantum observable is measured. A token physical event e_S occurs just in case in a measurement M of observable O the outcome is o_i, an eigenvalue of $\hat{O}$ with eigenspace S. Here M is a token physical procedure corresponding to the mathematical object $\mathcal{B}_M$, a Boolean subalgebra of a lattice $\mathcal{L}(H)$ of subspaces including S, where $\hat{O}$ is a self-adjoint operator on H.

When a quantum gamble is defined over a set $\mathcal{M}$ of incompatible measurements, each of these is characterized by the corresponding Boolean subalgebra of $\mathcal{L}(H)$. A token event e_S settles a gamble if it is the outcome of an actual measurement procedure M corresponding to the mathematical object $\mathcal{B}_M \in \mathcal{M}$, where the bookie chose to perform M rather than some other measurement in the gamble. Following the subjectivist tradition pioneered by Ramsey, Pitowsky stressed that, for a quantum gamble to reveal an agent's degrees of belief, the outcome of whatever measurement is chosen by the bookie must be settleable.

> A proposition that describes a possible event in a probability space is of a rather special kind. It is constrained by the requirement that there should be a viable procedure to determine whether the event occurs, so that a gamble that involves it can be unambiguously decided. This means that we exclude many propositions. For example, propositions that describe past events of which we have only a partial record, or no record at all. (Pitowsky 2003, p.217)
>
> Recall that in section 2.2 we restricted "matters of fact" to include only observable records. Our notion of "fact" is analytically related to that of "event" in the sense that a bet can be placed on x_1 only if its occurrence, or failure to occur, can be unambiguously recorded. (*op. cit.*, p.231)

In the previous section I questioned the force of Pitowsky's argument that a rational agent's degrees of belief in the outcomes of quantum gambles will be representable as a quantum probability measure on $\mathcal{L}(H)$. But even if that argument were sound it would not extend to all uses of probability in applications of quantum theory. The Born rule may be legitimately applied to propositions concerning events whose outcomes are unknown or even unknowable. For such events there is no viable procedure to determine what the outcome is, so there is no settleable quantum gamble involving these events. An agent acting according to the standards of rationality associated with a quantum gamble need not assign *any* probabilities to the possible outcomes of such an event: rules 1 and 2 need not apply. Such a rational agent may have degrees of belief concerning these outcomes that cannot be represented as a quantum measure on a corresponding $\mathcal{L}(H)$ even though each outcome has a well-defined Born probability.

Quantum theory is often applied to occurrences in distant spacetime regions from which no observable records are accessible. These include processes in the center of stars (including the cores of neutron stars) and quantum field fluctuations in the early universe. There is no restriction on the application of the Born rule to calculate probabilities of outcomes associated with such processes. Of course these cannot be understood as the outcomes of experiments since no experimenters could have been present in those regions. But the Born rule is commonly applied to yield probabilities of outcomes of processes in which no experiment is involved, whether or not these are referred to as measurements. The notion of measurement is notoriously obscure in quantum mechanics, as Bell forcefully pointed out in his article "Against 'Measurement'". But even there he posed the rhetorical question

> If the theory is to apply to anything but highly idealized laboratory operations, are we not obliged to admit that more or less 'measurement-like' processes are going on more or less all the time, more or less everywhere? (Bell 2004, p.216)

These days references to measurement are often replaced by talk of decoherence, though it is now generally acknowledged that if there is a serious measurement problem then appeals to decoherence will not solve it. While agreeing with Pitowsky that there is no BIG measurement problem, in the next section I will indicate how quantum models of decoherence may be used as guides to the legitimate application of the Born rule in a pragmatist view of quantum mechanics. But in this view a proposition to which the Born rule assigns a probability does not describe the outcome of a measurement but the physical event of a magnitude taking on a value, whether or not this is precipitated by measurement.

Even when a magnitude (a.k.a. observable) takes on a value recording the outcome of a quantum measurement, that record may be subsequently lost, at which point Pitowsky's view excludes any event described by a proposition about that outcome from a quantum probability space. While such scruples may not be out of place in the context of a quantum gamble, only an indefensible form of verificationism would prohibit retrospective application of the Born rule to that event. So quantum mechanics permits application of the concept of probability in circumstances in which it cannot be understood to be defined on a space of events consistent with Pitowsky's view.

One might object that no record of a measurement outcome is ever irretrievably lost. But recent arguments (Healey 2018; Leegwater 2018) challenge such epistemic optimism in the context of *Gedankenexperimenten* based on developments of the famous Wigner's friend scenario. These arguments threaten the objectivity of measurement outcomes under the assumption that unitary, no–collapse single-world quantum mechanics is applicable at all scales, even when applied to one observer's measurement on another observer's lab (including any devices in that lab recording the outcomes of prior quantum measurements). It is characteristic of these arguments that an observation by one observer on the lab of another completely erases all of the latter's records of the outcome of his or her prior quantum measurement. As Leegwater (2018, p.13) put it

> such measurements in effect erase the previous measurement's result; they must get rid of all traces from which one can infer the outcome of the first measurement.

In this situation, the first observer's measurement outcome was a real physical occurrence, observable by and (then) known to that observer. But the second observer's measurement on the first observer's lab erased all records of that outcome, and neither of these observers, nor any other observer, can subsequently verify what it was. Indeed, even the first observer's memory of his own result has been erased. This is clearly a situation in which Pitowsky would have excluded the event corresponding to the first observer's measurement outcome from any probability space. For in no sense does a true proposition describing that outcome state (what he called) a "matter of fact".

I wonder what Pitowsky would have made of these recent arguments. Leegwater (2018) presents his argument as a "no-go" result for unitary, no–collapse single-outcome (relativistic) quantum mechanics, while Healey (2018) takes the third argument he considers to challenge the objectivity of measurement outcomes. Brukner (2018) formulates a similar argument in a paper entitled "A no-go theorem for observer-independent facts", in which he says

> We conclude that Wigner, even as he has clear evidence for the occurrence of a definite outcome in the friend's laboratory, cannot assume any specific value for the outcome to coexist together with the directly observed value of his outcome, given that all other

> assumptions are respected. Moreover, there is no theoretical framework where one can assign jointly the truth values to observational propositions of different observers (they cannot build a single Boolean algebra) under these assumptions. A possible consequence of the result is that there cannot be facts of the world per se, but only relative to an observer, in agreement with Rovelli's relative-state interpretation, quantum Bayesianism, as well as the (neo)-Copenhagen interpretation.

Pitowsky founded his Bayesian approach to quantum probability on the notion of a quantum gamble, from which he excluded propositions about measurement outcomes that describe past events of which *we* have only a partial record, or no record at all. This strongly suggests that he would be unwilling to countenance applications of quantum theory to propositions about measurement outcomes stating observer-dependent facts. His tolerance of agent-dependent probabilities did not extend to acquiescence in non-objective facts about their subject matter. The paper by Hemmo and Pitowsky (2007) criticized the way many-worlds interpretations of quantum mechanics understood the use of probability in quantum mechanics. So Pitowsky would almost certainly have rejected the option of evading the conclusion of recent non-go arguments by countenancing multiple outcomes of a single quantum measurement. I speculate that his response to these arguments would have been to follow von Neumann by rejecting the assumption that unitary quantum mechanics may be legitimately applied to the measurement process itself.

14.5 A Pragmatist View of Quantum Probability

As we saw in Sect. 14.3, Pitowsky's Bayesian view of quantum probability relied on his Rule 1, which requires an agent's degrees of belief in the possible outcomes of a quantum measurement characterized by a Boolean algebra $\mathcal{B}$ to be a (finitely additive) probability measure on the associated algebra of sets. Here he followed in the tradition of Ramsey and de Finetti, who first employed synchronic Dutch book arguments in support of the probability laws as standards of synchronic coherence for degrees of belief. But each of them also took an agent's betting behavior as a *measure* of his or her degrees of belief, giving operational significance to the numerical probabilities by which these could be represented. Pitowsky quoted Ramsey as follows:

> "The old-established way of measuring a person's belief " by proposing a bet, and seeing what are the lowest odds which he will accept, is "fundamentally sound". (Pitowsky 2006, p.223)

If degrees of belief are to be behaviorally defined in terms of betting behavior, then it is essential that these bets be settleable—as Pitowsky insisted. But operational definitions of individual cognitive states like beliefs and desires or preferences are now commonly regarded as inadequate. These are better understood within a broadly functionalist approach to the mind in which the behavioral manifestations of an individual belief are a function of many, if not all, an agent's other cognitive states. Even if there are such things as degrees of belief, these cannot be reliably measured by an agent's betting behavior of the kind that figures in an argument intended to show that a (prudentially) rational agent's degrees of belief will be representable as probabilities.

A Dutch book argument may still be used to justify Bayesian coherence of a set of degrees of belief as a normative condition on epistemic rationality, analogous to the condition that a rational agent's full beliefs be logically compatible. The view that ideally rational degrees of belief must be representable as probabilities has been called *probabilism* (Christensen 2004, p.107): such degrees of belief are known as credences. The force of a Dutch book argument for probabilism is independent of whether there are any bookies or whether bets in a book are settleable. Understood this way, Rule 1 is justified as a condition of epistemic rationality on degrees of belief in a quantum event, whether or not a gamble that involves it can be unambiguously decided.

Probabilism places only minimal conditions on the degrees of belief of an individual cognitive agent, just as logical compatibility places only minimal conditions on his or her full beliefs. Pitowsky sought to justify additional conditions, sufficient to establish the result that a rational agent's credences in quantum events involving a system be representable by a quantum measure on the lattice $\mathcal{L}(H)$ of subspaces of a Hilbert space H associated with that system (provided that $dim(H) \geqslant 3$). This was the key result he took to justify his claim that the Hilbert space formalism of quantum mechanics is a new theory of probability.

Achieving this result depended on Rule 2: the further condition that a rational agent's credence in a quantum event represented by subspace $S \in \mathcal{L}(H)$ be the same, no matter whether it occurs as an outcome of measurement M_1 (represented by Boolean sub-lattice $\mathcal{B}_1 \subset \mathcal{L}(H)$) or M_2 (represented by $\mathcal{B}_2$). Pitowsky took Rule 2 to follow from the identity between events, and the principle that identical events in a probability space have equal probabilities. As embodied in Eq. (14.2), Pitowsky took these judgments to be natural extensions of the classical case (as embodied in (14.1)): and he said of these judgments "a posteriori, they are all justified empirically".

This justification for Rule 2 is quite different from the justification offered for Rule 1, which was justified not empirically but as a normative requirement on epistemic rationality. As such, Rule 1 showed why an epistemically rational agent should adopt degrees of belief representable as probabilities over a classical probability space of events associated with each Boolean subalgebra $\mathcal{B}$ of $\mathcal{L}(H)$: in that sense, it justified considering $\mathcal{B}$ as a probability space. It is because no analogous normative requirement justifies taking $\mathcal{L}(H)$ itself to be a probability space that Pitowsky appealed instead to empirical considerations. But while such empirical considerations may justify the introduction of a quantum measure μ over $\mathcal{L}(H)$ as a convenient device for generating a posteriori justified objective probability measures over Boolean subalgebras of $\mathcal{L}(H)$, this does not show that μ is itself a probability, or $\mathcal{L}(H)$ a probability space. Only by appeal to some additional normative principle of epistemic rationality could a Bayesian in the tradition of Ramsey and de Finetti attempt to show that.

The bearing of empirical considerations on credences has been a controversial issue among Bayesians. De Finetti maintained that there is nothing more to probability than each agent's actual credences. Ramsey allowed that probability in physics may require more, and in this he has been followed by the physicist Jaynes (2003) and other objective Bayesians. Contemporary QBists portray the Born rule as an empirically motivated additional normative constraint on credences (Fuchs and Schack 2013), while most physicists still follow Feynman in seeking an objective correlate of probability in stable frequencies of outcomes in repeated experimental trials.

Lewis believed that objective probability required the kind of indeterminism generally thought to be manifested by radioactive decay. He took orthodox quantum mechanics to involve such indeterminism, with objective probabilities supplied by the Born rule serving as paradigm instances of what he called *chance*. Lewis (1980) formulated the Principal Principle he took to state all we know about chance by linking this to credence. He later proposed a modification to square it with his Humean metaphysics. But Ismael (2008) argued that the modification was a mistake, and essentially restated his original principle. In Lewis's (1994, pp.227–8) words,

> If a rational believer knew that a chance of [an event *e*] was 50%, then almost no matter what he might or might not know as well, he would believe to degree 50% that [*e*] was going to occur. Almost no matter, because if he had reliable news from the future about whether *e* would occur, then of course that news would legitimately affect his credence.

There is now a large literature on how a Principal Principle should be stated, and how, if at all, it may be justified. This includes Ismael's helpful formulation as an implicit definition of a notion of chance:

> The [modified Lewisian] chance of A at p, conditional on any information H_p about the contents of p's past light cone satisfies: $Cr_p(A/H_p) =_{df} Ch_p(A)$.

Lewis was right to add to credence a second, more objective, concept of probability: but he was wrong to restrict its application to indeterministic contexts. The physical situation of a localized agent frequently imposes limitations on access to relevant information about an event about whose occurrence he or she wishes to form reasonable credences. This occurs, for example, in so-called games of chance and in classical statistical mechanics as well as in the quantum domain. In such situations general principles or a physical theory may provide a reliable way of "packaging" the accessible information in a way that permits generally reliable inferences and appropriate actions. That is how I understand the role of the Born rule in quantum mechanics. To accept quantum mechanics is to grant the Born rule objective authority over one's credences in quantum events and thereby regard it as epistemic expert in this context.

The Born rule associates a set of general probabilities to events of certain types involving a kind of physical system assigned a quantum state. The physical situation of an actual or merely hypothetical agent gives that agent access to information about the surrounding circumstances that may be sufficient to assign a specific quantum state to one or more individual systems. Differently situated agents should sometimes correctly assign different quantum states because different information is accessible to each. This may occur just because the agents do not share a single spacetime location, in conformity to Ismael's modified Lewisian chance: Born probabilities supply many examples of such objective probabilities. After a specific quantum state is assigned to an individual system, the Born rule yields a probability measure over events of certain types involving it. By instantiating the general Born rule, the agent may then derive an objective chance distribution over possible events.

This is a classical probability distribution over an event space with the structure of a Boolean algebra $\mathcal{B}$, not a quantum measure over a non-Boolean lattice: in a legitimate application of the Born rule no probability is assigned to events that are not elements of $\mathcal{B}$. Events assigned a probability in this way are given canonical descriptions of the form $Q_s \in \Delta$, where Q is a dynamical variable (observable), s is a quantum system, and Δ is a Borel set of real numbers: I call $Q_s \in \Delta$ a *magnitude claim*. Some of these events are appropriately redescribed as measurement outcomes: for others, this redescription is less appropriate.

Probability theory had its origins in a dispute concerning dice throws, so it may be helpful to draw an analogy with applications of probability theory to throws of a die. The faces of a normal die are marked in such a way that the point total of opposite faces sums to 7, and the faces marked 4, 6 meet on one edge of the die. Consider an evenly weighted die that includes a small amount of magnetic material carefully distributed throughout its bulk. This die may be thrown onto a flat, level surface in one or other of two different kinds of experiments: a magnetic field may be switched on or off underneath the surface as the die is thrown. In an experiment with the magnetic field off, a throw of the die is fair, so there is an equal chance of $1/6$ that each face will land uppermost. (Notice that, though natural, this use of the term 'chance' does not conform to Lewis's Principal Principle—even in Ismael's modified formulation—since it does not presuppose that dice throws are indeterministic processes.) In an experiment with the magnetic field on, the die is biassed so that some faces are more likely to land uppermost than others. But because of the careful placement of the magnetic material within the die, the chance is still $1/3$ that a face marked 4 or 6 lands uppermost.

In each kind of experiment, a probabilistic model of dice-throwing may be tested against frequency data from repeated dice throws of that kind. This model includes general probability statements of the form $P_X(E) = p$ specifying the probability of an event of type E describing a possible outcome of a throw of the die in an experiment of type X. Consider the situation of an actual or hypothetical agent immediately prior to an individual throw of the die. The information accessible to this agent does not include the outcome of that throw: nor could it include all potentially relevant microscopic details of the initial and boundary conditions present in that throw, even if it were a deterministic process. But it does include knowledge of whether the magnetic field is on or off: failure to obtain this relevant, accessible information would be an act of epistemic irresponsibility.

Having accepted a probabilistic model on the evidence provided by relative frequencies of different outcome types in repeated throws of both kinds, this actual or hypothetical agent has reason to instantiate the general probability statement of the form $P_X(E) = p$ for each possible event e of type E in experiment x of kind X. The result is the *chance* of e in x—that to which an epistemically rational agent should match his or her credence concerning event e in experiment x. What that chance is may depend on the experiment x. If e is an event of face 1 landing uppermost, then the chance of e may be less if the magnetic field is on in x than it would have been if the field had been off. But if e' is the event of a prime-numbered face landing uppermost, then the chance of e' is $2/3$ both in experiment x (magnetic field off) and in experiment x' (magnetic field on).

Here is the analogy between this dice-throwing example and quantum probabilities. In both cases there are general probability rules that may be instantiated to give chances to which an (actual or hypothetical) rational agent who accepts those rules should match his or her credences in circumstances of the kind specified by the rule. In both cases these circumstances may be described as those of an experiment, but in neither case is this description essential—no experimenter need be present if the circumstances occur naturally.

In both cases a possible event of a certain type may be assigned the same chance in different circumstances, describable as experiments of different kinds. Both cases feature non-contextual probability assignments to events of the same type. There is no temptation to combine the classical probability spaces corresponding to different kinds of experiment in the dice-throwing example into a single non-classical event space. The rich formal structures to which Pitowsky appealed in the quantum case have no analog in the dice-throwing example. But I remain unpersuaded that the absence of any corresponding structures in the dice-throwing example undermines the force of the analogy.

It is the physical circumstances in which a system finds itself that determine which applications of the Born rule are legitimate (if any), and which are not. Quantum mechanics itself does not specify these circumstances, but physicists have developed reliable practical knowledge of when they obtain. Quantum models of decoherence can provide useful guidance in judging whether and which application of the Born rule is legitimate, by selecting an appropriate Boolean algebra $\mathcal{B}$ of events corresponding to a so-called "pointer basis" of subspaces of the Hilbert space H_s of s. Here s will typically differ from the target system t to which a quantum state has been assigned in order to apply the Born rule.

Models of decoherence are neither sufficient nor necessary to solve any BIG measurement problem: there is no such problem. That problem would arise only on the mistaken assumption that a quantum state specifies all the physical properties of the system to which it is assigned. But I agree with Pitowsky that a quantum state does not do this: instead, it acts as a book-keeping device for a rational agent's credences by requiring them to conform to the associated Born rule probabilities. Quantum mechanics cannot explain why measurements have definite outcomes, since application of its Born rule presupposes that they do. But since a quantum state does not specify its physical condition, there is no tension between a system's being assigned a superposed state and the truth of a magnitude claim that it has a particular eigenvalue of the corresponding operator.

The Born rule is legitimately applied only to significant magnitude claims. The significance of a magnitude claim is not certifiable in terms of truth-conditions in some kind of "quantum semantics". In my pragmatist view, the significance of any claim arises from its place in an inferential web of statements, ultimately linked to perception and action through practical rather than theoretical inference. This inferentialist "metasemantics" supports an analog rather than digital view of the content of each magnitude claim as coming in degrees rather than making the claim simply meaningful or meaningless. Though extraordinarily rapid, complete, and practically irreversible, decoherence is also a matter of degree in applicable quantum models. It follows that there can be no sharp line dividing meaningful from meaningless magnitude claims, or legitimate from illegitimate applications of the Born rule.

This sheds light on the import of the arguments discussed in the previous section purporting to show that the universal applicability of unitary quantum mechanics is incompatible with the assumption that quantum measurements always have unique, objective outcomes. In my pragmatist view, the outcome of a quantum measurement can be stated in a magnitude claim $Q_s \in \Delta$ about a system s that may be thought of as a measuring apparatus. The claim $\mathbf{Q}_s \in \Delta$ is true at a time if and only if (the value of) Q_s is then an element of Δ. But $\mathbf{Q}_s \in \Delta$ derives its content not through this (trivial) truth-condition, but through the reliable inferences into which the claim then enters.

In ordinary circumstances we assume that observation of an experimenter's laboratory records is a reliable way for anyone to determine the outcome of a quantum measurement in that laboratory. Indeed, that is a common understanding of what it is for the measurement to have a unique, objective outcome. But this assumption breaks down in the extreme circumstances of *Gedankenexperimenten* like those that figure in arguments considered in Healey (2018) and Leegwater (2018). In such circumstances an observer in an initially physically isolated laboratory will correctly report the outcome of his experiment even though another observer may report a different outcome after subsequently entering that laboratory and making her own observations of its records. Moreover these observers will not then register any disagreement, since all of the first observer's initial records (including memories) will have been erased by the time of the second observation.

In such cases, processes modeled by decoherence confined to the first observer's laboratory rendered reliable a host of inferences based on magnitude claims stating records of his quantum measurement. This endowed these claims with a high degree of significance, so his observations warranted him in taking them truly to state the unique, physical outcome of his measurement. But the extraordinary circumstances of the *Gedankenexperiment* in fact restrict the domain of reliability to the context internal to the laboratory within which these processes were confined. There is a wider context that includes subsequent physical processes coupling that laboratory to a second observer and her laboratory. In that wider context, inferences based on the magnitude claims stating records of his quantum measurement cease to be reliable, thereby curtailing the significance of these claims, which were nevertheless true in the context in which he made them.

In my pragmatist view, quantum measurements would have unique, physical outcomes even in extraordinary circumstances like those described in the

Gedankenexperimenten that figure in the arguments considered in Healey (2018) and Leegwater (2018). The outcomes could be described by true magnitude claims about individual experimenters' physical records of them. Some such claims made by different experimenters may seem inconsistent. But the inconsistency is merely apparent. A correct understanding of these claims requires that their content be relativized to the physical context to which they pertain. So contextualized, each experimenter in one of these *Gedankenexperimenten* can allow the objective truth of the others' reports of each of their unique measurement outcomes while consistently stating the outcome of his or her own measurement.

This implicit limitation on the content of observation reports of the outcomes of quantum measurements may usually be neglected. Quantum decoherence is so pervasive that we will never be able to realize the extraordinary circumstances required by the *Gedankenexperimenten* that figure in the arguments considered in Healey (2018) and Leegwater (2018): even a powerful quantum computer would not constitute an agent capable of performing and reporting the outcome of a quantum measurement in a physically isolated laboratory. Because we all inevitably share a single decoherence context, true reports of the unique, physical outcomes of quantum measurements provide the objective data that warrant acceptance of the theory we use successfully to predict their objective probabilities.

14.6 Conclusion

Quantum mechanics is not a new theory of probability. On the contrary, it constitutes perhaps our most successful deployment of classical probability theory in physics. It is not only the mathematics of probability that are classical here: the concept itself functions in basically the same way in quantum mechanics that it always has. This function is as a source of expert, "pre-packaged" advice to an actual or merely hypothetical situated agent on how strongly to believe statements whose truth-value that agent is not in a position to determine from the accessible information.

Quantum probabilities may be represented by means of quantum measures on non-Boolean lattices, in which case Gleason's theorem offers an elegant characterization of the range of quantum probabilities yielded by applications of the Born rule. But a quantum measure is not a probability measure, and the lattice of closed subspaces of a Hilbert space is not a probability space, despite the non-contextuality of quantum probabilities.

Dutch book arguments for synchronic coherence support an epistemic norm that an agent's degrees of belief in a set of propositions should be representable by a probability measure. But there is a more objective notion of probability than that of an individual agent's actual credences. An agent's credences become subject to additional norms through acceptance of general probabilities: these importantly include the Born probabilities prescribed by a legitimate application of the Born rule. Coherence is an epistemic norm even for beliefs concerning unsettleable events, including the outcomes of quantum 'measurements' that no-one knows, and those not everyone *can* know—but even these outcomes are as objective as science needs them to be.

In all these ways I have come to disagree with Itamar's view of probability in quantum mechanics. I only wish he could now reply to show me why I am wrong: we could all learn a lot from the ensuing debate. Let me finish by reiterating two important points of agreement. There is no BIG quantum measurement problem: there is only the small problem of physically modeling actual measurements within quantum theory and showing why some are much easier than others. A quantum state is not an element of physical reality ($|\psi\rangle$ is not a beable); it is a book-keeping device for updating an agent's credences as time passes or in the light of new information (even if there is no actual agent or bookie to keep the books)!

References

Bell, J. S. (1966). On the problem of hidden variables in quantum mechanics. *Reviews of Modern Physics, 38*, 447–52.
Bell, J. S. (2004). *Speakable and unspeakable in quantum mechanics* (2nd ed.). Cambridge: Cambridge University Press.
Brukner, Č. (2018). A no-go theorem for observer-independent facts. *Entropy, 20*, 350.
Christensen, D. (2004). *Putting logic in its place*. Oxford: Oxford University Press.
De Finetti, B. (1937). La prévision: ses lois logiques, ses sources subjectives. *Annales de l'Institute Henri Poincaré, 7*, 1–68.
Earman, J. (2018). The relation between credence and chance: Lewis' "Principal Principle" is a theorem of quantum probability theory. http://philsci-archive.pitt.edu/14822/. Accessed 6 May 2019.
Feynman, R. P. (1951). The concept of probability in quantum mechanics. In *Second Berkeley Symposium on Mathematical Statistics and Probability, 1950* (pp. 533–541). Berkeley: University of California Press.
Fine, A. (1982). Joint distributions, quantum correlations, and commuting observables. *Journal of Mathematical Physics, 23*, 1306–1310.
Fuchs, C., & Schack, R. (2013). Quantum-Bayesian coherence. *Reviews of Modern Physics, 85*, 1693–1715.
Gleason, A. M. (1957). Measures on the closed subspaces of a Hilbert space. *Journal of Mathematics and Mechanics, 6*, 885–893.
Healey, R. A. (2012). Quantum theory: a pragmatist approach. *British Journal for the Philosophy of Science, 63*, 729–71.
Healey, R. A. (2017). *The quantum revolution in philosophy*. Oxford: Oxford University Press.
Healey, R. A. (2018). Quantum theory and the limits of objectivity. *Foundations of Physics, 48*, 1568–89.
Hemmo, M., & Pitowsky, I. (2007). Quantum probability and many worlds. *Studies in History and Philosophy of Modern Physics, 38*, 333–350.
Ismael, J. T. (2008). Raid! Dissolving the big, bad bug. *Nous, 42*, 292–307.
Jaynes, E. T. (2003). In G. Larry Bretthorst (Ed.), *Probability theory: The logic of science*. Cambridge: Cambridge University Press.
Kochen, S., & Specker, E. (1967). The problem of hidden variables in quantum mechanics. *Journal of Mathematics and Mechanics, 17*, 59–87.
Leegwater, G. (2018). When Greenberger, Horne and Zeilinger meet Wigner's friend. arXiv:1811.02442v2 [quant-ph] 7 Nov 2018. Accessed 6 May 2019.

Lewis, D. K. (1980). A subjectivist's guide to objective chance. In R. C. Jeffrey (Ed.), *Studies in inductive logic and probability* (Vol. II, pp. 263–293). University of Berkeley: California Press.
Lewis, D. K. (1994). Humean supervenience debugged. *Mind, 103*, 473–90.
Pitowsky, I. (2003). Betting on the outcomes of measurements: A Bayesian theory of quantum probability. *Studies in History and Philosophy of Modern Physics, 34*, 395–414.
Pitowsky, I. (2006). Quantum mechanics as a theory of probability. In W. Demopoulos & I. Pitowsky (Eds.), *Physical theory and its interpretation* (pp. 213–240). Dordrecht: Springer.
Ramsey, F. P. (1926). Truth and probability. Reprinted In D. H. Mellor (Ed.), *F. P. Ramsey: Philosophical Papers*. Cambridge: Cambridge University Press. 1990.
Ruetsche, L. (2011). *Interpreting quantum theories*. Oxford: Oxford University Press.
Ruetsche, L., & Earman, J. (2012). Infinitely challenging: Pitowsky's subjective interpretation and the physics of infinite systems. In Y. Ben-Menahem & M. Hemmo (Eds.), *Probability in physics*. Berlin: Springer.

Chapter 15
Quantum Mechanics as a Theory of Probability

Meir Hemmo and Orly Shenker

Abstract We examine two quite different threads in Pitowsky's approach to the measurement problem that are sometimes associated with his writings. One thread is an attempt to understand quantum mechanics as a probability theory of physical reality. This thread appears in almost all of Pitowsky's papers (see for example 2003, 2007). We focus here on the ideas he developed jointly with Jeffrey Bub in their paper 'Two Dogmas About Quantum Mechanics' (2010) (See also: Bub (1977, 2007, 2016, 2020); Pitowsky (2003, 2007)). In this paper they propose an interpretation in which the quantum probabilities are objective chances determined by the *physics of a genuinely indeterministic* universe. The other thread is sometimes associated with Pitowsky's earlier writings on quantum mechanics as a Bayesian theory of quantum probability (Pitowsky 2003) in which the quantum state seems to be a credence function tracking the experience of agents betting on the outcomes of measurements. An extreme form of this thread is the so-called Bayesian approach to quantum mechanics. We argue that in both threads the measurement problem is solved by implicitly adding structure to Hilbert space. In the Bub-Pitowsky approach we show that the claim that decoherence gives rise to an effective Boolean probability space requires adding structure to Hilbert space. With respect to the Bayesian approach to quantum mechanics, we show that it too requires adding structure to Hilbert space, and (moreover) it leads to an extreme form of idealism.

Keywords Bub-Pitowsky's approach · Bayesian approach · measurement problem · preferred basis problem

---

M. Hemmo (✉)
Department of Philosophy, University of Haifa, Haifa, Israel
e-mail: meir@research.haifa.ac.il

O. Shenker
Edelstein Center for History and Philosophy of Science, The Hebrew University of Jerusalem, Jerusalem, Israel
e-mail: Orly.shenker@mail.huji.ac.il

M. Hemmo, O. Shenker (eds.), *Quantum, Probability, Logic*,
Jerusalem Studies in Philosophy and History of Science,
https://doi.org/10.1007/978-3-030-34316-3_15

15.1 Introduction

The measurement problem in quantum mechanics is the problem of explaining how individual measurement outcomes are brought about, given the Hilbert space structure of the quantum state space and the unitary dynamics (the Schrödinger equation in the non-relativistic case). Influential interpretations of quantum mechanics such as Bohm's (1952) 'hidden variables' theory and the dynamical collapse theory by Ghirardi, Rimini and Weber (1986; see also Bell 1987a) propose solutions to this problem by adding assumptions about the structure of physical reality and the dynamical equations of motion over and above the Hilbert space structure, respectively, Bohmian trajectories and GRW collapses (or flashes; see Tumulka 2006). By contrast, Pitowsky's approach to quantum mechanics as a (non-classical) theory of probability aims at solving (or dissolving) the measurement problem, without adding anything to Hilbert space.[1]

In this paper we examine two quite different threads in Pitowsky's approach to the measurement problem that are sometimes associated with his writings. One thread is an attempt to understand quantum mechanics as a theory of probability which is about physical reality. This thread appears in almost all of Pitowsky's papers (see for example 2003, 2007). We shall focus here on the ideas he developed jointly with Jeffrey Bub in their paper 'Two Dogmas About Quantum Mechanics' (2010).[2] In this paper they propose an interpretation in which the quantum probabilities are *objective chances* determined by the *physics of a genuinely indeterministic* universe. The other thread is sometimes associated with Pitowsky's earlier writings on quantum mechanics as a Bayesian theory of quantum probability (Pitowsky 2003) in which the quantum state seems to be a credence function tracking the *experience* of agents betting on the outcomes of measurements. An extreme form of this thread is the so-called Bayesian approach to quantum mechanics, which as we read it addresses only the experience of agents.[3] We argue that in both approaches the measurement problem is solved by implicitly adding structure to Hilbert space. In the Bub-Pitowsky approach we show that the claim that decoherence gives rise to an effective Boolean probability space requires adding structure to Hilbert space. With respect to the Bayesian approach to quantum mechanics, we show that it too requires adding structure to Hilbert space, and (moreover) it leads to an extreme form of idealism.

The paper is structured as follows. In Sect. 2 we briefly describe the Bub-Pitowsky approach. In Sect. 3 we describe the role of decoherence in their approach and how it gives rise to an effectively classical probability space of the events associated with our familiar experience. But we argue that this falls short of providing what they call a 'consistency proof' that the quantum (unitary) dynamics

[1] Everett's (1957) interpretation of quantum mechanics has a similar attempt in this respect.

[2] See also: Bub (1977, 2007, 2016, 2020); Pitowsky (2003, 2007).

[3] For versions of the Bayesian approach to quantum mechanics, see e.g., Caves et al. (2002a, b, 2007); Fuchs and Schack, (2013); Fuchs et al. (2014); Fuchs and Stacey (2019).

is consistent with our experience in a way that is analogous to special relativity. Finally, in Sect. 4 we criticise the Bayesian approach to quantum mechanics.

15.2 The Bub-Pitowsky Approach

According to Bub and Pitowsky (2010), quantum mechanics is primarily a theory of non-classical probability describing a genuinely indeterministic universe in which "probabilities (objective chances) are uniquely given from the start" by the geometry of Hilbert space (Bub and Pitowsky 2010, p. 444; see also von Neumann 2001, p. 245). The event space in this theory of probability is the projective geometry of closed subspaces of a Hilbert space, which form an infinite family of intertwined Boolean algebras, but cannot be embedded into a single Boolean algebra. This space imposes built-in objective constrains on the *possible* configurations of individual events (see Pitowsky 2003, 2007; Bub and Pitowsky 2010). The unitary evolution of the quantum state evolves the *entire* structure of correlations between events in the Hilbert space, but it does *not* describe the transition from one configuration of individual events to another (a special case of which is measurement). By Gleason's theorem, the Hilbert space structure determines unique probabilities of individual events (e.g., measurement outcomes) which are encoded in the quantum state. They take this to imply that the quantum state is a credence function (Bub and Pitowsky 2010, p. 440):

> [T]he quantum state is a credence function, a bookkeeping device for keeping track of probabilities – *the universe's objective chances*... (Bub and Pitowsky 2010, p. 441; our emphasis)

In their approach, von Neumann's projection postulate turns out to stand for a rule of updating probabilities by conditionalisation on a measurement outcome, that is the von Neumann-Lüders rule. The fact that the updating is non-unitary expresses in their view the 'no cloning' principle in quantum mechanics, according to which there is no universal cloning machine capable of copying the outputs of an arbitrary information source. This principle entails, via Bub and Pitowsky's (2010, p. 455) "information loss theorem", that a measurement necessarily results in the loss of information (e.g., the phase in the pre-measurement state). Since the 'no cloning' principle is implied by the Hilbert space structure, they argue, the loss of information in measurement is *independent* of how the measurement process might be implemented dynamically. By this they mean that if no structure is added to Hilbert space, any dynamical quantum theory of matter and fields that will attempt to account for the occurrence of *individual* events must satisfy this principle (and other constraints imposed by the quantum event space, such as 'no signaling'[4]), *regardless* of its details.

[4]The 'no signaling' principle asserts, roughly, that the marginal probabilities of events associated with a quantum system are independent of the particular set of mutually exclusive and collectively exhaustive events associated with any *other* system.

This idea, that quantum mechanics is a *theory of probability* in which the event space is Hilbert space, leads them to reject what they identify as two 'dogmas' of quantum mechanics: that measurement is a process that, as a matter of principle, should always be open to a complete dynamical analysis; and that a quantum pure state is analogous to the classical state given by a point in phase space, that is the quantum state is a description (or representation) of physical reality. On the view they propose, the quantum state is a credence function, which keeps track of the objective chances in the universe, it is not the 'truthmaker' of propositions about the occurrence or non-occurrence of individual events; and measurement is primitive in the sense that the no-cloning principle follows from the non-Boolean structure of the event space, and if one does not add structure to Hilbert space (like Bohmian trajectories, GRW collapses), then there is no further deeper story to be told about how measurement outcomes are brought about dynamically. In this sense the measurement problem, the 'big' problem as they call it, is a *pseudo* problem.

But why is there no deeper story to be told? After all, Bub and Pitowsky themselves realise that the quantum event space imposes constrains on the 'objective chances' of all *possible* configurations of individual events, and the unitary dynamics evolves this entire structure: it does not describe the evolution from *one* configuration of the universe to *another*. But if the theory does not describe the transition from one configuration to the next, what are the 'objective chances' chances of? And in what sense is this quantum theory of probability a physical theory of an indeterministic universe? In fact, in what sense is the universe indeterministic at all, given the determinism of the unitary dynamics (which is all there is)?

To these questions, Bub and Pitowsky reply by an analogy from special relativity:

> If we take special relativity as a template for the analysis of quantum conditionalization and the associated measurement problem, the information-theoretic view of quantum probabilities as 'uniquely given from the start' by the structure of Hilbert space as a kinematic framework for an indeterministic physics is the proposal to interpret Hilbert space as a constructive theory of information-theoretic structure or probabilistic structure, part of the kinematics of a full constructive theory of the constitution of matter, where the corresponding principle theory includes information-theoretic constraints such as 'no signaling' and 'no cloning.' Lorentz contraction is a physically real phenomenon explained relativistically as a kinematic effect of motion in a non-Newtonian spacetime structure. Analogously, the change arising in quantum conditionalization that involves a real loss of information is explained quantum mechanically as a kinematic effect of *any* process of gaining information of the relevant sort in the non-Boolean probability structure of Hilbert space (irrespective of the dynamical processes involved in the measurement process). Given 'no cloning' as a fundamental principle, there can be no deeper explanation for the information loss on conditionalization than that provided by the structure of Hilbert space as a probability theory or information theory. The definite occurrence of a particular event is constrained by the kinematic probabilistic correlations encoded in the structure of Hilbert space, and only by these correlations – it is otherwise 'free'. (Bub and Pitowsky, p. 448)

On their account, the Hilbert space structure is analogous to the Minkwoski spacetime[5]; and the quantum conditionalisation is analogous to the physically real

[5]Or Einstein's principle of relativity and the light postulate.

Lorentz contraction (the contraction is different in different Lorentz frames, but all the frames are equally real). Since a conditionalisation is expressed in terms of some specific basis states in the Hilbert space, all choices of basis are (prima facie) equally real, presumably in an analogous sense to the equal status of all Lorentz frames in special relativity. It seems to us that in the quantum case, the deep explanation of the conditionalisation in measurement – which, by the Bub-Pitowsky analogy, is also *physically real* – is the non-Boolean probabilistic structure of the Hilbert space, which includes the 'no cloning' principle. By the analogy, dynamical analyses of the information loss in measurement are to be seen as attempts to solve a consistency problem: that is, to find a dynamical analysis of the conditionalisation upon measurement that is consistent with the Hilbert space structure. We shall come back to this point later.

Bub and Pitowsky argue that dynamical analyses of the breaking of the thread in Bell's (1987b) spaceships scenario (an example which Bub and Pitowsky 2010 consider) in terms of Lorentz covariant forces is a consistency proof that a relativistic dynamics compatible with the geometrical structure of Minkowski spacetime is *possible*.[6] The breaking of the thread in this scenario is a real fact about which all the Lorentz frames agree (although the fact is explained differently in different frames). Bub and Pitowsky's claim is that the deep explanation for the breaking of the thread is the Lorentz contraction, which is a feature of the geometry of Minkowski spacetime that is independent of the material constitution of the thread and the kind of interactions involved.

What is the quantum analogue of the breaking of the thread? There are three candidates that come to mind (perhaps there are more): (i) The fact that a measurement has a definite outcome in our experience. Bub and Pitowsky do not attempt to explain this fact, so we set this issue aside. (ii) The fact that the structure of the probability space of the events in our experience seems to be classical (Boolean). Bub and Pitowsky set this fact (ii) as their desideratum. We shall criticise their approach on this particular point in the next section. Another candidate is: (iii) The objective basis-independent fact that the Hamiltonian of the universe includes decoherence interactions.

In the literature, there are quantum mechanical theories that provide dynamical analyses of measurement, such as Bohm's (1952) theory and the GRW (1986) theory. But these theories add structure to Hilbert space, and in virtue of this structure they solve the measurement problem. For this reason Bub and Pitowsky (2010, p. 454) take Bohm's theory and the GRW theory as "'Lorentzian' interpretations of quantum mechanics", which assume the Newtonian spacetime structure and add to it the ether as an additional structure for the propagation of electromagnetic effects.' Because they add structure, these 'Lorentzian' theories are *not* solutions to the consistency problem (of finding a dynamical analysis that is *compatible* with the Hilbert space structure (they call this problem, the 'small' measurement problem).

---

[6]See Brown (2005) for a different view about the role of dynamical analyses in special relativity; and Brown and Timpson (2007) for a criticism of the analogy with special relativity.

What is the solution they propose to the consistency problem in quantum mechanics? They appeal to decoherence theory which analyses quantum mechanically the interaction between macroscopic systems and the environment.[7] Consider a typical quantum measurement situation in which an observer (call her Alice) measures (say, by means of a Stern-Gerlach device; omitted in our description), the z-spin of an electron, which is prepared in some superposition of the up- and down-eigenstates of the z-spin. At the 'initial' time t_0 the quantum state of this universe is this[8]:

$$| \Psi (t_0)\rangle = 1/\sqrt{2}\left(| \uparrow_z\rangle + | \downarrow_z\rangle\right) \otimes | \psi_0\rangle \tag{15.1}$$

where $|\uparrow_z\rangle$ and $|\downarrow_z\rangle$ are (respectively) the up- and down-eigenstates of the z-spin of the electron; and $| \psi_0\rangle$ is Alice's brain state in which she is ready to measure the z-spin of the electron. The Schrödinger equation (in the non-relativistic case) maps the state in (1) to the following superposition state at the end of the measurement at some time t:

$$| \Psi(t)\rangle = 1/\sqrt{2} \mid \uparrow_z\rangle \otimes |\psi_\uparrow\rangle + 1/\sqrt{2} \mid \downarrow_z\rangle \otimes |\psi_\downarrow\rangle \tag{15.2}$$

where the $| \psi_{\uparrow\downarrow}\rangle$ are the brain states of Alice in which she would believe that the outcome of the z-spin measurement is up (or down, respectively), *if* the initial state of the electron were only the up-state (or only the down-state). We assume here that Alice's brain states $| \psi_{\uparrow\downarrow}\rangle$ include *all* the degrees of freedom that are correlated with whatever *macro*scopic changes occurred in Alice's brain due to this transition. For simplicity, the description in (2) omits many details of the microphysics, but nothing in our argument hinges on this. This is the generic case in which the measurement problem arises in which Alice's brain states and the spin states of the electron are entangled.

The essential idea of decoherence (for our purposes) is very briefly this. Since Alice is a relatively massive open system that continuously interacts with a large number of degrees of freedom in the environment, the quantum state of the universe is not (2), but rather this state:

$$|\Psi_t\rangle = 1/\sqrt{2} \mid \uparrow_z\rangle \otimes |\psi_\uparrow\rangle \otimes |E_+(t)\rangle + 1/\sqrt{2}\, |\downarrow_z\rangle \otimes |\psi_\downarrow\rangle \otimes |E_-(t)\rangle \tag{15.3}$$

in which the $| E_\pm(t)\rangle$ are the collective (relative-)states of all the degrees of freedom in the environment[9]. According to decoherence theory, the Hamiltonian that governs

[7]For decoherence theory, and references, see e.g., Joos et al. (2003).

[8]Interactions never begin, so state (1) (which is a product state) should be replaced with a state in which Alice + electron are (weakly) entangled; otherwise the dynamics would not be reversible.

[9]In discrete models, both $| \psi_{\uparrow\downarrow}\rangle$ and $| E_\pm(t)\rangle$ are states of the form: $\prod_{i=1}^{N} \mid \mu_i\rangle$ defined in the corresponding multi-dimensional Hilbert spaces.

the interaction with the environment *commutes* (or approximately commutes) with some observables of the macroscopic system, for example the position of its center of mass. In our example, Alice's brain states $|\psi_{\uparrow\downarrow}\rangle$ denote the eigenstates of the observables on which the interaction Hamiltonian depends. These states are called in the literature the 'decoherence basis'. The interaction couples (approximately) the states of the environment $|E_{\pm}(t)\rangle$ to the states $|\psi_{\uparrow\downarrow}\rangle$ in the decoherence basis. Since the environment consists of a large number of dynamically independent degrees of freedom (this is expressed by assuming a suitable 'random' probability distribution over the states of the different degrees of freedom in the environment), the $|E_{\pm}(t)\rangle$ are (highly likely to become) extremely quickly *almost* orthogonal to each other. Because of the large number of degrees of freedom in the environment, this behaviour is likely to persist for very long times (in some models, for times comparable with the age of the universe) for exactly the same reasons.

As long as decoherence persists, that is the $|E_{\pm}(t)\rangle$ remain nearly orthogonal to each other, the interference terms in the superposition (3) in the decoherence basis are very small. This means that at all these times, the off-diagonal elements of Alice's *reduced* density matrix (calculated by partial tracing over the environmental degrees of freedom) in the decoherence basis $|\psi_{\uparrow\downarrow}\rangle$ are very small. Since the interference small the branches defined in the decoherence basis $|\psi_{\uparrow\downarrow}\rangle$ evolve in time in the course of the Schrödinger evolution of (3) *almost separately* of each other. That is, given the Schrödinger evolution of the universal state (3), which results in decoherence, it turns out that (in the decoherence basis) each of the branches in (3) evolves in a way that is almost independent of the other branches. In these senses, the decoherence basis is special: It has the above formal features when decoherence obtains, were in other bases in which state (3) may be written, the expansion is much more cumbersome.

Notice that in our outline of decoherence we have defined the branches by Alice's brain states; but this is just an artefact of our example. The states of any decohering system (such as a macroscopic body in a thermal or radiation bath, a Brownian particle in suspension in a fluid, a pointer of a measuring device, Schrödinger's cat) are equally suitable to define the branches. It might even be that Alice's relevant brain states are not subject directly to decoherence (whether or not this is the case is the subject matter of brain science after all). All one needs is that Alice's relevant brain states (whatever they are) that give rise to her experience be coupled to the states of a decohering system. In state (3) we use Alice's brain states to define the decoherence basis only for simplicity.

The essential point for Bub and Pitowsky is that, since the interference terms are very small in the decoherence basis, the dynamical process of decoherence gives rise to an effectively classical (Boolean) probability space in which the probabilities for the events are given by the Born probabilities. The macroscopic events in this space are given by the projections onto the eigenstates of the decohering variables (in our example, the projections onto the states in the decoherence basis $|\psi_{\uparrow\downarrow}\rangle$) and these turn out to be the macroscopic variables that feature in our experience, and more generally in classical mechanics. And the probability space (approximately) satisfies classical probability theory as long as decoherence endures. For example, the conditional probabilities defined by sequences of such Boolean spaces (associated with sequences of measurements) are additive. Since the process of decoherence is a consequence of the unitary dynamics, Bub and Pitowsky take this analysis to provide the consistency proof they seek, that is, a proof that shows how a classical probability space of the kind of events we experience emerges in decoherence situations, without (allegedly) adding anything to Hilbert space.

'The analysis shows that a quantum dynamics, consistent with the kinematics of Hilbert space, suffices to underwrite the emergence of a classical probability space for the familiar macroevents of our experience, with the Born probabilities for macroevents associated with measurement outcomes derived from the quantum state as a credence function. The explanation for such non-classical effects as the loss of information on conditionalization is not provided by the dynamics, but by the kinematics, and given 'no cloning' as a fundamental principle, there can be no deeper explanation. In particular, there is no dynamical explanation for the definite occurrence of a particular measurement outcome, as opposed to other possible measurement outcomes in a quantum measurement process – the occurrence is constrained by the kinematic probabilistic correlations encoded in the projective geometry of Hilbert space, and only by these correlations.' (Bub and Pitowsky, p. 454).

However, this doesn't work. In special relativity 'consistency proofs' show that Lorentz covariant forces 'track' the structure of Minkowski spacetime of which Lorentz contraction and time dilation are real physical features. So, by the analogy, the quantum consistency proof should prove – without adding structure to Hilbert space – that decoherence gives rise to the emergence of an effectively classical probability space of the events in our experience and that the von-Neumann-Lüders conditionalisation 'tracks' real physical features of events occurring in the universe (the objective chances). But we shall argue that this cannot be done. The analogy with special relativity breaks down, because even in decoherence situations the quantum (unitary) dynamics of the universe does not result in the emergence of an effectively classical (Boolean) probability space. In the next section we explain why.

15.3 The Role of Decoherence in the Consistency Proof

It is true that the decoherence process is basis-independent and that the interaction Hamiltonian picks out the decoherence basis as *dynamically* preferred (given a factorisation of the Hilbert space into system and environment), as we explained above. In fact, since the process of decoherence is continuous, there is a *continuous* infinity of basis states in Alice's state space *other* than the $|\psi_{\uparrow\downarrow}\rangle$ which satisfy to the *same* degree the conditions of decoherence (namely the near orthogonality of the relative states of the environment and the fact that this behaviour is stable over long time periods). It is an objective fact that as long as decoherence endures the interference between the branches defined in any such basis (and the off-diagonal elements in the reduced state of the decohering system) are very small and can hardly be detected by measurements. In many cases in physics it is practically useful to ignore small terms which make no detectible difference for calculating empirical predictions. Here are two examples.

In probability theory it is sometimes assumed that events with very small probability will not occur with *practical* (or moral) certainty, or equally, that events with very high probability which is not strictly one, will occur with practical certainty (see Kolmogorov 1933). In these cases, the identification of very small probability with zero probability is practically useful whenever one assumes that cases with very small probability occur with very small relative frequency. Although such cases will appear in the long run of repeated trials, they are unlikely to appear in our relatively short-lived experience. Another example is the Taylor series of a function which is used to approximate the value of a function at the expansion point by the function's derivatives at that point (when the derivatives are defined and continuous around the point). Although the Taylor series is an infinite sum of terms, it is true in many cases (but not always) that one can approximate the function to a sufficient degree of accuracy by truncating the Tailor series and taking only the first finite number of the terms in the series. So effectively in these cases the rest of the terms in the series which typically decay as the order of the derivatives increases, but do not strictly vanish, are set to zero. One can obtain a bound via Taylor's theorem, on the size of the error in the approximation depending on the length of the truncated series which might be much below the degree of accuracy that one can measure. Although in general the Tailor infinite series is not equal to the function (even if the series converges at every point), using the truncated series gives accurate empirical predictions about the values of the function near the expansion points and its behavior in various limits. But no claim is made that the truncated small terms do not exist, the claim is only that we are not able to detect their existence in our experience.

It seems to us that the role of decoherence in the foundations of quantum mechanics should be understood along similar lines to the above two examples. Suppose that the Hamiltonian of the universe includes interactions that we call measurements of certain observables and decoherence interactions (as in our spin example above). The evolution of the universal state can be written in infinitely many bases in the Hilbert space of the universe. Different bases pick out different events, in some bases the interference terms between the components describing the events are small, and in others they are large. In some intuitive sense, the bases in which the interference terms are small seem to be esthetically more suitable or more convenient for us to describe the quantum state. But given the Hilbert space structure these bases have no other special status, in particular they are not more physically real (as it were) than the other bases. In this sense the analogy of Hilbert space bases with Lorentz frames in special relativity is in place. It turns out that the events that feature in our experience correspond to the projections onto the decoherence basis. If we take this fact into account without a deeper explanation of how it comes about, then it is only natural that we write the quantum state describing our measurements in the decoherence basis. And then for matters of calculating our empirical predictions the small interference terms in the decoherence basis make no detectable difference to what we can measure in our experience, and for this purpose these terms can be set to zero. To detect the superposition in (3), by means of measurements which are described in the decoherence basis, one needs to manipulate the relative states of the environment in a way that results in significant interference. But this is practically impossible, due to the large number of degrees of freedom in the environment and the fact that they are much beyond our physical reach. In this practical sense, setting the interference terms in the decoherence basis to effectively zero is a perfectly useful approximation.

But in order for Bub and Pitowsky's consistency proof to be analogous to the dynamical proofs in special relativity, these effective considerations are immaterial. It is an objective fact about the universe that decoherence obtains (given a factorisation). This is our candidate (iii) for the analogy with the breaking of the thread in special relativity. But what about Bub and Pitowsky's target (candidate (ii)) for the analogy, namely the fact that the structure of the probability space of the events in our experience seems to be classical (Boolean)? The decoherence basis is analogous only to *certain* Lorentz frames, and the decoherence interaction explains why the interference terms of state (3), described in this basis, are small; the decoherence interaction equally explains why in other bases of the Hilbert space of the universe (which are analogous to other Lorentz frames) the interference terms of state (3) are large. Consequently, the decoherence interaction explains why a probability space of events picked out by the decoherence basis is effectively classical (Boolean); the decoherence interaction equally explains why the probability space of events picked out by some other bases are *not* effectively classical. But the decoherence interaction does *not* explain why events picked out by the decoherence basis, rather than by these other bases, feature in our experience.

For this reason, the analogy with special relativity breaks down. In special relativity one *can* explain why the experiences of certain (physical) observers are associated with certain Lorentz frames by straightforward physical facts about these observers, for example, their relative motion.[10] But in quantum mechanics, if we do not add structure to Hilbert space, one cannot explain by appealing to physical facts why the experiences of certain observers, for example, human observers, are associated with the events picked out by the decoherence basis. The emergence of the probability space of these events in our experience is *not* a basis-independent

[10]In statistical mechanics, here is an analogous scenario: the phase space of the universe can be partitioned into infinitely many sets corresponding to infinitely many *macrovariables*. Some of these partitions exhibit certain regularities, for example, the partition to the thermodynamic sets which human observers are sensitive to, exhibit the thermodynamic regularities; while other partitions to which human observers are not sensitive, although they are *equally real* exhibit other regularities and may even be anomalous (that is, exhibit no regularities at all). We call this scenario in statistical mechanics *Ludwig's problem* (see Hemmo and Shenker 2012, 2016). But one *can* explain by straightforward physical facts why (presumably) human observers experience the thermodynamic sets and regularities but not the equally existing other sets (although it might be that in our experience some other sets also appear). This case too is dis-analogous to the basis-symmetry in quantum mechanics, from which the problem of *no* preferred basis follows.

fact in quantum mechanics. And if quantum mechanics is a complete theory, as Bub and Pitowsky assume, then it follows that components of the universal state in some or other basis do *not* pick out anything that is physically 'more real' (as it were) than components in any other basis. In particular, the decoherence basis has absolutely no preferred ontological status by comparison to any other basis in the Hilbert space of the universe, despite the fact that it is special in that the interference terms of states like (3) described in this basis are small. So: what is the explanation of the fact that events picked out by the decoherence basis (for which the probability space is effectively classical) feature in our experience? On the Bub-Pitowsky approach this cannot be explained by appealing to facts about, say, Alice's brain and its interactions, for the same reason that (as we see from our example), Alice's brain, even if it is subject to decoherence, can be equally described by infinitely many bases all of which are on equal status. The only way to explain why we experience events that are picked out by the decoherence basis, rather than events picked out by other bases, is to add laws, or structure, to Hilbert space, from which it will follow that the decoherence basis is physically preferred, in the sense that it picks out the actually occurring events in our experience.[11]

What about the quantum conditionalisation (and von Neumann's projection postulate)? One might argue that the von Neumann-Lüders conditionalisation does not amount to adding structure to Hilbert space, since it expresses the information loss in measurement, which is itself entailed by the 'no cloning' theorem. But this is beside the point, since the conditionalisation depends not only on the individual outcome that actually occurs (a desideratum that Bub and Pitowsky do not attempt to explain), but also on a choice of a Hilbert space basis. As we argued above, the Hilbert space structure, however, says *nothing* about the preference, or choice, of any basis, let alone the occurrence (or non-occurrence) of individual events. This is why the additional laws (or structure) are required. This is why decoherence cannot provide the consistency proof Bub and Pitowsky seek by their analogy with relativity theory.

15.4 From Subjectivism to Idealism

In the Bayesian approach to probability, probability theory is about how to make rational decisions in the face of uncertainty, and therefore the starting point is that probability is a measure of ignorance (a credence function describing degrees of belief) in a single trial, and no a priori connection is assumed between the probability and limiting frequencies, except as derived by conditionalisation on the *experienced* frequencies, and adding inductive generalisations.

[11] A preferred basis in exactly the same sense is also required in all the contemporary versions of the Everett (1957) interpretation of quantum mechanics, so that one must add a new law of physics to the effect that worlds emerge relative to the decoherence basis; see (Hemmo and Shenker 2019).

In the face of the measurement problem, the Bayesian approach to quantum mechanics says this.[12] The quantum state is *just* a credence function, tracking the limiting frequencies *in experience*, it is *not* a bookkeeping device for tracking the 'universe's objective chances' (as it is in the Bub-Pitowsky approach). The credence function changes in accordance with the unitary dynamics in the absence of measurements, but in measurement contexts it changes by quantum conditionalisation (the projection postulate or the von Neumann-Lüders rule), which is the quantum analogue of the classical Bayesian updating. Both laws are seen as inductive generalisations from experience. Decoherence in this approach is *not* a physical process but some sort of a mental process that is subject to a rationality constraint on the agent's experiences, in particular, belief states, which is related to so-called Dutch-book consistency.[13] In this approach no attempt is made at explaining how our experience comes about.

On this approach quantum mechanics is a theory about the *mental states (experiences, beliefs) of observers*. Each observer describes the rest of the world by the unitary dynamics, and applies the quantum conditionalisation law only to the evolution of its *own* mental state. This extreme form of subjectivism is necessary to avoid contradictions in the 'internal scenarios' (or the experience) of each observer.[14] This is because, if there are two observers or more, nothing prevents a case in which different observers will experience different outcomes for the same measurement. But this does not lead to a contradiction in the internal scenario (as it were) of each observer separately: as is well known, quantum mechanics ensures that it will always seem to *me* that the experiences reported by other observers are the *same* as mine. On this picture, quantum mechanics is not about anything external to experience. Whether or not the experience is somehow correlated with anything external (and whether there is anything external) is beyond quantum mechanics.

We wish to point out two consequences of this approach. First, as in the Bub-Pitowsky approach, the Bayesian approach adds structure to Hilbert space. In implementing the quantum conditionalisation law, *experience* plays two crucial roles. The first role of experience is *to choose one basis* in the Hilbert space of the measured system, the basis formed by the set of eigenstates of the observable measured, with respect to which the agent applies the von Neumann-Lüdres rule[15]; and the *second role* is to *choose one of the eigenstates* in this basis as the experienced

[12]Recent variations are, for example, Caves et al. (2002a, b, 2007); Fuchs and Schack (2013); Fuchs et al. (2014); Fuchs and Stacey (2019).

[13]For the role of decoherence in the Bayesian approach and its relation to Dutch-book consistency, see Fuchs and Scack (2012).

[14]See Hagar and Hemmo (2006); Hagar (2003).

[15]Bohm's (1952) theory and the GRW theory (Ghirardi et al. 1986) assume a preferred basis from the start. Our point is that the structure added by the Bayesian approach is no less than e.g., Bohmiam trajectories or GRW flashes.

measurement outcome, where the probability for the choice is dictated by the Born rule.[16]

The second consequence of the Bayesian approach is that as a *matter of principle*, not of ignorance about the external physical world, mental states (and experience in general) *cannot* be accounted for by physics. Here is why. Contemporary science (brain and cognitive science) is far from explaining how mental experience comes about. But the standard working hypothesis in these sciences – which is fruitful and enjoys vast empirical support – is that our mental experience is to be explained (somehow, by and large) by features of our brains, perhaps together with some other parts of the body; for simplicity we call all of them 'the brain'. How and why we can treat biological systems in effectively classical terms is part of the questioin of the classical limit of quantum mechanics. But here our specific question is this. Let us suppose that quantum mechanics does yield in some appropriate limit classical behavior of (say) our brains. Can the Bayesian approach accept the standard working hypothesis that mental states are to be accounted for by brain states (call it the 'brain hypothesis')?

If the brain hypothesis is true, there is some physical state of my brain which underlies my credence function. Since according to contemporary physics quantum mechanics is our best physical theory, this physical state (whatever its details are) should be some quantum state of my brain. But on the Bayesian approach, the quantum state of my brain is a credence function of some observer, not the physical state of anything; so *a fortiori* it can't be the physical state of my brain. It follows that, as a matter of principle, mental states cannot be accounted for by brain states: they *are not physical even fundamentally*.

In classical physics the Bayesian theory of probability is not meant to replace the physical theory (i.e., classical relativistic physics) describing how the relative frequencies of events are brought about. After all, as Pitowsky (2003, p. 400) has put it: "the theory of probability, even in its most subjective form, associates a person's degree of belief with the objective possibilities in the physical world." In the classical case the subjective interpretation of probability (de Finetti 1970; Ramsey 1929) may assume in the background that – as a matter of principle – not only the frequencies but also the credences of agents and their evolution over time might be ultimately accounted for by physics. In this sense, the subjectivist approach to classical probability is *consistent* with the *physicalist* thesis in cognition and philosophy of mind. But the Bayesian approach to quantum mechanics, is not only inconsistent with physicalism: it is a form of idealism.

Acknowledgement We thank Guy Hetzroni and Cristoph Lehner for comments on an earlier draft of this paper. This research was supported by the Israel Science Foundation (ISF), grant number 1114/18.

[16]These two roles are carried out by the mind at one shot (as it were), but analytically they are different.

References

Bell, J. S. (1987a). Are there quantum jumps? In J. Bell (Ed.), *Speakable and unspeakable in quantum mechanics* (pp. 201–212). Cambridge: Cambridge University Press.

Bell, J. S. (1987b). How to teach special relativity. In J. Bell (Ed.), *Speakable and unspeakable in quantum mechanics* (pp. 67–80). Cambridge: Cambridge University Press.

Brown, H. (2005). *Physical relativity: Spacetime structure from a dynamical perspective*. Oxford: Oxford University Press.

Brown, H. R., & Timpson, C. G. (2007). Why special relativity should not be a template for a fundamental reformulation of quantum mechanics. In Pitowsky (Ed.), *Physical theory and its interpretation: Essays in honor of Jeffrey Bub, W. Demopoulos and I.* Berlin: Springer.

Bub, J. (1977). Von Neumann's projection postulate as a probability conditionalization rule in quantum mechanics. *Journal of Philosophical Logic, 6*, 381–390.

Bub, J. (2007). Quantum probabilities as degrees of belief. *Studies in History and Philosophy of Modern Physics, 38*, 232–254.

Bub, J. (2016). *Bananaworld: Quantum mechanics for primates*. Oxford: Oxford University Press.

Bub, J. (2020). 'Two Dogmas' Redux. In Hemmo, M., Shenker, O. (eds.) *Quantum, probability, logic: Itamar Pitowsky's work and influence*. Cham: Springer.

Bub, J., & Pitowsky, I. (2010). Two dogmas about quantum mechanics. In S. Saunders, J. Barrett, A. Kent, & D. Wallace (Eds.), *Many worlds? Everett, quantum theory, and reality* (pp. 431–456). Oxford: Oxford University Press.

Bohm, D. (1952). A suggested interpretation of the quantum theory in terms of "hidden" variables. *Physical Review, 85*(166–179), 180–193.

Caves, C. M., Fuchs, C. A., & Schack, R. (2002a). Unknown quantum states: The quantum de Finetti representation. *Journal of Mathematical Physics, 43*(9), 4537–4559.

Caves, C. M., Fuchs, C. A., & Schack, R. (2002b). Quantum probabilities as Bayesian probabilities. *Physical Review A, 65*, 022305.

Caves, C. M., Fuchs, C. A., & Schack, R. (2007). Subjective probability and quantum certainty. *Studies in History and Philosophy of Modern Physics, 38*, 255–274.

de Finetti, B. (1970). *Theory of probability*. New York: Wiley.

Everett, H. (1957). 'Relative state' formulation of quantum mechanics. *Reviews of Modern Physics, 29*, 454–462.

Fuchs C. A. & Schack, R. (2012). Bayesian conditioning, the reflection principle, and quantum decoherence. In Ben-Menahem, Y., & Hemmo, M. (eds.), *Probability in physics* (pp. 233–247). The Frontiers Collection. Berlin/Heidelberg: Springer.

Fuchs, C. A., & Schack, R. (2013). Quantum Bayesian coherence. *Reviews of Modern Physics, 85*, 1693–1715.

Fuchs, C. A., Mermin, N. D., & Schack, R. (2014). An introduction to QBism with an application to the locality of quantum mechanics. *American Journal of Physics, 82*, 749–754.

Fuchs, C. A., & Stacey, B. (2019). QBism: Quantum theory as a hero's handbook. In E. M. Rasel, W. P. Schleich, & S. Wölk (Eds.), *Foundations of quantum theory: Proceedings of the International School of Physics "Enrico Fermi" course 197* (pp. 133–202). Amsterdam: IOS Press.

Ghirardi, G., Rimini, A., & Weber, T. (1986). Unified dynamics for microscopic and macroscopic systems. *Physical Review, D, 34*, 470–479.

Hagar, A. (2003). A philosopher looks at quantum information theory. *Philosophy of Science, 70*(4), 752–775.

Hagar, A., & Hemmo, M. (2006). Explaining the unobserved – Why quantum mechanics ain't only about information. *Foundations of Physics, 36*(9), 1295–1324.

Hemmo, M., & Shenker, O. (2012). *The road to Maxwell's demon*. Cambridge: Cambridge University Press.

Hemmo, M., & Shenker, O. (2016). *Maxwell's demon*. Oxford University Press: Oxford Handbooks Online.

Hemmo, M. & Shenker, O. (2019). *Why quantum mechanics is not enough to set the framework for its interpretation*, forthcoming.

Joos, E., Zeh, H. D., Giulini, D., Kiefer, C., Kupsch, J., & Stamatescu, I. O. (2003). *Decoherence and the appearance of a classical world in quantum theory*. Heidelberg: Springer.

Pitowsky, I. (2003). Betting on the outcomes of measurements: A Bayesian theory of quantum probability. *Studies in History and Philosophy of Modern Physics, 34*, 395–414.

Kolmogorov, A. N. (1933). *Foundations of the Theory of Probability*. New York: Chelsea Publishing Company, English translation 1956.

Pitowsky, I. (2007). Quantum mechanics as a theory of probability. In W. Demopoulos & I. Pitowsky (Eds.), *Physical theory and its interpretation: Essays in honor of Jeffrey Bub* (pp. 213–240). Berlin: Springer.

Ramsey, F. P. (1990). *Truth and Probability*. (1926); reprinted in Mellor, D.H. (ed.), F. P. Ramsey: Philosophical Papers. Cambridge: Cambridge University Press.

Tumulka, R. (2006). A relativistic version of the Ghirardi–Rimini–Weber model. *Journal of Statistical Physics, 125*, 821–840.

von Neumann, J. (2001). Unsolved problems in mathematics. In M. Redei & M. Stoltzner (Eds.), *John von Neumann and the foundations of quantum physics* (pp. 231–245). Dordrecht: Kluwer Academic Publishers.

Chapter 16
On the Three Types of Bell's Inequalities

Gábor Hofer-Szabó

Abstract Bell's inequalities can be understood in three different ways depending on whether the numbers featuring in the inequalities are interpreted as classical probabilities, classical conditional probabilities, or quantum probabilities. In the paper I will argue that the violation of Bell's inequalities has different meanings in the three cases. In the first case it rules out the interpretation of certain numbers as probabilities of events. In the second case it rules out a common causal explanation of conditional correlations of certain events (measurement outcomes) conditioned on other events (measurement settings). Finally, in the third case the violation of Bell's inequalities neither rules out the interpretation of these numbers as probabilities of events nor a common causal explanation of the correlations between these events – provided both the events and the common causes are interpreted non-classically.

Keywords Bell's inequalities · Conditional probability · Correlation polytope

16.1 Introduction

Ever since its appearance three decades ago, Itamar Pitowsky's *Quantum Probability – Quantum Logic* has been serving as a towering lighthouse showing the way for many working in the foundations of quantum mechanics. In this wonderful book Pitowsky provided an elegant geometrical representation of classical and quantum probabilities: a powerful tool in tackling many difficult formal and conceptual questions in the foundations of quantum theory. One of them is the meaning of the violation of Bell's inequalities.

However, for someone reading the standard foundations of physics literature on Bell's theorems, it is not easy to connect up Bell's inequalities as they are presented

---

G. Hofer-Szabó (✉)
Research Center for the Humanities, Budapest, Hungary
e-mail: szabo.gabor@btk.mta.hu

M. Hemmo, O. Shenker (eds.), *Quantum, Probability, Logic*,
Jerusalem Studies in Philosophy and History of Science,
https://doi.org/10.1007/978-3-030-34316-3_16

in Pitowsky's book with the inequalities presented by Bell, Clauser, Horne, Shimony etc. The main difference, to make it brief, is that Bell's inequalities in their original form are formulated in terms of *conditional* probabilities, representing certain measurement outcomes provided that certain measurement are performed, while Pitowsky's Bell inequalities are formulated in terms of *unconditional* probabilities, representing the distribution of certain underlying properties or events responsible for the measurement outcomes.

Let us see this difference in more detail. Probabilities enter into quantum mechanics *via* the trace formula

$$p = \mathrm{Tr}(\hat{\rho}\hat{A}) \tag{16.1}$$

where $\hat{\rho}$ is a density operator, $\hat{A}$ is the spectral projection associated to eigenvalue α of the self-adjoint operator $\hat{a}$, and Tr is the trace function. Let us call the probabilities generated by the trace formula (16.1) *quantum probabilities*.

Now, what is the physical interpretation of quantum probabilities? There are two possible answers to this question according to two different interpretations of quantum mechanics:

1. On the *operational (minimal) interpretation*, the density operator $\hat{\rho}$ represents the state or preparation s of the system; the self-adjoint operator $\hat{a}$ represents the measurement a performed on the system; and the spectral projection $\hat{A}$ represents the outcome A of the measurement a. On this interpretation the quantum probability is interpreted as the *conditional probability*

$$p = p_s(A|a) \tag{16.2}$$

   that is the probability of obtaining outcome A *provided* that the measurement a has been performed on the system previously prepared in state s. This interpretation is called *minimal* since the fulfillment of (16.2) is a necessary condition for the theory to be empirically adequate.
2. On the *ontological (deterministic hidden variable, property) interpretation*, the density operator $\hat{\rho}$ represents the distribution ρ of the ontological states $\lambda \in \Lambda$ in the preparation s; the operator $\hat{a}$ represents the physical magnitude a^*; and the projection $\hat{A}$ represents the event A^* that the value of a^* is A^*.[1] Denote by $\Lambda^{A^*} = \{\lambda^{A^*}\}$ the set of those ontological states for which the value a^* is A^*. On the ontological interpretation the quantum probability is (intended to be) interpreted as the *unconditional probability*

$$p = p_s(A^*) = \int_{\Lambda^{A^*}} \rho(\lambda)\, d\lambda \tag{16.3}$$

   that is the probability of the event A^* in the state s.

[1] I denote the event and the value by the same symbol.

The ontological and the operational definition are connected as follows. The measurement a is said to measure the physical magnitude a^* only if the following holds: the outcome of measurement a performed on the system will be A if and only if the system is in an ontological state for which the value of a^* is A^*:

$$p_\lambda(A|a) = 1 \qquad \text{if and only if} \qquad \lambda \in \Lambda^{A^*}$$

As is seen, the two interpretations differ in how they treat quantum probabilities. On the operational interpretation quantum probabilities are condition probabilities, while on the ontological interpretation they are unconditional probabilities. (Note that in both interpretations probabilities are "conditioned" on the preparation which will be dropped from the next section.)

Now, consider a *set* of self-adjoint operators $\{\hat{a}_i\}$ $(i \in I)$ each with two spectral projections $\{\hat{A}_i, \hat{A}_i^\perp\}$. Correspondingly, consider a set of measurements $\{a_i\}$ each with two outcomes $\{A_i, A_i^\perp\}$ and a set of magnitudes $\{a_i^*\}$ each with two values $\{A_i^*, A_i^{*\perp}\}$. If $\hat{A}_i$ and $\hat{A}_j$ are commuting, then the quantum probabilities

$$p_i = \text{Tr}(\hat{\rho}\hat{A}_i) \tag{16.4}$$

$$p_{ij} = \text{Tr}(\hat{\rho}\hat{A}_i\hat{A}_j) \tag{16.5}$$

can be interpreted either operationally:

$$p_i = p_s(A_i|a_i) \tag{16.6}$$

$$p_{ij} = p_s(A_i \wedge A_j|a_i \wedge a_j) \tag{16.7}$$

or ontologically:

$$p_i = p_s(A_i^*) \tag{16.8}$$

$$p_{ij} = p_s(A_i^* \wedge A_j^*) \tag{16.9}$$

Consider a paradigmatic Bell inequality, the Clauser-Horne inequalities:

$$\begin{aligned} -1 \quad &\leqslant p_{ij} + p_{i'j} + p_{ij'} - p_{i'j'} - p_i - p_j \leqslant 0 \\ &i, i' = 1, 2; \ \ j, j' = 3, 4; \ \ i \neq i'; \ \ j \neq j' \end{aligned} \tag{16.10}$$

The numbers featuring in (16.10) are probabilities. But which type of probabilities? Are they simply (uninterpreted) quantum probabilities of type (16.4)–(16.5), or condition probabilities of type (16.6)–(16.7), or unconditional probabilities of type (16.8)–(16.9)? Depending on how the probabilities in the Clauser-Horne inequalities are understood, the inequalities can be written out in the following three different forms:

$$-1 \leqslant \mathrm{Tr}(\hat{\rho}\hat{A}_i\hat{A}_j) + \mathrm{Tr}(\hat{\rho}\hat{A}_{i'}\hat{A}_j) + \mathrm{Tr}(\hat{\rho}\hat{A}_i\hat{A}_{j'}) - \mathrm{Tr}(\hat{\rho}\hat{A}_{i'}\hat{A}_{j'}) - \mathrm{Tr}(\hat{\rho}\hat{A}_i) - \mathrm{Tr}(\hat{\rho}\hat{A}_j) \leqslant 0 \quad (16.11)$$

(where $\hat{A}_i\hat{A}_j$, $\hat{A}_{i'}\hat{A}_j$, $\hat{A}_i\hat{A}_{j'}$ and $\hat{A}_{i'}\hat{A}_{j'}$ are commuting projections); or

$$-1 \leqslant p_s(A_i \wedge A_j|a_i \wedge a_j) + p_s(A_{i'} \wedge A_j|a_{i'} \wedge a_j) + p_s(A_i \wedge A_{j'}|a_i \wedge a_{j'}) - p_s(A_{i'} \wedge A_{j'}|a_{i'} \wedge a_{j'}) - p_s(A_i|a_i) - p_s(A_j|a_j) \leqslant 0 \quad (16.12)$$

or

$$-1 \leqslant p_s(A_i^* \wedge A_j^*) + p_s(A_{i'}^* \wedge A_j^*) + p_s(A_i^* \wedge A_{j'}^*) - p_s(A_{i'}^* \wedge A_{j'}^*) - p_s(A_i^*) - p_s(A_j^*) \leqslant 0 \quad (16.13)$$

Thus, altogether we have three different types of Bell inequalities depending on whether and how the probabilities featuring in (16.10) are physically interpreted. The aim of the paper is to clarify as to what is exactly excluded if Bell's inequalities of type (16.11), (16.12) or (16.13) are violated. I will argue that the violation has three different meanings in the three different cases. In case of (16.13), when the probabilities are classical unconditional probabilities, it rules out the interpretation of certain numbers as probabilities of events or properties. These are the inequalities which Pitowsky identified and categorized. The violation of inequalities (16.12), when the probabilities are classical conditional probabilities, does not rule out the interpretation of certain numbers as conditional probabilities of events but only a common causal explanation of the conditional correlations between these events. These are the Bell inequalities used in the standard foundations of physics literature. Finally, the violation of Bell's inequalities (16.11) neither rules out the interpretation of certain numbers as probabilities of events nor a common causal explanation of the correlations between these events—provided that both events and common causes are interpreted non-classically. The violation of these Bell's inequalities is used for another purpose: it places a bound on the strength of correlations between these events.

I will proceed in the paper as follows. In Sect. 16.2 I analyze Bell's inequalities for classical probabilities, and in Sect. 16.3 for classical conditional probabilities. In Sect. 16.4 the two types will be compared. I turn to Bell's inequalities in terms of quantum probabilities in Sect. 16.5. In Sect. 16.6 I apply the results to the EPR-Bohm scenario and finally conclude in Sect. 16.7.

To make the notation simple, I drop both the hat and the asterisk from the next section on, that is I write A instead of both $\hat{A}$ and also A^*. The semantics of A will be clear from the context: in classical conditional probabilities A will refer to a measurement outcome, in classical unconditional probabilities it will refer to a property/event, and in quantum probabilities it will refer to a projection.

My approach strongly relies on Szabó (2008) distinction between Bell's original inequalities and what he calls the Bell-Pitowsky inequalities. For a somewhat parallel research see Gömöri and Placek (2017).

16.2 Case 1: Bell's Inequalities for Classical Probabilities

Consider real numbers p_i and p_{ij} in $[0, 1]$ such that $i = 1 \ldots n$ and $(i, j) \in S$ where S is a subset of the index pairs $\{(i, j)\,|i < j;\ i, j = 1 \ldots n\}$. When can these numbers be probabilities of certain events and their conjunctions? More precisely: given the numbers p_i and p_{ij}, is there a classical probability space (Ω, Σ, p) with events A_i and $A_i \wedge A_j$ in Σ such that

$$p_i = p(A_i)$$
$$p_{ij} = p(A_i \wedge A_j)$$

In brief, do the numbers p_i and p_{ij} admit a *Kolmogorovian representation*?

Itamar Pitowsky's (1989) provided an elegant geometrical answer to this question. Arrange the numbers p_i and p_{ij} into a vector

$$\vec{p} = \big(p_1, \ldots, p_n; \ldots, p_{ij}, \ldots\big)$$

called *correlation vector*. $\vec{p}$ is an element of an $n+|S|$ dimensional real linear space, $R(n, S) \cong \mathbb{R}^{n+|S|}$, where $|S|$ is the cardinality of S.

Now, we construct a polytope in $R(n, S)$. Let $\varepsilon \in \{0, 1\}^n$. To each ε assign a *classical vertex vector* (truth-value function): $\vec{u}^{\,\varepsilon} \in R(n, S)$ such that

$$\vec{u}_i^{\,\varepsilon} = \varepsilon_i \qquad i = 1 \ldots n$$
$$\vec{u}_{ij}^{\,\varepsilon} = \varepsilon_i \varepsilon_j \quad (i, j) \in S$$

Then we define the *classical correlation polytope* in $R(n, S)$ as the convex hull of the classical vertex vectors:

$$c(n, S) := \left\{ \vec{p} \in R(n, S) \,\middle|\, \vec{p} = \sum_{\varepsilon \in \{0,1\}^n} \lambda_\varepsilon \vec{u}^{\,\varepsilon}\,;\ \lambda_\varepsilon \geqslant 0;\ \sum_{\varepsilon \in \{0,1\}^n} \lambda_\varepsilon = 1 \right\}$$

The polytope $c(n, S)$ is a simplex, that is any correlation vector in $c(n, S)$ has a unique expansion by classical vertex vectors.

Pitowsky's theorem (Pitowsky 1989, p. 22) states that $\vec{p}$ admits a Kolmogorovian representation if and only if $\vec{p} \in c(n, S)$. That is numbers can be probabilities of certain events and their conjunctions if and only if the correlation vector composed of these numbers is in the classical correlation polytope.

Now, Bell's inequalities enter the scene as the *facet inequalities* of the classical correlation polytopes. The simplest such correlation polytope is the one with $n = 2$ and $S = \{(1, 2)\}$. In this case the vertices are: $(0, 0; 0)$, $(1, 0; 0)$, $(0, 1; 0)$ and $(1, 1; 1)$ and the facet inequalities (Bell's inequalities) are the following:

$$0 \leqslant p_{12} \leqslant p_1, p_2 \leqslant 1$$
$$p_1 + p_2 - p_{12} \leqslant 1$$

Another famous polytope is $c(n, S)$ with $n = 4$ and $S = \{(1, 3), (1, 4), (2, 3), (2, 4)\}$. The facet inequalities are then the following:

$$0 \leqslant p_{ij} \leqslant p_i, p_j \leqslant 1 \qquad i = 1, 2;\ \ j = 3, 4 \tag{16.14}$$
$$p_i + p_j - p_{ij} \leqslant 1 \qquad i = 1, 2;\ \ j = 3, 4 \tag{16.15}$$
$$-1 \leqslant p_{ij}+p_{i'j}+p_{ij'}-p_{i'j'}-p_i-p_j \leqslant 0 \qquad i, i'=1, 2;\ j, j'=3, 4;\ i \neq i';\ j \neq j' \tag{16.16}$$

The facet inequalities (16.16) are called the *Clauser-Horne inequalities*. They express whether $4+4$ real numbers can be regarded as the probability of four classical events and their certain conjunctions.

A special type of the correlation vectors are the *independence vectors* that is correlation vectors such that for all $(i, j) \in S$, $p_{ij} = p_i p_j$. It is easy to see that all independence vectors lie in the classical correlation polytope with coefficients

$$\lambda_\varepsilon = \prod_{i=1}^{n} p_i^*, \qquad \text{where } p_i^* = \begin{cases} p_i & \text{if } \varepsilon_i = 1 \\ 1 - p_i & \text{if } \varepsilon_i = 0 \end{cases} \tag{16.17}$$

Classical vertex vectors are independence vectors by definition; they are extremal points of the classical correlation polytope. For a classical vertex vector $p_i \in \{0, 1\}$ for all $i = 1 \ldots n$. Thus we will sometimes call a classical vertex vector a *deterministic* independence vector and an independence vector which is *not* a classical vertex vector an *indeterministic* independence vector.

Although correlation vectors in $c(n, S)$ has a unique convex expansion by classical vertex vectors, they (if not classical vertex vectors) can have many convex expansions by indeterministic independence vectors. Moreover, for a classical correlation vector $\vec{p} \in c(n, S)$ (which is not a classical vertex vector) there always exist many sets of indeterministic independence vectors $\{\vec{p}^{\,\varepsilon}\}$ such that

$$\vec{p} = \sum_{\varepsilon} \lambda_\varepsilon \vec{u}^{\,\varepsilon} = \sum_{\varepsilon} \lambda_\varepsilon\, \vec{p}^{\,\varepsilon}$$

that is the coefficients λ_ε of the expansion of $\vec{p}$ by classical vertex vectors and by indeterministic independence vectors are the same.[2]

Let us call a correlation vector which is not an independence vector a *proper correlation vector*. For a proper correlation vector $\vec{p}$ there is at least one pair $(i, j) \in S$ such that $p_{ij} \neq p_i p_j$. When lying in the classical correlation polytope $\vec{p}$ represents the probabilities of certain events and their conjunctions and the events associated to indices i and j are correlated:

$$p(A_i \wedge A_j) \neq p(A_i)\, p(A_j) \tag{16.19}$$

Now, we ask the following question: When do the correlations in the Kolmogorovian representation of a proper correlation vector in $c(n, S)$ have a common causal explanation?

A common cause in the Reichenbachian sense (Reichenbach 1956) is a screener-off partition of the algebra Σ (or an extension of the algebra; see Hofer-Szabó et al. 2013, Ch. 3 and 6). In other words, correlations (16.19) are said to have a *joint common causal explanation* if there is a partition $\{C_k\}$ $(k \in K)$ of Σ such that for any (i, j) in S and $k \in K$

$$p(A_i \wedge A_j | C_k) = p(A_i | C_k)\, p(A_j | C_k) \tag{16.20}$$

Introduce the notation

$$p_i^k = p(A_i | C_k)$$
$$c_k = p(C_k)$$

where $\sum_k c_k = 1$ and construct for each $k \in K$ a *common cause vector*

$$\vec{p}^{\,k} = \left(p_1^k, \ldots, p_n^k; \ldots, p_i^k p_j^k, \ldots\right)$$

---

[2]For example:

$$\begin{aligned}
\vec{p} &= \left(\frac{2}{5}, \frac{2}{5}; \frac{1}{5}\right) \\
&= \frac{1}{5}(1, 1; 1) + \frac{1}{5}(1, 0; 0) + \frac{1}{5}(0, 1; 0) + \frac{2}{5}(0, 0; 0) \\
&= \frac{1}{5}\left(\frac{1}{4}, \frac{1}{4}; \frac{1}{16}\right) + \frac{1}{5}\left(\frac{3+\sqrt{5}}{8}, \frac{3+\sqrt{5}}{8}; \frac{7+3\sqrt{5}}{32}\right) \\
&\quad + \frac{1}{5}\left(\frac{3-\sqrt{5}}{8}, \frac{3-\sqrt{5}}{8}; \frac{7--3\sqrt{5}}{32}\right) + \frac{2}{5}\left(\frac{1}{2}, \frac{1}{2}; \frac{1}{4}\right)
\end{aligned} \tag{16.18}$$

for the classical correlation vector $\vec{p}$. Due to (16.20) and the theorem of total probability

$$\vec{p} = \sum_k c_k \, \vec{p}^{\,k}$$

Since the common cause vectors are independence vectors lying in the classical correlation polytope $c(n, S)$, therefore their convex combination $\vec{p}$ also lies in the classical correlation polytope—which, of course, we knew since we assumed that $\vec{p}$ has a Kolmogorovian representation.

We call a common cause vector $\vec{p}^{\,k}$ *deterministic*, if $\vec{p}^{\,k}$ is a deterministic independence vector (classical vertex vector); otherwise we call it *indeterministic*. All classical correlation vectors have 2^n deterministic common cause vectors, namely the classical vertex vectors. In this cases $k = \varepsilon \in \{0, 1\}^n$ and the probabilities are:

$$p_i^{\varepsilon} = \varepsilon_i$$
$$c_{\varepsilon} = \lambda_{\varepsilon}$$

where λ_{ε} is specified in (16.17). Classical correlation vectors which are not classical vertex vectors also can be expanded as a convex sum of indeterministic common cause vectors in many different ways.

Conversely, knowing the k common cause vectors and the probabilities c_k alone, one can easily construct a common causal explanation of the correlations (16.19). First, construct the classical probability space Σ associated to the correlation vector $\vec{p}$ (for the details see Pitowsky 1989, p. 23). Then extend Σ containing the events A_i and $A_i \wedge A_j$ such that the extended probability space contains also the common causes $\{C_k\}$ (for such an extension see Hofer-Szabó et al. (1999)).

To sum up, in the case of classical probabilities the fulfillment of Bell's inequalities is a necessary and sufficient condition for a set of numbers to be the probability of certain events and their conjunctions. If these events are correlated, then the correlations will always have a common causal explanation. In other words, having a common causal explanation does not put a further constraint on the correlation vectors. Thus, Bell's inequalities have a double meaning: they test whether numbers can represent probabilities of events and at the same time whether the correlations between these events (provided they exist) have a joint common cause. These two meanings of Bell's inequalities will split up in the next section where we treat Bell's inequalities with classical *conditional* probabilities.

16.3 Case 2: Bell's Inequalities for Classical Conditional Probabilities

Just as in the previous section suppose that we are given real numbers p_i and p_{ij} in $[0, 1]$ with $i = 1 \ldots n$ and $(i, j) \in S$. But now we ask: when can these numbers be *conditional* probabilities of certain events and their conjunctions? More precisely: given the numbers p_i and p_{ij}, do there exist events A_i and a_i $(i = 1 \ldots n)$ in a classical probability space (Ω, Σ, p) such that p_i and p_{ij} are the following classical conditional probabilities:

$$p_i = p(A_i|a_i)$$
$$p_{ij} = p(A_i \wedge A_j|a_i \wedge a_j)$$

Or again in brief, do the numbers p_i and p_{ij} admit a *conditional Kolmogorovian representation*?

The answer here is more permissive. Except for some extremal values the numbers p_i and p_{ij} always admit a conditional Kolmogorovian representation: any correlation vector $\vec{p}$ admits a conditional Kolmogorovian representation if $\vec{p}$ has no $(i, j) \in S$ such that either (i) $p_i = 0$ or $p_j = 0$ but $p_{ij} \neq 0$; or (ii) $p_i = p_j = 1$ but $p_{ij} \neq 1$.

Obviously, any Kolmogorovian representations is a conditional Kolmogorovian representation with $a_i = \Omega$ for all $i = 1 \ldots n$. However, correlation vectors admitting a conditional Kolmogorovian representation are not necessarily in the classical correlation polytope.[3]

[3]For example consider the following correlation vector in $R(n, S)$ with $n = 2$ and $S = \{1, 2\}$:

$$\vec{p} = \left(\frac{2}{3}, \frac{2}{3}; \frac{1}{5}\right)$$

The vector $\vec{p}$ violates Bell's inequality

$$p_1 + p_2 - p_{12} \leqslant 1$$

hence it is not in $c(n, S)$ and, consequently, it does not admit a Kolmogorovian representation. However, $\vec{p}$ admits a conditional Kolmogorovian representation with the following atomic events and probabilities:

$$p(A_1 \wedge A_2 \wedge a_1 \wedge a_2) = \frac{1}{25}$$
$$p(A_1^\perp \wedge A_2^\perp \wedge a_1 \wedge a_2) = \frac{4}{25}$$
$$p(A_1^\perp \wedge A_2 \wedge a_1^\perp \wedge a_2) = p(A_1 \wedge A_2^\perp \wedge a_1 \wedge a_2^\perp) = \frac{9}{25}$$
$$p(A_1^\perp \wedge A_2^\perp \wedge a_1^\perp \wedge a_2) = p(A_1^\perp \wedge A_2^\perp \wedge a_1 \wedge a_2^\perp) = \frac{1}{25}$$

(The probability of all other atomic events is 0.)

Now, any correlation vector can be expressed as a convex combination of (not necessarily classical) vertex vectors. A *vertex vector* $\vec{u}$ is defined as follows: $\vec{u}_i, \vec{u}_{ij} \in \{0, 1\}$ for all $i = 1 \ldots n$ and $(i, j) \in S$. Obviously, classical vertex vectors are vertex vectors but not every vertex vector is classical. For example the vertex vectors $(0, 0; 1)$, $(1, 0; 1)$, $(0, 1; 1)$ or $(1, 1; 0)$ in $R(n, S)$ with $n = 2$ and $S = \{1, 2\}$ are not classical. There are $2^{n+|S|}$ different vertex vectors $\vec{u}^{\,k}$ in $R(n, S)$ and 2^n different classical vertex vectors $\vec{u}^{\,\varepsilon}$.

Denote the convex hull of the vertex vectors by

$$u(n, S) := \left\{ \vec{v} \in R(n, S) \,\middle|\, \vec{v} = \sum_{k=1}^{2^{n+|S|}} \lambda_k \vec{u}^{\,k} \; ; \; \lambda_k \geqslant 0; \; \sum_{k=1}^{2^{n+|S|}} \lambda_k = 1 \right\}$$

Contrary to $c(n, S)$, the polytope $u(n, S)$ is not a simplex, hence the expansion of the correlation vectors in $u(n, S)$ by vertex vectors is typically not unique.[4]

Now, the set of correlation vectors admitting a conditional Kolmogorovian representation is dense in $u(n, S)$: all interior points of $u(n, S)$ admit a conditional Kolmogorovian representation and also all surface points, except for which there is a pair $(i, j) \in S$ such that either (i) $p_i = 0$ or $p_j = 0$ but $p_{ij} \neq 0$; or (ii) $p_i = p_j = 1$ but $p_{ij} \neq 1$. Denote the set of correlation vectors admitting a conditional Kolmogorovian representation by $u'(n, S)$.

Now, let's go over to the common causal explanation of conditional correlations. Let $\vec{p}$ be a proper correlation vector in $u'(n, S)$. Lying in $u'(n, S)$ the correlation vector $\vec{p}$ represents the conditional probabilities of certain events and their conjunctions and the events associated to some pairs $(i, j) \in S$ are *conditionally correlated*:

$$p(A_i \wedge A_j | a_i \wedge a_j) \neq p(A_i | a_i)\, p(A_j | a_j) \tag{16.21}$$

Interpret now the events A_i and a_i in the context of physical experiments: Let a_i represent a possible *measurement* that an experimenter can perform on an object. Then the event $a_i \wedge a_j$ will represent the joint performance of measurements a_i and a_j. Let furthermore the event A_i represent an outcome of measurement a_i

[4]The correlation vector

$$\vec{p} = \left(\frac{2}{3}, \frac{2}{3}; \frac{1}{5} \right)$$

for example can be expanded in many different ways:

$$\begin{aligned} \vec{p} &= \frac{1}{3}(0, 1; 0) + \frac{1}{3}(1, 0; 0) + \frac{1}{5}(1, 1; 1) + \frac{2}{15}(1, 1; 0) \\ &= \frac{1}{3}(0, 0; 0) + \frac{1}{5}(1, 1; 1) + \frac{7}{15}(1, 1; 0) \end{aligned}$$

etc.

and $A_i \wedge A_j$ represent an outcome of the jointly performed measurement $a_i \wedge a_j$. Then (16.21) expresses a correlation between two measurement outcomes provided their measurements have been performed.

When do the conditional correlations in a conditional Kolmogorovian representation of $\vec{p}$ have a common causal explanation?

The set of conditional correlations are said to have a *non-conspiratorial joint common causal explanation* if there is a partition $\{C_k\}$ $(k \in K)$ of Σ such that for any (i, j) in S and $k \in K$

$$p(A_i \wedge A_j | a_i \wedge a_j \wedge C_k) = p(A_i | a_i \wedge C_k)\, p(A_j | a_j \wedge C_k) \tag{16.22}$$

$$p(a_i \wedge a_j \wedge C_k) = p(a_i \wedge a_j)\, p(C_k) \tag{16.23}$$

Equations (16.22) express the Reichenbachian idea that the common cause is to *screen off* all correlations. Equations (16.23) express the so-called *no-conspiracy*, the idea that common causes should be causally, and hence probabilistically, independent of the measurement choices. The common causal explanation is *joint* since all correlations (16.21) have the same common cause.

Now, suppose the correlations in the a given conditional Kolmogorovian representation of (16.21) of $\vec{p}$ in $u'(n, S)$ have a non-conspiratorial joint common causal explanation. Introduce again the notation

$$p_i^k = p(A_i | a_i \wedge C_k)$$

$$c_k = p(C_k)$$

and consider the k common cause vectors of the correlation vector $\vec{p}$:

$$\vec{p}^{\,k} = \left(p_1^k, \ldots, p_n^k; \ldots, p_i^k p_j^k, \ldots\right)$$

We call a non-conspiratorial joint common cause *deterministic* if $\vec{p}^{\,k}$ is a deterministic common cause vector for all $k \in K$; otherwise we call it *indeterministic*. We call a deterministic non-conspiratorial joint common cause $\{C_k\}$ $(k \in K)$ a *property*; and an indeterministic common cause a *propensity*.

Now, due to (16.22)–(16.23) and the theorem of total probability

$$\vec{p} = \sum_k c_k\, \vec{p}^{\,k}$$

Since common cause vectors are independence vectors lying in $c(n, S)$, their convex combination also lies in the classical correlation polytope $c(n, S)$. Thus, $\vec{p}$ being a classical correlation vector is a necessary condition for a conditional Kolmogorovian representation of $\vec{p}$ to have a non-conspiratorial joint common causal explanation.

Conversely, knowing the k common cause vectors and the probabilities c_k alone, one can construct the classical probability space Σ with the conditionally correlating events A_i and a_i and the common causes $\{C_k\}$.

Observe that the situation is now different from the one in the previous section: the numbers p_i and p_{ij} can be conditional probabilities of events and their conjunctions even if they violate the corresponding Bell inequalities. However, the conditional correlations between these events have a non-conspiratorial joint common causal explanation if and only if the correlation vector composed of the numbers p_i and p_{ij} lies in the classical correlation polytope. In other words, in case of classical *conditional* probabilities Bell's inequalities do not test the conditional Kolmogorovian representability but whether correlations can be given a common causal explanation.

16.4 Relating Case 1 and Case 2

How do the Kolmogorovian and the conditional Kolmogorovian representation relate to one another?

In this section I will show that (i) a conditional Kolmogorovian representation of a classical correlation vector has a property explanation only if the representation is non-signaling (see below); and (ii) a correlation vector $\vec{p}$ has a Kolmogorovian representation if and only if it has a property explanation for any non-signaling conditional Kolmogorovian representation.

Let $\vec{p}$ be a proper correlation vector in $c(n, S)$ and consider a conditional Kolmogorovian representation of $\vec{p}$. Obviously, there are many such representations of $\vec{p}$ depending on the measurement conditions a_i. Let's say, somewhat loosely, that a conditional Kolmogorovian representation has a property/propensity explanation if the conditional correlations in the representation have a property/propensity explanation.

First, we claim that a conditional Kolmogorovian representation of a correlation vector $\vec{p}$ in $c(n, S)$ has a property explanation only if the representation satisfies *non-signaling*:

$$p(A_i|a_i) = p(A_i|a_i \wedge a_j) \tag{16.24}$$

$$p(A_j|a_j) = p(A_j|a_i \wedge a_j) \tag{16.25}$$

for any $(i, j) \in S$. Note that non-signaling is not a feature of the correlation vector itself but of the representation. For the same correlation vector in $c(n, S)$ one can provide both non-signaling and also signaling representations.

Now, a conditional Kolmogorovian representation can have a property explanation only if it satisfies non-signaling. Recall namely that for a property $\{C_k\}$:

$$p(A_i|a_i \wedge C_k) \in \{0, 1\}$$

and hence

$$p(A_i|a_i \wedge C_k) = p(A_i|a_i \wedge a_j \wedge C_k) \tag{16.26}$$

for any $i, j = 1 \ldots n$ and $k \in K$. But (16.26) together with no-conspiracy (16.23) and the theorem of total probability imply non-signaling (16.24)–(16.25). Thus, satisfying non-signaling is a necessary condition for a conditional Kolmogorovian representation to have a property explanation. Signaling conditional Kolmogorovian representations do not have a property explanation.

Second, suppose that a conditional Kolmogorovian representation of $\vec{p}$ have a property explanation. That is $\vec{p}$ has a conditional Kolmogorovian representation in a classical probability space Σ and all the conditionally correlating event pairs have a non-conspiratorial deterministic joint common cause in Σ. Then these properties (deterministic common causes) provide a Kolmogorovian (unconditional) representation for $\vec{p}$.

Observe namely that

$$p_i = p(A_i|a_i) = \frac{p(A_i \wedge a_i)}{p(a_i)} = \frac{\sum_k p(A_i \wedge a_i \wedge C_k)}{p(a_i)} = \frac{\sum_k p(A_i|a_i \wedge C_k)p(a_i \wedge C_k)}{p(a_i)}$$
$$\stackrel{*}{=} \frac{\sum_k p(A_i|a_i \wedge C_k)p(a_i)p(C_k)}{p(a_i)} = \sum_k p(A_i|a_i \wedge C_k)p(C_k) = \sum_{k:\, p_i^k=1} p(C_k)$$

where the equation $\stackrel{*}{=}$ holds due to no-conspiracy (16.23) and the symbol $\sum_{k:\, p_i^k=1}$ means that we sum up for all k for which $p_i^k = 1$. Similarly, using (16.22)–(16.23) one obtains

$$p_{ij} = \sum_{k:\, p_i^k=1,\ p_j^k=1} p(C_k)$$

That is the events

$$C_i = \vee_{k:\, p_i^k=1} C_k$$
$$C_i \wedge C_j = \vee_{k:\, p_i^k=1,\ p_j^k=1} C_k$$

provide a Kolmogorovian representation for the numbers p_i and p_{ij}.

Third, conversely, if $\vec{p}$ admits a Kolmogorovian representation, then it also admits a property explanation for any non-signaling conditional Kolmogorovian representation.

Namely, if $\vec{p}$ admits a Kolmogorovian representation, then there is a partition $\{C_\varepsilon\}$ in Σ such that

$$p_i = p(C_i) = \sum_{\varepsilon:\, \varepsilon_i = i} p(C_\varepsilon) = \sum_{\varepsilon:\, \varepsilon_i = i} \lambda_\varepsilon \tag{16.27}$$

$$p_{ij} = p(C_i \wedge C_j) = \sum_{\varepsilon:\, \varepsilon_i = 1,\, \varepsilon_j = 1} p(C_\varepsilon) = \sum_{\varepsilon:\, \varepsilon_i = 1,\, \varepsilon_j = 1} \lambda_\varepsilon \tag{16.28}$$

Now, consider a non-signaling conditional Kolmogorovian representation of $\vec{p}$, that is let there be events A_i and a_i in Σ' such that

$$p_i = p(A_i | a_i)$$
$$p_{ij} = p(A_i \wedge A_j | a_i \wedge a_j)$$

and suppose that non-signaling (16.24)–(16.25) hold for any $(i, j) \in S$.

Now, the events C_ε provide a propensity explanation for the conditional Kolmogorovian representation of $\vec{p}$ in the following sense. First, let $\varepsilon, \varepsilon' \in \{0, 1\}^n$ and define the events $a_{\varepsilon'}$ in Σ' as follows:

$$a_{\varepsilon'} := \wedge_i \, a_{\varepsilon'_i} \qquad \text{where } a_{\varepsilon'_i} = \begin{cases} a_i \text{ if } \varepsilon'_i = 1 \\ \overline{a}_i \text{ if } \varepsilon'_i = 0 \end{cases}$$

Then, extend the algebra Σ' to Σ'' generated by the atomic events $D_{\varepsilon',\varepsilon}$ defined as follows:

$$\vee_\varepsilon \, D_{\varepsilon',\varepsilon} = a_{\varepsilon'} \tag{16.29}$$

$$\vee_{\varepsilon'} D_{\varepsilon',\varepsilon} = C_\varepsilon \tag{16.30}$$

$$p(D_{\varepsilon',\varepsilon}) = p(a_{\varepsilon'})\, p(C_\varepsilon) = p(a_{\varepsilon'})\, \lambda_\varepsilon \tag{16.31}$$

$$p(A_i | D_{\varepsilon',\varepsilon}) = \varepsilon'_i \, p_i^\varepsilon \tag{16.32}$$

$$p(A_i \wedge A_j | D_{\varepsilon',\varepsilon}) = \varepsilon'_i \, p_i^\varepsilon \varepsilon'_j \, p_j^\varepsilon \tag{16.33}$$

where the numbers $p_i^\varepsilon \in [0, 1]$ will be specified below. Now, using (16.29)–(16.33) one obtains

$$p(A_i \wedge A_j | a_i \wedge a_j \wedge C_\varepsilon) = p_i^\varepsilon \, p_j^\varepsilon = p(A_i | a_i \wedge C_\varepsilon)\, p(A_j | a_j \wedge C_\varepsilon)$$

$$p(a_i \wedge a_j \wedge C_\varepsilon) = \sum_{\varepsilon':\, \varepsilon'_i = 1,\, \varepsilon'_j = 1} p(D_{\varepsilon',\varepsilon}) = p(a_i \wedge a_j)\, p(C_\varepsilon)$$

That is the partition $\{C_\varepsilon\}$ provides a propensity explanation for the conditional Kolmogorovian representation of $\vec{p}$ with probabilities specified in (16.27)–(16.28). Now

$$p(A_i|a_{\varepsilon'}) = \frac{p(A_i \wedge a_{\varepsilon'})}{p(a_{\varepsilon'})} = \frac{\sum_\varepsilon p(A_i \wedge D_{\varepsilon',\varepsilon})}{p(a_{\varepsilon'})} = \frac{\sum_\varepsilon p(A_i|D_{\varepsilon',\varepsilon})\, p(D_{\varepsilon',\varepsilon})}{p(a_{\varepsilon'})}$$
$$= \frac{\sum_\varepsilon \varepsilon'_i\, p_i^\varepsilon\, p(a_{\varepsilon'})\lambda_\varepsilon}{p(a_{\varepsilon'})} = \varepsilon'_i \sum_\varepsilon p_i^\varepsilon\, \lambda_\varepsilon$$

that is

$$p_i = p(A_i|a_i) = \sum_\varepsilon p_i^\varepsilon\, \lambda_\varepsilon \tag{16.34}$$

and similarly

$$p_{ij} = p(A_i \wedge A_j|a_i \wedge a_j) = \sum_\varepsilon p_i^\varepsilon\, p_j^\varepsilon\, \lambda_\varepsilon \tag{16.35}$$

Composing 2^n independence vectors $\vec{p}^{\,\varepsilon}$ from the numbers p_i^ε:

$$\vec{p}^{\,\varepsilon} = \left(\ldots p_i^\varepsilon \ldots p_j^\varepsilon \ldots; \ldots p_i^\varepsilon p_j^\varepsilon \ldots\right)$$

(16.34)–(16.35) reads as follows:

$$\vec{p} = \sum_\varepsilon \lambda_\varepsilon\, \vec{p}^{\,\varepsilon}$$

This means that the numbers p_i^ε are to be taken from $[0, 1]$ such that the 2^n independence vectors $\vec{p}^{\,\varepsilon}$ provide a convex combination for $\vec{p}$ with the *same* coefficients λ_ε as the vertex vectors $\vec{u}^{\,\varepsilon}$ do. One such expansion always exists. Namely, when $\vec{p}^{\,\varepsilon} = \vec{u}^{\,\varepsilon}$. In this case $p_i^\varepsilon = \varepsilon_i$ for any $i = 1 \ldots n$ and $\varepsilon \in \{0, 1\}^n$ and (16.34)–(16.35) reads as follows:

$$p_i = p(A_i|a_i) = \sum_\varepsilon \varepsilon_i\, p(C_\varepsilon) = p(C_i)$$

$$p_{ij} = p(A_i \wedge A_j|a_i \wedge a_j) = \sum_\varepsilon \varepsilon_i \varepsilon_j\, p(C_\varepsilon) = p(C_i \wedge C_j)$$

That is we obtain a property explanation for the conditional Kolmogorovian representation of $\vec{p}$. However, for classical correlation vectors which are not classical vertex vectors one can also provide for $\vec{p}$ various convex combinations by indeterministic independence vectors with the coefficients λ_ε (see (16.18) as an example). In this case we obtain a propensity explanation for the given conditional Kolmogorovian representation of $\vec{p}$.

To sum up, a correlation vector $\vec{p}$ has a Kolmogorovian representation if and only if it has a property explanation for any non-signaling conditional Kolmogorovian representation. This equivalence justifies retrospectively why we used the term "property" both for a deterministic common cause in the conditional Kolmogorovian representation and also as a synonym of "event" in the unconditional Kolmogorovian representation.

16.5 Case 3: Bell's Inequalities for Quantum Probabilities

Again start with real numbers p_i and p_{ij} in $[0, 1]$ with $i = 1 \ldots n$ and $(i, j) \in S$. But now the question is the following: when can these numbers be *quantum* probabilities? That is, given the numbers p_i and p_{ij}, do there exist projections A_i $(i = 1 \ldots n)$ representing quantum events in a quantum probability space $(\mathcal{P}(H), \rho)$, where $\mathcal{P}(H)$ is the projection lattice of a Hilbert space H and ρ is a density operator on H representing the quantum state, such that p_i and p_{ij} are the following quantum probabilities:

$$
\begin{aligned}
p_i &= \mathrm{Tr}(\rho A_i) \\
p_{ij} &= \mathrm{Tr}(\rho(A_i \wedge A_j))
\end{aligned}
$$

(Here $A_i \wedge A_j$ denotes the projection projecting on the intersection of the closed subspaces of H onto which A_i and A_j are projecting.) Again in brief, do the numbers p_i and p_{ij} admit a *quantum representation*?

The answer is again given by Pitowsky (1989, p. 72). Introduce the notion of a quantum vertex vector. A quantum vertex vector is a vertex vector in $R(n, S)$ such that

$$
\begin{aligned}
\vec{u}_i^{\varepsilon} &= \varepsilon_i & i &= 1 \ldots n \\
\vec{u}_{ij}^{\varepsilon} &\leqslant \varepsilon_i \varepsilon_j & (i, j) &\in S
\end{aligned}
$$

Obvious, any classical vertex vector is a quantum vertex vector and any quantum vertex vector is a vertex vector, but the reverse inclusion does not hold. In $R(n, S)$ with $n = 2$ and $S = \{1, 2\}$ for example the vertex vector $(1, 1; 0)$ is quantum but not classical, and the vertex vectors $(0, 0; 1)$, $(1, 0; 1)$, $(0, 1; 1)$ are not even quantum.

Denote by $q(n, S)$ the convex hull of quantum vertex vectors. Pitowsky then shows that almost all correlation vectors in $q(n, S)$ admit a quantum representation. More precisely, Pitowsky shows that—denoting by $q'(n, S)$ the set of quantum correlation vectors that is the set of those correlation vectors which admit a quantum representation—the following holds:

(i) $c(n, S) \subset q'(n, S) \subset q(n, S)$;
(ii) $q'(n, S)$ is convex (but not closed);
(iii) $q'(n, S)$ contains the interior of $q(n, S)$

We can add to this our result in the previous section:

(iv) $q(n, S) \subset u'(n, S) \subset u(n, S)$

Thus, the set of numbers admitting a quantum representation is strictly larger than the set of numbers admitting a Kolmogorovian representation but strictly smaller than the set of numbers admitting a conditional Kolmogorovian representation.

Let us turn now to the question of the common causal explanation. Let $\vec{p}$ be a proper quantum correlation vector. $\vec{p}$ then represents the quantum probabilities of certain quantum events and their conjunctions and the events associated to some pairs $(i, j) \in S$ are *correlated*:

$$\mathrm{Tr}(\rho(A_i \wedge A_j)) \neq \mathrm{Tr}(\rho A_i)\,\mathrm{Tr}(\rho A_j) \tag{16.36}$$

When do the correlations (16.36) have a common causal explanation?

The set of quantum correlations has a *joint quantum common causal explanation* if there is a partition $\{C_k\}$ $(k \in K)$ in $\mathcal{P}(H)$ (that is a set of mutually orthogonal projection adding up to the unity) such that for any (i, j) in S and $k \in K$

$$\mathrm{Tr}(\rho_k(A_i \wedge A_j)) = \mathrm{Tr}(\rho_k A_i)\,\mathrm{Tr}(\rho_k A_j) \tag{16.37}$$

where

$$\rho_k := \frac{C_k \rho\, C_k}{\mathrm{Tr}(\rho\, C_k)}$$

is the density operator ρ after a selective measurement by C_k. If the partition $\{C_k\}$ is commuting with the each correlating pair (A_i, A_j), then we call the common causes *commuting*, otherwise *noncommuting*.

Now, suppose the correlations (16.36) have a joint quantum common causal explanation. Does it follow that $\vec{p}$ is in $c(n, S)$ that is Bell's inequalities are satisfied?

Introduce again the notation

$$p_i^k = \mathrm{Tr}(\rho_k A_i)$$
$$c_k = \mathrm{Tr}(\rho\, C_k)$$

and consider the k common cause vectors of the correlation vector $\vec{p}$:

$$\vec{p}^{\,k} = \left(p_1^k, \ldots, p_n^k; \ldots, p_i^k p_j^k, \ldots\right)$$

The common cause vectors are independence vectors. Hence, using (16.37) and the theorem of total probability, the convex combination of the common cause vectors

$$\vec{p}^{\,c} = \sum_k c_k\, \vec{p}^{\,k} \tag{16.38}$$

will be in $c(n, S)$. However $\vec{p}^{\,c}$ is not necessarily identical with the original correlation vector $\vec{p}$. More precisely, $\vec{p}^{\,c} = \vec{p}$ for any ρ if $\{C_k\}$ are commuting common causes. But if $\{C_k\}$ are noncommuting common causes, then $\vec{p}^{\,c}$ and $\vec{p}$ can be different and hence even if $\vec{p}^{\,c} \in c(n, S)$, $\vec{p}$ might be outside $c(n, S)$. In short, a quantum correlation vector can have a joint *noncommuting* common causal explanation even if it lies outside the classical correlation polytope. The correlation vector $\vec{p}$ is confined in $c(n, S)$ only if the common causes are required to be *commuting*.[5]

To sum up, in the case of quantum probabilities we found a scenario different from both previous cases. Here Bell's inequalities neither put a constraint on whether numbers can be quantum probabilities nor on whether the correlation between events with the prescribed probability can have a common causal explanation. Bell's inequalities constrain common causal explanations only if common causes are understood as commuting common causes.

Perhaps it is worth mentioning that in algebraic quantum field theory (Rédei and Summers 2007; Hofer-Szabó and Vecsernyés 2013) and quantum information theory (Bengtson and Zyczkowski 2006) the violation of Bell's inequalities composed of quantum probabilities is used for another purpose: it places a bound on the strength of correlations between certain events. Abstractly, one starts with two mutually commuting C^*-subalgebras $\mathcal{A}$ and $\mathcal{B}$ of a C^*-algebra $\mathcal{C}$ and defines a *Bell operator* R for the pair $(\mathcal{A}, \mathcal{B})$ as an element of the following set:

$$\mathbb{B}(\mathcal{A}, \mathcal{B}) := \Big\{ \frac{1}{2}\big(A_1(B_1 + B_2) + A_2(B_1 - B_2)\big) \,\big|\, A_i = A_i^* \in \mathcal{A}; \\ B_i = B_i^* \in \mathcal{B};\ -\mathbf{1} \leqslant A_i, B_i \leqslant \mathbf{1} \Big\}$$

where $\mathbf{1}$ is the unit element of $\mathcal{C}$. Then one can prove that for any Bell operator R, $|\phi(R)| \leqslant \sqrt{2}$ for any state ϕ; but $|\phi(R)| \leqslant 1$ for separable states (i.e. for convex combinations of product states).

In other words, in these disciplines one fixes a Bell operator (a witness operator) and scrolls over the different quantum states to see which of them is separable. In this reading, Bell's inequalities neither test Kolmogorovian representation nor common causal explanation but separability of certain normalized positive linear functionals. Obviously, this role of Bell's inequalities is completely different from the one analyzed in this paper.

16.6 The EPR-Bohm Scenario

Bell's inequalities in the EPR-Bohm scenario are standardly said to be violated. But which Bell inequalities and what does this violation mean?

[5] For the details and for a concrete example see Hofer-Szabó and Vecsernyés (2013, 2018).

Let us start in the reverse order, with quantum representation. Consider the EPR-Bohm scenario with pairs of spin-$\frac{1}{2}$ particles prepared in the singlet state. In quantum mechanics the state of the system and the quantum events are represented by matrices on $M_2 \otimes M_2$, where M_2 is the algebra of the two-dimensional complex matrices. The singlet state is represented as:

$$\rho^s = \frac{1}{4}\Big(\mathbf{1} \otimes \mathbf{1} - \sum_{k=1}^{3} \sigma_k \otimes \sigma_k\Big)$$

and the event that the spin of the particle is "up" on the left wing in direction $\vec{a}_i$ $(i = 1, 2)$; on the right wing in direction $\vec{b}_j$ $(i = 3, 4)$; and on both wings in directions $\vec{a}_i$ and $\vec{b}_j$, respectively, are represented as:

$$A_i = \frac{1}{2}\big((\mathbf{1} + \vec{a}_i \cdot \vec{\sigma}) \otimes \mathbf{1}\big)$$

$$A_j = \frac{1}{2}\big(\mathbf{1} \otimes (\mathbf{1} + \vec{b}_j \cdot \vec{\sigma})\big)$$

$$A_{ij} = \frac{1}{4}\big((\mathbf{1} + \vec{a}_i \cdot \vec{\sigma}) \otimes (\mathbf{1} + \vec{b}_j \cdot \vec{\sigma})\big)$$

where $\mathbf{1}$ is the identity matrix on M_2, $\vec{\sigma} = (\sigma_1, \sigma_2, \sigma_3)$ is the Pauli vector, and $\vec{a}_i$ $(i = 1, 2)$ and $\vec{b}_j$ $(j = 3, 4)$ are the spin measurement directions on the left and right wing, respectively. Furthermore, $A_{ij} = A_i A_j = A_i \wedge A_j$ since A_i and A_j are commuting.

The quantum probabilities are generated by the trace formula:

$$p_i = \mathrm{Tr}(\rho^s A_i) = \frac{1}{2} \tag{16.39}$$

$$p_j = \mathrm{Tr}(\rho^s A_j) = \frac{1}{2} \tag{16.40}$$

$$p_{ij} = \mathrm{Tr}(\rho^s A_{ij}) = \frac{1}{2}\sin^2\left(\frac{\theta_{ij}}{2}\right) \tag{16.41}$$

where θ_{ij} denotes the angle between directions $\vec{a}_i$ and $\vec{b}_j$. As it is well known, for the measurement directions

$$\vec{a}_1 = (0, 1, 0) \qquad \vec{b}_3 = \frac{1}{\sqrt{2}}(1, 1, 0)$$

$$\vec{a}_2 = (1, 0, 0) \qquad \vec{b}_4 = \frac{1}{\sqrt{2}}(-1, 1, 0)$$

the Clauser-Horne inequality

$$-1 \leqslant p_{13} + p_{23} + p_{14} - p_{24} - p_1 - p_3 = -\frac{1+\sqrt{2}}{2} \tag{16.42}$$

is violated.

What does the violation of the Clauser-Horne inequality (16.42) mean with respect to the quantum representation? As clarified in Sect. 16.5, it does not mean that the numbers (16.39)–(16.41) cannot be given a quantum mechanical representation. Equations (16.39)–(16.41) just provide one. On the other hand, the violation of (16.42) neither means that the EPR correlations

$$p_{ij} = \text{Tr}(\rho^s A_{ij}) \neq \text{Tr}(\rho^s A_i)\text{Tr}(\rho^s A_j) = p_i\, p_j \qquad i, j = 1, 2 \tag{16.43}$$

cannot be given a joint quantum common causal explanation. In Hofer-Szabó and Vecsernyés (2012, 2018) we have provided a partition $\{C_k\}$ such that the screening-off conditions (16.37) hold for all EPR correlations (16.43). That is $\{C_k\}$ is a joint common causal explanation for all the four EPR correlations.[6] As also shown in Sect. 16.5, these joint common causes need to be *noncommuting* common causes, since commuting common causes would imply the Clauser-Horne inequality (16.42) which is violated in the EPR-Bohm scenario. Thus, the violation of the Clauser-Horne inequality (16.42) does not exclude a common causal explanation of the EPR-Bohm scenario—as long as noncommuting common causes are tolerated in the explanation.[7]

To see the thrust of the violation of the Clauser-Horne inequalities, we have to go over to the interpretations of quantum mechanics. As for the operational interpretation, the violation of the Clauser-Horne inequalities, as clarified in Sect. 16.3, again does not mean that the numbers (16.39)–(16.41) cannot be given an operational interpretation. There *is* such an interpretation; otherwise quantum mechanics would not be empirically adequate. The operational interpretation is the following:

$$p_i = p(A_i|a_i)$$
$$p_j = p(B_j|b_j)$$
$$p_{ij} = p(A_i \wedge B_j|a_i \wedge b_j)$$

where a_i denotes the event that the spin on the left particle is measured in direction $\vec{a}_i$, and A_i denotes the event that the outcome in this measurement is "up". (Similarly, for b_j and B_j.) That is, the probabilities are conditional probabilities of certain measurement outcomes provided that certain measurements are performed. However, the violation of the Clauser-Horne inequality (16.42) does exclude that the conditionally correlating pairs of outcomes

[6]More than that, we have also shown that in an algebraic quantum field theoretic setting the common causes can even be localized in the common past of the correlating events.

[7]The problem with such noncommuting common causes, however, is that it is far from being clear how they should be interpreted.

$$p_{ij} = p(A_i \wedge A_j | a_i \wedge a_j) \neq p(A_i | a_i)\, p(A_j | a_j) = p_i\, p_j \qquad i, j = 1, 2$$

has a non-conspiratorial joint common causal explanation. Thus, in the operational interpretation, Bell's inequalities filter common causal explanations.

Finally, let us turn to the ontological interpretation. As is clarified by Pitowsky, the violation of the Clauser-Horne inequality (16.42) excludes the numbers (16.39)–(16.41) to be classical unconditional probability of certain events or properties and their conjunctions. That is, there are no such numbers A_i in a classical probability space such that

$$p_i = p(A_i)$$
$$p_j = p(B_j)$$
$$p_{ij} = p(A_i \wedge B_j)$$

Consequently, there are no correlations

$$p_{ij} = p(A_i \wedge A_j) \neq p(A_i)\, p(A_j) = p_i\, p_j \qquad i, j = 1, 2$$

and a fortiori no need to look for a common causal explanation.

As shown in Sect. 16.4, a correlation vector has a Kolmogorovian representation if and only if it has a property explanation for any non-signaling conditional Kolmogorovian representation. The probabilities in the EPR-Bohm scenario are non-signaling:

$$p_i = p(A_i | a_i) = p(A_i | a_i \wedge b_j)$$
$$p_j = p(A_i | b_j) = p(B_j | a_i \wedge b_j)$$

Hence, the violation of (16.42) again excludes a property explanation for any non-signaling conditional Kolmogorovian representation.

16.7 Conclusions

What does it mean that a set of numbers violates Bell's inequalities? One can answer this question in three different ways depending on whether the numbers are interpreted as classical unconditional probabilities, classical conditional probabilities, or quantum probabilities:

(i) In the first case, the violation of Bell's inequalities excludes these numbers to be interpreted as the classical (unconditional) probabilities of certain events/properties and their conjunctions. The satisfaction of Bell's inequalities does not only guarantee the existence of such events, but also the existence of a set of joint common causes screening off the correlations between these events.

(ii) In the second case, the violation of Bell's inequalities does not exclude that these numbers can be interpreted as the classical conditional probability of certain events (measurement outcomes) conditioned on other events (measurement settings). However, it does exclude a non-conspiratorial joint common causal explanation for the conditional correlations between these events.

(iii) Finally, the violation or satisfaction of Bell's inequalities has no bearing either on whether a set of numbers can be interpreted as quantum probability of certain projections, or whether there can be given a joint common causal explanation for the correlating projections—as long as noncommuting common causes are adopted in the explanation.

Acknowledgements This work has been supported by the Hungarian Scientific Research Fund, OTKA K-115593 and a Research Fellowship of the Institute of Advanced Studies Köszeg. I wish to thank Márton Gömöri and Balázs Gyenis for reading and commenting on the manuscript.

References

Bengtson, I., & Zyczkowski, K. (2006). *Geometry of quantum states: An introduction to quantum entanglement*. Cambridge: Cambridge University Press.

Gömöri, M., & Placek, T. (2017). Small probability space formulation of Bell's theorem. In G. Hofer-Szabó & L. Wronski (Eds.), *Making it formally explicit – Probability, causality and indeterminism* (European studies in the philosophy of science series, pp. 109–126). Dordrecht: Springer Verlag.

Hofer-Szabó, G., & Vecsernyés, P. (2012). Noncommuting local common causes for correlations violating the Clauser–Horne inequality. *Journal of Mathematical Physics, 53*, 12230.

Hofer-Szabó, G., & Vecsernyés, P. (2013). Bell inequality and common causal explanation in algebraic quantum field theory. *Studies in History and Philosophy of Science Part B – Studies in History and Philosophy of Modern Physics, 44*(4), 404–416.

Hofer-Szabó, G., & Vecsernyés, P. (2018). *Quantum theory and local causality*. Dordrecht: Springer Brief.

Hofer-Szabó, G., Rédei, M., & Szabó, L. E. (1999). On Reichenbach's common cause principle and on Reichenbach's notion of common cause. *The British Journal for the Philosophy of Science, 50*, 377–399.

Hofer-Szabó, G., Rédei, M., & Szabó, L. E. (2013). *The principle of the common cause*. Cambridge: Cambridge University Press.

Pitowsky, I. (1989). *Quantum probability – Quantum logic*. Dordrecht: Springer.

Rédei, M., & Summers, J. S. (2007). Quantum probability theory. *Studies in History and Philosophy of Modern Physics, 38*, 390–417.

Reichenbach, H. (1956). *The direction of time*. Los Angeles: University of California Press.

Szabó, L. E. (2008). The Einstein-Podolsky-Rosen argument and the Bell inequalities. *Internet encyclopedia of philosophy*, http://www.iep.utm.edu/epr/.

Chapter 17
On the Descriptive Power of Probability Logic

Ehud Hrushovski

Abstract In a series of imaginary conversations, a philosopher of physics and a mathematical logician discuss the investigation of a mathematical relational universe and the description of the world by physical theory, as metaphors for each other. They take for granted that basic events can be observed directly, as atomic formulas, and ask about the ability to interpret deeper underlying structures. This depends on the logic used. They are interested specifically in the role of classical probability logic in this endeavor. Within the model they adopt, as a case study, probability logic is highly competent in discovering and describing spatial structures; where a very general notion of space is taken to correspond to the logical notion of monadic, both with natural relativistic corrections. However, this logic can go no further in itself, quite unlike the unbounded abilities of first-order logic. In some cases new first-order structures emerge at the point of failure of probability logic.

Keywords Probability logic · Model theory · Expressive power

17.1 Introduction: Logic as a Descriptive Tool

Logic is often defined as the "systematic study of the form of valid inference".[1] "Let C be the planets, B to be near, A not to twinkle," wrote Aristotle; you can then study the implication from the premises $(\forall x)(A(x) \rightarrow B(x))$ and $(\forall x)(C(x) \rightarrow A(x))$, in modern notation, to the conclusion that the planets are near. It is the pride and weakness of his method that "no further term is required from without in order to

[1]From the wikipedia entry on Logic, retrieved June 2019.

E. Hrushovski (✉)
Mathematical Institute, Oxford, UK

M. Hemmo, O. Shenker (eds.), *Quantum, Probability, Logic*,
Jerusalem Studies in Philosophy and History of Science,
https://doi.org/10.1007/978-3-030-34316-3_17

make the consequence necessary" (Aristotle 1906). By the end of the argument, we may learn a new *fact* about B and C; we will not be introduced to any new idea D.

Modern logic has, it seems to me, an additional role: the introduction of new concepts. The primary use of the first-order relational calculus is not for judging the wisdom of asserting a certain hypothesis, whose meaning is assumed clear. It is first of all a tool for carving out new ideas and dividing lines, about which assertions can subsequently be formulated. An ancient example is the notion of parallel lines, defined from a basic relation of incidence between a point and a line. With the aid of a quantifier, concerning the (non-) existence of an intersection point, a new relation – parallelism – has been defined from one taken to be more basic (a point lying on a line). Taking such definitions further, Descartes shows how to interpret algebraic operations in geometry, in such a way that one can turn back and define the geometry in the algebra. Or to move to the nineteenth century, take the definition of *continuity* of a function F:

$$(\forall x)(\forall \epsilon > 0)(\exists \delta > 0)(\forall y)(d(x, y) < \delta \rightarrow d(F(x), F(y)) < \epsilon)$$

Continuity is clearly a new concept; yet it somehow emerges from a background consisting only of an idea of *distance* d, with ordered values, using the first-order logical symbols alone.

To be sure, a concept like continuity does not reduce to the formal definition; it continues to have associated pictures and ideas, without which we would drown in an exponentially growing set of potential formal proofs. But if mathematics as a human activity would make little sense without the conceptual imagery, the specifically mathematical nature of this activity does have much to do with the casting of these images in the clay of logic. Having proved itself for a certain time, the formal definition becomes the final arbiter; more than that, it informs and guides the future evolution of the initial idea. In any case, whatever one's position on this, at the very least we have here a formal avatar of concept creation.

It is the dynamic of alternation of quantifiers that has this dramatic creative power; it encodes an activity: given x and ϵ, search for a good δ, whose discovery may lead to questions about a new x requiring a new witness. Alternation is only possible if the relations we begin with are at least binary, and thus have n-ary boolean combinations for any n. The formula

$$(\forall y)(d(x, y) < \delta \rightarrow d(F(x), F(y)) < \epsilon)$$

relates 3 variables, x, ϵ, δ, in a new way, and we can now afford to quantify one out. By contrast, if we only have unary relations, quantifiers produce nothing new beyond *sentences*; there the monadic expressive power ends. We obtain assertions whose truth we can try to decide, but no new definitions, no new distinctions among elements populating the given world. This is the fundamental difference between Aristotle's logic, and the modern version.

It is easy to multiply examples of the birth of new concepts in first-order language; mathematics forms a vast web of fields of discourse, whose basic concepts are introduced and inter-related in this manner. Sets are sometimes used, but set theory itself is understood via first-order logic; the primary use of the assumption that Y consists of *all* subsets of X is the first-order comprehension axiom.

The interpretation of mathematics in set theory does show, in one way among many, that one binary relation – the smallest non-monadic relational language – suffices to interpret anything we like. And having properly started, Gödel's theorem tells us it would be impossible to set fixed limits to this process.

The mathematics used in physics is no exception. Science is only at the very end (if that) about checking the truth of universal sentences. There is first the step of recognizing or creating the notions, in terms of which the theory is formulated and may yield universal consequences. Regular viewings of a light point in the night sky are interpreted in terms of a celestial body; an orbital ellipse describes its movement; a differential equation predicts the ellipse. Newton's laws are phrased in terms of derivatives; when fully clarified, they are formulated in terms of tangent and cotangent spaces, vector fields and flows and symplectic forms. The definition of all of these, and their connection to the simpler mathematical objects that have a direct interpretation – the distance a stone will fly – is carried out using the first-order quantifiers.

It is in this light that I would like to consider probability logic, where the absolute quantifiers are replaced by numerical ones giving probabilities.[2] In the setting of Mendelian genetics, say, an ordinary quantifier may be used to ask whether a yellow pea *ever* descends from two orange ones. A probability quantifier rather asks: what proportion of such descendents is yellow? If we pick one at random, what is the likelihood that it is yellow? Robust logical calculi where such expressions are permitted have been constructed.[3] Such systems admit a completeness theorem, so that the (false) dichotomy between probabilistic and measure-theoretic language becomes a useful duality, with the measure theory on the semantic side complementing the random-variable viewpoint in the language. They are applicable when a natural measure forms part of the structure of the universe we are trying to describe.

In certain circumstances, probability quantifiers significantly enrich the power of the logic. A trivial way this may happen is at the very reading of the initial data. A graph on a large number n of points, obtained by a random choice of edges, looks the same to first-order eyes regardless of whether the edge probability is 1%, or 99%, provided n is big enough compared to the formulas considered. Probability quantifiers, by their design, will of course directly and immediately detect the difference.

But is probability logic of use beyond the simple access to the basic data? Does it also help in the formulation of new concepts, new relations, and if so, is it possible

---

[2] See Mumford (2000) for a contrasting view.

[3] We will work with a finitary first-order calculus close to that of Keisler (1985) Sect. 4.1; the importance of a theory of approximations and limiting behavior favors this over the infinitary versions.

to say in what way? This is the question we attempt to answer here. This is quite different from the obvious importance of probability or measure as mathematical tools in their own right, studied from the outside using ordinary quantifiers; we are interested in the creative power coded *within* that calculus.

Consider this thought-experiment. You are plunged into a world of which nothing is known; or perhaps you sit outside, but have the possibility of probing it. It is a Hilbert-style world, of the general kind of his axiomatization of geometry; with recognizable 'things' (perhaps points and lines, 'material particles' (Maxwell 1925), planets, packets of radiation, small colored pixels, $\cdots$.) Some basic relations among them – we will start with unary and binary relations – can be directly sensed. If you are unlucky, your world consists of Kolmogorovian noise; no pattern can be inferred, no means of deducing any additional observations from some given ones. But if there is a pattern, how much of it you will discern depends not only on your eyes, but on the logical language you have available. Quite possibly, the laws of the world in question can only be formulated in terms of concepts and structures that are not immediately given in the basic relations. If you use first-order logic, starting with what seem to be points and lines, interpreted in algebra and leading to number theory, such entities may be discovered and analyzed; there is no limit to the level of abstraction that may be needed. Is there a similar phenomenon if your logic is purely stochastic? What features will you be able to make out?

This essay is conceived as a dialogue between a mathematical logician (L), and a philosopher of physics (P), sometimes joined by P's brilliant student (S). The philosopher, over L's objections that he is engaged in purely technical lemmas, asks L to present his thoughts on the expressive power of probability logic. This leads to a number of conversations, in which the interlocutors attempt a philosophical interpretation of the lemmas in question, and their mathematical background.

To try to gain insight into this question, they try to study a purified extract; a logic with probability quantifiers *replacing* the usual ones entirely.

Their first conclusion is that the probability logic considered excels in characterizing, and recognizing, space-like features. If the given universe includes a metric giving a notion of distance, the metric space may be characterized by the stochastic logical theory (a theorem of Gromov.) Under certain finite-dimensionality assumptions, the logic is sometimes able to distinguish infinitesimal features, even from coarse-grained data: if the metric is not directly given at the smallest scale, but only some mid-scale relation depending on it is directly observable, nevertheless the probability quantifiers can penetrate into the small scale structure of the world. If we further assume that the universe (with respect to certain relations) has the same statistical properties at every point, it can be shown to approximate a Riemannian homogeneous space, a rather unique kind of mathematical object.

But there it ends. Probability logic can describe a locally compact space-like substrate of a given relational world; and color this space by the local correlation functions of some conjunctions of the basic relations. Beyond this, it can go no further on its own. Conditioned on this spatial structure, all (or almost all) events appear statistically independent; hidden structures beyond the monadic space may exist, but can not be detected with stochastic logic.

In some cases, hints of a deeper structure can be gleaned from the probability-zero sets where the statistical independence breaks down; they must be analyzed by other means. The general formulation of concepts in probability logic, with both ordinary and probability quantifiers present, can be viewed then as an iteration of a first-order layer, then a purely probabilistic one, and so on. The role of the probability quantifiers appears restricted to a determination of the space associated with the current conceptual framework, along with some correlations along it; certain measure zero sets suggest new phenomena to be explored, once again with first order logic.

Does the picture change if we use ternary or higher relations? Here a full mathematical description is not yet available; but it is clear that the expressive power of pure probability logic remains limited, and perhaps again restricted to specifying parameters for a somewhat more generalized notion of space; the new phenomena that occur will be briefly considered.

This essay is dedicated to the memory of my friend Itamar Pitowsky. The first lines of the dialogue really do represent a meeting with Itamar, in the academic year 2009/2010, on the Givat Ram campus. I was already thinking about related questions, and Itamar was – as always – interested. But we first diverged to different subjects, and the rest could never materialize in reality; these dialogues are purely imaginary.

Needless to say, P does not stand for Pitowsky; I could not possibly, and did not try, to guess Itamar's real reactions.

One of the points that arises is the distinction between monadic logic and full first order logic, and the apparent non-existence of any intermediate (beyond the relativistic correction that is discussed in the text.) There is at least an analogy, probably a deeper connection, between this and the question of existence of a mathematical world lying strictly between linear algebra (linear equations) and full algebraic geometry (all algebraic equations.) In this Itamar was interested over many years, for reasons stemming (in part) from his deep study of Descarte's geometry, and the significance of quadratic equations there.[4] I was likewise interested for quite different reasons – such a world flickers briefly, intriguingly, in and out of existence within Zilber's proof[5] (by contradiction) of the linearity of totally categorical structures. I am still not aware of such a world, or a framework in which it may be formulated.'

17.2 Continuous Logic

P. L! You are back. What are you thinking about these days?

L. Actually, it is related to probability logic; but purely classical, Kolmogorov-style probability, I'm afraid. A certain statement I can't quite put my hands on; perhaps it will end up requiring some higher-categorical language.

---

[4]See Pitowsky (1999).

[5]Zilber (1984), see also Pillay (1996).

P. I'd be very interested to hear.

L. It's a bit technical, I'd need to explain first a specific implementation of the logic, and preliminary to that, the continuous logic it is embedded in.

P. All the better! Let it be several strolls. But here is my student Sagreda, perhaps she will join us?

L. Let us start with continuous logic, then. Probabilities, like distances or masses, are real numbers. We could discuss possible probabilities one by one, with relation symbols R_α asserting: the probability of P is α. But that would be clumsy; it would not capture the basic feature of continuity of probabilities. We think of probability 0.5 as just barely distinguishable from 0.50000001; but if we set them up discretely, they appear to be just as different as 0 and 1. It is preferable to use a logic whose values admit a compatible notion of approximation. In real-valued continuous logic,[6] a basic relation symbol R comes with a possible range of values – say, a closed interval on the real line. To specify a structure, for each assignment of elements to the variables, we need to give a value for R. In two-valued logic, a typical predicate has two possible outcomes, in a given possible world: Socrates is mortal, or immortal; the mortality of Socrates is a fact, or a falsehood. In real-valued logic, such two-valued predicates are still allowed. But so are terms such as "height", or "mass" or "temperature"; in these cases the possible values are real numbers. We will take them to lie in some bounded range, known in advance; though 'unbounded' modifications of the logic also exist.[7] This is not the 'fuzzy logic' viewpoint; one should not think of the values as measures of the degree of truth of a predicate, but rather that one asks questions whose answers are not expected to be binary in the first place.

P. But in discrete first order logic, we allow in any case infinitely many relations. If we wish to have a relation R taking values between 0 to 1, say, couldn't we equally well dedicate infinitely many ordinary predicates R_n for the purpose, with $R_n(x)$ specifying the n'th binary digit of $R(x)$?

L. This is a very good way of understanding continuous logic formulas. The one difference lies in allowing for *connectedness* of the real line. Continuous logic treats a structure giving a reading of $0.99999\cdots$ as *identical* to one where the reading is $1.0000\cdots$.

This is not too important for a single structure. But many of the fundamental notions of logic concern the approximation of one structure by another. Recall for example that a substructure A of a structure B is called *an elementary submodel* if an observer living within A has no way of sensing that it is not the entire universe B; any formula $\phi(x)$ with parameters from A having a solution b in B, has a solution b' in A too.

[6] See Chang and Keisler (1966), Ben Yaacov et al. (2008), and Ben Yaacov and Usvyatsov (2010).

[7] Ben Yaacov (2008). One way to present such structures is to introduce a metric and use it at both ends, with 0 an ideal limit of increasing resolution, and ∞ an ideal upper limit of distances; so formulas, when considered up to a given precision, can only access a bounded region, as well as only distinguish points at some resolution bounded above 0.

Now suppose A is an elementary submodel in the usual 2-valued sense, and let us use your coding. If $R(b)$ takes a real value $\alpha = 0.abc\cdots$ and A is an elementary submodel, then one can find b' in A with value $0.abc\cdots$, agreeing with α for as many digits as we wish. In particular, $R(b')$ is arbitrarily close to α; this is indeed what we would want. But the fact that the specifically chosen binary expansions agree is very slightly stronger, and not a natural demand.

P. I see. So the right notion of approximation in continuous logic is simply this: in continuous logic, A is an elementary submodel of B if for any formula $\phi(x)$ with parameters from A, if $\phi(b)$ has value $r \in \mathbb{R}$, then for any $\epsilon > 0$ one can find b' in A with $\phi(b')$ of value between $r - \epsilon$ and $r + \epsilon$.

S. I think I can guess what connectives must be. Returning to sequences of formulas in ordinary logic, say $R_1, R_2, \cdots$ and $S_1, S_2, \cdots$, we can produce a sequence $T_1, T_2, \cdots$ where each T_n is a Boolean function of finitely many R_i and S_j; say $T_1 = R_1 \& S_1$, $T_2 = R_1 \vee S_3$, and so on. In terms of binary expansions of numbers, we obtain any function $C(x, y)$ such that the n'th binary digit of $C(x, y)$ depends only on the first m digits of x and of y, for some m. I suppose the continuous logic analogue would be any function from $\mathbb{R}^2$ to $\mathbb{R}$, whose value up to an ϵ-accuracy depends only on the inputs up to some δ?

L. Just so; in other words, you are simply describing a continuous function.

P. Isn't it excessive to allow any continuous function as a connective? True, in the two-valued case, too, we have an exponential number of n-place connectives. But we can choose a small basis, for instance conjunction and negation, that generate the rest.

L. In the continuous logic case, one can also approximate any connectives with arbitrary accuracy using a small basis, for instance just the binary connective giving the minimum of two values, multiplication by rationals as unary operators, and rational constants.

P. And what about quantifiers?

L. Here a natural basis consists of sup and inf; see Ben Yaacov and Usvyatsov (2010). There is an equivalent, perhaps more philosophically satisfying approach in Chang and Keisler (1966), where a quantifier is seen to be an operator from a *set* of values to a single value, continuous in the appropriate sense. Thus the existential quantifier is viewed as the operator telling us if a set includes 1 or not; $(\exists y)\phi(y)$ is true in M if $1 \in \{[\phi(a)] : a \in M\}$.

P. After describing the usual predicate calculus, one steps back into the metalanguage to define the notion of logical validity.

L. It looks the same here. One can ask about the set of possible truth values of a sentence σ in models M; or for simplicity, just whether that set reduces to a single element $\{\alpha\}$. This will be the case if and only if for any given $\epsilon > 0$, a syntactic proof can be found (in an appropriate system) that shows the value to be within ϵ of α.

In fact all the major theorems of first order logic persist, as if they were made for this generalization. Compactness, effectiveness statements remain the same; notions such as model completeness and criteria for quantifier elimination; stability and beyond. It took substantial work of a number of model theorists over several generations to get us used to the intrusion of real numbers; but once the right point of view was found, it became clear that continuous logic is a very gentle generalization of the old.

P. We have so far paid no attention to equality.

L. The analogue of equality in continuous logic can be taken to be a metric, the data of the distance between any two points. This leads to completions, which do not appear in the first-order case, and do sometimes require more care. But issues of this type will not concern us here.

S. And what is the gain in moving to continuous logic?

L. As you know, unlike the 'total' foundations of mathematics envisaged by Russell and Whitehead, or Hilbert, model theory has to tame one area at a time; using continuous logic permitted a great expansion of the repertoire. See for instance Ben Yaacov et al. (2008). Once one has continuous logic, Robinson's phrase, that model theory is the metamathematics of algebra, can be extended to meaningful parts of analysis. From another angle, constructions such as the monadic space of a theory, while first encountered in the discrete setting (Shelah 1990; Lascar 1982; Kim and Pillay 1997), are more naturally presented in the continuous logic setting.

P. It also makes mathematical logic look much more like the foundations of physics. In a structure one takes an atomic formula, and upon specifying parameters, obtains a value. In classical physics, we have 'observables', often depending on parameters (such as a test function supported in a small region of some point of space), and obtains a numerical answer. And a given experiment will confirm a theory, or measure the value of an observable or a fundamental constant, only to a given degree of approximation.

L. For many years I was puzzled by the fact that the simplest physical models, and with them many areas of mathematics, seem not to be approachable by first-order logic, without the intermediary of a large measure of set theory. The latter seems to be used only as a scaffolding, yet it makes model-theoretic decidability results impossible. It is surprising to what extent this impression was due to no more than a slight discrepancy in the formalization, correctible without any change of the superstructure it supports.

P. I wonder to what extent the role of Hilbert in the formalization of both mathematical logic and mathematical physics has to do with this parallelism, later broken and, I see, mended.[8]

[8]Corry (2004) has an interesting account of the proximity of the two in Hilbert's own thought. Hilbert's foundations of specific subjects looks on this account very much like a direct predecessor of the model-theoretic program of finding 'tame' axiomatizations of wide (but circumscribed) areas of mathematics; while his 'continuity axiom' in this context looks like a part of a description of continuous logic: If a sufficiently small degree of accuracy is prescribed in advance as our condition for the fulfilment of a certain statement, then an adequate domain may be determined, within which one can freely choose the arguments [of the function defining the statement], without however deviating from the statement, more than allowed by the prescribed degree. (Hilbert 1905, cited in Corry (2004).)

S. But still, in the specific setting of probability logic, can I just ignore this issue if I pretend that probabilities are meaningful only to a given precision (say to a single atom in a mole)? Can I then use a *finite* coding of probabilities by ordinary first order formulas?

L. Yes, if you make this assumption you can ignore today's discussion. When probabilities behave discretely, they can be treated using ordinary predicates. Conversely, ordinary quantifiers are clearly special cases of probability quantifiers, assuming there is a gap between 0 and the next highest possible probability for a given question. Yet the results we will discuss are already meaningful, even under such restrictions.[9]

P. But even if we ignore real-valued logic, taking all probabilities to be rational, this discussion was useful in order to remind us of the significance of the notion of approximation, which is surely just as important in discrete first order logic.

I think it underlies most achievements of modern logic, in Gödel's proof of the consistency of the continuum hypothesis no less than the Ax-Kochen asymptotic theory of p-adic fields, to choose two random examples.

L. I agree. Is this notion as pervasive in the philosophy of physics?

P. I think physicists never evoke a mathematical structure without being aware of the meaning of approximations to the same. They may exhibit this awareness in their practice, rather than in their definitions, which is why Dirac felt comfortable with his δ-functions long before Schwarz.

It is also the standard model of the history of physics, as opposed to physics proper. We have a succession of theories, approximating reality with respect to more and more classes of phenomena, and to better and better accuracy. Until perhaps one reaches a final theory, and needs only to fine-tune some parameters, approaching their physical value.

L. Sometimes it seems to go the other way: the perfect mathematical model is the theory; the universe is the rough approximation; it is understandable precisely because it *approximates* a beautiful theory. A hill of sand has the perfect shape predicted by a differential equation, whose terms are dictated by considerations of weight and forces of the wind. But we know that the grains are discrete; the differential equation (taken literally) is wrong. The right mathematical statement is that the actual beach lies on a sequence of ideal ones, with sand made of ever smaller grains; it is the limit shape of the various sand hills, that will really satisfy the differential equation; reality is just a very good approximation to the ideal.

S. What notion of 'limit shape' do you have in mind here?

P. Perhaps some variation on Gromov-Hausdorff distance. In a known ambient space, say 3-dimensional Euclidean space, Hausdorff showed how to view the space

[9]There are in fact interesting mathematical worlds where probabilities are always rational, and with bounded denominator in any given family. Thus for instance Chatzidakis et al. (1992) study 'nonstandard finite fields', infinite models of the theory of all finite fields; they carry a notion of probability deriving from the counting-based measure on finite fields. There, the probability of being a square is exactly $\frac{1}{2}$, as is easy to see; the probability of x and $x+1$ both being squares is a little more complicated to analyze, but still equals exactly $\frac{1}{4}$.

of all compact subsets as a metric space. Gromov defined a metric on abstract compact metric spaces, given without any embedding in an ambient space. For instance, any compact metric space X is the limit of finite metric spaces; one can pick a finite ϵ-dense of points, i.e. any point of the space is at distance at most ϵ from a point of the finite subset; as ϵ approaches zero, the finite subspace will approach X.

L. Gromov's distance agrees, incidentally, with the continuous-logic notion of approximation of structures that we discussed, with respect to existential and universal formulas.

On the other hand, individual compact metric spaces play a special role in continuous logic; they are the analogues of *finite sets* in the discrete case. To any given resolution, they do look finite.

17.3 Probability Logic

Probabilities behave formally like a quantifier. Given a formula ϕ, with free variable x and possibly some others y, one can consider another statement, asserting that the probability – if one chooses x at random – that $\phi(x, y)$ be true, is α. This can be viewed as a true/false statement about y and α, but is more conveniently construed (as discussed above) as a statement about y, whose value is α; we can denote it as $E_x\phi(x, y)$.

P. But how do you iterate quantifiers in this language? After all, the complexity of first-order logic is created by quantifier alternation. And here you started with a $0/1$-valued formula, and obtained a real-valued one.

L. It would be possible, though clumsy, to stick with $0/1$-valued statements, and then iterate them, in expressions such as: if we choose x at random, with probability $>.99$, a further random choice of y will have probability $<.01$ of making ϕ true. But it is more convenient to generalize probabilities to expectations. For a 0/1-valued statement, the expectation of the truth value is the same as the probability. But expectations make sense for a real-valued formula as well, and one can simply write $E_y(E_x\phi(x, y))$. These two formalisms convey the same information, one can translate one into the other.

P. I see; we have a formal operator taking a formula $\phi(x, y, z)$ (in free variables x, y, z say) to a formula $E_x\phi(x, y, z)$ with free variables y, z – indeed a kind of quantifier. We view $\phi(x, b, c)$ as describing an event relative to the variable x; depending on parameters b, c. Then $E_x\phi(x, b, c)$ gives the probability. Or more generally, if $\phi(x, b, c)$ denotes a 'weighted event', or random variable, then $E_x\phi(x, b, c)$ should be the expectation. The first axioms for "fields of probability" in Kolmogorov (1950) are entirely algebraic; they already occur in some form in Hilbert and earlier, cf. Corry (2004). It should be easy to formalize them here.

L. Just so. Generalized to expectation of 'weighted' events, the axioms read:

- $E_x(\phi + \psi) = E_x(\phi) + E_x(\psi)$
- $E_x(\psi(y) \cdot \phi(x, y)) = \psi(y) E_x(\phi(x, y))$
- $E_x(|\phi|) \geq 0$
- $E_x(1) = 1$

The first three axioms express *linearity* and *positivity* of expectation. The fourth is a normalization axiom, asserting that the probability of a tautologically true condition is 1 (here the first 1 denotes the constant function with value 1, in various sets of variables.)

P. But what about countable additivity, the axiom Kolmogorov adds in the second chapter? All of measure theory appears to depend on this.

L. In our setting, countable additivity comes of itself. Unlike the Kolmogorov setting where the field of events is an arbitrary Boolean algebra of sets, here the events are given by formulas. Using a syntax/semantics duality – Stone's or Gelfand's, according to whether the logic is discrete or 'continuous' – we can represent formulas as continuous functions on a compact Hausdorff space (the space of types). The operator E_x then induces a unique countably additive probability measure on the algebra of Borel subsets. (Riesz representation theorem.)

P. Great; so this amounts to a completeness theorem for the logic. A consistent theory has a model, which is a structure in the usual sense along with a probability measure on the σ-algebra generated by the definable sets.

S. But does it mean that Fubini's theorem comes for free, too? If we also have expectation operators for another variable y, is it automatic that $E_x E_y \phi(x, y) = E_y E_x \phi(x, y)$?

L. This is interesting; it speaks again to the difference between unary and relational. The usual formulation of Fubini's theorem concerns the *product measure*; it measures only sets (or for us formulas) of the form $\psi(x)$ or $\theta(y)$, and their combinations. But $\phi(x, y)$ may not arise from unary formulas at all. Thus while Fubini's theorem – being a theorem – does hold, it does not give the commutativity that you mentioned. If you are interested, see Simon (2015), Theorem 7.29 for example, for a place where such commutativity, or non-commutativity, has model-theoretic significance.

I realized just today the close connection of what I'm calling here *pure* probability logic, with the flag algebras of Razborov (2007). They include additional axioms – most likely Fubini will suffice- as they are aimed at *proving* statements of additive combinatorics. As our interest is in a different question – finding a ceiling on new concepts definable in this – we are best with the minimal set of axioms. I strongly recommend by the way the beautiful exposition (Razborov 2013).

P. It is interesting to note non-commutativity of operators, the source of the uncertainty principle in some presentations, can arise from classical logic for purely model theoretic reasons.

L. We agreed not to discuss quanta today; but since you raise the analogy, let me also indulge in a comment on the non-measurability hypothesis of Pitowsky (1982). I used to think that non-measurable sets are pathological and unlikely to occur in physics; the problem being not doubts about their existence, dependent on the axiom of choice, but the fact that this axiom, when used to show existence of one set, tends to bring with it a plethora of others, too similar to each other; such sets can rarely if ever be uniquely characterized, so that hopes of a meaningful physical theory using them seem forlorn in advance. But thinking about measures on Cartesian products suggests a quite different perspective. There, non-measurability only means, in usual usage: with respect to the product measure. It can simply indicate a binary relation that does not reduce to unary ones.

Consider a family of finite structures, F_n approaching a structure F_*. Each F_n carries a probability measure induced by counting – a subset X has measure $|X|/|F_n|$. As probability logic structures, they approach (F_*, μ_1) for some measure μ_1. Likewise, Subsets R of the Cartesian plane F_*^2 are *always* measurable for an appropriate limit measure μ_2 on F_*^2. But they are very rarely measurable for the product measure $\mu_1 \times \mu_1$.

S. Are these exotic sets that are not measurable? Or everyday ones like the diagonal?

L. The diagonal on F_* itself, the graph Δ of the equality relation, will be measurable since it has measure zero. But thickenings of the diagonal – pairs of elements at distance $\leq \epsilon$ from each other, for some ϵ and an appropriate notion of distance – will usually fail to be measurable. In case F_n is a group, there are natural neighborhoods Δ_U of Δ of the form $xy \in v \in U$, where U is a 'neighborhood' of 1, of small but positive measure on F_n. If F_n is a sufficiently noncommutative group,[10] such neighborhoods will not be measurable. But there is nothing pathological about them.

The specific case of non-measurable spin functions on a sphere may not be attainable in a pseudo-finite manner, as suggested by results in Farah and Magidor (2012). But the larger idea of a significant two-dimensional set, not measurable for the product measure but restricting to a measurable set in every linear one-dimensional section, attains in this light a different allure.

S. But let us go back to completeness. Is there also an incompleteness theorem? Of course if one throws in the usual quantifiers, the usual incompleteness theorems apply. But if we use exclusively probabilistic quantifiers, replacing "all" by"almost all" in the appropriate sense? Gödel showed that an alternation of n quantifiers does not reduce to an alternation of $n-1$; is it likewise true here that alternating "almost surely" and "almost never"?

P. I would guess not. It is often pointed out (by Laplace as well as Keynes, if memory serves me) that the knowledge of *highly probable* events – with probability very close to 1 – suffices to determine all probabilities; this, by the law of large numbers, provided we can discuss the probability outcome of a large number of choices. The same principle can be used to read out, by measuring a single probability, when (for instance) 99% of elements x there are about 3% of elements

[10] I.e. without a commutative subgroup of bounded index. The more precise statement is that the map $(x, y) \mapsto (x, yx)$ will not be measurable at the limit.

y satisfying some relation. L. It suffices to 'guess' a very large number of random elements, and see if this statement holds for *frequency* among them.

L. Indeed this is Hoover's normal form; see Keisler (1985). Any sequence of probability or expectation quantifiers, and connectives, is equivalent to a single string $Ex_1 Ex_2 \cdots Ex_n$ upfront, followed by a formula made up using connectives from atomic ones. It is a kind of universal quantifier-elimination result for probability logic. Quite amazing to a model theorist, who spends years studying some special theory of geometric provenance, and trying to eliminate its quantifiers based on laws specific to that case.

The obverse side of the coin, of course, is that this is a first hint of the limited expressive power of probability logic.

S. But it does still leave a block of expectation quantifiers at the front, of arbitrary length. And even with one quantifier, and one binary relation R, couldn't we define complicated, intrinsically n-ary relations, just as in the first-order case, by quantifying over a Boolean combination?

L. Before discussing this, we will need to refine our notion of *monadic*. But it is late; shall we continue tomorrow?

17.4 The Recognition of Space

P. You said probability logic is good at discerning spatial structures. Could you give some examples?

L. Let us start with Gromov's reconstruction theorem.[11] It applies to a language for probability logic with a distance predicate $d(x, y)$, satisfying the axioms of a metric space. A model for this is thus a complete metric space M carrying a measure; we assume moreover that all balls are measurable, with *positive* measure. If there are any measure zero balls, we discard them; probability logic will not notice the difference, since it is blind to measure zero phenomena. Likewise, if M is not complete, we can just pass to the completion. Then the theorem asserts that any such model is determined by its theory. It is usually presented in more concrete terms: if for each n we know the probability distribution for the distances between n randomly chosen points, then we know the space.

P. And if M carries additional structure?

L. In continuous logic, all predicates are assumed uniformly continuous with respect to the structural metric. With this assumption, the theorem remains true. The case of an additional constant is already interesting; it implies that if the same geometry is visible when you stand at a or at b inside M – for instance, in the ball of radius 1 around either of them, there is such-and-such a probability for a triangle

[11]Gromov (1998), see Vershik (2004).

to be obtuse,[12] the same probability for a and b; and ditto for all such questions – then there exists an automorphism of M carrying a to b.

P. So if the geometry looks the same from any vantage point on M, statistically speaking, then the space is homogeneous; there are automorphisms taking any given point to any other.

L. Yes. M will be compact then, and a whole body of theorems on the structure of compact and locally compact groups applies to it; in the present case, the Peter-Weyl theorem. It means in particular that with respect to any given formula, or finite number of formulas, M can be quite precisely approximated by finite unions of compact, finite dimensional, homogeneous Riemannian manifolds. Such manifolds are very special; with somewhat stronger homogeneity assumptions (on directions), they can be fully classified. Spheres and tori of various dimensions are good examples. It is really a vertiginous change of perspective, from the very abstract situation we started with to these classical geometries.

This phenomenon persists if the measure is infinite, but finite on any ball. In this case M is locally compact; Euclidean space, say $\mathbb{R}^3$, is the most classical example. Locally compact groups were the subject of Hilbert's 5th problem, resolved in the 1950s, and showing that they are quite close to Lie groups, finite dimensional groups with a differentiable structure; the simple Lie groups were classified in the nineteenth century.

S. Going back – can I summarize Gromov's theorem by saying that M can be identified internally – that a physicist in M, using probability logic, and able to sample distances between points, will be able to recognize the geometry M?

P. Prima facie, the theorem asserts only that the universe is determined by the answers to *all* stochastic experiments. So your physicist will get closer and closer to the full description, that does determine M. But there is no guarantee that at some given time, inside M, a physicist can guess at M as the right answer. At any point, there may be many other possibilities.

L. That's right; there need not exist a finite description of M at all, after all. But suppose M is in fact homogeneous. Our physicists can see that the formulas they measure have very close values at different points. They can then make the leap and assume it *is* homogeneous. Then I expect there will often (say when $Aut(M)$ is semisimple) be in fact be a *unique* homogeneous M where a certain finite number of stochastic questions have the experimentally found answers.

S. But how can the physicists of M guess that this particular sentence – expressing homogeneity – once gleaned from the statistics, would have such fruitful consequences? How to distinguish it from others, that presumably also exist, that lead to nothing special, or else to a completely wild, Gödelian domain?

L. If we knew that, we'd be able to answer the question: What is model theory?

P. Seriously, it is an interesting topic of mathematical philosophy of physics: which possible geometries are amenable to discovery by their denizens, in the sense

[12]Expressed via the sides, i.e. the square of one side is greater than the sum of squares of the other two.

that a finite number of statements, if seen to have high probability and assumed to hold absolutely, will characterize the world uniquely? I think physicists are well aware of such questions (and for related reasons, were disappointed to find large numbers of Calabi-Yau manifolds.)

L. This leap incidentally, from high probability to *always*, cannot be justified within the probability calculus; I don't know what it means to say that it is *likely to be true*, or becomes likelier with more observations; a condition like homogeneity would seem in any reasonable model to have probability zero, if it has one at all. Nevertheless this seems to be the kind of leap science is based on. And then, following the passage from probabilistic to definite rules, a long mathematical analysis using classical logic, before going back to probabilistic verification.

S. There is also an issue at the opening that we ignored. We assumed that the distance between two points is directly observable, and with arbitrary precision. But what if one only has access to everyday distances, not too small or big? Suppose these physicists have size on the order of 100 units. They can estimate distances up to the nearest integer unit; by weighing or other means, like Dalton, they can measure the number of particles in a region a few units wide ball, or at least the ratio of particle numbers between two such regions. But to describe the geometry – e.g. to see whether it is Riemannian – requires understanding distances at much smaller scales. How are they to get this information?

L. This is not unrelated to Hilbert's fifth problem, mentioned a few moments ago. The breakthrough on this problem was Gleason's construction, starting from a compact neighborhood of the identity in a group, of the *tangent space*, the infinitesimal vectors. A modern version with pseudo-finite ramifications is due to Breuillard et al. (2012). (and in fact this is how I got interested in such questions.) The description of the metric at small distances, given the statistics of multiplication at a medium range, is a step in this construction.

But this can be done without any homogeneity assumptions; the most direct description is due to Lovász and Szegedy (2010), though their setting was slightly different.[13] Suppose, in your scenario, that $R(x, y)$ denotes the relation: x, y are at most a unit distance from each other. The Lovász-Szegedy metric can then be written in one line as a probability logic formula:

$$d_{LS}(x, y) := E_z(|E_t(R(t, x)\&R(t, z)) - E_t(R(t, y)\&R(t, z))|)$$

The first thought would be that x, y are nearby if the unit balls around them are 99% equal, in terms of the number of particles they contain; there is of course a grain of truth in this. But what actually works is at one remove: intersect the unit balls around x and z, and compare this to the result for y and z. If for the vast majority of

[13]The theory of *graphons* includes ideas very close to those presented here. Lovász and his collaborators begin with measure-theoretic limits of sequences of graphs, then builds a topology. In the same years, unaware of this, model theorists added measures to their already topological type and strong type spaces. Many results are represented in the two theories, in different terms; but a systematic comparison has not been undertaken.

particles z, the two numbers are close, then x, y must be close. Even at the pseudo-finite limit, where the number of particles in a ball is infinite, this metric can be shown to be compact; and a unit distance ball, in naive units, is still contained in a finite ball of this metric. This indicates that the new metric must be seeing short-range distances fairly faithfully; if all distances below 1 were lumped together as in the original metric, the unit ball would be infinite and discrete, and not compact. The story goes further than that, but you see that probability quantifiers are a powerful tool in the recognition of space.

17.5 Unary and Monadic

P. We were able so far to talk without agreeing on the notion of *space*; just as mathematicians felt no need for a definition of algorithm, as long as they were showing many things were algorithmic. But if we are to discuss the capabilities of probability logic beyond the spatial, we should perhaps agree what we mean by the latter.

L. The most basic example is a 'space' allowing two positions. We could call one of them Q. the other then is $\neg Q$. A unary predicate. In, or out.

S. Isn't it a little artificial to call this a space? Do any mathematical definitions of 'space' apply here?

L. Well, you can think of it as a very basic, two-element topological space X_0, with points 0 and 1. Given a structure M, the interpretation of Q gives a map $M \to X_0$. We think of M as spread out over X, with Q lying above 1, and the complement above 0.

S. But X_0 is discrete; you are not really using the topology.

L. For one predicate, that is true. With more unary predicates, we can distinguish more and more regions in this primitive space, positions that elements may take; we can subdivide the house into rooms, or delineate additional areas outside it. As long as the set of possible positions is finite, it is discrete and the topology is not used. But generally there are infinitely many unary formulas to consider, and then the best way to keep track of the possible positions is a topology. For instance if we have infinitely many unary predicates $P_1, P_2, \cdots$ with no relations among them – all positions are possible – then X_0 would be the Cantor space of 0/1-sequences, with $01\cdots$ coding $P_1 \cap (\neg P_2) \cap \cdots$. If constraints exist, we will have a certain subspace of that.

S. And in continuous logic, I suppose, these positions may form a connected space; the range of values of a unary predicate can be an extended interval.

L. Quite so. In any case it is compact; up to a given resolution, there are finitely many possibilities.

P. Perhaps this is a good place to disambiguate the word 'finite'. When a combinatorialist like Erdős says 'let n be a number', n is a finite number; but one that we view as potentially increasing towards infinity. A theorem on all but finitely many n would often be satisfactory. In other cases, it is really viewed as fixed.[14]

L. Well, the space X_0, to any given resolution, is not pseudo-finite; it is not moving; it's dead finite. We may well visualize large finite structures, where Q has n elements and $\neg Q$ some much bigger number m perhaps; at the limit we talk of a pseudo-finite n. But X_0 will still have two.

S. So the position of an element a in X_0 is what is called in Chang and Keisler (1990) the type of a; it determines the value in M of all unary formulas concerning a. But given two elements a, b, and a relation R, the positions of a and of b in X_0 will not necessarily determine whether $R(a, b)$ holds. By 'spatial' you just mean 'unary', then?

P. The spatial configuration of n points should be determined by the position of each of them; in this sense, we do view space as monadic, or unary.

L. So far we took 'monadic' to be synonymous to 'unary'. But the literal notion of unary is not robust. Starting with a purely unary structure and *forgetting* some information, we can arrive at one that is not quite unary. Let us go back to our unary predicate Q; we think of it as specifying one side of some border, some distinction in the world. You might think a single unary predicate is the least possible structure a relational universe can have, but it is not: you can easily conceive of a language able to specify less: it gives the demarcation line, without breaking the symmetry by giving a name to either side. You simply need a relation $E(x, y)$, true if both x, y are in the extension of the predicate, or if both are out; an equivalence relation with two classes. Technically, $E(x, y)$ is a binary relation. But it could have been defined from the unary predicate Q, if we had it: $E(x, y) \iff (Q(x) \wedge Q(y)) \vee (\neg Q(x) \wedge \neg Q(y))$. A world with E as a basic relation carries no more structure than one with Q, and possibly less: there may be no way to tell apart the two classes.

S. What happens to X_0 in this case?

L. Given a model M, we can *define* X_0 as M/E, the set of classes; even without having a name for one of them. So we still have the map $p : M \to X_0$, giving what we take for the spatial location of an element. The elements of X_0 are called *finite imaginaries*; it was precisely for them, and quite in the present setting, that Shelah introduced his notion of imaginary elements (or T^{eq}; Shelah 1990).

S. But as a topological space, X_0 is just the same as in the case of unary predicates?

L. Yes, but there is some further information in it, of a different kind: about what we are *not* able to name. If E has n classes (n being a limited number), then X_0 has n points; some of them may have names, and some not; if we pick one, we may have a way of naming another. The right ideas to describe the situation were found by Galois: there is a finite group acting on the set of classes; all classes, or sets of

[14]Jordan (1878), introducing his theorem regarding *all but finitely many* matrix groups of a given dimension, attempted to introduce the term 'limited' (see p. 114), insisting that *limité* et *illimité* …ne sont pas synonymes de *fini* et *infini*", but unlike the concept and theorem, this linguistic advice was not fully taken up by later mathematicians.

classes, or even relations on the classes describable in the language are preserved by the group, and conversely any invariant relation can be described by a formula. It is exactly the situation with the roots of an irreducible polynomial; the Galois theory underlying Shelah's finite imaginaries directly generalizes the Galois's theory for solutions of algebraic equations.

P. And if we use continuous logic?

L. The need for a more general monadic space actually arose earlier, in the ordinary, two-valued first-order case, in Lascar (1982) and Kim and Pillay (1997). In place of maps $p : M \to X_0$ with X_0 a finite set, we have to take into account maps into compact spaces; but using the Peter-Weyl theorem, it suffices in fact to consider maps $p : M \to Y$ into finite dimensional Riemannian manifolds. The Galois theory persists: there is a canonical group H acting on Y, a subgroup of the orthogonal group of automorphisms of $\mathbb{R}^n$ preserving Euclidean distance. A Riemannian distance $\rho(y, y')$ on Y is preserved by H. And while the literal spatial position $p(m)$ in Y is not definable, the distance between two points – $\rho(p(m), p(m'))$- is definable in the sense of continuous logic. The full monadic space X is X is the projective limit of all such p; inasfar as any finite number of formulas is concerned, to a given approximation, it suffices to think of one $p : M \to Y$ as above.

P. The move you describe from unary to monadic closely parallels the relativistic correction of Galileo. The elements of the universe are mapped into a space Y, but only relative distances are meaningful. And for any given question, if you fix a sufficient number of reference points, spatial location becomes unary again; it is determined by the distances to each of these reference points.

S. But the Galilean group is not compact; it includes the translation group $\mathbb{R}^3$; like the space $\mathbb{R}^3$ itself, it is only locally compact.

L. We mentioned earlier are 'unbounded' generalizations of continuous logic; there is also a natural generalization of probability logic, where the measures are finite only locally. This should really be our framework, but as they do add a technical layer, it seemed better to describe the bounded case. In the full setting we do obtain locally compact groups and spaces; so $\mathbb{R}^3$ is included.

P. It is quite striking how Euclidean spaces, that one thinks of as essential but possibly contingent features of physics and geometry, arise of themselves in a first-order logic setting. Do you also see something resembling time?

L. Not so far. Nor Lorentzian metrics.

P. I think we can provisionally accept this as the monadic, and in a sense the *compact* or *locally compact* spatial part of the theory of M;

Now what about probability logic?

L. If we have probability quantifiers too, we obtain an induced measure on the space X; and we can ignore any open set of measure zero, as the probability quantifiers will not see it. This is quite close to the measured metric spaces we encountered yesterday.

17.6 Nothing But Space

S. Is it time now to repeat my question from day 3? Suppose we have at least one binary relation R in the language, two-valued if you like. Define

$$R^*(y_1, \ldots, y_n) = (Ex)(R(x, y_1) \wedge \cdots \wedge R(x, y_n))$$

Isn't this a new, essentially n-ary relation, obtained in probability logic?

L. Surprisingly, no; in pure probability logic – now including the Fubini axiom – we can actually write an identity giving the value of $R^*(b_1, \ldots, b_n)$ in terms of basic data. If we knew that the events $R(x, b_i)$ are statistically independent, of course it would just be the product of their probabilities. Now they may not be independent, since they are spread out differently over the monadic space X. For instance suppose X has just two points, as above; corresponding to two classes Q, Q'; say with probabilities $1/3, 2/3$, for definiteness. Possibly $R(x, b_1)$ is contained in Q, and the others are contained in Q'; then of course they are not independent. A better guess for the expectation of their conjunction would be $(1/3)E((R(x, b1)) + (2/3)E(R(x, b2) \wedge R(x, b3))$. The general expression uses a little more technique from measure theory, but comes to the same thing: we decompose the sets $R(x, b_i)$ along X, and assume that when conditioned on any given $x \in X$, they are statistically independent. Then we sum up the result.

S. And does this guess actually work?

L. Yes; provided $b_1, \ldots, b_n$ do not fork.

S. Do not what?

L. Forking is the fundamental relation found in Shelah (1990), on which the enormous edifice of stability theory rests. For instance in algebra, an algebraically independent n-tuple of numbers does not fork; a polynomial relation does cause forking. What matters to us is that the set of forking tuples has measure zero (in fact it has measure zero for *any* definable measure.) In particular the identity described above holds for almost all $b_1, \ldots, b_n$. In pure probability logic, this is indistinguishable from it holding for *all* $b_1, \ldots, b_n$.

Thus pure probability logic sees, very precisely, the monadic space of the theory; and conditional probabilities of basic relations as functions on this space. Beyond this, or relative to it, the universe is boring; everything is locally statistically independent, no special features can be made out, and no new relations can be described.

P. This has a Kantian flavor: with probabilistic eyes we see space, since that is all we are able to see. Classical physics does not take the worlds to be relational; The dynamics takes place on particles in or fields on this pre-conceived space. It is sometimes even seen as preliminary to the very language used to formulate queries, i.e. we may have observables (how many particles in such and such a location?) whose very definition depends on a point in space. But even if the world *were* relational, observed pure probabilistically, it would still appear to be space-based in a similar way; the binary relation will have no influence on the physical laws, and will not make itself felt beyond direct, empirical local correlations.

S. Do you know Gromov's (2008), where he speaks about the "reconstruction of the geometry of physical space" from statistical data?

P. Of course. He first evokes the first genetic map, obtained in Sturtevant (1913) long before the physical nature of a gene was understood, and confirming at the same time the one-dimensionality of the genetic code; a beautiful instance of the statistics-geometry connection. The key is an idea of Morgan's, that the probability of a break between A and B should depend on their physical distance. Gromov proposes a generalization to "geometric structures on a set L that are (...) encoded by probability measures on the set of all subsets K of L, or by something similar to such measures." All *definable* subsets K in our sense would perhaps be even closer to the discussion there, of *observed samples* and *arranged experiments*. As a model example, he takes a set L of "pixels (...) or light-sensitive cells in the retina of an eye (...) stripped of any structure." L is presented to us as a pure set, with n points; it comes as part of a two-sorted structure, where the second sort P is the set of all subsets of L, so of size 2^n, and the membership relation is part of the language. An element of P then codes a black and white picture on L. This is all generic; the key element is the interpretation of the expectation quantifiers; they reflect not the counting measure on all of P, but a measure that concentrates on images possible in a given world, according to their likelihood. Gromov suggests how the geometry on P may be reconstructed from these statistics: nearby points correlate well, faraway points not at all. He asks: " Is there, yet unknown, a mathematical theory (...)incorporating these ideas and being useful not only for "specifying parameters" in a (given class of) structures but for also for predicting and/or generating new (classes of) structures?

S. The theory of the monadic space does seem to be a reconstruction of geometry, incorporating these ideas systematically.

P. Yes; and on it, in this case, the Lovász-Szegedy metric would be natural. But Gromov may have meant more than this. Measured metric spaces are well-understood structures; indeed by Gromov's reconstruction theorem, they can be parameterized; perhaps we should view the recognition of space, once it is known that a space is what we are looking for, as merely a form of 'specifying parameters"?

L. In this case our conclusion is that a mathematical theory such as Gromov wishes, cannot be found within pure probability logic.

P. But perhaps at the limits of this logic. The forking relation we encountered earlier is invisible to the probability logic, but is found precisely by tracing the limits of validity of this logic. Finite improbabilities were incorporated in the monadic space X, and influenced its form; notably they were used to recognize new imaginaries, forming another layer of X. Now relative to the full X, none remain. If any surprises are left – failures of statistical independence relative to X – they are an infinitely improbable thing; having measure zero they may hint at a new structure.[15] Presumably, the next step, for the physicists of L, would be trying to test the new relation for universal laws, in first order logic again.

[15] The discontinuity points of the correlation functions, also of measure zero, may be another such hint.

L. Let us take a simple example, consisting of addition alone (whereas the relational language associated with Euclidean geometry, according to Descartes, has multiplication too.) The set of pixels L, of illimited size $n = |L|$, is presented to us 'stripped of all structure'; but it secretly carries the structure of an abelian group. To simplify further, assume $(L, +)$ has exponent 2, so it can be viewed as a vector space over the two-element field $\mathbb{F}_2 = \{0, 1\}$. Let P, the set of pictures, be the dual group of characters of L; in our case it is the dual space L^* of homomorphisms from L to the 2-element field. Each character $a \in P$ is thus a black-and-white picture on L (with $a(l) = 1$ denoting that l is black.) We treat P using probability quantifiers, with the unbiased measure on P. As for L, we have a choice, either using probability quantifiers (with the counting measure) or ordinary ones. In the former case, we will be unable to discern any structure whatever on L. If we choose any $a_1, a_2, a_3 \in L$, for instance, then with probability 1, all 8 possible pictures on the 3-element set $\{a_1, a_2, a_3\}$ will be equally likely, with expectation 2^{-3} each; nothing to report. But if we use first-order logic on L, we notice some exceptions. For a small number of triples – a measure zero set – the probabilities are radically different: the probability of $x(a_1) = x(a_2) = x(a_3) = 0$ is $1/4$, while $x(a_1) = x(a_2) = x(a_3) = 1$ has probability 0. An external observer aware of the addition that secretly rules this world will see that this happens just when $a_1 + a_2 = a_3$. Addition involves forking, and is inaccessible to the probability logic. A denizen of L, becoming aware of such exceptions, will have to begin by noting the non-probabilistic associative law, and analyze what follows in first-order logic, including perhaps the construction of a new measure appropriate to the graph of addition.

Perhaps a mathematical investigation is warranted, as to *what kind* of algebraic structure can be encountered in this way.

S. And what about higher-order relations? Could it simply be that binary relations, for pure probability logic, form a limited world, but unlimited expressibility appears once past this barrier?

L. No. By the methods of Towsner (2011), one can see that all formulas defined in pure probability logic have the general feature we noticed for monadic ones: to any given approximation, if you fix a sufficient number of reference points as parameters, ϕ becomes unary; the value of $\phi(a_1, \ldots, a_m)$ is determined by certain unary formulas regarding the a_i, using the chosen points as parameters. This does not reveal the precise geometry of the situation as we have in the binary case; but it does impose a clear limit.

P. We should call it a day. But before we break off, can you say a little more about the new issues that are seen when the language includes, say, ternary relations?

L. At least two refinements of the notion of space are needed.

The first is simple. In the ternary case already, the additional structure required involves an inner space associated to each object, with a compact group acting on it transitively; only relations invariant under the group action are meaningful, so that this new space contributes nothing at the level of a single object; but still the full description, involving relations among several, requires it.

S. This is reminiscent of the basic idea of gauge theories in physics, there too the first thing beyond space!

P. But the details appear different – in physics a bundle is induced on the ambient space, here it appears as still relational and on the particles. This too might be worth looking into.

L. The second issue is equality, and higher imaginaries; it is probably cognate to what the higher topos theory of Lurie (2009) is after. We already had to consider *imaginary elements*, introduced by Shelah into model theory in precisely this setting. They are just ordinary elements (or tuple of elements) a, but with a new notion of equality, provided by a definable equivalence relation. The imaginary is an ideal element a/E, representing one or any of the elements of a given equivalence class C. Now suppose the elements a themselves carry some structure $\Sigma(a)$, and all the $\Sigma(a)$ are isomorphic for $a \in C$. We would like to be able to speak of an ideal Σ, isomorphic to any concrete embodiment $\Sigma(a)$. This cannot be done with ordinary imaginaries, but does form a part of the space that probability quantifiers can discern. A special case, when $G = Aut(N/M)$, N algebraic over a sort M, $R(g, a, b)$ if g fixes the fibers about (a, b), was treated in Hrushovski (2012).

It is conceivable that these two considerations are the only new ones needed.

Acknowledgements Warm thanks to Dan Drai for his reading and comments on the text; to Martin Guttman for a conversation, explaining to me inter alia that monads have no windows; to the editors, Meir and Orly, for their encouragement and patience.

References

Aristotle. (1906). *Prior analytics* (A. J. Jenkinson, Trans.) and *Posterior analytics* (G. R. G. Mure, Trans.). Oxford: Oxford Translation. Available from: http://classics.mit.edu/Aristotle/prior.html, http://classics.mit.edu/Aristotle/posterior.html.

Ben Yaacov, I. (2008). Continuous first order logic for unbounded metric structures. *Journal of Mathematical Logic, 8*(2), 197–223.

Ben Yaacov, I., & Usvyatsov, A. (2010). Continuous first order logic and local stability. *Transactions of the American Mathematical Society, 362*, 5213–5259.

Ben Yaacov, I., Berenstein, A., Henson, C. W., & Usvyatsov, A. (2008). Model theory for metric structures. In *Model theory with applications to algebra and analysis* (Vol. 2, pp. 315–427). London mathematical society lecture note series (Vol. 350). Cambridge: Cambridge University Press.

Breuillard, E., Green, B., & Tao, T. (2012). The structure of approximate groups. *Publications Mathematiques- Institut Des Hautes Etudes Scientifiques, 116*, 115–221.

Chang, C.-C., & Keisler, H. J. (1966). *Continuous model theory*. Annals of mathematics studies (Vol. 58, pp. xii+166). Princeton: Princeton University Press.

Chang, C. C., & Keisler, H. J. (1990). *Model theory* (3rd ed.). Studies in logic and the foundations of mathematics (Vol. 73). Amsterdam: North-Holland Publishing Co.

Chatzidakis, Z., van den Dries, L., & Macintyre, A. (1992). Definable sets over finite fields. *Journal für Die Reine und Angewandte Mathematik, 427*, 107–135.

Corry, L. (2004). David Hilbert and the axiomatization of physics (1898–1918). From Grundlagen der Geometrie to Grundlagen der Physik. Archimedes: New studies in the history and philosophy of science and technology (Vol. 10, pp. xvii+513). Dordrecht: Kluwer Academic Publishers.

Farah, I., & Magidor, M. (2012). Independence of the existence of Pitowsky spin models. arXiv:1212.0110.

Gromov, M. (1998). *Metric structures for riemannian and non-riemannian spaces*. Boston: Birkhuaser.

Gromov, M. (2008). Mendelian dynamics and Sturtevant's paradigm. Geometric and probabilistic structures in dynamics. Contemporary mathematics (Vol. 469, pp. 227–242). Providence: American Mathematical Society.

Hrushovski, E. (2012). Groupoids, imaginaries and internal covers. *Turkish Journal of Mathematics, 36*(2), 173–198. (Arxiv: math.LO/0603413.)

Jordan, C. (1878). Mémoire sur les équations différentielles linéaires intégrale algébriques. *Journal für Die Reine und Angewandte mathematik, 84*, 89–215.

Keisler, H. J. (1985). Probability quantifiers. In J. Barwise & S. Feferman (Eds.), *Model-theoretic logics* (pp. 509–556). Perspectives in mathematical logic. New York: Springer.

Kim, B., & Pillay, A. (1997). Simple theories. Joint AILA-KGS model theory meeting (Florence, 1995). *Annals of Pure and Applied Logic, 88*(2–3), 149–164.

Kolmogorov, A. N. (1950). Grundbegriffe der Wahrscheinlichkeitrechnung. Ergebnisse der Mathematik (1933); English translation: New York: Chelsea.

Lascar, D. (1982). On the category of models of a complete theory. *Journal of Symbolic Logic, 47*(2), 249–266.

Lovász, L., & Szegedy, B. (2010). Regularity partitions and the topology of graphons. *An irregular mind, Szemerédi is 70* (pp. 415–446). Berlin/Heidelberg: János Bolyai Mathematical Society and Springer-Verlag.

Lurie, J. (2009). *Higher topos theory*. Annals of mathematics studies (Vol. 170). Princeton/Oxford: Princeton University Press.

Maxwell, J. C. (1925). *Matter and motion, 1877*. London: Sheldon Press.

Mumford, D. (2000). The dawning of the age of stochasticity. *Mathematics: Frontiers and perspectives* (pp. 197–218). Providence: American Mathematical Society.

Pillay, A. (1996). *Geometric stability theory*. Oxford logic guides (Vol. 32). New York: The Clarendon Press/Oxford University Press. x+361. Oxford Science Publications.

Pitowsky, I. (1982). Resolution of the Einstein-Podolsky-Rosen and Bell paradoxes. *Physical Review Letters, 48*(19), 1299–1302.

Pitowsky, I. (1999). *The limits of computability and the limits of formalization (hebrew)* (Vol. 48, pp. 209–229). Iyyun: The Jerusalem Philosophical Quarterly.

Razborov, A. (2007). Flag algebras. *Journal of Symbolic Logic, 72*, 1239–1282.

Razborov, A. (2013). What is a flag algebra? *Notices of the AMS, 60*(10), 1324–1327.

Shelah, S. (1990). *Classification theory and the number of nonisomorphic models* (2nd ed.). Studies in logic and the foundations of mathematics (Vol. 92). Amsterdam: North-Holland Publishing Co.

Simon, P. (2015). *A guide to NIP theories*. Lecture notes in logic (Vol. 44, pp. vii+156). Cambridge: Cambridge Scientific Publishers. Association for symbolic logic, Chicago.

Sturtevant, A. H. (1913). The linear arrangement of six sex-linked factors in Drosophila, as shown by their mode of association. *Journal of Experimental Zoology, 14*, 43–59.

Towsner, H. (2011). A model theoretic proof of Szemerédi?s Theorem. arXiv:1002:4456v3, January 2011.

Vershik, A. M. (2004). Random and universal metric spaces. In A. Maass, S. Martínez, & J. San Martín (Eds.), *Dynamics and randomness II*. Nonlinear phenomena and complex systems (Vol. 10, pp. 199–228). Dordrecht: Kluwer Academic Publishers.

Zilber, B. I. (1984). Strongly minimal countably categorical theories. II. *Sibirskii matematicheskii zhurnal, 25*(3), 71–88.

Chapter 18
The Argument Against Quantum Computers

Gil Kalai

Dedicated to the memory of Itamar Pitowsky

My purpose in this paper is to examine, in a general way, the relations between abstract ideas about computation and the performance of actual physical computers.

— *Itamar Pitowsky,* "The physical Church thesis and physical computational complexity" (1990)

Abstract We give a computational complexity argument against the feasibility of quantum computers. We identify a very low complexity class of probability distributions described by noisy intermediate-scale quantum computers, and explain why it will allow neither good-quality quantum error-correction nor a demonstration of "quantum supremacy." Some general principles governing the behavior of noisy quantum systems are derived. Our work supports the "physical Church thesis" studied by Pitowsky (Iyuun Jerus Philos Q 39:81–99, 1990) and follows his vision of using abstract ideas about computation to study the performance of actual physical computers.

Keywords Quantum computation · Noise sensitivity and stability · Quantum error-correction · Physical Church-Turing thesis

18.1 Introduction

My purpose in this paper is to give an argument against the feasibility of quantum computers. In a nutshell, our argument is based on the following statement:

G. Kalai (✉)
Einstein Institute of Mathematics, The Hebrew University of Jerusalem, Jerusalem, Israel

Efi Arazy School of Computer Science, Interdisciplinary Center Herzliya, Herzliya, Israel

M. Hemmo, O. Shenker (eds.), *Quantum, Probability, Logic*,
Jerusalem Studies in Philosophy and History of Science,
https://doi.org/10.1007/978-3-030-34316-3_18

> Noisy quantum systems will not allow building quantum error-correcting codes needed for quantum computation.

To support this statement we study the relations between abstract ideas about computation and the performance of actual physical computers, which are referred to as noisy intermediate-scale quantum computers or NISQ computers for short. We need to explain why it is the case that quantum error-correcting codes are needed to gain the computational advantage of quantum computing, and we also need to explain why quantum error-correcting codes are out of reach. The crux of the matter lies in a parameter referred to as the rate of noise. We will explain why reducing the rate of noise to the level needed for good quantum error-correction will already enable NISQ computers to demonstrate "quantum supremacy." We will also explain why NISQ computers represent a very low level computational complexity class which cannot support quantum supremacy. This provides a good, in-principle, argument for the infeasibility of both good quantum error-correcting codes and quantum supremacy.

We will show where the argument breaks down for classical computation. Rudimentary classical error-correcting codes *are supported* by the low-level computational complexity class that describes NISQ computers.

Pitowsky (1990) formulates and studies a strong form of the "physical" Church–Turing thesis (referred to later as the extended Church–Turing thesis (ECTT))[1], namely

> whether we can invoke physical laws to reduce a computational problem that is manifestly or presumably of exponential complexity, and actually complete it in polynomial time.

and considers the then recently discovered model of quantum computers[2] as a possible counterexample to this newly formulated physical Church–Turing thesis. Very strong support for "quantum supremacy" – the ability of quantum computers to perform certain tasks that classical computers cannot perform, – came four years after Pitowsky's paper, with Peter Shor's discovery (Shor 1994, 1999) that quantum computers can factor integers efficiently.

Our argument, which is in agreement with quantum mechanics, supports the validity of the extended Church–Turing thesis but does not rely on this thesis. Our argument is based on computational complexity considerations and thus manifests the vision of Pitowsky of using abstract ideas about computation to draw conclusions on the performance of actual physical computers, and of integrating computational complexity into theories of nature. Our computational complexity considerations refer to very low-level computational classes and rely on definite mathematical theorems; they do not depend on the unproven conjectures underlying our view of the computational complexity world, such as the $\mathbf{P} \neq \mathbf{NP}$ conjecture. The novelty as well as a certain weakness of my argument is that it uses asymptotic insights (with unspecified constants) to draw conclusions on the behavior of small-

[1] Pitowsky's paper attributes the physical Church–Turing thesis to Wolfram (1985).

[2] Quantum computers were proposed by Feynman (1982), Deutsch (1985), and others. The idea can be traced to Feynman's 1959 visionary lecture, "There is Plenty of Room at the Bottom."

and intermediate-scale systems and, in particular, on the value of constants that naively appear to reflect just engineering capabilities.

The argument against quantum computers leads to various predictions on near-term experiments, and to general principles governing the behavior of noisy quantum systems. It may shed light on a variety of questions about quantum systems and computation. We emphasize that the argument predicts the failure of near-term experimental goals of many groups around the world to demonstrate on NISQ computers quantum supremacy and good-quality quantum error-correcting codes.

In Sect. 18.2 we will consider basic models of computation and basic insights about computational complexity. Our argument against quantum computers is presented and discussed in Sect. 18.3, along with concrete predictions on near-term experimental efforts. The argument crucially relies on the study by Kalai and Kindler (2014) of noise stability and sensitivity for systems of non-interacting bosons. This study is built on the theory of noise stability and noise sensitivity of Boolean functions developed by Benjamini et al. (1999). Section 18.4 discusses underlying principles behind the failure of quantum computers and some consequences. Section 18.5 concludes and Sect. 18.6 is about Itamar. This paper complements, updates, and presents in nontechnical language some of my earlier works on the topic of quantum computers; see Kalai (2016a,b, 2018), and earlier papers cited there.

Note added in proof (Nov. 2019). The very recent paper Arute et al. (2019), claims experimental demonstration of "quantum supremacy" on a 53-qubit quantum computer. In contrast, the argument of our paper predicts that this supremacy claim will not stand.

18.2 Basic Models of Computation

Pitowsky (1990) described three types of computers where each type is more restrictive than the previous one: "the Platonist computer," "the constructivist computer," and "the finitist computer." We will consider versions of the last two types. Our definitions rely on the notion of Boolean circuits.[3]

18.2.1 Pitowsky's "Constructivist Computer," Boolean Functions, and Boolean Circuits

> The 'constructivist computer' is just a (deterministic) Turing machine, and associated with this notion is the class of all Turing machine computable, i.e., recursive functions. Can a physical machine compute a non-recursive function? Itamar Pitowsky (1990)

[3]The precise relation between the Turing machine model (considered in Pitowsky 1990) and the circuit model is interesting and subtle but we will not discuss it in this paper.

Model 1 – Computers (circuits) A computer (or circuit) has n bits, and it can perform certain logical operations on them. The NOT gate, acting on a single bit, and the AND gate, acting on two bits, suffice for the full power of classical computing.

The outcome of the computation is the value of certain k output bits. When $k = 1$ the circuit computes a *Boolean function*, namely, a function $f(x_1, x_2, \ldots, x_n)$ of n Boolean variables ($x_i = 1$ or $x_i = 0$) so that the value of f is also 0 or 1.

18.2.2 Easy and Hard Problems and Pitowsky's "Finitist Computer"

> The 'finitist computer,' is a deterministic, polynomial-time Turing machine (with a fixed finite number of tapes). Associated with this notion is the class of all functions which are computable in a number of steps bounded by a polynomial in the size of the input. As is well known, many important computational problems seem to lie outside the class **P**, typically the so-called **NP**-hard problem. Itamar Pitowsky (1990)

While general circuits can compute arbitrary Boolean functions, matters change when we require that the size of the circuit be at most a polynomial in the number of input variables. Here the size of the circuits is the total number of gates.

Model 2 – Polynomial size circuit A circuit with n input bits of size at most An^c. Here A and c are positive constants.

The complexity class **P** refers to problems that can be solved using a polynomial number of steps in the size of the input. The complexity class **NP** refers to problems whose solution can be verified in a polynomial number of steps. Our understanding of the computational complexity world depends on a whole array of conjectures: **NP** $\neq$ **P** is the most famous one. Many natural computational tasks are **NP**-complete. (A task in **NP** is **NP**-complete if a computer equipped with a subroutine for solving this task can solve, in a polynomial number of steps, *every* problem in **NP**.)

Let me mention two fundamental examples of computational "easiness" and "hardness." The first example is: multiplication is easy, factoring is hard![4] Multiplying two n-digit numbers requires a simple algorithm with a number of steps that is quadratic in n. On the other hand, the best-known algorithm for factoring an n-digit number to its prime factors is (we think) exponential in $n^{1/3}$ (Fig. 18.1).

The second fundamental example of complexity theory is that "determinants are easy but permanents are hard." Recall that the permanent of an n-by-n matrix M has a similar (but simpler) formula compared to the more famous determinant. The difference is that the signs are eliminated, and this difference accounts for a huge computational difference. Gaussian elimination gives a polynomial-time algorithm

[4]Factoring belongs to **NP** $\cap$ **coNP** and therefore, according to the basic conjectural view of computational complexity, it is not an **NP**-complete problem. The hardness of factoring is a separate important computational complexity conjecture.

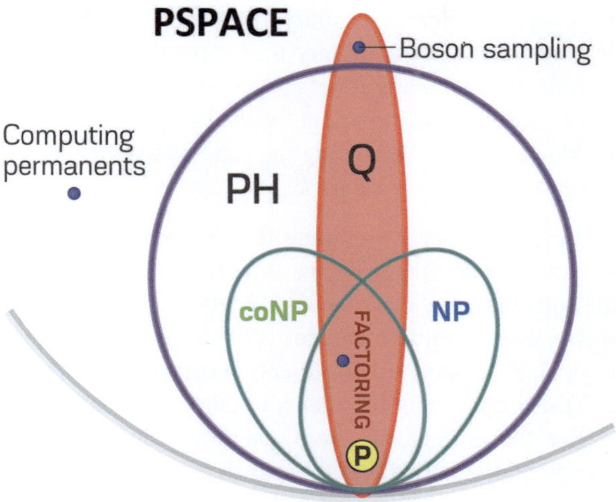

Fig. 18.1 The (conjectured) view of some main computational complexity classes. The red ellipse represents efficient quantum algorithms. Note the position of three important computational tasks: factoring, computing permanents, and boson sampling

to compute determinants. By contrast, computing permanents is harder than solving **NP**-complete problems.[5]

Classical circuits equipped with random bits lead to randomized algorithms, which are both practically useful and theoretically important.

Model 3 – Randomized circuit A circuit with n bits of input and with one additional type of gate that provides a random bit. Again we assume that the circuit is of size at most An^c (here, again, A and c are positive constants). This time the output is a sample from a probability distribution on strings of length k of zeroes and ones.[6]

18.2.3 Quantum Computers

We will now give a brief description (based on circuits) of quantum computers.

Model 4 – Quantum computers

- A *qubit* is a piece of quantum memory. A qubit is a unit vector in $\mathbb{C}^2$, and the state of a qubit is a unit vector in a two-dimensional complex Hilbert space $H = \mathbb{C}^2$. The memory of a quantum computer (quantum circuit) consists of n qubits and the state of the computer is a unit vector in the 2^n-dimensional Hilbert space, i.e., $(C^2)^{\otimes n}$.

[5]The hardness of computing permanents defines an important computational class **#P** (in words, number-P or sharp-P) that is larger than **PH** (the entire "polynomial hierarchy").

[6]For simplicity, we use in this paper the term **P** (rather than **BPP**) to describe the computational power of polynomial-time computers (and polynomial-size circuits) with randomization. We also use the term **Q** rather than **BQP** (for decision problems) and **Quantum Sampling** (for sampling problems) to describe the computational power of polynomial-time quantum computers (and polynomial-size quantum circuits).

- *A Quantum gate* is a unitary transformation. We can put one or two qubits through gates representing unitary transformations acting on the corresponding two- or four-dimensional Hilbert spaces. As for classical computers, there is a small list of gates that are sufficient for the full power of quantum computing.
- *Measurement*: Measuring the state of k qubits leads to a probability distribution on 0–1 vectors of length k.

Here and later we assume that the number of gates is at most polynomial in n. Quantum computers allow sampling from probability distributions well beyond the capabilities of classical computers (with random bits). Shor's famous algorithm shows that quantum computers can factor n-digit integers efficiently, in roughly n^2 steps! It is conjectured that quantum computers cannot solve **NP**-complete problems and this implies that they cannot compute permanents.

For the notions of noisy quantum computers and for fault-tolerant quantum computation that we discuss next, it is necessary to allow measurement throughout the computation process and to be able to feed the measurement results back into later gates.

18.2.4 Noisy Quantum Circuits

Next we consider noise. Quantum systems are inherently noisy; we cannot accurately control them, and we cannot accurately describe them. In fact, every interaction of a quantum system with the outside world amounts to noise.

Model 5 – Noisy quantum computers A noisy quantum circuit has the property that every qubit is corrupted in every "computer cycle" with a small probability t, and every gate is t-imperfect. Here, t is a small constant called the rate of noise.

Here, in a "computer cycle" we allow several non-overlapping gates to perform in parallel. We do not specify the precise technical meaning of "corrupt" and "t-imperfect," but the following intuitive explanation (for a restricted form of noise called depolarizing noise) could be useful. When a qubit is corrupted then its state is replaced by a uniformly distributed random state on the same Hilbert state. A gate is t-imperfect if with probability t the state of the qubits that are involved in the gate is replaced by a uniformly random state in the associated Hilbert space.

Theorem 1 ("The threshold theorem") *If the error rate is small enough, noisy quantum circuits allow the full power of quantum computing.*

The threshold theorem was proved around 1995 by Aharonov and Ben-Or (1997), Kitaev (1997), and Knill et al. (1998). The proof relies on quantum error-correcting codes first introduced by Shor (1995) and Steane (1996).

A common interpretation of the threshold theorem is that it shows that large-scale quantum computers are possible in principle. A more careful interpretation is that if we can control noisy intermediate-scale quantum systems well enough, then we can build large-scale universal quantum computers. As we will see, there are good reasons for why we *cannot* control the quality of noisy intermediate-scale quantum systems well enough.

18.2.5 Quantum Supremacy and NISQ Devices

Model 6 – Noisy intermediate-scale quantum (NISQ) computers These are simply noisy quantum circuits with at most 500 qubits.

The number 500 sounds arbitrary but there is a reason to regard NISQ circuits and other NISQ devices as an important separate computation model: NISQ circuits are too small to apply quantum error-correction for quantum fault-tolerance, since a single good-quality "logical" qubit requires a quantum error-correcting code described by a NISQ circuit with more than a hundred qubits.

A crucial theoretical and experimental challenge is to understand NISQ computers.[7] Major near-future experimental efforts are aimed at demonstrating quantum supremacy using NISQ circuits and other NISQ devices, and are also aimed at using NISQ circuits to build high-quality quantum error-correcting codes.

For concreteness we mention three major near-term experimental goals.

Goal 1: Demonstrate quantum supremacy via systems of non-interacting bosons. (See Section 18.3.3.)

Goal 2: Demonstrate quantum supremacy on random circuits (namely, circuits that are based on randomly choosing the computation process, in advance), with 50–100 qubits.

Goal 3: Create distance-5 surface codes on NISQ circuits that require a little over 100 qubits.

In our discussion we will refer to the task of demonstrating with good accuracy boson sampling for 10–20 bosons as **baby goal 1**, to the task of building with good accuracy random circuits with 10–30 qubits, as **baby goal 2**, and to the task of creating distance-3 surface codes on NISQ circuits that require a little over 20 qubits, as **baby goal 3**.

18.3 The Argument Against Quantum Computers

18.3.1 The Argument

In brief, the argument against quantum computers is based on a computational complexity argument for why NISQ computers cannot demonstrate quantum supremacy and cannot create the good-quality quantum error-correcting codes needed for quantum computation. We turn now to a detailed description of the argument, which is based on the following assertions on NISQ devices.

---

[7] John Preskill coined the terms "quantum supremacy" (in 2012) and "NISQ systems" (in 2018). Implicitly, these notions already play a role in Preskill (1998). The importance of multi-scale analysis of the quantum computer puzzle is emphasized in Kalai (2016a).

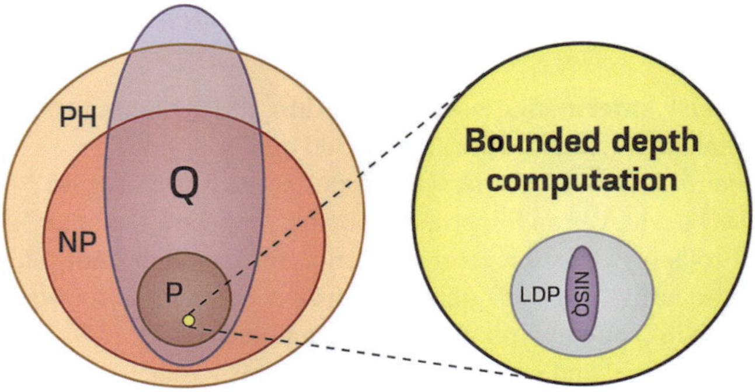

Fig. 18.2 NISQ circuits are computationally very weak and therefore unlikely to create quantum error-correcting codes needed for quantum computers

(A) Probability distributions described (robustly) by NISQ devices can be described by low-degree polynomials (LDP). LDP-distributions represent a very low-level computational complexity class well inside bounded-depth (classical) computation.
(B) Asymptotically low-level computational devices cannot lead to superior computation.
(C) Achieving quantum supremacy is easier than achieving quantum error-correction.

Concretely, we argue that Goals 1 and 2 both represent superior computation that cannot be reached by the low-level computational NISQ devices and that Goal 3 (creating distance-5 surface codes) is even harder than Goal 2 (Fig. 18.2).

We will discuss in Sects. 18.3.3 and 18.3.4 our argument for assertion (A) based on the theory of noise sensitivity and noise stability, and, in particular, a detailed study of noisy systems of noninteracting bosons. It would be interesting to explore other theoretical arguments for (A).

Assertion (B) requires special attention and it can be regarded as both a novel and a weak link of our argument. There is no dispute that we can apply asymptotic computational insights to the behavior of computing devices in the small and intermediate scale when we know or can estimate the constants involved. This is not the case here. The constants depend on (unknown) engineering abilities to control the noise. I claim that (even when the constants are unknown) the low-level asymptotic behavior implies or strongly suggests limitations on the computing power and hence on the engineering ability. Moreover, these limitations already apply for the intermediate scale, namely, for NISQ circuits, already when the number of qubits is rather small. In my view, this type of reasoning is behind many important successes regarding the interface between the theory of computing and the practice of computing.[8] Several researchers disagree with this part of my analysis and find it unconvincing.

There is substantial empirical evidence for (C) and experimentalists often refer to achieving quantum supremacy as a step toward achieving quantum error-correction. Quantum circuits are expected to exhibit probability distributions beyond the power of classical computers already for 70–100 qubits, while quantum error-correcting codes of the quality needed for quantum fault-tolerance will require at least 150–500 qubits. It will be interesting to conduct a further theoretical study of the hypothesis that the ability to build good-quality quantum error correction codes already leads to quantum systems that demonstrate quantum supremacy. There is good theoretical evidence for a weaker statement (C'). (Claims (A), (B), and (C') already support a weaker form of the argument against quantum computers.)

(C') High-quality quantum error-correcting codes are not supported by the very low-level computational power **LDP** of NISQ systems.

18.3.2 Predictions on NISQ Computers

The quality of individual qubits and gates is the major factor for the quality of the quantum circuits built from them. One consequence of our argument is that there is an upper bound on the quality of qubits and gates, which is quite close, in fact, to what people can achieve today. This consequence seems counterintuitive to many researchers. As a matter of fact, the quantum-computing analogue of Moore's law, known as "Schoelkopf's law," asserts, to the contrary, that roughly every three years, quantum decoherence can be delayed by a factor of ten. Our argument implies that Schoelkopf's law will be broken before we reach the quality needed for quantum supremacy and quantum fault-tolerance. (Namely, now!)[9] Our first prediction is thus that

[8]A nice example concerns the computation of the Ramsey number $R(k, r)$, which is the smallest integer n such that in every party with n participants you can find either a set of k participants where every two shook hands, or a set of r participants where no two shook hands. This is an example where asymptotic computational complexity insights into the large-scale behavior (namely, when k and r tend to infinity) explain the small-scale behavior (namely, when k and r are small integers). It is easy to verify that $R(3, 3) = 6$, and it was possible with huge computational efforts (and human ingenuity) to compute the Ramsey number $R(4, 5) = 25$ (McKay and Radziszowski 1995). It is commonly believed that computing $R(7, 7)$ and certainly $R(10, 10)$ is and will remain well beyond our computational ability. Building gates that are two orders of magnitude better than the best currently existing gates can be seen as analogous to computing $R(10, 10)$: we have good computational complexity arguments that both these tasks are beyond reach.

[9]Schoelkopf's law was formulated by Rob Schoelkopf's colleagues based on Schoelkopf's experimental breakthroughs, and only later was adopted as a prediction for the future. An even bolder related prediction by Neven asserts that quantum computers are gaining computational power relative to classical ones at a "doubly exponential" rate. Our argument implies that Neven's law does not reflect reality.

(a) The quality of qubits and gates cannot be improved beyond a certain threshold that is quite close to the best currently existing qubits and gates. This applies also to topological qubits (Sect. 18.3.5).

Our argument leads to several further predictions on near-term experiments, for example, on distributions of 0–1 strings based on a random quantum circuit (Goal 2), or on circuits aimed at creating good-quality quantum error-correcting codes such as distance-5 and even distance-3 surface codes (Goal 3).

(b) For a larger amount of noise, robust experimental outcomes are possible but they will represent probability distributions that can be expressed in terms of low-degree polynomials (**LDP**-distributions, for short) that are far away from the desired noiseless distributions.
(c) For a wide range of smaller amounts of noise, the experimental outcomes will not be robust. This means that the resulting probability distributions will strongly depend on fine properties of the noise and hence the outcomes will be chaotic.

    In other words, in this range of noise rate the probability distributions are still far away from the desired noiseless distributions and, moreover, running the experiment twice will lead to very different probability distributions.

Following predictions (b) and (c) we further expect that

(d) The effort required to control k qubits to allow good approximations of the desired distribution will increase exponentially and will fail in a fairly small number of qubits. (My guess is that the number will be ≤ 20.)

18.3.3 Non-interacting Bosons

Caenorhabditis elegans (or C. elegans for short) is a species of a free-living roundworm whose biological study shed much light on more complicated organisms. Our next model, *boson sampling* can be seen as the C. elegans of quantum computing. It is both technically and conceptually simpler than the circuit model and yet it allows definite and clear-cut insights that extend to more general models and more complicated situations.

Model 7 – Boson Sampling (Troyansky and Tishby 1996; Aaronson and Arkhipov 2013) Given a complex n-by-m matrix X with orthonormal rows, sample subsets of columns (with repetitions) according to the absolute value-squared of permanents. This task is referred to as boson sampling.

Boson sampling represents a very limited form of quantum computers based on non-interacting bosons. Quantum circuits can perform boson sampling efficiently on the nose! There is a good theoretical argument by Aaronson–Arkhipov (2013) that these tasks are beyond the reach of classical computers. If we consider non-interacting fermions rather than bosons, we obtain a similar sampling task, called fermion sampling, with determinants instead of permanents. This sampling task can be performed by a classical randomized computer (Model 3).

Model 8 – Noisy boson sampling (Kalai and Kindler 2014) Let G be a complex Gaussian n-by-m noise matrix (normalized so that the expected row norm is 1). Given an input matrix A, we average the boson sampling distributions over $\sqrt{(1-t)}A + \sqrt{t}G$. Here, t is the rate of noise.

To study noisy boson sampling we expand the outcome distribution in terms of Hermite polynomials in nm variables which correspond to the entries of the input matrix A. The effect of the noise is exponential decay of high-degree terms and, as it turns out, the Hermite expansion for the boson sampling model is very simple and quite beautiful (much like the simplicity of the graph of synaptic connectivity of C. elegans). This analysis leads to the following theorems:

Theorem 2 (Kalai and Kindler 2014) *When the noise level is constant, distributions given by noisy boson sampling are well approximated by their low-degree Fourier–Hermite expansion. (Consequently, these distributions can be approximated by bounded-depth polynomial-size circuits.)*

Theorem 3 (Kalai and Kindler 2014) *When the noise level is higher than $1/n$, noisy boson sampling is very sensitive to noise, with a vanishing correlation between the noisy distribution and the ideal distribution.*

Theorem 2 asserts that noisy intermediate-scale systems of non-interacting bosons are, for every fixed level of noise, "classical computing machines"; namely, they can be simulated by classical computers. We do not expect *classical* computing devices to exhibit quantum supremacy. In this case, the argument is especially strong in view of the very low-level computational class described by Theorem 2, as well as in view of Theorem 3 that asserts that in a wide range of subconstant levels of noise the experimental outcomes will be unrelated to the noiseless outcomes. Furthermore, Theorem 3 strongly suggests that in this wide range of subconstant noise levels, experimental outcomes will be chaotic; namely, the correlation between the outcomes of two different runs of the experiment will also tend to zero (Fig. 18.3).

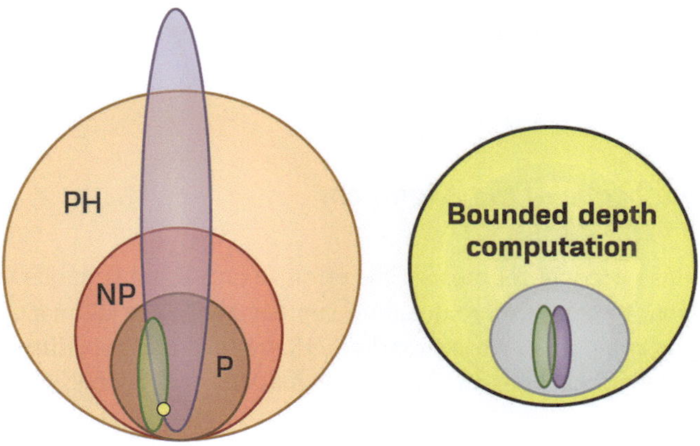

Fig. 18.3 The huge computational gap (left) between boson sampling (purple) and fermion sampling (green) vanishes in the noisy versions (right)

18.3.4 From Boson Sampling to NISQ Circuits

The following open-ended mathematical conjecture is crucial to part (A) of our analysis, as well as to predictions (b)–(d) in Sect. 18.3.2.

Conjecture 4 Theorems 2 and 3 both extend to all NISQ computers (in particular, to noisy quantum circuits) and to all realistic forms of noise.

(i) When the noise level is constant, distributions given by NISQ systems are well approximated by their low-degree Fourier expansion. In particular, these distributions are very low-level computationally.
(ii) For a wide range of lower noise levels, NISQ systems are very sensitive to noise, with a vanishing correlation between the noisy distribution and the ideal distribution. In this range, the noisy distributions depend on fine parameters of the noise.

Assertion (A) of our argument is supported by Conjecture 4(i). The conjecture asserts that, for every fixed level of noise, probability distributions described by NISQ circuits can be simulated by classical computers. As in the case of boson sampling, the argument is especially strong in view of the very low-level computational class described by Conjecture 4(i), and in view of Conjecture 4(ii) that asserts that in a wide range of subconstant levels of noise the experimental outcomes will be unrelated to the noiseless outcomes and, furthermore, the correlation between the outcomes of two different runs of the experiment will also tend to zero.

One delicate conceptual aspect of the argument is that it is a multi-scale argument. The threshold theorem asserts that noisy quantum circuits for a small enough level of noise *support* universal computation! However, in the NISQ regime the situation is different and Conjecture 4(i) asserts that NISQ circuits are "classical" and of a very low level of computational complexity even in the classical computational kingdom. It follows that in the NISQ regime quantum supremacy (e.g., Goal 2) is out of reach, and high-quality quantum error-correction (e.g. Goal 3) is therefore out of reach as well. This behavior of NISQ circuits implies lower bounds on the rate of noise that show that, in larger scales, the threshold theorem is not relevant.

18.3.5 The Scope of the Argument

NISQ computers account for most of the effort to create stable qubits via quantum error-correction, to demonstrate quantum supremacy, and to ultimately build universal quantum computers. We expect (Sect. 18.4.1) that noise stability and the very low-level computational class LDP that we identified for noisy NISQ systems apply in greater generality.

Specifically, it is plausible that the failure of NISQ computers to achieve stable logical qubits extends to systems, outside the NISQ regime, that do not apply quantum error-correction. It is plausible that Google's attempts to build "surface codes" with quantum circuits of 50–1000 qubits will meet the same fate as Microsoft's attempts to build a single topological qubit based on non-Abelian anyons. The microscopic quantum process for Microsoft's attempts is not strictly in the NISQ regime, but it is plausible that it represents the same primitive computational power of noisy NISQ circuits.

In brief, the extended argument against quantum computers is based on a paradox. On the one hand, you cannot achieve quantum supremacy without quantum error-correction. On the other hand, if you could build quantum error-correcting codes of the required quality then you could demonstrate quantum supremacy directly on smaller and simpler quantum systems that involve no quantum error-correction. To sum

> Noisy quantum systems that refrain from using error correction represent a low-level computational ability that will not allow building quantum error-correcting codes needed for quantum computation.

It is not easy to define quantum devices that "refrain from using error correction," but this property holds for NISQ circuits that simply do not have enough qubits to use quantum error correction, and appears to hold for all known experimental processes aimed at building high quality physical qubits including toplogical qubits. What we see next is that the low level complexity class of NISQ circuits does allow building basic forms of classical error-correcting codes.

18.3.6 Noise Stability, Noise Sensitivity, and Classical Computation

The study of noisy boson sampling relies on a study of Benjamini et al. (1999) of noise stability and noise sensitivity of Boolean functions. See Kalai (2018) where both these theories and their connections to statistical physics, voting rules, and computation are discussed. A crucial mathematical aspect of the theory is that the noise has a simple description in terms of Fourier expansion: it reduces "high frequency" coefficients exponentially and, therefore, noise stability reflects "low frequency" (low-degree) Fourier expansion.

The theory of noise stability and noise sensitivity posits a clear distinction between classical information and quantum information. The class **LDP** of probability distributions that can be approximated by low-degree polynomials does not support quantum supremacy and quantum error-correction but it still supports robust *classical* information, and with it also classical communication and computation. The ("majority") Boolean function, defined by $f(x_1, x_2, \ldots, x_n) = 1$ if more than half the variables have a value of 1, allows for very robust bits based on a large number of noisy bits or qubits and admits excellent low-degree approximations. It can be argued that *every form* of robust information, communication, and computation in nature is based on a rudimentary form of classical error-correction where information is encoded by repetition (or simple variants of repetition) and decoded by the majority function (or a simple variant of it). In addition to this rudimentary form of classical error-correction, we sometimes witness more sophisticated forms of classical error-correction.

18.3.7 The Extended Church–Turing Thesis Revisited

The extended Church–Turing thesis asserts (informally) that

(ECTT) It is impossible to build computing devices that demonstrate computational supremacy.

Here, computational supremacy is the ability to perform certain computations far beyond the power of digital computers.

A *classical computing device* (or a **P**-device) is a computing device for which we can show or have good reason to believe that it can be modeled by polynomial-size Boolean circuits.

We now formulate the weak extended Church–Turing thesis that reads

(WECTT) It is impossible to build *classical* computing devices that demonstrate computational supremacy.

WECTT seems almost tautological and it is in wide agreement. A popular example of an application of WECCT is to the human brain. As many argue, it is unlikely that quantum effects in the human brain lead to some sort of quantum computational supremacy. Another implication of WECCT is that the recent proposal by Johansson and Larsson (2017) to demonstrate Shor's factoring algorithm on certain classical analog devices cannot be realistic.

The argument given in this section asserts that for constant level of noise, NISQ devices are **P**-devices. Consequently, WECTT suffices to rule out quantum supremacy of NISQ devices. What makes this argument stronger is that we have good evidence that

1. Probability distributions described by NISQ devices for constant error rate are actually modeled by **LDP**, a computational class well below bounded-depth computation, which is well below **P**.
2. The outcomes of the computation process are chaotic for a wide range of subconstant levels of errors.

We argue that NISQ devices are, in fact, low-level classical computing devices, and I regard it as strong evidence that engineers will not be able to reach the level of noise required for quantum supremacy. Others argue that the existence of low-level simulation (in the NISQ regime) for every fixed level of noise does not have any bearing on the engineering question of reducing the level of noise. These sharply different views will be tested in the next few years.

In his comments on this paper, physicist and friend Aram Harrow wrote that my predictions are so sweeping that they say that hundreds of different research teams will all stop making progress. Indeed, my argument and predictions say that the goals of these groups (such as Goals 1–3 and other, more ambitious, goals) will not be achieved. Rather, if I am correct, these research teams will make important progress in our understanding of the inherent limitations of computation in nature, and the limitations of human ability to control quantum systems.

Aram Harrow also wrote: "The Extended Church–Turing Thesis (ECTT) is a major conjecture, implicating the work of far more than quantum computing. Long before Shor's algorithm it was observed that quantum systems are hard in practice to simulate because we do not know algorithms that beat the exponential scaling and work in all cases, or even all practical cases. Simulating the 3 or so quarks (and gluons) of a proton is at the edge of what current supercomputers can achieve despite decades of study, and going to larger nuclei (say atomic weight 6 or more) is still out of reach. If the extended Church–Turing thesis is actually true then there is a secret poly-time simulation scheme here that the lattice QCD community has overlooked. Ditto for chemistry where hundreds of millions of dollars are spent each year simulating molecules. Thus the ECTT is much stronger than just saying we cannot build quantum computers."

I agree with the spirit of Harrow's comment, if not with the specifics, and the comment can serve as a nice transition to the next section that discusses a few underlying principles and consequences that follow from the failure of quantum computing.

18.4 The Failure of Quantum Computers: Underlying Principles and Consequences

In this section we will propose some general underlying principles for noisy quantum processes that follow from the failure of quantum computers. (For more, see Kalai 2016b and Kalai 2018.)

18.4.1 Noise Stability of Low-Entropy States

Principle 1: Probability Distributions Described by Low-Entropy States Are Noise Stable and Can Be Expressed by Low-Degree Polynomials

The situation for probability distributions described by NISQ systems appears to be as follows: noise causes the terms, in a certain Fourier-like expansion, to be reduced exponentially with the degree. Noise-stable states, namely, those states for which the effect of the noise is small, have probability distributions expressed by low-degree terms. Probability distributions realistically (and robustly) obtained by NISQ devices represent a very low-level computational complexity class, **LDP**, the class of probability distributions that can be approximated by low-degree polynomials. In the absence of quantum fault-tolerance this description extends to (low-entropy) quantum states that arbitrary quantum circuits and (other quantum devices) can reach.

Noise stability and descriptions by low-degree polynomials might be relevant to some general aspects of quantum systems in nature. In the previous section we considered one such aspect: the feasibility of classical information and computation. The reason that our argument against quantum computation does not apply for classical computation is that noise stability still allows basic forms of classical error-correction. In fact, every form of robust information, communication, and computation in nature is based on some rudimentary form of classical error-correction that is supported by low-degree polynomials.

We will now consider two additional aspects: the possibility of efficiently learning quantum systems, and the ability of physical systems to reach ground states. Before we proceed I would like to make the following two remarks. The first remark is that the specific "Fourier-like expansion" required for the notions of noise stability and noise sensitivity is different for different situations. (The computational class **LDP** applies to all these situations.) An interesting direction for future research would be to extend the noise stability/noise sensitivity dichotomy and to find relevant Fourier-like expansions and definitions of noise for mathematical objects of high-energy physics. The second remark is that the study of noise stability is important in other scientific disciplines, and the assumption that systems are noise stable (aka "robust") may provide useful information for studying them. Robustness is an important ingredient in the study of chemical and biological networks; see, e.g., Barkai and Leibler (1997).

18.4.1.1 Learnability

It is a simple and important insight of machine learning that the approximate value of a low-degree polynomial can efficiently be learned from examples. This offers an explanation for our ability to understand natural processes and the parameters defining them. Rudimentary physical processes with a higher entropy rate may represent simple extensions of the class of **LDP** for which efficient learnability is still possible, and this is an interesting topic for further research.

18.4.1.2 Reaching Ground States

Reaching ground states is computationally hard (**NP**-hard) for classical systems, and even harder for quantum systems. So how does nature reach ground states so often? The common answer relies on two ingredients: the first is that physical systems operate in positive temperature rather than zero temperature, and the second is that nature often reaches meta-stable states rather than ground states. However, these explanations are incomplete: we have good theoretical reasons to think that, for general processes in positive temperature, reaching meta-stable states is computationally intractable as well. First, for general quantum or classical systems, reaching a meta-stable state can be just as computationally hard as reaching a ground state. Second, one of the biggest breakthroughs in computational complexity, the "PCP theorem," asserts (in physics language) that positive temperature offers no computational complexity relief for general (classical) systems.

Quantum evolutions and states approximated by low-degree polynomials represent severe computational restrictions and the results on the hardness of reaching ground states no longer apply. This may give a theoretical support to the natural phenomenon of easily reached ground states (Fig. 18.4).

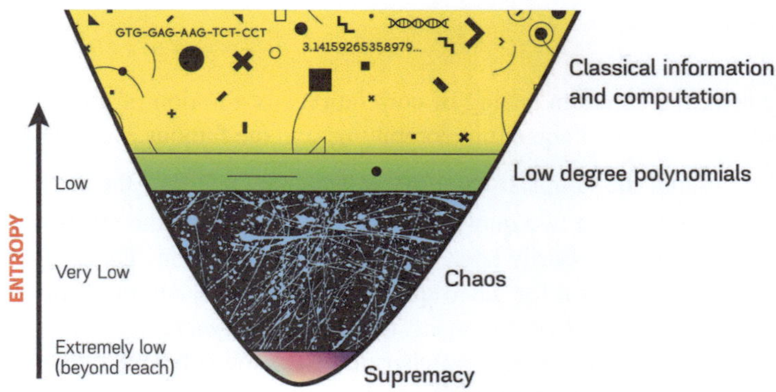

Fig. 18.4 Low-entropy quantum states give probability distributions described by low degree polynomials, and very low-entropy quantum states give chaotic behavior. Higher entropy enables classical information

18.4.2 Noise and Time-Dependent Evolutions

Principle 2: Time-Dependent (Local) Quantum Evolutions Are Inherently Noisy

It is an interesting challenge to give a satisfactory mathematical description of what "time dependent" means. We proposed in Kalai (2016a) to base a lower bound on the rate of noise on a measure of the non-commutativity of unitary operations. This measure of non-commutativity can also serve as an "internal" clock of the noisy quantum evolution. We also proposed in Kalai (2016a) a restricted class of noisy evolutions called "smoothed Lindblad evolutions" defined via a certain smoothing operation that mathematically expresses the idea that in the absence of quantum error-correction, noise accumulates. An important consequence of Principle 2 is that there are inherent lower bounds on the quality of qubits, or, more concretely, an inherent upper bound on the number of non-commuting Pauli operators you can apply to a qubit before it get destroyed. We note that lower bounds on the rate of noise in terms of a measure of non-commutativity arise also in Polterovich (2014) in the study of quantizations in symplectic geometry.

18.4.3 Noise and Correlation

Principle 3: Entangled Qubits Are Subject to Positively Correlated Noise

Entanglement, the quantum analog of correlations, is a notion of great importance in quantum physics and in quantum computing, and the famous *cat state* $\frac{1}{\sqrt{2}}\left|00\right\rangle + \frac{1}{\sqrt{2}}\left|11\right\rangle$ represents the simplest example of entanglement and the strongest form of entanglement between two qubits. Principle 3 asserts that the errors for the two qubits of a cat state necessarily have large positive correlation. This is an observed and accepted phenomenon for gated qubits and without quantum error-correction it is inherited to all pairs of entangled qubits. *Error synchronization* refers to a substantial probability that a large number of qubits, much beyond the average rate of noise, are corrupted. An important consequence of Principle 3 is that complicated states and evolutions lead to error-synchronization. Principle 3 leads to additional predictions about NISQ systems that can be tested for current experiments.

(e) Every pair of gated qubits will be subject to errors with large positive correlation.
(f) Every pair of qubits in a cat state will be subject to errors with large positive correlation.

(g) All experimental attempts to achieve Goals 2 and 3 will lead to a strong effect of error-synchronization.

We emphasize that our argument against quantum computers in Sect. 18.3 does not depend on Principle 3 and, moreover, the main trouble with quantum computers is that the noise rate of qubits and gates cannot be reduced toward the threshold for fault-tolerance. Correlation is part of this picture since, as is well known, the value of the threshold becomes smaller when there is a larger positive correlation for 2-qubit gated errors. We note (Kalai 2016a) that Prediction (f) implies that the quality of logical qubits via quantum error-correction is limited by the quality of gates. See also Chubb and Flammia (2018) for an interesting analysis of the effect of correlated errors on surface codes.

A reader may question how it is possible that entangled states will experience correlated noise, even long after they have been created and even if the constituent particles are far apart and not directly interacting. We emphasize that Principle 3 does not reflect any form of "non local" noise and that no "additional" correlated noise on far apart entangled qubits is ever assumed or conjectured. Principle 3 simply reflects two facts. The first fact is that errors for gated qubits are positively correlated from the time of creation. The second fact is that unless the error rate is small enough to allow quantum error-correction (and this may be precluded by the argument of the previous section), cat states created indirectly via a sequence of computational steps will inherit errors with large positive correlation.

Remark Error-synchronization is related to an interesting possibility (Kalai 2016b) that may challenge conventional wisdom and can also be checked for near term NISQ (and other) experiments. A central property of Hamiltonian models of noise in the study of quantum fault-tolerance (see, e.g., Preskill 2013), is that error fluctuations are sub-Gaussian. Namely, when there are N qubits, the standard deviation for the number of qubit errors behaves like $\sqrt{N}$ and the probability of more than $t\sqrt{N}$ errors decays as it does for Gaussian distribution. Many other models in mathematical physics have a similar property. However, it is possible that fluctuations for the total amount of errors for noisy quantum circuits and other quantum systems with interactions are actually super-Gaussian and perhaps even linear.

18.4.4 A Taste of Other Consequences

The failure of quantum computers and the infeasibility of high-quality quantum error-correction have several others far-reaching consequences imposing severe restrictions of humans' and nature's ability to control quantum states and evolutions. Following the tradition of using cats for quantum thought experiments, consider an ordinary living cat.

If quantum computation is not possible, it will be impossible to teleport the cat; it will be impossible to reverse time in the life of the cat; it will be impossible to

Fig. 18.5 Human imagination allows for a cat with an unusual geometry. Drawing by Netta Kasher (a tribute to Louis Wain)

implement the cat on a very different geometry; it will be impossible to superpose the lives of two distinct cats; and, finally, even if we place the cat in an isolated and monitored environment, the life of this cat cannot be predicted.

These restrictions will already be in place when our "cat" represents realistic (but involved) quantum evolutions described by quantum circuits in the small scale, namely, with less than 50 qubits (baby goals 1–3 can serve as good candidates for such "cats"). See Kalai (2016b) for a detailed discussion (Fig. 18.5).

18.5 Conclusion

> The intrinsic study of computation transcends human-made artifacts, and its expanding connections and interactions with all sciences, integrating computational modeling, algorithms, and complexity into theories of nature and society, marks a new scientific revolution! Avi Wigderson – *Mathematics and Computation* 2019.

Understanding noisy quantum systems and potentially even the failure of quantum computers is related to the fascinating mathematical theory of noise stability and noise sensitivity and its connections to the theory of computing. Our study shows how insights from the computational complexity of very low-level computational classes support the extended Church–Turing thesis. Exploring this avenue may have important implications for various areas of quantum physics. These connections between mathematics, computation, and physics are characterized by a rich array of conceptual and technical methods and points of view. In this study, we witness a delightfully thin gap between fundamental and philosophical issues on the one hand and practical and engineering aspects on the other.

18.6 Itamar

This paper is devoted to the memory of Itamar Pitowsky. Itamar was great. He was great in science and great in the humanities. He could think and work like a mathematician, and like a physicist, and like a philosopher, and like a philosopher of science, and probably in various additional ways. And he enjoyed the academic business greatly, and took it seriously, with humor. Itamar had an immense human wisdom and a modest, level-headed way of expressing it.

Itamar's scientific path interlaced with mine in many ways. Itamar made important contributions to the foundations of the theory of computation, probability theory, and quantum mechanics, all related to some of my own scientific endeavors and to this paper. Itamar's approach to the foundation of quantum mechanics was that quantum mechanics is a theory of non-commutative probability that (like classical probability) can be seen as a mathematical language for the other laws of physics. (My work, to a large extent, adopts this point of view. But, frankly, I do not quite understand the other points of view.) (Fig. 18.6)

In the late 1970s when Itamar and I were both graduate students I remember him enthusiastically sharing his thoughts on the Erdős–Turán problem with me. This is a mathematical conjecture that asserts that if we have a sequence of integers $0 < a_1 < a_2 < \cdots < a_n < \cdots$ such that the infinite series $\sum \frac{1}{a_n}$ diverges then we can find among the elements of the sequence an arithmetic progression of length k, for every k. Over the years both of us spent some time on this conjecture without much to show for it. This famous conjecture is still open even for $k = 3$. Some years later both Itamar and I became interested, for entirely different reasons, in remarkable geometric objects called cut polytopes. Cut polytopes are obtained by taking the convex hull of characteristic vectors of edges in all cuts of a graph. Cut polytopes arise naturally when you try to understand correlations in probability theory and Itamar wrote several fundamental papers studying them; see Pitowsky (1991). Cut polytopes came into play, in an unexpected way, in the work of Jeff Kahn and myself where we disproved Borsuk's conjecture (Kahn and Kalai 1993). Over the years, Itamar and I became interested in Arrow's impossibility theorem (Arrow 1950), which Itamar regarded as a major twentieth-century intellectual achievement. He gave a course centered around this theorem and years later so did I. The study of the noise stability and noise sensitivity of Boolean functions has close ties to Arrow's theorem.

The role of skepticism in science and how (and if) skepticism should be practiced is a fascinating issue. It is, of course, related to this paper that describes skepticism over quantum supremacy and quantum fault-tolerance, widely believed to be possible. Itamar and I had long discussions about skepticism in science, and about the nature and practice of scientific debates, both in general and in various specific cases. This common interest was well suited to the Center for the Study of Rationality at the Hebrew University that we were both members of, where we and other friends enjoyed many discussions, and, at times, heated debates.

A week before Itamar passed away, Itamar, Oron Shagrir, and I sat at our little CS cafeteria and talked about probability. Where does probability come from? What does probability mean? Does it just represent human uncertainty? Is it just an emerging mathematical concept that is convenient for modeling? Do matters change when we move from classical to quantum mechanics? When we move to quantum physics the notion of probability itself changes for sure, but is there a change in the interpretation of what probability is? A few people passed by and listened, and it felt like this was a direct continuation of conversations we had while we (Itamar and I; Oron is much younger) were students in the early 1970s. This was to be our last meeting.

Acknowledgements Work supported by ERC advanced grants 320924, & 834735, NSF grant DMS-1300120, and BSF grant 2006066. The author thanks Yuri Gurevich, Aram Harrow, and Greg Kuperberg for helpful comments, and Netta Kasher for drawing Figs. 18.1, 18.2, 18.3, 18.4, and 18.5.

Fig. 18.6 Itamar Pitowsky around 1985

References

Aaronson, S., & Arkhipov, A. (2013). The computational complexity of linear optics. *Theory of Computing, 4*, 143–252.

Aharonov, D., & Ben-Or, M. (1997). Fault-tolerant quantum computation with constant error. In *STOC'97* (pp. 176–188). New York: ACM.

Arrow, K. (1950). A difficulty in the theory of social welfare. *Journal of Political Economy, 58*, 328–346.

Arute, F., et al. (2019). Quantum supremacy using a programmable superconducting processor. *Nature, 574*, 505–510.

Barkai, N., & Leibler, S. (1997). Robustness in simple biochemical networks. *Nature, 387*, 913.

Benjamini, I., Kalai, G., & Schramm, O. (1999). Noise sensitivity of Boolean functions and applications to percolation. *Publications Mathématiques de l'Institut des Hautes Études Scientifiques, 90*, 5–43.

Chubb, C. T., & Flammia, S. T. Statistical mechanical models for quantum codes with correlated noise, arXiv:1809.10704.

Deutsch, D. (1985). Quantum theory, the Church–Turing principle and the universal quantum computer. *Proceedings of the Royal Society of London A, 400*, 96–117.

Feynman, R. P. (1982). Simulating physics with computers. *International Journal of Theoretical Physics, 21*, 467–488.

Johansson, N., & Larsson, J.-A. Realization of Shor's algorithm at room temperature, arXive:1706.03215.

Kahn, J., & Kalai, G. (1993). A counterexample to Borsuk's conjecture. *Bulletin of the American Mathematical Society, 29*, 60–62.

Kalai, G. (2016a). The quantum computer puzzle. *Notices of the American Mathematical Society, 63*, 508–516.

Kalai, G. (2016b). The quantum computer puzzle (expanded version), arXiv:1605.00992.

Kalai, G. (2018). Three puzzles on mathematics, computation and games. In *Proceedings of the International Congress of Mathematicians*, Rio de Janeiro (Vol. I, pp. 551–606).

Kalai, G., & Kindler, G. Gaussian noise sensitivity and BosonSampling, arXiv:1409.3093.

Kitaev, A. Y. (1997). Quantum error correction with imperfect gates. In *Quantum communication, computing, and measurement* (pp. 181–188). New York: Plenum Press.

Knill, E., Laflamme, R., & Zurek, W. H. (1998). Resilient quantum computation: Error models and thresholds. *Proceedings of the Royal Society of London A, 454*, 365–384.

McKay, B. D., & Radziszowski, S. P. (1995). R(4,5)=25. *Journal of Graph Theory, 19*, 309–322.

Pitowsky, I. (1990). The physical Church thesis and physical computational complexity. *Iyuun, A Jerusalem Philosophical Quarterly, 39*, 81–99.

Pitowsky, I. (1991). Correlation polytopes: Their geometry and complexity. *Mathematical Programming, A50*, 395–414.

Polterovich, L. (2014). Symplectic geometry of quantum noise. *Communications in Mathematical Physics, 327*, 481–519.

Preskill, J. (1998). Quantum computing: Pro and con. *Proceedings of the Royal Society of London A, 454*, 469–486.

Preskill, J. (2013). Sufficient condition on noise correlations for scalable quantum computing. *Quantum Information and Computing, 13*, 181–194.

Shor, P. W. (1995). Scheme for reducing decoherence in quantum computer memory. *Physical Review A, 52*, 2493–2496.

Shor, P. W. (1999). Polynomial-time algorithms for prime factorization and discrete logarithms on a quantum computer. *SIAM Review, 41*, 303–332. (Earlier version, *Proceedings of the 35th Annual Symposium on Foundations of Computer Science*, 1994.)

Steane, A. M. (1996). Error-correcting codes in quantum theory. *Physical Review Letters, 77*, 793–797.

Troyansky, L., & Tishby, N. (1996). Permanent uncertainty: On the quantum evaluation of the determinant and the permanent of a matrix. In *Proceedings of the 4th Workshop on Physics and Computation.*

Wigderson, A. (2019). *Mathematics and computation*. Princeton University Press. Princeton, New Jersey.

Wolfram, S. (1985). Undecidability and intractability in theoretical physics. *Physical Review Letters, 54*, 735–738.

Chapter 19
Why a Relativistic Quantum Mechanical World Must Be Indeterministic

Avi Levy and Meir Hemmo

Abstract We present the framework of Empirical Models and Hidden Variables Models in the foundations of Quantum Mechanics and its relation to Special Relativity. In this framework we discuss the definitions of properties such as non-signaling, strong and weak determinism, locality and, especially contextuality and show that (under mild assumptions) contextuality is equivalent to non-locality and to weak determinism, and non-contextuality is equivalent to locality and strong determinism. The central derivation of this paper is that a contextual Empirical Model cannot be accounted for by a Lorentz invariant (in the broad sense; see Sect. 19.1) and deterministic (strong or weak) theory. Hence, a relativistic theory which yields the empirical predictions of quantum mechanics *must be genuinely stochastic* (i.e., cannot be deterministic at its fundamental level). Finally, we discuss certain debates about the existence of relativistic versions of Bohm's and the GRW interpretations of quantum mechanics.

Keywords Quantum contextuality · Quantum probability · Weak-determinism · Strong-determinism · Locality · Special relativity

19.1 Introduction

It is well known, as demonstrated by Bell (1964), that quantum probability theory differs from classical probability theory (see for example Pitowsky 1986, 1989, 2007, for explicit arguments in this direction). In fact, there is a continuum of probability theories in which the classical non-contextual theory of probability (as described by Kolmogorov's 1933 axioms) stands at one end and highly contextual theories (e.g., the PR box; see Popescu and Rohrlich 1994) stand at the other end. The contextual quantum probability stands somewhere in between these two

A. Levy · M. Hemmo (✉)
Department of Philosophy, University of Haifa, Haifa, Israel
e-mail: levya@technion.ac.il; meir@research.haifa.ac.il

M. Hemmo, O. Shenker (eds.), *Quantum, Probability, Logic*, Jerusalem Studies in Philosophy and History of Science,
https://doi.org/10.1007/978-3-030-34316-3_19

ends.[1] One of the aims of this paper (see more details below) is to give a general characterization of contextuality in the context of Quantum Mechanics (QM) and its relation to other concepts such as locality and determinism.

Empirical Models (EM) (defined in Brandenburger and Yanofsky 2008; Abramsky and Brandenburger 2011) are probabilistic frameworks for analyzing the properties of measurement scenarios that arise in both classical and contextual experiments. One may view the EM framework as a general mathematical framework in which the full scope of probability theories can be analyzed. An EM consists of a set of admissible measurement sequences and the space of their corresponding outcomes together with a probability structure defined on these spaces. In simple terms, an EM is a "table" of measurements and their outcomes with the relative frequency at which these outcomes are observed.

The more commonly used term Hidden Variables Model (HVM) is an extension of the EM where a "state" space is appended to the measurement and outcome spaces. In a HVM the state of the measured system is often taken to provide an "explanation" for the frequencies of the measurement outcomes captured in the corresponding EM. The EM is the phenomenological description of the experimental results while the HVM describes a deeper structure that realizes the observed measurement outcomes and their relative frequencies as depicted by the EM. Since in general there may be several HVMs, that differ in their properties, all of which give rise to the same EM, one may conclude that the EM is the more fundamental layer of describing the physical reality in the sense that it gives the empirically compelling constraints on the HVM.

The EM and HVM are very general structures that are not restricted specifically to quantum experiments and can describe any reasonable experimental setting. As demonstrated beautifully in Brandenburger and Yanofsky (2008) (see also Abramsky and Brandenburger 2011; Brandenburger and Keisler 2016) the HVM framework allows defining, in an accurate mathematical way, interesting properties of physical systems such as Determinism (Strong and Weak), Locality, No-Signaling and Contextuality and deriving relations between these properties. (We use these terms now without explicit definitions; we shall give rigorous mathematical definitions as we proceed.)

Our first aim in this paper is to extend the definitions of the physical properties of Determinism, Locality and Non-Signaling from the HVM framework to the EM framework. Such extension is partially proposed in the above-mentioned references, but we would like to complete it here. If one looks only at the EM, the relations between these properties can be attributed to the actual "physical reality" rather than to a specific HVM theory. Indeed, this is exactly the way in which Bell's Theorem (1964) is understood (see e.g. Pitowsky 1986; Albert 1992; Maudlin 1994), namely that any violation of the Bell inequality (given Bell's assumptions) entails that the

[1]The "degree" of contextuality can be measured by the value of certain bounds on inequalities pertaining to the probability distribution that are violated for contextual (non-classical) probability theories.

physical reality itself is nonlocal. We shall define the above properties in terms of EM, examine their relations and compare them to the corresponding properties and relations of the HVM.

The second aim of this paper is to characterize the contextuality property which as we will show is relevant to EM only. There seems to be some confusion in the literature about how precisely to understand contextuality and we would like to clarify it. This discussion will lead us to propose that under reasonable assumptions Non-Contextuality is in a sense equivalent to Locality and to Strong Determinism, and Contextuality is equivalent to Non-Locality and to Weak (that is, not Strong) Determinism.

Our third aim is to consider the EM and HVM framework in the context of Special Relativity (SR). The No-Signaling property is an implicit requirement stemming from the principles of SR, which implies that no physical signal can be transmitted in a superluminal speed. Here we shall add explicitly the postulate according to which a HVM compatible with SR must be *genuinely* Lorentz invariant, where we understand this last requirement in a *broad* sense as follows: the laws of physics should be the same in every Lorentz frame of reference, and moreover there is no (absolute) temporal order (and therefore no preferred Lorentz reference frame) that fixes the temporal direction of the cause and effect[2] relation. We shall take it that these conditions should hold in any relativistic theory *regardless* of whether or not a violation of them may be empirically detected. We shall call this condition from now on *Lorentz invariance*. We will show (in Sect. 19.5) that requiring Lorentz invariance in this sense implies the No-Signaling property. In fact, assuming Lorentz invariance, we will prove the following general result:

> A contextual Empirical Model does not have a Strongly Deterministic (Local) or a Weakly Deterministic (Non-Local) Hidden Variable Model.

Since the quantum mechanical probabilities of measurement outcomes that have been confirmed by actual experiments match a *contextual* empirical model (i.e. the quantum probabilities are non-classical; see Sect. 19.4), our proof of the above proposition implies that under these assumptions (including our additional requirement of the existence of *no* preferred Lorentz frame) *the physical reality must be genuinely stochastic*. In other words, God must play dice in a relativistic quantum world.

Bell (1964) was the first to prove that the EPR setting cannot be realized by a local and (strongly) deterministic HVM. Results pertaining to non-local (weak) determinism have been proposed in certain other contexts, e.g., Hardy and Squires (1992), Clifton et al. (1992), Conway and Kochen (2006, 2009), Gisin (2011). But our theorem above is obtained in a very general context which is explicitly independent of quantum theory, and which is constrained only by the contextuality of measurement outcomes.

---

[2]Regardless of how precisely should one understand the relation of causation.

The paper is organized as follows. In Sect. 19.2 we briefly give the basic definitions, examples and the main results in the EM and HVM framework. In Sect. 19.3 we present the definitions of the physical properties mentioned above and their logical relations in the EM framework. In Sect. 19.4 we analyze the property of contextuality. In Sect. 19.5 we extend the discussion of EM and HVM to the framework of Special Relativity and we prove our general theorem, namely that there is no deterministic theory that is consistent with both the predictions of quantum mechanics and with the relativistic postulate of no preferred Lorentz frame of reference. In Sect. 19.6 we briefly draw some implications of our proof with regard to current interpretations of quantum mechanics. Section 19.7 is the conclusion.

19.2 The Hidden Variable Model (HVM) Framework

This section is rather technical and uses probability theory for the definitions part. We put it forward for the sake of completeness of the paper. The attached examples and the results pertaining to the relations between the properties can be understood without going into the formal details of all the definitions. Unless otherwise stated, all the definitions and results, but not the examples, are based on Brandenburger and Yanofsky (2008), Abramsky and Brandenburger (2011) and Brandenburger and Keisler (2016). However, our definitions of Empirical Model and Hidden Variable Model are closer to those in Abramsky and Brandenburger (2011) than to the definitions in the other two references. Hence there are certain differences in the main formulations which we shall discuss at the end of Sect. 19.2.2. When appropriate we add explicitly in comments the physical motivation and interpretation of the formal definitions, as we understand them.

19.2.1 Basic Definitions

Definition: Empirical Model (EM) Let X be a finite set of *Measurements* $\{A, B, C, \dots\}$ and let M be a *Measurement Context* i.e. a set of subsets of X that constitutes "admissible series" of measurements.

For example, if $X = \{A, B, C\}$ then M may be $\{(A, B), (B, C)\}$ where (A, C) and (A, B, C) are not included in M.

Remark the "admissible series" of measurement concept originated from QM where certain series of measurements may be *inadmissible* as they cannot be performed simultaneously, since the corresponding operators do not commute.

For each measurement $x \in X$ there exists a finite outcome set O_x and hence for each admissible series $m = (m_1, m_2, \dots) \in M$, where $m_i \in X$, there is an outcome space $O_m = O_{m_1} \times O_{m_2} \times \ldots$. For example, consider the experiment of tossing two dice independently from a set of 3 dice, then the set of measurements may be denoted $X = \{A, B, C\}$ and for each measurement the outcome set is $O_x = O = \{1, 2, 3, 4, 5, 6\}$.

Let the physical situation be such that it is possible to toss only two dice simultaneously: A, B or A, C together, but not B, C.

Then the *measurement context* is $M = \{(A, B), (A, C)\}$ and $O_m = \{1, \ldots, 6\} \times \{1, \ldots, 6\}$ for both members of M.

The *Experiments Space E* is the set of all admissible measurement series together with the sets of all their possible outcomes $E = \{E_m = \{(m, O_m)\} : m \in M\}$.

An *Empirical Model* is the experiments space E together with a family of probability distributions Q_m on the set of possible outcomes of each admissible series $m \in M$.

In the example of the dice, given that all dice are fair, the experiments space E is the union of two disjoint subsets:

$E_{AB} = \{(a, b) : a \in \{1, \ldots, 6\}, b \in \{1, \ldots, 6\}\}$ and $E_{AC} = \{(b, c) : a \in \{1, \ldots, 6\}, c \in \{1, \ldots, 6\}\}$.

And the probabilities family is correspondingly.

$Q_{AB} = \left\{Q(a, b) = {}^{1}\!/_{36}\right\}$, $Q_{AC} = \left\{Q(a, c) = {}^{1}\!/_{36}\right\}$ for $a \in \{1, \ldots, 6\}, b \in \{1, \ldots, 6\}, c \in \{1, \ldots, 6\}$.

Definition: Hidden Variable Model (HVM) A S*tate Space* is a set Λ whose elements are denoted by λ.

A *hidden variable model* is a pair $(V, \{P_m\})$ where $V = E \times \Lambda$ is the product of an Experiments Space E with a State Space Λ, and the product is equipped with a family of probability distributions P_m on the pairs $(E_m \times \Lambda)$.

Definition: A Consistent HVM An HVM $(E \times \Lambda, \{P_m\})$ is said to be consistent with the empirical model $(E, \{Q_m\})$ if $Q_{A,B,\ldots}(a, b, \ldots) = P_{A,B,\ldots}\{(a, b, \ldots) \times \Lambda\}$ for every admissible series of measurements $(A, B, \ldots) \in M$ and every outcome $(a, b, \ldots) \in O_{A,B,\ldots}$.

That is, a given HVM is consistent with the EM if one can describe the probabilities of the measurements outcomes by summing over all the possible system states λ in the model.

For example, a consistent HVM for the EM presented before of tossing two dice can be constructed by defining the state space of the three dice as the set of all three-tuples $\lambda = (\alpha, \beta, \gamma)$, where $\alpha, \beta, \gamma \in \{1, 2, 3, 4, 5, 6\}$.

The probability distribution that ensures the consistency is given by setting $p(\lambda) = \frac{1}{6^3}$ uniformly and defining:

$P_{A,B}(a, b, \lambda) = P_{A,B}(a, b|\,\lambda)p(\lambda)$, $P_{A,C}(a, c, \lambda) = P_{A,C}(a, c|\,\lambda)p(\lambda)$ where $P_{A,B}(a, b|\,(\alpha, \beta, \gamma)) = \delta(a = \alpha)\delta(b = \beta)$, $P_{A,C}(a, c|\,(\alpha, \beta, \gamma)) = \delta(a{=}\alpha)\delta(c{=}\gamma)$.

This is because, as we have seen, $Q_{A,B}(a,b) = \frac{1}{36}$ and direct computation yields $\sum_{\lambda\in\Lambda} P_{A,B}\{(a,b)\times\Lambda\} = \sum_{\alpha,\beta,\gamma} P_{A,B}(a,b|(\alpha,\beta,\gamma))\,p(\alpha,\beta,\gamma) = \sum_{\alpha,\beta,\gamma} \delta(a=\alpha)\,\delta(b=\beta)\frac{1}{6^3} = \frac{1}{6^2}$.

(and similarly for $Q_{AC}(a,c) = {}^{1}\!/_{36}$).

Comments

- Given an EM $(E, \{Q_m\})$, the HVM defined by $(E \times \Lambda, \{P_m\})$, where Λ is a singleton state space $\Lambda = \{\lambda\}$, and $P_{A,\,B,\,\ldots}(a, b, \ldots, \lambda) = Q_{A,\,B,\,\ldots}(a, b, \ldots)$ is consistent with $(E, \{Q_m\})$. Hence every EM has a "trivial HVM" that is consistent with it.
- Given an HVM $(E \times \Lambda, \{P_m\})$, one can define an EM $(E, \{Q_m\})$ where the probability measures are given by $Q_{A,B,\ldots}(a, b, \ldots) = \sum_{\lambda\in\Lambda} P_{A,B,\ldots}(a, b, \ldots, \lambda)$. Then, by definition, the given HVM is consistent with this "derived" EM.

19.2.2 Definitions of Properties of Hidden Variable Models (HVM)

Strong Determinism (SD)
An HVM is *Strongly Deterministic* if the state of the system determines uniquely the outcome of every measurement.

More formally, for every state λ the probability of observing each outcome for every measurement is either 0 or 1: for every $A \in M, \lambda \in \Lambda\ \ \exists\, a$ s. t. $P_A(a|\,\lambda) = 1$.

Weak Determinism (WD)
An HVM is *Weakly Deterministic* if the state of the system determines uniquely the outcomes of every admissible series of measurements.

Formally, for every state λ and for every admissible measurement series A, B, C, ... the probability of observing, a certain set of outcomes is either 0 or 1, i.e.

$$\exists\,(a, b, c, \ldots) \text{ s.t.} P_{A,B,C,\ldots}(a, b, c, \ldots|\lambda) = 1.$$

Note that in Strong Determinism the determinism applies to every measurement individually, but in Weak Determinism it applies only to admissible *sequences* of measurements. Hence SD implies WD but the converse is not true as can be seen from the following examples.

Example 2.1 Assume a HVM with three measurements A, B, C each with a binary outcome, two admissible measurement series (A, B), (A, C) and a two states $\lambda = 0$, $\lambda = 1$ with equal probabilities.

The HVM depicted by the following table

	0, 0	0, 1	1, 0	1, 1
$A, B \mid \lambda = 0$	1	0	0	0
$A, C \mid \lambda = 0$	1	0	0	0
$A, B \mid \lambda = 1$	0	0	0	1
$A, C \mid \lambda = 1$	0	0	0	1

is Strongly Deterministic since the state $\lambda = 0$ implies $a = b = c = 0$ and $\lambda = 1$ implies $a = b = c = 1$.

Example 2.2 However, a similar setting with the table:

	0, 0	0, 1	1, 0	1, 1
$A, B \mid \lambda = 0$	0	1	0	0
$A, C \mid \lambda = 0$	0	0	0	1
$A, B \mid \lambda = 1$	0	0	1	0
$A, C \mid \lambda = 1$	1	0	0	0

is Weakly Deterministic since given the state λ the outcome of the two admissible measurements is uniquely determined, but it is not Strongly Deterministic since given $\lambda = 0$ (and also $\lambda = 1$) the outcome of the measurement of A can be either 0 or 1 depending on whether it was measured together with B or with C. This counter intuitive property which characterizes certain quantum systems is called contextuality and will be investigated later in great details.

No-Signaling (Parameter Independence)[3] (NS)
An HVM is *No-Signaling* if for every state λ, the outcome of a measurement does not depend on the other measurements performed simultaneously with it.

Formally: for every λ: $P_{A,B,\ldots}(a|\lambda) = P_{A,B',\ldots}(a|\lambda) = P_A(a|\lambda)$.

The No-Signaling condition is associated with Special Relativity, since a HVM that fails NS may allow superluminal signaling if the measurements are performed in space-like separated locations.

Example 2.1 is No-Signaling but Example 2.2 is not as indicated before.

---

[3]The term ‘Parameter Independence’ is more common in the literature, but due to our focus on the relation between EM and SR we prefer the term ‘No-Signaling’, which emphasizes the physical implication of this property. Also note that we use the term ‘signaling’ in a *neutral* way, disregarding completely whether or not the signals can be noticed or readout, unless we say otherwise. See Maudlin (1994) for a broader discussion of signaling.

Comment

Even when an HVM does not satisfy the No-Signaling property, it may still be impossible to transmit superluminal signals. This is the case when it is not feasible[4] to prepare the system at those states where violation of the NS property takes place. Bohm's (1952) theory is an example for this situation (see discussions of this point and its implications in Pitowsky 1989; Dürr, Goldstein and Zanghi 1992; Albert 1992; Maudlin 1994; Hemmo and Shenker 2013).

Outcome Independence (OI)

An HVM is *Outcome independent* if for every state λ, the outcome of a measurement does not depend on the outcomes of other measurements performed simultaneously with it.

Formally: $P_{A,\,B,\,C,\,\ldots}(a, b, c, \ldots \mid \lambda) = P_{A,\,B,\,C,\,\ldots}(A = a \mid \lambda) P_{A,\,B,\,C,\,\ldots}$ $(B = b \mid \lambda) P_{A,\,B,\,C,\,\ldots}(C = c \mid \lambda) \cdots$, for every λ.

Example 2.3 In tossing two independent binary coins with measurements A, B respectively a state parameter $\lambda \in \{1, 2, 3, 4\}$ characterized the 4 possible outcomes. Let $P(\lambda) = {}^{1}/_{4}$ for every λ then the HVM described by the following table satisfied the Outcome Independence property:

	0, 0	0, 1	1, 0	1, 1
$A, B \mid \lambda = 0$	1	0	0	0
$A, B \mid \lambda = 1$	0	1	0	0
$A, B \mid \lambda = 2$	0	0	1	0
$A, B \mid \lambda = 3$	0	0	0	1

Example 2.4 The following "two balls in two jars" scenario is an example of HVM that is not Outcome-Independence.

A red ball and a blue ball are placed in two jars with equal probabilities. The measurement of A detects the color of the ball in the first jar and B in the second jar. Assume a singleton state space Λ and a probability table defined by:

	(r, r)	(r, b)	(b, r)	(b, b)
(A, B)	0	${}^{1}/_{2}$	${}^{1}/_{2}$	0

It is clearly No-Signaling but not Outcome Independence since the outcomes of the two measurements are dependent. Note that for the derived EM (which is

[4]We don't address here the question of why in a given theory the preparation of a system in a certain state λ is not feasible.

identical in this case to the HVM since Λ is a singleton) there is a consistent HVM (with two states of equal probabilities) that is No-Signaling and Outcome Independence which is given by

	(r, r)	(r, b)	(b, r)	(b, b)
$(A, B) \mid \lambda = 0$	0	1	0	0
$(A, B) \mid \lambda = 1$	0	0	1	0

Locality

The locality property is the conjunction of No-Signaling and Outcome Independence properties, as defined by Bell (1964):

$$P_{A,B,C\ldots}(a, b, c, \ldots \mid \lambda) = P_A(a \mid \lambda) P_B(b \mid \lambda) P_C(c \mid \lambda) \cdots \quad \text{for every } \lambda.$$

λ – Independence (LI)

Brandenburger and Yanofsky (2008) define an additional property of HVM, which they call λ-Independence (LI). This property entails that the state of the measured system as determined by the value of the hidden variable is independent of the measurement performed on the system and vice versa. However, in our settings of HVM this definition is meaningless since we did not assume a probabilistic structure on the admissible series of measurements. In fact, we tacitly assumed that there can be *no* dependence between the set of admissible measurements and the state of the measured system. Assuming the converse means that there is some sort of a "universal conspiracy" that correlates between the choices of the experimenter and the state of the measured system. For this reason, the λ-Independence property is sometimes associated with lack of 'free will' (we don't address this issue here). As is well known, Bell (1964) has assumed λ-Independence. Although retro-causal approaches to QM reject λ-Independence and admit correlations between the choices of the experimenter and the state of the measured system (see e.g. Price and Wharton 2013), we shall assume, as we said above, the more commonly accepted assumption that λ-Independence holds.

19.2.3 Existence Theorems and Relations Between HVM Properties

The following theorems and propositions are stated without their proofs. The interested reader can find the proofs in Brandenburger and Yanofsky (2008) and Brandenburger and Keisler (2016).

Existence Theorem
For every Empirical Model there exists a consistent Weakly Deterministic Hidden Variable Model.

Note that this theorem does not guarantee that there is a consistent Strongly Deterministic HVM. The most remarkable example for a Weakly Deterministic HVM theory is Bohm's interpretation of QM. As we shall see later it's Weakly Deterministic (but not SD) nature is intimately related to the fact that the values assigned to all quantum mechanical observables other than position are contextual. It follows from the above existence theorem that the Weakly Deterministic property of Bohm's theory is not an exception. However, it is still remarkable of course that Bohm's theory provides dynamics for the hidden variables that solves the measurement problem in QM in a way that is compatible with the Schrödinger dynamics of the quantum wavefunction, while this existence theorem refers only to measurement outcomes without mentioning dynamics at all.

However, the generality of this theorem may come as a surprise since in the history of QM it took a few decades to realize that von Neumann's "no-go proof," according to which hidden variables theories consistent with the predictions of QM do not exist, presupposed that the hidden variables have pre-measurement values that can be measured as quantum observables (see Bub (2010) for a discussion of this point). In the terminology of the EM and HVM framework, von Neumann's assumption is equivalent to requiring that the HVM satisfies Strong Determinism, and indeed such an HVM does *not* exist for a contextual Empirical Model as we shall explain later.

Comment on Strongly Deterministic Explanation
Brandenburger and Yanofsky (2008) consider another existence theorem:

Existence Theorem
For every Empirical Model there exists a consistent Strongly Deterministic HVM (but not necessarily λ- independent).

Note that although this theorem guarantees that every Empirical Model has a Strongly Deterministic HVM, this comes with the price that this model may not satisfy λ – Independence (described at the end of Sect. 19.2.2). As we said, in our setting we presuppose independence between the admissible measurements set and the state of the measured system. Hence, in this setting λ – Independence is guaranteed and therefore a Strongly Deterministic explanation may fail. We will see in Sect. 19.4 that if the Empirical Model is classical (non-contextual; see below) then it does have a consistent Strongly Deterministic HVM even in our setting (i.e. an HVM that satisfies λ – Independence), but if it is contextual then a consistent Strongly Deterministic HVM does not exist.

Relations Between HVM Properties ('$\rightarrow$' Means Material Implication)
Let us state the following relations between properties of HVM without explicit proofs:

1. Strong Determinism → Weak Determinism → Outcome Independence
2. Strong Determinism → No signaling, hence
3. Strong Determinism → Locality (= Outcome Independence & No signaling)
4. Strong determinism ↔ Weak Determinism and No signaling, hence
5. Strong determinism ↔ Weak Determinism and Locality

Example 2.5 Consider two spin-half particles denoted 1, 2 that are in the singlet state $\psi = {}^{1}\big/_{\sqrt{2}}\,(|\uparrow\rangle_1|\downarrow\rangle_2 + |\downarrow\rangle_1|\uparrow\rangle_2)$.

Let A, A' denote the spin measurements of particle 1 in two different directions and B, B' the spin measurements of particle 2 in the same two directions. The outcomes of the measurements are $a \in \{\pm 1\}$ for A, A' and $b \in \{\pm 1\}$ for B, B'. Finally, let the state space be $\Lambda = \{0, 1\}$ with uniform distribution and define the measurement outcomes deterministically by the table below.

One can notice that this model is Weakly Deterministic since for every (λ, m_1, m_2) the tuple (a, b) is determined uniquely, but it is not Strongly Deterministic as for $(\lambda = 0, m_1 = A, m_2 = B)$, $b = -1$ but for $(\lambda = 0, m_1 = A', m_2 = B)$, $b = 1$ and hence the outcome of measurement B is not determined by the state alone. The model is Outcome Independent but is not No-Signaling which can be verified directly or concluded from relations (1), (4) above, and therefore this model is not local.

λ	m_1	m_2	a	b
0	A	B	1	−1
1	A	B	−1	1
0	A'	B	1	1
1	A'	B	−1	−1
0	A	B'	1	1
1	A	B'	−1	−1
0	A'	B'	1	−1
1	A'	B'	−1	1

19.3 Properties of Empirical Models and Their Relations

The properties of Outcome Independence (OI), No-Signaling (NS), Locality, Strong Determinism (SD) and Weak Determinism (WD) for HVM have been defined and analyzed in Brandenburger and Yanofsky (2008), Brandenburger and Keisler (2016) and Abramsky (2013).

For EM however, No-Signaling is the only property defined (mistakenly as non-contextuality) in Brandenburger and Yanofsky (2008), correctly in Abramsky (2013) and more generally in Abramsky and Brandenburger (2011).[5]

An EM satisfies No-Signaling if the marginal distribution pertaining to a sub-sequence of an admissible sequence of measurements does not depend on the rest of the measurements in the sequence. More formally, for a sub-sequence of one measurement:

For every admissible measurement series $A, B, C, \ldots$ and $A, B', C', \ldots$, and every $a \in O_A$

$$Q_{A,B,C,\ldots}(a) = Q_{A,B',C',\ldots}(a) = Q_A(a).$$

We propose extending the definitions of all the HVM properties to Empirical Models as follows. An Empirical Model has a certain property just in case there exists a consistent HVM with this property. This extension is desirable since, as we said in the Introduction, we wish to formulate the abovementioned properties in a way that refers only to the phenomena and does not depend on a given theory.

Firstly let us show that our definition of No-Signaling is equivalent to the above given definition:

By proposition 2.2 from Branderburger and Yanofsky (2008) if a HVM satisfies No-Signaling and λ – independence then every equivalent EM satisfies No-Signalimg. In the other direction, if an EM is No-Signaling, then the trivial (i.e. with singleton state space) equivalent HVM is No-Signaling by definition.

We now investigate the relations between the EM properties and the way they differ from the HVM properties. Due to the existence theorem every Empirical Model is either Strongly Deterministic or else it is Weakly Deterministic but not Strongly Deterministic. Let us denote the second option (i.e. a WD model that is not SD) by WD_*. The significance of this distinction will be investigated in Sect. 19.4, which is devoted to the concept of contextuality. In contrast to the HVM case, an EM that fails to satisfy No-Signaling allows for superluminal signaling (for measurements performed at space-like separated locations) that can be empirically carried out. Hence, to comply with Special Relativity an EM must be No-Signaling, while it may have a consistent HVM that is not NS. We will show later that if an Empirical Model is WD_* it *must* have a consistent HVM that does not satisfy No-Signaling.

Example 3.1 The EM derived from the HVM in Example 2.2 in Sect. (19.2) satisfies No-Signaling, since the trivial HVM (singleton Λ) is NS by observing that$P_{A,\,B}(a) = P_{A,\,C}(a) = P_A(a)$ for $a \in \{0, 1\}$.

[5]In this reference a notion of compatibility between marginal probability distributions is defined as a generalization of the No-Signaling property which refers to one measurement marginal probability only.

$a, b =$	0, 0	0, 1	1, 0	1, 1
A, B	0	$^1/_2$	$^1/_2$	0
A, C	$^1/_2$	0	0	$^1/_2$

However, the original HVM defined in this Example (2.2) is not No-Signaling (since it is WD_*). This example demonstrates the fact that the phenomenal level EM may have properties that are different from a specific physical HVM that realizes it.

By contrast to the HVM case, the conjunction of Outcome Independence and No-Signaling for EM is not equivalent to the Locality property. This is due to the fact that there may be a consistent HVM that is Outcome Independence and another consistent HVM that is No-Signaling but there is no single consistent HVM that satisfies both OI and NS. Example 3.1 above demonstrate this situation since it has a Weakly Deterministic consistent HVM that is Outcome Independence (by relation (1)) and the trivial HVM is No-Signaling but it is not local (i.e. does not have a local HVM) since it is not classical. We will discuss this last point in Sect. 19.4.

Relations Between Properties of an Empirical Model

The first three relations pertaining to the properties of HVM properties (see Sect. 19.2) are satisfied also by EM:

1. Strong Determinism → Weak Determinism → Outcome Independence
2. Strong Determinism → No signaling, hence
3. Strong Determinism → Locality

However, relation (4): Weak Determinism and No signaling → Strong Determinism, fails because (as we mentioned above) there can be a consistent HVM which is WD and *another* consistent HVM which is No signaling but there is no *single* consistent HVM with both properties. This situation is demonstrated by the following example:

Example 3.2 Let there be four measurements A, A', B, B' each with outcomes in {0, 1} and the following conditional probability table:

	(0, 0)	(0, 1)	(1, 0)	(1, 1)
$A\ B$	$^1/_2$			$^1/_2$
$A'\ B$	$^1/_2$			$^1/_2$
$A\ B'$	$^1/_2$			$^1/_2$
$A'\ B'$		$^1/_2$	$^1/_2$	

This EM is No-Signaling (the trivial HVM is NS) and it is Weakly Deterministic, since every EM has a WD consistent HVM, but it is not Strongly Deterministic as we explain in the next section.

Relation (5): Strong Determinism $\nleftrightarrow$ Weak Determinism and Locality, has in fact a stronger version for Empirical Models. For an EM, Strong Determinism is equivalent to Locality, a result that is proved in Abramsky and Brandenburger (2011, Theorem 8.1). We discuss this result in Sect. 19.4.

Finally, note that if an HVM is No-Signaling then the derived (consistent) EM is also No-Signaling. This is shown by Brandenburger and Yanofsky (2008).

19.4 Contextuality

Non-Contextuality is defined by Peres (1995) as the property that "the outcome of experiments performed at a given location in space be independent of the arbitrary choice of other experiments that can be performed, simultaneously, at distant locations" (p. 187).

A common schematic example for contextuality (see e.g. Peres 1995) is the following. Consider a measurement of an observable A (represented by a Hermitian operator) that is performed simultaneously with a set of observables $B_1, B_2, \ldots$, all of which pairwise commute with A. Then, consider another measurement of A simultaneously with another set $C_1, C_2, \ldots$ that also commute pairwise with A. Suppose that the operators $B_1, B_2, \ldots$ do not commute with $C_1, C_2, \ldots$ This means that the outcome of the measurement of A may depend on the (measurement) context in which it is performed.

The subtle point in this definition and example is that if 'independence' is understood as "stochastic independence" (by which we mean that the distribution of the probabilities of the outcomes depends on the measurement context), then contextuality will allow superluminal signaling, which we assume here is impossible. This point has been noted by, for example Badzіąg et al. (2009) and this prompted them to distinguish between what they call "contextuality of outcomes" and "contextuality of probabilities."

In the spirit of the above definition, Brandenburger and Yanofsky (2008) define a non-contextual Empirical Model as one for which the probability of obtaining a particular outcome of a measurement does not depend on the other measurements performed. More formally:

An EM satisfies "*non-contextuality*" if for all $a, A, B, B', C, C', \ldots$, $Q_{A,B,C,\ldots}(a) = Q_{A,B',C',\ldots}(a) = Q_A(a)$.

Here we point out that this definition is *identical* to our definition of the No-Signaling property for EM, given in Sect. 19.3 (it is also identical to the definition of NS in Abramsky and Brandenburger 2011). Since No-Signaling must be satisfied for every EM in order to be compatible with Special Relativity, using the contextuality of probabilities property may not be a useful definition for differentiating "classical" scenarios from quantum mechanical ones.

For this reason, we propose that non-contextuality should mean something similar to the idea of "non-contextuality of outcomes" (as in Badzią̧g et al. 2009). Here the idea is that the outcome of the measurement of A should be independent of which additional measurements, $B_1, B_2, \ldots$ or $C_1, C_2, \ldots$, are performed with it. However, this intuition cannot be directly captured by the EM framework in which one only has *probabilities* for outcomes, not individual outcomes. This intuition leads us to join Abramsky and Brandenburger (2011) and suggest that non-contextuality for an Empirical Model is satisfied if there exists a consistent "classical" HVM, which in our terminology means that it is Strongly Deterministic. We argue that this is a useful definition since it sharply distinguishes between a "classical" and a "quantum" EM (see below, for explicit definitions of these notions), while No-Signaling is common to both.

Before presenting our definition of non-contextuality for EM we comment that defining a similar property for HVM is superfluous since it will coincide with previously defined terms. In the context of HVM, No-Signaling is equivalent to non-contextuality of probabilities and Strong Determinism is equivalent to non-contextuality of outcomes.

Characterization of a Classical Empirical Model

An EM is *classical* if its probability model is classical, i.e. it belongs to the convex hull of the extreme probability distributions (those with zero or one probabilities only).[6]

Therefore, it is possible to write:

$$Q_{A,B,C,\ldots}(a,b,c,\ldots) = \sum_{\lambda} \delta_{A,B,C,\ldots,\lambda}(a,b,c,\ldots)\,\mu(\lambda),$$

where $\delta_{A,B,C,\ldots,\lambda}$ are the extreme probability distribution indexed by a parameter $\lambda \in \Lambda$ for some finite set Λ and $\sum_{\lambda} \mu(\lambda) = 1$. Hence, the sum in the right-hand side of this formula is a convex combination and $\mu(\lambda)$ constitutes a probability measure on the set of indices.

This characterization means that there exists an HVM, consistent with the given EM, such that the state space of the hidden variables consists precisely of the parameter set Λ and it is Strongly Deterministic. The reason is that the $\delta_{A,B,C,\ldots,\lambda}$ are extreme probabilities and hence can be written as $\delta_{A,B,C,\ldots,\lambda}(a,b,c,\ldots) = \prod_{\substack{x=a,b,c,\ldots \\ X=A,B,C,\ldots}} \delta_X(x|\lambda)$, that is, every separate measurement has exactly a *single* outcome that has probability 1.

[6]See Pitowsky (1986) for characterizations of classical and quantum probability spaces as convex subsets of the space of phenomenal distributions.

We now turn to the main definition of this section.

Definition: Non-Contextual Empirical Model An EM is non-*contextual* if and only if it has a consistent Strongly Deterministic HVM, i.e.:

$$P_{A,B,C\ldots}(a,b,c,\ldots,\lambda) = P(\lambda) \prod_{\substack{x=a,b,c,\ldots \\ X=A,B,C,\ldots}} \delta_X(x|\lambda) \text{ and}$$

$$Q_{A,B,C\ldots}(a,b,c,\ldots) = \sum_{\lambda} P(\lambda) \prod_{\substack{x=a,b,c,\ldots \\ X=A,B,C,\ldots}} \delta_X(x|\lambda)$$

Now, in Abramsky and Brandenburger (2011, Theorem 8.1), and earlier in Fine (1982) it is proved that an EM has a consistent Local (factorizable) HVM if and only if it has a Strongly Deterministic HVM. Hence, we conclude that for an Empirical Model the three properties: Strong Determinism, Locality and Non-Contextuality are *equivalent* properties.

Therefore, there are only two possible scenarios for an Empirical Model:
It is either

- Classical (=Local = Strongly Deterministic = Non-Contextual), i.e., it has a Strongly Deterministic consistent HVM.

Or else:

- It is Non-Classical (=Non-Local = Contextual), i.e., it has a Weakly Deterministic consistent HVM, but not a Strongly Deterministic one.

It seems to us that this result in a sense vindicates the arguments in Pitowsky (2007) and Bub and Pitowsky (2010) that the essential character of quantum mechanics is fixed by the non-classical, that is, the contextual nature of the quantum probability.

Before concluding this section, we would like to mention the interesting question of how to decide if a given EM is contextual or non-contextual. This question can be formulated equivalently by asking if there exists a common "classical" probability distribution Q for the outcomes of the entire set of measurements $\{A, B, C, \ldots\}$ such that the family of probability distributions Q_m defined each on one admissible series of measurements are marginal distributions of Q.

It turns out that for a finite EM this question can be answered by constructing a system of linear equations on the probability distributions that characterized the model in such a way that it has a solution if and only if this EM is classical, namely if it has a Strongly Deterministic HVM (see Abramsky and Brandenburger 2011). Going into the details of this topic is beyond the scope of this paper, but let us only say that this result allows us to show that, on the one hand, Examples 2.2, 2.5, 3.1 and 3.2 (the famous PR box model) are contextual, and hence do not have a Strongly Deterministic HVM; on the other hand, Examples 2.1, 2.3, 2.4 are classical. Even without resorting to this method we have been able to distinguish the two cases by showing that the classical EM has a consistent Strongly Deterministic HVM and the contextual EM has a consistent Weakly Deterministic, but not Strongly Deterministic HVM.

One last comment here which is important for the next section. If one wants that a Contextual Empirical Model be compatible with Special Relativity, the model must satisfy No-Signaling. Let there be a consistent Weakly Deterministic HVM (which is guaranteed by the existence theorem). Then, this HVM cannot satisfy No-Signaling, since if it did, it would be Strongly Deterministic (by property (4) of Sect. 19.2), which implies that the Empirical Model is *not* contextual (contrary to our assumption). We conclude that *every deterministic HVM consistent with a contextual EM must have a state that allows superluminal signaling!* Therefore, in order to comply with Special Relativity, it is either impossible to prepare a system in such a state, or else a Weakly Deterministic HVM cannot exist. In the next section we will present a strong argument in support of the second option.

19.5 Empirical Models and Special Relativity

The interplay between the non-locality of QM and the relativistic constraints imposed by the Special theory of Relativity (SR) has drawn much attention in the literature (see e.g. Redhead 1987 and references therein). A number of results support the intuition that if one requires genuine adherence to Lorentz invariance,[7] then even a nonlocal but deterministic explanation of certain quantum mechanical scenarios leads to a contradiction (see e.g. Hardy and Squires 1992; Clifton et al. 1992; Conway and Kochen 2006, 2009; Gisin 2011; Suarez 2010). Rohrlich and Aharonov (2013, p. 206), for example, ask how can action at a distance be compatible with relativistic causality, and their answer is that: "God plays dice, because it is the only way for nonlocality and relativistic causality to coexist."[8] Similarly, Ben Menahem (2012, p. 164) argued that: "Once we distinguish between nonlocality and no signaling, the lesson of QM is that a combination of nonlocality and no signaling is made possible by indeterminism." Here we shall generalize the arguments from Gisin (2011) to the framework of Empirical Models and Hidden Variable Models and consider the implications of this approach with respect to the existence of Weakly Deterministic and Lorentz invariant HVM that is consistent with a contextual EM.

Let us start by reiterating the roles played by EM and HVM as representations of physical systems at different levels. An EM describes a physical system at the phenomenological level, i.e., it depicts the distribution of outcomes of a given set of measurements performed on the system. The HVM adds another layer to the description, namely the physical *state* of the system. It may be impossible to

[7]As we said in Sect. 19.1, by Lorentz invariance we mean that there is no preferred Lorentz frame of reference and therefore no absolute time order and no absolute direction of causation.

[8]This point has been stressed by Yakir Aharanov, since (as far as we know) the mid 1980s.

measure directly the value of this state, but when it is known, the probability distribution over the measurement outcomes described by the EM can be explained. This explanation may be deterministic (Strongly Deterministic or Weakly Deterministic HVM) or otherwise stochastic. Therefore, the HVM plays the role of the physical theory that allows making predictions about, or producing explanations for, the results of measurements performed on the system.

Although the EM and HVM framework is aimed at analyzing physical systems, it does *not* refer explicitly to *space and time*. Yet, the properties of No-Signaling and Locality originated from physical systems that are positioned in space-time and hence it is only natural to introduce these aspects to the EM and HVM framework.

A simple way of doing this is to assign to each measurement a space-time label and then every pair of measurements can be in space-like, time-like or light-like relation. And under these relations one can require, for example, that properties such as No-Signaling or Outcome Independence will apply only to measurements that are in the space-like relation. One can conclude (from special relativity) that an EM that does not satisfy No-Signaling with respect to two space-like related measurements cannot represent a physical system, since it will allow transmitting information in superluminal speed.

A HVM, on the other hand (as explained in Sect. 19.2.2), may violate No-Signaling but nevertheless it may constitute a deeper physical structure that gives rise to the EM, since the HVM might also rule out the possibility of preparing a system in a specific state, so that transmitting superluminal signals is not physically feasible.

However, as we shall see below, requiring that a HVM is Lorentz Invariant in the sense that it predicts consistent measurement outcomes in different inertial frames, implies that the HVM must be No-Signaling, and from this fact we shall prove a *no-go* theorem according to which a Weakly Deterministic HVM realizing a contextual EM *does not exist.*

We stress that this no-go theorem is strictly stronger than the exclusion of a strongly deterministic HVM for a contextual EM presented in Sect. 19.4, since the existence theorem of Sect. 19.2.3, guarantees the existence of a Weakly Deterministic HVM for *every* EM, including a *contextual* one. On the other hand, our no-go theorem does *not* contradict this existence theorem, since it is the Lorentz Invariance additional requirement that excludes the existence of the Weakly Deterministic HVM. So, as we shall see below, our no-go theorem does not imply that there is no Weakly Deterministic HVM consistent with a contextual EM. It does, however, show that such a HVM will be *inconsistent* with the fact that the time order of space-like measurements may be reversed for certain inertial frames.

To establish this no-go theorem let's start with proving the following Lemma.

Lemma 5.1 An HVM that is consistent with Lorentz Invariance must be No-Signaling.

By consistency with Lorentz Invariance we mean that given two measurements in space-like locations the HVM yields identical probabilistic predictions for every inertial frame of reference.

Proof We generalize the argument in Gisin (2011), which is explicitly formulated with respect to a Weakly Deterministic account of a Bell type scenario, to *any stochastic* HVM pertaining to a measurement scenario with two space-like related measurements.

Suppose that there are two observers Alice and Bob, in space-like locations, that can perform measurements A, A' and B, B', respectively, each with binary outcomes, where all four combinations of measurements (A, B), (A', B), (A, B'), (A', B') are admissible series of measurements. Since the two measurement locations are in space-like relation, there are two inertial Lorentz reference frames, such that in one of them the measurement of A or A' takes place *before* the measurement of B or B', whereas in the other inertial frame the temporal order of the measurements is reversed. We denote these two inertial frames by $\mathbb{AB}$ and $\mathbb{BA}$, respectively, to indicate the temporal order in which the measurements are performed (see Fig. 19.1 below).

Let us denote the outcomes of the measurements A, A' by a, the outcomes of the measurements B, B' by b, and the parameter space of the HVM by Λ.

For each admissible measurement $(X, Y) \in \{(A, B), (A', B), (A, B'), (A', B')\}$ and each $\lambda \in \Lambda$ the HVM defines the conditional probabilities: $P^{\mathbb{AB}}_{X,Y,\ldots}(X = a|\lambda)$, $P^{\mathbb{AB}}_{X,Y,\ldots}(Y = b|\lambda)$ and $P^{\mathbb{BA}}_{X,Y,\ldots}(X = a|\lambda)$, $P^{\mathbb{BA}}_{X,Y,\ldots}(Y = b|\lambda)$. Now, in the $\mathbb{AB}$ Lorentz frame, the probability $P^{\mathbb{AB}}_{X,Y,\ldots}(X = a|\lambda)$ cannot depend on Y, since the

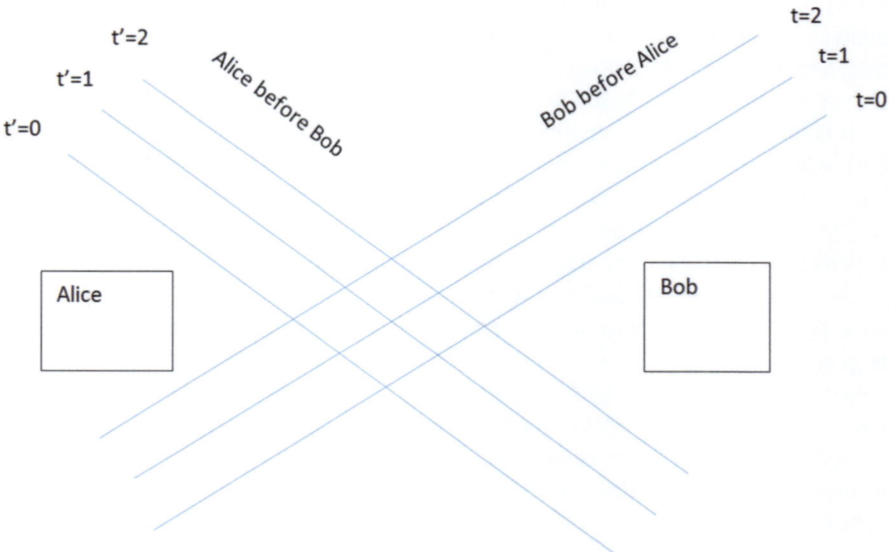

Fig. 19.1 Spacetime diagram of Bell type scenario

choice of measurement B or B' may be taken *after* the measurement in Alice's location.[9] Hence $P^{\mathbb{AB}}_{X,Y,\ldots}(X=a|\lambda) = P^{\mathbb{AB}}_{X}(X=a|\lambda)$.

And similarly in the $\mathbb{BA}$ frame: $P^{\mathbb{BA}}_{X,Y,\ldots}(Y=b|\lambda) = P^{\mathbb{BA}}_{Y}(Y=b|\lambda)$.

In addition, the two Lorentz frames must agree on the probability distributions of every event and therefore one concludes that:

$$P^{\mathbb{AB}}_{X,Y,\ldots}(X=a|\lambda) = P^{\mathbb{BA}}_{X,Y,\ldots}(X=a|\lambda) = P^{\mathbb{AB}}_{X}(X=a|\lambda)$$

$$P^{\mathbb{BA}}_{X,Y,\ldots}(Y=b|\lambda) = P^{\mathbb{AB}}_{X,Y,\ldots}(Y=b|\lambda) = P^{\mathbb{BA}}_{Y}(Y=b|\lambda)$$

$$\ldots$$

This implies that the HVM is No-Signaling.

Corollary A contextual EM has no Lorentz Invariant Weakly Deterministic consistent HVM.

Proof Assume that such a consistent Weakly Deterministic HVM exists. Then, since it is also No-Signaling (being Lorentz Invariant), it is in fact Strongly Deterministic by relation (4) of Sect. 19.2.3. But this is a *contradiction*, since a contextual EM cannot have a Strongly Deterministic consistent HVM.

Although, as we have seen, there is no deterministic (Weak or Strong) HVM of a contextual EM, there *always exists a probabilistic Lorentz Invariant and consistent* HVM, since the trivial consistent HVM has this property. So, in this sense our analysis vindicates the above arguments (at the beginning of this section) that the quantum mechanical nonlocal correlations are compatible with special relativity *only* in a genuinely stochastic world.

It is important to mention that Abramsky and Brandenburger (2011) proved that QM, which can be seen as a *genuine stochastic* HVM, with projective measurements and quantum states playing the role of the λ parameter, is indeed No-Signaling! If QM would not have this property it would be physically possible to transmit superluminal signals, since quantum states can actually be prepared!

As we said, similar results have been derived in the literature for specific QM scenarios.[10] But, in contrast to these specific scenarios, the argument presented here is *general*. It pertains to *any contextual* EM and does not depend on the details of a specific theory. Even if QM would turn out to be incomplete or even wrong, the framework of EM and HVM would remain intact. The proof above that there is no deterministic HVM for contextual and Non-Signaling EM depends *only* on the results obtained in experiments and on the requirement of Lorentz invariance of the laws of physics.

[9]We assume here λ-independence according to which there are no correlations between the choice of measurement and the outcome of *other* measurements through the parameter λ.

[10]See e.g., the two Mach-Zehnder type interferometers by Hardy and Squires (1992), the Kochen-Specker set up by Conway and Kochen (2006, 2009), Bell's EPR scenario by Gisin (2011), and the Suarez-Scarai model by Suarez (2010).

19.6 Application of Our No-Go Proof to Interpretations of QM

Let us now briefly show how using the EM and HVM frameworks may be helpful in dissolving certain debates that have been raised recently concerning the possibility of constructing interpretations of QM that are Lorentz invariant. An argument appearing in Conway and Kochen (2009) claims that for a contextual EM it is possible to exclude also the existence of a Lorentz invariant consistent *probabilistic* HVM. Conway and Kochen (2009) apply an analysis of the probability distribution functions of a contextual EM which is similar to Gisin's argument. However, their conclusion is *unsound*, as indicated also in Goldstein et al. (2010) for a specific scenario, since given a contextual, No-Signaling EM, the trivial (singleton Λ) consistent HVM is a No-Signaling Lorentz invariant probabilistic explanation for this EM.

Our analysis shows that it is impossible to construct *any deterministic* Lorentz invariant (and hence No-Signaling) HVM that yields the quantum mechanical predictions. Since, for example, Bohm's (1952) theory is Weakly Deterministic, it must fail to be Lorentz invariant, and in this sense Bohm's theory is *incompatible* with Special Relativity (SR).[11] The attempts to overcome this problem resort to postulating that there is a preferred Lorentz reference frame which describes the absolute temporal order of events (see Dürr et al. 2013). This absolute temporal order circumvents our proof above exactly where we assume that a later event in some reference frame cannot have an impact on an earlier event in the same frame.

The case of stochastic theories (of the kind originally proposed by Ghirardi et al. 1986, GRW) is subtler, since prima facie a genuinely stochastic theory may have a Lorentz invariant relativistic version compatible with QM, as we have shown above. In Tumulka (2006) it is suggested that the flash theory denoted by rGRWf is fully compatible with SR, in particular in the sense that it requires no preferred Lorentz frame. Tumulka's (2006) proposal has been criticized by Esfeld and Gisin (2014) who claim that "neither rGRWf, nor any other theory, can account for the occurrences of the Alice-flash and the occurrences of the Bob-flash in a Lorentz-invariant manner". This conclusion is based on the argument presented originally in Gisin (2011) which we also used in this section to derive our no-deterministic HVM result. Let us now examine Esfeld and Gisin's (2014) argument in the context of the rGRWf theory and the EM and HVM framework.

The core of Esfeld and Gisin's (2014) argument (on our understanding) is that the probabilities observed at space-like separation by Alice and Bob are "classical",

[11] For explicit proofs that Bohm's theory requires an *absolute* temporal order (i.e., a preferred Lorentz frame), see Albert (1992); Hemmo and Shenker (2013). Of course, as we said above, Bohm's theory does not allow to use its No-Signaling property, since its additional dynamical law for the motion of particles together with its statistical postulate concerning the distribution of particles entail that preparing a system at a *known* position that is more exact than the system's prepared wavefunction is physically *impossible*.

i.e. they can be "computed and simulated on a classical computer." Hence, they conclude that there exists a classical variable $\lambda \in \Lambda$ (that can be stored on a computer's hard disc) such that the outcomes on both sides of the EPR experiment are given by the four deterministic functions that depend on this variable and on the settings of the measurements as presented in our proof above. From this assumption, they derive their conclusion according to which the rGRWf (and any other theory) cannot provide a genuinely relativistic (i.e., Lorentz invariant) account of an EPR experiment.

However, as we have just seen in this section, there always *exists* a probabilistic HVM that is consistent with EPR-type experiments (or every other contextual experiment) and this fact obviously disagrees with Esfeld and Gisin's conclusion. We will now argue that Esfeld and Gisin's argument implicitly *presupposes* an assumption which in fact need *not* hold in a stochastic theory, and specifically does *not* hold in rGRWf.

Esfeld and Gisin (2014) are correct in arguing that the marginal probabilities of an EPR empirical model, namely those computed by Alice or Bob, are classical.[12] However, as explained in Sect. 19.4 above, this does *not* imply that there is a *common classical probability model* for *all* possible combinations of measurements on the two wings in the EPR setting. Assuming that the classical marginal probabilities imply the existence of a common classical probability is *equivalent* to assuming that the EM is *non*-contextual, which is not satisfied by the EM of the EPR experiment.

So perhaps Esfeld and Gisin (2014) mean that there exists a consistent Weakly Deterministic (non-classical) HVM for the EPR probabilities, which is indeed true, as shown by Existence Theorem in Sect. 19.2. In this case one can derive a contradiction with Lorentz invariance, as we have just done in this section. However, even this assumption will not make their argument against the rGRWf theory valid, unless one can prove that the rGRWf theory is Weakly Deterministic. But we do not see how this can be done, because both the original GRW theory and the rGRWf theory are *manifestly genuinely stochastic* theories. Of course, our point here should not be taken as a positive proof that rGRWf is a genuinely relativistic interpretation of QM. Such a proof seems to go beyond the framework of EM and HVM.

19.7 Conclusion

We saw in this paper that using the EM and HVM mathematical framework allows us to clarify certain relations between significant physical properties such as Determinism, Locality and Contextuality. Moreover, we were able to have an insight into the debate concerning the prospects of interpretations of QM that are

[12] A non-classical EM may have *many* classical marginal probabilities; this is the core of Abramsky and Brandenburger (2011).

compatible with Special Relativity. In particular we have proved here using the EM and HVM framework that the physical reality of our world (assuming that there is no absolute time order) *must be stochastic*.[13]

Viewed from this framework one can even think of standard QM (in von Neumann's formulation) as a stochastic HVM (in which the quantum state plays the role of the hidden variable) that is consistent with the observed frequencies of measurement outcomes. As we saw in Sect. 19.5, from a mathematical point of view, Lorentz invariance (in the sense that there is no preferred Lorentz frame)[14] by itself is consistent with a genuinely stochastic HVM that yields the predictions of QM. The genuine stochastic nature of QM allows it to have the property of No-Signaling even at the HVM level.

However, the EM and HVM framework cannot really settle the question of the compatibility of QM and SR because the dynamical aspect of physics is missing in this framework. And obviously, one should not expect to find in this framework a resolution to the measurement problem in QM. But it is nonetheless encouraging that there is no mathematical barrier to finding a stochastic theory that is compatible with both QM and SR. Whether or not rGRWf or other attempts should be considered satisfactory from a physical point of view is still to be settled.

Acknowledgement We thank Orly Shenker and Yemima Ben-Menahem for discussions of issues related to this paper. We also thank Ori Feiglin, Aharon Gill, Zohar Maliniak and Moti Suess, from the research group in philosophy of physics at the University of Haifa for valuable comments on this paper. Finally, we thank Samson Abramsky for very valuable comments on an earlier draft of this paper. This research has been supported by the Israel Science Foundation (ISF), grant number 1148/18.

References

Abramsky, S. (2013). Relational hidden variables and non-locality. *Studia Logica, 101*(2), 411–452.

Abramsky, S., & Brandenburger, A. (2011). The sheaf-theoretic structure of non-locality and contextuality. *New Journal of Physics, 13*, 113036.

Albert, D. (1992). *Quantum mechanics and experience*. Cambridge, MA: Harvard University Press.

Badziąg, P., Bengtsson, I., Cabello, A., & Pitowsky, I. (2009). Universality of state-independent violation of correlation inequalities for noncontextual theories. *Physical Review Letters, 103*, 050401.

Bell, J. (1964). On the Einstein-Podolsky-Rosen paradox. *Physics, 1*, 195–200.

Ben-Menahem, Y. (2012). Locality and determinism: The odd couple. In Y. Ben-Menahem & M. Hemmo (Eds.), *Probability in physics* (pp. 149–166). Berlin/Heidelberg: Springer.

[13]We did not go into the question why God's dice is quantum rather than, say, a PR dice. This will involve adding extra assumptions that inhibit the appearance of super-quantum theories (see Rohrlich and Aharonov 2013 for a recent survey).

[14]Of course, in a relativistic quantum theory, the Schrödinger dynamics is replaced with a relativistic equation of motion.

Bohm, D. (1952). A suggested interpretation of the quantum theory in terms of "hidden" variables. *Physical Review, 85*(166–179), 180–193.

Brandenburger, A., & Keisler, H. J. (2016). Fiber products of measures and quantum foundation. In J. Chubb, A. Eskandarian, & V. Harizanov (Eds.), *Logic and algebraic structures in quantum computing, lecture notes in logic* (Vol. 45, pp. 71–87). Cambridge: Cambridge University Press.

Brandenburger, A., & Yanofsky, N. (2008). A classification of hidden variable properties. *Journal of Physics A: Mathematical and Theoretical, 41*, 425302.

Bub, J. (2010). Von Neumann's 'no hidden variables' proof: A re-appraisal. *Foundations of Physics, 40*(9), 1333–1340.

Bub, J., & Pitowsky, I. (2010). Two dogmas about quantum mechanics. In S. Saunders, J. Barrett, A. Kent, & D. Wallace (Eds.), *Many worlds? Everett, quantum theory and reality* (pp. 433–459). Oxford: Oxford University Press.

Clifton, R., Pagonis, C., & Pitowsky, I. (1992). Relativity, quantum mechanics and EPR. In D. Hull, M. Forbes, & K. Okruhlik (Eds.), *PSA 1992* (Vol. 1, pp. 114–128). East Lansing: Philosophy of Science Association.

Conway, J., & Kochen, S. (2006). The free will theorem. *Foundations of Physics, 36*(10), 1441–1473.

Conway, J., & Kochen, S. (2009). The strong free will theorem. *Notices of the American Mathematical Society, 56*(2), 226–232.

Dürr, D., Goldstein, S., & Zanghi, N. (1992). Quantum equilibrium and the origin of absolute uncertainty. *Journal of Statistical Physics, 67*(5/6), 843–907.

Dürr, D., Goldstein, S., & Norsen, T. (2013). Can Bohmian mechanics be made relativistic? Proceedings of the Royal Society A, 470, No. 2162, 20130699.

Esfeld, M., & Gisin, N. (2014). The GRW flash theory: A relativistic quantum ontology of matter in space-time? *Philosophy of Science, 81*(2), 248–264.

Fine, A. (1982). Hidden variable, joint probability and the Bell inequalities. *Physical Review Letters, 48*(5), 291.

Ghirardi, G., Rimini, A., & Weber, T. (1986). Unified dynamics for microscopic and macroscopic systems. *Physical Review, D, 34*, 470–479.

Gisin, N. (2011). Impossibility of covariant deterministic nonlocal hidden-variable extensions of quantum theory. *Physical Review A, 020102*(R), 83.

Goldstein, S., Tausk, D. V., Tumulka, R., & Zanghi, N. (2010). What does the free will theorem actually prove? *Notices of the American Mathematical Society, 57*, 1451–1453.

Hardy, L., & Squires, E. J. (1992). On the violation of Lorentz-invariance in deterministic hidden-variable interpretations of quantum theory. *Physics Letters A, 168*(3), 169–173.

Hemmo, M., & Shenker, O. (2013). Probability zero in Bohm's theory. *Philosophy of Science, 80*, 1148–1158.

Kolmogorov, A. N. (1933). *Foundations of the theory of probability*. New York: Chelsea Publishing Company. English translation 1956.

Maudlin, T. (1994). *Quantum non-locality and relativity: Metaphysical intimations of modern physics*. Oxford: Wiley-Blackwell Press.

Peres, A. (1995). *Quantum theory: Concepts and methods*. Dordrecht: Kluwer Academic Press.

Pitowsky, I. (1986). The range of quantum probability. *Journal of Mathematical Physics, 27*, 1556.

Pitowsky, I. (1989). *Quantum probability, quantum logic*. University of California, Springer-Verlag.

Pitowsky, I. (2007). Quantum mechanics as a theory of probability. In: W. Demopoulos & I. Pitowsky (Eds.), *Physical theory and its interpretation, essays in honor of Jeffrey Bub, the Western Ontario series in philosophy of science* (Vol. 72, pp. 213–240). Springer.

Popescu, S., & Rohrlich, D. (1994). Quantum nonlocality as an axiom. *Foundations of Physics, 24*(3), 379–385.

Price, H., & Wharton, K (2013). Dispelling the quantum spooks – a clue that Einstein missed? arXiv 1307.7744 [physics.hist-ph].

Redhead, M. (1987). *Incompleteness, nonlocality and realism*. Oxford: Clarendon.

Rohrlich, D., & Aharonov, Y. (2013). PR-box correlations have no classical limit. In D. C. Struppa & J. M. Tollaksen (Eds.), *Quantum theory: A two-time success story (Yakir Aharonov festschrift)* (pp. 205–211). New York: Springer.
Suarez, A. (2010). The general free will theorem. arXiv: 1006.2485 [quant-ph].
Tumulka, R. (2006). A relativistic version of the Ghirardi–Rimini–Weber model. *Journal of Statistical Physics, 125*, 821–840.

Chapter 20
Subjectivists About Quantum Probabilities Should Be Realists About Quantum States

Wayne C. Myrvold

Abstract There is a significant body of literature, which includes Itamar Pitowksy's "Betting on the outcomes of measurements," that sheds light on the structure of quantum mechanics, and the ways in which it differs from classical mechanics, by casting the theory in terms of agents' bets on the outcomes of experiments. Though this approach, by itself, is neutral as to the ontological status of quantum observables and quantum states, some, notably those who adopt the label "QBism" for their views, take this approach as providing incentive to conclude that quantum states represent nothing in physical reality, but, rather, merely encode an agent's beliefs. In this chapter, I will argue that the arguments for realism about quantum states go through when the probabilities involved are taken to be subjective, if the conclusion is about the agent's beliefs: an agent whose credences conform to quantum probabilities should believe that preparation procedures with which she associates distinct pure quantum states produce distinct states of reality. The conclusion can be avoided only by stipulation of limitations on the agent's theorizing about the world, limitations that are not warranted by the empirical success of quantum mechanics or any other empirical considerations. Subjectivists about quantum probabilities should be realists about quantum states.

Keywords Quantum mechanics · Quantum ontology · Pusey-Barrett-Rudolph theorem · QBism · Subjective probability

20.1 Introduction

There is a conception of probability, which often goes by the name of *subjective probability*, that takes probability to have to do with an agent's degrees of belief. An important step in the development of this conception was its integration with

W. C. Myrvold (✉)
Department of Philosophy, The University of Western Ontario, London, ON, Canada
e-mail: wmyrvold@uwo.ca

M. Hemmo, O. Shenker (eds.), *Quantum, Probability, Logic*,
Jerusalem Studies in Philosophy and History of Science,
https://doi.org/10.1007/978-3-030-34316-3_20

decision theory, pioneered by Ramsey and de Finetti, and developed more fully by Savage. On this view, an agent's preferences between acts (often illustrated by choices of bets to make or accept) are taken to be indications of the agent's degrees of belief in various propositions.

It is an attractive idea to extend this conception to quantum probabilities. One systematic development of this idea can be found in Pitowsky's "Betting on the outcomes of measurements: a Bayesian theory of quantum probability" (Pitowsky 2003). The idea also forms the basis of the position (or family of positions) known as QBism (Caves et al. 2002, 2007; Fuchs and Schack 2013; Fuchs et al. 2014; Fuchs 2017; Fuchs and Stacey 2019). Proponents of such views tend to also take quantum states to be *nothing more* than codifications of an agent's beliefs. But the argument for this conclusion is never spelled out with sufficient clarity.

On the other hand, we find arguments—of which the theorem of Pusey et al. (2012) is the best known—for the opposite conclusion, namely, that quantum states should be taken to represent elements of physical reality. If interpreting quantum probabilities as subjective leads one inexorably to rejecting an ontic view of quantum states, then there must be something about the reasoning behind these arguments that doesn't go through on a subjective reading of probability. In this chapter, I will argue that this is not the case. Suitably interpreted, a version of the PBR argument leads to the conclusion that an agent who takes quantum mechanics as a guide for setting her subjective degrees of belief about the outcomes of future experiments ought to believe that preparations with which she associates distinct pure states result in ontically distinct physical states of affairs (this terminology will be explained in Sect. 20.3, below). Note that this conclusion is *entirely about the agent's beliefs*. We are not presuming, as part of the argument, that the agent's beliefs in any way reflect physical reality. That is a matter for the agent's own judgment (though, of course, there is an implication that if *you* take quantum mechanics as a guide for setting your degrees of belief about the outcomes of future experiments, you should take distinct pure states to be ontically distinct).

The argument, of course, involves assumptions about the agent's credences. It is also assumed that the agent may entertain theories about the way the world is, and have degrees of belief in such theories. It is, of course, *logically possible* to reject all such theorizing. If such a move is proposed in connection with quantum theory, then the question arises whether the empirical evidence that led the community of physicists to accept quantum theory over classical theory provides any incentive for the move. I will argue that we have no reason whatsoever to make such a move. Anyone is, of course, free to make such a move without reason, as a free choice. Dissuasion of someone so inclined will not be the topic of this chapter; what concerns us here is the evidential situation.

20.2 Credence and Action: Some Preliminary Considerations

In this section I would like to bring to the reader's mind some considerations of the sort that may be called "pre-theoretical"—the sort of considerations that tend to be taken for granted, prior to any physical theorizing. They are, for that reason, to be thought of as neither particularly classical nor quantum. I do not take myself to be saying anything new, radical, or contentious. Of course, one may start a scientific investigation with certain presuppositions about the way the world is, and end up concluding, as a result of one's investigations, that some of those presuppositions are false. If any readers are inclined to dispute some assertion made in this section, then I will ask that reader to provide evidence for the falsity of the assertion. I hope that it goes without saying that the mere fact that some proposition leads to an unwelcome conclusion does not suffice as evidence that the proposition is false.

Consider the following situation. You are attending a lecture, and you are taking notes on a laptop. This process involves a sequence of choices: you are choosing, at every moment, which, if any, keys to hit. Why are you doing this? If you are taking notes for your own benefit, as an aid to memory, then, presumably, you believe that, at some later date, you will be able to see or hear or feel (depending on the device used for readout) something that will be informative about your prior choices of key-strokes. If you are taking notes for someone else's benefit, then, presumably, you believe that, at a later date, that other agent will be able to see or hear or feel something that will be informative about your earlier choices of key-strokes.[1]

Why are you using a laptop, rather than, say, a kumquat? Presumably, you believe that there is something about the internal structure of the laptop that makes it suitable for the purpose. Suppose that you can use the readout to distinguish between two alternative choices of things to write today. For concreteness, consider a choice between writing

A 'Twas brillig, and the slithy toves did gyre and gimble in the wabe.
B Now is the winter of our discontent made glorious summer by this sun of York.

If you believe that the readout tomorrow will allow you to distinguish between having made choice *A* and choice *B* today (in case you're forgetful)—that is, if you believe that the laptop can serve as an auxiliary memory—this is to attribute to the laptop a certain role in a causal process linking your choice today to your experience tomorrow. Even if you have little or no idea of the internal workings of the laptop, attribution of this causal role to the laptop involves an assumption that your action will have an effect on the internal state of the laptop, and that the effect your action had on the internal state of the laptop will, in turn, have an effect on what it is that you experience tomorrow.

Now, an extreme version of operationalism might forbid you to entertain conjectures about the internal state of the laptop, and speak only of the relations between your act today and the readout tomorrow, with no mention of intervening variables. But the fact is that you are using a laptop rather than a kumquat, and, in fact, would not attempt to use a kumquat for this purpose, and that means that

[1]One might also consider the following situation. You are reading a text of which you are not the author. Presumably, you take what you are reading as informative about what choices of keystrokes the author made while composing it. If not, then why are you still reading?

you believe that there is something about the physical constitution of the laptop that makes it suitable for the role you wish it to play in the causal chain between your act today and the readout tomorrow. Caution should be exercised in theorizing about the internal workings of the laptop, but, as long as we are careful not to be committed to anything more than what we have good evidence for, there is no harm, and some potential benefit, in theorizing about the processes that mediate your actions today and experiences tomorrow.

Suppose, then, that you begin—tentatively, and with all due caution—to form a theory about the inner workings of the laptop. What constraints do your beliefs about the input-output relations of the laptop place on the sorts of theories you should entertain about its inner workings?

It seems that there are a few general things that can be said. If you believe that the laptop can serve as a means for discriminating (with certainty) between act A and act B, you should believe that there are distinct sets of internal states corresponding to these two acts. That is, the set of states that the laptop could end up in, as a result of your performing act A, has no overlap with the set of states that it could end up in as a result of your performing act B. Moreover, these two sets of states are distinguishable upon readout—the readout mechanism produces outputs, corresponding to the two sets of internal states associated with acts A and B, that are perceptually distinguishable by you. The only alternative to believing this is to believe in a direct causal effect of your choice today on your experience tomorrow, unmediated by any effect on the state of the world in the interim.

Of course, discrimination with absolute certainty is too much to ask. You might not regard your laptop as a *completely* reliable means for obtaining information about your past choices of key-strokes. That's okay; it can still be informative, as long as if you take some perceived outputs to be more likely, given some choices of keystrokes, than others. We can cash this out in terms of your conditional degrees of belief.

All of this will be formalized in the next section.

20.3 Constructing a Framework

Let us begin to formalize the informal considerations of the previous section.[2]

To simplify matters, we will imagine a Bayesian agent, Alice, whose gradations of strength of belief in various propositions can be represented by a real-valued function satisfying the axioms of probability. We will call this function the agent's *credence function*, and the value that it takes on a given proposition p, the agent's *credence* in p. We will avoid talking of "probability," in case that, for some, the word has connotations of something non-subjective.

[2]The framework in this section is based on the ontological models framework of Harrigan and Spekkens (2010).

We consider some physical system (you may think of the laptop as a running example), with which our agent will engage. She has some set of operations $\{\phi_1, \ldots, \phi_n\}$ that can be performed on the system at time t_0 (think choices of key-strokes). We will also consider situations in which an act is chosen without Alice's knowledge of which one; she has an n-sided die that she can toss (and when she does, her credence is equally divided among the possible outcomes), and a gadget that will choose among the operations $\{\phi_1, \ldots, \phi_n\}$, depending on the outcome of the die-throw.

At some later time t_1, she performs an operation R (a "readout" operation), which will result in one of a set of perceptually distinguishable results $\{r_1, \ldots r_m\}$. (We can, of course, generalize, and give her a choice of operations to perform, but for our purposes, one will suffice.)

We assume that our agent has conditional credences, $cr^A(r_k|\phi_i)$, representing credence in result r_k on the supposition of act ϕ_i. Upon learning the result of the operation R, she uses these to update, via conditionalization, her credences about which act was performed.

Alice does not believe that the acts $\{\phi_i\}$ have a direct influence on the outcomes $\{r_k\}$, unmediated by any effect of those acts on the system S. We will permit her to entertain various theories about the workings of the system. Any theory T of that sort will involve a set Ω_T of possible physical states. At the level of generality at which we are operating, we will assume nothing about the structure of these state-spaces Ω_T, and, in particular, do not assume that they are either classical or quantum in structure. We do assume that, on the supposition of theory T, Alice can form credences in propositions about the state of the system S to the effect that the system's state is in a subset Δ of Ω_T, and that, for each theory T, there is an algebra $\mathcal{A}_T$ of subsets of Ω_T deemed suitable for credences of that sort.

Suppose, now that Alice regards the state of the system to be relevant to the outcome of the readout operation. What she thinks about that relevance may, of course, depend on the theory T. To represent these judgments, we suppose her to have conditional credences of the form $cr^A(r_k|\phi_i, T, \omega)$. This is her conditional credence, on the supposition of theory T and the supposition that the state of the system is $\omega \in \Omega_T$, that the result of the readout will be r_k, if act ϕ_i was performed. It should be stressed that these are *Alice's* conditional credences; there is no assumption that these conditional credences mirror any causal structure out there in the rest of the world. Alice may hope that her conditional credences reflect something of the causal structure of the world (or rather, the causal structure, according to theory T), but no assumption to the effect that they do reflect the causal structure of the world will form any part of our argument, because our conclusions will only be about what Alice should believe.

We assume, also, that, on the supposition of T, Alice has credences about the states produced by the acts ϕ_i. For any subset Δ of Ω_T that is in $\mathcal{A}_T$, let $cr^A(\Delta|T, \phi_i)$ be Alice's conditional credence, on the supposition of theory T, that performing act ϕ_i would put the system into a state in Δ. Her credences about the state should mesh with her conditional credences $cr^A(r_k|\phi_i, T, \omega)$ to yield her act-outcome conditional credences, on the supposition of theory T:

$$cr^A(r_k|\phi_i, T) = \langle cr^A(r_k|\phi_i, T, \omega)\rangle, \tag{20.1}$$

where the angle-brackets $\langle \cdot \rangle$ indicate expectation value taken with respect to her credences $cr^A(\Delta|T, \phi_i)$ about the state, given act ϕ_i. If $\{T_j\}$ is the set of theories that Alice takes seriously enough to endow with nonzero credence, then her act-outcome conditional credences $cr^A(r_k|\phi_i)$ should satisfy

$$cr^A(r_k|\phi_i) = \sum_j cr^A(r_k|\phi_i, T_j)\, cr^A(T_j). \tag{20.2}$$

As mentioned, Alice does not believe that the acts $\{\phi_i\}$ have a direct influence on the outcomes $\{r_k\}$, unmediated by any effect of those acts on the system S. This will be reflected in her conditional credences; her belief can be captured by the condition that $cr^A(r_k|\phi_i, T, \omega)$ be independent of which act is performed. That is, for all i, j,

$$cr^A(r_k|\phi_i, T, \omega) = cr^A(r_k|\phi_j, T, \omega), \tag{20.3}$$

for all states $\omega \in \Omega_T$. This is the condition that her choice of act has an influence on the later outcome *only* via the influence of this choice on the state of the system.

Now, part of the point of taking notes on a laptop is that what you see later will permit you to distinguish between choices of key-strokes you made earlier. There may be a limit to such discrimination; if you type something and then delete it, the later readout might fail to distinguish between alternatives for the deleted text. Some pairs of acts will be readout-distinguishable; others will not. We will say that Alice takes two acts $\{\phi_i, \phi_j\}$ to be *R-distinguishable* if, for every outcome r_k of the operation R, either $cr^A(r_k|\phi_i)$ or $cr^A(r_k|\phi_j)$ is zero.

If Alice takes two acts to be R-distinguishable, then, provided she takes the influence of acts performed on the system on the later readout to be mediated by a change in the state of the system, then she should also take it that the sets of states that can result from the two acts are distinct. We can formalize this; say that Alice takes the effects of acts $\{\phi_i, \phi_j\}$ to be *ontically distinct*, on the supposition of theory T, if there exists $\Delta \subseteq \mathcal{A}_T$ such that $cr^A(\Delta|T, \phi_i) = 1$ and $cr^A(\Delta|T, \phi_j) = 0$. We have the following simple theorem.

Theorem 1 *If an agent's credences satisfy* (20.1), (20.2) *and* (20.3)*, then, if she takes two acts to be R-distinguishable for some operation R, then she also takes the effects of those acts to be ontically distinct on any theory T in which she places nonzero credence.*

We emphasize: this is a theorem about Alice's credences, not about actual states of the system. Alice could be wrong in her judgment that the effects on her laptop of different choices of keystrokes are ontically distinct. It might be that her keystrokes have no effect whatsoever on the internal state of the laptop (in which case she would be misguided in taking those acts to be readout-distinguishable).

All of this is very general, and no assumptions have been made about the sorts of theories that Alice might entertain. Now, suppose we apply it to quantum mechanics, with quantum probabilities construed subjectively. Suppose that Alice's credences about the results of future readout operations performed on system S can be represented by quantum states, and that which quantum state it is that represents her credences depends on her choice of act. If two acts are distinguishable by some experiment that can be done on the system, the corresponding quantum states are orthogonal. If Alice places nonzero credences only in theories for which her conditional credences satisfy (20.3), then she should also believe that the effects of acts with which she associates orthogonal quantum states are ontically distinct.

20.4 Ontic Distinctness of Non-orthogonal States

We have come to the conclusion that Alice should take the effects of preparation-acts that she regards as R-distinguishable for some operation R to be ontically distinct. This includes acts with which she associates orthogonal quantum states. But what about non-orthogonal quantum states?

Let $\{\phi_1, \ldots, \phi_n\}$ be a set of preparation-acts that are such that Alice's credences about the results of future experiments, upon performance of these acts, can be represented by pure quantum states $\{|\phi_1\rangle, \ldots, |\phi_n\rangle\}$. If Alice takes the effects of these acts to be ontically distinct, this means that she believes that the physical state carries a trace of which preparation was performed and, that, if she knew the physical state, this would be enough to specify which of the quantum states $\{|\phi_1\rangle, \ldots, |\phi_n\rangle\}$ she deems appropriate to use to set her credences about the results of future experiments. That is, she believes that, for each physical state that could arise from one of these preparations, there is a unique quantum state that "corresponds" to it as the state to use in setting her credences. The correspondence might not be one-one, as there might be a multiplicity of physical states that correspond to the same quantum state.

If Alice believes that the effects of any pair of preparation-acts with which she associates distinct pure quantum states are ontically distinct, then Alice believes that there is something in physical reality corresponding to these quantum states. We will say, in such a case, that *Alice is a realist about pure quantum states.*

The PBR theorem can be adapted to provide a set of conditions on Alice's credences sufficient for her to be a realist about pure quantum states.

Following the terminology of Leifer (2014), we will say that Alice takes a set $\{\phi_1, \ldots, \phi_n\}$ of preparation-acts to be *antidistinguishable* if there is an operation R such that, for each result r_k, $cr^A(r_k|\phi_i)$ is zero for at least one ϕ_i. Though Alice might not be able to uniquely decide which of the acts was performed, she will always be able to rule at least one out, whatever the result of operation R is.

The setup of the PBR theorem is as follows. Suppose that we have two systems, S_1 and S_2, of the same type, and suppose that each of them can be subjected to either of two preparation-acts, $\{\phi, \psi\}$. We assume that each choice of preparation act performed on one system is compatible with both choices of act performed on the other. This gives us four possibilities of preparation-acts for the joint system, which we will designate by $\{\phi_1 \otimes \phi_2, \phi_1 \otimes \psi_2, \psi_1 \otimes \phi_2, \psi_1 \otimes \psi_2\}$.

Suppose, now, that Alice regards these four preparation-acts as antidistinguishable. Under certain further conditions, which we will now specify, she should then take the effects of the preparation-acts ϕ and ψ on the two systems to be ontically distinct.

A theory about the physical states of the two systems, presumably, would be able to treat S_1 and S_2 on their own, and also to regard the composite system that has S_1 and S_2 as parts as a system in its own right. It will, therefore, include state-spaces Ω_1 and Ω_2 for the individual systems, and a state space Ω_{12} for the composite system. In a classical theory, to specify the physical state of the composite system $S_1 + S_2$, it suffices to specify the states of the component parts, and so the state of the composite can be represented as an ordered pair $\langle \omega_1, \omega_2 \rangle$ of states of the components, and Ω_{12} can be taken to be the set of all such ordered pairs. That is, Ω_{12} can be taken to be the cartesian product of Ω_1 and Ω_2. For other theories, such as quantum mechanical theories, the cartesian product $\Omega_1 \times \Omega_2$ might be included in Ω_{12} without exhausting it. In quantum mechanics, for any states ψ_1, ψ_2 of two disjoint systems, there is a corresponding product state $\psi_1 \otimes \psi_2$ of the composite, but not all pure states of the composite system are of this form, as there also pure entangled states.

We will say that a theory T satisfies the *Cartesian Product Assumption* (CPA) if $\Omega_1 \times \Omega_2$ is a subset of Ω_{12}. Given a theory satisfying the CPA, Alice's credences regarding a pair of preparations ϕ_1, ϕ_2 that can be performed on the two subsystems will be said to satisfy the *No Correlations Assumption* (NCA) if her credences about the joint state of the two systems, on the supposition of T and the preparations ϕ_1, ϕ_2, are concentrated on the cartesian product $\Omega_1 \times \Omega_2$ and if her credences are such that information about the state of one system would be completely uninformative about the state of the other. The conjunction of the Cartesian Product Assumption and the No Correlations Assumption is called the *Preparation Independence Postulate* (PIP). Note that this is a combination of an assumption about the structure of the state space of a theory T and an assumption about Alice's credences, conditional on the supposition of T and on the preparations ϕ_1, ϕ_2.

Now suppose that we have a pair of systems S_1, S_2, a theory T about them that satisfies the CPA, and, for each system, a pair of preparation-acts ϕ_i, ψ_i, which are such that:

1. There is an operation R that can be performed on the pair of systems, such that, for each outcome r_k of the operation, at least one of

$$\{cr^A(r_k|\phi_1 \otimes \phi_2), cr^A(r_k|\phi_1 \otimes \psi_2), cr^A(r_k|\psi_1 \otimes \phi_2), cr^A(r_k|\psi_1 \otimes \psi_2)\}$$

   is equal to zero. That is, Alice regards these preparation-acts as antidistinguishable.
2. Alice's credences about the states of the systems S_1 and S_2, conditional on the supposition of theory T and each of the preparations mentioned, satisfy the NCA.

It then follows, by the argument of Pusey et al. (2012), that Alice should regard the effects of these preparations to be ontically distinct, conditional on the supposition of theory T.

Now, of course, Alice might entertain theories for which the CPA does not hold, and, among those for which it does, her conditional credences might satisfy the NCA for some but not for others. In this case, she should not be certain that the effects of these preparations are ontically distinct. However, her credence that they are must be at least as high as her total credence in the class of theories for which the above-listed conditions are true. Another way to put this is: in order to be *certain* that the effects of these preparations are not ontically distinct—that is, in order to categorically deny onticity of quantum states—Alice's credences in all such theories must be strictly zero.

Though it seems natural from a classical standpoint, quantum mechanics gives us incentive to question the Cartesian Product Assumption. In the set of pure states of a bipartite system, product states are rather special, and every neighbourhood of any product state contains infinitely many entangled states. If both systems have Hilbert spaces of infinite dimension, then the entangled states are dense in the set of all states, pure or mixed (see Clifton and Halvorson 2000, 2001). We should, therefore, take seriously the cases of theories whose state spaces either lack a Cartesian product component, or are such that product states cannot reliably be prepared exactly, although we may be able to approximate them arbitrarily closely. A modification of the PBR argument that does not employ the Cartesian Product Assumption is therefore desirable.

In Myrvold (2018, 2020), a substitute for the PIP is proposed, which is strictly weaker than it and dispenses with the Cartesian Product Assumption. This substitute is called the Preparation Uninformativeness Condition (PUC).

To state the assumption, we consider the following set-up. Suppose that, for systems S_1, S_2, we have some set of possible preparation-acts that can be performed on the individual systems. Suppose that the choice of preparation for each of the subsystems is made independently, say, by rolling two separate dice. Following the preparation of the joint system, which consists of individual preparations on the subsystems, you are not told which preparations have been performed, but you are given a specification of the ontic state of the joint system. On the basis of this information, you form credences about which preparations were performed. In the case of preparations whose effects you take to be ontically distinct, you will be certain about what preparation has been performed; otherwise, you may have less than total information about which preparations were performed.

We ask: under these conditions, if you are now given information about which preparation was performed on one system, is this informative about which preparation was performed on the other? The Preparation Uninformativeness Condition is the condition that it is not. This condition is satisfied whenever the PIP is. It is also satisfied if you regard the effects of the preparations to be ontically distinct: in such a case, given the ontic state of the joint system, you are already certain about which preparations have been performed, and being told about the preparation on one system will not shift your credences.

The PUC is implied by the PIP, but it is strictly weaker. Even if the CPA is assumed, it is possible to construct models for the PBR setup, outlined in the previous section, in which the PUC is satisfied but the PIP is not. See Myrvold (2018) for one such construction.

On the assumption of the Preparation Uninformativeness Condition—which, in the current context, is a condition on the agent's credences, independent of the structure of the state space of the theory considered—one can get a ψ-ontology proof for a class of nonorthogonal states. Let $\{\phi_i, \psi_i\}$ $(i = 1, 2)$ be preparation-acts such that Alice's credences about the results of future experiments can be represented by quantum states $\{|\phi_i\rangle, |\psi_i\rangle\}$. If $|\langle\phi_i|\psi_i\rangle| \leq 1/\sqrt{2}$, then $\{|\phi_1\rangle|\phi_2\rangle, |\phi_1\rangle|\psi_2\rangle, |\psi_1\rangle|\phi_2\rangle, |\psi_1\rangle|\psi_2\rangle\}$ is an antidistinguishable set. From these assumptions, together with a mild side-assumption called the *principle of extendibility*,[3] it follows that, for any theory T such that Alice's credences satisfy the PUC for each of the preparation-acts $\{\phi_1 \otimes \phi_2, \phi_1 \otimes \psi_2, \psi_1 \otimes \phi_2, \psi_1 \otimes \psi_2\}$, Alice takes the effects of the preparation acts $\{\phi_i, \psi_i\}$ to be ontically distinct.

A key feature of quantum mechanics used in these proofs is the fact that, if $\{|\phi_i\rangle, |\psi_i\rangle\}$ are any state vectors with $|\langle\phi_i|\psi_i\rangle| \leq 1/\sqrt{2}$, then $\{|\phi_1\rangle|\phi_2\rangle, |\phi_1\rangle|\psi_2\rangle, |\psi_1\rangle|\phi_2\rangle, |\psi_1\rangle|\psi_2\rangle\}$ is an antidistinguishable set. What is needed for the argument is that there exist some preparations and an operation such that Alice's preparation-response conditional credences match the Born-rule probabilities yielded by these states for the outcomes of the experiment envisaged. This is not a necessity or probabilistic coherence. What we will assume is that Alice accepts quantum mechanics, in the following sense: for any quantum state ρ of a system, and any complete set of mutually orthogonal projections $\{P_1, \ldots, P_n\}$ on the Hilbert space of the system, there is *some* combination of preparation-act and readout-operation such that Alice's credences in the possible results $\{r_1, \ldots, r_n\}$ match (or at least closely approximate) the Born-rule probabilities $\text{Tr}(\rho P_k)$.[4]

The upshot of these theorems is that, for any theory T for which the assumptions (which may include both assumptions about the theory and about Alice's credences conditional on the supposition of the theory), Alice's credences, conditional on T, should take the effects of the preparation-acts considered to be ontically distinct. Thus, her credence in the ontic view of quantum states should be at least as high as her total credence in the class of all such theories. The only way that she can categorically deny that preparation-acts corresponding to distinct quantum states yield ontically distinct states of reality is to attach credence zero to the class of all such theories.

[3]This says that any system composed of N subsystems of the same type can be regarded as a part of a larger system consisting of a greater number of systems of the same type.

[4]Of course, we could reasonably expect something stronger, that this holds, not just for projections, but also for any positive-operator valued measure (POVM). But the weaker condition is all that we need.

20.5 The QBist Response

Advocates of QBism will not accept the conclusion that a subjectivist about quantum probabilities should be a realist about quantum states. Yet the argument so far has not violated any of the core tenets of QBism. To avoid the conclusion, QBists must add to its main tenets a further prohibition, logically independent of the explicitly stated core tenets.

In section 2, we first argued for what, in my opinion, ought to be an uncontentious claim: that an agent who regards a pair of preparation-acts to be distinguishable ought to regard the effects of these acts to be ontically distinct. This has the consequence that, if an agent associates orthogonal quantum states with a pair of preparation-acts, she ought to take their effects to be ontically distinct. The extension of this conclusion to nonorthogonal quantum states requires additional assumptions, which might be questioned. But the QBist will want to resist the very first step, having to do with orthogonal quantum states.

Nonetheless, the reasoning involved does not violate what proponents of QBism have identified as its fundamental tenets. These are given in different versions in different venues, but this is true for any of these versions.

Fuchs, Schack, and Mermin (2014) present the following as the "three fundamental precepts" of QBism.

1. A measurement outcome does not preexist the measurement. An outcome is created for the agent who takes the measurement action only when it enters the experience of that agent. The outcome of the measurement is that experience. Experiences do not exist prior to being experienced.
2. An agent's assignment of probability 1 to an event expresses that agent's personal belief that the event is certain to happen. It does not imply the existence of an objective mechanism that brings about the event. Even probability-1 judgments are judgments. They are judgments in which the judging agent is supremely confident.
3. Parameters that do not appear in the quantum theory and correspond to nothing in the experience of any potential agent can play no role in the interpretation of quantum mechanics.

Regarding Precept 1: We have not assumed that the result of a readout-operation (we have refrained from using the misleading term "measurement") preexists the operation, or that experiences exist prior to being experienced. Nor have we presupposed that the result of such an operation merely indicates a preexisting element of reality.

Regarding Precept 2: What we have concluded is that, if an agent takes two preparation-acts to be distinguishable with certainty, the agent should take the effects of those acts to be ontically distinct. The agent may be mistaken about this, of course. For example: I may be convinced that I will be able to learn tomorrow about my choices of key-strokes today, because I regard different choices of what to write today as readout-distinguishable tomorrow. Accompanying this belief is a belief that distinct choices of keystrokes today, if I save the file and don't erase it, have the effect of putting my laptop's internal mechanism into physically distinct states, which will have observationally distinguishable effects tomorrow. I could be wrong about this, of course. It might be that my laptop is malfunctioning and that there are, in fact, no lasting effects of my choices of keystrokes today. But—and this is the crucial point—if I were to revise my judgment about the ontic effects of my keystrokes, I would also revise my assessment of the readout-distinguishability of my choices tomorrow. What I would *not* do is simultaneously maintain that my choices today have no effect on the internal state of the laptop and that the result of the readout operation tomorrow will be informative about those choices.

Regarding Precept 3: The conclusion that an agent who takes two preparation-acts to be distinguishable should regard the effects of these acts to be ontically distinct is quite general, and is not specific to quantum mechanics. It invokes the sort of considerations that can be brought to bear on any sort of theorizing about the world. It should also be emphasized that the mere introduction of some sort of physical state-space Ω does not automatically consign us to the realm of hidden-variables interpretations of quantum mechanics. The structure of the physical state-space is left completely open, and our conclusion that one should take there to be distinctions in physical reality corresponding to distinctions between pure quantum state-preparations does not, by itself, commit us to anything else.

Fuchs (2017) presents the "three tenets" of QBism as

1. The Born rule is a normative statement.
2. All probabilities, including all quantum probabilities, are so subjective they never tell nature what to do. This includes probability-1 assignments. Quantum states thus have no "ontic hold" on the world.
3. Quantum measurement outcomes just are personal experiences for the agent gambling on them.

Regarding Tenet 1: Neither the Born rule, nor anything at all quantum-mechanical, was invoked in the argument for the conclusion that an agent should take the effects of preparation-acts that she regards as readout-distinguishable to be ontically distinct. A special case of this is a pair of preparations that are such that her credences can be represented by a orthogonal quantum states. In the extension to non-orthogonal quantum states, Born-rule probabilities were invoked only in consideration of what sorts of credences an agent who accepts quantum mechanics should have.

Regarding Tenet 2: Again, we have not assumed anything at all about what nature does. Our conclusion has only to do with the meshing of an agent's act-outcome conditional credences with her credences about the states of the physical system that mediates between the acts and the outcomes. As we have already mentioned in connection with the malfunctioning laptop, that system is under no obligation to conform to her credences, and, she may, indeed, find her expectations disappointed. If the last sentence in the statement of Precept 2 is supposed to indicate a denial of an ontic construal of quantum states, then, despite the "thus," it does not follow from the first. As we have seen, one can begin with the assumption that quantum probabilities are subjective and end with a conclusion that an agent whose credences conform to quantum mechanics ought to take quantum states to be ontic.

Regarding Tenet 3: Nothing in the argument depends on whether the results $\{r_k\}$ of the operation are taken to be things like states of the laptop screen, or states of the agent's nervous system, or her conscious experience upon looking at the screen.

Given that we have not, in our arguments, violated the fundamental precepts or tenets of QBism, why would, and how could, a QBist resist the conclusion? The only option open is to reject the claim that there is a causal relation between my choices of preparation-acts today and my experiences of the results of readout operations tomorrow that is mediated by the effects of my actions on the state of the physical system on which I act. There are two ways to do this. One is to maintain that there is a direct causal link between my present actions and future experiences, unmediated by any effect of my actions on the physical system on which I act. The other is to reject all talk of causal links between my present actions and future experiences.

There are passages in some writings by QBists that suggest the latter.

> There is a sense in which this unhinging of the Born Rule from being a "law of nature" in the usual conception ... i.e., treating it as a normative statement, rather than a descriptive one — makes the QBist notion of quantum *indeterminism* a far more radical variety than anything proposed in the quantum debate before. It says in the end that nature does what it wants, without a mechanism underneath, and without any "hidden hand" (Fine 1989) of the likes of Richard von Mises's *Kollektiv* or Karl Popper's *propensities* or David Lewis's objective chances, or indeed any conception that would diminish the autonomy of nature's events. Nature and its parts do what they want, and we as free-willed agents do what we can get away with. Quantum theory, on this account, is our best means yet for hitching a ride with the universe's pervasive creativity and doing what we can to contribute to it (Fuchs 2017, 272).

The suggestion is that the physical world obeys neither deterministic nor stochastic laws.

What sort of physical laws, if any, nature conforms to is not something that can be known a priori. If someone proposes a physical theory according to which one should expect to observe certain statistical regularities, that theory can be subjected to empirical test according to the usual Bayesian methods. This will never produce certainty that the theory is correct, but can result in high credence that the theory, or something like it, is at least approximately correct within a certain domain of application. In particular, if someone were to propose a theory according to which a given sort of preparation-act (say, passing a particle of a certain type through a Stern-Gerlach apparatus oriented vertically, and selecting the + beam) bestowed determinate chances on certain sorts of subsequent experiments, we could test that assertion by repeating the preparation-experiment sequence, and looking to see whether the relative frequencies converged to a stable value, in accordance with the weak law of large numbers. If we routinely find the theoretical expectation fulfilled, this should boost credence in the theory; if the outcomes of the experiment are not at all like what the theory leads us to expect, this counts as evidence against the theory. One could imagine that persistent failure to find *any* sort of regularity at all might boost credence in the proposition that there are no regularities to be found—

if the agent could survive that long! (It may be a requirement for the existence of agents that could have credences that there be some modicum of predictability in the world.)

Despite passages like the one quoted above, I am skeptical that any QBist actually rejects the claim that there is a causal link between their present actions and their future experiences, mediated by an effect of their actions on the physical systems with which they interact. I find it difficult to imagine how someone whose beliefs were like that would act. I am certain, however, that someone with beliefs like that would not judge a laptop that was advertised as having more memory—that is, more readout-distinguishable internal states that the user could choose between via choice of actions performed on the machine—to be worth more money than one with less memory. Indeed, it is hard to see why an agent like that could prefer any laptop at all to a kumquat as an implement for taking notes.[5]

Fuchs does not explain what, if anything, he takes to be evidence for the radical indeterminism of which he speaks. Such evidence *cannot* have come from the results of quantum experiments, which, on the contrary, seem to indicate law-like connections, best expressed in terms of probabilistic laws, between preparation-acts and experimental outcomes. At any rate, the evidence *against* the claim that the world is so completely devoid of law-like regularities that theorizing about it is pointless seems to be overwhelming.

There is, it seems, one possible response for the QBist. This would be to say that he accepts the thesis of radical indeterminism, of a world subject neither to deterministic nor stochastic laws, not on the basis of evidence, but as a leap of faith, or an expression of temperament (see Fuchs 2017, 251–253). Though he believes in no causal connection between his actions on the world and future experiences, as a psychological weakness he is unable to set his credences about the future at will, and can only affect his credences about the future by undergoing a ritual manipulation of the physical objects around him. He finds himself in the position of someone who does not believe that a horseshoe hung on the door will bring good luck, but hangs it anyway because he finds that doing so makes him more optimistic about the future. If that is the view—namely, that a QBist can offer no reason whatsoever to accept the radical renunciation of physical theorizing about the world that acceptance of the view entails—then we are in full agreement on that point.

[5]It is worth re-emphasizing: I am *not* claiming that anything about the laptop or the kumquat follows from assertions about the agent's credences about these systems. The agent may prefer the laptop to the kumquat as a note-taking device because of a *false* belief that the laptop is more suitable to play that role in a causal chain between the agent's actions and future experiences. The point is merely that the agent's belief, true or false, is a belief about the physical workings of the laptop.

20.6 An Argument from Locality?

In addition to the tenets and precepts listed above, QBism also involves a severe restriction on the application of quantum mechanics. An agent is required to only apply quantum probabilities to her own future experiences, and not to events (including the experiences of other agents) that she herself will not experience.

The severity of this restriction of scope is not always sufficiently emphasized. Though some of the ideas adopted by QBists are inspired by quantum information theory, the restriction of credences to one's own experiences eliminates vast swaths of quantum information theory, including everything to do with communication and cryptography. The point of a communication is to influence the credences of another agent, and, in designing a communication protocol, one must consider probabilities of changes to the recipient's belief-state.[6]

There may seem to be an advantage to this restriction: one thereby avoids quantum nonlocality. If an agent uses quantum mechanics to assign probabilities to events along her own worldline, and never to events at a distance from herself, all the events to which she assigns probabilities are timelike-related. There can be no question of action-at-a-distance because there are no distances between the events she assigns credences to.

This advantage obtains only if one refrains from doing any theorizing, quantum or otherwise, about events at a distance. Any theory, quantum or otherwise, that attempts to account for outcomes of tests of Bell inequalities—if we mean by 'outcomes' what is usually meant, namely, detector-registrations that occur at spacelike separation from each other—will have to violate at least one of any set of conditions sufficient to yield Bell inequalities.

The seeming advantage is illusory. If one refrains from giving an account of goings-on in the world that occur at a distance from each other *in order to avoid giving an account of nonlocal goings-on*, and if it is held that such a renunciation is *necessary* to avoid some sort of nonlocality that is thought to be objectionable, this is tantamount to saying that *if one were to give an account of goings-on in the world that occur at a distance from each other, such an account would involve the objectionable sort of nonlocality*. If this differs from acknowledging that the supposedly objectionable nonlocality is a feature of the world, I don't know how. It is as if one declined a request to evaluate a co-worker by saying that one is declining

[6]It might be claimed, in response, that Alice, in sending a signal to Bob, is concerned about the impact on Bob only insofar as it will later impinge on her. This, as everyone knows, is false. Some of the intended results of a communication are ones that the sender will never be aware of. As a particularly vivid example, consider a spy whose cover has been blown, who sends out one last encrypted message before taking a cyanide pill to avoid capture. A QBist must regard this as a misuse of quantum mechanics if the encryption scheme is a quantum scheme. It is not! Nor is it a misuse of quantum mechanics for an engineer designing a nuclear waste storage facility to use quantum mechanics to compute half-lives with the aim of constructing a facility that will be safe for generations to come.

the request in order to avoid saying that the co-worker is incompetent. In giving such a response, one has not refrained from conveying one's judgment of the co-worker's competence.

20.7 Conclusion

Subjectivitism about probabilities is not, by itself, an escape from ψ-ontology theorems. The theorems go through with all probabilities interpreted subjectively, with a conclusion that an agent whose credences satisfy the conditions of the proof should believe that preparation procedures that she associates with distinct pure quantum states leave the systems on which those procedures are performed in distinct ontic states. If this conclusion is to be avoided, some other stricture must be placed on the agent's credences.

The chief condition that underlies the proof is the assumption of the possibility of performing a preparation procedure that effectively screens off correlations between the system on which it is performed and the rest of the world, rendering a further specification of its physical state uninformative about the states of other systems. This sort of assumption lies deep at the heart of virtually all experimental science. None of the empirical evidence that motivates a shift from classical to quantum theory provides grounds for rejecting assumptions of this sort. If we ever come to a point at which we have reasons to doubt this sort of assumption, this will not come about as a result of experiments that presuppose it. And if we were presented with a reason to doubt this sort of assumption, it is hard to see how this doubt could be sufficiently contained so as not to undermine all of experimental science. Fortunately, we are not in such a position.

References

Caves, C. M., Fuchs, C. A., & Schack, R. (2002). Quantum probabilities as Bayesian probabilities. *Physical Review A, 65*, 022305.

Caves, C. M., Fuchs, C. A., & Schack, R. (2007). Subjective probability and quantum certainty. *Studies in History and Philosophy of Modern Physics, 38*, 255–274.

Clifton, R., & Halvorson, H. (2000). Bipartite-mixed-states of infinite-dimensional systems are generically nonseparable. *Physical Review A, 61*, 012108.

Clifton, R., & Halvorson, H. (2001). Entanglement and open systems in algebraic quantum field theory. *Studies in History and Philosophy of Modern Physics, 32*, 1–13.

Fine, A. (1989). Do correlations need to be explained? In J.T. Cushing and E. McMullin (Eds.), Philosophical Consequences of Quantum Theory: Reflections on Bell's Theorem. *Notre Dame, IN: University of Notre Dame Press, Notre Dame*, 175–194.

Fuchs, C. A. (2017). Notwithstanding Bohr, the reasons for QBism. *Mind & Matter, 15*, 245–300.

Fuchs, C. A., & Schack, R. (2013). Quantum Bayesian coherence. *Reviews of Modern Physics, 85*, 1693–1715.

Fuchs, C. A., & Stacey, B. (2019). QBism: Quantum theory as a hero's handbook. In E. M. Rasel, W. P. Schleich, & S. Wölk (Eds.) *Foundations of quantum theory: Proceedings of the international school of physics "Enrico Fermi" course 197*, pp. 133–202. Amsterdam: IOS Press.

Fuchs, C. A., Mermin, N. D., & Schack, R. (2014). An introduction to QBism with an application to the locality of quantum mechanics. *American Journal of Physics, 82*, 749–754.

Harrigan, N., & Spekkens, R. W. (2010). Einstein, Incompleteness, and the Epistemic View of Quantum States. *Foundations of Physics, 40*, 125–157.

Leifer, M. S. (2014). Is the quantum state real? An extended review of ψ-ontology theorems. *Quanta, 3*, 67–155.

Myrvold, W. C. (2018). ψ-ontology result without the Cartesian product assumption. *Physical Review A, 97*, 052109.

Myrvold, W. C. (2020). On the status of quantum state realism. In S. French & J. Saatsi (Eds.) *Scientific realism and the quantum*. Oxford: Oxford University Press, 229–251.

Pitowsky, I. (2003). Betting on the outcomes of measurements: A Bayesian theory of quantum probability. *Studies in History and Philosophy of Modern Physics, 34*, 395–414.

Pusey, M. A., Barrett, J., & Rudolph, T. (2012). On the reality of the quantum state. *Nature Physics, 8*, 475–478.

Chapter 21
The Relativistic Einstein-Podolsky-Rosen Argument

Michael Redhead

Abstract We present the possibility of a relativistic formulation of the Einstein-Podolsky-Rosen argument. We pay particular attention to the need for a reformulation of the so-called reality criterion. We introduce such a reformulation for the reality criterion due to Ghirardi and Grassi and show how it applies to the nonrelativistic EPR argument. We elaborate on Ghirardi and Grassi's proof and explain why it cannot be circumvented. Also we draw a new conclusion by assuming locality. Finally, we review and summarise our own views. This is a continuation of the paper by myself and Patrick La Rivière whose defects are exposed, and corrected, in the present work. In the Appendix locality principles are conveniently spelled out.

Keywords New critique of Ghirardi and Grassi's paper assuming locality · New refutation of Redhead and Rivière's paper · Nonlocality issues revisited

21.1 Preliminaries

At first glance, the realist interpretations of quantum mechanics such as Bohm's (see Bohm and Hiley 1993) offer many advantages over standard interpretations of the theory. In particular, they give a clear, intuitive picture of many potentially paradoxical, physical situations, such as the two-slit experiment and the phenomenon of barrier penetration. At the same time, their chief drawback—a form of nonlocality, as we have claimed, to conflict with the constraints of relativity theory – is apparently shared by the standard 'anti-realist' interpretations that reject hidden variables and assume completeness, as was demonstrated by the original Einstein-Podolsky-Rosen (EPR) argument (see Einstein et al. 1935).

M. Redhead (✉)
Centre for Philosophy of Natural and Social Science, London School of Economics and Political Science, London, UK
e-mail: mlr1000@cam.ac.uk

M. Hemmo, O. Shenker (eds.), *Quantum, Probability, Logic*,
Jerusalem Studies in Philosophy and History of Science,
https://doi.org/10.1007/978-3-030-34316-3_21

While the Bell argument that establishes nonlocality for realistic interpretations such as Bohm's has been formulated in an experimental relativistic context (see Bohm and Hiley, Chap. 12, 1993) there is no well-established relativistic formulation of the EPR argument. In the absence of such a formulation, it seems hasty to conclude that the tension between the standard interpretations and relativity theory is just as great as that between detailed Bohmian interpretations and relativity. Clearly, if a relativistic formulation of EPR could be given that did not entail nonlocality, antirealist interpretations would have an advantage over the Bohmian interpretation. (To be sure anti-realist interpretations that do not require nonlocality include Healey's pragmatism (see Healey 2017), Qbism (see Caves et al. 2002) and epistemicism (see Williamson 1994).

Let us now investigate the possibility of a relativistic formulation of the EPR argument. First, the standard nonrelativistic version of the EPR argument is reviewed and the problematics of translating it into a relativistic context are considered, paying particular attention to the need for a reformulation of the so-called reality criterion. Then, we introduce one such reformulated reality criterion, due to GianCarlo Ghirardi and Renata Grassi (see Ghirardi and Grassi 1994) and show how it is applied to the nonrelativistic EPR argument. Next, we discuss the application of the new reality criterion in a relativistic context.

A relativistic version of the EPR argument must differ from the nonrelativistic version in two principal ways. First, the particle states must be described by a relativistic wave-function. The details don't concern us here; we need only require that the wave-function preserve the maximal, mirror-image correlations of the nonrelativistic singlet state. And indeed, the existence of maximal correlations in the vacuum state of relativistic algebraic quantum field theory has recently been demonstrated (see Redhead 1995). Second, the argument must not depend on the existence of absolute time ordering between the measurement events on the left and right wings of the system, for in the relativistic argument these may be spacelike separated. As it turns out, the nonrelativistic version of the argument does invoke absolute time ordering. To see how to get around this problem, we must briefly review the standard formulation of the incompleteness argument.

For EPR, a necessary condition for the completeness of a theory is that every element of physical reality must have a counterpart in the theory. To demonstrate that quantum mechanics is incomplete, EPR need simply point to an element of physical reality that does not have a counterpart in the theory. In this vein, they consider measurements on a pair of scattered particles with correlated position and momentum, but the formulation of the argument in Bohm (see Bohm 1951) in terms of a pair of oppositely moving, singlet-state, spin-1/2 decay products of a spin-0 particle, is conceptually simpler. In this case, the formalism of quantum mechanics demands a strict correlation between the spin components of the two spatially separated particles, such that a measurement of, say, the z-component of spin of one particle allows one to predict with certainty the outcome of the same measurement on the distant particle . This ability to predict with certainty, or at least probability one, the outcome of a measurement is precisely the EPR criterion for the existence of an element of reality at the as-yet-unmeasured particle. By invoking one final assumption, a locality assumption stating that elements of reality pertaining to one system cannot be affected by measurements performed 'at a distance' on another system, EPR can establish that the element of reality at the unmeasured particle must have existed even before the measurement was performed at the distant particle . But the quantum formalism describes the particles at this point with the singlet state, and thus has no counterpart for the element of reality at the unmeasured particle. It follows that the quantum description was incomplete. Schematically,

$$\text{Quantum Formalism} \wedge \text{Locality} \rightarrow \sim \text{Completeness.} \tag{21.1}$$

Alternatively, if one assumes completeness, the argument may be rearranged as a proof of nonlocality:

$$\text{Quantum Formalism} \wedge \text{Completeness} \rightarrow \sim \text{Locality.} \tag{21.2}$$

The problematic assumption of absolute time ordering entered the argument in the reality criterion, which turns on the possibility of predicting with certainty the outcome of a measurement along one wing *subsequent* to having obtained the result of a measurement along the other. Of course, for spacelike separated events, notions like precedence and subsequence are reference-frame dependent, not absolute. So to translate the EPR argument to a relativistic context requires a modified criterion for the attribution of elements of reality that is not contingent on the time ordering of the measurement events. Ghirardi and Grassi have undertaken to formulate just such a criterion. For the sake of clarity, we shall first describe how this criterion applies to the nonrelativistic version of the argument.

Ghirardi and Grassi's criterion rests on the truth of certain classes of counterfactual statements-statements of the form 'if φ were true, then ψ would be true' where the antecedent φ is in general known to be false. In particular, they wish to link the attribution at a time t of the property corresponding to [observable **a** having value a] to the truth of the counterfactual assertion: if measurement of **a** were performed at time t, then the outcome would be a (they use David Lewis's analysis of counterfactuals here. See Lewis 1986).

With this criterion in hand, Ghirardi and Grassi can now run the *nonrelativistic* EPR argument essentially as before. They assume a measurement of property **a** is performed on the right-hand particle at time t_R, yielding a specific result a. To ascertain whether an element of reality corresponding to property $a = a'$ exists at the left-hand particle, they must assess the truth of the counterfactual assertion: 'if I were to perform a measurement of property **a** at the left-hand particle at time t_L, I would obtain the result a$'$.' In the nonrelativistic case, the truth of this counterfactual assertion follows naturally from the presence of absolute time ordering. For if $t_R < t_L$, then the outcome of the right-hand measurement can be assumed to be the same in all of the 'accessible' (most similar) worlds used to evaluate the counterfactual, because it is strictly in the past of the counterfactual's antecedent. The strict correlation laws of quantum mechanics, also assumed to hold in all accessible worlds, then demand that the result of a measurement on the left wing also be fixed in all possible worlds (specifically, the laws require that $a' = -a$). Thus the counterfactual is true, and an element of reality can be said to exist at the left-hand particle. From here, the argument unrolls in the usual way, and by supplementing this reality criterion with a locality assumption (they call it G-Loc, after Galileo), Ghirardi and Grassi can deduce that quantum mechanics is incomplete. Once again, we can represent their argument schematically by

$$\text{Quantum Formalism} \wedge \text{G-Loc} \rightarrow \sim \text{Completeness} \tag{21.3}$$

or

$$\text{Quantum Formalism} \wedge \text{Completeness} \rightarrow \sim \text{G-Loc}. \tag{21.4}$$

21.2 The Crux of the Argument

'A system cannot be affected by actions on a system from which it is isolated. In particular, elements of physical reality of a system cannot be influenced by actions on systems from which it is isolated.' An examination of the structure of Ghirardi and Grassi's argument reveals that they make use not of the general principle stated but of a special case of this general principle, namely that elements of reality cannot be brought into existence 'at a distance.' It is this special case of G-Loc, call it ER-Loc (for elements of reality) that enters toward the end of the argument to establish that the measurement at the right wing could not have created an element of reality at the left wing and thus it must have existed prior to the measurement at the right wing, when the quantum formalism said the particles were in the singlet state. Thus they conclude that quantum mechanics is incomplete. All is well so far, but when one turns the argument around, assuming completeness and dispensing with locality, one must ask, can one be more precise as to which locality principle should be given up: the principle they label G-Loc, or the special case ER-Loc? Indeed it is the latter, for only it entered into the argument. As it turns out the distinction between G-Loc and ER-Loc does not affect their conclusions in the nonrelativistic case, because the conclusion they choose to highlight- the creation of elements of reality at a distance- is precisely one that does follow from dispensing only with ER-Loc.

In the relativistic case they want to claim peaceful coexistence between nonlocality and special relativity. However, greater care must be taken with the statement of the locality principle, this time called L-Loc (after Lorentz by Ghirardi and Grassi), because a locality principle must enter at the very beginning of the argument as well as in the usual way at the end. The argument begins in the same way as in the nonrelativistic case, with the occurrence of a measurement on the right-hand side, but now the absence of absolute time ordering means the result of this measurement can no longer tacitly be assumed to be the same in all the accessible worlds used to evaluate the element-of-reality counterfactual at the lefthand side. Locality must be invoked to establish the independence of the outcome of the righthand measurement from the occurrence of the measurement at the left. This done, Ghirardi and Grassi then demonstrate the existence of an element of reality at the left-hand side following the same reasoning as above. From here, the argument unrolls once again in the usual way and locality makes a second appearance in its familiar place at the end of the argument. In this way, Ghirardi and Grassi can again prove that standard quantum mechanics plus 'locality' implies incompleteness.

But there are two quite distinct cases of L-Loc that are actually being employed, one used in getting the argument started and the other appearing in the conclusion. Ghirardi and Grassi define L-Loc as the following: 'An event cannot be influenced by events in spacelike separated regions. In particular, the outcome obtained in a measurement cannot be influenced by measurements performed in spacelike separated regions; and analogously, possessed elements of physical reality referring to a system cannot be changed by actions taking place in spacelike separated regions.' As in the nonrelativistic case, it is not the general principle but rather the two special cases, call them OM-Loc (for outcome of measurement) and ER-Loc (again for element of reality), that are doing the logical work in their argument. OM-Loc affirms that the outcome of a measurement cannot be influenced by performing another measurement at a spacelike separation, while ER-Loc affirms that elements of reality cannot be created by performing a measurement at spacelike separation. Ghirardi and Grassi invoke OMLoc at the beginning of the argument while applying the counterfactual reality criterion, as discussed above, and they invoke ER-Loc at the end of the argument, as they did in the nonrelativistic case. So if we write L-Loc = OM-Loc ^ ER-Loc, then, schematically, their argument looks like this:

$$\text{Quantum Formalism} \wedge \text{OM-Loc} \wedge \text{ER-Loc} \rightarrow \sim \text{Completeness} \tag{21.5}$$

or

$$\text{Quantum Formalism} \wedge \text{Completeness} \rightarrow \sim \text{OM-Loc} \vee \sim \text{ER-Loc}. \tag{21.6}$$

Ghirardi and Grassi now argue, in effect, as follows. Assuming OM-Loc we can again demonstrate from Completeness a violation of ER-Loc, i.e. Einstein's *spooky* action-ata-distance *creating* elements of reality at a distance. But if we don't assume OM-Loc, then we cannot deduce a violation of ER-Loc. All this is quite correct, but the price we have to pay for *not* being able to demonstrate a violation of ER-Loc is precisely that we have to accept a violation of OM-Loc!

In other words, the relativistic formulation of the EPR argument does not help with the thesis of peaceful coexistence between quantum mechanics and special relativity, unless one argues that violating ER-Loc is more serious than violating OM-Loc from a relativistic point of view. This is hard to maintain since violating OM-Loc involves a case-by-case version of what Shimony (see Shimony 1984) refers to as violating parameter independence, i.e. the independence of the probability of the outcome on one wing of the EPR experiment with respect to the parameters controlling the type of experiment being performed on the other wing. By analogy, violating ER-Loc is also a form of parameter dependence.

So to return again, Ghirardi and Grassi's claim of peaceful coexistence is already justified, with a novel idea of how it works out, a violation of OM-Loc.

21.3 The Thesis of Myself and La Rivière's Paper

However they say they need to identify an additional assumption omitted from (21.5) and (21.6), which, if challenged, could undermine the inference. Recall that to run the argument in either the nonrelativistic or relativistic case, Ghirardi and Grassi must establish that the outcome of, say, the right-hand measurement is the same in all accessible worlds. With this established, the correlation laws of quantum mechanics imply that the outcome of the left-hand measurement is the same in all accessible worlds, and hence establish the truth of the counterfactual assertion about the left-hand measurement result that permits the attribution of an element of reality to the left hand particle. In the nonrelativistic case, the constancy of the right-hand result is a natural consequence of the absolute time ordering as discussed above; in the relativistic case. A premise akin to one that Michael Redhead labels the Principle of Local Counterfactual Definiteness (PLCD) (see Redhead, p. 92, 1989) is needed to do this sort of work.

In the present case, PLCD may be taken to assert that the result of an experiment that *could* be performed on a microscopic system has a definite value that does not depend on the occurrence of a measurement at a distant apparatus. Ghirardi and Grassi implicitly assume that PLCD is licensed by their locality principle, for they invoke only OM-Loc to establish the constancy of the right-hand outcome in all accessible worlds. But we argue that PLCD does *not* follow directly from any typical locality principle, certainly not from one like OM-Loc, which asserts that the outcome obtained in a measurement cannot be influenced by measurements in spacelike separated regions. The reason is quite simple: while invoking locality may prevent measurements on the left hand particle from influencing the result at the right and from breaking the constancy of the accessible worlds as far as the right-hand result is concerned, it does not prevent *indeterminism* from wreaking that sort of havoc. Intuitively, we can imagine that we run the world over again, this time performing the measurement on the left-hand particle. If we consider this left-hand measurement schematically as a point event with a backward light cone identical to that in the actual world, we are concerned with what will happen in the complement of the forward and backward light cones. Under indeterminism the events in this complement (the absolute elsewhere) simply cannot be assumed to remain the same.

Thus it is maintained that Ghirardi and Grassi need both OM-Loc and an assumption of determinism to get their argument off the ground.

We need, schematically, to replace (21.6) by

$$\text{Quantum Formalism} \wedge \text{Completeness} \wedge \text{Determinism} \rightarrow \sim \text{OM-Loc v} \sim \text{ER-Loc.} \tag{21.7}$$

Thus,

$$\text{Determinism} \rightarrow \text{OM-Loc} (\rightarrow \text{PLCD})$$

and hence by the disjunctive syllogism, we can infer ~ER-Loc,
So text of myself and La Rivière (see Redhead and La Rivière 1997) ends here.

21.4 But there Is a Catch . . .

But there is a catch. We note first that Determinism ➔ Bell Inequality (assuming randomness assumption is assumed to be true, for which independent engagement can be given). But this contradicts the conclusion that the Bell Inequality is false. So we cannot infer ~ER-Loc.

But, according to Lewis (private communication), OM-Loc ➔ PLCD and hence allows the Bell Inequality, independently of Determinism. The argument here appears to be the following. For possible worlds we allow so-called quasi-miracles to happen, but also they might not happen. To prove the counterfactual we need to close down worlds where miracles happen, leaving the counterfactual to succeed.

We disagree with this claim since it involves discussion of possible worlds, whereas in our world the counterfactual comes out as false.

Put succinctly, the argument is the following. A counterfactual supposition about an event can, and indeed should, be taken to preserve as much as possible of the actual facts, compatible with the truth of the supposition. Since in a relativistic setting, a counterfactual supposition about an event is compatible with preserving the the actual facts that are separated from it. It seems to us, under indeterminism, one's treatment of a counterfactual supposition should 'respect' the indeterminism by allowing the ways that history could play out, compatibly with the counterfactual supposition to be considered.

But if we follow Lewis we again allow the Bell Inequality so on both counts the attempt to rescue the justification of ~ER-Loc fails.

We make a remark on Ghirardi and Grassi's proof on peaceful existence. For Ghirardi and Grassi there are two sorts of argument: (1) EPR argument assuming locality; (2) argument showing, correctly, how nonlocality works in violating Bell. I confused the two sorts of arguments, starting with (1) and trying to prove peaceful existence according to (2)! I learned from my mistakes as Popper would have said!

But notice how we have shown, using (1), that peaceful coexistence actually works out.

21.5 Conclusion

So what conclusion concerning nonlocality can we draw? Taking first the realist option, in which all observables have sharp values at all times, if we assume determinism or restrict the discussion to cases of strict correlation or anticorrelation where we can derive determinism on plausible assumptions, there seems very little scope for avoiding the conclusion of nonlocality for any 'realist' reconstruction of quantum mechanics. But experimentally we never observe absolutely *strict* correlations, so it may be argued that discussions of nonlocality should not deal with this ideal case, but should be based upon the experimentally realistic nonideal case. We are not convinced by this argument, since idealization is an essential aspect of any scientific theorizing. Be that as it may, the assumption of determinism for the nonideal case where correlations are less than perfect, may well be suspect. In such cases one is forced, as we have seen, to go to the stochastic hidden-variable framework, and much of the recent discussion has focused on the significance of violating the locality assumptions involved in this framework. Following Jon Jarrett (Jarrett 1984) these break down into two classes.

1. Independence of the probability for a particular outcome of measurement on one particle, conditionalized on the collective hidden variables, from the *outcome* of measurement carried out on a remote particle. This is what Shimony calls parameter independence, which is already referred to above.
2. Independence of the probability for a particular outcome of measurement on one particle conditionalized on the collective hidden variables, from the type of measurement being carried out on a remote particle. This is what Shimony calls outcome independence. (Shimony 1984).

Violation of (1) has been generally held to mark probabilistic causal dependence between the outcome of a local measurement and the 'setting' of a remote piece of apparatus, in prima facie conflict with standard interpretations of special relativity as prohibiting causal links between spacelike separated events. Violation of (2) has led to more controversy. It seems to demonstrate that the precise specification of the state of the source, or more generally of the particles before the measurements are undertaken, cannot be regarded as a common cause of the measurement outcomes. Indeed a direct causal connection exists between the events consisting of the measurement results on distant particles obtaining in the way they actually do. We have challenged this view (Redhead, pp.90–96, pp.98–107, 1989) by proposing to interpret violation of (2), in the way proposed, but in terms of a noncausal direct dependence between measurement outcomes that lacks the necessary 'robustness' in respect of how the measurement results are brought about, to merit description as a causal connection. This approach gives up the stochastic hidden-variable framework in favour of trying to understand the correlations between distant events in terms of a harmony-at-a distance, or, as Shimony describes it, a passionat-a distance (Shimony 1984).

This idea ties in with, but is really quite distinct from, the fact that violation of (2) cannot be used to transmit signals between distant locations (Redhead, b p.106, 1989) since the necessity to recover quantum probability distributions by averaging over the hidden variables, means that we have no way of controlling the local marginal distributions for the results of measurement on one particle, by varying the type of measurement performed on distant particles. The no-signalling result is often cited to defuse the tension between nonlocality in quantum mechanics and the constraints of special relativity, but, at a deeper level, the 'nonrobustness' argument may be preferable.

Turning to the anti-realist option, we have examined Ghirardi and Grassi's attempt to reformulate the EPR argument in a relativistic context and argued that peaceful coexistence does arise.

There remains of course the question of how to interpret the violation of outcome independence. Assuming completeness, outcome dependence famously follows. In the EPR set-up this means that when measurements are performed at spacelike separation on the two wings of the experiment, the results are mirror-image correlated. As one potentiality gets actualized on the left, say, how does this happen exactly in tandem with the opposite result on the right? Are we faced with a causal effect, namely result-to-result causation, so that peaceful coexistence with relativity is still challenged? Similar answers involving no-signalling or nonrobustness may be provided as in the discussion above of the 'realist' option.

But it must be stressed that the mysterious harmony of the result-to-result correlations remains arguably 'spooky' even if it does not involve causal dependence.

Another distinguishing feature of this harmony is its symmetrical character, quite unlike the asymmetry that one would normally want to ascribe to a causal connection. Shimony's phrase 'passion-at-a -distance' seems exactly the right one to capture what is going on, even if one concedes that the mystery of the EPR correlations is not eliminated merely by introducing an apt nomenclature.

For the anti-realist the role of measurement is to actualize potentialities. If there are no measurements, then there are no actualities! This observation is particularly relevant to cosmological applications of quantum mechanics, where there is nothing 'outside' the universe to serve as a measuring device! So, from the cosmological perspective, one can argue that the realist option is after all to be preferred. This assumes the universe is finite in spacial extent. For an infinite universe the argument fails. I owe this comment to my erstwhile student Tian Yu Cao.

Here one can argue that the nonseparability approach, as exemplified in blocking the nonlocality proof by denying the 0-Loc principle, may in the end be the best way of understanding the peculiar features of entangled states in quantum mechanics.

Many people find it difficult to see the distinction between violating 0-Loc (nonseparability) and violating E-Loc (actionat-distance). A simple example due to David Lewis (private communication) may help here. Consider someone with what she calls a bilocal hand. There is in reality just one hand that is manifested in two different places. If you shake hands with this curiously disabled person, her other hand will move in synchrony with the one you are shaking, not because there is an interaction between the two hands, but just because, in reality, there is only one hand, bilocally located! According to Lewis, this is an example of nonseparability as distinct from action-at-distance.

Acknowledgements I acknowledge Jeremy Butterfield and Bryan Roberts for reading the manuscript, and for helpful comments.

A.1 Appendix

A.1.1 Realist Interpretations

E-Loc: Local elements of reality cannot be affected by changes in a distant, spacelike separated environment.

O-Loc: Local elements of reality can be specified independently of a holistic context.

A.1.2 Anti-Realist Interpretations

G-Loc: Events cannot be affected by a distant measurement performed simultaneously.

L-Loc: Same as for G-Loc but with 'simultaneously' replaced by 'at spacelike separation'.

ER-Loc: Elements of reality cannot be created by distant measurements performed simultaneously/at spacelike separation – a special case of G-Loc/L-Loc.

OM-Loc: The outcome of a measurement at one location cannot be affected by a distant measurement performed at spacelike separation – another special case of L-Loc.

PLCD: The outcome of a measurement that could be performed has a definite value independent of whether or not another measurement takes place at a distant spacelike separated location. OM-Loc implies PLCD under an assumption of determinism, but not of indeterminism.

Parameter independence: The *statistics* of measurement results at one location is independent of the parameters defining a measurement procedure at a distant spacelike separated location.

Outcome independence: The *statistics* of measurement results at one location is independent of the outcome of measurements performed at a distant spacelike separated location.

References

Bohm, D. (1951). *Quantum theory*. Englewood Cliffs: Prentice-Hall.

Bohm, D., & Hiley, B. J. (1993). *The undivided universe*. Routledge: Taylor & Francis.

Caves, C. M., Fuchs, C. A., & Shack, R. (2002). Unknown quantum states: The de Finnetti representation. *Journal of Mathematical Physics, 43*, 4537–4559.

Einstein, A., Podolsky, B., & Rosen, R. (1935). Can quantum-mechanical description of physical reality be considered complete? *Physical Review, 47*, 777–800.

Ghirardi, G., & Grassi, R. (1994). Outcome predictions and property attribution: The EPR argument reconsidered. *Studies in History and Philosophy of Science, 25*, 397–423.

Healey, R. (2017). *The quantum revolution in philosophy*. Oxford: Oxford University Press.

Jarrett, J. (1984). On the physical significance of the locality conditions in the bell arguments. *Nous, 18*, 569–589.

Lewis, D. (1986). *Philosophy papers* (Vol. 2). Oxford: Oxford University Press.

Redhead, M. L. G. (1989). *Incompleteness, nonlocality and realism*. Oxford: Clarendon Press.

Redhead, M. L. G. (1995). More ado about nothing. *Foundations of Physics, 25*, 123–137.

Redhead, M. L. G., & La Rivière, P. (1997). The EPR relativistic argument. In R. Cohen, M. Horne, & J. Stachel (Eds.), *Potentiality, entanglement and passion-at-a-distance. Quantum Mechanical Studies for Abner Shimony* (Vol. 2, pp. 207–215). Dordrecht: Kluwer.

Shimony, A. (1984). Controllable and uncontrollable non-locality. In S. Kamafuchi (Ed.), *Proceedings of the international symposium: Foundations of quantum mechanics in the light of new technology* (pp. 225–230). Tokio: Physical Society of Japan.

Williamson, T. (1994). *Vagueness*. London: Routledge.

Chapter 22
What Price Statistical Independence? How Einstein Missed the Photon

Simon Saunders

Abstract Einstein's celebrated 1905 argument for light quanta was phrased in terms of the concept of 'statistical independence'– wrongly. It was in fact based on the notion of local, non-interacting entities, and as such was neutral on the question of statistical independence. A specific kind of statistical dependence had already been introduced by Gibbs, in connection with the Gibbs paradox, but went unrecognised by Einstein. A modification of Einstein's argument favours the latter. This kind of statistical dependence, as eventually attributed by Einstein to Bose, although expressed, in quantum mechanics, by the (anti)symmetrisation of the wave-function, does not require 'genuine' entanglement. This chapter demystifies quantum indistinguishability; its novel features derive from the finiteness of state-space in quantum mechanics not indistinguishability.[1]

Keywords Einstein · Boson · Photon · Gibbs paradox

22.1 Overview

According to the physicist and historian Abraham Pais, the photon was the first particle to be predicted theoretically. It was also, he noted, met with scepticism on an epic scale: '[i]ts assimilation came after a struggle more intense and prolonged than for any other particle ever postulated'.[2] Yet it was also, in a certain sense, *derived* from empirical data – specifically, derived by Einstein from the Wien limit of Planck's black-body spectral distribution, taken as a phenomenological given

[1]This chapter is based on Saunders (2018, 2020), but it's focus is different from either.

[2]Pais (1982, p. 361).

S. Saunders (✉)
Faculty of Philosophy, University of Oxford, Oxford, UK
e-mail: simon.saunders@philosophy.ox.ac.uk

M. Hemmo, O. Shenker (eds.), *Quantum, Probability, Logic*,
Jerusalem Studies in Philosophy and History of Science,
https://doi.org/10.1007/978-3-030-34316-3_22

(it was by then accurately confirmed at high frequencies and low temperatures). The derivation was what John Norton has called Einstein's 'miraculous argument', contained in the most revolutionary of all of Einstein's great papers of 1905.[3]

If it was derived from the data, how was it opposed? The answer in part is that it was incomplete: whatever light was really made of, it could not *only* be Einstein's light quanta, because if it were, it would contravene the Planck distribution in the low frequency high-temperature regime (the Rayleigh-Jeans regime, also by then accurately confirmed). What was the nature of the struggle, that was so intense and prolonged? The answer in part was the difficulty of reconciling a particulate structure to radiation with Maxwell's equations (or any wave equation, needed to explain interference effects), but another part was down to Einstein's insistence on statistical independence. It was *Einstein's* struggle; and it was only concluded (and then only temporarily) with his quantum theory of the ideal monatomic gas, published in three instalments in the period 1924–1925.

To see the nature of the difficulty, consider that radiation in thermal equilibrium behaves as a gas of indistinguishable particles, photons, satisfying the Bose-Einstein statistics; and that the latter, as Einstein admitted in his second paper of his gas trilogy, implies that particles are not statistically independent of each other; yet Einstein's 'miraculous argument' of 1905 had shown that thermal radiation, in the Wien regime, is made of statistically independent particles. Contradiction.

An easy way out is to suppose that in the Wien regime photons just are statistically independent of each other; and it is true that in this regime they satisfy Maxwell-Boltzmann statistics. However, the latter is not the signature of statistical independence: classical indistinguishable particles, in Gibbs sense, obey Maxwell-Boltzmann statistics, but they are not statistically independent in the requisite sense.

The better answer is that Einstein did *not* show that radiation, in the Wien regime, is composed of statistically independent particles. His argument showed that in the Wien regime thermal radiation is made of *non-interacting, localised* entities; it left open the question of whether they were indistinguishable, in Gibbs' sense, or statistically independent, in Einstein's. However, as we shall see, a slightly different kind of fluctuation than the one Einstein considered resolves the under-determination in favour of the Gibbs concept – for roughly the reasons Gibbs himself advocated it, namely, as a solution to what is now called the Gibbs paradox.

Had Einstein in 1905 understood this point, the history of the development of quantum theory would have been significantly different, for it would have been a relatively short step to Einstein's quantum gas theory. It is therefore something of an embarrassment that Gibbs' concept of indistinguishability, introduced in terms of the notion of 'generic phase' (as opposed to 'specific phase', as had hitherto been implicitly assumed), was to be found in his *Elementary Principles in Statistical Mechanics*, published in 1902, and indeed already published in German translation in 1904.

[3]Norton (2006).

It seems that Einstein knew nothing of Gibbs' work at this time, but we know that he had read at least parts of the *Principles* by 1910 (and that it impressed him greatly)[4]; however, we also know that he was sceptical of Planck's claims, in the period 1912–1918, on the extensivity of the entropy function of a material gas, that cited Gibbs[5] (the failure of extensivity is one formulation of the Gibbs paradox). When Einstein embraced essentially the same solution to the extensivity puzzle in the second of his papers on gas theory, it was in the name of what he called 'Bose's statistical approach', not Gibbs'.

Bose himself, by his own admission, was unaware that his approach had differed from the usual one – and, indeed, his procedure so obscured the underlying statistical assumptions that Einstein, when he took it over in his first paper on gas theory, did not realise (or at any rate did not mention) that it implied the failure of statistical independence. In his second paper, published at the beginning of 1925, Einstein presented the expression for the entropy in a different form, from which this failure was 'easily recognised'; but he still attributed the approach to Bose. That expression for the entropy, I am claiming, is the natural transposition of Gibbs' notion of generic phase to the case where the one-particle state space has a finite number of states (when it is divided into 'elementary cells', of size h^3) – the central assumption of the 'old' quantum theory.

No more did Schrödinger identify Gibbs as the progenitor of 'Bose's approach', although he came close. He had publicly criticised Planck 's arguments for an extensive entropy, that drew on Gibbs', and now found fault with Einstein's new gas theory precisely on the point that, following Bose's method, particles could not be statistically independent of each other. Einstein, by then, was prepared to grant that it was for nature to decide; not so Schrödinger, who went on (in a paper immediately preceding the first of his historic papers on wave mechanics) to propose a new quantisation method for the quantum gas that maintained statistical independance. Neither Einstein nor Schrödinger mentioned Gibbs by name, nor the concept of generic phase. The following year, in 1926, Dirac laid down the definitive treatment of particle indistinguishability in quantum mechanics, again with no reference to Gibbs (or to Schrödinger on gas theory) – and few physicists looked back to its precursors since.

Historians, meanwhile, seem to have taken Einstein, Schrödinger, and Dirac rather too much at face value; since none of them mentioned Gibbs or the generic phase, the equivalence of the latter with 'Bose's approach' has not been recognised, or even seriously entertained.

There are other reasons why the equivalence was not recognised. One is because Bose-Einstein statistics is directly related to the symmetrisation of the state in quantum mechanics (Dirac cited Bose and Einstein). But the latter, by wide consensus, is thought to involve entanglement, a quintessentially quantum concept. That it is in fact entanglement of a particularly trivial kind – that does not, for

---

[4]As recounted by Pais (1982, p. 55).

[5]From correspondence with Ehrenfest in 1918, quoted in Darrigol (1991, pp. 288–89).

example, support the violation of any Bell inequality – has only recently been shown. Another reason the equivalence has not been recognized comes from the classical side, where it is thought that the concept of particle indistinguishability cannot apply to classical mechanics 'because classical particles can always be distinguished by their trajectories'. That phrase, or one like it, recurs again and again in the subsequent literature.[6] But it meets the counter-argument that *quantum* particles can also have trajectories – either approximately, for sufficiently massive and weakly-interacting particles in well-localised states, or exactly, in the case of non-interacting particles, taking 'trajectories' as orbits of orthogonal one-particle states. In neither case do the particles cease to be indistinguishable, in the sense that their total state remains symmetrised[7] We shall see a simple example in the concluding section.

It appears that the struggle initiated by Einstein with the photon concept is so intense, and so prolonged, that it continues unabated to this day. I can offer up nothing quite as remarkable as Itamar Pitowsky's discovery, in George Boole's 'impossibilities of experience', of Bell's inequalities. His beautiful theory of correlation polytopes has long been a source of inspiration to me, as has his independence of thought and temperament. But it is my hope that he would have found pleasing the discovery in Gibbs' paradox of the kind of failure of statistical independence on which quantum indistinguishability is based – as within the realm of possible experience. Jon Dorling, in his acute historical and theoretical observations, was a similar inspiration, and himself wrote on my topic, arguing that in the limit of the Wein regime Einstein's argument showed light quanta were truly point-like. Their friendship and support has always meant much to me. This essay is dedicated to Itamar and to Jon.

22.2 The Gibbs Paradox and 'Mutual Independence'

Boltzmann's equation

$$S = k \ln W + \text{const}$$

was first written down in this notation by Planck, but it was introduced by Boltzmann in his 1877 memoire, often accounted his most important contribution to statistical mechanics. It was derived anew by Einstein in 1905 on the basis of what he called 'Boltzmann's principle', namely the principle that *the entropy of a system*

[6]See, for example, Epstein (1936a, p. 557), Darrigol (1986, pp. 205–9), Bach (1997, p. 8), Dieks (2014). (Bach's defence of the concept of classical particle indistinguishability extends only to the notion of probability distributions that give no information on individual trajectories.)

[7]Saunders (2006, pp. 199–200, 207).

is a function of the probability W of its instantaneous state. His argument was as follows[8]:

If it makes sense to talk about the probability of a state of his system, and if, further, each entropy increase can be conceived as a transition to a more probable state, then the entropy S_A of a system A is a function of the probability W_A of its instantaneous state. Therefore, if we have two systems S_A and S_B that do not interact with each other, we can put

$$S_A = \varphi_A(W_A),\, S_B = \varphi_B(W_B).$$

If these two systems are viewed as a single system of entropy S and probability W, we have

$$S = S_A + S_B = \varphi(W) \tag{22.1}$$

and

$$W = W_A . W_B.$$

This last relation tells us that the states of the two systems are mutually independent events.

From these equations it follows that

$$\varphi(W_A . W_B) = \varphi_A(W_A),\, \varphi_B(W_B),$$

and from this we get, finally,

$$\varphi_A(W_A) = C \ln(W_A) + \text{const}$$

$$\varphi_A(W_A) = C \ln(W_A) + \text{const}$$

$$\varphi(W) = C \ln(W) + \text{const.}$$

The quantity C is thus a universal constant; it follows from the kinetic theory of gases that its value is R/N [Boltzmann's constant k, in Planck's notation].

The argument has power and beauty, but we must look to its premises. The left-hand equality of (22.1) is nowadays called the *additivity* of the entropy. In thermodynamics, it applies to non-interacting systems A and B each in thermal equilibrium (not necessarily with each other), but Einstein went on to use it to broader effect. Let there be z one-particle states, all equiprobable, so that the probability that a particle will be found in a given state is $1/z$, and suppose there are n non-interacting particles. They will be 'mutually independent' in Einstein's sense if the probability that they will all be found in that state is $1/z^n$. The joint probability, in

[8]Einstein (1905, p. 94–95); I follow Einstein's notation, save for replacing '1' and '2' by '*A*' and '*B*'.

other words, should factorise, as the product of probabilities for individual particles considered separately. In this sense, the probability of a set of outcomes on throwing of n dice should be the same, whether the dice are thrown sequentially or all at once, so long as they are thrown independently and do not influence each other.

If thermodynamic probabilities are to factorise in this way when particles are non-interacting, and if they are to connect in a simple way with the available phase space volume,[9] then the volume measure on the many-particle phase space had better factorise; and that in turn requires that the phase space itself factorises, i.e. that it is the Cartesian product of 1-particle phase-spaces γ:

$$\Gamma = \gamma \times \ldots . \times \gamma \tag{22.2}$$

with volume measure:

$$W_V = V^n.$$

The dependence of the entropy of n particles on the spatial volume V is then of the form

$$S_V = k \ln V^n. \tag{22.3}$$

There is the inevitable attendant failure of extensivity: S_V as given by (22.3) does not scale with the extensive quantities, n and V. Double the latter and the entropy is not only doubled, but picks up an additional term $2nk \ln 2$. Of course it is true that extensivity is something of an approximation, as the molecules making up real material gases are at least weakly interacting and there may be boundary effects, but it is an *extremely* good approximation for thermal radiation, so long as the temperature is not too high, and it follows, or *should* follow, for an ideal material gas as well.

In terms of the set-up of the Gibbs paradox, suppose two samples of the same ideal gas are confined to volume V_A (n_A particles) and V_B (n_B particles), at equal pressure and temperature. The initial total entropy is then (from additivity):

$$S_{A+B} = S_A + S_B = k \ln V_A^{n_A} + k \ \ln V_B^{n_B}. \tag{22.4}$$

Let the two volumes be contiguous and let the two gases subsequently diffuse into each other. Then the final entropy will be:

$$S_{A \cup B} = k \ln (V_A + V_B)^{(n_A + n_B)}. \tag{22.5}$$

[9] I shall use the terms 'probability' and 'phase space volume' more or less interchangeably, and lazily use the same symbol V for both spatial volume and volume of γ; the context should make clear which is intended.

The result is an increase in entropy; in the simplest case $V_A = V_B$ the difference between (22.4) and (22.5) is $2nk \ln 2$. It is the same term spoiling extensivity that we encountered before, where now it reflects the two possibilities for each particle after the partition is removed, whether located in *A* or in *B*, where before there was only one. This is the form of the Gibbs paradox that we shall be concerned with, where there is no exchange of particles with outside systems, but only between particles in the two volumes V_A and V_B.

In order to obtain no entropy of mixing, and thereby an extensive entropy function, Gibbs proposed to use the quotient of Γ as given by (22.2) under the permutation group Π instead,[10] to use the space of what he called *generic* phases (now usually called *reduced* phase space):

$$\tilde{\Gamma} = \gamma \times \ldots . \times \gamma/\Pi.$$

It is the space arrived at by identification of points in Γ that are related by a permutation of particles (a permutation with respect to which particle has which position and momentum; a permutation of sequence-position in the Cartesian product). The phase space (22.2) was called by Gibbs the space of *specific* phases. As Gibbs' remarked, the reduced phase space has a considerably more complicated topology, but the volume measure has a simple expression:

$$\tilde{W}_V = \frac{V^n}{n!}. \tag{22.6}$$

The latter does not factorise, and particles described in this way *fail* Einstein's test of statistical independence. For suppose particles n_A are statistically independent from particles n_B, both occupying the same region *V*. Then the probability of the instantaneous state of the composite $\tilde{W}_{A\cup B}$ should be the product of the probabilities of the state of each, $\tilde{W}_A \cdot \tilde{W}_B$. However, by inspection:

$$\tilde{W}_{A\cup B} = \frac{V^{(n_A+n_B)}}{(n_A+n_B)!} \not\approx \frac{V^{n_A}}{n_A!} \cdot \frac{V^{n_B}}{n_B!} = \tilde{W}_A \cdot \tilde{W}_B \tag{22.7}$$

and the probability does not factorise. Contrast the specific probabilities, that do:

$$W_{A\cup B} = V^{(n_A+n_B)} = V^{n_A} \cdot V^{n_B} = W_A \cdot W_B.$$

Returning to the generic phase, our argument used '*A*' and '*B*' as names for the two collections of particles, dispersed through the same volume, when they are supposed to be indistinguishable; is this move legitimate? But surely it is if

[10] Gibbs did not of course use the concept 'group', and gave no independent motivation for the generic phase other than that it gave no entropy of mixing. It was seen as ad hoc by Lorentz and Ehrenfest; see Saunders (2020), Darrigol (1991) for further discussion.

'indistinguishable' is merely to renounce factor position in the Cartesian product Eq. (22.3) – what I shall call the *realist* reading of the Gibbs' generic phase. The n-tuple of pairs in a definite sequence:

$$\langle\langle q_a, p_a\rangle, \langle q_b, p_b\rangle, \ldots, \langle q_d, p_d\rangle\ \rangle \in \Gamma$$

where $\langle q_a, p_a\rangle \in \gamma$, etc., is a specific phase (specific phase space point). The identification of all such points related by a permutation of entries in sequence position, is to pass to a space the points of which are represented by unordered collections:

$$\{\langle q_a, p_a\rangle, \langle q_b, p_b\rangle, \ldots, \langle q_d, p_d\rangle\} \in \tilde{\Gamma}.$$

There is no longer a sequence position to pick out a particle, independent of its position and momentum, but the latter can be specified as such directly, and thereby the particle that has that position and momentum, supposing there are no repetitions.[11] The identification of points in Γ related by permutations does not mean that the resulting space $\tilde{\Gamma}$ does not have points. Nor does it mean that we must interpret those points as an equivalence class of specific phases (sequences); but as Gibbs remarked, the latter is useful from the point of doing analysis on manifolds.[12]

There is nevertheless a fundamental distinction between this way of specifying the partition of the gas of n particles into two collections, each scattered throughout the same spatial volume, and specifying the partition in terms of disjoint volumes instead (and indeed disjoint regions of the one-particle phase space, regions in both momentum space and position space). Specify the sub-systems A and B in the latter way, and we are back to the Gibbs set-up. The generic probability now *does* factorise[13]:

$$\tilde{W}_{A+B} = \tilde{W}_A . \tilde{W}_B = \frac{V^{n_A}}{n_A!} \cdot \frac{V^{n_B}}{n_B!}.$$

If the two gases are allowed to mix, they are no longer statistically independent of each other, as we have seen, and the joint probability is given by the left-hand side of Eq. (22.7). Using the Sterling approximation:

---

[11] The assumption is innocent in a continuum theory of point-like particles, or for extended but impenetrable molecules, but for a finite state space it is the difference between fermions (no repetitions) and bosons (repetitions).

[12] Gibbs (1902, p. 188). Whether my realist reading of the generic phase should be attributed to Gibbs is a delicate question, for Gibbs writings were so brief. For further discussion, see my (2020).

[13] If this is in doubt, consider the analogous question using a discrete state-space, and apply Eq. (22.16) to find the number of ways of distributing n_A indistinguishable particles over z_A states, and likewise for n_B particles over z_B disjoint states; the product of the two numbers is the same as the number of ways of distributing $n = n_A + n_B$ indistinguishable particles over $z = z_A + z_B$ states, under the constraint that n_A are in V_A and n_B are in V_B (with no regard to which is in which).

$$\ln n! \approx n \ln n - n$$

and bearing in mind that since the two gases prior to mixing had the same temperature and pressure, their densities are equal as well, $n_A/V_A = n_b/V_B$; it follows after a short calculation:

$$\tilde{W}_{A+B} \approx \tilde{W}_{A\cup B}. \qquad (22.8)$$

The reason the equality is only approximate in (22.8) is because after the gases mix the states available include those in which numbers of particles different from n_A and n_B may be found in the regions A and B respectively (but not of course by virtue of which particles are found in each of the two regions). After mixing, the available number of microstates is thus increased, but by a number that can be neglected in the Sterling approximation.

The lesson of the Gibbs paradox is that indistinguishable particles, in Gibbs' sense, even if non-interacting, are not statistically independent in Einstein's sense. To put the matter vividly: if there are n_A particles alone, initially confined to volume V_A, expanding into volume V, the entropy increases. If there are n_B particles alone, initially confined to volume V_B, with the same density, expanding into volume V, the entropy increases. Yet let n_A, n_B particles be present in each of the two volumes, of the same density, and allowed to expand into volume V, then there is no entropy increase, even though the particles are non-interacting, and even though the trajectory of every particle is the same as if all the others were not there. The entropy of the joint system is not the sum of the entropies of the two subsystems, and the entropy *change* of the joint system is not the sum of the entropy changes of the two subsystems. Einstein's criterion fails for both, and with it mutual independence.

22.3 Einstein's 'Miraculous Argument'

Consider now Einstein's celebrated argument from the Wien black-body distribution formula. This claimed to show that insofar as that equation is obeyed, light behaves as if made of elements that were 'mutually independent' and, moreover, that were 'moveable points'.

First 'mutually independent'. We must take Einstein to mean what he meant earlier in the same paper, in the argument for the universality of the entropy as a function of W just reviewed. Mutual independence, then, means the factorising of probabilities. Is it the same as the condition that the particles are non-interacting? It has been interpreted in this way,[14] but that would be a mistake: even if Einstein did ultimately believe that statistical correlations must be explained by some

[14]Most recently by Norton (2006), according to whom 'independence is just the simplest case of no inter-molecular forces' (p. 93).

dynamical coupling, he consistently used the phrase 'mutual independence' and later 'statistical independence' (and not 'non-interacting').

Given all this, and given that the Wien limit to the radiation law just is the limit in which Gibbs' measure for the generic phase is valid, there is the obvious difficulty: the probabilities as defined by Gibbs' measure do *not* factorise, as we have just seen. How then, working from the Wien regime, could Einstein have concluded the probabilities factorise, and hence that light quanta were mutually independent of each other? Indeed, on a careful reading, this was not exactly his claim, nor an accurate reading of his argument. He actually showed that *if* thermal radiation consisted of mutually independent and localised entities, *then* its spectral distribution would be the Wien distribution. From this we infer that thermal radiation, in the Wien limit, behaves *as if* made of mutually independent and localised entities, not that it is made of such entities.

Einstein himself, in his summary conclusion, put the onus entirely on mutual independence:

> Monochromatic radiation of low density (within the range of validity of Wien's radiation formula) behaves thermodynamically as if it consisted of mutually independent energy quanta of magnitude $h\nu$.

The omission of locality may have been no more than a slip; does he commit the fallacy of affirming the consequent as well? He treads a fine line: light behaves as if it is so constituted, granted; but it so behaves if it is constituted in other ways as well – in particular, if the energy quanta are non-interacting and localised but *lack* statistical independence, in the very particular way that they are correlated when described by the generic phase.

We clearly need to see the details. From the first law of thermodynamics:

$$dU = TdS + pdV$$

it follows:

$$\left.\frac{dS}{dU}\right|_{V=const} = \frac{1}{T}.$$

Einstein worked directly with the spectral entropy density φ of black-body radiation (the entropy per unit frequency range and volume), as a function of the spectral energy density ρ (the energy per unit frequency and volume), to derive what he called 'the law of black bodies':

$$\frac{\partial\varphi}{\partial\rho} = \frac{1}{T}. \tag{22.9}$$

The Wien spectral distribution law, as first proposed by Wilhelm Wien in 1896, is

$$\rho = \alpha v^3 e^{-\beta v/T} \tag{22.10}$$

where α and β were new and unknown constants. It was well-confirmed by experiment for $v/T \gg 1$, as Einstein noted. Using (22.10) to write the inverse temperature as a function of ρ and the frequency, and inserting in (22.9), Einstein obtained a differential equation for φ. It could readily be integrated to obtain:

$$\varphi = -\frac{\rho}{\beta v}\left(\ln \frac{\rho}{\alpha v^3} - 1\right). \tag{22.11}$$

Planck had found this equation in 1898, likewise working from the Wien distribution (and he had used the same method again in 1900, to obtain an equation for the entropy as a function of ρ, working from the Planck distribution). Einstein's originality lay in what came next.

Let $S_0 = V_0\varphi$ be the entropy of the radiation in volume V_0 per unit frequency range and let E be the energy $V_0\rho$. Multiplying both sides of (22.11) by V_0, it follows, writing ρ in terms of E and V_0:

$$S_0 = -\frac{E}{\beta v}\left(\ln \frac{E}{\alpha v^3 V_0} - 1\right).$$

Suppose now that this volume of radiation (in the given frequency range) undergoes a spontaneous fluctuation, and is contained instead in the sub-volume V; then it will have the entropy:

$$S = -\frac{E}{\beta v}\left(\ln \frac{E}{\alpha v^3 V} - 1\right).$$

The entropy change is therefore:

$$S - S_0 = \frac{E}{\beta v} \ln \frac{V}{V_0}$$

or more suggestively:

$$S - S_0 = k \ln \left(\frac{V}{V_0}\right)^{\frac{E}{k\beta v}}. \tag{22.12}$$

It is the same volume-dependence of the entropy function as for an ideal gas of $n = E/k\beta v$ particles; in Planck's notation, for $n = E/hv$ particles; in Einstein's words, as for a gas of *n independently movable points*.

Here is Einstein's argument in full:

> Let us consider a part of the volume V_0 of magnitude V and let all n moveable points be transferred into the volume V without any other change in the system. It is obvious that this state has a different value of entropy (S), and we now wish to determine the entropy difference with the aid of Boltzmann's principle.

We ask: How great is the probability of the last-mentioned state relative to the original one? Or: How great is the probability that at a randomly chosen instant of time all n independently movable points in a given volume V_0 will be contained (by chance) in volume V?

Obviously, for this probability, which is a 'statistical probability',[15] one obtains the value[16]

$$W^{rel} = \left[\frac{V}{V_0}\right]^n; \tag{22.13}$$

from this, by applying Boltzmann's principle, one obtains

$$S - S_0 = k \ln \left(\frac{V}{V_0}\right)^n. \tag{22.14}$$

By 'probability ... .relative to the original' we take it Einstein means the *ratio* of the probabilities of the two states, so that, in accordance with (22.14), its logarithm is the difference in the entropies of the two states.[17] And whether or not (22.13) really is obvious: as before, we take Einstein to mean that if independent, the probability for the joint state of all n moveable points is the product of the probabilities for each; or at least that this is true of the relative probabilities. The relative probability that a single particle, initially in volume V_0, is subsequently in the sub-volume V, is V/V_0; hence the relative probability that this is true of all n particles is the product of the relative probabilities of each, Eq. (22.13).

This is, 'obviously', consistent with the volume dependence of the probability on the total phase space of the n particles of the form already considered:

$$W = V^n.$$

It reproduces the required relative probability, in that

$$W^{rel} = \frac{W}{W_0} = \left(\frac{V}{V_0}\right)^n.$$

It is the same as the ratio of probabilities for Gibbs' specific phase. But the ratio of probabilities for Gibbs' *generic* phases is exactly the same:

$$\tilde{W}^{rel} = \frac{\tilde{W}}{\tilde{W}_0} = \left(\frac{V}{V_0}\right)^n$$

even though $\tilde{W}$, as given by Eq. (22.6), cannot be factorised. Non-interacting moveable points that are indistinguishable in Gibbs sense are not mutually independent in

[15]The reference is to his preference for an ergodic interpretation of probabilities, rather than Boltzmann's combinatorial method, that he had emphasised in his early papers of 1903 and 1904.

[16]To avoid confusion, we write W^{rel} where Einstein wrote W.

[17]This is consonant with e.g. Pais (1982, p. 72); I am not aware of any alternative reading.

Einstein's sense, yet yield the same relative probability and have the same volume dependence of their entropy as do mutually independent movable points. The Wien law, concerning this kind of fluctuation, does not discriminate between them.

Would it be better to read Einstein's notion of independence as requiring the factorising only of *relative* probabilities? In that case indistinguishable particles in Gibbs' sense may come out as statistically independent after all. But that contravenes what we have learned from Gibbs' paradox in Sect. 22.2. Indeed, consider a fluctuation of a rather different sort than the one Einstein imagined, in which n_A particles initially occupying volume V are subsequently found in the sub-volume V_A, and the remaining n_B particles, initially in volume V, are subsequently found in the disjoint sub-volume V_B, where $V = V_A + V_B$. Suppose first that the n_A particles are statistically independent of the n_B particles. Then the relative probability of the fluctuation should factorise into the product of the relative probability of each considered independently:

$$W^{rel} = W_A^{rel}.W_B^{rel} = \left(\frac{V_A}{V}\right)^{n_A}.\left(\frac{V_B}{V}\right)^{n_B}.$$

It is the same, whether calculated as the ratio of the probability of the joint final state to the initial, or as the product of the ratios of final and initial states of each separately. Suppose now that the fluctuation does not change the density, so that:

$$\frac{(n_A + n_B)}{(V_A + V_B)} \approx \frac{n_A}{V_A} \approx \frac{n_B}{V_B}.$$

The fluctuation is then the time reverse of a process of mixing of gases, as in the Gibbs set-up. It follows there is a decrease in entropy (for $V_A = V_B$, a decrease equal to $nk \ln 2$), which is as to be expected, given that the reverse process yields an increase in entropy.

If now we use generic phases instead, there is no such decrease, for the same reason that the time-reverse process gives no increase. The relative probability as given by the ratio of the probability of the joint final state to the probability of the joint initial state is:

$$\tilde{W}^{rel} = \frac{V_A{}^{n_A}/n_A! \cdot V_B{}^{n_B}/n_B!}{V^N/n!}$$

which, as we have already seen, is approximately unity in the Sterling approximation. It is not the product of the relative probability for a fluctuation in the state of n_A particles, initially occupying volume V, by which they are found in volume V_A, with the relative probability for a fluctuation in the state of the remaining n_B particles, from V to V_B. Eq. (22.7) holds here just as it did for the Gibbs paradox; therefore the relative probability does not factorise.

Our earlier conclusion therefore stands: indistinguishable classical particles in Gibbs' sense are not statistically independent in Einstein's sense, not even if

Einstein's factorizability condition is only required for relative probabilities. Since photons in the Wien limit have the same functional dependence on the volume and particle number as do indistinguishable classical particles, in Gibbs' sense, the result applies as much to photons. Photons are not statistically independent, in Einstein's sense, not even in the Wien regime.

22.4 Gibbs' Generic Phase as the Limit of Bose-Einstein Statistics

There are of course other solutions, or purported solutions, to the Gibbs paradox, or rather paradoxes (for it has been taken to mean many different things), than the one we have considered; and there are other criticisms, or purported criticisms, of Gibbs' concept of generic phase. But there is a clear structural correspondence between statistics as made out in terms of the generic phase, vs the specific phase, and statistics following what Einstein in 1924 called 'Bose's statistical approach', vs that for 'statistically independent molecules', following Boltzmann's. Bose's method, in the Wien limit, is the same as Gibbs', using the generic phase.

Alas, neither this structural correspondence, nor the connection with Gibbs paradox, was so much as mentioned by any of the key players, nor were they for some years.[18] The first to consider the matter in any detail appears to have been Paul Epstein, in a contribution to the *Commentary on the Scientific Writings of J. Willard Gibbs* published by Yale University, Gibbs' alma mater, in 1936. Epstein highlighted the relationship between specific vs generic phase, in Gibbs' sense, and statistical independence vs Bose's method, in Einstein's; but he also maintained, as did virtually everyone following Dirac's 1926 intervention, that the absence of trajectories plays an essential role in quantum indistinguishability, which therefore has no analogue in classical statistical mechanics (whether using specific or generic phases)[19] – precisely the opposite conclusion to the one advanced here.

Epstein's intervention, whatever its merits, came too late for historians of the indistinguishability concept. It came too late for Gibbs, for by then photons were a species of bosons; gibbsons were not to be. Yet the connection was there for the taking in Einstein's discussion. The key expression for the thermodynamic probability, the signature of the new statistics, had been at the heart of the quantum enigma since it first appeared in Planck's publication of 1900 (where Planck's symbol 'h' appeared for the first time as well):

---

[18]As remarked, Schrödinger (1925) came the closest, but did not mention Gibbs or the generic phase. For further reading, see Darrigol (1991, pp. 293–8), Mehra and Rechenberg (1987, pp. 361–66), Saunders (2020).

[19]Epstein (1936a, p. 557). Epstein also argued that a discrete state space was inconsistent with a continuous dynamics, in the case of generic phase, in (1936a, p. 564, p. 527) and (1936b, pp. 485–91).

$$\frac{(n+z-1)!}{(z-1)!n!}. \tag{22.15}$$

It has a simple combinatorial meaning: it is the number of ways of distributing n objects, among z cells, where it makes no difference as to which object is assigned each cell, but only how many. It is the count of microstates, where each is defined as a specification, of each cell, of the number of objects it contains, but not of *which* of the n objects is assigned each cell. This was well-understood by 1911 if not already by Boltzmann in 1877 (who was the first to write down (22.15), as the number of ways that 'energy elements', meaning integral multiples of a unit ε could be distributed among n objects). It may well have been known to Gibbs, although Gibbs probably had no reason to take seriously the idea that z might have a well-defined physical meaning (for that was provided by Planck's discovery in 1900, putting $\varepsilon = h\nu$, and as later developed in terms of the number of elementary cells to phase space of volume h^3). If Gibbs read Planck's publications in 1900 and 1901, he did not comment on them; he died unexpectedly in 1903.

Planck's notation did not make evident that the numbers n and z were tied to a range of frequency and position, denote $[s, s + \Delta s]$. Einstein's notation, 25 years later, did, and following Bose, generalised it from a range of frequency to a range of momentum (as well as spatial volume). Such a range Einstein called an *elementary region* s of the 1-particle state space, of the form $V \times [p, p + \Delta p] \subset \gamma$ (provided by what we would today call a *coarse-graining* of the one-particle state-space). Dividing by h^3 gave the number of elementary cells z_s. For the number of microstates, if there are n_s particles distributed over the z_s cells of region s, and if a microstate only says how many are in each cell ('Bose's method'), Einstein wrote down the expression:

$$\frac{(z_s+n_s-1)!}{(z_s-1)!n_s!} \tag{22.16}$$

immediately remarking that 'it is easily recognized' that the distribution of molecules over cells is not statistically independent. We shall denote it $\widetilde{Z}_s$, as the analog of $\widetilde{W}_s$, the continuous measure on the reduced phase space of the region s. The total probability of a macrostate is then the total number of microstates, consistent with the specification of the numbers n_s, for all the regions s, i.e. the product[20]

$$\widetilde{Z} = \prod_s \frac{(z_s+n_s-1)!}{(z_s-1)!n_s!}. \tag{22.17}$$

[20]The probability factorises for the elementary regions, in the same way and for the same reasons, as it does for the subsystems defined by V_A and V_B in the Gibbs set-up (see fn.13).

Einstein contrasted this expression with that following from 'the hypothesis of statistical independence of the molecules'. There 'a state is defined microscopically by the fact that for each molecule one indicates in which cell it is sitting (complexion)',[21] where 'complexion' was Boltzmann's 1877 term for microstates. In that case the number of microstates of n_s particles in region s is:

$$z_s{}^{n_s},$$

denote Z_s, and across all the regions for given occupation numbers $n_1, n_2 \ldots$ it is

$$\frac{n!}{n_1!n_2!\ldots\ldots}\prod_s z_s{}^{n_s} \tag{22.18}$$

where n is the total number of particles, and where the pre-factor

$$\frac{n!}{n_1!n_2!\ldots\ldots}$$

is Boltzmann's famous 'permutability', as he called it. It reflected the additional multiplicity in microstates, introduced when partitioning the total number of molecules n among the elementary regions, so that n_1 are in region 1, n_2 in region 2, and so on.

The different expressions (22.17) and (22.18) yield different entropy functions. Only one can be correct – and on this point Einstein thought the question was an empirical question, that 'nature must decide'. But of course, in the case of thermal radiation, nature's decision was already known, favouring the Bose-Einstein entropy and the new statistics.

And there, more or less, Einstein left the matter. He did not comment on the Wien limit; neither did Bose (and nor, in 1926, did Dirac). We should verify that it is indeed the limit of the Gibbs generic phase. The Planck distribution is:

$$\rho = \frac{8\pi v^2}{c^3}\frac{hv}{e^{hv/kT}-1}.$$

It goes over to the Wien distribution (22.10) in the limit $hv \gg kT$. But that is the limit in which very few modes of the electromagnetic field are excited, for the minimum required is much greater than the thermal energy available (of order kT), and those that are excited are at the lowest excitation (so the occupation numbers are all 0's and 1's). Since z_s is the number of modes, it follows the Wien limit is the limit $z_s \gg n_s$. In that limit it can easily be seen:

[21] Einstein (1925); 'each molecule' appears to be functioning as a proper name. See Saunders (2006, 2013, 2016) for more on philosophy of language and indistinguishability.

$$\frac{(z_s + n_s - 1)!}{(z_s - 1)!n_s!} \approx \frac{{z_s}^{n_s}}{n_s!}. \tag{22.19}$$

The total number of microstates for all elementary regions s for which $z_s \gg n_s$ is then approximately:

$$\prod_{s;z_s \gg n_s} \frac{{z_s}^{n_s}}{n_s!} \tag{22.20}$$

recovering the Maxwell-Boltzmann probability Eq. (22.18), save for the overall factor $n!$ (where now n is the total number of particles in the Wien regime). In terms of the continuous phase space measure $\widetilde{W}_s$ for n_s particles in region s, $\widetilde{Z}_s$ and $\widetilde{W}_s$ are proportional only in the Wien regime. In contrast the equation

$$Z_s = \frac{1}{\tau^{n_s}} W_s$$

holds for every region s and numbers n_s, where τ is the unit of phase space volume γ. Classically τ can always be chosen as sufficiently small to ensure the Wien limit is enforced (only the size of the elementary regions must be kept large, so as to contain many particles, so that the Sterling approximation applies). Boltzmann's combinatorial method then goes over unchanged (apart from the overall factor $n!$), using Gibbs' generic phase, on replacing 'cells' by 'elementary regions', the same as using the specific phase. In the quantum case $\tau = h^3$, where h is Planck's constant, whereupon the number of cells z_s cannot be taken as arbitrarily large, and the approximation (22.19) fails for sufficiently large n_s.

None of this was mentioned by Einstein. The correspondence between (22.18 and 22.20) should have been obvious much earlier, for example, to the Ehrenfests, at the time of writing their famous Encyclopaedia article in 1912, for they knew of the distinction between elementary regions and cells (which indeed goes back to Boltzmann in 1877). They too did not mention it.[22]

Fast-forwarding to Dirac, all of the foregoing applies to equilibrium distributions of photons. In the Wien regime, the total number of microstates in each elementary region is well-approximated by (22.19), so photons are not statistically independent, in Einstein's sense. The exact expression for every frequency and temperature (so long as the temperature is not so high that pair-creation of electrons becomes possible) is $\widetilde{Z}_s$, as given by (22.16). Indeed, $\widetilde{Z}_s$ is exactly the dimension of the subspace of the Hilbert space spanned by symmetrised wave-functions in the n_s-fold

[22]They assigned $z_s = \omega$ in Eq. (22.18), with ω an arbitrary constant independent of s; whereupon the comparison of (22.20) with (22.18) is all but invisible (Ehrenfest and Ehrenfest 1912, p. 27). Ehrenfest and Kamerlingh-Onnes in their (1914) came very close to the identification, but also missed it; see my (2020) for further discussion.

tensor product of the one-particle Hilbert space $\mathcal{H}_s$, corresponding to the range of frequency (momentum) $[s, s + \Delta s]$ in spatial volume V_s, where $\mathcal{H}_s$ has dimension z_s.

22.5 Locality and Entanglement

If Einstein was wrong to think that in the Wien limit light was made up of statistically independent entities, his argument surely did show that in that regime it has a discrete structure, and consists of non-interacting localised entities[23] – in that nobody has been able to come up with another interpretation of the fluctuation formula (22.12) that says any different. He was right in another respect as well: in recommending that light quanta be thought of as *movable*. That might seem an obvious presupposition of his thought-experiment (how else was the fluctuation in volume supposed to happen unless the particles were 'movable'?), but it was a first step towards investing the photon, as the concept was eventually named, with different properties at different times – in short, to considering it as the sort of entity that might have different states at different times, and hence that might have a state-space at all; whereas 'energy elements', of the sort that had been discussed by Boltzmann, Planck, and Natanson, were never thought of like that. The latter were alike in *all* of their properties; they were hardly thought of as capable of change at all, unless it be the transfer of energy, emission and absorption, creation and annihilation.

Did it matter, historically, that light quanta were wrongly thought to be statistically independent in the Wien regime? For those in the critical period 1905–1925 prepared to extend the concept of light quanta to the Rayleigh-Jeans regime, it led to the idea that correlations *only* came into play in that regime. This encouraged the picture of a cooperative behaviour ('the light molecule') of quanta all in the same cell, or all in the same normal mode of cavity radiation. Einstein was sympathetic to these ideas. They were also entertained, in one of his early papers, by Louis de Broglie. Einstein warmed to this theme in his second paper on gas theory in 1925. He recommended de Broglie's newly published PhD thesis, as taken up, among others, by Schrödinger. In these respects the confusion was positively beneficial, for it led Schrödinger to de Broglie's concept of matter wave and to his concept of quantisation conditions as defined by solutions to wave equations; but confusion it was.

The story did not stop there, however. Einstein called the failure of statistical independence 'an initially completely puzzling mutual influence of the molecules' (ibid, p.374). He also spoke of 'spooky' action-at-a-distance, a line of inquiry that eventually led to the EPR paradox. The latter, we know, exploits entanglement.

[23] Whether the 'movable points' must really be *points* will depend, presumably, on the precision with which the probability of the fluctuation is defined (perhaps only in the limit $\nu \to \infty$). I refer to Dorling (1971) for discussion, already mentioned.

(Anti-)symmetrisation of the wave-function also involves entanglement, and surely explains the failure of statistical independence. The two, it might be thought, are at bottom the same, an argument developed at length by Don Howard (and recently endorsed by Norton).[24] Yet there are important differences: as already remarked, there is a growing body of literature to show that the particular kind of entanglement required by symmetrisation is essentially trivial. Not only is it consistent with the existence of well-defined particle trajectories, it is also insufficient to violate any Bell inequality.[25] It is, I suggest, just what is required, no more and no less, to ensure that factor-position, on taking the tensor-product of one-particle Hilbert spaces, can play no role in the definition of any physical quantity. Thus the partial trace of the state, as defined by factor-position, with respect to one or more of all other particles, delivers exactly the same density matrix whatever the factor-position (a result that has led some madmen to conclude that every electron is in exactly the same state).

Then what does explain the failure of statistical independence? An alternative but still somewhat tentative answer is that microstates should not be considered as built up out of particle motions taken one at a time; that a microstate at an instant of time supervenes globally, rather than locally, on the states of individual particles. The idea that the dice thrown all at once should yield outcomes with the same probabilities, if non-interacting, as were each die thrown sequentially in time, may be the culprit; for if n dice are thrown sequentially in time, each with z equiprobable outcomes, there will be

$$\frac{n!}{n_1!n_2!\dots n_z!}$$

different possible temporal sequences, all with the same final set of occupation numbers $n_1, \dots n_z$, with each sequence equiprobable – and we are back to Maxwell-Boltzmann statistics.

22.6 Particle Trajectories and the Gibbs Paradox

There is more to learn from the failure of statistical independence apparent in the Gibbs paradox. The Gibbs paradox is properly speaking a cluster of puzzles, where the one so-far considered made direct contact with Einstein's idea of statistical independence. Here is another formulation.

---

[24] Howard (1990), Norton (2006). Howard – and Norton – may well be right that the two were connected in Einstein's thinking; whether Einstein was right to connect them is the point I am contesting.

[25] For triviality, see Ghirardi et al. (2002), Ghirardi and Marinatto (2004), Penrose (2004); for quantum trajectories, see Saunders (2006, p. 199–200, 2016, p. 169–70, 2018, p. 9–10); for Bell inequalities, see Caulton (2020).

If there is to be no entropy of mixing in the Gibbs set-up, there can be no new (or comparatively few) microstates available after mixing, when the partition is removed, than were not already present before. Yet it seems there are such states available. Thus Versteegh and Dieks have recently argued[26]:

> Consider again two gas-filled chambers, each containing n identical particles. Before the partition is removed the number of available states per particle is $\frac{1}{2}V$. After the partition has been removed, the number of available states is V. The reason is that after the partition's removal it has become possible for the particles to move to the other chamber. The doubling of the number of available microstates thus expresses a physical freedom that was not present before the partition was taken away. Trajectories in space-time have become possible from the particles' initial states to states in the other chamber. Particles from the left and right sides can physically exchange their states.

Their conclusion is that according to classical theory an entropy of mixing is unavoidable, and that an extensive entropy function, in classical theory, is no more than a useful fiction. More recently, Dieks has argued indistinguishability in the classical case amounts to invariance under particle relabelling, with no physical consequences[27]:

> For example, two widely separated electrons, following distinct trajectories and labelled a and b, still represent exactly the same physical situation after we have exchanged labels (in the sense that the electron at the first position is now called electron b, and *vice versa*). But this example also immediately demonstrates the physical insignificance of this type of indistinguishability: in spite of the exchangeability of the *labels*, the *particles* themselves can clearly be distinguished physically—we could easily tell them apart by a detection device that is sensitive to positions. So it seems that this permutability argument, and division by $n!$ on its basis, only relates to our mode of description and does not possess physical import. This division by $n!$ cannot possibly lead to different predictions; which makes it highly implausible that it could be relevant for a solution of the Gibbs paradox.

It appears that for indistinguishability of particles to have physical import it must be impossible to tell them apart: it is indistinguishability in the sense of Boltzmann, Planck, and Natanson, not Gibbs, Bose, and Einstein. But Dieks is right that division by $n!$ relates to our 'mode of description', specifically, to the choice of state space, whether, in the classical case, it is the reduced state space $\tilde{\Gamma}$ or the specific state space Γ, and in the quantum case, whether it is the symmetrised subspace or the full tensor product of Hilbert spaces. In all cases, with the entropy given by the volume or dimensionality of the state space, it is *highly plausible* that this choice will be relevant to Gibbs paradox. For consider: is it true, as Versteegh and Dieks claim, that the number of available microstates doubles on removal of the partition? Figure 22.1 is a space-time diagram for the Gibbs set-up. In Fig. 22.1b, the partition remains in place throughout; in Fig. 22.1b, at time t', it is removed. New trajectories are then possible after t'. But does it follow that new *states* are made available at t_2?

We may simplify further. Figure 22.2 is a spacetime diagram for the 2-particle Gibbs paradox. Figure 22.2a shows an illustrative pair of trajectories when the

[26] Versteegh and Dieks (2011, p. 742) (I have made minor changes of notation).

[27] Dieks (2014, p. 1308).

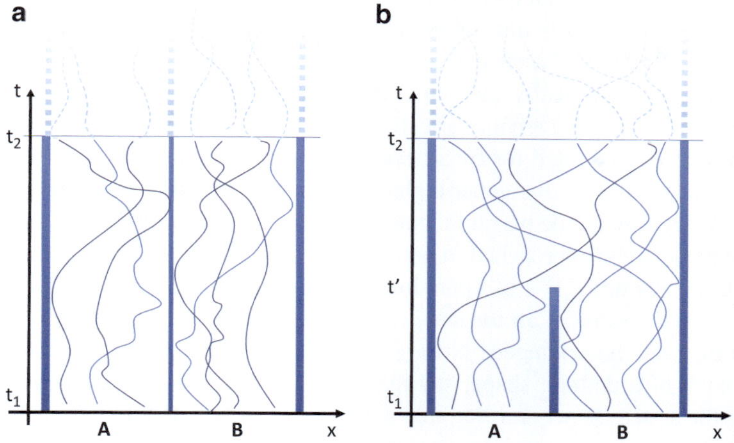

Fig. 22.1 Space-time diagram for the mixing of gases. In (**a**) the partition remains in place, whilst in (**b**) it is removed at time t'

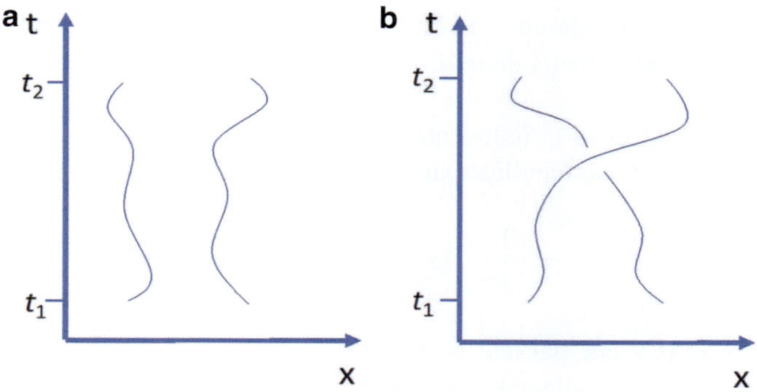

Fig. 22.2 The 2-particle Gibbs paradox

partition is not removed, and Fig. 22.2b shows the kind of new trajectory possible when it is, chosen to yield the same final positions and momenta. The intuition to which Versteegh and Dieks appeal suggests that the two final microstates in Fig. 22.2a, b are different because they differ as to which particle has which position and momentum. This is, fairly obviously, consistent with ordinary notions of macroscopic objects, for macroscopic objects differ in their intrinsic, position- and momentum-independent properties; but for microscopic particles, whose intrinsic properties are exactly the same?

To insist that there is a difference even in this case is to insist on the use of the specific phase, in Gibbs sense. In the space of specific phases of the two particles, a continuous curve connecting two points related by particle interchange is a physical process by which each particle acquires the position and velocity of the other. In the space of specific states, it is an open curve whose end points are distinct. In the space of generic states, the initial and end points are one and the same. But the continuous curve connecting them is not obliterated – the exchange has still taken place – it is merely that it is now a closed curve in (reduced) phase space, with beginning and end points numerically the same. (Of course these states obtain at different times; as *time slices* they are numerically distinct; as such each may have a different past, but that is accounted for by a change in the Hamiltonian, as time-dependent, not the state.) On using the generic phase, it remains true that new trajectorie*s* are possible, on removing the partition, but not new states; it is only that the same states (microstates) can be approached in new ways.[28]

The Gibbs paradox, if it is a paradox, now reduces to this: are the final states of Fig. 22.2a, b the same? Can the state of two particles be unchanged, when each of the two particles has changed? No *logical* contradiction is involved. Imagine two things, each of a definite shape, say that one is round and one is square; we can imagine the square slowly turning into a circle, and the circle into a square, in such a way that each occupies the same position as the other at the end.

It points, metaphysically, to the idea that the right notion of how the instantaneous state of a joint system depends on the instantaneous state of its subsystems is global. The state of the joint system supervenes on both – it is not built up from the state of each subsystem considered individually. It may be this idea is more at home in quantum theory, and perhaps there it is forced; but it may be the right metaphysics more generally.

To see that it is forced in (conventional) quantum mechanics, consider a totally symmetrised state of two identical particles, of the form:

$$|\Psi\rangle = \frac{1}{\sqrt{2}}\Big(|\varphi\rangle \otimes |\psi\rangle + |\psi\rangle \otimes \Big|\varphi\rangle\Big)$$

where $|\varphi\rangle \in \mathcal{H}$ is a one-particle state (and similarly $|\psi\rangle$, where $|\varphi\rangle$and $|\psi\rangle$ are orthogonal). Let the particles be non-interacting, and suppose the unitary evolution smoothly switches the two:

$$\hat{U}_t\ |\varphi\rangle = |\psi\rangle\,;\ \hat{U}_t\ |\psi\rangle = |\varphi\rangle\,. \tag{22.21}$$

Then the final state is unchanged:

$$\hat{U}_t \otimes \hat{U}_t\,|\Psi\rangle = |\Psi\rangle\ .$$

Each of $|\varphi\rangle$ and $|\psi\rangle$ is (continuously) changed, yet the initial and final state is the same.

The same question can be posed in terms of the Feynman sum-over-paths formalism, using the same Fig. 22.2. In computing the transition amplitude between

[28]Discounting the new states in which particle numbers on each side are changed, as already mentioned.

initial and final states: do we sum the amplitudes for 2a and 2b coherently? We should if the final state in the two cases is the same.

It was the same question posed, and answered, by Dirac in 1926, in terms of the matrix mechanics, supplemented with Schrödinger's new concept of state. He concluded

> If the positions of two of the electrons are interchanged, the new state of the atom is physically indistinguishable from the original one. In such a case one would expect only symmetrical functions of the coordinates of all the electrons to be capable of being represented by matrices. It is found that this allows one to obtain two solutions of the problem satisfying all the necessary conditions, and the theory is incapable of deciding which is the correct one. One of the solutions leads to Pauli's principle that not more than one electron can be in any given orbit, and other, when applied to the analogous problem of the ideal gas, leads to the Einstein-Bose statistical mechanics.[29]

Dirac explicitly considered the interchange of electrons in terms of a physical process connecting the original state and the interchanged state. The two are the same using Gibbs' concept of generic phase, just as they are the same using Dirac's concept of symmetrised states; and indeed, only symmetric functions of the specific phases can be represented by functions on the space of generic phases.

There is no question that in standard quantum mechanics, for identical particles, initial and final states are identified, even if the process connecting them is a switch of the form Eq. (22.21). If there is a rationally compelling argument to the contrary, it awaits to be stated. But it may have seemed the onus was on the other side to Einstein, Ehrenfest, and Lorentz, in that fateful first quarter of the last century. In my opening remarks I suggested that there may have been a genuine misconception at issue in their rejection of Gibbs' concept of generic phase. Perhaps it was this: they thought it meant, inter alia, that whether or not an interchange of like particles had actually happened was somehow a *meaningless* question, according to Gibbs. As realists, and given the existence of particle trajectories, they thought (rightly) that that made no sense.

The view that quantum particle indistinguishability involves the renunciation of particle trajectories has this in its favour. *Departures* from classical statistics involves the absence of particle trajectories, understood as well-defined orbits of 1-particle states, for those departures involve more than one particle being found in the same 1-particle state – prohibited if each has a well-defined and non-overlapping trajectory.

But in another respect the verdict is quite different. The attribution of histories to indistinguishable particles – or more precisely, the attribution of histories, or changes of states, to light quanta, indeed, to an individual light quantum – the idea of a light quantum as having a state-space at all – was Bose's essential innovation in 1924. It was this that broke the conceptual deadlock over the interpretation of thermal radiation in terms of light quanta for every range of frequency and

[29]Dirac (1926, p. 662). These are the two possibilities already remarked on in fn.11. Note that parastatistics do not arise, unless we admit microstates that differ by particle permutation alone.

temperature, and not just in the Wien regime. Had light quanta already been recognised as non-independent in the Wien limit, the Rayleigh-Jeans regime might better have been recognised for what it was: the regime in which the discrete structure to radiation was no longer point-like, rather than non-independent; but still *movable*.

References

Bach, A. (1997). *Classical indistinguishable particles*. Berlin: Springer.

Boltzmann, L. (1877). On the relationship between the second fundamental theorem of the mechanical theory of heat and probability calculations regarding the conditions for thermal equilibrium (K. Sharp & F. Matschinky, Trans.). *Entropy* 17 (2015) pp. 1971–2009. https://doi.org/10.3390/e17041971.

Caulton, A. (2020). *Entanglement by (anti-)symmetrisation does not Violate Bell's Inequalities*, forthcoming.

Darrigol, O. (1986). The origin of quantized matter waves. *Historical Studies in the Physical and Biological Sciences, 16*, 197–253.

Darrigol, O. (1991). Statistics and combinatorics in early quantum theory, II: Early symptoma of indistinguishability and holism. *Historical Studies in the Physical and Biological Sciences, 21*, 237–298.

Dieks, D. (2014). The logic of identity. Distinguishability and indistinguishability in classical and quantum physics. *Foundations of Physics, 44*, 1302–1316.

Dirac, P. (1926). On the theory of quantum mechanics. *Proceedings of the Royal Society of London, A112*, 661–677.

Dorling, J. (1971). Einstein's introduction of photons: Argument by analogy or deduction from the phenomena? *British Journal for the Philosophy of Science, 22*, 1–8.

Ehrenfest, P., & Ehrenfest T. (1912). *The conceptual foundations of the statistical approach in mechanics* (M. J. Moravcsik, Trans.). New York: Dover, 1990.

Ehrenfest, P., & Kamerlingh-Onnes, H. (1914). Simplified deduction of the formula from the theory of combinations which Planck uses as the basis of his radiation theory. *Proceedings of the Amsterdam Acadamie, 17*, 870–872. Reprinted in *Paul Ehrenfest, Collected Scientific Papers*, Klein, M. (ed).; North-Holland, 1959, pp.353–56.

Einstein, A. (1905). On a heuristic point of view concerning the production and transformation of light. In J. Stachel, D. Cassidy, J. Renn, & R. Schulmann (Eds.), [1990]: *The Collected Papers of Albert Einstein, Vol.2. The Swiss Years: Writings, 1900–1909. English Translation*: Trans. A. Beck, Princeton: Princeton University Press (1989).

Einstein, A. (1925). Quantum theory of the ideal monatomic gas. Second part. In D. K. Buchwald, J. Illy, Z. Rosenkranz, T. Sauer, & O. Moses (Eds.), [2015]: *The collected papers of Albert Einstein, Vol. 14. English Translation*. Princeton: Princeton University Press.

Epstein, P. (1936a). Gibbs' methods in quantum statistics. In A. Haas (Ed.), *A commentary on the scientific writings of J. Willard Gibbs Vol. I: Theoretical physics* (pp. 521–584). New Haven: Yale University Press.

Epstein, P. (1936b). Critical appreciation of Gibbs' statistical mechanics. In A. Haas (Ed.), *A commentary on the scientific writings of J. Willard Gibbs Vol. I: Theoretical physics* (pp. 461–518). New Haven: Yale University Press.

Ghirardi, G., & Marinatto, L. (2004). General criterion for the entanglement of two indistinguishable states. *Physical Review A, 70*, 012109.

Ghirardi, G., Marinatto, L., & Weber, Y. (2002). Entanglement and properties of composite quantum systems: A conceptual and mathematical analysis. *Journal of Statistical Physics, 108*, 112.

Gibbs, J. W. (1902). *Elementary principles in statistical mechanics*. New York: Scribner.
Howard, D. (1990). "Nicht sein kann was nicht sein darf" or the prehistory of EPR, 1909–1935: Einstein's early worries about the quantum mechanics of composite systems. In A. I. Miller (Ed.), *Sixty-two years of uncertainty*. New York: Plenum.
Mehra, J., & Reckenberg, H. (1987). *The historical development of quantum theory: Vol. 5*. New York: Springer.
Norton, J. D. (2006). Atoms, entropy, quanta: Einstein's miraculous argument of 1905. *Studies in History and Philosophy of Modern Physics, 37*, 71–100.
Pais, A. (1982). *Subtle is the Lord: The science and the life of Albert Einstein*. Oxford: Oxford University Press.
Penrose, R. (2004). *The Road to Reality*. London: Jonathan Cape.
Saunders, S. (2006). On the explanation of quantum statistics. *Studies in the History and Philosophy of Modern Physics, 37*, 192–111. Available online at arXiv:quant-ph/0511136.
Saunders, S. (2013). Indistinguishability. In R. Batterman (Ed.), *The Oxford handbook of philosophy of physics* (pp. 340–380). Oxford: Oxford University Press. http://philsci-archive.pitt.edu/12448/.
Saunders, S. (2016). On the emergence of individuals in physics. In A. Guay & T. Pradeu (Eds.), *Individuals across the sciences*. New York: Oxford University Press.
Saunders, S. (2018). The Gibbs paradox. *Entropy, 20*(8), 552. https://doi.org/10.3390/e20080552.
Saunders, S. (2020). The concept 'indistinguishable', forthcoming. *Studies in the History and Philosophy of Modern Physics*.
Schrödinger, E. (1925). Bemerkungen über die statistische Entropiedefinition beim idealen Gas. *Preussische Akademie der Wissenschaften (Berlin), Proceeding* (pp. 434–441).
Versteegh, M., & Dieks, D. (2011). The Gibbs paradox and the distinguishability of identical particles. *American Journal of Physics, 79*, 741–746.

Chapter 23
How (Maximally) Contextual Is Quantum Mechanics?

Andrew W. Simmons

Abstract Proofs of Bell-Kochen-Specker contextuality demonstrate that there exists sets of projectors that cannot each be assigned either 0 or 1 such that each basis formed from them contains exactly one 1-assigned projector. Instead, at least some of the projectors must have a valuation that depends on the context in which they are measured. This motivates the question of how many of the projectors must have contextual valuations. In this paper, we demonstrate a bound on what fraction of rank-1 projective measurements on a quantum system must be considered to have context-dependent valuations as a function of the quantum dimension, and show that quantum mechanics is not as contextual, by this metric, as other possible physical theories. The figure of merit is a property only of the orthogonality graph, or event structure, of a quantum scenario, which meshes with Pitowskys suggestion that all relevant quantum behaviour can be captured by finite orthogonality graphs, and raises questions about the same link Pitowsky explored between contextuality and restrictions on quantum probabilities in the manner of Gleason's theorem.

Keywords Quantum foundations · Quantum logic · Quantum contextuality · Maximal contextuality · Kochen-Specker contextuality

23.1 Introduction

The Bell-Kochen-Specker theorem tells us that there exist sets of projectors onto states $\{|\psi_i\rangle\}$, such that it is impossible to assign to each of them a valuation of 0 or 1 in a context-independent way, and have there be exactly one 1-valued projector in every PVM context. We might ask the following question: under an assumption of outcome-definiteness, how many such projectors *can* we give

A. W. Simmons (✉)
Department of Physics, Imperial College London, London, UK
e-mail: andrew@simmons.co.uk

M. Hemmo, O. Shenker (eds.), *Quantum, Probability, Logic*,
Jerusalem Studies in Philosophy and History of Science,
https://doi.org/10.1007/978-3-030-34316-3_23

a noncontextual valuation to, and how many projectors (or more robustly, what *fraction* of projectors), must be considered to have context-dependent valuations?

Historically, Kochen-Specker style proofs of contextuality, also known as strong or maximal contextuality, have tended to have a "lynch pin" quality; removal of any part of their structure causes the proof to fall apart in its entirety. In fact, this fact has been exploited in order to tame the maximal contextuality demonstrated by the Peres-Mermin magic square for all qubit states Bermejo-Vega et al. (2016), as we need only remove the ability to measure one context before it fails to be a proof of contextuality at all. For such proofs, then, since removing one context is sufficient to remove all contextuality, certainly allowing a projector to vary in its value assignment by context (by assigning it a C) will. This minimalism is partially motivated by the fact that since the original paper by Kochen and Specker, there has been interest in trying to find examples of small Kochen-Specker sets (Arends et al. 2011). When only a single projector need be considered to take a contextual valuation, finding a small such set is the only way of attaining a higher fraction of contextual projectors. We will be looking for an assignment to all projection operators onto a Kochen-Specker set $\{|\psi_i\rangle\langle\psi_i|\} \rightarrow \{0, C, 1\}$ such that the following rules hold: in any basis formed from the projectors onto $\{|\psi_i\rangle\langle\psi_i|\}$, there must be at least one with a C- or 1-valuation, and no more than one with a 1-valuation. The fraction which must be given a C-valuation is the key figure-of-merit that will be explored in this paper. This figure of merit, then, can be seen as a generalisation or successor to the search for small Kochen-Specker sets. Pitowsky (1998, 2005) explored the idea that Gleason's theorem, viewed as a proof of the Bell-Kochen-Specker theorem, can be thought of as providing more quantitative, rather than qualitative, view on the Kochen-Specker theorem when applied to quantum subtheories. Given any set of measurements, it is possible to calculate the maximum deviation from the Born rule that the scenario permits; thinking of a noncontextual 0,1-valuation as representing the maximum possible deviation from the continuous Born rule probabilities. From this perspective, the problem approached in this paper can be viewed as a complementary approach to that of Pitowsky; we find our quantitative figure of merit by seeing how much we have to relax the noncontextuality assumption to permit a classical model for a certain compatibility structure, rather than how much the event structure can tolerate deviations from quantum-like probabilities given by the Born rule.

In this paper, following Pitowsky's approach (Pitowsky 1998), we will use graph-theoretic methods as a computational tool to investigate how robust, in this sense, quantum-mechanically accessible Kochen-Specker proofs can be. We will demonstrate that it is possible that other physical theories can be more contextual, given this definition, than quantum mechanics. We can consider this as a stronger analogue than the observation that a PR-box is more nonlocal with respect to the CHSH inequality than any quantum-mechanically accessible scenario. The figure of merit in this paper can be defined referring only to this orthogonality graph, which agrees with Pitowsky's suggestion that all quantum behaviour should be viewed as a product of the compatibility structure of quantum measurements, *via* the restrictions imposed on possible probability rules by Gleason's theorem. The central question we ask and partially resolve in this paper explores the limits of Pitowsky's ideas and work on logical indeterminacy (Pitowsky 1998); unlike in the situation Pitowsky considered, we see that allowing some observables to have valuations that contextually vary allows many others to be deterministic. We are then able to query to what extent the logical indeterminacy principle remains true, even in the presence of explicit contextuality.

One metric for contextuality is the *contextual fraction* Abramsky et al. (2017), in which we consider our probability distribution to be a convex mixture between a noncontextual theory and a contextual one. This notion is what prompts the term "maximal contextuality"; if a scenario is maximally contextual then it has a contextual fraction that is the highest possible, *viz* 1. It is independent under a re-labelling of the measurement effects, but the contextual fraction is hard to calculate when the number of corners of the noncontextual polytope is large, which is true in many natural situations such as in nonlocality scenarios in which one party has access to measurements with three or more outcomes Simmons (2018). The figure-of-merit explored in this paper acts as a complementary measure of contextuality that can distinguish different scenarios which have the maximum possible contextual fraction of 1. However, it is likely to be hard to calculate in many circumstances, and in the case in which one is using the entirety of Hilbert space as one's vector set, we shall explore the connection to open problems in mathematics known as Witsenhausen's problem (Witsenhausen 1974) and the Hadwiger-Nelson problem (Soifer 2008).

23.2 How Robust Can Quantum Maximal Contextuality Be?

Consider the orthogonality graph $\mathcal{G}$ of a maximal nonlocality scenario, in which we identify projectors with vertices, and two vertices g_1, g_2, identified with $|\psi_1\rangle, |\psi_2\rangle$ are joined by an edge iff $\langle\psi_1|\psi_2\rangle = 0$. Following Renner and Wolf Renner and Wolf (2004), we will restrict our attention to outcome assignments in which we do not ever assign a "1" valuation to two orthogonal vectors; this is known as a *weak Kochen-Specker set*. A weak Kochen-Specker set can be extended to a full Kochen-Specker set with the addition of $O(||\psi_i\rangle|^2 d)$ extra vectors, or we can consider the incomplete bases as being completed by additional projectors, whose valuations are allowed to be contextual as they are outside the scope of our scenario. In fact, this distinction will not be crucial for any of the main results; it does however make certain definitions conceptually clearer. An *independent set* is a subset of vertices, none of which are joined by an edge. Any noncontextual assignment of "1"s, then, will form an independent set since we cannot have two orthogonal vectors, (which correspond to two vertices joined by an edge) in our set of vectors which are noncontextually given "1" assignments. The size of the largest independent set of a graph $\mathcal{G}$ is denoted $\alpha(\mathcal{G})$ and is called the *independence number*. Another important concept will be that of a clique:

Definition 1 (Clique) For a graph $\mathcal{G} = \{V, E\}$, vertex subset $C \subset V$ is a *Clique* if the subgraph induced from $\mathcal{G}$ on C is *complete*, that is, every vertex in C is connected to every other; and C is inclusion-maximal.

We say that a clique is a *maximum clique* if there is no clique in $\mathcal{G}$ with more vertices. The size of a maximum clique in $\mathcal{G}$ is denoted $\omega(\mathcal{G})$, and so we note that for a d-dimensional quantum system, as long as the set of vectors we can measure includes at least one basis, we will have $\omega(\mathcal{G}) = d$. In a hidden variable model for a scenario, we require that in every context, there be at least one vector that can take a "1" value either contextually or noncontextually. This motivates the following definition:

Definition 2 (Maximum-clique hitting-set) For a graph $\mathcal{G} = \{V, E\}$, a vertex subset $T \subset V$ is a *maximum-clique hitting-set* or *maximum-clique transversal* if for all cliques C, $|C| = \omega(\mathcal{G})$, $T \cap C \neq \emptyset$.

In any contextual hidden-variable model, the set of vertices associated with a set of vectors given either noncontextual or contextual "1" assignments is a Maximum-clique hitting set, since every context needs an outcome. This allows us to define our first figure-of-merit for a set of vectors forming a Kochen-Specker type proof of the Bell-Kochen-Specker theorem.

$$q_s(\mathcal{G}) = \min_{T} \min_{A \subset T} |T| - |A|, \tag{23.1}$$

where T is a maximum-clique hitting-set. We see that this is a graph-theoretic representation of the figure-of-merit we posited above: we take a set T of vectors which will recieve a "1" valuation, either contextually or noncontextually, and subtract from it the number of those vectors that can support a noncontextual valuation. This quantity then represents the number of vectors that must receive a contextual valuation in any model for the scenario. However, we can note that it is possible to achieve an arbitrarily high value of q_s just by taking an equal number of noninteracting copies of, for example, the Peres-Mermin magic square. In order that we might more directly compare this quantity between graphs of different sizes, we can consider its graph density:

$$q(\mathcal{G}) = \frac{q_s(\mathcal{G})}{|V(\mathcal{G})|}. \tag{23.2}$$

We now have a graph-theoretic formulation of our quantity of interest. What values of this quantity are obtainable, either by quantum-mechanically realisable scenarios, or in general probabilistic theories? We note the following trivial lower bound:

$$q_s(\mathcal{G}) \geq \min |T| - \alpha(\mathcal{G}). \tag{23.3}$$

We note that in general, it is **NP**-complete to calculate the value of $\min |T|$ if we know $\omega(\mathcal{G})$, since for triangle-free graphs, this is equivalent to finding a vertex cover,

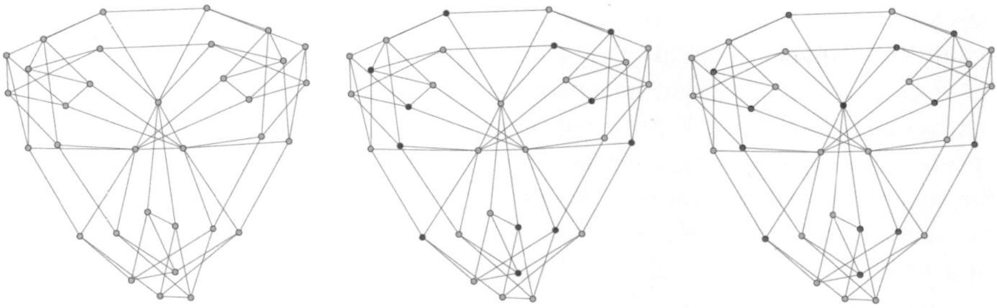

Fig. 23.1 Three images illustrating the calculation of the figure of merit. The first is the orthogonality graph for the 33-ray proof of the Kochen-Specker theorem by Peres (1991). Highlighted in the second in red is a maximal independent set, that represents a maximal assignment of noncontextual 1-valuations. In fact, only a single context lacks a 1-valued vector, visible in the centre of the graph. Highlighted in the third in blue is a projector that can be given a contextual valuation in order that every context contain a ray assigned either a 1 or a contextual valuation; it will take a 1 valuation in the "central" context, and a 0 in all others

which is **NP**-complete (Poljak 1974). Also, it is **NP**-complete to calculate $\alpha(\mathcal{G})$ (Garey and Johnson 1979). This suggests that calculation of this quantity may be computationally difficult. A tension exists in trying to optimise $q(\mathcal{G})$. Particularly well-connected graphs may achieve an exponentially small independence density – this is true for a few of the families of graphs considered in this paper – so few vectors can receive a noncontextual assignment of 1, but this same connectedness means that allowing a single vector's valuation to vary contextually "fixes" many contexts (Fig. 23.1).

We will now demonstrate an upper bound on the quantity $q(\mathcal{G})$ for any quantum-mechanically accessible $\mathcal{G}$ in terms of the quantum dimension of the vectors that give rise to it.

Theorem 1 *Any orthogonality graph $\mathcal{G}$ induced by a set of vectors $|\psi_i\rangle$ in a d-dimensional Hilbert space has*

$$q(\mathcal{G}) \leq \left(1 - \frac{1}{d}\right)^{d-1} - \frac{1}{2^{d-1}}. \qquad (23.4)$$

Proof The proof proceeds by demonstrating an explicit procedure for finding a T and an $A \subset T$, for any $\{|\psi_i\rangle\}$, with upper-bounded worst-case performance.

The set of vectors $|\phi\rangle \in \mathcal{H}_d$ such that $|\langle\psi|\phi\rangle|^2 \geq t$ for some $t \in \mathbb{R}$ forms a closed complex hyperspherical cap centred at $|\psi\rangle$. Consider a basis for $\mathcal{H}_d$. Without loss of generality, we can take this to be the standard basis for the set. By symmetry concerns, the quantity $\min_i |\langle\psi|i\rangle|$ is maximised (albeit nonuniquely) by $|\psi\rangle = d^{-1/2} \sum_i |i\rangle$. This puts an upper bound on the size of the largest hyperspherical angle that can separate an arbitrary vector from its closest basis element. A closed complex hyperspherical cap subtending this angle, $\{|\psi\rangle \big| |\langle\psi|\phi\rangle|^2 \geq 1/d\}$, will then necessarily capture at least one element from the basis, and so any set of vectors within such a hyperspherical cap C construes a maximum-clique hitting-set, T_C.

Since columns of a Haar-random unitary matrix over $\mathbb{C}^d$ are uniformly random unit vectors, the volume V of this hyperspherical cap as a proportion of that of the sphere is given by $1 - F(t)$, where $F(t)$ is the cumulative distribution function of the squared modulus of an entry in such a Haar-random unitary. This quantity is well known in the field of random matrix theory and it can be found, for example, as a special case of a result by Życzkowski and Sommers (2000), and will be demonstrated in Sect. 23.3.

$$\frac{V}{V_S} = (1-t)^{d-1}. \tag{23.5}$$

Clearly, the value of $q(\mathcal{G})$ for a quantum graph is upper bounded by the size of any maximum-clique hitting set, so if we can bound the minimum number of vectors we must intersect with such a hyperspherical cap, this also constitutes a bound on the value $q(\mathcal{G})$. However, we would not expect this bound to be tight, since we have ignored the independent set subtrahend entirely. We would like to demonstrate an independent set within our maximum-clique hitting-set T_C, and therefore give a better upper bound on the quantity $q_s(\mathcal{G}) = \min_T \min_{A \subset T} |T| - |A|$, giving an explicit construction procedure for a T and $A \subset T$.

We note that a hyperspherical cap containing the vectors $|\phi\rangle$ such that $|\langle\psi|\phi\rangle|^2 > 1/2$ cannot contain two orthogonal vectors as a simple application of the triangle inequality. We can place this smaller hyperspherical cap anywhere within the larger hyperspherical cap, and this will form an independent set A that is a subset of T_C. For the rest of this proof, we will assume that the two hyperspherical caps are centred on the same point; this will make no difference to the worst case scenario which forms the bound. The region of Hilbert space, then, in which vectors count towards the figure of merit, is a member of a two-parameter family of *annuli* on the complex hypersphere, illustrated in Fig. 23.2 and defined by

$$t_1 \leq |\langle\psi|\phi\rangle|^2 \leq t_2. \tag{23.6}$$

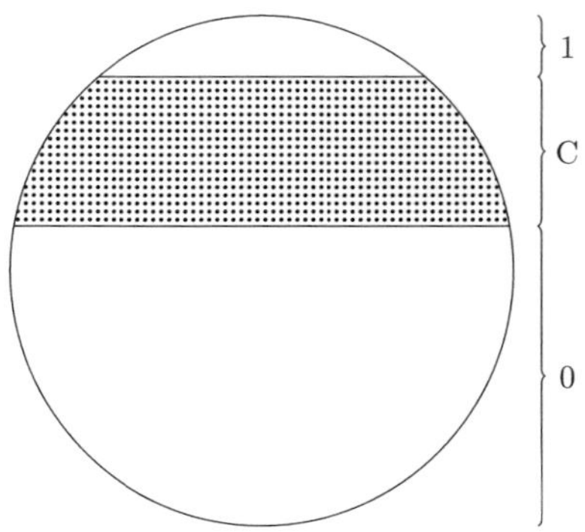

Fig. 23.2 An illustration of how this construction splits up the Hilbert space into different regions, that can be associated with projectors given noncontextual 1-assignments, contextual projectors, and projectors given noncontextual 0-assignments

Such an annulus takes up a proportion

$$p(t_1, t_2) = (1 - t_1)^{d-1} - (1 - t_2)^{d-1} \tag{23.7}$$

of the Hilbert space. Our construction has $t_1 = 1/d$, and $t_2 = 1/2$, giving us a hyperspherical annulus defined by $1/d \leq |\langle\psi|\phi\rangle|^2 \leq 1/2$. It now remains to be proved that no matter the positions of the set of vectors $\{\psi\}$, there must be some place in which we can place this annulus such that we only intersect strictly less than $p(t_1, t_2)$ of them.

Consider the canonical action of $SU(d)$ on the unit hypersphere in $\mathbb{C}^d$, which we shall denote $S_{\mathbb{C}}^{d-1}$. We define a space $SU(d) \times S_{\mathbb{C}}^{d-1}$, where $SU(d)$ is equipped with the Haar measure, and $S_{\mathbb{C}}^{d-1}$ is equipped with the uniform measure. Equivalently, we could have taken H_d rather than $S_{\mathbb{C}}^{d-1}$, in which case it would be equipped with the Fubini-Study metric.

At each $|\psi_i\rangle$, we place a bump function $f^{\epsilon}_{|\psi_i\rangle}$ with unit integral and support only within a disc of radius ϵ around $|\psi_i\rangle$. Let A_{t_1,t_2} denote the hyperspherical annulus $\{|\phi\rangle \mid t_1 \leq |\langle 0|\phi\rangle|^2 \leq t_2\}$. By Fubini's theorem, we have:

$$\int_{SU(d)} \int_{S_{\mathbb{C}}^{d-1}} \mathbb{1}_{g.z \in A_{t_1,t_2}} \sum_i f^{\epsilon}_{|\psi_i\rangle}(z)\, \mathrm{d}z\, \mathrm{d}g = \int_{S_{\mathbb{C}}^{d-1}} \int_{SU(d)} \mathbb{1}_{g.z \in A_{t_1,t_2}} \sum_i f^{\epsilon}_{|\psi_i\rangle}(z)\, \mathrm{d}g\, \mathrm{d}z \tag{23.8}$$

We note that if we extend our annulus's extent by ϵ, so that we have $t_1 \to t_1 - \epsilon$ and $t_2 \to t_2 + \epsilon$, we capture the entirety of the measure of the bump functions centred inside the original annulus. Hence, for all $g \in SU(d)$ we have $\left|\{|\psi_i\rangle \middle| |\psi_i\rangle \in g^{-1} A_{t_1,t_2}\}\right| \leq \int_{S_{\mathbb{C}}^{d-1}} \mathbb{1}_{g.z \in A_{t_1-\epsilon,t_2+\epsilon}} \sum_i f^{\epsilon}_{|\psi_i\rangle}(z)\, \mathrm{d}z$, and so we have

$$\int_{SU(d)} \left|\left\{|\psi_i\rangle \middle| |\psi_i\rangle \in g^{-1} A_{t_1,t_2}\right\}\right| \mathrm{d}g \leq \int_{S_{\mathbb{C}}^{d-1}} \int_{SU(d)} \mathbb{1}_{g.z \in A_{t_1-\epsilon,t_2+\epsilon}} \sum_i f^{\epsilon}_{|\psi_i\rangle}(z)\, \mathrm{d}g\, \mathrm{d}z, \tag{23.9}$$

Rearranging:

$$\int_{SU(d)} \left|\left\{|\psi_i\rangle \middle| |\psi_i\rangle \in g^{-1} A_{t_1,t_2}\right\}\right| \mathrm{d}g \leq \sum_i \int_{S_{\mathbb{C}}^{d-1}} f^{\epsilon}_{|\psi_i\rangle}(z) \int_{SU(d)} \mathbb{1}_{g.z \in A_{t_1-\epsilon,t_2+\epsilon}}\, \mathrm{d}g\, \mathrm{d}z \tag{23.10}$$

Twirling the indicator function $\mathbb{1}_{g.z \in A_{t_1-\epsilon,t_2+\epsilon}}$ with respect to the Haar measure over $SU(n)$ reduces it to a constant function with value $p(t_1 - \epsilon, t_2 + \epsilon)$, the proportion of the Hilbert space taken up by the annulus.

$$\int_{SU(d)} \left|\left\{|\psi_i\rangle \middle| |\psi_i\rangle \in g^{-1} A_{t_1,t_2}\right\}\right| \mathrm{d}g \leq p(t_1 - \epsilon, t_2 + \epsilon) \sum_i \int_{S_{\mathbb{C}}^{d-1}} f^{\epsilon}_{|\psi_i\rangle}(z)\, \mathrm{d}z \tag{23.11}$$

$$\leq p(t_1 - \epsilon, t_2 + \epsilon) \left|\left|\psi_i\right\rangle\right| \tag{23.12}$$

Since the left hand side of Equation 23.11 represents an average over the rotation group, there must be some $g \in SU(d)$ such that $\left|\{|\psi_i\rangle \big| |\psi_i\rangle \in g^{-1}A_{t_1,t_2}\}\right| \leq p(t_1 - \epsilon, t_2 + \epsilon)\left|\{|\psi_i\rangle\}\right|$, and this is true for all ϵ. Therefore:

$$q(\mathcal{G}) \leq \lim_{\epsilon \to 0} p(t_1 - \epsilon, t_2 + \epsilon) \tag{23.13}$$

$$\leq \left(1 - \frac{1}{d}\right)^{d-1} - \frac{1}{2^{d-1}}. \tag{23.14}$$

□

Corollary 1 *Any quantum mechanically accessible* $\mathcal{G}$ *has* $q(\mathcal{G}) \leq$ $^{4251920575}/_{11019960576} \sim 0.385838$.

Proof Calculating the derivative of $\left(1 - \frac{1}{d}\right)^{d-1} - \frac{1}{2^{d-1}}$ with respect to d shows that it takes a maximum between $d = 9$ and $d = 10$. Evaluation of the quantity at these points reveal the maximum to be at $d = 9$, when we get $(1 - {}^1/_9)^8 - {}^1/_{2^8} = {}^{4251920575}/_{11019960576}$. This, then, forms a hard limit of $q(\mathcal{G})$ for any quantum mechanically achievable $\mathcal{G}$. □

We note that as the quantum dimension approaches infinity, we achieve a limiting value of $^1/_e$, so quantum systems with very high dimension are viable candidates for providing robust contextuality scenarios. In Fig. 23.3, we see the bound on $q(\mathcal{G})$ plotted as a function of d.

Above, it was shown that a spherical cap taking up a proportion of $(1 - {}^1/_d)^{d-1}$ of Hilbert space must capture at least one vector from every orthonormal basis. However, if for many bases this spherical cap captures more than one basis element, this might suggest that this spherical cap was not the most efficient way of choosing a vector from each basis. We note, though, that the exact geometry of the set

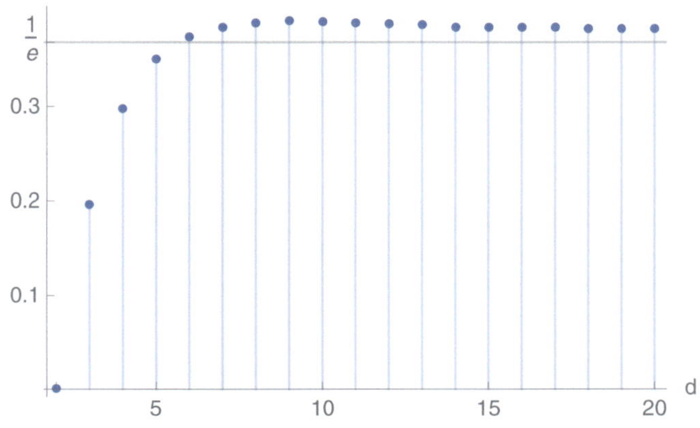

Fig. 23.3 A chart showing the value of the bound on $q(\mathcal{G})$ for quantum-mechanically accessible $\mathcal{G}$ as a function of the Hilbert space dimension d

of projectors will have a large effect on the efficiency of this construction. One one hand, an increase in the quantum dimension allows for more complicated structures and interplay of the permitted projectors that, as we shall demonstrate, seems to allow for construction of quantum graphs with very small independence number. However, it also has a dampening affect on the quantity $|T|$ since the bases themselves have more geometrical constraints.

In fact such an annulus can in the worst case can contain every vector from a basis. Below, we will see a specific example of a set of vectors providing a nonlocality scenario, in which *every* projector is captured by a pessimally-chosen spherical cap position. In d-dimensional Hilbert space, a spherical cap of proportion $(1-1/n)^{d-1}$ can capture n vectors out of a basis. So for any chosen n', as d increases, we see that the fraction of the spherical cap in the construction that cannot capture n' of a basis tends to 0. However, if we consider the fraction of the spherical cap that can capture a constant proportion of basis elements c, we get the following behaviour under increasing d:

$$\lim_{d\to\infty}\left(1-\frac{1}{cd}\right)^{d-1}=e^{-\frac{1}{c}}. \tag{23.15}$$

This might imply that the quality of the bound does not degrade as the dimension increases (Fig. 23.4).

Name	Quantum dimension	Transversal size	Independence number	Figureof merit
ABK triangle-free graphs (Alon et al. 2010)	✗	$n - c\sqrt{n\log n}$	$c\sqrt{n\log n}$	$1 - O\left(\frac{\log n}{n}\right)$
Quantum upper bound	d	$\leq n\left(1-\frac{1}{d}\right)^{d-1}$	$\geq \frac{n}{2^{d-1}}$	≤ 0.385838
E_8 root system	8	$\leq \frac{11n}{60}$	$\frac{n}{15}$	$\leq 0.11\dot{6}$
Two-qubit stabiliser quantum mechanics	4	$\frac{3n}{10}$	$\frac{n}{5}$	0.1
Peres-Mermin magic square (Peres 1991)	4	$\frac{7n}{24}$	$\frac{5n}{24}$	$0.08\dot{3}$
Cabello's 18-ray proof (Cabello 1997)	4	$\frac{5n}{18}$	$\frac{4n}{18}$	$0.0\dot{5}$
Peres's 33-ray proof (Peres 1991)	3	$\frac{9n}{33}$	$\frac{12n}{33}$	$0.\dot{0}\dot{3}$

Fig. 23.4 A table comparing the graph constructions in this paper; each concrete example represents the graph union of multiple noninteracting copies of the KS proof. Two-qubit stabiliser quantum mechanics forms a KS proof using 60 rays and 105 contexts; the E_8 root system forms a KS proof using 120 rays and 2025 contexts. The quantities related to each were calculated using a proof by exhaustion, or are the best found during a non-exhaustive search, such as is the case for the E_8 system. It is perhaps of note that the lower bound of the size of a traversal less the independence number forms a trivial bound only for Peres's 33-ray proof, and in fact is optimal for Cabello's 18-ray proof and the Peres-Mermin magic square

23.2.1 Triangle-Free Graphs

In triangle-free graphs, the maximum cliques are merely edges, and so each context consists of two measurement objects. Quantum mechanically, the only such graphs are uninteresting: the disjoint graph union of K_2 graphs. These do not have enough structure to form a proof of the Bell-Kochen-Specker theorem; this reflects the fact that each projector appears in only a single context. Many such graphs, then, are not quantum-mechanically accessible. However, they do allow us to prove a concrete bound on what values of $q(\mathcal{G})$ are achievable within the framework of generalised probabilistic theories. In such graphs, the concept of a maximum-clique hitting-set reduces to that of a covering of edges by vertices, known as a *vertex cover*. The size of the minimal vertex cover of $\mathcal{G}$ is denoted $\tau(\mathcal{G})$. This reduction of our figure-of-merit allows us to make use of a bound derived for Ramsey type problems.

In the triangle-free case, we have a powerful relation between maximum-clique hitting-sets and independent sets; namely that they are graph complements of each other. A maximum-clique hitting-set, or vertex cover, is a set of vertices T such that for every edge, at least one of that edge's vertices is in T. We can see, then, that the set $V-T$ must be independent, since if there were two vertices in $V-T$ connected by an edge, then neither of those vertices would be in T, a contradiction. Conversely, if we have an independent set A, then $V-A$ must be a vertex cover; if there were an edge with neither vertex in $V-A$, then both vertices are in A, a contradiction. This means that we can bound $q_s(\mathcal{G})$ as

$$q_s(\mathcal{G}) \geq \min_{\mathcal{G}} |T| - \alpha(\mathcal{G}) = |V(\mathcal{G})| - 2\alpha(\mathcal{G}). \tag{23.16}$$

In other words, to be able to lower bound $q_s(\mathcal{G})$, we need only consider the independence number, $\alpha(\mathcal{G})$. We seek, therefore, triangle-free graphs with low independence number. We can invoke here a theorem due to Alon et al. (2010).

Theorem 2 *There exists a constant c such that for all $n \in \mathbb{N}$ there exists a regular triangle-free graph $\mathcal{G}_n$, with $V(\mathcal{G}_n) = n$, and $\alpha(\mathcal{G}_n) \leq c\sqrt{n \log n}$.*

Hence, we have

$$\lim_{n\to\infty} q(\mathcal{G}_n) \geq \lim_{n\to\infty} \frac{n - 2c\sqrt{n\log n}}{n} \tag{23.17}$$

$$\geq 1 - 2c \lim_{n\to\infty} \sqrt{\frac{\log n}{n}} = 1. \tag{23.18}$$

Therefore, $q(\mathcal{G}_n) = 1$ is approachable in the limit of $n \to \infty$. We see then that quantum mechanics does not display maximal contextuality as robustly, given this definition of robustness, as other possible GPTs. This could potentially have an impact on attempts to recreate quantum mechanics as a principle theory.

23.3 Higher-Rank PVMs

We now extend the result to apply to projective valued measures including projectors of rank greater than 1, although we are still considering the case in which contexts are made up of a set of projectors which sum to the identity, rather than the more general case of a set of commuting projectors. We will allow a rank-k projector P to "inherit" a labelling as a function of the assignments to some $\left\{\left|\psi_P^{(i)}\right\rangle\right\}$, which form a specially chosen decomposition of P as $P = \sum_i \left|\psi_P^{(i)}\right\rangle\!\left\langle\psi_P^{(i)}\right|$.

Given a variable assignment to states $f : \mathcal{H}_d \to \{0, C, 1\}$, we can define a variable assignment to all projection operators on $\mathcal{H}_d$, $g : \mathcal{P}(\mathcal{H}_d) \to \{0, C, 1\}$ by the following:

$$g(P) = \begin{cases} 0 & \text{if } \exists \left\{\psi_P^{(i)}\right\},\ P = \sum_i \left|\psi_P^{(i)}\right\rangle\!\left\langle\psi_P^{(i)}\right|,\ f\left(\left|\psi_P^{(i)}\right\rangle\right) = 0\ \forall i, \\ 1 & \text{elif } \exists \left\{\psi_P^{(i)}\right\},\ P = \sum_i \left|\psi_P^{(i)}\right\rangle\!\left\langle\psi_P^{(i)}\right|,\ f\left(\left|\psi_P^{(i)}\right\rangle\right) \in \{0, 1\}\ \forall i, \\ C & \text{o/w.} \end{cases} \tag{23.19}$$

The motivation for this notion of value inheritance is as follows: in any context containing P, we can consider replacing it with one of its maximally fine-grained decompositions, which then inherits value assignments from f derived *via* the annulus method. Since the assignment of values to this rank-1 decomposition of P is necessarily consistent with the restrictions of the scenario, we can consider a post-processing that consists of a coarse-graining of those fine-grained results, "forgetting" which one of them occurred. This process must also be consistent. Hence, as long as there exists some decomposition of P into rank-1 projectors, each of which is assigned a noncontextual valuation, then P can inherit a noncontextual valuation. However if there must be a contextual vector in any such decomposition, we must treat P as having a contextual valuation.

The challenge, then, is to characterise which projectors P must be given a contextual valuation under this schema, and then prove a bound analogous to that of Theorem 1.

Theorem 3 *In a scenario in which the available measurements are rank-r projectors* $\{P_i\}$*, performed on a quantum system of dimension d, then at most* $(I_{1/2}(r, d-r) - I_{r/d}(r, d-r))|\{P_i\}|$ *are given a contextual valuation, where* $I_x(\alpha, \beta)$ *is the* regularised incomplete beta function.

Proof The proof proceeds similarly to the proof of Theorem 1. We will identify a generalised annulus within the space of rank-r projectors on $\mathcal{H}_d$ which can be associated with contextual valuations, and then use its volume in proportion to that of the overall Hilbert space to form a bound on how many must be given contextual valuations.

Lemma 1 *Under the extension of a valuation of rank-1 projectors given by the annulus method from Theorem* 1 *centred on* $|\phi\rangle$, *a rank-r projector* P_r *is contextual only if* ${}^r/_d \leq \mathrm{Tr}(P_r|\phi\rangle\langle\phi|) \leq {}^1/_2$.

Proof of Lemma First, consider the case that $\mathrm{Tr}(P_r|\phi\rangle\langle\phi|) > {}^1/_2$. We could equivalently write this as $\langle\phi|P_r|\phi\rangle = \langle\phi|P_r P_r|\phi\rangle > {}^1/_2$. We note that $P_r|\phi\rangle$ is a vector in the eigenvalue-1 subspace of P_r. As such, we can find a basis for P_r that includes $P_r|\phi\rangle$, up to a normalisation constant. By design, each of these other basis elements $|i\rangle$ will have $\langle i|\phi\rangle = \langle i|P_r|\phi\rangle = 0$. Hence, this corresponds to a decomposition of P_r in which every vector would be given either a 1 or a 0 valuation, and so P_r inherits a 1 valuation.

Next, we consider the case that $\mathrm{Tr}(P_r|\phi\rangle\langle\phi|) = {}^r/_d - \epsilon < {}^r/_d$. We wish to decompose P_r in such a way that each of the elements $|i\rangle$ of the decomposition have $|\langle i|\phi\rangle| < {}^1/_d$. Again, we consider the vector $P_r|\phi\rangle$ and choose a basis for the 1-eigenspace of P_r so that we have

$$P_r|\phi\rangle = \sum_{i=0}^{r-1}\left(\sqrt{\frac{1}{d} - \frac{\epsilon}{r}}\right)|i\rangle. \tag{23.20}$$

Each of these elements in the decomposition would recieve a noncontextual 0 valuation, so P_r inherits a 0 valuation. This concludes the proof. □

Using the result of the lemma, we can apply a proof method identical to that in Theorem 1 if we can calculate the proportion of the Hilbert space taken up by the set $\{P_r|{}^r/_d \leq \mathrm{Tr}(P_r|\phi\rangle\langle\phi|) < {}^1/_2\}$.

As before, entries of a uniformly random unit vector in $\mathbb{C}^d$ have the same distribution as the entries of a column in a Haar-random $d \times d$ unitary matrix. Applying a result of Życzkowski and Sommers (2000), if U is a Haar-random $d \times d$ unitary, then defining $Y = \sqrt{\sum_{k=1}^{r}|U_{k,1}|^2}$, we have

$$P^Y_{d,r}(y) = c_{d,r}\, y^{2r-1}(1-y^2)^{d-r-1}, \tag{23.21}$$

where $P^Y_{d,r}(y)$ is the probability density function for the random variable Y, and

$$c_{d,r} = \frac{2}{B\,(r, d-r)} = \frac{2\Gamma(d)}{\Gamma(r)\Gamma(d-r)}, \tag{23.22}$$

in which B is the Euler beta function. and Γ is the gamma function. Performing a change of variables, then, in which $T = Y^2$, we get:

$$P^T_{d,r}(t) = \left|\frac{\mathrm{d}}{\mathrm{d}t}\left(\sqrt{t}\right)\right| P^Y_{d,r}\left(\sqrt{t}\right) \tag{23.23}$$

$$= \frac{t^{r-1}(1-t)^{d-r-1}}{B(r, d-r)} \tag{23.24}$$

This is exactly a Beta distribution with shape parameters $\alpha = r$ and $\beta = d - r$; we have derived that $T \sim \text{Beta}(r, d-r)$. We note taking $r = 1$, and integrating, we recover our earlier result used in Theorem 1. Since the CDF for a Beta distributed random variable is given by the regularised incomplete beta function, $I_x(\alpha, \beta)$, the proportion of the Hilbert space taken up by the set $\{P_r | r/d \leq \text{Tr}(P_r |\phi\rangle\langle\phi|) \leq 1/2\}$ is given by

$$I_{1/2}(r, d-r) - I_{r/d}(r, d-r). \tag{23.25}$$

To complete the proof we once again apply the argument from Fubini's theorem used in the proof of Theorem 1. □

Corollary 2 *For any choice of d, r, the proportion of projectors requiring a contextual valuation is always less than 1/2.*

Proof We have for any d, r, that the proportion of projectors requiring a contextual valuation is bounded above by $I_{1/2}(r, d-r) - I_{r/d}(r, d-r)$. We wish, then, to bound this quantity above by $1/2$.

Since $I_x(\alpha, \beta)$ is the cumulative distribution function for a random variable with distribution $\text{Beta}(\alpha, \beta)$, a trivial upper bound for this quantity is given by

$$I_{1/2}(r, d-r) - I_{r/d}(r, d-r) < 1 - I_{r/d}(r, d-r). \tag{23.26}$$

In fact for large d, this is a reasonable approximation, since we can apply Chebyshev's inequality to show that $I_{1/2}(r, d-r) \geq 1 - O(d^{-2})$. We wish, then, to lower bound the value of $I_{r/d}(r, d-r)$. The mean-median-mode inequality (Kerman 2011) for the Beta distribution with $\alpha \leq \beta$, which is met here since r must be a nontrivial factor of d, is given by

$$\frac{\alpha - 1}{\alpha + \beta - 2} \leq m(\alpha, \beta) \leq \frac{\alpha}{\alpha + \beta}, \tag{23.27}$$

where $m(\alpha, \beta)$ represents the median of a random variable with a $\text{Beta}(\alpha, \beta)$ distribution. This is an equality only when $\alpha = \beta$; although this corresponds to a trivial contextuality scenario. Importantly, we have

$$m(r, d-r) \leq \frac{r}{d}. \tag{23.28}$$

Hence by the monotonicity of $I_x(\alpha, \beta)$, we have $I_{r/d}(r, d-r) \geq I_{m(r,d-r)}(r, d-r) = 1/2$, and the result follows.

□

23.4 Conclusion and Future Work

There is a large gulf between the best-known realisation of a quantum-mechanically accessible BKS proof and the bounds on the robustness of such a proof that we have derived in this article. However, it may be that to prove a tighter bound on what is quantum-mechanically possible in a way similar to that of this paper is mathematically difficult; this is because the method requires that we find a set that is independent, regardless of the specific structure of the set of vectors in question. This implies bounds on the independence number of the associated orthogonality graph for all finite contextuality scenarios, and by extension also the contextuality scenario that includes every quantum state. This problem, to choose the largest set of points on a hypersphere such that no two points in the set are orthogonal, is essentially a problem known as Witsenhausen's problem (Witsenhausen 1974) after the author of the paper that formulated this problem, and proved the trivial bound of $1/d$. It has been conjectured that the spherical cap approach is optimal for both the real case, as conjectured by DeCorte and Pikhurko (2016) by Gil Kalai and in the complex case by Montina. However, this is not even known for the original three-real-dimensional case considered by Kochen and Specker. In turn, the solution to Witsenhausen's problem implies lower bounds for the Hadwiger-Nelson problem (DeCorte and Pikhurko 2016): the number of colours needed to colour Euclidean spaces if no two points distance 1 away from each other can have the same colour. The solution to this problem is thought to depend on the specific model of ZF set theory adopted (Soifer 2008). Additionally, if a better construction is found, then we might also have to vary the shape of the area designed to capture at least one element from each basis, so that this latter area is a strict superset of the former, meaning that this efficiency may not be directly capturable by a variation of this proof. However, a more efficient construction would lead to a better bound in the high-dimensional limit.

In this paper, we considered contexts formed of projective matrices that sum to the identity. Recently, attention has been drawn to an analogous problem defined for sets of mutually commuting Hermitian matrices, and so an extension of variant of this result in that case would be helpful for bounding the amount of classical memory needed for an unbiased weak simulation of quantum subtheories (Karanjai et al. 2018).

A comparison between this problem and that explored by Pitowsky reveals potential connections that could be investigated, the most obvious of which is to investigate the extent to which the two figures of merit capture the same information about the contextuality of a quantum subtheory. They have similar properties, such as being defined as a function of just the compatibility structure of a scenario given by orthogonality graphs; and taking noncontextual scenarios as an extreme value. Such a comparison would be restricted only to connected graphs since Pitowsky's figure of merit is not naturally invariant under the addition of a rotated version of a scenario to itself, when the two copies have no commensurability.

Acknowledgements Special thanks to Mordecai Waegell, who suggested using the E_8 root system and two-qubit stabiliser quantum mechanics to demonstrate lower bounds on what values of the figure of merit in Sect. 23.2 were quantum-mechanically accessible, as well as his help calculating this lower bound. Special thanks also to Terry Rudolph for asking this question in the first place. I am grateful to Angela Xu, Angela Karanjai and Baskaran Sripathmanathan for their helpful discussions which have guided this project. I acknowledge support from EPSRC, *via* the CDT in Controlled Quantum Dynamics at Imperial College London; and from Cambridge Quantum Computing Limited. This research was supported in part by Perimeter Institute for Theoretical Physics. Research at Perimeter Institute is supported by the Government of Canada through the Department of Innovation, Science and Economic Development and by the Province of Ontario through the Ministry of Research and Innovation.

References

Abramsky, S., Barbosa, R. S., & Mansfield, S. (2017). The contextual fraction as a measure of contextuality. arXiv:1705.07918.

Alon, N., Ben-Shimon, S., & Krivelevich, M. (2010). A note on regular Ramsey graphs. *Journal of Graph Theory, 64*(3), 244–249.

Arends, F., Ouaknine, J., & Wampler, C. W. (2011). *On searching for small Kochen-Specker vector systems* (pp. 23–34). Berlin/Heidelberg: Springer.

Bermejo-Vega, J., Delfosse, N., Browne, D. E., Okay, C., & Raussendorf, R. (2016). Contextuality as a resource for qubit quantum computation. arXiv:1610.08529.

Cabello, A. (1997). *A proof with 18 vectors of the Bell-Kochen-Specker theorem* (pp. 59–62). Dordrecht: Springer.

DeCorte, E., & Pikhurko, O. (2016). Spherical sets avoiding a prescribed set of angles. *International Mathematics Research Notices, 2016*(20), 6096–6117.

Garey, M. R., & Johnson, D. S. (1979). *Computers and intractability; a guide to the theory of NP-completeness*. W.H. Freeman and Company. New York.

Karanjai, A., Wallman, J., & Bartlett, S. (2018). Contextuality bounds the efficiency of classical simulation of quantum processes. arXiv:1802.07744.

Kerman, J. A closed-form approximation for the median of the beta distribution. arXiv:1111.0433, November 2011.

Peres, A. (1991). Two simple proofs of the Kochen-Specker theorem. *Journal of Physics A: Mathematical and General, 24*, 175–178.

Pitowsky, I. (1998). Infinite and finite Gleason's theorems and the logic of indeterminacy. *Journal of Mathematical Physics, 39*, 218.

Pitowsky, I. (2005). Quantum mechanics as a theory of probability. In W. Demopoulos & I. Pitowsky (Eds.) *Physical theory and its interpretation* (The western Ontario series in philosophy of science, Vol. 72). Dordrecht: Springer.

Poljak, S. (1974). A note on stable sets and colorings of graphs. *Commentationes Mathematica Universitatis Carolinae, 15*(2), 307–309.

Renner, R., & Wolf, S. (2004). Quantum pseudo-telepathy and the Kochen-Specker theorem. In *International Symposium on Information Theory, 2004. ISIT 2004. Proceedings* (pp. 322–322).

Simmons, A. W. (2018, February). On the computational complexity of detecting possibilistic locality. *Journal of Logic and Computation, 28*(1), 203–217. https://doi.org/10.1093/logcom/exx045.

Soifer, A. (2008). *The mathematical coloring book: Mathematics of coloring and the colorful life of its creators*. New York/London: Springer.

Witsenhausen, H. S. (1974). Spherical sets without orthogonal point pairs. *American Mathematical Monthly, 81*, 1101–1102.

Życzkowski, K., & Sommers, H.-J. (2000). Truncations of random unitary matrices. *Journal of Physics A: Mathematical and General, 33*(10), 2045.

Chapter 24
Roots and (Re)sources of Value (In)definiteness *Versus* Contextuality

Karl Svozil

Abstract In Itamar Pitowsky's reading of the Gleason and the Kochen-Specker theorems, in particular, his Logical Indeterminacy Principle, the emphasis is on the value indefiniteness of observables which are not within the preparation context. This is in stark contrast to the prevalent term contextuality used by many researchers in informal, heuristic yet omni-realistic and potentially misleading ways. This paper discusses both concepts and argues in favor of value indefiniteness in all but a continuum of contexts intertwining in the vector representing a single pure (prepared) state. Even more restrictively, and inspired by operationalism but not justified by Pitowsky's Logical Indeterminacy Principle or similar, one could identify with a "quantum state" a single quantum context – aka the respective maximal observable, or, in terms of its spectral decomposition, the associated orthonormal basis – from the continuum of intertwining context, as per the associated maximal observable actually or implicitly prepared.

Keywords Logical indeterminacy principle · Contextuality · Conditions of possible experience · Quantum clouds · Value indefiniteness · Partition logic

24.1 Introduction

An upfront *caveat* seems in order: The following is a rather subjective narrative of my reading of Itamar Pitowsky's thoughts about classical value indeterminacy on quantum logical structures of observables, amalgamated with my current thinking on related issues. I have never discussed these matters with Itamar Pitovsky explicitly; therefore the term "my reading" should be taken rather literally; namely

K. Svozil (✉)
Institute for Theoretical Physics, Vienna University of Technology, Vienna, Austria
e-mail: svozil@tuwien.ac.at; http://tph.tuwien.ac.at/~svozil

M. Hemmo, O. Shenker (eds.), *Quantum, Probability, Logic*,
Jerusalem Studies in Philosophy and History of Science,
https://doi.org/10.1007/978-3-030-34316-3_24

as taken from his publications. In what follows classical value indefiniteness on collections of (intertwined) quantum observables will be considered a consequence, or even a synonym, of what he called indeterminacy. Whether or not this identification is justified is certainly negotiable; but in what follows this is taken for granted.

The term value indefiniteness has been stimulated by recursion theory (Rogers, Jr. 1967; Odifreddi 1989; Smullyan 1993), and in particular by *partial functions* (Kleene 1936) – indeed the notion of partiality has not diffused into physical theory formation, and might even appear alien to the very notion of functional value assignments – and yet it appears to be necessary (Abbott et al. 2012, 2014, 2015) if one insists (somewhat superficially) on classical interpretations of quantized systems.

Value indefiniteness/indeterminacy will be contrasted with some related interpretations and approaches, in particular, with contextuality. Indeed, I believe that contextuality was rather foreign to Itamar Pitowsky's thinking: the term *"contextuality"* appears marginally – as in "a different context" – in his book *Quantum Probability – Quantum Logic* (Pitowsky 1989b), nowhere in his reviews on Boole-Bell type inequalities (Pitowsky 1989a, 1994), and mostly with reference to *contextual quantum probabilities* in his late writings (Pitowsky 2006). The emphasis on value indefiniteness/indeterminacy was, I believe, independently shared by Asher Peres as well as Ernst Specker.

I met Itamar Pitowsky (Bub and Demopoulos 2010) personally rather late; after he gave a lecture entitled *"All Bell Inequalities"* in Vienna (ESI – The Erwin Schrödinger International Institute for Mathematical Physics 2001) on September 6th, 2000. Subsequent discussions resulted in a joint paper (Pitowsky and Svozil 2001) (stimulating further research (Sliwa 2003; Colins and Gisin 2004)). It presents an application of his correlation polytope method (Pitowsky 1986, 1989a,b, 1991, 1994) to more general configurations than had been studied before. Thereby semi-automated symbolic as well as numeric computations have been used.

Nevertheless, the violations of what Boole called (Boole 1862, p. 229) *"conditions of possible experience,"* obtained through solving the hull problem of classical correlation polytopes, was just one route to quantum indeterminacy pursued by Itamar Pitowsky. One could identify at least two more passages he contributed to: One approach (Pitowsky 2003, 2006) compares differences of classical with quantum predictions through conditions and constraints imposed by certain intertwined configurations of observables which I like to call *quantum clouds* (Svozil 2017b). And another approach (Pitowsky 1998; Hrushovski and Pitowsky 2004) pushes these predictions to the limit of logical inconsistency; such that any attempt of a classical description fails relative to the assumptions. In what follows we shall follow all three pursuits and relate them to new findings.

24.2 Stochastic Value Indefiniteness/Indeterminacy by Boole-Bell Type Conditions of Possible Experience

The basic idea to obtain all classical predictions – including classical probabilities, expectations as well as consistency constraints thereof – associated with (mostly complementary; that is, non-simultaneously measurable) collections of observables is quite straightforward: Figure out all "extreme" cases or states which would be classically allowed. Then construct all classically conceivable situations by forming suitable combinations of the former.

Formally this amounts to performing the following steps (Pitowsky 1986, 1989a,b, 1991, 1994):

- Contemplate about some concrete structure of observables and their interconnections in intertwining observables – the quantum cloud.
- Find all two-valued states of that quantum cloud. (In the case of "contextual inequalities" (Cabello 2008) include all variations of true/1 and false/0, irrespective of exclusivity; thereby often violating the Kolmogorovian axioms of probability theory even within a single context.)
- Depending on one's preferences, form all (joint) probabilities and expectations.
- For each of these two-valued states, evaluate the joint probabilities and expectations as products of the single particle probabilities and expectations they are formed of (this reflects statistical independence of the constituent observables).
- For each of the two-valued states, form a tuple containing these relevant (joint) probabilities and expectations.
- Interpret this tuple as a vector.
- Consider the set of all such vectors – there are as many as there are two-valued states, and their dimension depends on the number of (joint) probabilities and expectations considered – and interpret them as vertices forming a convex polytope.
- The convex combination of all conceivable two-valued states yields the surface of this polytope; such that every point inside its convex hull corresponds to a classical probability distribution.
- Determine the conditions of possible experience by solving the hull problem – that is, by computing the hyperplanes which determine the inside–versus–outside criteria for that polytope. These then can serve as necessary criteria for all classical probabilities and expectations considered.

The systematic application of this method yields necessary criteria for classical probabilities and expectations which are violated by the quantum probabilities and expectations. Since I have reviewed this subject exhaustively (Svozil 2018c, Sect. 12.9) (see also Svozil 2017a) I have just sketched it to obtain a taste for its relevance for quantum indeterminacy. As is often the case in mathematical physics the method seems to have been envisioned independently a couple of times. From its (to the best of my knowledge) inception by Boole (1862) it has been discussed in the measure theoretic context by Chochet theory (Bishop and Leeuw 1959) and by Vorob'ev (1962). Froissart (Froissart 1981; Cirel'son (=Tsirel'son) 1993) might have been the first explicitly proposing it as a method to generalized Bell-type inequalities. I suggested its usefulness for non-Boolean cases (Svozil 2001) with "enough" two-valued states; preferable sufficiently many to allow a proper distinction/separation of all observables (cf. Kochen and Specker's Theorem 0 (Kochen and Specker 1967, p. 67)). Consideration of the pentagon/pentagram logic – that is, five cyclically intertwined contexts/blocks/Boolean subalgebras/cliques/orthonormal bases popularized the subject and also rendered new predictions which could be used to differentiate classical from quantized systems (Klyachko 2002; Klyachko et al. 2008; Bub and Stairs 2009, 2010; Badziąg et al. 2011).

A *caveat:* the obtained criteria involve multiple mutually complementary summands which are not all simultaneously measurable. Therefore, different terms, when evaluated experimentaly, correspond to different, complementary measurement configurations. They are obtained at different times and on different particles and samples.

Explicit, worked examples can, for instance, be found in Pitowsky's book (Pitowsky 1989b, Section 2.1), or papers (Pitowsky 1994) (see also Froissart's example (Froissart 1981)). Empirical findings are too numerous to even attempt a just appreciation of all the efforts that went into testing classicality. There is overwhelming evidence that the quantum predictions are correct; and that they violate Boole's conditions of possible classical experience (Clauser 2002) relative to the assumptions (basically non-contextual realism and locality).

So, if Boole's conditions of possible experience are violated, then they can no longer be considered appropriate for any reasonable ontology forcing "reality" upon them. This includes the realistic (Stace 1934) existence of hypothetical counterfactual observables: "unperformed experiments seem to have no consistent outcomes" (Peres 1978). The inconsistency of counterfactuals (in Specker's scholastic terminology *infuturabilities* (Specker 1960, 2009)) provides a connection to value indefiniteness/indeterminacy – at least, and let me again repeat earlier provisos, relative to the assumptions. More of this, piled higher and deeper, has been supplied by Itamar Pitowsky, as will be discussed later.

24.3 Interlude: Quantum Probabilities from Pythagorean "Views on Vectors"

Quantum probabilities are vector based. At the same time those probabilities mimic "classical" ones whenever they must be classical; that is, among mutually commuting observables which can be measured simultaneously/concurrently on the same particle(s) or samples – in particular, whenever those observables correspond to projection operators which are either orthogonal (exclusive) or identical (inclusive).

At the same time, quantum probabilities appear "contextual" (I assume he had succumbed to the prevalent nomenclature at that late time) according to Itamar Pitowsky's late writings (Pitowsky 2006) if they need not be classical: namely among non-commuting observables. (The term "needs not" derives its justification from the finding that there exist situations (Moore 1956; Wright 1990) involving complementary observables with a classical probability interpretation (Svozil 2005)).

Thereby, classical probability theory is maintained for simultaneously co-measurable (that is, non-complementary) observables. This essentially amounts to the validity of the Kolmogorov axioms of probability theory of such observables within a given context/block/Boolean subalgebra/clique/orthonormal basis, whereby the probability of an event associated with an observable

- is a non-negative real number between 0 and 1;
- is 1 for an event associated with an observable occurring with certainty (in particular, by considering any observable or its complement); as well as
- additivity of probabilities for events associated with mutually exclusive observables.

Sufficiency is assured by an elementary geometric argument (Gleason 1957) which is based upon the Pythagorean theorem; and which can be used to explicitly construct vector-based probabilities satisfying the aforementioned Kolmogorov axioms within contexts: Suppose a pure state of a quantized system is formalized by the unit state vector $|\psi\rangle$. Consider some orthonormal basis $\mathscr{B} = \{|\mathbf{e}_1\rangle, \dots, |\mathbf{e}_n\rangle\}$ of $\mathscr{V}$. Then the square $P_\psi(\mathbf{e}_i) = |\langle\psi|\mathbf{e}_i\rangle|^2$ of the length/norm $\sqrt{\langle\psi|\mathbf{e}_i\rangle\langle\mathbf{e}_i|\psi\rangle}$ of the orthogonal projection $(\langle\psi|\mathbf{e}_i\rangle)\,|\mathbf{e}_i\rangle$ of that unit vector $|\psi\rangle$ along the basis element $|\mathbf{e}_i\rangle$ can be interpreted as the probability of the event associated with the $0-1$-observable (proposition) associated with the basis vector $|\mathbf{e}_i\rangle$ (or rather the orthogonal projector $\mathbf{E}_i = |\mathbf{e}_i\rangle\langle\mathbf{e}_i|$ associated with the dyadic product of the basis vector $|\mathbf{e}_i\rangle$); given a quantized physical system which has been prepared to be in a pure state $|\psi\rangle$. Evidently, $1 \leq P_\psi(\mathbf{e}_i) \leq 1$, and $\sum_{i=1}^{n} P_\psi(\mathbf{e}_i) = 1$. In that Pythagorean way, every context, formalized by an orthonormal basis $\mathscr{B}$, "grants a (probabilistic) view" on the pure state $|\psi\rangle$.

It can be expected that these Pythagorean-style probabilities are different from classical probabilities almost everywhere – that is, for almost all relative measurement positions. Indeed, for instance, whereas classical two-partite correlations are linear in the relative measurement angles, their respective quantum correlations follow trigonometric functions – in particular, the cosine for "singlets" (Peres 1993). These differences, or rather the vector-based Pythagorean-style quantum probabilities, are the "root cause" for violations of Boole's aforementioned conditions of possible experience in quantum setups.

Because of the convex combinations from which they are derived, all of these conditions of possible experience contain only *linear* constraints (Boole 1854, 1862; Fréchet 1935; Hailperin 1965, 1986; Ursic 1984, 1986, 1988; Beltrametti and Mączyński 1991, 1993, 1994, 1995; Pykacz and Santos 1991; Sylvia and Majernik 1992; Dvurečenskij and Länger 1994; Beltrametti et al. 1995; Del Noce

1995; Länger and Mączyński 1995; Dvurečenskij and Länger 1995a,b; Beltrametti and Bugajski 1996; Pulmannová 2002). And because linear combinations of linear operators remain linear, one can identify the terms occurring in conditions of possible experience with linear self-adjoint operators, whose sum yields a self-adjoint operator, which stands for the "quantum version" of the respective conditions of possible experience. This operator has a spectral decomposition whose min-max eigenvalues correspond to the quantum bounds (Filipp and Svozil 2004a,b), which thereby generalize the Tsirelson bound (Cirel'son (=Tsirel'son) 1980). In that way, every condition of possible experience which is violated by the quantum probabilities provides a direct criterium for non-classicality.

24.4 Classical Value Indefiniteness/Indeterminacy by Direct Observation

In addition to the "fragmented, explosion view" criteria allowing "nonlocality" *via* Einstein separability (Weihs et al. 1998) among its parts, classical predictions from quantum clouds – essentially intertwined (therefore the Hilbert space dimensionality has to be greater than two) arrangements of contexts – can be used as a criterium for quantum advantage over (or rather "otherness" or "distinctiveness" from) classical predictions. Thereby it is sufficient to observe of a single outcome of a quantized system which directly contradicts the classical predictions.

One example of such a configuration of quantum observables forcing a "one-zero rule" (Svozil 2009b) because of a true-implies-false set of two-valued classical states (TIFS) (Cabello et al. 2018) is the "Specker bug" logic (Kochen and Specker 1965, Fig. 1, p. 182) called "cat's cradle" (Pitowsky 2003, 2006) by Itamar Pitowsky (see also Belinfante (1973, Fig. B.l. p. 64), Stairs (1983, p. 588–589), Clifton (1993, Sects. IV, Fig. 2) and Pták and Pulmannová (1991, p. 39, Fig. 2.4.6) for early discussions), as depicted in Fig. 24.1.

For such configurations, it is often convenient to represent both its labels as well as the classical probability distributions in terms of a partition logic (Svozil 2005) of the set of two-valued states – in this case, there are 14 such classical states. Every maximal observable is characterized by a context. The atoms of this context are labeled according to the indices of the two-valued measure with the value 1 on this atom. The axioms of probability theory require that, for each two-valued state, and within each context, there is exactly one such atom. As a result, as long as the set of two-valued states is separating (Kochen and Specker 1967, Theorem 0), one obtains a set of partitions of the set of two-valued states; each partition corresponding to a context.

Classically, if one prepares the system to be in the state $\{1, 2, 3\}$ – standing for any one of the classical two-valued states 1, 2 or 3 or their convex combinations – then there is no chance that the "remote" target state $\{7, 10, 13\}$ can be observed. A direct observation of quantum advantages (or rather superiority in terms of the

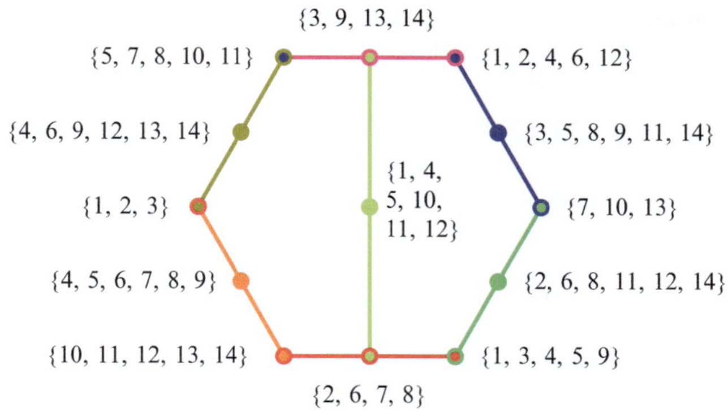

Fig. 24.1 The convex structure of classical probabilities in this (Greechie) orthogonality diagram representation of the Specker bug quantum or partition logic is reflected in its partition logic, obtained through indexing all 14 two-valued measures, and adding an index $1 \le i \le 14$ if the ith two-valued measure is 1 on the respective atom. Concentrate on the outermost left and right observables, depicted by squares: Positivity and convexity requires that $0 \le \lambda_i \le 1$ and $\lambda_1 + \lambda_2 + \lambda_3 + \lambda_7 + \lambda_{10} + \lambda_{13} \le \sum_{i=1}^{14} \lambda_i = 1$. Therefore, if a classical system is prepared (a generalized urn model/automaton logic is "loaded") such that $\lambda_1 + \lambda_2 + \lambda_3 = 1$, then $\lambda_7 + \lambda_{10} + \lambda_{13} = 0$, which results in a TIFS: the classical prediction is that the latter outcome never occurs if the former preparation is certain

frequencies predicted with respect to classical frequencies) is then suggested by some faithful orthogonal representation (FOR) (Lovász et al. 1989; Parsons and Pisanski 1989; Cabello et al. 2010; Solís-Encina and Portillo 2015) of this graph. In the particular Specker bug/cats cradle configuration, an elementary geometric argument (Cabello 1994, 1996) forces the relative angle between the quantum states $|\{1, 2, 3\}\rangle$ and $|\{7, 10, 13\}\rangle$ in three dimensions to be not smaller than $\arctan\left(2\sqrt{2}\right)$, so that the quantum prediction of the occurrence of the event associated with state $|\{7, 10, 13\}\rangle$, if the system was prepared in state $|\{1, 2, 3\}\rangle$ is that the probability can be at most $|\langle\{1, 2, 3\}|\{7, 10, 13\}\rangle|^2 = \cos^2\left[\arctan\left(2\sqrt{2}\right)\right] = \frac{1}{9}$. That is, on the average, if the system was prepared in state $|\{1, 2, 3\}\rangle$ at most one of 9 outcomes indicates that the system has the property associated with the observable $|\{7, 10, 13\}\rangle\langle|\{7, 10, 13\}|$. The occurrence of a single such event indicates quantum advantages over the classical prediction of non-occurrence.

This limitation is only true for the particular quantum cloud involved. Similar arguments with different quantum clouds resulting in TIFS can be extended to arbitrary small relative angles between preparation and measurement states, so that the relative quantum advantage can be made arbitrarily high (Abbott et al. 2015; Ramanathan et al. 2018). Classical value indefiniteness/indeterminacy comes naturally: because – at least relative to the assumptions regarding non-contextual value definiteness of truth assignments, in particular, of intertwining, observables – the existence of such definite values would enforce non-occurrence of outcomes which are nevertheless observed in quantized systems.

Very similar arguments against classical value definiteness can be inferred from quantum clouds with true-implies-true sets of two-valued states (TITS) (Stairs 1983; Clifton 1993; Johansen 1994; Vermaas 1994; Belinfante 1973; Pitowsky 1982; Hardy 1992, 1993; Boschi et al. 1997; Cabello and García-Alcaine 1995; Cabello et al. 1996, 2013, 2018; Cabello 1997; Badziąg et al. 2011; Chen et al. 2013). There the quantum advantage is in the non-occurrence of outcomes which classical predictions mandate to occur.

24.5 Classical Value Indefiniteness/Indeterminacy Piled Higher and Deeper: The Logical Indeterminacy Principle

For the next and final stage of classical value indefiniteness/indeterminacy on quantum clouds (relative to the assumptions) one can combine two logics with simultaneous classical TIFS and TITS properties at the same terminals. That is, suppose one is preparing the same "initial" state, and measuring the same "target" observable; nevertheless, contemplating the simultaneous counterfactual existence of two different quantum clouds of intertwined contexts interconnecting those fixated "initial" state and measured "target" observable. Whenever one cloud has the TIFS and another cloud the TITS property (at the same terminals), those quantum clouds induce contradicting classical predictions. In such a setup the only consistent choice (relative to the assumptions; in particular, omni-existence and context independence) is to abandon classical value definiteness/determinacy. Because the assumption of classical value definiteness/determinacy for any such logic, therefore, yields a complete contradiction, thereby eliminating prospects for hidden variable models (Abbott et al. 2012, 2015; Svozil 2017b) satisfying the assumptions.

Indeed, suppose that a quantized system is prepared in some pure quantum state. Then Itamar Pitowsky's (Pitowsky 1998; Hrushovski and Pitowsky 2004) *indeterminacy principle* states that – relative to the assumptions; in particular, global classical value definiteness for all observables involved, as well as context-independence of observables in which contexts intertwine – any other distinct (non-collinear) observable which is not orthogonal can neither occur nor not occur. This can be seen as an extension of both Gleason's theorem (Gleason 1957; Zierler and Schlessinger 1965) as well as the Kochen-Specker theorem (Kochen and Specker 1967) implying and utilizing the non-existence of any two-valued global truth assignments on even finite quantum clouds.

For the sake of a concrete example consider the two TIFS and TITS clouds – that is, logics with 35 intertwined binary observables (propositions) in 24 contexts – depicted in Fig. 24.2 (Svozil 2018b). They represent quantum clouds with the same terminal points $\{1\} \equiv \{1'\}$ and $\{2, 3, 4, 5, 6, 7\} \equiv \{1', 2', 3', 4', 5'\}$, forcing

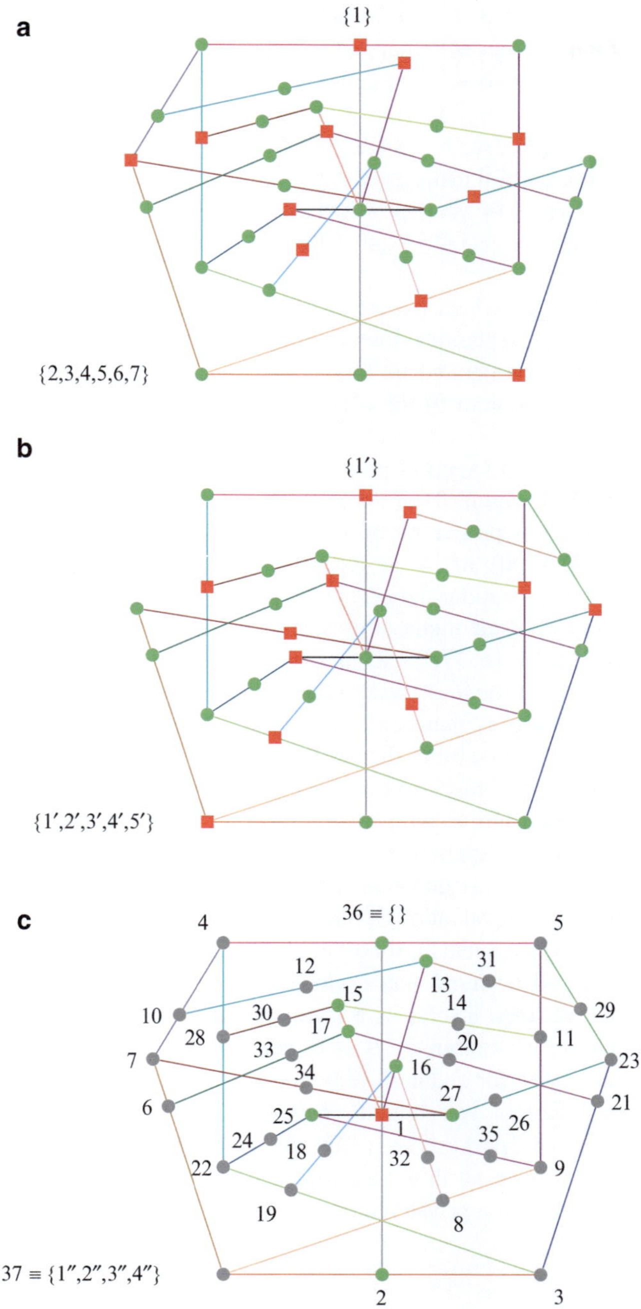

Fig. 24.2 (**a**) TIFS cloud, and (**b**) TITS cloud with only a single overlaid classical value assignment if the system is prepared in state $|1\rangle$ (Svozil 2018b). (**c**) The combined cloud from (**a**) and (**b**) has no value assignment allowing $36 = \{\}$ to be true/1; but still allows 8 classical value assignments enumerated by Table 24.1, with overlaid partial coverage common to all of them. A faithful orthogonal realization is enumerated in Abbott et al. (2015, Table. 1, p. 102201–7)

the latter ones (that is, $\{2, 3, 4, 5, 6, 7\}$ and $\{1', 2', 3', 4', 5'\}$) to be false/0 and true/1, respectively, if the former ones (that is, $\{1\} \equiv \{1'\}$) are true/1.

Formally, the only two-valued states on the logics depicted in Fig. 24.2a and b which allow $v(\{1\}) = v'(\{1'\}) = 1$ requires that $v(\{2, 3, 4, 5, 6, 7\}) = 0$ but $v'(\{1', 2', 3', 4', 5'\}) = 1 - v(\{2, 3, 4, 5, 6, 7\})$, respectively. However, both these logics have a faithful orthogonal representation (Abbott et al. 2015, Table. 1, p. 102201–7) in terms of vectors which coincide in $|\{1\}\rangle = |\{1'\}\rangle$, as well as in $|\{2, 3, 4, 5, 6, 7\}\rangle = |\{1', 2', 3', 4', 5'\}\rangle$, and even in all of the other adjacent observables.

The combined logic, which features 37 binary observables (propositions) in 26 contexts has no longer a classical interpretation in terms of a partition logic, as the 8 two-valued states enumerated in Table 24.1 cannot mutually separate (Kochen and Specker 1967, Theorem 0) the observables 2, 13, 15, 16, 17, 25, 27 and 36, respectively.

It might be amusing to keep in mind that, because of non-separability (Kochen and Specker 1967, Theorem 0) of some of the binary observables (propositions), there does not exist a proper partition logic. However, there exist generalized urn (Wright 1978, 1990) and finite automata (Moore 1956; Schaller and Svozil 1995, 1996) model realisations thereof: just consider urns "loaded" with balls which have no colored symbols on them; or no such balls at all, for the binary observables (propositions) 2, 13, 15, 16, 17, 25, 27 and 36. In such cases it is no more possible to empirically reconstruct the underlying logic; yet if an underlying logic is assumed then – at least as long as there still are truth assignments/two-valued states on the logic – "reduced" probability distributions can be defined, urns can be loaded, and automata prepared, which conform to the classical predictions from a convex combination of these truth assignments/two-valued states – thereby giving rise to "reduced" conditions of experience *via* hull computations.

For global/total truth assignments (Pitowsky 1998; Hrushovski and Pitowsky 2004) as well as for local admissibility rules allowing partial (as opposed to total, global) truth assignments (Abbott et al. 2012, 2015), such arguments can be extended to cover all terminal states which are neither collinear nor orthogonal. One could point out that, insofar as a fixed state has to be prepared the resulting value indefiniteness/indeterminacy is state dependent. One may indeed hold that the strongest indication for quantum value indefiniteness/indeterminacy is the *total absence/non-existence* of two-valued states, as exposed in the Kochen-Specker theorem (Kochen and Specker 1967). But this is rather a question of nominalistic taste, as both cases have no direct empirical testability; and as has already been pointed out by Clifton in a private conversation in 1995: *"how can you measure a contradiction?"*

Table 24.1 Enumeration of the 8 two-valued states on 37 binary observables (propositions) of the combined quantum clouds/logics depicted in Fig. 24.2a and b. Row vector indicate the state values on the observables, column vectors the values on all states per the respective observable

#	1	2	3	4	...											...										...							34	35	36	37
1	1	0	0	1	0	0	0	0	0	0	1	1	0	0	0	0	0	0	1	0	1	0	0	1	0	1	0	0	1	1	0	1	1	1	1	0	1
2	1	0	0	1	0	0	0	0	0	0	1	1	0	0	0	0	0	0	1	1	0	0	1	1	0	0	0	0	0	1	1	1	1	1	1	0	1
3	1	0	0	0	1	0	0	0	0	1	0	0	0	1	0	0	0	0	1	0	1	0	0	1	0	1	0	1	0	0	1	1	1	1	1	0	1
4	1	0	0	0	1	0	0	0	0	1	0	0	0	1	0	0	0	1	0	0	1	1	0	0	0	1	0	0	0	1	1	1	1	1	1	0	1
5	1	0	1	1	0	1	0	1	0	0	1	1	0	0	0	0	0	1	0	1	0	0	0	1	0	1	0	0	1	1	0	0	0	1	1	0	0
6	1	0	1	1	0	1	0	0	1	0	0	1	0	1	0	0	0	1	0	1	0	0	0	1	0	1	0	0	1	1	0	1	0	1	0	0	0
7	1	0	1	0	1	1	0	1	0	1	0	0	0	1	0	0	0	1	0	1	0	0	0	1	0	1	0	1	0	0	1	0	0	1	1	0	0
8	1	0	1	0	1	0	1	1	0	0	0	1	0	1	0	0	0	1	0	1	0	0	0	1	0	1	0	1	0	0	1	0	1	0	1	0	0

24.6 The "Message" of Quantum (In)determinacy

At the peril of becoming, as expressed by Clauser (2002), "evangelical," let me "sort things out" from my own very subjective and private perspective. (Readers adverse to "interpretation" and the semantic, "meaning" aspects of physical theory may consider stop reading at this point.)

Thereby one might be inclined to follow Planck (against Feynman (Clauser 2002; Mermin 1989a,b)) and hold it as being not too unreasonable to take scientific comprehensibility, rationality, and causality as a (Planck 1932, p. 539) (see also Earman 2007, p. 1372) *"heuristic principle, a sign-post ... to guide us in the motley confusion of events and to show us the direction in which scientific research must advance in order to attain fruitful results."*

So what does all of this – the Born rule of quantum probabilities and its derivation by Gleason's theorem from the Kolmogorovian axioms applied to mutually comeasurable observables, as well as its consequences, such as the Kochen-Specker theorem, the plethora of violations of Boole's conditions of possible experience, Pitowsky's indeterminacy principle and more recent extensions and variations thereof – "try to tell us?"

First, observe that all of the aforementioned postulates and findings are (based upon) assumptions; and thus consequences of the latter. Stated differently, these findings are true not in the absolute, ontologic but in the epistemic sense: they hold relative to the axioms or assumptions made.

Thus, in maintaining rationality one needs to grant oneself – or rather one is forced to accept – the abandonment of at least some or all assumptions made. Some options are exotic; for instance, Itamar Pitowsky's suggestions to apply paradoxical set decompositions to probability measures (Pitowsky 1983, 1986). Another "exotic escape option" is to allow only unconnected (non-intertwined) contexts whose observables are dense (Godsil and Zaks 1988, 2012; Meyer 1999; Havlicek et al. 2001). Some possibilities to cope with the findings are quite straightforward, and we shall concentrate our further attention to those (Svozil 2009b).

24.6.1 Simultaneous Definiteness of Counterfactual, Complementary Observables, and Abandonment of Context Independence

Suppose one insists on the simultaneous definite omni-existence of mutually complementary, and therefore necessarily counterfactual, observables. One straightforward way to cope with the aforementioned findings is the abandonment of context-independence of intertwining observables.

There is no indication in the quantum formalism which would support such an assumption, as the respective projection operators do not in any way depend on the contexts involved. However, one may hold that the outcomes are context dependent as functions of the initial state and the context measured (Svozil 2009a, 2012; Dzhafarov et al. 2017); and that they actually "are real" and not just "idealistically occur in our imagination;" that is, being *"mental through-and-through"* (Segal and Goldschmidt 2017/2018). Early conceptualizations of context-dependence aka contextuality can be found in Bohr's remark (in his typical Nostradamus-like style) (Bohr 1949) on *"the impossibility of any sharp separation between the behavior of atomic objects and the interaction with the measuring instruments which serve to define the conditions under which the phenomena appear."* Bell, referring to Bohr, suggested (Bell 1966), Sec. 5) that *"the result of an observation may reasonably depend not only on the state of the system (including hidden variables) but also on the complete disposition of the apparatus."*

However, the common, prevalent, use of the term "contextuality" is not an explicit context-dependent form, as suggested by the realist Bell in his earlier quote, but rather a situation where the classical predictions of quantum clouds are violated. More concretely, if experiments on quantized systems violate certain Boole-Bell type classical bounds or direct classical predictions, the narratives claim to have thereby "proven contextuality" (e.g., see Hasegawa et al. (2006), Cabello et al. (2008), Cabello (2008), Bartosik et al. (2009), Amselem et al. (2009), Bub and Stairs (2010) and Cabello et al. (2013) for a "direct proof of quantum contextuality").

What if we take Bell's proposal of a context dependence of valuations – and consequently, "classical" contextual probability theory – seriously? One of the consequences would be the introduction of an *uncountable multiplicity* of counterfactual observables. An example to illustrate this multiplicity – comparable to de Witt's view of Everett's relative state interpretation (Everett III 1973) – is the uncountable set of orthonormal bases of $\mathbb{R}^3$ which are all interconnected at the same single intertwining element. A continuous angular parameter characterizes the angles between the other elements of the bases, located in the plane orthogonal to that common intertwining element. Contextuality suggests that the value assignment of an observable (proposition) corresponding to this common intertwining element needs to be both true/1 and false/0, depending on the context involved, or whenever some quantum cloud (collection of intertwining observables) demands this through consistency requirements.

Indeed, the introduction of multiple quantum clouds would force any context dependence to also implicitly depend on this general perspective – that is, on the respective quantum cloud and its faithful orthogonal realization, which in turn determines the quantum probabilities *via* the Born-Gleason rule: Because there exist various different quantum clouds as "pathways interconnecting" two observables, context dependence needs to vary according to any concrete connection between the prepared and the measured state.

A single context participates in an arbitrary, potentially infinite, multiplicity of quantum clouds. This requires this one context to "behave very differently" when it comes to contextual value assignments. Alas, as quantum clouds are hypothetical constructions of our mind and therefore *"mental through-and-through"* (Segal and Goldschmidt 2017/2018), so appears context dependence: as an idealistic concept, devoid of any empirical evidence, created to rescue the *desideratum* of omni-realistic existence.

Pointedly stated, contextual value assignments appear both utterly ad hoc and abritrary – like a *deus ex machina* "saving" the *desideratum* of a classical omni-value definite reality, whereby it must obey quantum probability theory without grounding it (indeed, in the absence of any additional criterium or principle there is no reason to assume that the likelihood of true/1 and false/0 is other than 50:50); as well as highly discontinuous. In this latter, discontinuity respect, context dependence is similar to the earlier mentioned breakup of the intertwine observables by reducing quantum observables to disconnected contexts (Godsil and Zaks 1988, 2012; Meyer 1999; Havlicek et al. 2001).

It is thereby granted that these considerations apply only to cases in which the assumptions of context independence are valid throughout the entire quantum cloud – that is, uniformly: for all observables in which contexts intertwine. If this were not the case – say, if only a single one observable occurring in intertwining contexts is allowed to be context-dependent (Svozil 2012; Simmons 2020) – the respective clouds taylored to prove Pitowsky's Logical Indeterminacy Principle and similar, as well as the Kochen-Specker theorems do not apply; and therefore the aforementioned consequences are invalid.

24.6.2 Abandonment of Omni-Value Definiteness of Observables in All But One Context

Nietzsche once speculated (Nietzsche 1887, 2009–,,) that what he has called *"slave morality"* originated from superficially pretending that – in what later Blair (aka Orwell 1949) *"doublespeak"* – weakness means strength. In a rather similar sense the lack of comprehension – Planck's *"sign-post"* – and even the resulting inconsistencies tended to become reinterpreted as an asset: nowadays consequences of the vector-based quantum probability law are marketed as *"quantum supremacy"* – a *"quantum magic"* or *"hocus-pocus"* (Svozil 2016) of sorts.

Indeed, future centuries may look back at our period, and may even call it a second "renaissance" period of scholasticism (Specker 1960). In years from now historians of science will be amused about our ongoing queer efforts, the calamities and "magic" experienced through our painful incapacity to recognize the obvious – that is, the non-existence and therefore value indefiniteness/indeterminacy of certain counterfactual observables – namely exactly those mentioned in Itamar Pitowsky's indeterminacy principle.

This principle has a positive interpretation of a quantum state, defined as the maximal knowledge obtainable by simultaneous measurements of a quantized system; or, conversely, as the maximal information content encodable therein. This can be formalized in terms of the *value definiteness* of a single (Zeilinger 1999; Svozil 2002, 2004, 2018b; Grangier 2002) context – or, in a more broader (non-operational) perspective, the continuum of contexts intertwined by some prepared pure quantum state (formalized as vector or the corresponding one-dimensional orthogonal projection operator). In terms of Hilbert space quantum mechanics this amounts to the claim that the only value definite entity can be a single orthonormal basis/maximal operator; or a continuum of maximal operators whose spectral sum contain proper "true intertwines." All other "observables" grant an, albeit necessarily stochastic, value indefinite/indeterministic, view on this state.

If more than one context is involved we might postulate that all admissable probabilities should at least satisfy the following criterium: they should be classical Kolmogorov-style *within* any single particular context (Gleason 1957). It has been suggested (Auffèves and Grangier 2017, 2018) that this can be extended and formalized in a quantum multi-context environment by a double stochastic matrix whose entries $P(\mathbf{e}_i, \mathbf{f}_j)$, with $1 \leq i, j \leq n$ (n is the number of distinct "atoms" or exclusive outcomes in each context) are identified by the conditional probabilities of one atom $\mathbf{f}_j$ in the second context, relative to a given one atom $\mathbf{e}_i$ in the first context. The general multi-context case yields row stochastic matrices (Svozil 2018a). Various types of decompositions of those matrices exist for particular cases:

- By the Birkhoff-von Neumann theorem double stochastic matrices can be represented by the Birkhoff polytope spanned by the convex hull of the set of permutation matrices: let $\lambda_1, \ldots, \lambda_k \geq 0$ such that $\sum_{l=1}^{k} \lambda_l = 1$, then $P(\mathbf{e}_i, \mathbf{f}_j) = \left[\sum_{l=1}^{k} \lambda_l \Pi_l\right]_{ij}$. Since there exist $n!$ permutations of n elements, k will be bounded from above by $k \leq n!$. Note that this type of decomposition may not be unique, as the space spanned the permutation matrices is $\left[(n-1)^2+1\right]$-dimensional; with $n! > (n-1)^2+1$ for $n > 2$. Therefore, the bound from above can be improved such that decompositions with $k \leq (n-1)^2+1 = n^2-2(n+1)$ exist (Marcus and Ree 1959). Formally, a permutation matrix has a quasi-vectorial (Mermin 2007) decomposition in terms of the canonical (Cartesian) basis, such that, $\Pi_i = \sum_{j=1}^{n} |\mathbf{e}_j\rangle\langle\mathbf{e}_{\pi_i(j)}|$, where $|\mathbf{e}_j\rangle$ represents the n-tuple associated with the jth basis vector of the canonical (Cartesian) basis, and $\pi_i(j)$ stands for the ith permutation of j.
- Vector based probabilities allow the following decomposition (Auffèves and Grangier 2017, 2018): $P(\mathbf{e}_i, \mathbf{f}_j) = \text{Trace}\left(\mathbf{E}_i\mathbf{R}\mathbf{F}_j\mathbf{R}\right)$, where $\mathbf{E}_i$ and $\mathbf{F}_i$ are elements of contexts, formalized by two sets of mutually orthogonal projection operators, and $\mathbf{R}$ is a real positive diagonal matrix such that the trace of $\mathbf{R}^2$ equals the dimension n, and $\text{Trace}\left(\mathbf{E}_i\mathbf{R}^2\right) = 1$. The quantum mechanical Born rule is recovered by identifying $\mathbf{R} = \mathbb{I}_n$ with the identity matrix, so that $P(\mathbf{e}_i, \mathbf{f}_j) = \text{Trace}\left(\mathbf{E}_i\mathbf{F}_j\right)$.
- There exist more "exotic" probability measures on "reduced" propositional spaces such as Wright's 2-state dispersion-free measure on the pentagon/pentagram (Wright 1978), or another type of probability measure based on a discontinuous 3(2)-coloring of the set of all unit vectors with rational coefficients (Godsil and Zaks 1988, 2012; Meyer 1999; Havlicek et al. 2001) whose decomposition appear to be ad hoc; at least for the time being.

Where might this aforementioned type of stochasticism arise from? It could well be that it is introduced by interactions with the environment; and through the many uncontrollable and, for all practical purposes (Bell 1990), huge number of degrees of freedom in unknown states.

The finiteness of physical resources needs not prevent the specification of a particular vector or context. Because any other context needs to be operationalized within the physically feasible means available to the respective experiment: it is the measurable coordinate differences which count; not the absolute locatedness relative to a hypothetical, idealistic absolute frame of reference which cannot be accessed operationally.

Finally, as the type of context envisioned to be value definite can be expressed in terms of vector spaces equipped with a scalar product – in particular, by identifying a context with the corresponding orthonormal basis or (the spectral decomposition of) the associated maximal observable(s) – one may ask how one could imagine the origin of such entities? Abstractly vectors and vector spaces could originate from a great variety of very different forms; such as from systems of solutions of ordinary linear differential equations. Any investigation into the origins of the quantum mechanical Hilbert space formalism itself might, if this turns out to be a progressive research program (Lakatos 1978), eventually yield to a theory indicating operational physical capacities beyond quantum mechanics.

24.7 Biographical Notes on Itamar Pitowsky

I am certainly not in the position to present a view of Itamar Pitowsky's thinking. Therefore I shall make a few rather anecdotal observations. First of all, he seemed to me as one of the most original physicists I have ever met – but that might be a triviality, given his *opus*. One thing I realized was that he exhibited a – sometimes maybe even unconscious, but sometimes very outspoken – regret that he was working in a philosophy department. I believe he considered himself rather a mathematician or theoretical physicist. To this I responded that being in a philosophy department might be rather fortunate because there one could "go wild" in every direction; allowing much greater freedom than in other academic realms. But, of course, this had no effect on his uneasiness.

He was astonished that I spent a not so little money (means relative to my investment capacities) in an Israeli internet startup company which later flopped, depriving me of all but a fraction of what I had invested. He told me that, at least at that point, many startups in Israel had been put up intentionally only to attract money from people like me; only to collapse later.

A late project of his concerned quantum bounds in general; maybe in a similar – graph theoretical and at the time undirected to quantum – way as Grötschel, Lovász and Schrijver's theta body (Grötschel et al. 1986; Cabello et al. 2014). The idea was not just deriving absolute (Cirel'son (=Tsirel'son) 1980) or parameterized, continuous (Filipp and Svozil 2004a,b) bounds for existing classical conditions of possible experience obtained by hull computations of polytopes; but rather genuine quantum bounds on, say, Einstein-Podolsky-Rosen type setups.

Acknowledgements I kindly acknowledge enlightening criticism and suggestions by Andrew W. Simmons, as well as discussions with Philippe Grangier on the characterization of quantum probabilities. All remaining misconceptions and errors are mine. The author acknowledges the support by the Austrian Science Fund (FWF): project I 4579-N and the Czech Science Foundation: project 20-09869L.

References

Abbott, A. A., Calude, C. S., Conder, J., & Svozil, K. (2012). Strong Kochen-Specker theorem and incomputability of quantum randomness. *Physical Review A, 86*, 062109. https://doi.org/10.1103/PhysRevA.86.062109, arXiv:1207.2029

Abbott, A. A., Calude, C. S., & Svozil, K. (2014) Value-indefinite observables are almost everywhere. *Physical Review A, 89*, 032109. https://doi.org/10.1103/PhysRevA.89.032109, arXiv:1309.7188

Abbott, A. A., Calude, C. S., & Svozil, K. (2015). A variant of the Kochen-Specker theorem localising value indefiniteness. *Journal of Mathematical Physics, 56*(10), 102201. https://doi.org/10.1063/1.4931658, arXiv:1503.01985

Amselem, E., Rådmark, M., Bourennane, M., & Cabello, A. (2009). State-independent quantum contextuality with single photons. *Physical Review Letters, 103*(16), 160405. https://doi.org/10.1103/PhysRevLett.103.160405

Auffèves, A., & Grangier, P. (2017). Recovering the quantum formalism from physically realist axioms. *Scientific Reports, 7*(2123), 43365 (1–9). https://doi.org/10.1038/srep43365, arXiv:1610.06164

Auffèves, A., & Grangier, P. (2018). Extracontextuality and extravalence in quantum mechanics. *Philosophical Transactions of the Royal Society A: Mathematical, Physical and Engineering Sciences, 376*(2123), 20170311. https://doi.org/10.1098/rsta.2017.0311, arXiv:1801.01398

Badziąg, P., Bengtsson, I., Cabello, A., Granström, H., & Larsson, J. A. (2011). Pentagrams and paradoxes. *Foundations of Physics, 41*. https://doi.org/10.1007/s10701-010-9433-3.

Bartosik, H., Klepp, J., Schmitzer, C., Sponar, S., Cabello, A., Rauch, H., & Hasegawa, Y. (2009). Experimental test of quantum contextuality in neutron interferometry. *Physical Review Letters, 103*(4), 040403. https://doi.org/10.1103/PhysRevLett.103.040403, arXiv:0904.4576.

Belinfante, F. J. (1973). *A survey of hidden-variables theories* (International series of monographs in natural philosophy, Vol. 55). Oxford/New York: Pergamon Press/Elsevier. https://doi.org/10.1016/B978-0-08-017032-9.50001-7.

Bell, J. S. (1966). On the problem of hidden variables in quantum mechanics. *Reviews of Modern Physics, 38*, 447–452. https://doi.org/10.1103/RevModPhys.38.447.

Bell, J. S. (1990). Against 'measurement'. *Physics World, 3*, 33–41. https://doi.org/10.1088/2058-7058/3/8/26.

Beltrametti, E. G., & Bugajski, S. (1996). The Bell phenomenon in classical frameworks. *Journal of Physics A: Mathematical and General Physics, 29*. https://doi.org/10.1088/0305-4470/29/2/005.

Beltrametti, E. G., & Mączyński, M. J. (1991). On a characterization of classical and nonclassical probabilities. *Journal of Mathematical Physics, 32*. https://doi.org/10.1063/1.529326.

Beltrametti, E. G., & Mączyński, M. J. (1993). On the characterization of probabilities: A generalization of Bell's inequalities. *Journal of Mathematical Physics, 34*. https://doi.org/10.1063/1.530333.

Beltrametti, E. G., & Mączyński, M. J. (1994). On Bell-type inequalities. *Foundations of Physics, 24*. https://doi.org/10.1007/bf02057861.
Beltrametti, E. G., & Mączyński, M. J. (1995). On the range of non-classical probability. *Reports on Mathematical Physics, 36*, https://doi.org/10.1016/0034-4877(96)83620-2.
Beltrametti, E. G., Del Noce, C, & Mączyński, M. J. (1995). Characterization and deduction of Bell-type inequalities. In: C. Garola & A. Rossi (Eds.), *The foundations of quantum mechanics – Historical analysis and open questions*, Lecce, 1993 (pp. 35–41). Dordrecht: Springer. https://doi.org/10.1007/978-94-011-00298_3.
Bishop, E., & Leeuw, K. D. (1959). The representations of linear functionals by measures on sets of extreme points. *Annals of the Fourier Institute, 9*, 305–331. https://doi.org/10.5802/aif.95.
Bohr, N. (1949). Discussion with Einstein on epistemological problems in atomic physics. In P. A. Schilpp (Ed.), *Albert Einstein: Philosopher-Scientist, the library of living philosophers*, Evanston (pp. 200–241). https://doi.org/10.1016/S1876-0503(08)70379-7.
Boole, G. (1854). An investigation of the laws of thought. http://www.gutenberg.org/ebooks/15114.
Boole, G. (1862). On the theory of probabilities. *Philosophical Transactions of the Royal Society of London, 152*, 225–252. https://doi.org/10.1098/rstl.1862.0015.
Boschi, D., Branca, S., De Martini, F., & Hardy, L. (1997). Ladder proof of nonlocality without inequalities: Theoretical and experimental results. *Physical Review Letters, 79*, 2755–2758. https://doi.org/10.1103/PhysRevLett.79.2755.
Bub, J., & Demopoulos, W. (2010). Itamar Pitowsky 1950–2010. *Studies in History and Philosophy of Science Part B: Studies in History and Philosophy of Modern Physics, 41*(2), 85. https://doi.org/10.1016/j.shpsb.2010.03.004.
Bub, J., & Stairs, A. (2009). Contextuality and nonlocality in 'no signaling' theories. *Foundations of Physics, 39*. https://doi.org/10.1007/s10701-009-9307-8.
Bub, J., & Stairs, A. (2010). Contextuality in quantum mechanics: Testing the Klyachko inequality. https://arxiv.org/abs/1006.0500, arXiv:1006.0500.
Cabello, A. (1994). A simple proof of the Kochen-Specker theorem. *European Journal of Physics, 15*(4), 179–183. https://doi.org/10.1088/0143-0807/15/4/004.
Cabello, A. (1996). Pruebas algebraicas de imposibilidad de variables ocultas en mecánica cuántica. PhD thesis, Universidad Complutense de Madrid, Madrid. http://eprints.ucm.es/1961/1/T21049.pdf.
Cabello, A. (1997). No-hidden-variables proof for two spin- particles preselected and postselected in unentangled states. *Physical Review A, 55*, 4109–4111. https://doi.org/10.1103/PhysRevA.55.4109.
Cabello, A. (2008). Experimentally testable state-independent quantum contextuality. *Physical Review Letters, 101*(21), 210401. https://doi.org/10.1103/PhysRevLett.101.210401, arXiv:0808.2456.
Cabello, A., & García-Alcaine, G. (1995). A hidden-variables versus quantum mechanics experiment. *Journal of Physics A: Mathematical and General Physics, 28*. https://doi.org/10.1088/0305-4470/28/13/016.
Cabello, A., Estebaranz, J. M., & García-Alcaine, G. (1996). Bell-Kochen-Specker theorem: A proof with 18 vectors. *Physics Letters A 212*(4), 183–187. https://doi.org/10.1016/0375-9601(96)00134-X, arXiv:quant-ph/9706009.
Cabello, A., Filipp, S., Rauch, H., & Hasegawa, Y. (2008). Proposed experiment for testing quantum contextuality with neutrons. *Physical Review Letters, 100*, 130404. https://doi.org/10.1103/PhysRevLett.100.130404.
Cabello, A., Severini, S., & Winter, A. (2010). (Non-)contextuality of physical theories as an axiom. https://arxiv.org/abs/1010.2163, arXiv:1010.2163
Cabello, A., Badziag, P., Terra Cunha, M., & Bourennane, M. (2013). Simple Hardy-like proof of quantum contextuality. *Physical Review Letters, 111*, 180404. https://doi.org/10.1103/PhysRevLett.111.180404.

Cabello, A., Severini, S., & Winter, A. (2014). Graph-theoretic approach to quantum correlations. *Physical Review Letters, 112*, 040401. https://doi.org/10.1103/PhysRevLett.112.040401, arXiv:1401.7081.

Cabello, A., Portillo, J. R., Solís, A., & Svozil, K. (2018). Minimal true-implies-false and true-implies-true sets of propositions in noncontextual hidden-variable theories. *Physical Review A, 98*, 012106. https://doi.org/10.1103/PhysRevA.98.012106, arXiv:1805.00796.

Chen, J. L., Cabello, A., Xu, Z. P., Su, H. Y., Wu, C., & Kwek, L. C. (2013) Hardy's paradox for high-dimensional systems. *Physical Review A, 88*, 062116. https://doi.org/10.1103/PhysRevA.88.062116.

Cirel'son (=Tsirel'son), B. S. (1980). Quantum generalizations of Bell's inequality. *Letters in Mathematical Physics, 4*(2), 93–100. https://doi.org/10.1007/BF00417500.

Cirel'son (=Tsirel'son), B. S. (1993). Some results and problems on quantum Bell-type inequalities. *Hadronic Journal Supplement, 8*, 329–345. http://www.tau.ac.il/~tsirel/download/hadron.pdf.

Clauser, J. (2002). Early history of Bell's theorem. In R. Bertlmann & A. Zeilinger (Eds.), *Quantum (un)speakables: From Bell to quantum information* (pp. 61–96). Berlin: Springer. https://doi.org/10.1007/978-3-662-05032-3_6.

Clifton, R. K. (1993). Getting contextual and nonlocal elements-of-reality the easy way. *American Journal of Physics, 61*(5), 443–447. https://doi.org/10.1119/1.17239.

Colins, D., & Gisin, N. (2004). A relevant two qbit Bell inequality inequivalent to the CHSH inequality. *Journal of Physics A: Mathematical and General, 37*, 1775–1787. https://doi.org/10.1088/0305-4470/37/5/021, arXiv:quant-ph/0306129.

Del Noce, C. (1995). An algorithm for finding Bell-type inequalities. *Foundations of Physics Letters, 8*. https://doi.org/10.1007/bf02187346.

Dvurečenskij, A., & Länger, H. (1994). Bell-type inequalities in horizontal sums of boolean algebras. *Foundations of Physics, 24*. https://doi.org/10.1007/bf02057864.

Dvurečenskij, A., & Länger, H. (1995a). Bell-type inequalities in orthomodular lattices. I. Inequalities of order 2. *International Journal of Theoretical Physics, 34*. https://doi.org/10.1007/bf00671363.

Dvurečenskij, A., & Länger, H. (1995b). Bell-type inequalities in orthomodular lattices. II. Inequalities of higher order. *International Journal of Theoretical Physics, 34*. https://doi.org/10.1007/bf00671364.

Dzhafarov, E. N., Cervantes, V. H., & Kujala, J. V. (2017). Contextuality in canonical systems of random variables. *Philosophical Transactions of the Royal Society A: Mathematical, Physical and Engineering Sciences, 375*(2106), 20160389. https://doi.org/10.1098/rsta.2016.0389, arXiv:1703.01252.

Earman, J. (2007). Aspects of determinism in modern physics. Part B. In J. Butterfield & J. Earman (Eds.), *Philosophy of physics, handbook of the philosophy of science* (pp. 1369–1434). Amsterdam: North-Holland. https://doi.org/10.1016/B978-044451560-5/50017-8.

ESI – The Erwin Schrödinger International Institute for Mathematical Physics. (2001). Scientific report for the year 2000. https://www.esi.ac.at/material/scientific-reports-1/2000.pdf, eSI-Report 2000.

Everett III, H. (1973). *The many-worlds interpretation of quantum mechanics* (pp. 3–140). Princeton: Princeton University Press.

Filipp, S., & Svozil, K. (2004a). Generalizing Tsirelson's bound on Bell inequalities using a min-max principle. *Physical Review Letters, 93*, 130407. https://doi.org/10.1103/PhysRevLett.93.130407, arXiv:quant-ph/0403175.

Filipp, S., & Svozil, K. (2004b). Testing the bounds on quantum probabilities. *Physical Review A, 69*, 032101. https://doi.org/10.1103/PhysRevA.69.032101, arXiv:quant-ph/0306092.

Fréchet, M. (1935). Généralisation du théorème des probabilités totales. *Fundamenta Mathematicae, 25*(1), 379–387. http://eudml.org/doc/212798.

Froissart, M. (1981). Constructive generalization of Bell's inequalities. *Il Nuovo Cimento B (1971–1996), 64*, 241–251. https://doi.org/10.1007/BF02903286.

(aka George Orwell), E. A. B. (1949). Nineteen Eighty-Four (aka 1984). Twentieth century classics, Secker & Warburg, Cambridge. http://gutenberg.net.au/ebooks01/0100021.txt.

Gleason, A. M. (1957). Measures on the closed subspaces of a Hilbert space. *Journal of Mathematics and Mechanics (now Indiana University Mathematics Journal), 6*(4), 885–893. https://doi.org/10.1512/iumj.1957.6.56050.

Godsil, C. D., & Zaks, J. (1988, 2012) Coloring the sphere, https://arxiv.org/abs/1201.0486, University of Waterloo research report CORR 88-12, arXiv:1201.0486.

Grangier, P. (2002). Contextual objectivity: A realistic interpretation of quantum mechanics. *European Journal of Physics, 23*(3), 331–337. https://doi.org/10.1088/0143-0807/23/3/312, arXiv:quant-ph/0012122.

Grötschel, M., Lovász, L., & Schrijver, A. (1986). Relaxations of vertex packing. *Journal of Combinatorial Theory, Series B, 40*(3), 330–343. https://doi.org/10.1016/0095-8956(86)90087-0.

Hailperin, T. (1965) Best possible inequalities for the probability of a logical function of events. *The American Mathematical Monthly, 72*(4), 343–359, https://doi.org/10.2307/2313491.

Hailperin, T. (1986). *Boole's logic and probability: Critical exposition from the standpoint of contemporary algebra, logic and probability theory* (Studies in logic and the foundations of mathematics, Vol. 85, 2nd ed.). Elsevier Science Ltd. https://www.elsevier.com/books/booles-logic-and-probability/hailperin/978-0-444-87952-3.

Hardy, L. (1992). Quantum mechanics, local realistic theories, and lorentz-invariant realistic theories. *Physical Review Letters, 68*, 2981–2984. https://doi.org/10.1103/PhysRevLett.68.2981.

Hardy, L. (1993). Nonlocality for two particles without inequalities for almost all entangled states. *Physical Review Letters, 71*, 1665–1668. https://doi.org/10.1103/PhysRevLett.71.1665.

Hasegawa, Y., Loidl, R., Badurek, G., Baron, M., & Rauch, H. (2006). Quantum contextuality in a single-neutron optical experiment. *Physical Review Letters, 97*(23), 230401. https://doi.org/10.1103/PhysRevLett.97.230401.

Havlicek, H., Krenn, G., Summhammer, J., & Svozil, K. (2001). Colouring the rational quantum sphere and the Kochen-Specker theorem. *Journal of Physics A: Mathematical and General, 34*, 3071–3077. https://doi.org/10.1088/0305-4470/34/14/312, arXiv:quant-ph/9911040.

Hrushovski, E., & Pitowsky, I. (2004). Generalizations of Kochen and Specker's theorem and the effectiveness of Gleason's theorem. *Studies in History and Philosophy of Science Part B: Studies in History and Philosophy of Modern Physics, 35*(2), 177–194. https://doi.org/10.1016/j.shpsb.2003.10.002, arXiv:quant-ph/0307139

Johansen, H. B. (1994). Comment on "getting contextual and nonlocal elements-of-reality the easy way". *American Journal of Physics, 62*. https://doi.org/10.1119/1.17551.

Kleene, S. C. (1936). General recursive functions of natural numbers. *Mathematische Annalen, 112*(1), 727–742. https://doi.org/10.1007/BF01565439.

Klyachko, A. A. (2002). Coherent states, entanglement, and geometric invariant theory. https://arxiv.org/abs/quant-ph/0206012, arXiv:quant-ph/0206012.

Klyachko, A. A., Can, M. A., Binicioğlu, S., & Shumovsky, A. S. (2008). Simple test for hidden variables in spin-1 systems. *Physical Review Letters, 101*, 020403. https://doi.org/10.1103/PhysRevLett.101.020403, arXiv:0706.0126

Kochen, S., & Specker, E. P. (1965). Logical structures arising in quantum theory. In The theory of models, proceedings of the 1963 international symposium at Berkeley (pp. 177–189). Amsterdam/New York/Oxford: North Holland. https://www.elsevier.com/books/the-theory-of-models/addison/978-0-7204-2233-7, reprinted in Specker (1990, pp. 209–221).

Kochen, S., & Specker, E. P. (1967). The problem of hidden variables in quantum mechanics. *Journal of Mathematics and Mechanics (now Indiana University Mathematics Journal), 17*(1), 59–87. https://doi.org/10.1512/iumj.1968.17.17004.

Lakatos, I. (1978). *Philosophical papers. 1. The methodology of scientific research programmes.* Cambridge: Cambridge University Press.
Länger, H., Mączyński, M. J. (1995). On a characterization of probability measures on boolean algebras and some orthomodular lattices. *Mathematica Slovaca, 45*(5), 455–468. http://eudml.org/doc/32311.
Lovász, L., Saks, M., & Schrijver, A. (1989). Orthogonal representations and connectivity of graphs. *Linear Algebra and Its Applications, 114–115*, 439–454. https://doi.org/10.1016/0024-3795(89)90475-8, special Issue Dedicated to Alan J. Hoffman.
Marcus, M., & Ree, R. (1959). Diagonals of doubly stochastic matrices. *The Quarterly Journal of Mathematics, 10*(1), 296–302. https://doi.org/10.1093/qmath/10.1.296.
Mermin, D. N. (1989a). Could Feynman have said this? *Physics Today, 57*, 10–11. https://doi.org/10.1063/1.1768652.
Mermin, D. N. (1989b) What's wrong with this pillow? *Physics Today, 42*, 9–11. https://doi.org/10.1063/1.2810963.
Mermin, D. N. (2007). *Quantum computer science*. Cambridge: Cambridge University Press. https://doi.org/10.1017/CBO9780511813870.
Meyer, D. A. (1999). Finite precision measurement nullifies the Kochen-Specker theorem. *Physical Review Letters, 83*(19), 3751–3754. https://doi.org/10.1103/PhysRevLett.83.3751. arXiv:quant-ph/9905080.
Moore, E. F. (1956). Gedanken-experiments on sequential machines. In C. E. Shannon, & J. McCarthy (Eds.), *Automata studies, (AM-34)* (pp. 129–153). Princeton: Princeton University Press. https://doi.org/10.1515/9781400882618-006.
Nietzsche, F. (1887, 2009–). Zur Genealogie der Moral (On the Genealogy of Morality). http://www.nietzschesource.org/#eKGWB/GM, digital critical edition of the complete works and letters, based on the critical text by G. Colli & M. Montinari, Berlin/New York, de Gruyter 1967-, edited by Paolo D'Iorio.
Nietzsche, F. W. (1887, 1908; 1989, 2010). On the genealogy of morals and Ecce homo. Vintage, Penguin, Random House. https://www.penguinrandomhouse.com/books/121939/on-the-genealogy-of-morals-and-ecce-homo-by-friedrich-nietzsche-edited-with-a-commentary-by-walter-kaufmann/9780679724629/, translated by Walter Arnold Kaufmann.
Odifreddi, P. (1989). *Classical recursion theory* (Vol. 1). Amsterdam: North-Holland.
Parsons, T., & Pisanski, T. (1989) Vector representations of graphs. *Discrete Mathematics, 78*, https://doi.org/10.1016/0012-365x(89)90171-4.
Peres, A. (1978). Unperformed experiments have no results. *American Journal of Physics, 46*, 745–747, https://doi.org/10.1119/1.11393.
Peres, A. (1993). *Quantum theory: Concepts and methods*. Dordrecht: Kluwer Academic Publishers.
Pitowsky, I. (1982). Substitution and truth in quantum logic. *Philosophy of Science, 49*. https://doi.org/10.2307/187281.
Pitowsky, I. (1983). Deterministic model of spin and statistics. *Physical Review D, 27*, 2316–2326. https://doi.org/10.1103/PhysRevD.27.2316.
Pitowsky, I. (1986). The range of quantum probabilities. *Journal of Mathematical Physics, 27*(6), 1556–1565.
Pitowsky, I. (1989a). From George Boole to John Bell: The origin of Bell's inequality. In M. Kafatos (Ed.), *Bell's theorem, quantum theory and the conceptions of the universe* (Fundamental theories of physics, Vol. 37, pp. 37–49). Dordrecht: Kluwer Academic Publishers/Springer. https://doi.org/10.1007/978-94-017-0849-4_6.
Pitowsky, I. (1989b). *Quantum probability – Quantum logic* (Lecture notes in physics, Vol. 321). Berlin/Heidelberg: Springer. https://doi.org/10.1007/BFb0021186.
Pitowsky, I. (1991). Correlation polytopes their geometry and complexity. *Mathematical Programming, 50*, 395–414. https://doi.org/10.1007/BF01594946.
Pitowsky, I. (1994). George Boole's 'conditions of possible experience' and the quantum puzzle. *The British Journal for the Philosophy of Science, 45*, 95–125. https://doi.org/10.1093/bjps/45.1.95.

Pitowsky, I. (1998). Infinite and finite Gleason's theorems and the logic of indeterminacy. *Journal of Mathematical Physics, 39*(1), 218–228. https://doi.org/10.1063/1.532334.

Pitowsky, I. (2003). Betting on the outcomes of measurements: A Bayesian theory of quantum probability. *Studies in History and Philosophy of Science Part B: Studies in History and Philosophy of Modern Physics, 34*(3), 395–414. https://doi.org/10.1016/S1355-2198(03)00035-2, quantum Information and Computation, arXiv:quant-ph/0208121.

Pitowsky, I. (2006). Quantum mechanics as a theory of probability. In W. Demopoulos & I. Pitowsky (Eds.), *Physical theory and its interpretation* (The western Ontario series in philosophy of science, Vol. 72, pp. 213–240). Springer Netherlands. https://doi.org/10.1007/1-4020-4876-9_10, arXiv:quant-ph/0510095.

Pitowsky, I., & Svozil, K. (2001). New optimal tests of quantum nonlocality. *Physical Review A, 64*, 014102. https://doi.org/10.1103/PhysRevA.64.014102, arXiv:quant-ph/0011060.

Planck, M. (1932). The concept of causality. *Proceedings of the Physical Society, 44*(5), 529–539. https://doi.org/10.1088/0959-5309/44/5/301.

Pták, P., & Pulmannová, S. (1991). Orthomodular structures as quantum logics. Intrinsic properties, state space and probabilistic topics (Fundamental theories of physics, Vol. 44). Dordrecht: Kluwer Academic Publishers/Springer.

Pulmannová, S. (2002). Hidden variables and Bell inequalities on quantum logics. *Foundations of Physics, 32*, https://doi.org/10.1023/a:1014424425657.

Pykacz, J., & Santos, E. (1991). Hidden variables in quantum logic approach reexamined. *Journal of Mathematical Physics, 32*. https://doi.org/10.1063/1.529327.

Ramanathan, R., Rosicka, M., Horodecki, K., Pironio, S., Horodecki, M., & Horodecki, P. (2018). Gadget structures in proofs of the Kochen-Specker theorem. https://arxiv.org/abs/1807.00113, arXiv:1807.00113

Rogers, Jr H. (1967). *Theory of recursive functions and effective computability*. New York/Cambridge, MA: MacGraw-Hill/The MIT Press.

Schaller, M., & Svozil, K. (1995). Automaton partition logic versus quantum logic. *International Journal of Theoretical Physics, 34*(8), 1741–1750. https://doi.org/10.1007/BF00676288.

Schaller, M., & Svozil, K. (1996). Automaton logic. *International Journal of Theoretical Physics, 35*, https://doi.org/10.1007/BF02302381.

Segal, A., & Goldschmidt, T. (2017/2018). The necessity of idealism. In *Idealism: New essays in metaphysics* (pp. 34–49). Oxford: Oxford University Press. https://doi.org/10.1093/oso/9780198746973.003.0003.

Simmons, A. W. (2020). How (maximally) contextual is quantum mechanics? This volume, pp. 505–520.

Sliwa, C. (2003). Symmetries of the Bell correlation inequalities. *Physics Letters A, 317*, 165–168. https://doi.org/10.1016/S0375-9601(03)01115-0, arXiv:quant-ph/0305190.

Smullyan, R. M. (1993). *Recursion theory for metamathematics* (Oxford logic guides, Vol. 22). New York/Oxford: Oxford University Press.

Solís-Encina, A., & Portillo, J. R. (2015). Orthogonal representation of graphs. https://arxiv.org/abs/1504.03662, arXiv:1504.03662.

Specker, E. (1960). Die Logik nicht gleichzeitig entscheidbarer Aussagen. *Dialectica, 14*(2–3), 239–246. https://doi.org/10.1111/j.1746-8361.1960.tb00422.x, arXiv:1103.4537.

Specker, E. (1990). *Selecta*. Basel: Birkhäuser Verlag. https://doi.org/10.1007/978-3-0348-9259-9.

Specker, E. (2009). Ernst Specker and the fundamental theorem of quantum mechanics. https://vimeo.com/52923835, video by Adán Cabello, recorded on June 17, 2009.

Stace, W. T. (1934). The refutation of realism. *Mind, 43*(170), 145–155. https://doi.org/10.1093/mind/XLIII.170.145.

Stairs, A. (1983). Quantum logic, realism, and value definiteness. *Philosophy of Science, 50*, 578–602. https://doi.org/10.1086/289140.

Svozil, K. (2001). On generalized probabilities: Correlation polytopes for automaton logic and generalized urn models, extensions of quantum mechanics and parameter cheats. https://arxiv.org/abs/quant-ph/0012066, arXiv:quant-ph/0012066.

Svozil, K. (2002). Quantum information in base n defined by state partitions. *Physical Review A, 66*, 044306. https://doi.org/10.1103/PhysRevA.66.044306, arXiv:quant-ph/0205031.

Svozil, K. (2004). Quantum information via state partitions and the context translation principle. *Journal of Modern Optics, 51*, 811–819. https://doi.org/10.1080/09500340410001664179, arXiv:quant-ph/0308110.

Svozil, K. (2005). Logical equivalence between generalized urn models and finite automata. *International Journal of Theoretical Physics, 44*, 745–754. https://doi.org/10.1007/s10773-005-7052-0, arXiv:quant-ph/0209136.

Svozil, K. (2009a). Proposed direct test of a certain type of noncontextuality in quantum mechanics. *Physical Review A, 80*(4), 040102. https://doi.org/10.1103/PhysRevA.80.040102.

Svozil, K. (2009b). Quantum scholasticism: On quantum contexts, counterfactuals, and the absurdities of quantum omniscience. *Information Sciences, 179*, 535–541. https://doi.org/10.1016/j.ins.2008.06.012.

Svozil, K. (2012). How much contextuality? *Natural Computing, 11*(2), 261–265. https://doi.org/10.1007/s11047-012-9318-9, arXiv:1103.3980.

Svozil, K. (2016). Quantum hocus-pocus. *Ethics in Science and Environmental Politics (ESEP), 16*(1), 25–30. https://doi.org/10.3354/esep00171, arXiv:1605.08569.

Svozil, K. (2017a). Classical versus quantum probabilities and correlations. https://arxiv.org/abs/1707.08915, arXiv:1707.08915.

Svozil, K. (2017b). Quantum clouds. https://arxiv.org/abs/1808.00813, arXiv:1808.00813.

Svozil, K. (2018a). Kolmogorov-type conditional probabilities among distinct contexts. https://arxiv.org/abs/1903.10424, arXiv:1903.10424.

Svozil, K. (2018b). New forms of quantum value indefiniteness suggest that incompatible views on contexts are epistemic. *Entropy, 20*(6), 406(22). https://doi.org/10.3390/e20060406, arXiv:1804.10030.

Svozil, K. (2018c). *Physical [a]causality. Determinism, randomness and uncaused events.* Cham/Berlin/Heidelberg/New York: Springer. https://doi.org/10.1007/978-3-319-70815-7.

Sylvia, P., & Majernik, V. (1992). Bell inequalities on quantum logics. *Journal of Mathematical Physics, 33*. https://doi.org/10.1063/1.529638.

Ursic, S. (1984). A linear characterization of NP-complete problems. In R. E. Shostak (Ed.), *7th international conference on automated deduction*, Napa, 14–16 May 1984 Proceedings (pp. 80–100). New York: Springer. https://doi.org/10.1007/978-0-387-34768-4_5.

Ursic, S. (1986). Generalizing fuzzy logic probabilistic inferences. In *Proceedings of the second conference on uncertainty in artificial intelligence, UAI'86* (pp. 303–310). Arlington: AUAI Press. http://dl.acm.org/citation.cfm?id=3023712.3023752, arXiv:1304.3114.

Ursic, S. (1988). Generalizing fuzzy logic probabilistic inferences. In J. F. Lemmer & L. N. Kanal (Eds.), *Uncertainty in artificial intelligence 2 (UAI1986)* (pp. 337–362), Amsterdam: North Holland.

Vermaas, P. E. (1994). Comment on "getting contextual and nonlocal elements-of-reality the easy way". *American Journal of Physics, 62*. https://doi.org/10.1119/1.17488.

Vorob'ev, N. N. (1962). Consistent families of measures and their extensions. *Theory of Probability and Its Applications, 7*. https://doi.org/10.1137/1107014.

Weihs, G., Jennewein, T., Simon, C., Weinfurter, H., & Zeilinger, A. (1998). Violation of Bell's inequality under strict Einstein locality conditions. *Physical Review Letters, 81*, 5039–5043. https://doi.org/10.1103/PhysRevLett.81.5039.

Wright, R. (1978). The state of the pentagon. A nonclassical example. In A. R. Marlow (Ed.), *Mathematical foundations of quantum theory* (pp. 255–274). New York: Academic Press. https://www.elsevier.com/books/mathematical-foundations-of-quantum-theory/marlow/978-0-12-473250-6.

Wright, R. (1990). Generalized urn models. *Foundations of Physics, 20*(7), 881–903. https://doi.org/10.1007/BF01889696.

Zeilinger, A. (1999). A foundational principle for quantum mechanics. *Foundations of Physics, 29*(4), 631–643. https://doi.org/10.1023/A:1018820410908.

Zierler, N., & Schlessinger, M. (1965). Boolean embeddings of orthomodular sets and quantum logic. *Duke Mathematical Journal, 32*, 251–262. https://doi.org/10.1215/S0012-7094-65-03224-2, reprinted in Zierler and Schlessinger (1975).

Zierler, N., & Schlessinger, M. (1975). Boolean embeddings of orthomodular sets and quantum logic. In C. A. Hooker (Ed.), *The logico-algebraic approach to quantum mechanics: Volume I: Historical evolution* (pp. 247–262). Dordrecht: Springer Netherlands. https://doi.org/10.1007/978-94-010-1795-4_14.

Chapter 25
Schrödinger's Reaction to the EPR Paper

Jos Uffink

Abstract I discuss Schrödinger's response to the Einstein-Podolsky-Rosen (EPR) paper of 1935. In particular, it is argued, based on an unpublished notebook, that while Schrödinger sympathized with the EPR argument he worried about its lack of emphasis on the role of biorthogonal expansions and on the distinction between conclusions dependent on the particular outcome obtained in a single measurement, versus those that quantify over all possible outcomes of a measurement procedure. I also discuss the different views between Schrödinger and Einstein on the issue of the completeness of quantum mechanics. Finally, I discuss the question whether Schrödinger borrowed from his correspondence with Einstein to come up with his famous Cat Paradox.

Keywords Schrödinger · Quantum mechanics · Einstein-Podolski-Rosen argument · Completeness · Separability · Schrödinger's cat

25.1 Introduction

On March 25, 1935, the *Physical Review* received a paper by Einstein, Podolsky and Rosen (EPR), arguing that the quantum description of physical reality was incomplete. It was published on May 15 of the same year. It will be superfluous, on this occasion, to stress the importance of this paper for the subsequent discussions of the foundations of quantum theory in the 1930s and beyond, in particular by directing these discussions to focus on the phenomenon of entanglement, eventually opening up, in recent years, investigations into quantum information and computation. It is a pleasure to dedicate this paper to the memory of Itamar

J. Uffink (✉)
Philosophy Department and Program in History of Science, Technology and Medicine, University of Minnesota, Minneapolis, MN, USA
e-mail: jbuffink@umn.edu

M. Hemmo, O. Shenker (eds.), *Quantum, Probability, Logic*, Jerusalem Studies in Philosophy and History of Science, https://doi.org/10.1007/978-3-030-34316-3_25

Pitowsky, who did so much conceptual work to understand quantum entanglement and to distinguish it from mere classical correlations.

Actually, the term 'entanglement' does not appear in the EPR paper. This term (or its German equivalent '*Verschränkung*') was coined by Schrödinger in two papers he wrote during the following summer (1935a, 1935b) and finished nearly simultaneously around 11 August. One paper, for a general audience, was published in *Die Naturwissenschaften* and the second, more technical, in the *Mathematical Proceedings of the Cambridge Philosophical Society*, where he described entanglement as "not one but *the* characteristic trait of quantum mechanics that enforces its entire departure from classical lines of thought."

It is commonly assumed in the historical literature that Schrödinger's papers were triggered by the EPR paper, and indeed, Schrödinger acknowledged as much in a footnote in (1935b, p. 845), where, referring to the EPR paper, he wrote

> The appearance of this work gave the incentive for the present — shall I say seminar paper or general report?

However, although EPR obviously *stimulated* Schrödinger to publish his thoughts, this is not to say that he *developed* these thoughts only after he had learned of the EPR paper. Elsewhere, Christoph Lehner and I will present evidence from Schrödinger's unpublished notebooks, and (partly published) letters that detail Schrödinger's role in the development of the EPR argument. In particular, these notebooks and letters show how Schrödinger struggled with the concept of entanglement already from 1926-7 onwards, and how he essentially developed what we now call the EPR argument already in November 1931, $3\frac{1}{2}$ years before EPR, when he and Einstein were still colleagues in Berlin. Here, however, I will focus on Schrödinger's immediate reactions to the appearance of the EPR paper, and his subsequent correspondence with Einstein in 1935. Schrödinger wrote to Einstein on June 7, 1935, to express his joy at the appearance of the EPR paper, "about what we discussed so much in Berlin", probably referring to discussions in November 1931. However, Schrödinger's notebook entitled *Übungen*[1] contains notes which he took while preparing this letter. These notes clearly reveal that Schrödinger was not entirely happy with the presentation of the argument in the EPR paper. He also expressed his worries in the letter to Einstein:

> Can I add a few words? They may look at first sight like objections, but they are just points that I would like to have formulated more clearly (von Meyenn 2011, p. 527).

The points that Schrödinger worried about are subtle. They concern both the role of biorthogonal decompositions of bipartite wave functions and the crucial dependence on the measurement outcomes in the EPR argument. To understand his worries, it might be well to first recall the structure of the argument in the EPR paper.

---

[1]This notebook is accessible at https://fedora.phaidra.univie.ac.at/fedora/objects/o:159711/methods/bdef:Book/view. The relevant text is on pp. 35–41.

25.2 The Argumentation of the EPR Paper

EPR present a necessary Criterion for Completeness of a physical theory,[2] and a sufficient Criterion for Reality for what counts as "an element of physical reality".[3] After rehearsing some well-known aspects of quantum mechanics in their section 1, their section 2 introduces the example of two systems that have interacted in some period in the past, but between which any interaction has ceased since then.[4]

EPR then present a wave function for such a joint system, which can be expressed in different forms[5]:

$$|\Psi\rangle = \sum_n c_n |a_n\rangle_1 |\psi_n\rangle_2 = \sum_n c'_n |a'_n\rangle_1 |\phi_n\rangle_2 \tag{25.1}$$

Here, $\{|a_n\rangle\}$ denotes an orthonormal basis of eigenvectors of some physical quantity A of system 1 and $\{|a'_n\rangle\}$ another orthonormal basis of eigenvectors of quantity A' (which need not commute with A) of system 1. $\{|\psi_n\rangle\}$ and $\{|\phi_n\rangle\}$ denote two arbitrary collections of (linearly independent) unit vectors for system 2, neither of which needs to be mutually orthogonal nor complete. Note that the expansions of the wave function in (25.1) are not biorthogonal.

EPR argue that if we make a measurement of quantity A on system 1, say with result a_k, the second system is to be assigned a wave function corresponding to the vector $|\psi_k\rangle$, by the process of "the reduction of the wave packet" (or in more modern terms: by applying the projection postulate). However, if we decide instead to make a measurement of quantity A' on system 1, and obtain an eigenvalue a'_r, the corresponding wave function for system 2 is $|\phi_r\rangle$. But, as EPR argue, what is physically real for system 2 should not depend on what is done to system 1, since the two systems no longer interact. Thus, they draw the following, what I will refer to as their 'general conclusion' from their argument:

---

[2]"[E]very element of the physical reality must have a counterpart in the physical theory." (Einstein et al. 1935, p. 777)

[3]"If, without in any way disturbing a system, we can predict with certainty (i.e., with probability equal to unity) the value of a physical quantity, then there exists an element of reality corresponding to that quantity." (Einstein et al. 1935, p. 777)

[4]Incidentally, the EPR paper never refers to spatial separation between the two systems as a condition that would guarantee the absence of such interaction, unlike Schrödinger's argument in 1931.

[5]Here and in further quotations, I have not followed original notation. In particular, wave functions are presented here in the now common Dirac notation for perspicuity, even though, strictly speaking, symbols like $|\Psi\rangle$ are vectors in a Hilbert space, and the corresponding wave function $\Psi(x_1, x_2) = \langle x_1, x_2|\Psi\rangle$ is a representation of this vector in the position eigenbasis. Also, I abstain from refering to such entities as 'states', because it is an issue of contention between Einstein and Schrödinger whether wave functions represent states of a system.

[GENERAL CONCLUSION:]"Thus, *it is possible to assign two different wave functions* (in our example $|\psi_k\rangle$ and $|\phi_r\rangle$) *to the same reality* (the second system after its interaction with the first)." (original italicization)

After reaching this general conclusion, EPR proceed to discuss a special case, in which it happens that ψ_k and ϕ_r above are eigenfunctions of the non-commuting quantities P and Q, respectively, of system 2. For this purpose, they employ the wave function

$$\Psi(q_1, q_2) = \frac{1}{\sqrt{2\pi}} \int_{-\infty}^{+\infty} e^{i(q_1 - q_2 + q_0)p} dp \tag{25.2}$$

which, in Dirac notation, corresponds to

$$|\Psi\rangle = \frac{1}{\sqrt{2\pi}} \int e^{iq_0 p} |p\rangle_1 |-p\rangle_2 dp = \frac{1}{\sqrt{2\pi}} \int |q\rangle_1 |q - q_0\rangle_2 dq \tag{25.3}$$

where q_0 is a real-valued constant.

In this special case, measuring the momentum P_1 on system 1, with an outcome p, say, leads one to assign the wave function $|-p\rangle_2$ to system 2, i.e. an eigenfunction of the momentum P_2 of system 2 with eigenvalue $-p$. Hence, we can infer the prediction that, with 100% certainty, any measurement of the momentum of 2 will provide an outcome $-p$, and this value, by the Criterion of Reality, should therefore correspond to an element of physical reality. Similarly, if we were to measure the position Q_1 on system 1, and obtain an outcome q_1, we can not only assign the the wave function $|q_1 - q_0\rangle_2$ to system 2 but also infer that any measurement of the position Q_2 on system 2 will 100% certainly lead to the value $q_1 - q_0$. Hence by the Criterion of Reality, in this case the position of system 2 corresponds to an element of reality.

The conclusion from this special case can therefore be formulated in terms of different physical quantities of system 2, depending on what we measure on system 1, instead of just different wave functions as in the general conclusion. Yet, just like that previous conclusion, it is argued that the elements of reality for system 2 should not depend on what is done to system 1 (i.e.: the choice of measuring either P_1 or Q_1), so that the quantities for system 2 must represent simultaneous elements of reality:

[SPECIAL CONCLUSION:]" [...] we arrived at the conclusion that two physical quantities, with non-commuting operators, can have simultaneous reality."

And this conclusion is then shown to lead to a contradiction with the assumption that the wave function gives a complete description of reality, because quantum mechanics does not allow any wave function for system 2 in which both P_2 and Q_2 are both elements of reality.

It is worth emphasizing that the argument for the special conclusion, unlike the general conclusion, makes essential use of the circumstance that the special wave function (25.3) possesses multiple *biorthogonal* decompositions.

25.3 Schrödinger's Objections to the EPR Argument

The notes in the undated folder *Übungen* relating to the EPR paper start off with 2 pages entitled "Entwicklung e. Funktion zweier Variablen in eine biorthogonale Reihe" (Expansion of a function of two variables in a biorthogonal series). This is a rehearsal of a proof of the now well-known biorthogonal decomposition theorem,— a proof Schrödinger had in fact already worked out earlier, in the folder *Supraleitung 1931*.[6] He included a proof of the theorem along the lines of these notes in his June 7 letter to Einstein, and also in Schrödinger (1935a). In *Übungen*, he did not work out all the details of the theorem, but only to the point that he refreshed his knowledge of the topic enough to continue. He writes:

> The main issue established. To be worked out later.

Schrödinger then immediately turns to the EPR paper (ignoring Einstein's coauthors) with the remark:

> Einstein does *not* write of a complete biorthogonal decomposition.

And his argument ends with the conclusion:

> Thus, I have to assume a biorthogonal decomposition.

Even before we go into the details of his argument, these remarks show two points: he is contemplating what I called the general conclusion of the EPR paper, based on the non-biorthogonal decompositions (25.1), rather than the special conclusion, where (25.3) of course *does* provide biorthogonal decompositions. Secondly: Schrödinger thought that this general conclusion was problematic precisely because it did not assume a biorthogonal decomposition. In more detail, his argument in *Übungen* initially worries about the question what the non-biorthogonal decompositions in (25.1) imply for the application of the projection postulate after a measurement on system 1. But these initial worries do not materialize, the text breaks off, and he then switches to a different problem:

> However, the problem is different. From the mere fact that after one measurement [i.e. of A on system 1] we are stuck with one [wave function ψ_k], but after another measurement [of A' on system 1] we are stuck with another [wave] function [ϕ_r], where these functions correspond to contradictory properties of "the other" system [2], one cannot fabricate a noose by which to hang quantum mechanics (läßt sich die Qu. mech. noch kein Strick drehen). Because even one and the same method of measurement can and must, for identically prepared systems, provide sometimes this, sometimes another outcome, which lead to inferences about the "other" system that will likewise be contradictory.

Clearly, Schrödinger sees the general conclusion in the EPR paper as insufficient to produce a paradox. He continues:

> It is therefore also not a contradiction when the function [$\Psi(q_1, q_2)$] includes the possibility that *on one occasion* a measurement delivers a determinate value of position, and on *another*

[6] https://fedora.phaidra.univie.ac.at/fedora/objects/o:259940/methods/bdef:Book/view, pp.95–6.

occasion *another* measurement a determinate value of momentum. Because, indeed, it could be that *really* in one case the position, in the other case the momentum of the "other" system was sharp.

The contradiction only arises when I recognize the following:

1. There is *one* measurement arrangement that *always* lets me infer a determinate position for the other system. And
2. There is *another* measurement arrangement that at least *sometimes* provides a value for a quantity of the other system which is incompatible with determinateness of position.

Thus, I have to assume a biorthogonal decomposition:

$$\Psi(q_1, q_2) = \sum c_i g_i(q_1) u_i(q_2) \tag{25.4}$$

What was he thinking here? Let me try to reconstruct the argument by breaking the problem down in a piecemeal manner, proceeding from simple cases to more general ones. There are two main themes throughout this reconstruction attempt: the first is the distinction between biorthogonal and non-biorthogonal decompositions of the joint wave function $|\Psi\rangle$, as already indicated. The second theme is the distinction between inferences based on just a single performance of a measurement (*Einzelmessung*), which will naturally be conditional on the particular outcome obtained in this single measurement, versus conclusions that can be stated by quantifying over all possible outcomes of the measurement procedure, and therefore are not conditional on any particular outcome.

The simplest case is when the state vector $|\Psi\rangle$ for the pair of systems has a *unique* biorthogonal decomposition, say:

$$|\Psi\rangle = \sum_i c_i |a_i\rangle |b_i\rangle \tag{25.5}$$

Indeed, according to the Biorthogonal Decomposition Theorem, this is the generic case for almost all $|\Psi\rangle \in \mathcal{H}_1 \otimes \mathcal{H}_2$, i.e., as long as the absolute values of all coefficients c_i are different. If we measure quantity A on system 1, for this state, and obtain the outcome a_k, we assign wave function $|b_k\rangle$ to system 2. But if we obtain a different outcome, a_r, one should assign the wave function $|b_r\rangle$ to system 2. Thus, depending on which outcome we obtain, we should assign different wave functions to system 2. But could we argue in this case, that because system 2 does not interact with system 1, these different wave functions refer to the same physical reality for system 2? Surely not, because because even if one agrees to EPR's claim that "no real change can take place in the second system in consequence of anything that may be done to the first system", in the case at hand we are not *doing* different things on the first system at all (we measure the same quantity A in both cases, i.e. we are executing "one and the same method of measurement" as Schrödinger writes in the quote above.). Rather, the question is here whether the physical reality of system 2 should be independent of the *outcome* obtained in this measurement of A on system

1.[7] And the answer to that question is generally negative, even in classical physics, because two systems, even when they do not interact, might be correlated, so that the outcome obtained by a measurement on 1 can still be informative about what is real for system 2. Indeed, one of the main themes that surface again and again in Schrödinger's writings on entanglement and the EPR problem is that they express an intermixture of 'mental' (where today we would perhaps say 'epistemic') and 'physical' (where today we would perhaps say 'ontic') aspects of the wave function.

Therefore, in the special case of (25.5), the fact that, after the same measurement yields different outcomes on system 1, different wave functions are assigned to system 2, depending on the outcomes obtained on system 1, does not justify EPR's general conclusion that those wave functions represent the *same* physical reality for system 2.

Let us proceed to a more general case, where the decomposition is not biorthogonal, i.e.:

$$|\Psi\rangle = \sum_n c_n |a_n\rangle |\psi_n\rangle \tag{25.6}$$

where, as before, $|\psi_n\rangle$ denote linearly independent wave functions for system 2, not necessarily orthogonal or complete. However, for a given $|\Psi\rangle$, and a given orthonormal basis $\{|a_n\rangle\}$, this set is uniquely determined (up to inessential phase factors). In fact, the set $\{|\psi_n\rangle\}$ forms what Everett in 1957 called the relative states of system 2, relative to the choice of $|\Psi\rangle$ and the choice of basis $\{|a_n\rangle\}$ on $\mathcal{H}_1$.

The main difference with the previous case (25.5) is just that now the inference to "elements of physical reality" is not so simple. Wave functions are not the kind of entities that can be predicted with 100% certainty, especially not when one has to decide between a non-orthogonal set. However, this problem is alleviated by noting that every wave function, say $|\psi_i\rangle$, is an eigenfunction of some observable (or set of observables). In particular, one could consider the projectors $\boldsymbol{P}_i := |\psi_i\rangle\langle\psi_i|$ as physical quantities, each of which will have the property that when the wave function of system 2 is $|\psi_i\rangle$, one can predict with 100% probability that the physical quantity $\boldsymbol{P}_i$ takes the value 1. So, here too, one can argue that, conditional on a measurement of quantity A on system 1 yielding an outcome a_k, there is a quantity $\boldsymbol{P}_k$ of the second system that will have the value 1, even without actually intervening on system 2 and therefore, by the EPR reality criterion, to the inference that the quantity $\boldsymbol{P}_k$ must correspond to an element of reality for system 2.

However, in this case too, we are still left with the fact that this inference is conditional on the outcome obtained in the measurement on system 1. And given that the particular outcome obtained in such a measurement might be informative about what is real about system 2, in the simple case (25.5), it would be hard to deny a similar possibility too in the case of (25.6): different possible outcomes of the A-measurement on system 1, say a_k and a_r ($k \neq r$), are mutually exclusive data

[7]This point is reminiscent of the distinction between parameter independence and outcome independence in later discussions of the Bell inequalities, cf. Jarrett, Shimony.

obtained from system 1. If these outcomes are held to be informative about system 2, they might represent mutually exclusive pieces of information about system 2. But then, how could one argue that the inferences about system 2, i.e. that either of the quantities $\boldsymbol{P}_k$ or $\boldsymbol{P}_r$ represents an element of reality by having the value 1, refer to the *same* reality? Accordingly, Schrödinger does not regard it as paradoxical by itself that, conditional on different outcomes obtained in measurements on the first system, one should be able to predict (and infer the reality of) values of non-commuting quantities of the second system.

He expresses this worry in his June 7 letter to Einstein:

> In order to construct a contradiction, it is in my view not sufficient that for the same preparation of the system pair the following can happen: one concrete single measurement on the first system leads to a determinate value of a quantity $[B]$ of the second system, any other value of $[B]$ or $[B']$ is excluded for general reasons. I believe this is not sufficient. Indeed, the fact that same preparations do not always give the same measurement outcomes is something we have to concede anyway. So when on one occasion the measurement gives the value $[b_k]$, and on another occasion the value $[b_r]$ to quantity B, why would it not give on a third time, not the quantity B but the quantity B' reality, and provide it with a value $[b'_s]$?

The point that Schrödinger raises in the the last sentence, namely that the measurement on system 1 could lead to *different* (in particular non-commuting) quantities obtaining values, would be hard to appreciate without the context of his notes in the folder *Übungen*, because Schrödinger does not explicitly mention the issue of biorthogonality in this passage. But one can infer from these notes that the non-biorthogonality of the EPR wave function (25.1) was his main concern, and since my reconstruction argues that for a non-biorthogonal decomposition, like (25.6), one would indeed end up assigning reality to values of non-commuting quantities like $\boldsymbol{P}_k$ and $\boldsymbol{P}_r$, depending on the outcome obtained in the measurement on system 1, this last sentence just quoted makes sense.

But of course, the two cases just reviewed still do not cover the case treated by EPR's general conclusion, i.e. the state (25.1). Here, we consider a measurement of A, *or* a measurement of B, on system 1 with outcome a_k or b_r respectively, and contemplate their implications for system 2. Both EPR and Schrödinger agree that this choice of what is *done* to system 1 should not affect what is real on system 2. (As he said in 1931: what we choose to measure depends on an arbitrary choice (*Willesentscheidung*).) But what we infer about system 2 depends not only on what we *do* to system 1, it depends also on the outcome we obtain. Now, it would be hard to deny that, just as a_1 and a_2 are mutually incompatible outcomes for the measurement of A on system 1, that allow different inferences about what is real for system 2, in example (25.6), the outcomes a_k and b_r obtained in incompatible measurements on system 1 also provide incompatible data that enable different inferences about what is real for system 2. And so one may well ask whether this should be in any way more paradoxical than the case already discussed under (25.6). Schrödinger's answer seems to be negative.

To summarize, what I have been arguing so far is that Schrödinger's misgivings about the EPR argument focus on their general conclusion, and particularly on the fact that, at this stage of their argument, biorthogonality of the decomposition of the wave function is not assumed. He argues that any inference about the second system, based on a measurement of the first, depends not only on the quantity measured on the first system, but also on the outcome obtained in such a measurement. In the case of a non-biorthogonal decomposition of the wave function, he does not regard it as paradoxical that, depending on the outcome of a measurement on system 1, we should infer that the values of one or another of non-commuting quantities for system 2 constitute an element of reality.

Schrödinger also describes in his June 7 letter to Einstein what *he* thinks is a *satisfactory* way to argue for a contradiction. The argument is very similar to what he wrote in his *Übungen* notebook:

> In order to construct a contradiction, the following seems necessary to me. There should be a wave function for the pair of systems and there should be two quantities A and $[A']$, whose simultaneous reality is excluded for general reasons, and for those, the following should hold:
>
> 1. There exists a method of measuring, which for this wave function *always* gives the quantity A a sharp value (even if not always the same value), so that I can say, even without actually performing the measurement: in the case of such a wave function, A possesses reality — I don't care about which value it has. (welchen Wert es hat ist mir Wurst.)
> 2. Another method of measurement must provide, at least *sometimes*, a sharp value for the quantity $[A']$ (always for the same wave function, of course).

Schrödinger then points out that the maximally entangled wave function used in the example by EPR in their special conclusion does indeed satisfy these conditions, (and in fact always provides a sharp value in both measurements), but that this is due to the fact that their example wave function allows multiple biorthogonal decompositions. In fact, this example wave function is exceptionally special in the sense that the absolute values of the coefficients (e^{ipx_0}) are *all* the same, and this implies that by performing an arbitrary unitary basis transformation on both bases in the Hilbert spaces of the two systems, one can get another biorthogonal decomposition. He then expresses a secondary worry that a conclusion as far-reaching as EPR, should not depend on such an exceptional example. But he also points out that if one does not share this secondary worry, the example is just fine in demonstrating the argument. His letter ends with an outline of a proof of the biorthogonal decomposition theorem.

In this attempt at reconstructing Schrödinger's thinking, the theme of the distinction between conclusions based single measurements versus conclusions based on general conclusions that are not conditional on a particular, given, measurement outcome plays center stage, whereas the distinction between biorthogonal versus non-biorthogonal decompositions has dropped to an auxiliary role.

So *like* Schrödinger, one could argue as follows: suppose we have the wave function (25.1), considered in their general argument. Suppose we make a measurement of quantity A on system 1, whose eigenfunctions are $|a_i\rangle$, and eigenvalues a_i. If this measurement yields the outcome a_k we infer that the corresponding wave function for sytem 2 is ψ_k. This leads us to predict with 100% probability, before making any measurement on system 2 that the physical quantity $\boldsymbol{P}_k$ has the value 1. By the "criterion of reality", this entails that in this case $\boldsymbol{P}_k$ is an element of reality. However, this conclusion is conditional on the outcome obtained on system 1, i.e. the value a_k obtained in a measurement of A on system 1. And, even if system 2 is non-interacting with, or even spatially separated from system 1, we cannot conclude that the entire variety (for $k = 1, 2, \ldots$ of wave functions $|\psi_k\rangle$ or associated quantities $\boldsymbol{P}_k$ refer to the *same* physical reality of system 2, precisely because they do not depend only on what measurement is performed on system 1 but *also* on the outcome obtained.

However, *contra* Schrödinger, one could continue as follows: suppose the procedure mentioned in the previous paragraph is in place.

> We perform the measurement A on system 1 and obtain one of its possible outcomes a_k. For each of those specific outcomes one of the quantities $\boldsymbol{P}_k$ represents an element of reality for system 2. So, if we do not care what this outcome might be, we can infer that at least one of the quantities $\boldsymbol{P}_1, \boldsymbol{P}_2, \ldots$ represents a element of physical reality.

In other words, by quantifying over all possible outcomes of the A measurement, we obtain a logical disjunction of

$$[\boldsymbol{P}_1 = 1] \vee [\boldsymbol{P}_2 = 1] \cdots \vee [\boldsymbol{P}_n = 1] \tag{25.7}$$

and one could regard this description as representing an element of reality for the second system. Now compare this to another measurement on system 1, of the quantity A'. This would lead similarly to a logical disjunction of

$$[\boldsymbol{Q}_1 = 1] \vee [\boldsymbol{Q}_2 = 1] \cdots \vee [\boldsymbol{Q}_n = 1] \tag{25.8}$$

where $\boldsymbol{Q}_i = |\phi_i\rangle\langle\phi_i|$.

But in general, of course, these two disjunctions will be mutually exclusive, and indeed, there will be no description within quantum mechanics for a state in which both disjunctions are correct descriptions. So, assuming with EPR that what is real for system 2 cannot depend on the choice of measurement on system 1 we do reach their conclusion that QM is incomplete, even for non-biorthogonal wave functions.

What emerges from this discussion is thus that Schrödinger's insistence that the reliance on single measurements in EPR's general argument is not sufficient to establish their conclusion is well-founded. However, his insistence that biorthogonal decompositions are needed for this purpose seems too strong. What I have tried to argue is that, while they of course *simplify* the argument quite considerably, they are not essential to their argument.

25.4 Einstein's Reply to Schrödinger

Einstein's June 19 1935 reply to Schrödinger's letter has become rather well-known, since it has been quoted and discussed by several commentators (Fine 1986,

2017; Howard 1985, 1990; Bacciagaluppi and Crull 2020). In this letter, Einstein expresses his dissatisfaction with the formulation of the EPR paper, blaming Podolsky for "smothering it in erudition". It is also the first time he explicitly puts forward a 'Separation Principle' (*Trennungsprinzip*), which might be paraphrased by saying that for two spatially separated systems, each has its own physical state, independently of what is done to the other system.

Einstein first presents a simple classical illustration: a ball is always found in either of two boxes, and we describe this situation by saying that the probability of finding it in the first or second box is $\frac{1}{2}$. He asks whether this description could be considered as a complete specification, and contemplates two answers:

> **No:** A complete specification is: the ball is in the first box (or not). This is how the characterization of the state should look in the case of a complete description.
>
> **Yes:** Before I open the lid of a box, the ball is not at all in *one* of the two boxes. Its being in one determinate box only comes about because I open the lid. Only that brings about the statistical character of the world of experience, [...]
>
> We have analogous alternatives when we want to interpret the relation between quantum mechanics and reality. In the ball-system, the second "spiritist" or Schrödinger interpretation is, so to say, silly, and only the first "Born" interpretation would be taken seriously.[8] But the talmudic philosopher whistles at "reality" as a sock puppet of naivety and declares both views as only different in their ways of expression. [...]
>
> In quantum theory, one describes the real state of a system by means of a normalized function ψ of its coordinates (in configuration space).[...] One would now like to say the following: ψ is correlated one-to-one with the real state of the real system. [...] If this works, I speak of a complete description of reality by the theory. But if such an interpretation fails I call the theoretical description 'incomplete'. [...]
>
> Let us describe the joint system consisting of subsystems [1] and [2] by a ψ-function [$|\Psi\rangle$]. This description refers to a moment in time when the interaction has practically ceased. This ψ of the joint system can be built up from the normalized eigen-ψ $|a_i\rangle$, $|b_j\rangle$ that belong to the eigenvalues of the "observables" (respectively sets of commuting observables) [A] and [B] respectively. We can write:
>
> $$|\Psi\rangle = \sum_{ij} c_{ij}|a_i\rangle|b_j\rangle \tag{25.9}$$
>
> If we now make an [A]-measurement on 1, this expression reduces to
>
> $$|\psi\rangle = \sum_{j} c_{ij}|b_j\rangle \tag{25.10}$$
>
> This is the ψ-function of the subsystem [2] in the case I have made an [A]-measurement.
>
> Now instead of the eigenfunctions of the observables [A] and [B], I can also decompose in the eigenfunctions [of] [A'], and [B], where [A'] is another set of commuting observables:
>
> $$|\Psi\rangle = \sum_{ij} c'_{ij}|a'_i\rangle|b_j\rangle \tag{25.11}$$

[8]Here, the reference to the "Schrödinger" and "Born" interpretation refer back to the year 1926, when Schrödinger originally interpreted the wave function as corresponding to the physical state of the system, whereas Born proposed a statistical interpretation.

So that after measurement of $[A']$ one obtains

$$|\psi'\rangle = \sum_j c'_{ij} |b_j\rangle \tag{25.12}$$

Ultimately, it is only essential that $|\psi\rangle$ and $|\psi'\rangle$ are at all different. I hold that this variety is incompatible with the hypothesis that the ψ-description is correlated in a one-to-one fashion with physical reality.

This reply is striking, in several ways. First of all, as has been noted by previous authors, because the argument is significantly different from the EPR argument: Einstein does not mention the Reality Criterium, and employs a *very* different notion of completeness. The Separability Principle, although appearing here for the first time, is arguably implicitly present in the EPR paper (Fine 1986, 2017; Howard 1985, 1990). A second reason this reply is striking is that it seems to ignore the remarks that Schrödinger put forward in his previous letter, and with regard to Schrödinger's contention that biorthogonal decompositions are essential, Einstein's only remark is "Das ist mir wurst" (I could not care less).

Let me try to spell this out in more detail. The conclusion of Einstein's argument is that depending on which measurement we perform on system 1, one should assign different wave functions to system 2, even though the real physical state of system 2 should —according to the Separation Principle— be the same, regardless of what we choose to measure on system 1. This conclusion just reiterates the general conclusion of the EPR paper, which was the focus of Schrödinger's worries in the first place. But where, for EPR, this conclusion was only an initial step —and did not by itself imply the incompleteness of the quantum description of reality—, Einstein argues that it is all that is ultimately essential for incompleteness.

This contrast between Einstein's June 19 argument and the EPR argument is due, of course, to their different construals of the notion of completeness. EPR provide only a necessary condition for when a theoretical description is complete. On the other hand, Einstein, in the quote above, describes a necessary and sufficient condition in terms of a one-to-one correspondence between real states and their quantum description by means of a wave function.

Lehner (2014) has proposed an apt terminology for dissecting notions of completeness employed by EPR and in Einstein's letter to Schrödinger. Failures of a one-to-one correspondence between "real states" and a "theoretical description" may come in at least two ways:

Overcompleteness Several theoretical descriptions are compatible with the same real state.

Incompleteness Several real states are compatible with the same theoretical description.

Einstein in his letter to Schrödinger is content with showing that QM is overcomplete: depending on what measurement one performs on the first subsystem, a different wave function is attributed to the second system, even though (given Separability) the real state of the second system should be one and the same.

Indeed, one main reason why EPR need to continue their argument to establish their special conclusion is just that they aim to establish incompleteness rather than overcompleteness. (Another reason is of course their reliance on the Criterion of Reality— I will come back to that below.)

The surprising point, here, is that, cases of overcompleteness are ubiquitous within physics. Indeed, even in quantum theory, it is well-known that one can multiply a wave function ψ with an arbitrary phase factor $e^{i\alpha}$ ($\alpha \in \mathbb{R}$) to obtain another wave function $\psi' = e^{i\alpha}\psi$, where it is generally assumed that ψ and ψ' describe the same physical situation. More general examples of overcompleteness can be found in gauge theories. Indeed one might wonder whether overcompleteness is a worrisome issue at all in theoretical physics. Now of course, in the examples just alluded to, the gauge freedom in the theoretical description of a physical state can usually be expresses in terms of equivalence relations and be expelled by 'dividing out' the corresponding equivalence classes in the theoretical descriptions. No such equivalence relations exist in the case discussed by Einstein in the June 19 letter, and so the case of overcompleteness presented there in his is not so simple as in the examples of gauge freedom (cf. Howard 1985).

But apart from the distinctions between Einstein (on June 19) and EPR, let me return to Schrödinger. As we have seen, Schrödinger had misgivings precisely about the EPR general conclusion, and tried to point out to Einstein that the wave function assigned to system 2 depends not only on what measurement is performed on system 1, but also on its outcome.

Einstein's letter, however, does not contain any acknowledgement of this issue. For example, he argues that the wave function (25.10) "is the ψ function of the subsystem [2] in the case I have made an [A]-measurement". But it is not, or at the very least, Einstein's notation is ambiguous here. Inspection of the right-hand side of (25.10) reveals that this wave function depends on i, which labels the outcome obtained in the measurement of A. So a proper formulation, instead of (25.10) should have read something like

$$|\psi_i\rangle = \sum_j c_{ij}|b_j\rangle \tag{25.13}$$

This is the ψ-function of subsystem 2 in case I have made an A-measurement *and* obtained the outcome a_i.

But this notation makes it transparent that there will be different wave functions for system 2 for different outcomes a_k and a_r, and, as Schrödinger argued, this need not be paradoxical. A similar remark holds for the wave function (25.12).

Now, one might argue for a more charitable reading of Einstein's response of June 19 to Schrödinger. One such reading[9] is that Einstein may have wanted to point out that that if we perform measurement A on system 1 the wave function of system 2 would be one of the *form*

[9]I am grateful to Guido Bacciagaluppi for suggesting this reading.

$$|\psi_i\rangle = \sum_j c_{ij}|b_j\rangle \tag{25.14}$$

or, more precisely, that it would be one of the wave functions in the set

$$\{|\psi_1\rangle, \ldots |\psi_n\rangle\} \tag{25.15}$$

On the other hand, if we perform a measurement of A' on system 1, the second system will be left in one of the wave functions

$$\{|\psi'_1\rangle \ldots |\psi'_n\rangle\} \tag{25.16}$$

and in general, these two sets will be disjoint. So, if there were a wave function to describe the real physical state of system 2, independently of what measurement we perform on system 1, it would have to be a member of two disjoint sets. In other words, there would not even be any wave function to assign to system 2. And so, we would recover Einstein's conclusion, i.e. a failure of a one-to-one correspondence between between physical states and wave functions, not in terms of Lehner's incompleteness (many-to-one) nor overcompleteness (one-to-many) but rather as a one-to-none correspondence. Be that as it may, it would of course still be failure of one-to-one correspondence.

However, this still leaves another vulnerability in Einstein's argument to be discussed. This vulnerability is that the argument will only convince those who share Einstein's premise that the wave function is to be interpreted as a direct description of the real state of a system. In the June 19 letter, he ascribes this view to Schrödinger, but Schrödinger sets him right in a response of August 19:

> I have long passed the stage when I thought that one could see the ψ-function as a direct description of reality.

Indeed, there are numerous passages in Schrödinger's post-1927 work that indicate he came to acknowledge that the wave function contained epistemic factors, pertaining to our knowledge about the system, as well as objective factors pertaining to its physical state. But perhaps worse than Schrödinger's rejection of this premise is that adherents of the Copenhagen interpretation, and in particular the 'talmudic philosopher' (Bohr), i.e. the main target of the argument, will also reject it.

In this regard, the EPR paper itself fares much better. The reason is that it does not attempt to establish a link between physical reality and a wave function, but only between elements of physical reality and measurement outcomes of physical quantities, at least when they are fully predictable. How does this change the argument? In the case of their special conclusion, one could argue as follows: assume the Separability Principle. But depending on what quantity we measure on system 1 (either P or Q), we can infer with certainty the values of the corresponding measurements on system 2. They should therefore represent elements of physical reality, indeed simultaneously so, because what is real on system 2 should, by Separability, not depend on what measurement we perform on system 1. But there is no quantum description of a state with definite real values for both position an momentum for system 2. Therefore the quantum description of physical reality is incomplete (in the sense of a one-to-none correspondence).

The reason why the EPR special conclusion is more robust against objections, compared to Einstein's June 19 argument, is precisely because it avoids the premise that wave functions directly describe physical reality. But this comes at a price: the EPR special conclusion employs the maximally entangled EPR state with it its multiple biorthogonal decompositions, i.e. the assumption that Einstein said he could not care less about. However, as I have argued in the previous section, this conclusion can be generalized to a case without such biorthogonal ecompositions.

25.5 What Did Schrödinger Make of the EPR Argument?

Schrödinger was in 1935 no stranger to the EPR argument—having worked out the basics of the example in 1931—but unlike EPR, or Einstein in his 1935 correspondence to Schrödinger, he never expressed his conclusions to the argument in terms of (in)completeness.

25.5.1 *Schrödinger's Interpretation of the Wave Function*

Einstein, in his June 19 letter ascribes the view that the wave function presents a complete description of physical reality to Schrödinger. This view is indeed one that Schrödinger used in his original works on wave mechanics in the first half of 1926 (Joas and Lehner 2009). However, he abandoned this view in several letters written in November 1926, precisely because of the phenomenon we would now call entanglement.

Indeed, in his (1935b), he writes that the wave function is only a "representative" of the state of the system, and that under the proviso that he is describing the current view (Lehrmeinung) of quantum mechanics. More specifically, he characterizes a wave function as a mere catalogue of expectation values, and compares it to a travel guide (Baedeker). All this underlines that Schrödinger did not commit to a view that the wave function was in one-to-one correspondence with the real state of a system.

25.5.2 *Schrödinger's Take on the EPR Problem*

Einstein persistently and consistently saw the EPR problem as an objection to the claim that the quantum description of physical systems by means of a wave function was complete, and also persistently argued that this problem would disappear if only one would take the view of the wave function as a statistical ensemble (thus providing an incomplete description of a physical system). He made this argument in another letter to Schrödinger of August 8:

> My solution to the paradox presented in our [EPR] paper is this: the ψ-function does not describe the state of a *single* system, but rather gives a (statistical) description of an ensemble of systems.

Schrödinger, by contrast, did not harbour any such hopes. For Schrödinger, the argument pointed to a deeper problem, that would not be resolved by merely adopting an ensemble interpretation. Thus, in response to Einstein on August 19, he points out that Einstein's solution would not work. He provides an argument, using the example of the operator $\frac{P^2}{a^2} + a^2 Q^2$ which had been presented to Schrödinger for a similar purpose by Pauli in a letter of July 9 (von Meyenn et al. 1985). His (1935b) contains a similar argument. Einstein's response (on September 4) to this argument is

> If I claimed that I have understood your last letter, that would be a fat lie (von Meyenn 2011, p. 569).

In reply, (October 4) Schrödinger apologizes and uses the example of the operator for angular momentum, also suggested to him by Pauli on July 9, and also discussed in (1935b). Both arguments are actually versions of Von Neumann's argument against a hidden variables interpretation.

But if Schrödinger did not believe in Einstein's favorite way to resolve the paradox, what alternative did he have? It seems to me that Schrödinger did not see a clear way out. For example in his letter to Sommerfeld (December 1931) Schrödinger concluded

> One can hardly help oneself in any other way than by the following 'emergency decree': in quantum mechanics it is forbidden to make statements about what is 'real', about the object, such statements deal only with the relation object–subject, and apparently in a much more restrictive way than in any other description of nature. (von Meyenn 2011, Vol. I, p. 490)

This suggests that Schrödinger saw the only rescue in a non-realist interpretation of quantum mechanics. In the *Übungen* notebook, his conclusion was a rejection of the Copenhagen interpretation;

> Therefore: This interpretational scheme does not work. One has to look for another one.

But in his June 7 letter to Einstein, he put blame on the non-relativist nature of QM:

> The lesson (Der Vers) that I have drawn for myself until now about this whole business is this. We do not possess a quantum mechanics compatible with relativity theory, that is, amongst others, a theory that takes account of the finite speed of propagation of all interactions. [...] When we want to separate two systems, their interaction will, long before it vanishes, fail to approximate the absolute Coulomb or similar force. And that is where our understanding ends. The process of separation cannot at all be grasped in the orthodox scheme.

Similarly, his final remarks in (1935b) point out the non-relativistic nature of QM, and the obstacles towards a fully relativistic theory, and conjecture that non-relativistic QM might merely be a convenient calculational artifice. For Schrödinger, the conclusions to be drawn form the EPR argument were therefore not so clear-cut. On some occasions he would be satisfied by just claiming that it showed an inadequacy of the Copenhagen interpretation of quantum mechanics, on other occasions he hoped that a future relativistic version of quantum theory would make the problem disappear, but without having concrete proposals about how this would work.

25.6 The Origins of the Cat Paradox

Schrödinger's long essay, *Die gegenwärtige Situation in der Quantenmechanik* appeared in *die Naturwissenschaften* in three installments in subsequent issues in November and December of 1935. His famous cat paradox made its entry in the second installment, published on November 29, 1935. There is a widespread view in the literature (Fine 1986, 2017; Moore 1992; Gribbin 2012; Gilder 2008; Kaiser 2016) that Schrödinger only conceived of his famous cat paradox in response to his correspondence with Einstein over the summer of 1935. Indeed, Einstein wrote to Schrödinger on August 8 about an example of an explosive device, containing an amount of gunpowder in a chemically unstable state. Its wave function would then develop in the course of time to contain parts that represent an exploded and un unexploded state. This example is indeed, in all intent and purposes, very similar to Schrödinger's cat. Given the lapse between August 8 and November 29, it might seem, on first sight, very well possible that Schrödinger adapted this example of Einstein into his own paper. For example, in his authoritative biography of Schrödinger, (Moore 1992, p. 305) comments that "Einstein's gunpowder would soon reappear in the form of Schrödinger's cat." Similarly, (Gribbin 2012, p. 179–180) writes: "The ideas encapsulated in the famous 'thought experiment' involving Schrödinger's cat actually came in no small measure from Einstein, in the extended correspondence between the two triggered by the EPR paper."

However, if we look a bit closer into the genesis of Schrödinger (1935b), this view becomes more problematic. Arnold Berliner, founder and editor of *die Naturwissenschaften* invited several authors in June 1935 to comment on the EPR paper, amongst them Pauli, Heisenberg, and Schrödinger. And although Pauli and Heisenberg did consider this invitation seriously, and Heisenberg indeed wrote a manuscript for this purpose, (Bacciagaluppi and Crull 2020), only Schrödinger submitted a contribution. Berliner wrote to Schrödinger on July 1, to express his joy that Schrödinger had consented to contribute to a discussion about EPR in *die Naturwissenschaften*. He addresed the question about whether "the name" could be mentioned in Schrödinger's contribution.[10] Berliner emphasizes that Schrödinger should feel free to mention the *Innominato* in a footnote or otherwise would be

[10]As Von Meyenn (p. 548) notes, this issue concerns the fact that under the Nazi regime, the mention of Einstein's name in German scientific literature had been forbidden.

entirely up to his discretion. In a letter of July 13 to Einstein, Schrödinger mentions that he had begun writing such a paper:

> The great difficulty for me to reach even only a mutual understanding with the orthodox people, has led me to make an attempt to analyse the present interpretational situation *ab ovo* in a longish essay. Whether and what I will publish of this, I do not know.

The letter continues by describing some of the themes that did indeed end up in the pubished (Schrödinger 1935b).

On July 25, Schrödinger wrote to Berliner that his manuscript was finished in handwriting, and that a citation of the work from the Physical Review would only appear after three quarters of his manuscript, at the beginning of the final quarter. He mentions that it would take him some more time to produce a type-written version. He writes to Berliner again on August 11, saying he did not reply earlier:

> because the manuscript, which you so kindly agreed to publish, was doubly burning under my fingers. I have now put it all together, and will send it off tomorrow together with this letter.

Apparently, his paper, eventually published as (1935b) in November and December, was finished in typewriting on August 11, and sent off the next day.

Now, it is obvious that Einstein's August 8 letter with the gun-powder example, written from Old Lyme (Connetticut), could not have crossed the ocean to reach Schrödinger, in Oxford, before he sent off his manuscript on August 12. Indeed, in his reply (August 19) to Einstein's August 8 letter, Schrödinger states that in a long essay he had just sent off a week ago he had made use of an example, very similar to Einstein's gun powder example and describes his cat example to Einstein. Thus, the only sensible conclusion at this point can be that Einstein and Schrödinger conceived of their strikingly similar examples independently, around the same time, cf. Ryckman (2017).

Still, the most elaborate study and commentary of the Schrödinger-Einstein correspondence in 1935 to date is (Fine 1986) which includes a chapter called "Schrödinger's Cat and Einstein's: the genesis of a paradox", in which he discussed the question how Schrödinger might have conceived of his famous cat paradox. This work is presumably the origin for the wide-spread conception that that the Cat Paradox borrowed and built upon Einstein's exploding gunpowder example of August 8.

Indeed, Fine claims that

> The cat was put into (Schrödinger 1935b) in response to ideas and distinctions generated during the correspondence [with Einstein] over EPR (p. 73).

But Fine's argument differs in detail from Moore and Gribbin. He suggests that section 4 and also section 5 (containing the Cat) of (1935b) were *not* part of the paper sent to Berliner on August 12, but later additions:

> The conclusion seems inescapable that section 4 was put together only after the August 19 letter [by Schrödinger], and was not part of the original manuscript [Schrödinger] sent to Berliner. (p. 81)

> But if section 4 was only written after August 19, then most likely section 5 is a later addition as well. (p. 82)

There are several problems with this hypothesis. I will deal with its first claim (i.e. that section 4 of Schrödinger's (1935b) could only have been written after Schrödinger's August 19 letter) below. Let me first mention problems with the conjecture that section 5, containing the Cat Paradox, was composed later as well. The first problem is that it would imply that Schrödinger is literally lying to Einstein when he wrote on August 19, in response to Einstein's exploding gunpowder example, that he had already presented a quite similar example in the paper he just sent off a week earlier, and described the cat example. This is all the more improbable given that Schrödinger seems to be generally generous in issues of priority, and did not even emphasize in his June 7 letter that the EPR paper "about that which we discussed so much in Berlin" is in fact indebted to his own formulation of the EPR problem in 1931.

Of course, Fine's hypothesis could be tested, e.g. if a draft version of the Naturwissenschaften paper surfaces. I could not find any in the Schrödinger Nachlass, (but who knows what might emerge from archives of *Die Naturwissenschaften*, if they survived the war). However, one can rely on what is known from Schrödinger's correspondence in this period with Berliner.

When Schrödinger wrote to Berliner on July 25 that he had finished a handwritten draft of the paper, and would still need time to type it. He reported a word count of 10230 words, and estimated that this would fill 13.3 to 14 pages of *die Naturwissenschaften* not counting the paragraph headings. (von Meyenn 2011, p. 555). He suggested to that this "extraordinary length" probably required it to be cut into two or three installments, (but advising against the latter option).

On July 29, Berliner replied that he was looking forward to the submission, and that he would happily allot all the space Schrödinger needed, and promised that it would appear in two subsequent issues of the journal.

On August 11, Schrödinger wrote to Berliner that his typescript was finished and to be mailed the next day. On August 14, Berliner responded that he had suddenly been removed from his position as editor of *die Naturwissenschaften*, but implored Schrödinger that his paper should still be published in that journal, and that he had proposed Walther Matthée as his successor.[11]

On August 19, Schrodinger wrote Einstein, as mentioned before, and told him also that his first impulse was to withdraw the paper, but didn't because Berliner himself had insisted on its publication.

On August 24, he received a letter from Mathée, telling Schrödinger he was the new editor of *Die Naturwissenschaften*, thanking him for the manuscript, and mentions including the proofs of the paper. Mathée informed him that the

[11]Berliner was removed from the journal because he was Jewish (von Meyenn 2011, Vol. 2, pp. 546–7).

manuscript was to be split into three parts, to appear in consecutive issues, and asked Schrödinger to indicate in these proofs where the cuts were to be made.[12]

Schrödinger must have replied to Mathée soon afterwards, and there is a draft of this letter in the selection by Von Meyenn, but erroneously appended to his letter to Berliner from July 25. (von Meyenn 2011, Vol. 2, p. 557). Here, Schrödinger mentions that he had just sent a telegram with the message "Sorry, tripartion impossible" and maintains that the manuscript should only be split into two parts, or otherwise returned to him. He mentions that this division could well take place after section 9, which is "very exactly half of the manuscript". He ends, however, on a more reconciliatory note, saying that he trusts Mathée's judgment, who should not feel bound by an earlier promise, and that Schrödinger would feel the same way if Berliner had told him that the promise of a bipartition was not possible. There is no mention at all of a revised version with later additions to the paper at this stage, and clearly there would have been very little time to do so, if this revision is supposed to have occurred after August 19.

In fact, the paper was published in three installments in Schrödinger (1935b). In total, they comprise about 12270 words (not counting footnotes, references, figure captions, tables of contents and section titles)[13] and span about 15.5 pages. This implies that Schrödinger did expand his manuscript by about 2000 extra words, or about 2 pages, after July 25. This could actually fit Fine's hypothesis that its sections 4 and 5 (which together comprise just less than 2 pages, and about 1600 words) could be later additions — but not necessarily additions from after August 19.

The second problem for Fine's hypothesis concerns Schrödinger's estimate on July 25 that the first mention of Einstein's name would occur after three quarters of the paper, as well as his estimate, presumably a few days after August 24, that the end of section 9 would " very exactly" divide the manuscript into two equal halves. Both these estimates are actually quite accurate also for the published version. This suggests that, more likely, he made additions while typing his draft of July 25, more or less equally spaced throughout the manuscript. In any case, if Schrödinger added two whole new sections before section 9 in Fine's hypothetical revision one must surmise him to have also made additions of the same size to the second half of his manuscript as well in order to keep the end of section 9 at exactly half of the paper. But that clearly did not happen, because that would take the size of the manuscript beyond what was actually published.

Given this evidence, it seems quite unlikely that Schrödinger only thought of the cat paradox after he read Einstein's exploding device example. The more simple and straightforward explanation, as stated earlier, is that these two authors, independently, thought of very similar examples at roughly the same time.

Next, let me discuss Fine's argument for the "inescapable" claim that section 4 of Schrödinger (1935b) is a later addition to the manuscript Schrödinger sent to

[12]This letter is not included in the selection by Von Meyenn.

[13]I am indebted to Guido Bacciagaluppi for this careful word count.

Berliner on August 12. This claim is based on the fact that Einstein writes in a letter to Schrödinger on August 9, that he believes a mere ensemble interpretation of the wave function (which, of course would concede that the description by a wave function is incomplete (in Einstein's sense as well as the EPR sense of the term) would suffice to solve the paradox. Schrödinger responds on August 19 that this argument will not work. He proceeds to provide a counter-argument, using the Hamiltonian of the harmonic oscillator as an example. Section 4 of Schrödinger (1935b) contains a very similar argument.

Fine assumes that Schrödinger only could have developed this argument in *response* to Einstein's letter, and thus after August 19. However, it seems to me that this assumption overlooks Schrödinger's wider correspondence in this period. The very idea that an interpretation of the wave function as incomplete, i.e., as merely describing a statistical inhomogeneous ensemble is irreconcilable with the quantum predictions was mentioned to Schrödinger by Pauli in a letter of July 9 1935, specifically pointing out that the discrete spectrum that quantum mechanics requires for operators like the Hamiltonian $P^2 + \omega^2 Q^2$ for a harmonic oscillator or the angular momentum (in a 3-dimensional case) $P_x Q_y - P_y Q_x$ would be incompatible with this view. Schrödinger apparently took Pauli's remarks to heart. Both the harmonic oscillator Hamiltonian and the angular momentum examples are discussed in Schrödinger (1935b), with his own twist, adding an arbitrary multiplicative constant to the last term of the harmonic oscillator Hamiltonian. Moreover, he presents this argument in a version adapted to the EPR case in a letter to von Laue (July 25 1935). Again, I believe there seems little ground for the hypothesis that Schrödinger could only have composed this argument after he responded to Einstein on August 19, and even less for the claim that, therefore, section 4 of (1935b) must have been a later addition to the manuscript he submitted on August 12 of that year.

Indeed, it seems to me that Fine's analysis of Schrödinger's *Naturwissenschaften* paper, vis-a-vis his correspondence with Einstein in 1935, paints him as a passive receiver of Einstein's ideas, and as composing his own (Schrödinger 1935b) paper merely in response to those ideas by Einstein. I believe that we should redress this picture. Schrödinger knew about the essential aspects of the EPR argument $3\frac{1}{2}$ years before it was published, and might well have influenced Einstein in his transition from the photon box argument to the clearer and simpler version that is EPR. There was no need for Schrödinger to learn about EPR from Einstein. Instead, as I have argued, he had a lesson to teach Einstein, although he failed to get across.

Acknowledgements I am indebted to The Max Planck Institute for History of Science in Berlin for enabling this research, as well as the Vossius Institute in Amsterdam and the Descartes Centre in Utrecht for providing the opportunity to finish it. I also thank Christoph Lehner for many valuable discussions, and deciphering Schrödinger's handwriting, and Guido Bacciagaluppi for many helpful and insightful comments. I am also grateful to Alexander Zartl for assistance in browsing through the Schrödinger Nachlass in the Zentralbibliothek für Physik at the University of Vienna.

References

Bacciagaluppi, G., & Crull E. (2020). *The Einstein paradox: The debate on nonlocality and incompleteness in 1935*. Cambridge: Cambridge University Press.

Einstein, A., Podolsky, B., & Rosen, N. (1935). Can quantum mechanical description of reality be considered complete? *Physical Review, 47*, 777–780.

Fine, A. (1986). *The Shaky game: Einstein, realism and the quantum theory*. Chicago: University of Chicago Press.

Fine, A. (2017). The Einstein-Podolsky-Rosen argument in quantum theory. In E. N. Zalta (Ed) *The Stanford encyclopedia of philosophy* (Winter 2017 Edition). https://plato.stanford.edu/archives/win2017/entries/qt-epr/.

Gilder, L. (2008). *The age of entanglement*. New York: Knopf.

Gribbin, J. (2012). *Erwin Schrödinger and the quantum revolution*. London: Bantam Press.

Howard, D. (1985). Einstein on locality and separability. *Studies in History and Philosophy of Science, 16*, 171–201.

Howard, D. (1990). Nicht sein kann was nicht sein darf, or the Prehistory of EPR, 1909–1935: Einstein's early worries about the Quantum Mechanics of Composite Systems. In A. I. Miller (Ed.) *Sixty-two years of uncertainty: Historical, philosophical and physical inquiries in the foundations of quantum mechanics* (NATO ASI series, Vol. 226, pp. 61–111). New York: Plenum Press.

Joas, C. & Lehner, C. (2009). The classical roots of wave mechanics: Schrödinger's transformations of the optical-mechanical analogy. *Studies in the History and Philosophy of Modern Physics, 40*, 338–351.

Kaiser, D. (2016). How Einstein and Schrödinger conspired to kill a cat. *Nautilus*, Issue 041.

Lehner, C. (2014). Einstein's realism and his critique of quantum mechanics. In M. Janssen & C. Lehner (Eds) *The Cambridge companion to Einstein* (pp. 306–353). Cambridge: Cambridge University Press.

Moore, W. (1992). *Schrödinger, life and thought*. Cambridge: Cambridge University Press.

Ryckman, T. (2017). *Einstein*. Milton Park: Routledge.

Schrödinger, E. (1935a). Discussion of probability relations between separated systems. *Mathematical Proceedings of the Cambridge Philosophical Society, 31*, 555–563.

Schrödinger, E. (1935b). Die gegenwärtige Situation in der Quantenmechanik. *Die Naturwissenschaften, 23*, 807–812, 823–828, 844–848.

von Meyenn, K. (2011). *Eine entdeckung von ganz außerordentliche Tragweite; Schrödinger's Briefwechsel zur Wellenmechanik und zum Katzenparadoxon* (Vols. 1 and 2). Heidelberg: Springer.

von Meyenn, K., Hermann, A., & Weiskopf, V. F. (1985). *Wolfgang Pauli: Wissenschaftlicher Briefwechsel mit Bohr, Einstein, Heisenberg u.a. Vol II: 1930–1939*. Heidelberg: Springer.

Chapter 26
Derivations of the Born Rule

Lev Vaidman

Abstract The Born rule, a cornerstone of quantum theory usually taken as a postulate, continues to attract numerous attempts for its derivation. A critical review of these derivations, from early attempts to very recent results, is presented. It is argued that the Born rule cannot be derived from the other postulates of quantum theory without some additional assumptions.

Keywords Quantum probability · Born rule · Many-worlds interpretation · Relativistic causality

26.1 Introduction

My attempt to derive the Born rule appeared in the first memorial book for Itamar Pitowsky (Vaidman 2012). I can only guess Itamar's view on my derivation from reading his papers (Pitowsky 1989, 2003, 2006; Hemmo and Pitowsky 2007). It seems that we agree which quantum features are important, although our conclusions might be different. In this paper I provide an overview of various derivations of the Born rule. In numerous papers on the subject I find in depth analyses of particular approaches and here I try to consider a wider context that should clarify the status of the derivation of the Born rule in quantum theory. I hope that it will trigger more general analyses which finally will lead to a consensus needed for putting foundations of quantum theory on solid ground.

The Born rule was born at the birth of quantum mechanics. It plays a crucial role in explaining experimental results which could not be explained by classical physics. The Born rule is known as a statement about the probability of an outcome of a quantum measurement. This is an operational meaning which corresponds to

L. Vaidman (✉)
Raymond and Beverly Sackler School of Physics and Astronomy, Tel-Aviv University, Tel-Aviv, Israel
e-mail: vaidman@post.tau.ac.il

M. Hemmo, O. Shenker (eds.), *Quantum, Probability, Logic*, Jerusalem Studies in Philosophy and History of Science,
https://doi.org/10.1007/978-3-030-34316-3_26

numerous very different statements about ontology in various interpretations of quantum theory. Von Neumann's description of quantum measurement includes, at some stage, a collapse of the quantum state corresponding to a particular result of the measurement and the Born rule provides the probability of getting this result. In this framework there is no definition of when exactly it happens (where is the quantum-classical boundary?), so the Copenhagen interpretation and its recent development in the form of QBism (Caves et al. 2002) do not specify the ontology, leaving the definition of the principle on the operational level. In the framework of the Bohmian interpretation (Bohm 1952), which is a deterministic theory given the initial Bohmian positions of all particles, the Born rule is a postulate about the probability distribution which governs these random initial Bohmian positions. It is a postulate about the genuinely random stochastic collapse process in the framework of physical collapse theories (Ghirardi et al. 1986). In Aharonov's solution to the measurement problem (Aharonov et al. 2014), it is a postulate about the particular form of the backward evolving wavefunciton. In the framework of the many-worlds interpretation (MWI) (Everett III 1957), it is a postulate about experiences of an observer in a particular world (Vaidman 2002). So, in all interpretations, the Born rule is postulated, but the question of the possibility of its derivation is considered to be of interest, and it is still open (Landsman 2009).

A rarely emphasized important fact about the Born rule is that it might be even more relevant for explaining physics phenomena in which the probability is not explicitly manifested. Quantum statistical mechanics which leads to quantum thermodynamics heavily uses the Born rule for explaining everyday observed phenomena. There is nothing random when we observe a blue sky or reddish sun at sunset. The explanation includes scattered photons of various colors absorbed by cones in the eye with their color depended efficiency of the absorption. The ratio of the number of events of photon absorption in different cones corresponds to different experiences of the color of the sky and the sun, and we have to use the Born rule to explain our visual experience (Li et al. 2014). In this explanation we consider the cone photoreceptor in an eye as a single-photon detector and light from the sun scattered by molecules of air as a collection of photons. The quantum nature of light coming from modern artificial light sources is even more obvious. An observer looks on a short flash of a fluorescent soft white bulb and announces the color she has seen. The spectrum of the light from this bulb consists of red, green and blue photons, but nobody would say that she saw red light or that she saw green light from the fluorescent bulb. The Born Rule is needed to calculate the ratio of the signals from the cones. The large number of events of photon detection by cones explains why nobody would say they saw a color different from white, since the Born rule provides an almost vanishing probability for such event. The Born rule also tells us that there is an astronomically small probability that we will see a red sky and a blue sun, but it is not different from other quantum tiny tails which we neglect when we explain the classical world we observe through underlying quantum reality.

26.2 Frequentist Approach

One of the early approaches relied on consideration of infinite series of repeated measurements. In the frequentist approach to probability, we consider the ratio of particular outcomes to the total number of measurements. The probability acquires its meaning only in the infinite limit. The important milestones were the works of Hartle (1968) and Farhi et al. (1989). Then the program was extended by replacing infinite tensor products of Hilbert spaces by continuous fields of C*-algebras (Van Wesep 2006; Landsman 2008). The core feature of these arguments involves taking a limit of an infinite number of quantum systems. Aharonov and his collaborators (Aharonov and Reznik 2002; Aharonov et al. 2017) presented, in my view, the simplest and the most elegant argument based on this type of infinite limit.

Consider a large number N of identical systems all prepared in the same state

$$|\Psi\rangle = \sum_i \alpha_i |a_i\rangle, \tag{26.1}$$

which is a superposition of nondegenerate eigenstates of a variable A. Consider the "average" variable $\bar{A} \equiv \frac{1}{N}\sum_{n=1}^{N} A_n$. Applying the universal formula (Aharonov and Vaidman 1990)

$$A|\Psi\rangle = \langle\Psi|A|\Phi\rangle\,|\Psi\rangle + \Delta A|\Psi_\perp\rangle, \tag{26.2}$$

where $\langle\Psi|\Psi_\perp\rangle = 0$, we obtain

$$\bar{A}\prod_{n=1}^{N}|\Psi\rangle_n = \langle\Psi|A|\Psi\rangle\prod_{n=1}^{N}|\Psi\rangle_n + \frac{\Delta A}{N}\sum_{k=1}^{N}\prod_{n\neq k}|\Psi\rangle_n|\Psi_\perp\rangle_k. \tag{26.3}$$

The amplitude of the first term in the right hand side of the equation is of order 1 while the amplitude of the second term (the sum) is proportional to $\frac{1}{\sqrt{N}}$, so in the limit as N tends to infinity, the second term can be neglected and the product state $\prod_{n=1}^{N}|\Psi\rangle_n$ can be considered an eigenstate of the variable $\bar{A}$ with eigenvalue $\langle\Psi|A|\Psi\rangle$.

Now consider the measurement of $\bar{A}$ followed by measurements of A of each of the individual systems. N_i is the number of outcomes $A = a_i$. The probability of outcome a_i is defined as the limit $p_i \equiv \lim_{N\to\infty}\frac{N_i}{N}$. To derive the Born rule we consider the shift of the pointer of the measuring device measuring $\bar{A}$, in two ways. First, since in the limit, the state is an eigenstate with eigenvalue $\langle\Psi|A|\Psi\rangle$, the pointer is shifted by this value. Second, consider the evolution backward in time given that we have the results of individual measurements of variable A of each system. Then the shift has to be $\sum_{i=1}^{N}\frac{a_i N_i}{N}$. In the limit we obtain $\langle\Psi|A|\Psi\rangle = \sum_i |\alpha_i|^2 a_i = \sum_{i=1}^{N} a_i p_i$. This equation can be generally true only if $p_i = |\alpha_i|^2$ for all eigenvalues a_i. This proves the Born rule.

The legitimacy of going to the limit $N \to \infty$ in the earlier proofs was questioned in Squires (1990), Buniy et al. (2006) and Aharonov's approach was analyzed in Finkelstein (2003). I am also skeptical about the possibility of arguments relying on the existence of infinities to shed light on Nature. Surely, the infinitesimal analysis is very helpful, but infinities lead to numerous very peculiar sophisticated features which we do not observe. I see no need for infinities to explain our experience. Very large numbers can mimic everything and are infinitely simpler than infinity. The human eye cannot distinguish between 24 pictures per second and continuous motion, but infinite information is required to describe the latter. There is no need for infinities to explain all what we see around.

Another reason for my skepticism about possibility to understand Nature by neglecting vanishing terms in the infinite limit is the following example in which these terms are crucial for providing common sense explanation. In the modification of the interaction-free measurement (Elitzur and Vaidman 1993) based on the Zeno effect (Kwiat et al. 1995) we get information about the presence of an object without being near it. The probability of success can be made as close to 1 as we wish by changing the parameters. Together with this, there is an increasing number of moments of time at which the particle can be absorbed by the object with decreasing probability of each absorption. In the limit, the sum of the probabilities of absorption at all different moments goes to zero, but without these cases the success of the interaction-free measurement seems to contradict common sense. These are the cases in which there is an interaction. Taking the limit in proving the Born rule is analogous to neglecting these cases.

The main reason why I think that this approach cannot be the solution is that I do not see what is the additional assumption from which we derive the Born rule. Consider a counter example. Instead of the Born rule, the Nature has "Equal rule". Every outcome a_i of a measurement of A has the same probability given that it is possible, i.e., $\alpha_i \neq 0$. Of course, this model contradicts experimental results, but it does not contradict the unitary evolution part of the formalism of quantum mechanics. I do not see how making the number of experiments infinite can rule out Equal rule. Note that an additional assumption ruling out this model is hinted in Aharonov and Reznik (2002) "the results of physical experiments are stable against small perturbations". A very small change of the amplitude can make a finite change in the probability in the proposed model. (This type of continuity assumption is present in some other approaches too.)

26.3 The Born Rule and the Measuring Procedure

The Born rule is intimately connected to the measurement problem of quantum mechanics. Today there is no consensus about its solution. The Schrödinger equation cannot explain definite outcomes of quantum measurements. So, if it does not explain the existence of a (unique) outcome, how can it explain its probability? It is the collapse process (which is not explained by the Schrödinger equation) that provides the unique outcome, so it seems hopeless to look for an explanation of the Born rule based on the Schrödinger equation.

What might be possible are consistency arguments. If we accept the Hibert space structure of quantum mechanics and we accept that there is probability for an outcome of a quantum measurement, what might this probability measure be? Itamar Pitowsky suggested to take it as the basis and showed how Gleason's theorem (Gleason 1957) (which has its own assumptions) leads to the Born rule (Pitowsky 1998). He was aware of "two conceptual assumptions, or perhaps dogmas. The first is J. S. Bell's dictum that the concept of measurement should not be taken as fundamental, but should rather be defined in terms of more basic processes. The second assumption is that the quantum state is a real physical entity, and that denying its reality turns quantum theory into a mere instrument for predictions" (Bub and Pitowsky 2010). In what followed, he recognized the problem as I do: "This last assumption runs very quickly into the measurement problem. Hence, one is forced either to adopt an essentially non-relativistic alternative to quantum mechanics (e.g. Bohm without collapse, GRW with it); or to adopt the baroque many worlds interpretation which has no collapse and assumes that all measurement outcomes are realized." The difference between us is that he viewed "the baroque many worlds interpretation" as unacceptable (Hemmo and Pitowsky 2007), while I learned to live with it (Vaidman 2002).

Maybe more importantly, we disagree about the first dogma. I am not ready to accept "measurement" as a primitive. Physics has to explain all our experiences, from observing results of quantum measurements to observing the color of the sun and the sky at sunset. I do believe in the ontology of the wave function (Vaidman 2016, 2019) and I am looking for a direct correspondence between the wave function and my experience considering quantum observables only as tools for helping to find this correspondence. I avoid attaching ontological meaning to the values of these observables. It does not mean that I cannot discuss the Born rule. The measurement situation is a well defined procedure and our experiences of this procedure (results of measurements) have to be explained.

The basic requirement of the measurement procedure is that if the initial state is an eigenstate of the measured variable, it should provide the corresponding eigenvalue with certainty. Any procedure fulfilling this property is a legitimate measuring procedure. The Born rule states that the probability it provides should be correct for any legitimate procedure, and this is a part of what has to be proved, but let us assume that the fact that all legitimate procedures provide the same probabilities is given. I will construct then a particular measurement procedure (which fulfills the property of probability 1 for eigenstates) which will allow me to explain the probability formula of the Born rule.

Consider a measurement of variable A performed on a system prepared in the state (26.1). The measurement procedure has to include coupling to the measuring device and the amplification part in which the result is written in numerous quantum systems providing a robust record. Until this happens, there is no point in discussing the probability, since outcomes were not created yet. So the measurement process is:

$$|\Psi\rangle \prod_m |r\rangle_m^{MD} \rightarrow \sum_i \alpha_i |a_i\rangle \prod_{m\in S_i} |\checkmark\rangle_m^{MD} \prod_{m\notin S_i} |r\rangle_m^{MD}, \qquad {}_m^{MD}\langle r|\checkmark\rangle_m^{MD} = 0 \;\forall m \tag{26.4}$$

where "ready" states $|r\rangle_m^{MD}$ of the numerous parts m of the measuring device, $m \in S_i$, are changed to macroscopically different (and thus orthogonal) states $|\checkmark\rangle_m^{MD}$, in correspondence with the eigenvalue a_i. For all possible outcomes a_i, the set S_i of subsystems of the measuring device which change their states to orthogonal states has to be large enough. This part of the process takes place according to all interpretations. In collapse interpretations, at some stage, the state collapses to one term in the sum.

This schematic description is not too far from reality. Such a situation appears in a Stern-Gerlach experiment in which the atom, the spin component of which is measured, leaves a mark on a screen by exciting numerous atoms of the screen. But I want to consider a modified measurement procedure. Instead of a screen in which the hitting atom excites many atoms, we put arrays of single-atom detectors in the places corresponding to particular outcomes. The arrays cover the areas of the quantum uncertainty of the hitting atom. The arrays are different, as they have a different number N_i of single-atom detectors which we arrange according to equation $N_i = |\alpha_i|^2 N$. (We assume that we know the initial state of the system.)

In the first stage of the modified measuring procedure an entangled state of the atom and sensors of the single-atom detectors is created:

$$|\Psi\rangle \prod_n |r\rangle_n^{sen} \rightarrow \sum_i \frac{\alpha_i}{\sqrt{N_i}} |a_i\rangle \sum_{k_i} |\checkmark\rangle_{k_i}^{sen} \prod_{n\neq k_i} |r\rangle_n^{sen}, \tag{26.5}$$

where $|r\rangle_n^{sen}$ represents an unexcited state of the sensor with label n running over sensors of all arrays of single-photon detectors. N is the total number of detectors. For each eigenvalue a_i there is one array of N_i detectors with sensors in a superposition of entangled states in which one of the sensors k_i has an excited state $|\checkmark\rangle_{k_i}^{sen}$. At this stage the measurement has not yet taken place. The number of sensors with changed quantum state might be large, but no "branches" with many systems in excited states has been created. We need also the amplification process which consists of excitation of a large number of subsystems of individual detectors. In the modified measurement, instead of a multiple recording of an event specified by the detection of a_i, we record activation of every sensor k_i by excitation of a large (not necessarily the same) number of quantum subsystems m belonging to the set S_{i_k}. Including in our description these subsystems, the description of the measurement process is:

$$|\Psi\rangle \prod_n |r\rangle_n^{sen} \prod_m |r\rangle_m^{MD} \rightarrow \frac{1}{\sqrt{N}} \sum_i |a_i\rangle \sum_{k_i} |\checkmark\rangle_{k_i}^{sen} \prod_{n\neq k_i} |r\rangle_n^{sen} \prod_{m\in S_{i_k}} |\checkmark\rangle_m^{MD} \prod_{m\notin S_{i_k}} |r\rangle_m^{MD}. \tag{26.6}$$

Here we also redefined the states $|\checkmark\rangle_{k_i}^{sen}$ to absorb the phase of α_i to see explicitly that all terms in the superposition have the same amplitude. Every term in the superposition has macroscopic number of subsystems of detectors with states which are orthogonal to the states appearing in other terms. This makes all the terms separate. We have N different options. They consist of sets according to all different possible eigenvalues when the set corresponding to eigenvalue a_i has N_i elements. Assuming that all options are equiprobable, we obtain the Born rule. The probability of a reading corresponding to eigenvalue a_i is $p_i = \frac{N_i}{N} = |\alpha_i|^2$. And this procedure is a good measurement according to our basic requirement: If the initial state is an eigenstate, we will know it with certainty.

In Fig. 26.1, we demonstrate such a situation for a modified Stern-Gerlach experiment with the initial state $|\Psi\rangle = \sqrt{0}.4|\uparrow\rangle + \sqrt{0}.6|\downarrow\rangle$. There are $N = 5$ single-photon detectors. The description of the measurement process (represented for a general case by [26.6]) is now:

$$\left(\sqrt{0}.4|\uparrow\rangle + \sqrt{0}.6|\downarrow\rangle\right)\prod_{n=1}^{5}|r\rangle_n^{sen}\prod_m|r\rangle_m^{MD} \rightarrow$$

$$\frac{1}{\sqrt{5}}\left(|\uparrow\rangle|\checkmark\rangle_1^{sen}\prod_1 + |\uparrow\rangle|\checkmark\rangle_2^{sen}\prod_2 + |\downarrow\rangle|\checkmark\rangle_3^{sen}\prod_3 + |\downarrow\rangle|\checkmark\rangle_4^{sen}\prod_4 + |\downarrow\rangle|\checkmark\rangle_5^{sen}\prod_5\right), \tag{26.7}$$

where

$$\prod_i \equiv \prod_{n\neq i}|r\rangle_n^{sen}\prod_{m\in S_i}|\checkmark\rangle_m^{MD}\prod_{m\notin S_i}|r\rangle_m^{MD}.$$

We obtain a superposition of five equal-amplitude states, each corresponding to one detector clicks and others are not. It is natural to accept equal probabilities for clicks of all these detectors and since there are two detectors corresponding to the outcome 'up' and three detectors corresponding to the outcome 'down' we obtain the Born rule probabilities for our example.

An immediate question is: how can I claim to derive $p_i = |\alpha_i|^2$ when in my procedure I put in by hand $\frac{N_i}{N} = |\alpha_i|^2$? The answer is that making another choice would not lead to a superposition of orthogonal terms with equal amplitudes, so with another choice the derivation does not go through.

This derivation makes a strong assumption that in the experiment, the firing of each sensor has the same probability. It is arranged that all these events correspond to terms in the superposition with the same amplitude, so the assumption is that equal amplitudes correspond to equal probabilities. It is this fact that is considered to be the main part in the derivation of the Born rule. I doubt that the formalism of quantum mechanics by itself is enough to provide a proof for this statement, see also Barrett (2017). In the next section I will try to identify the assumptions added in various proofs of the Born rule.

Without the proof of the connection between amplitudes and probabilities, the analysis of the experiment I presented above is more of an explanation of the Born rule than its derivation. We also use an assumption that all valid measurement

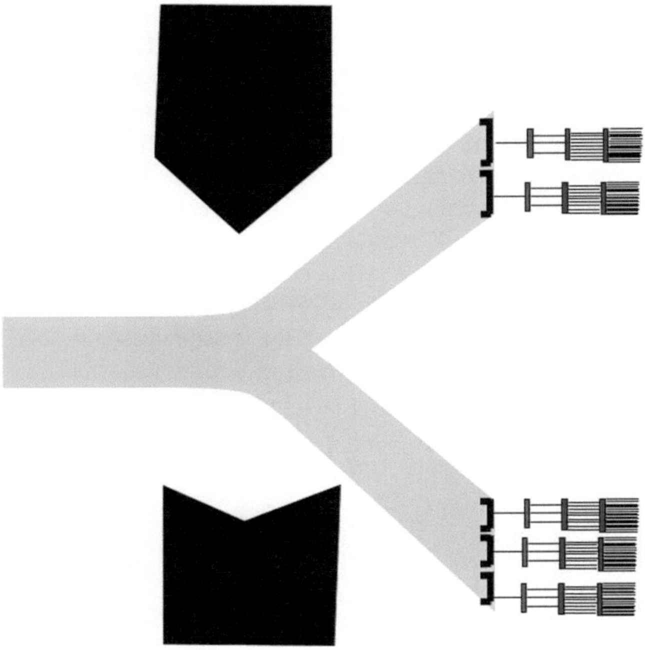

Fig. 26.1 Modified Stern-Gerlach experiment specially tailored for the state $\sqrt{0.4}|\uparrow\rangle + \sqrt{0.6}|\downarrow\rangle$. There are two detectors in the location corresponding to the result 'up' and three detectors in the location corresponding to the result 'down'

experimental procedures provide the same probabilities for outcomes. The modified procedure has a very natural combinatorial counting meaning of probability. It can be applied to the collapse interpretations when we count possible outcomes and in the MWI where we count worlds. The objection that the number of worlds is not a well defined concept (Wallace 2010) is answered when we put weights (measures of existence) on the worlds (Vaidman 1998; Greaves 2004; Groisman et al. 2013).

26.4 Symmetry Arguments

In various derivations of the Born rule, the statement that equal amplitudes lead to equal probabilities relies on symmetry arguments. The starting point is the simplest (sometimes named pivotal) case

$$|\Psi\rangle = \frac{1}{\sqrt{2}}(|a_1\rangle + |a_2\rangle). \tag{26.8}$$

The pioneer in attempting to solve this problem was Deutsch (1999) whose work was followed by extensive development by Wallace (2007), a derivation by Zurek

(2005), and some other attempts such as Sebens and Carroll (2016) and also my contribution with McQueen (Vaidman 2012; McQueen and Vaidman 2018). The key element of these derivations is the symmetry under permutation between $|a_1\rangle$ and $|a_2\rangle$. It is a very controversial topic, with numerous accusations of circularity for some of the proofs (Hemmo and Pitowsky 2007; Barnum et al. 2000; Saunders 2004; Gill 2005; Schlosshauer and Fine 2005; Lewis 2010; Rae 2009; Dawid and Thébault 2014).

In all these approaches there is a tacit assumption (which I also used above) that the probability of an outcome of a measurement of A does not depend on the procedure we use to perform this measurement. Another assumption is that probability depends only on the quantum state. In the Deutsch-Wallace approach some manipulations, swapping, and erasures are performed to eliminate the difference between $|a_1\rangle$ and $|a_2\rangle$, leading to probability half due to symmetry. If the eigenstates do not have internal structure except for being orthogonal states, then symmetry can be established, but it seems to me that these manipulations do not provide the proof for important realistic cases in which the states are different in many respects. It seems that what we need is a proof that all properties, except for amplitudes, are irrelevant. I am not optimistic about the existence of such a proof without adding some assumptions. Indeed, what might rule out the "Equal rule", a naive rule according to which probabilities for all outcomes corresponding to nonzero amplitudes are equal, introduced above?

The assumption of continuity of probabilities as functions of time rules Equal rule out, but this is an additional assumption. The Deutsch-Wallace proof is in the framework of the MWI, i.e. that the physical theory is just unitary evolution which is, of course, continuous, but it is about amplitudes as functions of time. The experience, including the probability of self-location of an observer, supervenes on the quantum state specified by these amplitudes, but the continuity of this supervenience rule is not granted.

Zurek made a new twist in the derivation of the Born rule (Zurek 2005). His key idea is to consider entangled systems and rely on "envariance" symmetry. A unitary evolution of a system which can be undone by the unitary evolution of the system it is entangled with. For the pivotal case, the state is

$$|\Psi\rangle = \frac{1}{\sqrt{2}}(|a_1\rangle|1\rangle + |a_2\rangle|2\rangle), \tag{26.9}$$

where $|1\rangle$, $|2\rangle$ are orthogonal states of the environment. The unitary swap $|a_1\rangle \leftrightarrow |a_2\rangle$ followed by the unitary swap of the entangled system $|1\rangle \leftrightarrow |2\rangle$ brings us back to the original state which, by assumption, corresponds to the original probabilities. Other Zurek's assumptions are that a manipulation of the second system does not change the probability of the measurement on the system, while the swap of the states of the systems swaps the probabilities for the two outcomes. This proves that the probabilities for the outcomes in the pivotal example must be equal.

In my view, the weak point is the claim that swapping the states of the system swaps the probabilities of the outcomes. This property follows from the quantum formalism when the initial state is an eigenstate, but in our case, when the mechanism for the choice of the outcome is unknown, we also do not know how it is affected by unitary operations. Note, that it is not true for the Stern-Gerlach experiment in the framework of Bohmian mechanics.

Zurek, see also Wallace (2010), Baker (2007), and Boge (2019), emphasises the importance of decoherence: entanglement of environment with eigenstates of the system. Indeed, decoherence is almost always present in quantum measurements and its presence might speed up the moment we can declare that the measurement has been completed, but, as far as I understand, decoherence is neither necessary, nor sufficient for completing a quantum measurement. It is not necessary, because it is not practically possible to perform an interference experiment with a macroscopic detector in macroscopically different states even if it is isolated from the environment. It is not sufficient, because decoherence does not ensure collapse and does not ensure splitting of a world. For a proper measurement, the measuring device must be macroscopic. It is true that an interaction of the system with an environment, instead of a macroscopic measuring device, might lead to a state similar to (26.4) with macroscopic number of microscopic systems of environment "recording" the eigenvalue of the observable. It, however, does not ensure that the measurement happens. It is not clear that macroscopic number of excited microsystem causes a collapse, see analysis of a such situation in the framework of the physical collapse model (Ghirardi et al. 1986) in Albert and Vaidman (1989) and Aicardi et al. (1991). In the framework of the many-worlds interpretation we need splitting of worlds. The moment of splitting does not have a rigorous definition, but a standard definition (Vaidman 2002) is that macroscopic objects must have macroscopically different states. Decoherence might well happen due to a change of states of air molecules which do not represent any macroscopic object.

What I view as the most problematic "symmetry argument proof" of probability half for the pivotal example is the analysis of Sebens and Carroll (2016), see also Kent (2015). Sebens and Caroll considered the measurement in the framework of the MWI and apply the uncertainty of self-location in a particular world as a meaning of probability (Vaidman 1998). However, in my understanding of the example they consider, this uncertainty does not exist (McQueen and Vaidman 2018). In their scenario, a measurement of A on a system in state (26.8) is performed on a remote planet. Sebens and Caroll consider a question: What is the probability of an observer who is here, i.e., far away from the planet, to be in a world with a particular outcome? This question is illegitimate, because he is certainly present in both worlds, there is no uncertainty here. This conclusion is unavoidable in the MWI as I understand it (Vaidman 2002), which is a fully deterministic theory without any room for uncertainty. However, uncertainty in the MWI is considered in Saunders (2004) and Saunders and Wallace (2008), so if this program succeeds (see however Lewis 2007), then the Sebens-Caroll proof might make sense. Another way to make sense of the Sebens-Caroll proof was proposed by Tappenden (2017) based on his unitary interpretation of mind, but I have difficulty accepting this metaphysical picture.

A scenario in which an observer is moved to different locations according to an outcome of a quantum measurement without getting information about this outcome (Vaidman 1998), allows us to consider the probability based on observer's ignorance about self-location and without uncertainty in the theory. This by itself, however, does not prove the probability half for the pivotal case. The proof (McQueen and Vaidman 2018), which is applicable to all interpretations, has two basic assumptions. First, it is assumed that space in Nature has symmetry, so we can construct the pivotal case with symmetry between the states $|a_1\rangle$ and $|a_2\rangle$. We do not rely on permutation of states, we rely on the symmetry of physical space and construct a symmetric state with identical wave packets in remote locations 1 and 2. The second assumption is that everything fulfills the postulate of the theory of special relativity according to which we cannot send signals faster than light. Changing probability by a remote action is sending signals. This proves that changing the shape or even splitting a remote state will not change the probability of finding a_1 provided its amplitude was not changed.

26.5 Other Approaches

Itamar Pitowsky's analysis of the Born rule on the basis of Gleason's theorem (Pitowsky 1998) was taken further to the case of generalized measurements (Caves et al. 2004). Galley and Masanes (2017) continued research which singles out the Born rule from other alternatives. Note that they also used symmetry ("bit symmetry") to single out the Born rule. Together with Muller, they extended their analysis (Masanes et al. 2019) and claimed to prove *everything* just from some "natural" properties of measurements which are primitive elements in their theory. So, people walked very far on the road paved by Itamars's pioneering works. I have to admit that I am not sympathetic to this direction. The authors of Masanes et al. (2019) conclude "Finally, having cleared up unnecessary postulates in the formulation of QM, we find ourselves closer to its core message." For me it seems that they go away from physics. Quantum mechanics was born to explain physical phenomena that classical physics could not. It was not a probability theory. It was not a theory of measurements, and I hope it will not end as such. "Measurements" should not be primitives, they are physical processes as any other, and physics should explain all of them.

Similarly, I cannot make much sense of claims that the Born rule appears even in classical systems presented in the Hilbert space formalism (Brumer and Gong 2006; Deumens 2019). Note that in the quantum domain, the Born rule appears even outside the framework of Hilbert spaces in the work of Saunders (2004), who strongly relies on operational assumptions such as a continuity assumption: "sufficiently small variations in the state-preparation device, and hence of the initial state, should yield small variations in expectation value." This assumption is much more physical than postulates of general probabilistic theories.

The dynamical derivation in the framework of the Bohmian interpretation championed by Valentini (Valentini and Westman 2005; Towler et al. 2011) who argued that under some (not too strong) requirements of complexity, the Born distribution arises similarly to thermal probabilities in ordinary statistical mechanics. See extensive discussion in Callender (2007) and recent analysis in Norsen (2018) which brings also similar ideas from Dürr et al. (1992). The fact that for some initial conditions of some systems relaxation to Born statistics does not happen is a serious weakness of this approach. What I find more to the point as a proof of the Born rule is that the Born statistical distribution remains invariant under time evolution in *all* situations. And that, under some very natural assumptions, is the only distribution with this strong property (Goldstein and Struyve 2007).

Wallace (2010) and Saunders (2010) advocate analyzing the issue of probability in the framework of the consistent histories approach. It provides formal expressions which fit the probability calculus axioms. However, I have difficulty seeing what these expressions might mean. I failed to see any ontological meaning for the main concept of the approach "the probability of a history", and it also has no operational meaning apart from the conditional probability of an actually performed experiment (Aharonov et al. 1964), while the approach is supposed to be general enough to describe evolution of systems which were not measured at the intermediate time.

26.6 Summary of My View

I feel that there is a lot of confusion in the discussions of the subject and it is important to make the picture much more clear. Even if definite answers might not be available now, the question: What are the open problems? can be clarified. First, it is important to specify the framework: collapse theory, hidden variables approach or noncollapse theory. Although in many cases the "derivation of the Born rule" uses similar structure and arguments in all frameworks, the conceptual task is very different. I believe that in all frameworks there is no way to prove the Born rule from other axioms of standard quantum mechanics. The correctly posed question is: What are the additional assumptions needed to derive the Born rule?

Standard quantum mechanics tells us that the evolution is unitary, until it leads to a superposition of quantum states corresponding to macroscopically different classical pictures. There is no precise definition of "macroscopically different classical pictures" and this is a very important part of the measurement problem, but discussions of the Born rule assume that this ambiguity is somehow solved, or proven irrelevant. The discussions analyze a quantitative property of the nonunitary process which happens when we reach this stage assuming that the fact that it happens is given. I see no possibility to derive the quantitative law of this nonunitary process from laws of unitary evolution. It is usually assumed that the process depends solely on the quantum state, i.e. that the probability of an outcome of a measurement of an observable does not depend on some hidden variables and does not depend on the way the observable is measured. The process also should not alter the unitary evolution when a superposition of states corresponding to macroscopically different classical pictures was not created. This, however, is not enough to rule out proposals different from the Born rule, e.g., Equal rule described above. We have to add something to derive the Born rule. We are not supposed to rely on experimental results, they do single out the Born rule, but this is not a "derivation". Instead, if we take some features of observed results as the basis, it is considered as a derivation. I am not sure that it is really better, unless these features are considered not as properties of Nature, but as a basic reason for Nature to be as it is. Then the Born rule derivations become a part of the program to get quantum mechanics from simple axioms (Popescu and Rohrlich 1994; Hardy 2001; Chiribella et al. 2011). In these derivations, quantum mechanics is usually considered as a general probability theory and the main task is to derive the Born rule.

In McQueen and Vaidman (2018) the program is more modest. Unitary quantum mechanics is assumed and two physical postulates are added. First, that there are symmetries in space and second that there is no superluminal signalling. The first principle allows us to construct a pivotal example described by (26.8) in which there is symmetry between states $|a_1\rangle$ and $|a_2\rangle$. The second principle allows us to change one of the eigenstates in the pivotal state without changing the probability to find the other eigenvalue. This is the beginning of the procedure, first shown by Deutsch in (1999), who pioneered these types of derivations.

The situation in the framework of the MWI is conceptually different. The physical essence of the MWI is: unitary evolution of a quantum state of the universe is all that there is. There is no additional process of collapse behaviour which should be postulated. So it seems that here there is no room for additional assumptions and that the Born rule must be derived just from the unitary evolution.

However, the MWI has a problem with probability even before we discuss the quantitative formula of the Born rule. The standard approach to the probability of an event requires that there to be a matter of fact about whether this event and not the other takes place, but in the MWI all events take place. On the other hand, we do have experience of one particular outcome when we perform a measurement. My resolution of this problem (Vaidman 1998) is that indeed, there is no way to ask what is the probability of what will happen, because all outcomes will be actual. The "probability" rule is still needed to explain statistics of observed measurements in the past. There are worlds with all possible statistics, but we happen to observe Born rule statistics. The "probability" explaining these statistics is the probability of self-location in a particular world. In Vaidman (1998) I constructed a scenario with quantum measurements in which the observer is split (and together with him, his world) according to the outcome of the measurement without being aware of the result of the measurement. This provides the ignorance probability of the observer about the world specified by the outcome of the measurement he is a part of. Tappenden (2010) argues that merely considering such a construction allows us to discuss the Born rule. These are supporting arguments of the solution: there is no probabilistic process in Nature: with certainty all possible outcomes of a quantum measurement will be realized, but an observer, living by definition in one of the worlds, can consider the question of probability of being located in a particular world.

All that is in Nature, according to MWI, is a unitary evolving quantum state of the universe and observers correspond to parts of this wave function (Vaidman 2016, 2019). So, there is a hope that the experience of observers, after constructing the theory of observers (chemistry, biology, psychology, decision theory, etc.) can be, in principle, explained solely from the evolution of the quantum state. Apparently, experiences of an observer can be learned from his behavior which is described by the evolution of the wave function. Then, it seems that the Born rule should be derivable from the laws of quantum mechanics. However, I believe that this is not true.

Consider Alice and Bob at separate locations and they have a particle in a state (26.8) where $|a_1\rangle$ corresponds to a particle being at Alice's site and $|a_2\rangle$ corresponds to a particle being at Bob's site. Now assume that instead of the Born rule, which states that the probability of self-location in a world is proportional to the square of the amplitude, Nature has the Equal rule which yields the same probability of self location in all the worlds, i.e. probability $\frac{1}{N}$, where N is the number of worlds. Equal rule allows superluminal signaling. Alice and Bob agree that at a particular time t Alice measures the presence of the particle at her site, i.e. she measures the projection on state $|a_1\rangle$. To send bit 0, Bob does nothing. Alice's measurement splits the world into two worlds: the one in which she finds the particle and the other, in which she does not. Then she has equal probability to find herself in each of the worlds, so she has probability half to find herself in the world in which she finds the particle. For sending bit 1, just before time t, Bob performs a unitary operation on the part of the wave at his site splitting it to a hundred orthogonal states

$$|a_2\rangle \rightarrow \sum_{k=1}^{100} |b_k\rangle, \tag{26.10}$$

and immediately measures operator B which tells him the eigenvalue b_k. This operation splits the initial single world with the particle in a superposition into hundred and one worlds: hundred worlds with one of Bob's detectors finding the particle, and one world in which the particle was not found by Bob's detectors. Prior to her measurement, Alice is present in all these worlds. Her measurement tests if she is in one particular world, so she has only probability of $\frac{1}{101}$ to find the particle at her site.

Bob's unitary operation and measurement change the probability of Alice's outcome. With measurements on a single particle, the communication is not very reliable, but using an ensemble will lead to only very rare cases of an error. The Equal rule will ensure that Alice and Bob meeting in the future will (most probably) verify correctness of the bit Bob has sent.

We know that unitary evolution does not allow superluminal communication. (When we consider a relativistic generalisation of the Schrödinger equation.) Can, a supertechnology, capable of observing superposition of Alice's and Bob's worlds, given that the actual probability rule of self-location is the Equal rule (and not the Born rule), use the above procedure for sending superluminal signals? No! Only Alice and Bob, inside their worlds, have the ability of superluminal communication. It does not contradict relativistic properties of physics describing unitary evolution of all worlds together.

What I argue here, is that the situation in the framework of MWI is not different from collapse theories. There is a need for an independent probability postulate. In collapse theory it is a physical postulate telling us about the dynamics of the ontology, dynamics of the quantum state of the universe describing the (single) world. In MWI, the postulate belongs to the part connecting observer's experiences with the ontology. In the MWI, as in a collapse theory, the experiences supervene on the ontology, the quantum state. The supervenience rule is the same when the quantum state corresponds to a single world, but it has an additional part regarding the probability of self-location, when the quantum state of the universe corresponds to more than world. The postulate describes this supervenience rule.

We can justify the Born rule postulate of self-location by experimental evidence, or by requiring the relativistic constraint of superluminal signaling also within worlds. I find a convincing explanation in the concept of the measure of existence of a world (Vaidman 1998; Groisman et al. 2013). While there is no reason to postulate that the probability of self-location in every world is the same, it is natural to postulate that the probability of self-location in worlds of equal existence (equal square of the amplitude) is the same. Adding another natural assumption that probability of self-location in a particular world should be equal to the sum of the probability of self-location in all the worlds which split from the original one, provides the Born rule.

My main conclusion is that there is no way to derive the Born rule without additional assumptions. It is true both in the framework of collapse theories and, more surprisingly, in the framework of the MWI. The main open question is not the validity of various proofs, but what are the most natural assumptions we should add for proving the Born rule.

Acknowledgements This work has been supported in part by the Israel Science Foundation Grant No. 2064/19.

References

Aharonov, Y., & Reznik, B. (2002). How macroscopic properties dictate microscopic probabilities. *Physical Review A, 65*, 052116.

Aharonov, Y., & Vaidman, L. (1990). Properties of a quantum system during the time interval between two measurements. *Physical Review A, 41*, 11–20.

Aharonov, Y., Bergmann, P. G., & Lebowitz, J. L. (1964). Time symmetry in the quantum process of measurement. *Physical Review, 134*, B1410–B1416.

Aharonov, Y., Cohen, E., Gruss, E., & Landsberger, T. (2014). Measurement and collapse within the two-state vector formalism. *Quantum Studies: Mathematics and Foundations, 1*, 133–146.
Aharonov, Y., Cohen, E., & Landsberger, T. (2017). The two-time interpretation and macroscopic time-reversibility. *Entropy, 19*, 111.
Aicardi, F., Borsellino, A., Ghirardi, G. C., & Grassi, R. (1991). Dynamical models for state-vector reduction: Do they ensure that measurements have outcomes? *Foundations of Physics Letters, 4*, 109–128.
Albert, D. Z., & Vaidman, L. (1989). On a proposed postulate of state-reduction. *Physics letters A, 139*, 1–4.
Baker, D. J. (2007). Measurement outcomes and probability in Everettian quantum mechanics. *Studies in History and Philosophy of Science Part B: Studies in History and Philosophy of Modern Physics, 38*, 153–169.
Barnum, H., Caves, C. M., Finkelstein, J., Fuchs, C. A., & Schack, R. (2000). Quantum probability from decision theory? *Proceedings of the Royal Society of London. Series A: Mathematical, Physical and Engineering Sciences, 456*, 1175–1182.
Barrett, J. A. (2017). Typical worlds. *Studies in History and Philosophy of Science Part B: Studies in History and Philosophy of Modern Physics, 58*, 31–40.
Boge, F. J. (2019). The best of many worlds, or, is quantum decoherence the manifestation of a disposition? *Studies in History and Philosophy of Science Part B: Studies in History and Philosophy of Modern Physics, 66*, 135–144.
Bohm, D. (1952). A suggested interpretation of the quantum theory in terms of "hidden" variables. I. *Physical Review, 85*, 166.
Brumer, P., & Gong, J. (2006). Born rule in quantum and classical mechanics. *Physical Review A, 73*, 052109.
Bub, J., & Pitowsky, I. (2010). Two dogmas about quantum mechanics. In S. Saunders, J. Barrett, A. Kent, & D. Wallace (Eds.), *Many Worlds?: Everett, quantum theory, & reality* (pp. 433–459). Oxford: Oxford University Press.
Buniy, R. V., Hsu, S. D., & Zee, A. (2006). Discreteness and the origin of probability in quantum mechanics. *Physics Letters B, 640*, 219–223.
Callender, C. (2007). The emergence and interpretation of probability in Bohmian mechanics. *Studies in History and Philosophy of Science Part B: Studies in History and Philosophy of Modern Physics, 38*, 351–370.
Caves, C.M., Fuchs, C.A., & Schack, R. (2002). Quantum probabilities as Bayesian probabilities. *Physical Review A, 65*, 022305.
Caves, C. M., Fuchs, C. A., Manne, K. K., & Renes, J. M. (2004). Gleason-type derivations of the quantum probability rule for generalized measurements. *Foundations of Physics, 34*, 193–209.
Chiribella, G., D'Ariano, G. M., & Perinotti, P. (2011). Informational derivation of quantum theory. *Physical Review A, 84*, 012311.
Dawid, R., & Thébault, K. P. (2014). Against the empirical viability of the Deutsch–Wallace–Everett approach to quantum mechanics. *Studies in History and Philosophy of Science Part B: Studies in History and Philosophy of Modern Physics, 47*, 55–61.
Deumens, E. (2019). On classical systems and measurements in quantum mechanics. *Quantum Studies: Mathematics and Foundations, 6*, 481–517.
Deutsch, D. (1999). Quantum theory of probability and decisions. *Proceedings of the Royal Society of London. Series A: Mathematical, Physical and Engineering Sciences, 455*, 3129–3137.
Dürr, D., Goldstein, S., & Zanghi, N. (1992). Quantum equilibrium and the origin of absolute uncertainty. *Journal of Statistical Physics, 67*, 843–907.
Elitzur, A. C., & Vaidman, L. (1993). Quantum mechanical interaction-free measurements. *Foundations of Physics, 23*, 987–997.
Everett III, H. (1957). "Relative state" formulation of quantum mechanics. *Reviews of Modern Physics, 29*, 454–462.
Farhi, E., Goldstone, J., & Gutmann, S. (1989). How probability arises in quantum mechanics. *Annals of Physics, 192*, 368–382.

Finkelstein, J. (2003). Comment on "How macroscopic properties dictate microscopic probabilities". *Physical Review A, 67*, 026101.
Galley, T. D., & Masanes, L. (2017). Classification of all alternatives to the Born rule in terms of informational properties. *Quantum, 1*, 15.
Ghirardi, G.C., Rimini, A., & Weber, T. (1986). Unified dynamics for microscopic and macroscopic systems. *Physical Review D, 34*, 470–491.
Gill, R. (2005). On an argument of David Deutsch. In M. Schürmann & U. Franz (Eds.), *Quantum probability and infinite dimensional analysis: From foundations to applications* (QP-PQ Series, Vol. 18, pp. 277–292). Singapore: World Scientific.
Gleason, A. M. (1957). Measures on the closed subspaces of a Hilbert space. *Journal of Mathematics and Mechanics, 6*, 885–893.
Goldstein, S., & Struyve, W. (2007). On the uniqueness of quantum equilibrium in Bohmian mechanics. *Journal of Statistical Physics, 128*, 1197–1209.
Greaves, H. (2004). Understanding Deutsch's probability in a deterministic multiverse. *Studies in History and Philosophy of Science Part B: Studies in History and Philosophy of Modern Physics, 35*, 423–456.
Groisman, B., Hallakoun, N., & Vaidman, L. (2013). The measure of existence of a quantum world and the Sleeping Beauty Problem. *Analysis, 73*, 695–706.
Hardy, L. (2001). Quantum theory from five reasonable axioms. arXiv preprint quant-ph/0101012.
Hartle, J. B. (1968). Quantum mechanics of individual systems. *American Journal of Physics, 36*, 704–712.
Hemmo, M., & Pitowsky, I. (2007). Quantum probability and many worlds. *Studies in History and Philosophy of Science Part B: Studies in History and Philosophy of Modern Physics, 38*, 333–350.
Kent, A. (2015). Does it make sense to speak of self-locating uncertainty in the universal wave function? Remarks on Sebens and Carroll. *Foundations of Physics, 45*, 211–217.
Kwiat, P., Weinfurter, H., Herzog, T., Zeilinger, A., & Kasevich, M. A. (1995). Interaction-free measurement. *Physical Review Letters, 74*, 4763–4766.
Landsman, N. P. (2008). Macroscopic observables and the Born rule, I: Long run frequencies. *Reviews in Mathematical Physics, 20*, 1173–1190.
Landsman, N. P. (2009). Born rule and its interpretation. In D. Greenberger, K. Hentschel, & F. Weinert (Eds.), *Compendium of Quantum Physics: Concepts, Experiments, History and Philosophy* (pp. 64–70). Heidelberg: Springer.
Lewis, P. J. (2007). Uncertainty and probability for branching selves. *Studies in History and Philosophy of Science Part B: Studies in History and Philosophy of Modern Physics, 38*, 1–14.
Lewis, P. J. (2010). Probability in Everettian quantum mechanics. *Manuscrito: Revista Internacional de Filosofía, 33*, 285–306.
Li, P., Field, G., Greschner, M., Ahn, D., Gunning, D., Mathieson, K., Sher, A., Litke, A., & Chichilnisky, E. (2014). Retinal representation of the elementary visual signal. *Neuron, 81*, 130–139.
Masanes, L., Galley, T. D., & Müller, M. P. (2019). The measurement postulates of quantum mechanics are operationally redundant. *Nature Communications, 10*, 1361.
McQueen, K. J., & Vaidman, L. (2018). In defence of the self-location uncertainty account of probability in the many-worlds interpretation. *Studies in History and Philosophy of Science Part B: Studies in History and Philosophy of Modern Physics, 66*, 14–23.
Norsen, T. (2018). On the explanation of Born-rule statistics in the de Broglie-Bohm pilot-wave theory. *Entropy, 20*, 422.
Pitowsky, I. (1989). *Quantum probability-quantum logic*. Berlin: Springer.
Pitowsky, I. (1998). Infinite and finite Gleason's theorems and the logic of indeterminacy. *Journal of Mathematical Physics, 39*, 218–228.
Pitowsky, I. (2003). Betting on the outcomes of measurements: A Bayesian theory of quantum probability. *Studies in History and Philosophy of Science Part B: Studies in History and Philosophy of Modern Physics, 34*, 395–414.

Pitowsky, I. (2006). Quantum mechanics as a theory of probability. In *Physical theory and its interpretation* (pp. 213–240). Berlin/Heidelberg: Springer.

Popescu, S., & Rohrlich, D. (1994). Quantum nonlocality as an axiom. *Foundations of Physics, 24*, 379–385.

Rae, A. I. (2009). Everett and the Born rule. *Studies in History and Philosophy of Science Part B: Studies in History and Philosophy of Modern Physics, 40*, 243–250.

Saunders, S. (2004). Derivation of the Born rule from operational assumptions. *Proceedings of the Royal Society of London. Series A: Mathematical, Physical and Engineering Sciences, 460*, 1771–1788.

Saunders, S. (2010). Chance in the Everett interpretation. In S. Saunders, J. Barrett, A. Kent, & D. Wallace (Eds.), *Many worlds?: Everett, quantum theory, & reality* (pp. 181–205). Oxford: Oxford University Press.

Saunders, S., & Wallace, D. (2008). Branching and uncertainty. *The British Journal for the Philosophy of Science, 59*, 293–305.

Schlosshauer, M., & Fine, A. (2005). On Zurek's derivation of the Born rule. *Foundations of Physics, 35*, 197–213.

Sebens, C. T., & Carroll, S. M. (2016). Self-locating uncertainty and the origin of probability in Everettian quantum mechanics. *The British Journal for the Philosophy of Science, 69*, 25–74.

Squires, E. J. (1990). On an alleged "proof" of the quantum probability law. *Physics Letters A, 145*, 67–68.

Tappenden, P. (2010). Evidence and uncertainty in Everett's multiverse. *British Journal for the Philosophy of Science, 62*, 99–123.

Tappenden, P. (2017). Objective probability and the mind-body relation. *Studies in History and Philosophy of Science Part B: Studies in History and Philosophy of Modern Physics, 57*, 8–16.

Towler, M., Russell, N., & Valentini, A. (2011). Time scales for dynamical relaxation to the Born rule. *Proceedings of the Royal Society A: Mathematical, Physical and Engineering Sciences, 468*, 990–1013.

Vaidman, L. (1998). On schizophrenic experiences of the neutron or why we should believe in the many-worlds interpretation of quantum theory. *International Studies in the Philosophy of Science, 12*, 245–261.

Vaidman, L. (2002). Many-Worlds interpretation of Quantum mechanics. In E. N. Zalta (Ed.), *The Stanford encyclopedia of philosophy*. Metaphysics Research Lab, Stanford University. https://plato.stanford.edu/cgi-bin/encyclopedia/archinfo.cgi?entry=qm-manyworlds

Vaidman, L. (2012). Probability in the many-worlds interpretation of quantum mechanics. In Y. Ben-Menahem & M. Hemmo (Eds.), *Probability in physics* (pp. 299–311). Berlin: Springer.

Vaidman, L. (2016). All is ψ. *Journal of Physics: Conference Series, 701*, 012020.

Vaidman, L. (2019). Ontology of the wave function and the many-worlds interpretation. In O. Lombardi, S. Fortin, C. López, & F. Holik (Eds.), *Quantum worlds: Perspectives on the ontology of quantum mechanics*. (pp. 93–106). Cambridge: Cambridge University Press.

Valentini, A., & Westman, H. (2005). Dynamical origin of quantum probabilities. *Proceedings of the Royal Society A: Mathematical, Physical and Engineering Sciences, 461*, 253–272.

Van Wesep, R. A. (2006). Many worlds and the appearance of probability in quantum mechanics. *Annals of Physics, 321*, 2438–2452.

Wallace, D. (2007). Quantum probability from subjective likelihood: Improving on Deutsch's proof of the probability rule. *Studies In History and Philosophy of Science Part B: Studies In History and Philosophy of Modern Physics, 38*, 311–332.

Wallace, D. (2010). How to prove the Born rule. In S. Saunders, J. Barrett, A. Kent, D. Wallace (Eds.) *Many worlds?: Everett, quantum theory, & reality* (pp. 227–263). Oxford University Press.

Zurek, W. H. (2005). Probabilities from entanglement, Born's rule $p_k = |\psi_k|^2$ from envariance. *Physical Review A, 71*, 052105.

Chapter 27
Dynamical States and the Conventionality of (Non-) Classicality

Alexander Wilce

Abstract Itamar Pitowsky, along with Jeff Bub and others, long championed the view that quantum mechanics (QM) is best understood as a non-classical probability theory. This idea has several attractions, not least that it allows us even to pose the question of why quantum probability theory has the rather special mathematical form that it has—a problem that has been very fruitful, and to which we now have several rather compelling answers, at least for finite-dimensional QM. Here, however, I want to offer some modest caveats. One is that QM is best seen, not as a new probability theory, but as something narrower, namely, a class of probabilistic models, selected from within a much more general framework. It is this general framework that, if anything, deserves to be regarded as a "non-classical" probability theory.

Secondly, however, I will argue that, for individual probabilistic models, and even for probabilistic theories (roughly, classes of such models), the distinction between "classical" and a "non-classical" is largely a conventional one, bound up with the question of what one means by the state of a system. In fact, for systems with a high degree of symmetry, including quantum mechanics, it is possible to interpret general probabilistic models as having a perfectly classical probabilistic structure, but an additional dynamical structure: states, rather than corresponding simply to probability measures, are represented as certain probability measure-valued functions on the system's symmetry group, and thus, as fundamentally dynamical objects. Conversely, a classical probability space equipped with reasonably well-behaved family of such "dynamical states" can be interpreted as a generalized probabilistic model in a canonical way. It is noteworthy that this "dynamical" representation is not a conventional hidden-variables representation, and the question of what one means by "non-locality" in this setting is not entirely straightforward.

A. Wilce (✉)
Department of Mathematical Sciences, Susquehanna University, Selinsgrove, PA, USA
e-mail: wilce@susqu.edu

M. Hemmo, O. Shenker (eds.), *Quantum, Probability, Logic*,
Jerusalem Studies in Philosophy and History of Science,
https://doi.org/10.1007/978-3-030-34316-3_27

Keywords Probabilistic theories · Classical representations · Dynamical states · Simplices · Test spaces

27.1 Introduction

Itamar Pitowsky long championed the idea (also associated with Jeff Bub, among others) that quantum theory is best viewed as a non-classical probability theory (Pitowsky 2005). This position is strongly motivated by the fact that the mathematical apparatus of quantum mechanics can largely be reduced to the statement that the binary observables—the physically measurable, {0, 1}-valued quantitites—associated with a quantum system are in one-to-one correspondence with the projections in a von Neumann algebra, in such a way that commuting families of projections can be jointly measured and mutually orthogonal projections are mutually exclusive.[1] In other words, mathematically, quantum theory is what one gets when one replaces the Boolean algebras of classical probability theory with projection lattices. It is very natural to conclude from this that quantum theory simply *is* a "non-classical" or "non-commutative" probability theory.

This point of view has many attractions, not least that it more or less dissolves what Pitowsky called the "big" measurement problem (Bub and Pitowsky 2012; Pitowsky 2005). It also raises the question of why quantum mechanics has the *particular* non-classical structure that it does—that is, why projection lattices of von Neumann algebras, rather than something more general? The past decade has seen a good deal of progress on this question, at least as it pertains to finite-dimensional quantum systems (Hardy 2001; Masanes and Müller 2011; D'Ariano et al. 2011; Mueller et al. 2014; Wilce 2012).

Here, I want to offer some modest caveats. The first is that how we *understand* Pitowski's thesis depends on how we understand probability theory itself. On the view that I favor, quantum mechanics (henceforth, QM) is best seen, not as replacement for general classical probability theory, but rather, as a specific class of probabilistic models—a *probabilistic theory*—defined within a much broader framework, in which individual models are defined by essentially arbitrary sets of basic measurements and states, with the latter assigning probabilities (in an elementary sense) to outcomes of the former. It is this framework that, if anything, counts as a "non-classical" probability theory—but in fact, I think we should regard, and refer to, this general framework, simply as *probability theory*, without any adjective. Since it outruns what's usually studied under that heading, I settle here

[1]Given this, Gleason's Theorem, as extended by Christensen and Yeadon, identifies the system's states with states (positive, normalized linear functionals) on that same von Neumann algebra, and Wigner's Theorem then identifies the possible dynamics with one-parameter groups of unitary elements of that algebra. For the details, see Varadarajan (1985).

for the term *general probability theory*.[2] Within this framework, in addition to QM, we can identify classical probability theory, in the strict Kolmogorovian sense, as a special, and, I will argue, equally *contingent*, probabilistic theory.

However—and this is my second point—even as applied to individual probabilistic models, or, indeed, individual probabilistic theories, the distinction between "classical" and "non-classical" is not so clear cut. One standard answer to the question of what makes the probabilistic apparatus of QM "non-classical" is that its state space (the set of density operators on a Hilbert space, or the set of states on a non-commutative von Neumann algebra, depending on how general we want to be) is not a simplex: the same state can be realized as a mixture—a convex combination—of pure states in many different ways. Another, related, answer is that, whereas in classical probability theory any two measurements (or, if you prefer, experiments) can effectively be performed jointly, in QM this is not the case.

Neither of these answers is really satisfactory. In the first place, it is perfectly within the scope of standard classical probability theory to consider restricted sets of probability measures and random variables on a given measurable space: the former need not be simplices, and the latter need not be closed under the formation of joint random variables. We might also wish to allow "unsharp" random variables having some intrinsic randomness (represented, depending on one's taste, by Markov kernels or by response-function valued measures). We might try to identify as "broadly classical" just those probabilistic models arising in this way, from a measurable space plus some selection of probability measures and (possibly unsharp) random variables. However, once we open the door this far, it's difficult to keep it on its hinges. This is because there exist straightforward, well known, and in a sense even canonical, recipes for representing essentially *any* "non-classical" probabilistic model in terms of such a broadly classical one, either by treating the state space as a suitable quotient of a simplex (in the manner of the Beltrametti-Bugajski representation of quantum mechanics), or by introducing contextual "hidden variables". In fact, for systems with a high degree of symmetry, including finite-dimensional quantum systems, there is another, less well known, but equally straightforward classical representation: the probabilistic structure is simply that of a single reference observable, and hence, entirely classical; states, however, rather than corresponding simply to probability measures, are represented by probability measure-valued functions on the system's symmetry group, and thus, as fundamentally dynamical objects. Conversely, a classical probability space equipped with a family of suitably covariant such "dynamical states"—what I call a *dynamical classical model*—can be interpreted as a generalized probabilistic model in a canonical way.

At the level of individual probabilistic models, then, it seems that the distinction between classical and non-classical is, at least mathematically, more or less one of convention (and, pragmatically, of convenience). There is, of course, another

[2]The framework in question is broadly equivalent to that of "generalized probabilistic theories". Luckily, both phrases can go by the acronym GPT.

consideration in play: that of locality. For probabilistic theories involving a notion of composite systems that admits entangled states—including, of course, quantum theory—the classical representations alluded to above are all manifestly non-local in one sense or another. One reading of Bell's Theorem is that, for such probabilistic theories, there is a necessary tension between "classicality" and locality. Indeed, one standard response to the existence of various classical representations is to dismiss them precisely *because* they are non-local. But locality is very much a *physical* constraint, rather than a "law of thought". Thus, we should be careful not to regard the probabilistic apparatus of quantum mechanics as a successor to classical probability theory as a *general* account of reasoning in the face of uncertainty. Rather, it is a particular, contingent, probabilistic *physical* theory, expressible in terms of a much broader generalized probability theory—a framework that we may or may not prefer to regard as non-classical, but which is in any event at most a very conservative generalization of classical probability theory. It is a collateral aim of this chapter to argue for this expansive view of what probability theory *is*.

A brief outline of this chapter is as follows. In Sect. 27.2, I give a condensed introduction to the framework—or rather, one particular version of the framework—of general(ized) probability theory, along the lines of Barnum and Wilce (2016). It is another collateral aim of this paper to advertise this framework as offering some additional measure of clarity to discussions of foundational issues in quantum theory. In Sect. 27.3, I ask how one ought to characterize *classical* probabilistic models in this framework, and review several more or less standard, or even canonical, ways in which an arbitrary probabilistic model can be interpreted in terms of a classical one. There is little here that is new, except possibly some streamlining of the material. In Sect. 27.4, I introduce the dynamical classical models alluded to above. In Sect. 27.5, I very briefly discuss some issues of locality, entanglement and so forth in this context, and in Sect. 27.6, I gather a few concluding thoughts. I have placed some technical material in a series of appendices. Appendix A establishes a folk-result, to the effect that a probabilistic model in which any two measurements are compatible in the sense of admitting a common refinement, is essentially classical. Appendix B gathers some straightforward observations about what are called semiclassical test spaces. Appendix C concerns a construction that can be used to generate large numbers of highly symmetric probabilistic models, or, equivalently, dynamical classical models. This follows, but in some respects generalizes, material in Wilce (2009). To keep measure-theoretic details to a minimum, I focus mainly (though not exclusively) on probabilistic models in which there is a finite upper bound on the number of distinct outcomes in a basic measurement.

Terminology: In the balance of this chapter, I will use the adjective *classical* sometimes in a broad and informal way, sometimes in a more technical sense, and sometimes ironically. I hope that context, occasionally aided by scare-quotes, will help make my meaning clear in each case. The term *classical representation* is also to be understood broadly and somewhat informally, to mean any mathematical representation of one probabilistic model or theory in terms of another such model or theory that is, in some reasonable sense, classical. At two points, my mathematical usage may be slightly nonstandard: first, by a *measurable space*, I will always mean a pair (S, Σ) where S is a set and Σ is an algebra, but not necessarily a σ-algebra, of subsets of S; and by a *measure* on S, unless otherwise specified I mean a finitely-additive measure. Secondly, by a *discrete Markov kernel* I mean a function $p : S \times T \to [0, 1]$, where S and T are sets, *possibly infinite*, such that $\sum_{y \in T} p(x, y) = 1$ for every $x \in S$, where the sum is meant in the unordered sense. Otherwise, my usage will be fairly standard. Terms likely to be unfamiliar to most readers are defined when introduced.

27.2 Probability Theory vs Probabilistic Theories

This section provides a brief but self-contained tutorial on the mathematical framework that I favor for a general probability theory. This is a slight elaboration of a formalism developed by D. J. Foulis and C. H. Randall; see, e.g., Foulis and Randall (1978, 1981). A more complete survey can be found in Barnum and Wilce (2016).

27.2.1 Test Spaces, Probability Weights, and Probabilistic Models

In many introductory treatments of probability theory, a *probabilistic model* is defined as a pair (E, p) where E is the outcome-set of some experiment, and p is a probability weight on E, meant to reflect the actual statistical behaviour of the experiment. An obvious generalization is to allow p to vary, i.e., to consider pairs (E, Ω) where Ω is some set of possible probability weights. For example, if $E = \{0, 1\}^n$, we might want to restrict attention to binomial probability weights with various probabilities of success.

An only slightly less obvious generalization is to allow E, too, to vary:

Definition 2.1 (Test spaces) A *test space* is a set $\mathcal{M}$ of non-empty sets $E, F, \ldots,$, understood as the outcome-sets of various experiments or, as we'll say, *tests*. If $X := \bigcup \mathcal{M}$ is the set of all outcomes of all tests in $\mathcal{M}$, a *probability weight* on $\mathcal{M}$ is a function $\alpha : X \to [0, 1]$ such that $\sum_{x \in E} \alpha(x) = 1$ for every $E \in \mathcal{M}$.

I will write $\mathbf{Pr}(\mathcal{M})$ for the set of all probability weights on a test space $\mathcal{M}$. Note that $\mathbf{Pr}(\mathcal{M})$ is a convex subset of $\mathbb{R}^X$: any weighted average $p\alpha + (1 - p)\beta$ of probability weights $\alpha, \beta \in \mathbf{Pr}(\mathcal{M})$ is again a probability weight on $\mathcal{M}$.

Definition 2.2 (Probabilistic models) A *probabilistic model* is a pair $A = (\mathcal{M}, \Omega)$ where $\mathcal{M}$ is a test space and $\Omega \subseteq \mathbf{Pr}(\mathcal{M})$ is a convex set of probability weights on $\mathcal{M}$, which we call the *states* of the model. Extreme points of Ω are called *pure states*.

It will be convenient to denote a probabilistic model by a single letter A, B, etc. When doing so, I write, e.g., $A = (\mathcal{M}(A), \Omega(A))$. The assumption that $\Omega(A)$ is convex is meant to reflect the possibility of randomizing the preparation of states; that is, if we have some procedure that produces state α and another that produces state β, we could flip a suitably biased coin to product the state $p\alpha + (1 - p)\beta$ for any $0 \leq p \leq 1$. It is also common to equip a probabilistic model A with a preferred group $G(A)$ of "physical transformations" (Hardy 2001; Dakič and Brukner 2011; Masanes and Müller 2011), and I will also do so in Sect. 27.4. Until then, however, this extra structure will not be necessary. Finally, I will suppose in what follows that Ω contains sufficiently many probability weights to separate points of X and that $\mathcal{M}$ contains sufficiently many tests (and hence X, sufficiently many outcomes) to distinguish the states in Ω (since otherwise, one would presumably just identify outcomes, respectively states, that are not distinguishable). Another condition that I will often (but not always) impose is that, for every outcome $x \in X(A)$, there exists at least one probability weight $\alpha \in \Omega(A)$ with $\alpha(x) = 1$. A model satisfying this condition is said to be *unital*.

Notice that we permit tests $E, F \in \mathcal{M}$ to intersect, that is, to share outcomes. From a certain point of view, this is absurd: when we perform a measurement, we usually retain a record of what measurement we've made! Nevertheless, we may have good reasons to want to *identify* outcomes from distinct tests, and to demand that physically meaningful operations on outcomes—symmetries, probability assignments, etc.—respect these identifications. The usual term for this is *non-contextuality*. In most of the probabilistic models one encounters in practice, such outcome-identifications are in fact made, and in any event, it is mathematically *more general* to allow tests to overlap than to forbid them to do so. Moreover, contextual probability weights can easily be accommodated by way of the following construction.

Definition 2.3 (Semi-classical test spaces) A test space $\mathcal{M}$ is *semi-classical* iff $E \cap F = \emptyset$ for all distinct tests $E, F \in \mathcal{M}$. The *semi-classical cover* of an arbitrary test space $\mathcal{M}$ is the test space

$$\widetilde{\mathcal{M}} = \{\, \widetilde{E} \mid E \in \mathcal{M} \,\},$$

where, for $E \in \mathcal{M}$, $\widetilde{E} = \{(x, E) | x \in E\}$.

The outcome-set of $\widetilde{\mathcal{M}}$ is thus $\widetilde{X} = \{(x, E) | x \in E \in \mathcal{M}\}$, i.e., the coproduct of the sets $E \in \mathcal{M}$. "Contextual" probability weights on $\mathcal{M}$ are best regarded as probability weights on $\widetilde{\mathcal{M}}$. In general $\mathbf{Pr}(\widetilde{\mathcal{M}})$ will be very much larger than $\mathbf{Pr}(\mathcal{M})$; indeed, if $\mathcal{M}$ contains infinitely many tests, $\mathbf{Pr}(\widetilde{\mathcal{M}})$ will be infinite dimensional, even if $\mathbf{Pr}(\mathcal{M})$ is finite-dimensional.

Definition 2.4 A probability weight α on $\mathcal{M}$ is *dispersion-free* or *deterministic* iff $\alpha(x) \in \{0, 1\}$ for every $x \in X = \bigcup \mathcal{M}$.

One of the most obvious distinctions between the probabilistic models associated with "classical" systems and those associated with quantum-mechanical systems is that the former have an abundance of dispersion-free states, while the latter, absent superselection rules, have none at all. Notice, however, that $\widetilde{\mathcal{M}}$ always has an abundance of dispersion-free probability weights (just choose an outcome from every test).

Definition 2.5 (Locally finite test spaces) A test space $\mathcal{M}$ is *locally finite* iff every test $E \in \mathcal{M}$ is a finite set. Every test space is associated with a locally finite test space $\mathcal{M}^{\#}$, consisting of finite partitions of tests in $\mathcal{M}$. That is, writing $\mathcal{D}(E)$ for the set of finite partitions of a set E,

$$\mathcal{M}^{\#} = \bigcup_{E \in \mathcal{M}} \mathcal{D}(E).$$

If $\mathcal{M}$ is a test space, an *event* for $\mathcal{M}$ is a subset of a test (that is, an event in the standard sense pertaining to that test). We write $\mathcal{E}v(\mathcal{M})$ for the set of all events of $\mathcal{M}$, so that $\mathcal{E}v(\mathcal{M}) = \bigcup_{E \in \mathcal{M}} \mathcal{P}(E)$. Thus, the outcome-space for $\mathcal{M}^{\#}$ is precisely the set $\mathcal{E}v(\mathcal{M}) \setminus \{\emptyset\}$ of non-empty events.

If α is a probability weight on $\mathcal{M}$, then we define the probability of an event $a \in \mathcal{E}v(\mathcal{M})$ by $\alpha(a) = \sum_{x \in a} \alpha(x)$. This evidently defines a probability weight on $\mathcal{M}^{\#}$. However, unless $\mathcal{M}$ is locally finite to begin with, $\mathcal{M}^{\#}$ will generally support many additional "singular" probability weights. (Indeed, if $\mathcal{M}$ is semiclassical, one can simply choose a non-principal ultrafilter on each $E \in \mathcal{M}$.)

Lemma 2.6 *Let $\mathcal{M}$ be locally finite with outcome-set X. Then the convex set* $\mathbf{Pr}(\mathcal{M})$ *of probability weights on $\mathcal{M}$ is compact as a subset of* $[0, 1]^X$ *(with the product topology).*

Proof Since each $E \in \mathcal{M}(A)$ is finte, the pointwise limit of a net of probability weights will continue to sum to unity over E. Thus, $\mathbf{Pr}(\mathcal{M})$ is closed, and hence, compact, in $[0, 1]^X$. □

It follows that if A is a probabilistic model with $\mathcal{M}(A)$ locally finite, the closure of $\Omega(A)$ in $\mathbf{Pr}(\mathcal{M}(A))$ will also be compact. Because it is both mathematically convenient and operationally reasonable to do so, I shall assume from now on that, unless otherwise indicated, ***all probabilistic models A under discussion have a locally finite test space $\mathcal{M}(A)$ and a closed, hence compact, state space $\boldsymbol{\Omega}(A)$.*** While I will make no direct use of it, it's worth noting that the Krein-Milman Theorem now tells us that every state in $\Omega(A)$ is a limit of convex combinations of pure states.

It is often useful to consider *sub-normalized* positive weights on a test space $\mathcal{M}$, that is, functions $\alpha : X := \bigcup \mathcal{M} \to \mathbb{R}_+$ such that $\sum_{x \in E} \alpha(x) =: r \leq 1$, independently of E. In fact, these are simply probability weights on a slightly larger test space:

Definition 2.7 (Adjoining a null-outcome) If $\mathcal{M}$ is a test space with outcome-space X, let $*$ be symbol not belonging to X, and define

$$\mathcal{M}_{+} = \{E \cup \{*\} | E \in \mathcal{M}\}.$$

We can think of $*$ as a "null outcome" representing the failure of a test, in a setting in which we regard this failure as meaning the same thing regardless of which test is performed. As an extreme example, $*$ might represent the destruction of the system under study. It should be clear that a probability weight on $\mathcal{M}_{+}$ corresponds exactly to a sub-normalized weight on $\mathcal{M}$.

27.2.2 Some Examples

Example 2.8 (Kolmogorovian models) We can treat standard measure-theoretic probability theory within the present framework in the following way. Let (S, Σ) be a measurable space and let $\mathcal{D}(S, \Sigma)$ be the set of finite partitions of S by elements of Σ, regarded as a test space. Note that $X = \bigcup \mathcal{D}(S, \Sigma)$ is simply $\Sigma \setminus \{\emptyset\}$. A probability weight on $\mathcal{D}(S, \Sigma)$, then, is a function $p : \Sigma \setminus \{\emptyset\} \to [0, 1]$ summing to unity on every finite partition of S by members of Σ. It is straightforward to show that such a function, extended to all of Σ by setting $p(\emptyset) = 0$, is a finitely additive probability measure on Σ, and conversely. We can now construct a probabilistic model $(\mathcal{D}(S, \Sigma), \Omega)$ in various ways, for instance, by taking Ω to consist of all such measures, or of all σ-additive probability measures, or of probability measures absolutely continuous with respect to some given measure, or of a single probability measure of interest, etc. I will refer to all models of the form $(\mathcal{D}(S, \Sigma), \Omega)$, where Ω is any convex set of probability measures on (S, Σ), as *Kolmogorovian*.[3]

Note that every point $s \in S$ defines a dispersion-free probability weight on $\mathcal{D}(S, \Sigma)$, namely, the point-mass $\delta_s(b) = 1$ if $b \in \Sigma$ with $s \in b$, and $\delta_s(b) = 0$ otherwise. (More generally, a dispersion-free probability weight on $\mathcal{D}(S, \Sigma)$ is an ultrafilter on Σ.)

Example 2.9 (Quantum models) If $\mathcal{H}$ is a Hilbert space, let $X(\mathcal{H})$ denote $\mathcal{H}$'s unit sphere and $\mathcal{M}(\mathcal{H})$, the set of maximal pairwise orthogonal subsets of $X(\mathcal{H})$, that is, the set of unordered orthonormal bases of $\mathcal{H}$. We shall call this the *quantum test space* associated with $\mathcal{H}$. Any density operator ρ on $\mathcal{H}$ gives rise to a probability weight on $\mathcal{M}(\mathcal{H})$, which I also denote by ρ, namely $\rho(x) = \langle \rho x, x \rangle$.[4] Let $\Omega(\mathcal{H})$ be the set of all probability weights on $\mathcal{M}(\mathcal{H})$ arising from density operators in this way: then $A(\mathcal{H}) := (\mathcal{M}(\mathcal{H}), \Omega(\mathcal{H}))$ is a probabilistic model representing the same

[3]This is one setting in which we would often *not* require Ω to be closed. For instance, if $\Omega = \Delta_\sigma(S, \Sigma)$, then Ω is not closed in the topology of event-wise convergence.

[4]Gleason's Theorem tells us that if $\dim(\mathcal{H}) > 2$, all probability weights arise in this way, but we will not make use of this fact.

quantum-mechanical system one associates with $\mathcal{H}$, but in a manner that puts its probabilistic structure in the foreground. Note that only if $\mathcal{H}$ is finite-dimensional, is $\Omega(\mathcal{H})$ compact.

While the model above is mathematically attractive, it is more usual to identify measurement outcomes with projection operators. The *projective* quantum test space consists of sets of projections p_i on $\mathcal{H}$ summing to $\mathbf{1}$. If ρ is a density operator on $\mathcal{H}$, we obtain a probability weight (also denoted by ρ) given by $\rho(p) = \mathrm{Tr}(\rho p)$. Letting Ω consist of all such pobability weights, we obtain what we can call the *projective* quantum model. More generally still, if $\mathfrak{A}$ is a von Neumann algebra (or, for that matter, any ring with identity), the collection $\mathcal{M}(\mathfrak{A})$ of sets E of projections $p \in \mathfrak{A}$ with $\sum_{p \in E} p = \mathbf{1}$ is a test space, on which every state on $\mathfrak{A}$ gives rise to a probability weight by evaluation. In this case, we might identify Ω with the state space (in the usual sense) of $\mathfrak{A}$, which is weak-$*$ compact, but which contains a host of non-normal states.

Two pathological examples The following simple examples are useful as illustrations of the range of possibilities comprended by the formalism sketched above.

Example 2.10 (Two bits) Let $\mathcal{M} = \{\{a, a'\}, \{b, b'\}\}$ be a test space consisting of two yes-no tests, one with outcomes a and a', the other with outcomes b and b'. The space of all probability weights on $\mathcal{M}$ is essentially the unit square, since such a weight α is determined by the pair $(\alpha(a), \alpha(b))$. While the state space is not a simplex, pure states *are* dispersion-free. The *square bit* B and *diamond bit* B' are the probabilistic models having the same test space, namely

$$\mathcal{M}(B) = \mathcal{M}(B') = \mathcal{M},$$

but the two different state spaces pictured below in Fig. 27.1.

The square bit figures, directly or indirectly, in the large literature on "PR boxes" inaugurated by Popescu and Rohrlich (1994). While the state space of B' is affinely isomorphic to that of B, it interacts with $\mathcal{M}$ differently. While both models are unital, in B', for each outcome x there is a *unique* state δ_x with $\delta_x(x) = 1$.

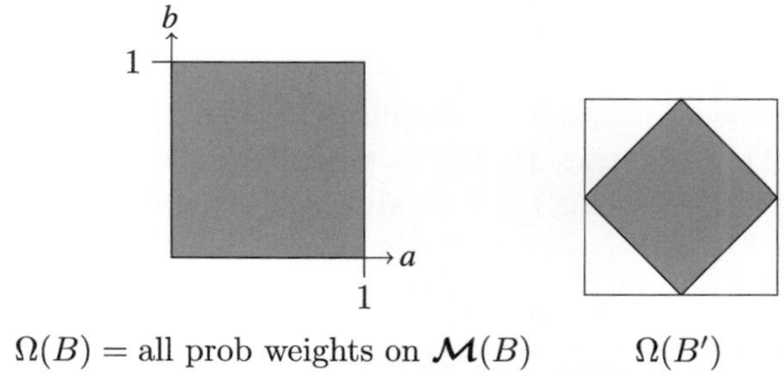

Fig. 27.1 Two bits

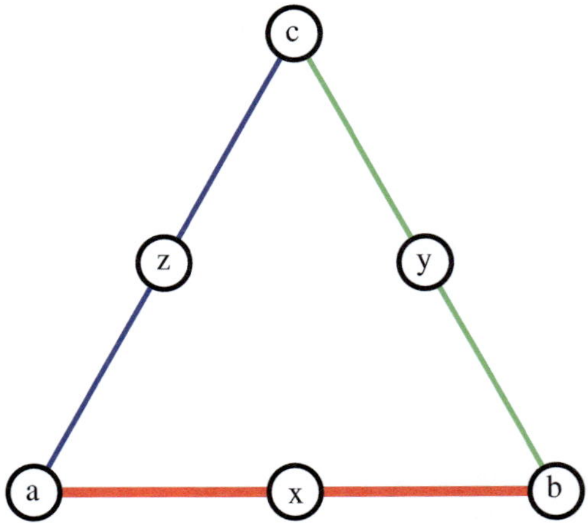

Fig. 27.2 A "Greechie diagram" of the Wright Triangle, in which outcomes belonging to a commont test lie along a smooth curve (here, a straight line)

Examples 2.11 (The Wright triangle[5]*)* A more interesting example consists of three overlapping, three-outcome tests pasted together in a loop, as in Fig. 27.2:

$$\mathbf{Pr}(\mathcal{M}) = \{\{a, x, b\}, \{b, y, c\}, \{c, z, a\}\}$$

The space $\mathbf{Pr}(\mathcal{M})$ of all probability weights on $\mathcal{M}$ is three-dimensional, since such a weight is uniquely determined by the triple $(\alpha(a), \alpha(b), \alpha(c))$. It is not difficult to check that is spanned by four dispersion-free states, determined by

$$\alpha(a) = \alpha(y) = 1, \beta(b) = \beta(z) = 1, \gamma(c) = \gamma(x) = 1 \text{ and } \delta(x) = \delta(y) = \delta(y) = 1,$$

and a fifth, non-dispersion free pure state

$$\epsilon(a) = \epsilon(b) = \epsilon(c) = 1/2, \ \epsilon(x) = \epsilon(y) = \epsilon(z) = 0.$$

In particular, Ω is not a simplex, and not all pure states are dispersion-free.

27.2.3 Probabilistic Models Linearized

It is often convenient to consider the vector space $\mathbf{V}(A) \leq \mathbb{R}^{X(A)}$ spanned by the state space $\Omega(A)$. This carries an obvious pointwise order. Writing $\mathbf{V}(A)_+$ for the cone of all $\alpha \in \mathbf{V}(A)$ with $\alpha(x) \geq 0$ for all $x \in X(A)$, it is not difficult to see that

[5]So well known in the quantum-logic community as to be almost a cliche, this gives rise to the simplest quantum logic that is an orthoalgebra but not an orthomodular poset. See Wilce (2017) for details. The example is named for R. Wright, one of Randall and Foulis's students.

each $\alpha \in \mathbf{V}(A)_+$ has the form $\alpha = r\beta$ where $\beta \in \Omega(A)$ and $r \geq 0$. Moreover, the coefficient r is unique, so $\Omega(A)$ is a *base* for the cone $\mathbf{V}(A)_+$.

An *effect* on A is a functional $f : \mathbf{V}(A) \to \mathbb{R}$ with $0 \leq f(\alpha) \leq 1$ for $\alpha \in \Omega(A)$. Every outcome $x \in X(A)$ determines an effect $\widehat{x} \in \mathbf{V}(A)^*$ by evaluation, that is, by $\widehat{x}(\alpha) = \alpha(x)$ for all $\alpha \in \mathbf{V}(A)$. We can regard an arbitrary f as representing an "in-principle" measurement outcome, not necessarily included among the outcomes in $X(A)$, but consistent with the convex structure of $\Omega(A)$: $f(\alpha)$ is the probability of the effect f occurring (if measured) in the state α. There is a unique *unit effect* u_A given by $u(\alpha) = 1$ for all $\alpha \in \Omega(A)$, representing the trivial outcome "something happened". If f is an effect, so is $u_A - f =: f'$; we interpret f' as representing the event that f did not occur. Thus, the pair $\{f, f'\}$ is an in-principle *test* (measurement, experiment) not necessarily included in $\mathcal{M}(A)$. More generally, if $f_1, \ldots, f_n$ is a set of effects with $\sum_i f_i = u_A$, we regard $\{f_1, \ldots, f_n\}$ as an "in principle" measurement or experiment—the standard term is "discrete observable"—not necessarily included in the catalogue $\mathcal{M}(A)$, but consistent with the structure of $\Omega(A)$. Each test $E \in \mathcal{M}(A)$ gives rise to a discrete observable $\{\widehat{x}|x \in E\}$ in this sense.

Examples 2.12

(a) As an illustration, let (S, Σ) be a measurable space, and let $\Delta = \Delta(S, \Sigma)$ be the space of finitely additive probability measures on Σ. Then $\mathbf{V}(\Delta)$ is the space of all signed measures $\mu = \mu_+ - \mu_-$ on Σ, where μ_+ and μ_- are positive measures. Any measurable function $\zeta : S \to [0, 1]$ defines an effect a through the recipe $a(\mu) = \int_S \zeta(s)d\mu(s)$. It is common to think of such an effect as a "fuzzy" or "unsharp" version of an indicator function (modelling, perhaps, a detector that responds with probability $\zeta(s)$ when the system is in a state s). The unit effect is the constant function 1. A discrete observable, accordingly, represents a generalization of a "fuzzy" or unsharp version of a finite measurable partition.

(b) In the case of a quantum-mechanical model, say $A = A(\mathcal{H})$, if we identify $\Omega(A)$ with the set of density operators on $\mathcal{H}$, then $\mathbf{V}(A)$ can be identified with the space of all compact self-adjoint operators on $\mathcal{H}$, ordered in the usual way; $\mathbf{V}(A)^*$ is then naturally isomorphic, as an ordered vector space, to the space of all bounded self-adjoint operators on $\mathcal{H}$, and an effect is an effect in the usual quantum-mechanical sense, that is, a positive self-adjoint operator a with $0 \leq a \leq \mathbf{1}$, where $\mathbf{1}$ is the identity operator on $\mathcal{H}$, which serves as the unit effect in this case. A discrete observable corresponds to a discrete positive-operator valued measure (POVM) with values in $\mathbb{N}$.

Unlike a test in $\mathcal{M}(A)$, a general discrete observable on A can have repeated values. Nevertheless, by considering their graphs, such observables can be organized into a test space, as follows. Here, $[0, u_A]$ denotes the set of all effects on A, and $(0, u_A]$ denotes the set of non-zero effects.

Definition 2.13 (Ordered tests) The space of *ordered tests* over the convex set Ω is the test space $\mathcal{M}^o(\Omega)$ with

(a) Outcome-space $X^o(\Omega) := \mathbb{N} \times (0, u]$;
(b) Tests, those finite sets $a \subseteq X^o$ that are graphs of mappings $a : I \to (0, u]$ with $\sum_{i \in I} a_i = u$, where $I \subseteq \mathbb{N}$ is finite.

In other words, tests in $\mathcal{M}^o(\Omega)$ are (graphs of) finite discrete observables with finite value-spaces $I \subseteq \mathbb{N}$. Every state $\alpha \in \Omega$ defines a probability weight α^o on $\mathcal{M}^o(\Omega)$ by $\alpha^o(i, a) = \alpha(a)$.[6] In effect, $\mathcal{M}^o(\Omega)$ is the largest test space supporting Ω as a separating set of probability weights.

27.2.4 Probabilistic Theories

A probabilistic *theory* ought to be more than a collection of probabilistic models. One also needs a collection of allowed mappings between such models turning $\mathcal{C}$ into a category. There are many plausible ways to define a morphism between probabilistic models, some more restrictive, some very general (see, e.g., Foulis and Randall 1978; Barnum and Wilce 2016). For the purposes of this chapter, it will be sufficient to consider the following, relatively strict definition (due to Foulis and Randall 1978). Recall that $\mathcal{E}v(\mathcal{M})$ denotes the set of events (subsets of tests) for a test space $\mathcal{M}$. Given two events $a, b \in \mathcal{E}v(\mathcal{M})$, we write $a \perp b$ to mean that $a \cap b = \emptyset$ and $a \cup b$ is again an event. That is, $a \perp b$ iff a and b are compatible (jointly testable), and mutually exclusive.

Definition 2.14 (Interpretations) Let $\mathcal{M}$ and $\mathcal{M}'$ be test spaces with outcome-sets X and X'. An *interpretation* from $\mathcal{M}$ to $\mathcal{M}'$ is a mapping $\phi : X \longrightarrow \mathcal{E}v(\mathcal{M}')$ such that if $x \perp y$ in X, then $\phi(x) \perp \phi(y)$ in $\mathcal{E}v(\mathcal{M}')$ and $\bigcup_{x \in E} \phi(x) \in \mathcal{M}'$ for all $E \in \mathcal{M}$.

In other words, an interpretation $\phi : \mathcal{M} \to \mathcal{M}'$ allows us to regard (to interpret!) each test $E \in \mathcal{M}$ as a coarse-grained version of a test in $\mathcal{M}'$, in a non-contextual way.

Definition 2.14 (Continued) The following terminology will be useful. An interpretation $\phi : \mathcal{M} \to \mathcal{M}'$ is *outcome-preserving* iff $\phi(x)$ is either empty or a singleton for every $x \in X$, in which case we can identify ϕ with the corresponding partial mapping from the outcome space $X = \bigcup \mathcal{M}$ to the outcome space $X' = \bigcup \mathcal{M}'$. If $\phi(x)$ is non-empty for every $x \in \bigcup \mathcal{M}$, then ϕ is *positive*. A positive, outcome-preserving interpretation $\phi : \mathcal{M} \to \mathcal{M}'$, amounts to a mapping $\phi : X \to X'$ with $\phi(\mathcal{M}) \subseteq \mathcal{M}'$. Where this mapping ϕ is injective, allowing us to identify $\mathcal{M}$ with a subset of $\mathcal{M}'$, we call it an *embedding*. Where $\phi(\mathcal{M}) = \mathcal{M}'$,

[6] It can be shown that, conversely, every probability weight on $\mathcal{M}^o(\Omega)$ is determined by a finitely-additive normalized measure on the effect algebra $[0, u]$, and hence, by a positive linear functional in $\mathbf{V}(A)^{**}$. Hence, probability weights on $\mathcal{M}^o(\Omega)$ that are continuous on $(0, u]$ in the latter's relative weak-$*$ topology, arise from elements of Ω.

we will say that ϕ is a *cover* of $\mathcal{M}'$ by $\mathcal{M}$. Where ϕ is bijective on both outcomes and tests, we call it an *isomorphism* of test spaces.

Examples 2.15

(a) If (S, Σ) and (T, Ξ) are measurable spaces, any measurable mapping $f : S \to T$ gives rise to an outcome-preserving interpretation $\phi : \mathcal{D}(T, \Xi) \to \mathcal{D}(S, \Sigma)$, given by $\phi(b) = \{f^{-1}(b)\}$ when $f^{-1}(b) \neq \emptyset$, and $\phi(b) = \emptyset$ otherwise.
(b) Suppose K_1 and K_2 are compact convex sets and $f : K_1 \to K_2$ is an affine mapping, we can define an outcome-preserving interpretation $\phi : \mathcal{M}^o(K_2) \to \mathcal{M}^o(K_1)$ by setting

$$\phi(i, b) = \begin{cases} \{(i, b \circ f)\} & b \circ f \neq 0 \\ \emptyset & b \circ f = 0 \end{cases}$$

(c) For an example of a positive but not outcome-preserving, interpretation, consider the inclusion mapping $\mathcal{E}v(\mathcal{M}) \setminus \{\emptyset\} \to \mathcal{E}v(\mathcal{M})$: we can understand this as an interpretation $\mathcal{M}^{\#} \to \mathcal{M}$ taking a non-empty event *qua* outcome of $\mathcal{M}^{\#}$ to the same event viewed as a set of outcomes of $\mathcal{M}$.

Notice that any interpretation $\phi : \mathcal{M}_1 \to \mathcal{M}_2$ gives rise to an affine mapping $\phi^* : \mathbf{Pr}(\mathcal{M}_2) \to \mathbf{Pr}(\mathcal{M}_1)$, defined for $\beta \in \Omega(\mathcal{M}_2)$ by $\phi^*(\beta) = \beta \circ \phi$.

Definition 2.16 (Interpretations of models) An interpretation $\phi : A \to B$ from a probabilistic model A to a probabilistic model B is an interpretation $\phi : \mathcal{M}(A) \to \mathcal{M}(B)$ such that $\phi^*(\Omega(B)) \subseteq \Omega(A)$. We say that ϕ is

(a) an *embedding* iff ϕ is an embedding of test spaces and ϕ^* is surjective;
(b) a *cover* iff ϕ is a cover of test spaces, in which case ϕ^* is injective.

The affine mapping $\phi^* : \Omega(B) \to \Omega(A)$ extends uniquely to a positive linear mapping $\phi^* : \mathbf{V}(B) \to \mathbf{V}(A)$, satisfying $u_A \circ \phi^* = u_B$. An apparently much more general notion of a morphism from a model A to a model B is a positive linear mapping $\Phi : \mathbf{V}(B) \to \mathbf{V}(A)$ with the feature that $u_A \circ \Phi \leq u_B$, that is, $u_B(\Phi(\beta)) \leq 1$ for all $\beta \in \Omega(B)$ (Barnum and Wilce 2016). Such a map represents a possibly "lossy" process that, given as input a state $\beta \in \Omega(B)$, produces an output state $\alpha := \Phi(\beta)/u_B(\Phi(\beta))$ with probability $u_B(\Phi(\beta))$—or, if $u_B(\Phi(\beta)) = 0$, simply destroys the system. However, given such a mapping Φ, and letting $\nabla(A)$ denote the convex set of sub-normalized states $r\alpha$ were $r \in [0, 1]$ and $\beta \in \Omega(B)$, we can define an interpretation $\phi : \mathcal{M}^o(\nabla(A)) \to \mathcal{M}^o(\Omega(B))$ as in Example 2.15 (b) above. Thus, we actually lose very little, if any, generality in restricting attention to categories of probabilistic models and interpretations. In the remainder of this chapter, a *probabilistic theory* will always mean such a category.

27.3 Classicality and Classical Representations

There are various ways of representing probabilistic models in terms of classical probabilistic models, most of them reasonably familiar in the context of QM. In this section, I want to discuss two of these that are rather canonical, both in the informal sense that they are more or less obvious, and apply to more or less *arbitrary* probabilistic models, and also in the technical sense that the constructions involved are actually functorial. Of these, it is my sense that only the first is really well known.

27.3.1 Classical Models and Classical Embeddings

Before we can discuss these representations, we need to address head on the central concern of this chapter: what does it mean (what *ought* it to mean) for a probabilistic model A to be "classical"? Whatever our final answer, presumably we would at least wish to consider models $A(S, \Sigma) := (\mathcal{D}(S, \Sigma), \Delta_\sigma(S, \Sigma))$ as classical. However, as discussed earlier, *even classically* it won't do to restrict attention to such models: any model of the form $(\mathcal{D}(S, \Sigma), \Omega)$ where Ω is *any* subset of $\Delta_\sigma(S, \Sigma)$, should almost certainly count as "classical", in the sense of falling within the scope of classical probability theory (Lindsay 1995).

We should also pause to ask whether or not the restriction to σ-additive measures should be regarded as one of the defining features of classicality. If μ is a finitely additive measure on an algebra Σ of sets, then letting $T = \beta(\Sigma)$ be the Stone space of Σ, regarded as a Boolean algebra, one finds that, owing to the fact that elements of Σ are clopen in T and T is compact, *no* countable disjoint family of sets in Σ has union equal to T. Hence, *every* measure on (T, Σ) is countably additive by default. In different language, whether or not a measure on a Boolean algebra Σ is countably additive depends on the particular representation of Σ as an algebra of sets.

For this reason, it seems prudent to allow any Kolmogorovian model, in the sense of Example 2.8—that is, any probabilistic model of the form $(\mathcal{D}(S, \Sigma), \Omega)$, where Σ is any algebra of subsets of S and Ω is any convex set of probability measures on Σ, countably additive or otherwise, to count as "classical". But if this is so, it seems equally reasonable to admit as classical models of the form $(\mathcal{M}, \Omega)$ as long as $\mathcal{M} \subseteq \mathcal{D}(S, \Sigma)$ and $\Omega \subseteq \Delta(S, \Sigma)$. That is, models admitting embeddings, in the sense of Definition 2.16, into Kolmogorovian models in the sense of Example 2.8, should themselves be considered "classical" in a broad sense.

When does an abstract model $(\mathcal{M}, \Omega)$ admit such a *classical embedding*? If $\phi : \mathcal{M}(A) \to \mathcal{D}(S, \Sigma)$ is an interpretation, then every point-mass δ_s in $\Delta(S, \Sigma)$ pulls back along ϕ to a dispersion-free probability weight on $\mathcal{M}(A)$. If ϕ is to be an embedding of test spaces, there must be enough of these dispersion-free states to separate outcomes in $X(A)$. Moreover, if ϕ is to be an embedding of probabilistic models, every state $\alpha \in \Omega(A)$ must belong to the closed convex hull of these dispersion-free weights.

Definition 3.1 (UDF models) A probabilistic model A is *unitally dispersion-free* or *UDF* iff every state $\alpha \in \Omega(A)$ lies in the closed convex hull of the set of dispersion-free probability weights on $\mathcal{M}(A)$.

Versions of the following can be found in many places in the quantum-logical literature.

Lemma 3.2 *A model admits a classical embedding iff it is UDF.*

Proof The "only if" direction is clear from the discussion above. For the converse, suppose $(\mathcal{M}, \Omega)$ is UDF, and let S be the set of dispersion-free probability weights on $\mathcal{M}$. For each $x \in X := \bigcup \mathcal{M}$, let

$$a_x \;:=\; \{s \in S | s(x) = 1\} \subseteq S.$$

Then $\{a_x | x \in E\}$ is a partition of S. Let Σ be the algebra of subsets of S generated by sets of the form a_x. Then the mapping $\phi : X \to \Sigma$ given by $\phi(x) = a_x$ gives rise to an outcome-preserving interpretation $\mathcal{M} \to \mathcal{D}(S, \Sigma)$, which is an embedding if and only if S separates points of X. The mapping $\phi^* : \mathbb{R}^\Sigma \to \mathbb{R}^X$ given by $\phi^*(\mu)(x) = \mu(a_x)$ is continuous with respect to the product topologies on these spaces, and for each $s \in S$, $\phi^*(\delta_s) = s$. It follows that $\overline{\mathrm{con}}(S) = \phi^*(\Delta(S, \Sigma))$. Hence, the probability weights in $\overline{\mathrm{con}}(S)$ have a common classical explanation. Given any closed convex set $\Omega \subseteq \overline{\mathrm{con}}(S)$, we then have an embedding of the model $(\mathcal{M}, \Omega)$ into the Kolmogorovian model $(\mathcal{D}(S, \Sigma), (\phi^*)^{-1}(\Omega))$. □

Let me stress that it is not sufficient for $\mathcal{M}(A)$ simply to have a large supply of dispersion-free probability weights: we also require that all of the allowed states $\alpha \in \Omega(A)$ arise as (limits of) averages of these. As an example, the Wright Triangle admits many dispersion-free states, but also a non-dispersion-free extreme state; this state can not be explained in terms of a classical embedding.

Of course, if $\mathcal{M}(A)$ has *no* dispersion-free states at all, then such a classical embedding is impossible in any case. Gleason's Theorem tells us that $\Omega(\mathcal{H})$ has no dispersion-free states if $\dim(\mathcal{H}) \geq 3$; this is the substance of one of the strongest "no-go" theorems for hidden variables. However, this is far from the end of the story. For one thing, the classical embeddings considered above are *non-contextual*. That is, if $x \in E \cap F$, where $E, F \in \mathcal{M}(A)$, then $\phi(x)$ does not depend on whether we regard x as an outcome of E or of F. As was realized very early on making $\phi(x)$ depend also on the choice of "measurement context" (e.g., E or F, above) allows for much more flexibility.

Besides the existence of an abundance of dispersion-free states, a distinctive feature of Kolmogorovian models is that any two partitions $E, F \in \mathcal{D}(S, \Sigma)$ can be regarded as coarse-grained versions of a *common refinement*: if

$$G = \{\, a \cap b \mid a \in E, b \in F,\ a \cap b \neq \emptyset \,\}$$

then $G \in \mathcal{D}(S, \Sigma)$, and every $a \in E$, and likewise every $b \in F$, is in an obvious sense equivalent to a subset of G. This allows us to make a simultaneous *joint* measurement of E and F, by performing a measurement of G. One can define a general notion of the refinement of one test by another in the context of an arbitrary probabilistic model, as follows:

Definition 3.3 (Refinement) A test $E \in \mathcal{M}(A)$ *refines* a test $F \in \mathcal{M}(A)$ iff there exists a surjection $f : E \to F$ such that $\alpha(y) = \alpha(f^{-1}(y))$ for every $y \in F$ and every state $\alpha \in \Omega(A)$.

Recall that a probabilistic model A is *unital* if every outcome $x \in X(A)$ has probability one in at least one state, i.e., there is at least one state $\alpha \in \Omega(A)$ with $\alpha(x) = 1$.

Theorem 3.4 *Let A be a unital model in which every pair of tests has a common refinement. Then A has a classical embedding.*

The proof is given in Appendix A.

Conversely, of course, if A has a classical embedding, then $\mathcal{M}(A)$ embeds in a test space $\mathcal{D}(S, \Sigma)$, in which every pair of tests has a common refinement. Thus, a necessary and sufficient condition for a unital model A to have a classical embedding is that it be possible to embed A in a model B in which every pair of tests has a common refinement. To this extent, one might reasonably say that it is the existence (or at least, the mathematical possibility) of joint measurements that is the hallmark of "classical" probabilistic theories. However, it is very important to note again at this point, that this is an *entirely contingent* feature of classical probability theory. That is, there is no "law of thought" that tells us that any two measurements or experiments can be performed jointly.

27.3.2 Classical Extensions

An embedding is not the only way of explaining one mathematical object in terms of another. One can also consider representing the object to be explained as a *quotient* of a more familiar object. In connection with probabilistic models, this line of thinking was explored by Holevo (1982) and, slightly later, by Bugajsky and Beltrametti (Beltrametti and Bugajski 1995; see also Hellwig and Singer 1991; Hellwig et al. 1993.) In the literature, this sort of representation of a quantum model is usually called a *classical extension*. The basic idea is that any convex set can arise as a quotient of another under an affine surjection. As a simple example, a square can arise as the projection of a regular tetrahedron (a 4-simplex) on a plane.

To develop this idea further, we will need a bit of background on Choquet theory (Alfsen 1971). Recall that the σ-field of *Baire sets* of a compact Hausdorff space S is the smallest σ-algebra containing all sets of the form $f^{-1}(0)$ where f ranges over continuous real-valued functions on S. Let K be a compact convex set, that is, a compact, convex subset of a locally convex topological vector space V, and let $\Delta_o(K)$ be the set of Baire probability measures thereon. This is an infinite-dimensional simplex in the sense that $\mathbf{V}(\Delta_o(K))$ is a vector lattice (Alfsen 1971). For each $\mu \in \Delta(K)$, define the *barycenter* of μ, $\widehat{\mu} \in V^{**}$, by

$$\widehat{\mu}(f) = \int_K f d\mu$$

for all functionals $f \in V^*$. Using the Hahn-Banach Theorem, one can show that $\widehat{\mu} \in K$ (Alfsen 1971, I.2.1). Thus, we have an affine mapping $\mu \mapsto \widehat{\mu}$ from $\Delta_o(K)$ to K. The barycenter of the point-mass δ_α at $\alpha \in K$ is evidently α, so this mapping is surjective. In this sense, *every compact convex set is the image of a simplex* under an affine mapping.

Definition 3.5 (Classical extensions) A *classical extension* of a probabilistic model A consists of a measurable space (S, Σ) and an affine surjection $q : \Delta_\sigma(S, \Sigma) \to \Omega(A)$.

This makes no reference to $\mathcal{M}(A)$. However, the affine surjection $q : \Delta_o(S) \to \Omega(A)$ can be dualized to yield a mapping $q^* : X(A) \to \text{Aff}(\Delta_o(S))$ taking every outcome $x \in X(A)$ to the functional $q^*(x)(\mu) = \widehat{\mu}(x)$, which is evidently an effect on $\Delta(S)$. Now, such an effect can be understood as a "unsharp" version of an indicator function, as discussed in Example 2.12(a) Accordingly, since $\sum_{x \in E} q^*(x)$ is identically 1, q^* is an effect-valued weight on $\mathcal{M}(A)$, representing each test as a partition of 1 by such fuzzy indicator functions, that is, $q^*(E)$ is a (discrete) "unsharp random variable" on S. For more on this, see Bacciagaluppi (2004). In such a representation, in other words, the state space is essentially classical (it's simply a set of probability measures), while outcomes and tests become "unsharp". While this may represent a slight extension of the apparatus of standard, Kolmogorovian probability theory, it is certainly within the scope of classical probability theory in the somewhat wider sense that concerns us here.

Another way to put all this is that a classical extension comes along with (and determines) an interpretation $\mathcal{M}(A) \to \mathcal{M}^o(\Delta(S, \Sigma))$, where the latter is the space of ordered tests on $\Delta(S, \Sigma)$, in the sense of Definition 2.13. This then gives us an interpretation of models $A \to (\mathcal{M}^o(\Delta(S, \Sigma)), \Delta(S, \Sigma))$. In this sense, a classical extension is an "unsharp" version of a classical embedding. However, as the discussion above shows, *every* probabilistic model has a *canonical* classical extension obtained by taking $S = \Omega(A)$ and Σ the Baire field of Ω. This construction is even functorial, since the construction $A \mapsto \Omega(A)$ is (the object part of) a contravariant functor from the category of probabilistic models to the category of compact convex sets, while $K \mapsto \Delta_0(K)$ and $K \mapsto \mathcal{M}^o(K)$ are, respectively, a covariant endofunctor on the category of compact convex sets and continuous affine maps, and from this category to the category of test spaces and interpretations. Putting these together gives us a covariant functor $A \mapsto (\mathcal{M}^o(\Delta_0(\Omega(A))), \Delta_0(\Omega(A)))$ from arbitrary (locally finite) probabilistic models to unsharp Kolmogorovian models.

One can also construct a classical extension in which the carrier space is the set K_{ext} of extreme points of K. This need not be a Borel subset of K; however, it becomes a measurable space if we let Σ_{ext} be the *trace* of K's Baire field Σ_o on K_{ext}, i.e., $\Sigma_{\text{ext}} = \{ b \cap K_{\text{ext}} \mid b \in \Sigma \}$. If μ is a probability measure on Σ_{ext}, then we can pull this back along the boolean homomorphism $\phi : b \mapsto b \cap K_{\text{ext}}$ to a probability measure $\mu \circ \phi$ on Σ_o, the barycenter of which, as defined above, defines a barycenter for μ; that is, we take $\widehat{\mu} := \widehat{\mu \circ \phi}$. The *Bishop-deLeeuw Theorem* (see Alfsen 1971, I.4.14) asserts that every point of K can be represented as the barycenter of a σ-additive probability measure on Σ_{ext}.

We can regard probability measures μ in the simplex $\Delta(S)$, $S = \Omega(A)_{\text{ext}}$, as representing *ensembles* of pure states, that is, preparation procedures that produce a particular range of pure states, with prescribed probabilities. The quotient map $\Delta(S) \to \Omega(A)$ simply takes each such ensemble to its probability-weighted average. Where $\Omega(A)$ is not a simplex, many different ensembles will yield this same average; operationally, that is, using the measurements available in $\mathcal{M}(A)$, distinct ensembles averaging to the same state are indistinguishable, so treating a state $\alpha \in \Omega(A)$ as "really" arising from one, rather than another, such ensemble represents a kind of contextuality, what Spekkens (2005) calls *preparation contextuality*.

27.3.3 Semiclassical Covers

Recall from Sect. 27.2.1 that the *semiclassical cover*, $\widetilde{\mathcal{M}}$, of a test space $\mathcal{M}$ is given by

$$\widetilde{\mathcal{M}} = \{ \widetilde{E} \mid E \in \mathcal{M} \},$$

where, for $E \in \mathcal{M}$, $\widetilde{E} = \{(x, E) | x \in E\}$. The outcome-set of $\widetilde{\mathcal{M}}$ is thus $\widetilde{X} = \{(x, E) | x \in E \in \mathcal{M}\}$, i.e., the coproduct of the tests $E \in \mathcal{M}$. There is an obvious outcome-preserving interpretation $\pi : \widetilde{\mathcal{M}} \to \mathcal{M}$, given by $\pi(x, E) = \{x\}$. Any probability weight α on $\mathcal{M}$ pulls back along this to a probability weight $\widetilde{\alpha} := \pi^*(\alpha)$ on $\widetilde{\mathcal{M}}$, given by $\widetilde{\alpha}(x, E) := \alpha(x)$ for all $x \in E \in \mathcal{M}(A)$. Given a model A, we define the semiclassical cover of A to be the model $\widetilde{A} := (\widetilde{\mathcal{M}}(A), \widetilde{\Omega}(A)\}$, where $\widetilde{\mathcal{M}}(A)$ is the semiclassical cover of $\mathcal{M}(A)$ and $\widetilde{\Omega}(A) := \{ \widetilde{\alpha} \mid \alpha \in \Omega \}$. The mapping $\alpha \mapsto \widetilde{\alpha}$ is an affine injection, so $\widetilde{\Omega}(A)$ is isomorphic to $\Omega(A)$. Thus, $\widetilde{A}$ differs from A *only* in the structure (that is, the comparative lack of structure) of its test space.

Since $\widetilde{\mathcal{M}}(A)$ is semiclassical, its space $\mathbf{Pr}(\widetilde{\mathcal{M}})$ of probability weights is essentially the Cartesian product $\Pi_{E \in \mathcal{M}} \Delta(E)$ of the finite-dimensional simplices $\Delta(E)$ of probability weights on the various tests $E \in \mathcal{M}$. Indeed, we can represent $\alpha \in \mathbf{Pr}(\widetilde{\mathcal{M}}(A))$ uniquely as (α_E) where α_E is α's restriction to $E \in \mathcal{M}(A)$. Hence, $\widetilde{\mathcal{M}}$ supports a wealth of dispersion-free probability weights, namely, those weights obtained by selecting a dispersion-free probability weight—a point-mass—from each simplex $\Delta(E)$.

In what follows, I'll write $S(A)$ for the set of all such dispersion-free states on $\widetilde{\mathcal{M}}(A)$. It is easy to show that these are exactly the extreme probability weights on $\widetilde{\mathcal{M}}(A)$, and that $S(A)$ is closed in $\mathbf{Pr}(\widetilde{\mathcal{M}}(A))$ (see Appendix B). This allows us to represent every $\alpha \in \mathbf{Pr}(\widetilde{\mathcal{M}})$ as the barycenter of a Borel probability measure on $S(A)$ (as every Baire probability measure on a compact space has a unique Borel extension). Letting $\Delta(S(A))$ denote the simplex of Borel probability measures on $S(A)$, this gives us an affine surjection $\phi : \Delta(S(A)) \to \mathbf{Pr}(\widetilde{\mathcal{M}}(A))$. Also, $\psi : (x, E) \mapsto \{\alpha \in S|\alpha(x, E) = 1\}$ gives an interpretation of each test $\widetilde{E} \in \widetilde{\mathcal{M}}(A)$ as an element of $\mathcal{D}(S, \Sigma)$, where Σ is the Borel algebra of $S(A)$, and it is straightforward that $\phi = \psi^*$. We therefore have the following picture:

$$\begin{array}{ccc} & & \mathcal{D}(S(A), \Sigma) \\ & & \uparrow \psi \\ \mathcal{M}(A) & \xleftarrow[\pi]{} & \widetilde{\mathcal{M}}(A) \end{array} \qquad \begin{array}{ccc} & & \Delta(S(A)) \\ & & \downarrow \phi = \psi^* \\ \Omega(A) & \xrightarrow[\pi^*]{} & \mathbf{Pr}(\widetilde{\mathcal{M}}(A)) \end{array}$$

The dual mapping $\pi^*(\alpha)(x, E) = \alpha(x)$ is injective, while ψ^* is surjective: again, $\Omega(A)$ is a convex subset of a quotient of a simplex.

However, in this representation observables associated with tests $E \in \mathcal{M}(A)$ correspond to *sharp* classical random variables on $S(A)$: letting

$$\widetilde{a}_x = \{s \in S(A)|s(x, E) = 1\},$$

$\{a_x|x \in E\}$ is a partition of $S(\widetilde{A})$.

This gives us a very simple representation of an arbitrary model as a quotient of a sub-model of Kolmogorovian model. To the extent that a representation of a model A as a quotient of a model B allows us to "explain" the model A in terms of the model B, and to the extent that representing a model B as a submodel of a model C also allows us to explain B in terms of C, then—to the extent that "explains" is transitive—we see that every model can be explained in terms of a classical one. This construction is even functorial: if $\phi : A \to B$ is an interpretation from a model A to a model B, there is a canonical extension of ϕ to an interpretation $\widetilde{\phi} : \widetilde{A} \to \widetilde{B}$, given by $\phi(x, E) = (\phi(x), \phi(E))$. Hence, an entire probabilistic theory can be rendered essentially classical, if we are willing to embrace contextuality.

Hidden Variables This is as good a place as any to clarify how the foregoing discussion connects to traditional notions of "hidden variables" (or "ontological representations" (Spekkens 2005)). Briefly, a (not necessarily deterministic, not necessarily contextual) *hidden variables representation* of a probabilistic model A consists of a measurable space Λ and, for every test $E \in \mathcal{M}$, a conditional probability distribution $p(\,\cdot\,|\lambda, E)$ on E—that is, $p(x|\lambda, E)$ is a non-negative real number for each $x \in E$, and $\sum_{x \in E} p(\,x \mid \lambda, E\,) = 1$. It is also required that, for every state $\alpha \in \Omega(A)$, there exist a probability measure μ on Λ such that for every $x \in X(A)$ and every test E with $x \in E$,

$$\int_\Lambda p(x|\lambda, E)d\mu(\lambda) = \alpha(x). \tag{27.1}$$

Now, if we rewrite $p(x|\lambda, E)$ as $p_\lambda(x, E)$, we see that p_λ is simply a probability weight on $\widetilde{\mathcal{M}}(A)$. We can then write (27.1) as

$$\alpha(x) = \left(\int_\Lambda p_\lambda d\mu(\lambda) \right) (x, E) \tag{27.2}$$

which merely asserts that α is the μ-weighted average of a family of probability weights on $\widetilde{\mathcal{M}}(A)$, parametrized by Λ.

27.3.4 Discussion

In view of the classical representations discussed in Sects. 27.3.2 and 27.3.3, it seems that we can always understand a "generalized probabilistic model", or, indeed, a generalized probabilistic theory, in terms of a classical model or theory. The only departure from classical Kolmogorovian probability theory, in these representations, is a restriction—which one can regard as epistemic or as "physical", as the case warrants—on which ensembles of probability measures we can prepare, which (classical) measurements we can perform, and hence, on which measurement outcomes and which mixtures of states we can distinguish. To be sure, we are under no obligation to take ensembles in $\Delta(\Omega_{\text{ext}})$ or dispersion-free states on $\widetilde{\mathcal{M}}$ seriously as part of any given physical theory, and we may also prefer to do without them for reasons of mathematical economy; but both are mathematically meaningful, and certainly have a place in any developed *general* probability theory. To this extent, arbitrary probabilistic models, and even probabilistic theories, have fairly canonical "explanations" in Kolmogorovian terms.

In the next section I will go just a step further and point out that, for models with certain strong symmetry properties (including quantum models), there is available a rather different kind of "classical" representation: one in which the model's probabilistic apparatus is *entirely* classical—there is, in fact, only one basic measurement. In such a representation it is, rather, the interplay between a system's symmetries, its states, and this perfectly classical probabilistic apparatus, that is (in a sense I try to pin down) perhaps not quite "classical".

27.4 Dynamical Models and Dynamical States

One of the things that most conspicuously distinguishes quantum probabilistic models from more general ones is the their very high degree of symmetry: given any two pure quantum states, there is a unitary operator taking one to the other, and likewise, given any two projective orthonormal bases, there is a unitary mapping one to the other. In this section, I want to sketch a different kind of classical representation, available for locally finite models whenever such a great deal of symmetry is in play. This is mathematically similar to the representation in terms of the semiclassical cover; conceptually, however, it is leaner, in that it privileges a single observable, and encodes the structure of the model in the ways in which probabilities on this observable can change.

27.4.1 Models with Symmetry

In this section, we will consider probabilistic models A that, in addition to the test space $\mathcal{M}(A)$ and state space $\Omega(A)$, are equipped with a distinguished *symmetry group* $G(A)$ acting on $X(A)$, $\mathcal{M}(A)$ and $\Omega(A)$. In most applications, $G(A)$ will be a lie group and, in our finite dimensional setting, compact.

Definition 4.1 (Symmetries of test spaces) A *symmetry* of a test space $\mathcal{M}$ with outcome-set $X = \bigcup \mathcal{M}$ is an isomorphism $g : \mathcal{M} \to \mathcal{M}$, or, equivalently, a bijection $g : X \to X$ such that $\forall E \subseteq X$, $g(E) \in \mathcal{M}$ iff $E \in \mathcal{M}$. We write $\text{Aut}(\mathcal{M})$ for the group of all symmetries of $\mathcal{M}$ under composition.

Note that $\text{Aut}(\mathcal{M})$ also has a natural right action on $\text{Pr}(\mathcal{M})$, since if $g \in \text{Aut}(\mathcal{M})$ and $\alpha \in \mathbf{Pr}(\mathcal{M})$, then $\alpha \circ g$ is again a probability weight, and the mapping $g^* : \alpha \mapsto \alpha \circ g$ is an affine bijection $\text{Pr}(\mathcal{M}) \to \text{Pr}(\mathcal{M})$.

Definition 4.2 (Dynamical models) A *dynamical probabilistic model* is a probabilistic model A with a distinguished *dynamical group* $G(A) \leq \text{Aut}(\mathcal{M}(A))$ under which $\Omega(A)$ is invariant (Barnum and Wilce 2016).

I use the terms *dynamical model* and *dynamical group* because in situations in which $G(A)$ is a Lie group, a *dynamics* for (or consistent with) the model will be a choice of a continuous one-parameter group $g \in \text{Hom}(\mathbb{R}, G(A))$, that is, a mapping $t \mapsto g_t$ with $g_{t+s} = g_t g_s$, which we interpret at tracking the system's evolution over time: if α is the state at some initial time, $g_t(\alpha) = \alpha \circ g_{-t}$ is the state after the elapse of t units of time. That g *is* a homomorphism encodes a Markovian assumption about the dynamics, namely, that a system's later state depends only on its initial state and the amount of time that has passed, rather than on the system's entire history.

Definition 4.3 (Symmetric and fully symmetric models) A dynamical probabilistic model A is *symmetric* iff $G(A)$ acts transitively on the set $\mathcal{M}(A)$ of tests, and the stabilizer, $G(A)_E$, of a (hence, any) test $E \in \mathcal{M}$ acts transitively on E. Note that this implies that all tests have a common size. We say that A is *fully symmetric* iff, additionally, for any test $E \in \mathcal{M}(A)$ and any permutation $\sigma : E \to E$, there exists at least one $g \in G(A)$ with $gx = \sigma(x)$ for every $x \in E$.

An equivalent way to express full symmetry is to say that all tests $E, F \in \mathcal{M}(A)$ have the same cardinality, and, for every bijection $f : E \to F$, there exists some $g \in G(A)$ with $gx = f(x)$ for every $x \in E$.

The quantum probabilistic model $A(\mathcal{H})$ is fully symmetric under the unitary group $\mathcal{U}(\mathcal{H})$, since any bijection between two (projective) orthonormal bases for $\mathcal{H}$ extends to a unitary operator.

We can reconstruct a symmetric test space $\mathcal{M}(A)$ from any one of its tests, plus information about its symmetries (Wilce 2009). Suppose we are given, in addition to the group $G(A)$, a single test $E \in \mathcal{M}(A)$, an outcome $x_o \in E$, and the two subgroups

$$H := \{ g \in G(A) \mid gE = E \} \text{ and } K := \{ g \in G(A) \mid gx_o = x_o \}.$$

We then have a canonical bijection $G(A)/K \to X(A)$, sending gK to gx_o, and sending $\{hK|h \in H\}$ bijectively onto E; $\mathcal{M}(A)$ can then be recovered as the orbit of E in $X(A)$. Thus, more abstractly, one could start with a triple (G, H, K) where G is a group and H, K are subgroups of G, define $X := G/K$ and $E = \{hK|h \in H\}$, and set $\mathcal{M} = \{gE|g \in G\}$. If H acts as the full permutation group on E, $\mathcal{M}$ will be fully symmetric under G. The choice of a closed convex set $\Omega \subseteq \Pr(\mathcal{M})$, invariant under G, completes the picture and gives us a G-symmetric probabilistic model.

The simplest cases are those in which $H \simeq \mathfrak{S}(E)$. This holds for ordinary quantum test spaces, since every permutation of an orthonormal basis determines a unique unitary. In what follows, we concentrate on this case. (For a projective quantum test space, in which E is a maximal set of pairwise orthogonal rank-one projections, H is generated by permutation matrices acting on E, plus unitaries commuting with these projections.)

27.4.2 A Representation in Terms of Dynamical States

We now reformulate these ideas in a way that gives more prominence to the chosen test $E \in \mathcal{M}(A)$. In brief: if we hold the test E fixed, but retain control over both the state space $\Omega(A)$ and the dynamical group $G(A)$, we obtain a mathematically equivalent picture of the model, but one in which the probabilistic structure is, from a certain point of view, essentially classical.

Consider the following situation: one has some laboratory apparatus, defining an experiment with outcome-set E. This is somehow coupled to a physical system having a state-space Ω, governed by a Lie group G. That is, G acts on Ω in such a way that all possible evolutions of this system are described by an initial state $\alpha_o \in \Omega$ and a one-parameter subgroup $(g_t)_{t\in\mathbb{R}}$ of G. It is harmless, and will be convenient, to take Ω to be a *right* G-space, that is, to denote the image of state $\alpha \in \Omega$ under the action of $g \in G$ by αg. We may suppose that the way in which the apparatus is coupled with the system manifests itself probabilistically, by a function

$$p : \Omega \times E \to [0, 1]$$

giving the probability $p(\alpha, x)$ to obtain outcome $x \in E$ when the system is in state α.[7] I will assume in what follows that the probability weights $p(\alpha, \cdot)$ separate outcomes in E, that is, that for all outcomes $x, y \in E$,

$$(\forall \alpha \in \Omega \; p(\alpha, x) \; = \; p(\alpha, y) \,) \; \Rightarrow \; x = y. \tag{27.3}$$

(If not, we can factor out the obvious equivalence relation on E.) I will also suppose that the experiment E and the group G together separate states, that is, knowing how the probabilities $p(\alpha g, x)$ vary with $g \in G$, for all outcomes $x \in E$, is enough to determine the state α uniquely. In other words,

$$(\forall g \in G, \forall x \in E \; p(\alpha g, x) = p(\beta g, x) \,) \; \Rightarrow \; \alpha = \beta \tag{27.4}$$

Notice that for each state $\alpha \in \Omega$, we can define a mapping $\widehat{\alpha} : G \to \Delta(E)$, where $\Delta(E)$ is the simplex of probability weights on E, given by

$$\widehat{\alpha}(g)(x) = p(\alpha g, x).$$

Alternatively, we can treat $\widehat{\alpha}$ as a mapping $G \times E \to [0, 1]$ with $\sum_{x \in E} \widehat{\alpha}(g, x) = 1$, by writing $\widehat{\alpha}(g, x) := \widehat{\alpha}(g)(x)$. I will leave it to context to determine which of these representations is intended. In either case, I will refer to $\widehat{\alpha}$ as the *dynamical state* associated with state $\alpha \in \Omega$. The state-separation condition (27.4) tells us that $\alpha \mapsto \widehat{\alpha} \in \Delta(E)^G$ is injective, so if we wish, we can *identify* states with the corresponding dynamical states.

The mapping $\widehat{\alpha} \in \Delta(E)^G$ is a (discrete) random probability measure on G, while $\widehat{\alpha} : G \times E \to [0, 1]$ is a discrete Markov kernel on $G \times E$. Thus, nothing we have done so far takes us outside the range of classical probability theory. In fact, if G is compact, there is a natural measure on $G \times E$, namely the product of the normalized Haar measure on G and the counting measure on E, such that

$$\int_{G \times E} \widehat{\alpha}(g, x) d(g, x) = \int_G \left(\sum_{x \in E} \widehat{\alpha}(g, x) \right) dg = \int_G 1 dg = 1.$$

In other words, for each state $\alpha \in \Omega$, $\widehat{\alpha}(g, x)$ defines a probability density on $G \times E$. Moreover, as $\sum_{x \in E} \widehat{\alpha}(g, x) = 1$, the conditional density $\widehat{\alpha}(x|g)$ is exactly $\widehat{\alpha}(g) \in \Delta(E)$. Nevertheless, *any symmetric pobabilistic model* provides an example of this scenario: simply chose a test $E \in \mathcal{M}(A)$, and let $p(\alpha, x) = \alpha(x)$ and $\widehat{\alpha}(g)(x) = \alpha(gx)$ for all $\alpha \in \Omega(A)$ and any $x \in E$. Conversely, as we will see, one can reconstruct a symmetric probabilistic model A with $E \in \mathcal{M}(A)$ and $\Omega \simeq \Omega(A)$ from the "classical" data described above, that is, the test E, the state space Ω, the symmetry group G and the probabilistic coupling p.

[7] I prefer to write this as $p(\alpha, x)$ rather than as $p(x|\alpha)$ in order to make the covariance conditions below come out more prettily.

Definition 4.4 (Implementing a permutation) Let E, Ω, G and p be as above, and let σ be a permutation of the outcome-set E. Then we shall say that $g \in G$ *implements* σ iff $p(\alpha g, x) = p(\alpha, \sigma x)$ for all $\alpha \in \Omega$ and all $x \in E$.

Our outcome-separation assumption (2.7.3) implies that if $p(\alpha, \sigma x) = p(\alpha, \tau x)$ for all $\alpha \in \Omega$ and all $x \in E$, then $\sigma x = \tau x$, i.e., $\sigma = \tau$. Thus, if $g \in G$ implements a permutation $\sigma \in \mathfrak{S}(E)$, it implements only one such permutation, which we may denote by σ_g.

Lemma 4.5 *Let $\mathbb{H}$ be the set of all group elements $g \in G$ implementing permutations of E. Then $\mathbb{H}$ is a subgroup of G and $\sigma : \mathbb{H} \to \mathfrak{S}(E)$, $g \mapsto \sigma_g$, is a homomorphism.*

Proof Let $g, h \in \mathbb{H}$. Then for all $\alpha \in \Omega$ and all $x \in E$, we have

$$p(\alpha gh, x) = p(\alpha g, \sigma_h x) = p(\alpha, \sigma_g \sigma_h x).$$

Hence, $gh \in \mathbb{H}$ with $\sigma_{gh} = \sigma_g \sigma_h$. We also have

$$p(\alpha g^{-1}, x) = p(\alpha g^{-1}, \sigma_g \sigma_g^{-1} x) = p(\alpha g^{-1} g, \sigma_g^{-1} x) = p(\alpha, \sigma^{-1} x)$$

so $g^{-1} \in H$. □

Thus, E carries a natural $\mathbb{H}$ action. In what follows, I will simplify the notation by writing hx for $\sigma_h x$, where $h \in \mathbb{H}$ and $x \in E$. It is natural to consider cases in which E is *transitive* as an $\mathbb{H}$-set, meaning that for every $x, y \in E$, there exits at least one $h \in H$ with $hx = y$. This motivates the following

Definition 4.6 (Dynamical classical models) A *dynamical classical model* (DCM) consists of groups $H \leq G$, a transitive left H-set E, a convex right G-space Ω, and a Markov kernel $p : \Omega \times E \to [0, 1]$ such that $p(\alpha, hx) = p(\alpha h, x)$ for all $\alpha \in \Omega$, $h \in H$ and $x \in E$.

Given a DCM as defined above, for each state $\alpha \in \Omega$ define $\widehat{\alpha} : G \to \Delta(E)$ by

$$\widehat{\alpha}(g)(x) = p(\alpha g, x).$$

We shall say that a DCM is *state-determining* iff $\alpha \mapsto \widehat{\alpha}$ is injective, i.e., the probability weights $\widehat{\alpha}(g)$ on $\Delta(E)$, as g ranges over G, determine the state α. Henceforth, I will assume that all DCMs are state-determining.

Any DCM can be reinterpreted as a symmetric dynamical-model in a routine way, which we shall now describe. Let $x_o \in E$ and set

$$\mathbb{K} := \{ g \in G \mid p(\alpha g, x_o) = p(\alpha, x_o) \ \forall \alpha \in \Omega \}.$$

Arguing in much the same way as above, it is easy to see that $\mathbb{K}$, too, is a subgroup of G. The following slightly extends a result from Wilce (2005); see also Wilce (2009):

Theorem 4.7 *Let E be a set acted upon transitively by a group H, let G be any group with $H \leq G$, and let K be any subgroup of G with $K \leq \mathbb{K}$ and $K \cap H = H_{x_o}$. Then there is a well-defined H-equivariant injection $\phi : E \to X$ given by $\phi(hx_o) = hK$ for all $h \in H$. Identifying E with $\phi(E) \subseteq X$,*

(a) $\mathcal{M} := \{ g\phi(E) \mid g \in G \}$ is a fully G-symmetric test space, and
(b) For every $\alpha \in \Omega$,

$$\underset{\sim}{\alpha}(x_g) := \alpha(g, x_o)$$

is a well-defined probability weight on $\mathcal{M}$; and
(c) The mapping $\alpha \mapsto \underset{\sim}{\alpha}$ is a G-equivariant affine injection.

The proof is given in Appendix C.

Thus, the essentially classical picture presented above—a single test (or observable) E, interacting probabilistically with a system with a state-space Ω and symmetry group G by means of the function $p : \Omega \times E \to [0, 1]$—can be reinterpreted as a "non-classical" probabilistic model in which E appears as one of many possible tests, provided that a sufficiently large set of permutations of E can be implemented physically by the group G. What is more important for our purposes, however, is the (even more) trivial converse: Any symmetric model A, and in particular, any (finite-dimensional) quantum model, arises from this construction: simply take H to be the stabilizer in $G(A)$ of a chosen test $E \in \mathcal{M}(A)$ and K to be the stabilizer in $G(A)$ of some chosen outcome $x_o \in E$, and we recover $(\mathcal{M}(A), \Omega(A))$ from Theorem 4.7.

To summarize: in the representation above, we have reinterpreted the states as, in effect, *random probability measures* (or rather, weights) on a fixed test E, indexed by the dynamical group $G(A)$. Another way to view this is that each state determines a *family of trajectories* in the simplex $\Delta(E)$ of classical probability weights on E. Given a state $\alpha \in \Omega$ plus the system's actual dynamics, as specified by a choice of one-parameter group $g : \mathbb{R} \to G(A)$, we obtain a path $\widehat{\alpha}_t := \widehat{\alpha}(g_t)$ in $\Delta(E)$. In this representation, notice,

(a) States are not viewed as probability weights on a non-classical test space, but rather, specify how classical probabilities change over time, given the dynamics.
(b) In general, the trajectories in $\Delta(E)$ arising from states and one-parameter subgroups of $G(A)$ are not governed by flows, that is, there is no one-parameter group of affine mappings $T^s : \Delta(E) \to \Delta(E)$ such that $\widehat{\alpha}_{t+s} = T^s(\alpha_t)$. In particular, the observed evolution of probabilities on E is not Markov. In this respect, it is the *dynamical*, rather than the probabilistic, structure of the DCM that can be regarded as non-classical.

It is also important to note that the representation of a symmetric probabilistic model as a DCM is *not* in any usual sense a "hidden variables" one: it invokes no additional structure, but is simply a different, and mathematically equivalent, way of *viewing* the model (in particular, its states)—one that, I am urging, we might nevertheless want to count as *probabilistically* "classical": to whatever extent it departs from classicality, it does so with respect to its dynamical, and not with respect to its probabilistic, structure.

27.5 Composite Models, Entanglement and Locality

Thus far, we have been discussing probabilistic models largely in isolation, and largely in abstraction from physics. Once we start to consider probabilistic models in relation to one another, and as representing actual physical systems *localized in spacetime*, e.g., particles (or laboratories), we encounter a host of new questions regarding compound systems in which two or more component systems occupy causally separate regions of spacetime. In particular, we unavoidably encounter the concepts of entanglement and locality.

In this section, I briefly review how these notions unfold in the context of general probabilistic theories (following Barnum and Wilce (2016), Foulis and Randall (1981)), and how the various classical representations discussed above interact with such composites. As we'll see, entangled states arise naturally in this context. There is a sense in which the representations in terms of classical extensions and in terms of semiclassical covers are both *obviously* non-local, simply because in each case the "classical" state spaces associated with a composite system AB is typically much larger than the Cartesian product of those associated with A and B separately. Thus, for instance, if $S(A)$ stands for the set of dispersion-free probability weights on $\widetilde{\mathcal{M}}(A)$, then $S(AB)$ is going to be much larger than $S(A) \times S(B)$. Similarly, unless $\Omega(A)$ or $\Omega(B)$ is a simplex, $\Omega(AB)$ will generally allow for entangled pure states, which essentially just *means* that $\Omega(AB)_{\text{ext}}$ will be larger than $\Omega(A)_{\text{ext}} \times \Omega(B)_{\text{ext}}$. A more technical notion of "non-locality" in terms of hidden variables is discussed below in Sect. 27.5.2, while the more delicate question of whether composites of DCMs should be regarded as local or not is discussed in Sect. 27.5.4.

27.5.1 *Composites of Probabilistic Models*

Suppose A and B are probabilistic models. A *joint probability weight* on the test spaces $\mathcal{M}(A)$ and $\mathcal{M}(B)$ is a function $\omega : X(A) \times X(B) \to [0, 1]$ that sums to 1 on each *product test* $E \times F$, where $E \in \mathcal{M}(A)$ and $F \in \mathcal{M}(B)$. Certainly if α and β are probability weights on A and B, we can form a joint probability weight

$\alpha \otimes \beta$ on A and B, by the obvious recipe $(\alpha \otimes \beta)(x, y) = \alpha(x)\beta(y)$ for outcomes $x \in X(A)$ and $y \in X(B)$. Note that joint probability weights can be understood as probability weights on the test space

$$\mathcal{M}(A) \times \mathcal{M}(B) := \{ E \times F \mid E \in \mathcal{M}(A), F \in \mathcal{M}(B) \}$$

consisting of all product tests.

Definition 5.1 (Non-signaling joint probability weights) A joint probability weight ω on $\mathcal{M}(A)$ and $\mathcal{M}(B)$ is *non-signaling* iff its *marginal* (or *reduced*) probability weights, given by

$$\omega_2(y) := \omega(E \times \{y\}) \text{ and } \omega_1(x) := \omega(\{x\} \times F)$$

are well-defined, i.e., independent of the choice of $E \in \mathcal{M}(A)$ and $F \in \mathcal{M}(B)$, respectively.

To the extent to which we think of the tests $E \in \mathcal{M}(A)$ as representing the physical actions performable on the system represented by A (in the sense that, to perform a test is to *do something to* the system, and then observe a result), and similarly for B, the non-signaling condition on ω tells us that no action performable on A can have any statistically detectable influence on B, and vice versa. Thus, the no-signaling condition is the *locality condition* appropriate to states *qua* probability weights.

A non-signaling joint probability weight, having well-defined marginals, also has well-defined *conditional* probability weights given by

$$\omega_{2|x}(y) := \frac{\omega(x, y)}{\omega_2(y)} \text{ and } \omega_{1|y}(x) := \frac{\omega(x, y)}{\omega_1(x)}$$

(with, say, the convention that both are zero if their denominators are). The marginal states can be recovered as convex combinations of these conditional states, in a version of the law of total probability:

$$\omega_1 = \sum_{y \in F} \omega_2(y)\omega_{1|y} \text{ and } \omega_2 = \sum_{x \in E} \omega_1(x)\omega_{2|x}.$$

We should certainly want these conditional and (hence) marginal states to belong to the designated state-spaces of each component of a reasonable composite model.

Definition 5.2 Given probabilistic models A and B, let $A \otimes B$ denote the probabilistic model with test space $\mathcal{M}(A \otimes B) := \mathcal{M}(A) \times \mathcal{M}(B)$ and state-space $\Omega(A \otimes B)$ consisting of all non-signaling joint states ω on $\mathcal{M}(A) \times \mathcal{M}(B)$ having conditional (and hence, marginal) states belonging to $\Omega(A)$ and $\Omega(B)$.

This is essentially the *maximal (or injective) tensor product* of $\Omega(A)$ and $\Omega(B)$ (Barnum and Wilce 2016; Namioka and Phelps 1969). Since it does not allow for non-product outcomes, it is too small to be generally useful as a model of composite systems, but it does afford an easy way to define something more general:

Definition 5.3 (Non-signaling composites (Barnum and Wilce 2016)) A *(non-signaling) composite* of probabilistic models A and B is a model AB, together with

(a) an outcome-preserving interpretation $\phi : A \otimes B \to AB$, taking each pair of outcomes $x \in X(A)$ and $y \in X(B)$ to a *product outcome* $xy := \phi(x, y)$,[8] and
(b) a bi-affine mapping $\Omega(A) \times \Omega(B) \to \Omega(AB)$ taking states $\alpha \in \Omega(A)$ and $\beta \in \Omega(B)$ to a *product state* $\alpha \otimes \beta$ such that $(\alpha \otimes \beta)(xy) = \alpha(x)\beta(y)$ for all $x \in X(A)$, $y \in X(B)$.

Examples 5.4 Consider quantum probabilistic models $A(\mathcal{H})$ and $A(\mathcal{K})$. If we take AB to be $A(\mathcal{H} \otimes \mathcal{K})$, then we have a natural map $\phi : X(\mathcal{H}) \times X(\mathcal{K}) \to X(\mathcal{H} \otimes \mathcal{K})$, given by $\phi(x, y) = x \otimes y$. It is easy to check that this is an interpretation from $A(\mathcal{H}) \times A(\mathcal{K})$. We also have a natural bi-affine mapping $\Omega(\mathcal{H}) \times \Omega(\mathcal{K}) \to \Omega(\mathcal{H} \otimes \mathcal{K})$ given by, $(\alpha_T, \alpha_W) \mapsto \alpha_{T \otimes W}$ (where, recall, α_T represents the probability weight determined by the density operator T). It is easy to see that these mappings satisfy the conditions above.

If AB is a composite of models A and B as in Definition 5.3, every state $\omega \in \Omega(AB)$ pulls back along ϕ to a nonsignaling joint probability weight $\phi^*(\omega)(x, y) = \omega(xy)$ on $\mathcal{M}(A) \times \mathcal{M}(B)$ with conditional and marginal weights belonging to $\Omega(A)$ and $\Omega(B)$. In general, this joint probability weight does not determine the state ω.

Definition 5.5 (Local tomography) A composite $\phi : A \otimes B \to AB$ of probabilistic models A and B is *locally tomographic* iff the mapping $\phi^* : \Omega(AB) \to \Omega(A \times B)$ is injective.

It is easy to show that, for finite-dimensional probabilistic models A and B, AB is locally tomographic iff the mapping $\otimes : \Omega(A) \times \Omega(B) \to \Omega(AB)$ extends to a linear isomorphism $\mathbf{V}(A) \otimes \mathbf{V}(B) \simeq \mathbf{V}(AB)$. Mathematically, then, local tomography is a great convenience. However, given its failure for real and quaternionic QM, unless we are looking for an excuse to rule out these variants of QM (and see, e.g., (Baez 2012) for reasons why we might not want to do so), it is probably best to avoid taking local tomography as an axiom.

It is natural to ask that a probabilistic theory be closed under some rule of composition, in the sense of Definition 5.3, that is consonant with the theory's categorical structure. A way to make this precise is to ask that a probabilistic theory be a symmetric monoidal category. See Barnum and Wilce (2016) for further discussion.

[8]A more general definition would drop the requirement that ϕ be outcome-preserving, thus allowing for the possibility that for some outcomes x and y, $\phi(x, y) =: xy$ might be a non-trivial (and even possibly empty) *event* of AB. We will not need this generality here.

27.5.2 Entanglement

If AB is a composite of models A and B, then any convex combination of product states, say $\sum_i t_i \alpha_i \otimes \beta_i$ or, more generally,

$$\int_\Lambda (\alpha_\lambda \otimes \beta_\lambda) d\mu(\lambda)^9$$

is said to be *separable*. Such states can (in principle) be prepared by randomly selecting product states α_λ and β_λ for A and B, respectively, in a classically correlated way. States *not* of this form are said to be *entangled*. The existence, and the basic properties, of entangled states are rather generic features of probabilistic models having non-simplex state spaces. In particular, we have the following

Lemma 5.6 *Let ω be any non-signaling state on AB. Then*

(*a*) *If $\alpha \otimes \beta$ is pure, then so are α and β;*
(*b*) *If either of the marginal states $\omega_1 \in \Omega(A)$ or $\omega_2 \in \Omega(B)$ is pure, then $\omega = \omega_1 \otimes \omega_2$;*
(*c*) *Hence, if ω is entangled, then ω_1 and ω_2 are mixed.*

Proof See Barnum and Wilce (2016) (and note that the arguments are virtually identical to the familiar ones in the context of QM).[10]

When A and B are semiclassical, the test space $\mathcal{M}(A \otimes B)$ defined in Definition 5.2 is again semiclassical. In this situation, we have lots of dispersion-free joint states. However:

Lemma 5.7 *Let A and B be models with semiclassical test spaces $\mathcal{M}(A)$ and $\mathcal{M}(B)$, and let $\omega \in$ $\mathbf{Pr}(\mathcal{M}(A \times B))$. If ω is non-signaling and dispersion-free, then $\omega = \delta \otimes \gamma$ where δ and γ are dispersion-free.*

Proof Suppose ω is dispersion-free. Since ω is also non-signaling, it has well-defined marginal states, which must obviously also be dispersion-free, hence, pure. But by Lemma 5.6, a non-signaling state with pure marginals is the product of these marginals. □

It follows that any average of non-signaling, dispersion-free states is separable.

---

[9]Where Λ is a measurable space and μ is a probability measure thereon, and where the integral exists in the obvious sense that

$$\left(\int_\Lambda \alpha_\lambda \otimes \beta_\lambda) d\mu(\lambda)\right)(z) = \int_\Lambda (\alpha_\lambda \otimes \beta_\lambda)(z) d\mu(\lambda).$$

[10]As an historical note, both of these points were first noted in this generality, but without any reference to entanglement, in a pioneering paper of Namioka and Phelps (1969) on tensor products of compact convex sets. They were rediscovered, and connected with entanglement, by Kläy (1988).

27.5.3 Locality and Hidden Variables

In the literature, a (contextual) *local* hidden variable model for a bipartite probability weight ω is usually taken to consist of a measurable space Λ, a probability measure μ thereon, and a pair of *response functions* p_A and p_B such that, such that for all tests $E \in \mathcal{M}(A)$, $F \in \mathcal{M}(B)$ and all outcomes $x \in E$ and $y \in F \in \mathcal{M}(B)$, $p(x|E, \lambda)$ is the probability of x given the parameter λ and the choice of measurement E, and similarly for $p(y|F, \lambda)$. These are required to satisfy

$$\omega(x, y) = \int_\Lambda p_A(x|E, \lambda)p(y|F, \lambda)d\mu(\lambda)$$

so that we can interpret the joint probability $\omega(x, y)$ as resulting from a classical correlation (given by μ) together with the local response functions p_A and p_B.

As discussed in Sect. 27.3.3, it is straightforward to reinterpret the functions p_A and p_B in terms of probability weights on $\widetilde{\mathcal{M}}(A)$ and $\widetilde{\mathcal{M}}(B)$, respectively, by writing $p_A(x|E, \lambda)$ as $p_{A,\lambda}(x, E)$ and $p_B(y|F, \lambda)$ as $p_{B,\lambda}(y, F)$ (emphasizing that λ can be treated as merely an index). We then have

$$\omega(x, y) = \left(\int_\Lambda p_{A,\lambda} \otimes p_{B,\lambda} d\mu(\lambda) \right) ((x, E), (y, F)),$$

independently of E and F. In other words, ω has a local HV model if, and only if, it is *separable*, in the integral sense, when regarded as a joint probability weight on $\widetilde{\mathcal{M}}(A \otimes B)$ (using the composite of Definition 5.2).

Remark A first reaction to this observation might be that it must be wrong, as it's well-known that there exist entangled Werner states that nevertheless have local HV representations (Werner 1989). The subtlety here (such as it is) resides in the fact that whether a state is separable or entangled depends on what "local" state-spaces are in play. Here, we are expressing ω as a weighted average of products of "contextual" states, whereas in standard discussions of quantum entanglement, a state is separable iff it can be expressed as a weighted average of products of quantum (in particular, non-contextual) states of the component systems.

27.5.4 Composites of Dynamical Models

Since dynamical classical models are simply another way of looking at symmetric probabilistic models, in principle they can be composed in the same way as the latter. I have discussed this in some technical detail elsewhere (Wilce 2009). Without going into such detail, I want to make a few remarks on the sense in which composites of DCMs can be regarded as local. Suppose, then, that

$$A = (\Omega(A), G(A), H(A), E(A), p_A) \text{ and } B = (\Omega(B), G(B), H(B), E(B), p_B)$$

are two DCMs as defined in Sect. 27.4.2, and suppose that AB is a DCM serving as a composite of these. This will require, at a minimum that $E(AB) = E(A) \times E(B)$, and that we have functions

$$\otimes : \Omega(A) \times \Omega(B) \to \Omega(AB) \text{ and } \otimes : G(A) \times G(B) \to G(AB)$$

with the former bi-affine and the latter a group homomorphism, such that

$$(G(A) \otimes G(B)) \cap H(AB) \leq H(A) \otimes H(B)$$

and

$$(\alpha \otimes \beta)(g \otimes h) = \alpha g \otimes \beta h$$

for all $(\alpha, \beta) \in \Omega(A) \times \Omega(B)$ and all $(g, h) \in G(A) \times G(B)$. Additional conditions are necessary to ensure that a non-signaling condition is satisfied, but this is enough for present purposes.

Examples 5.8

(a) By the *minimal* classical dynamical model associated with a finite set E, I mean the DCM $A(E)$ with $\Omega(A(E)) = \Delta(E)$, $G(A(E)) = H(A(E)) = \mathfrak{S}(E)$ (the symmetric group of all bijections $E \to E$) and $p_{A(E)}(\alpha, x) = \alpha(x)$. Now given two finite sets, then the minimal DCM $A(E \times F)$ is a composite of $A(E)$ and $A(F)$ with the maps $\otimes : \Delta(E) \times \Delta(F) \to \Delta(E \times F)$ and $\otimes : \mathfrak{S}(E) \times \mathfrak{S}(F) \to \mathfrak{S}(E \times F)$ the obvious ones, that is,

$$(\alpha \otimes \beta)(x, y) = \alpha(x)\beta(y) \text{ and } (g, h)(x, y) = (gx, hy)$$

for all $(x, y) \in E \times F$, $(\alpha, \beta) \in \Delta(E) \times \Delta(F)$ and $(g, h) \in \mathfrak{S}(E) \times \mathfrak{S}(F)$.

(b) Let A and B be quantum-mechanical systems associated with Hilbert spaces $\mathcal{H}_A$ and $\mathcal{H}_B$. Fixing orthonormal bases $E \in \mathcal{M}(\mathcal{H}_A)$ and $F \in \mathcal{M}(\mathcal{H}_B)$, we can regard A and B as DCMs with $\Omega(A)$ and $\Omega(B)$ the spaces of density operators on $\mathcal{H}_A$ and $\mathcal{H}_B$, respectively, $G(A) = \mathcal{U}(\mathcal{H}_A)$ and $G(B) = \mathcal{U}(\mathcal{H}_B)$, and with

$$p_A(\rho, x) = \text{Tr}(\rho p_x) = \langle \rho x, x \rangle$$

and similarly for p_B. Identifying $E \times F$ with the product basis $E \otimes F$ in $\mathcal{M}(\mathcal{H}_A \otimes \mathcal{H}_B)$, we can regard the quantum system associated with $\mathcal{H}_A \otimes \mathcal{H}_B$ as a DCM in the same way. In this case the maps $\otimes$ referred to above are the obvious ones: given unitaries $U \in \mathcal{U}(\mathcal{H}_A)$ and $V \in \mathcal{U}(\mathcal{H}_B)$, we have a untary $U \otimes V$ on $\mathcal{H}_A \otimes \mathcal{H}_B$, and if ρ_A and ρ_B are density operators on $\mathcal{H}_A$ and $\mathcal{H}_B$ respectively, the $\rho_A \otimes \rho_B$ is a density operator on $\mathcal{H}_A \otimes \mathcal{H}_B$.

As discussed above, the prevailing technical definition of a local probabilistic theory is one in which composite systems admit local hidden-variable models (ideally, in a functorial way, though this point is rarely discussed). More generally, one would like to say that a probabilistic theory is local iff composite systems admit "local" classical representations, in some well-defined (but reasonably general) sense. As I've mentioned, the classical extensions and the classical representations associated with semiclassical covers, discussed in Sects. 27.3.2 and 27.3.3, are certainly not local in any reasonable sense. What about composites of DCMs? In what sense are, or aren't, these to be regarded as *local* classical representations of symmetric probabilistic models? As the second example above illustrates, the state space $\Omega(AB)$ will in general be significantly larger $\Omega(A) \times \Omega(B)$ in the sense that there will be entangled states. In this weak sense, such a composite will generally be "non-local". On the other hand, it is not entirely clear how to discuss the locality of a composite of dynamical classical models *as* a classical representation, since it involves only a single, classical measurement $E(A) \times E(B)$, and invokes no hidden variables, non-local or otherwise—unless, of course, we wish to regard the symmetries in $G(AB)$ as "hidden variables". In that case, the "non-locality" presumably rests in the fact that the group $G(AB)$ will typically be a great deal larger than $G(A) \times G(B)$. This is true, of course, for quantum models, where the unitary group $\mathcal{U}(\mathcal{H}_A \otimes \mathcal{H}_B)$ is larger than $\mathcal{U}(\mathcal{H}_A) \otimes \mathcal{U}(\mathcal{H}_B)$. But it is also true for minimal classical dynamical models, since for sets E and F, the symmetric group $\mathfrak{S}(E \times F)$ is larger than $\mathfrak{S}(E) \times \mathfrak{S}(F)$.

27.6 Conclusion

Many of the observations collected above are well known, at least folklorically. My aim in bringing them together, in the particular way that I have, has been to draw attention to some points that I believe follow from these observations collectively, and that I believe are somewhat less widely appreciated than they might be:

(a) Classicality in the strict (that is, unrestricted) Kolmogorovian sense, is manifestly a contingent matter, since it is not a point of logical necessity that all experiments should be compatible. To the extent that *probability theory* is meant to be a completely general and a priori study of reasoning in the face of uncertainty or chance, it *isn't* Kolmogorovian, nor is it quantum-mechanical: it is, rather, the study of what I've here called probabilistic theories generally (whether formulated as I have done here, or in some roughly similar way).
(b) Within such a framework, what we (ought to) mean by saying that a probabilistic model or probabilistic theory is "classical" is not entirely obvious. For some values of "classical", *every probabilistic theory* is, or can be represented as, a classical one, at the cost of contextuality and, where we are dealing with a non-signaling theory, also of locality. Moreover, for certain kinds of theories (fully symmetric ones), there is a broad sense of "classical" in which such a theory simply *is* classical, without any need to invoke hidden variables, contextual, non-local or otherwise. Or at least, this is true for the theory's probabilistic structure: any departure from "classicality" is dynamical. Finite-dimensional quantum mechanics is such a theory.

(c) The selection of a class of probabilistic models—a probabilistic theory—must rest on some contingent assumptions about the nature of the entities one seeks to model. Quantum mechanics, as a probabilistic theory, is contingent in this way; so is (unrestricted, Kolmogorovian) classical probability theory, and the two are not the same. In *this* limited and unremarkable sense, quantum mechanics is (of course) a non-classical probabilistic theory.

I don't know how much, if any, of this Pitowsky would have endorsed. Regarding (b) at least, he was very clear:

> We can always avoid the radical view of probability by adopting a non-local, contextual hidden variables theory such as Bohm's. But then I believe, the philosophical point is missed. It is like taking Steven Weinberg's position on space-time in general relativity: There is no non-flat Riemannian geometry, only a gravitational field defined on a flat space-time that appears as if it gives rise to geometry ... I think that Weinberg's point and also Bohm's theory are justified only to the extent that they yield new discoveries in physics (as Weinberg certainly hoped). So far they haven't.

I have no issue to take with this, as far as it goes: we are always free to reject a classical representation of a probabilistic theory by objecting to the additional "classical" structure as "non-physical". But I would also want to say that doing so makes the theory a *physical* theory, and not a theory *of* probability. Or, to put it differently, whether one wants to take, for example, a set of contextual hidden variables, or a privileged observable, seriously in formulating a given physical theory is a pragmatic question about the best (the most elegant, the most convenient, the most fruitful) way to formulate mathematical models of a particular sort of physical situation. It is also (in particular!) a metaphysical question about what ontology one wants to admit in framing such a model. Such decisions are important; but they are not decisions about what kind of probability theory to use.

Pitowsky also says (emphasis mine):

> In this paper, all we have discussed is the Hilbert space formalism. I have argued that it is a new kind of probability theory that is *quite devoid of physical content,* save perhaps the indeterminacy principle which is built into axiom H4. Within this formal context there is no explication of what a measurement is, only the identification of "observables" as Hermitian operators. In this respect the Hilbert space formalism is really just a syntax which represents the set of all possible outcomes, of all possible measurements.

Here, of course, I disagree. The "syntax" of quantum probability theory is very definitely *not* devoid of physical content. Reconstructions of QM from various sets of postulates, ranging from the Piron-Amemiya-Araki-Solèr reconstruction cited by Pitowsky (see Holland 1995), to the various more recent finite-dimensional reconstructions (e.g., Mueller et al. 2014; D'Ariano et al. 2011; Dakič and Brukner 2011; Masanes and Müller 2011; Wilce 2012) have helped us understand the physical content of quantum theory more clearly, by isolating operationally meaningful principles that dictate that syntax. These principles invariably include some that are manifestly contingent, at least to the same degree that the classical co-measurability principle is contingent.

The final lines of Pitowsky's paper crystalize what I find problematic in his proposal. (Again, the emphasis is mine.)

> However, there is a structure to the set of events. Not only does each and every type of measurement yield a systematic outcome; but also the set of all possible outcomes of all measurements — including those that have been realized by an actual recording — hang together tightly in the structure of $L(\mathcal{H})$. *This is the quantum mechanical structure of reality.*

Taken in the context of Pitowsky's claim that QM represents a new, non-classical theory of probability, this suggests a certain naturalism regarding probability theory (echoing von Neumann/Birkhoff's naturalism regarding logic): that there is a *true* probability theory, which is the one that *works* best, and most broadly, to describe the world we actually live in. If the world really is, at every scale and in every way, described by some version of quantum theory, so that there just *are no* actual events that can't be described as effects in (say) some grand von Neumann algebra, then quantum probability theory *is* this true, correct probability theory, of which classical probability theory is simply a limiting or special case (where $\hbar$ tends to zero, or where we restrict attention to a set of commuting observables).

But this view seems to me misleading: not because naturalism *per se* is wrong (even about mathematics), but because it forgets that probability theory has a methodological as well as a metaphysical importance: it is not simply part of our scientific description of the world, but is part of the framework within which we *think about*, *criticize*, and (often) *revise* that description. In order for it to play this role, it can not be tied to any particular physical theory. This is why I think it is vital to maintain the distinction between *probability theory* and the various *probabilistic theories* that it studies. There's no such thing as *a* probability theory, classical or otherwise: there is only *probability theory*. There *can* be a correct *probabilistic theory* of the world—and perhaps it's quantum.

Appendix A Common Refinements

Let A be a probabilistic model, and let $E, F \in \mathcal{M}(A)$. Recall that E *refines* F, or that F is *a coarsening of* E, if there exists a surjection $f : E \to F$ such that, for every $y \in F$ and every state $\alpha \in \Omega$,

$$\alpha(y) = \alpha(f^{-1}(y)).$$

When this is so, we write $E \preceq F$, and say that f is a *coarsening map*.

Definition A.1 A set Ω of probability weights on a test space $\mathcal{M}$ *separates compatible events* iff, for every test $E \in \mathcal{M}$ and any pair of distinct events $a, b \subseteq E$, there exists a weight $\alpha \in \Omega$ with $\alpha(a) \neq \alpha(b)$.

Note that if A is unital, $\Omega(A)$ separates compatible events. Indeed, it is sufficient that, for every $x \in E$, there exist a state $\alpha \in \Omega(A)$ with $\alpha(x) > 1/2$.

Lemma A.2 *Let* $\Omega(A)$ *separate compatible events. Then*

(a) *There exists at most one coarsening map* $f : E \to F$*;*
(b) *If there exist coarsening maps* $f : E \to F$ *and* $g : F \to E$*, then* f *and* g *are bijections and* $g = f^{-1}$*;*
(c) *If* $f : G \to E$ *and* $g : G \to F$ *are coarsening maps and* $x \in E \cap F$*, then* $f^{-1}(x) = g^{-1}(x) \subseteq G$.

Proof (a) If $f, g : E \to F$ are coarsening maps, where $E \preceq F$ in $\mathcal{M}(A)$, then for every $y \in F$ we have $\alpha(f^{-1}(y)) = \alpha(y) = \alpha(g^{-1}(y))$ for every $\alpha \in \Omega(A)$. Hence, $f^{-1}(y) = g^{-1}(y)$ for every $y \in F$, whence, $f = g$. Now (b) follows, since the composition of coarsening maps is a coarsening map. For (c), observe that if $x \in E \cap F$, then for every state α we have $\alpha(f^{-1}(x)) = \alpha(x) = \alpha(g^{-1}(x))$. Since Ω separates compatible events, $f^{-1}(x) = g^{-1}(x)$. □

From now on, assume $\Omega(A)$ separates compatible events. We will write $f_{E,F}$ for the unique coarsening map $f : E \to F$, if one exists.

Suppose now that $E, F \in \mathcal{M}(A)$ have a *common refinement*, that is, that there exists a test $G \in \mathcal{M}(A)$ with $G \preceq E$ and $G \preceq F$. Then we have a natural surjection $\phi : G \to E \times F$, namely

$$\phi(z) = (f_{G,F}(z), f_{G,E}(z)).$$

If $\alpha \in \Omega$, then we have a probability weight on $E \times F$ given by

$$\phi_*(\alpha) := \alpha \circ \phi^{-1}.$$

This assigns to $(x, y) \in E \times F$ the probability

$$\alpha(\phi^{-1}(x, y)) = \alpha(a_x \cap b_y), \quad \text{where} \quad a_x = f_{G,E}^{-1}(x) \text{ and } b_y = f_{G,F}^{-1}(y).$$

It is easy to check that $\sum_{y \in F} \phi_*(\alpha)(x, y) = \alpha(x)$ and $\sum_{x \in E} \phi_*(\alpha)(x, y) = \alpha(y)$, i.e., $\phi_*(\alpha)$ has the "right" marginals to explain the probabilities that α assigns to E and F. In this sense, G can be regarded as a joint measurement of E and F.

Definition A.3 A is a *refinement ideal* iff every pair of tests in $\mathcal{M}(A)$ has a common refinement.

In other words, A is a refinement ideal iff the preordered set $(\mathcal{M}, \preceq)$ is downwardly directed. Define $S \subseteq \Pi_{E \in \mathcal{M}} E$ to be the set

$$S = \{\boldsymbol{x} = (x_E) \in \Pi_{E \in \mathcal{M}(A)} \mid \forall E \preceq F \ f_{E,F}(x_E) = x_F\}, \tag{27.5}$$

i.e., the inverse limit of $\mathcal{M}$, regarded as a small category under coarsening maps. As long as $\mathcal{M}$ is locally finite, one can show that this is non-empty (a consequence of the compactness of $\Pi_{E\in\mathcal{M}}E$). For each test $E \in \mathcal{M}(A)$, define $f_E : S \to E$ by $f_E(x) = x_E$, and for any $x_0 \in X(A)$, let

$$[x] := \{\, \boldsymbol{x} \in S \mid \forall E \in \mathcal{M}(A)\; x_0 \in E \;\Rightarrow\; x_E = x_0 \,\}.$$

Note that, by Lemma A.2, if $x_0 \in E \cap F$, then for every $\boldsymbol{x} \in S$, $x_E = x$ iff $x_F = x_0$; hence, $[x_0] \neq \emptyset$. If $a \in \mathcal{E}v(A)$, let $[\![a]\!] := \{[x] | x \in a\}$. Notice that $\bigcup [\![a]\!] = f_E^{-1}(a)$ where E is any test with $a \subseteq E$.

Recall that, if E and F are partitions of a set S, E is said to *refine* F iff for every $y \in F$, there is some $a \subseteq E$ with $y = \cup a$. I will write $E \sqsubseteq F$ to indicate this. If $\mathcal{M}$(A) is a collection of partitions of a set S such that every pair of partitions in $\mathcal{M}$(A) has a common refinement in this sense, then I will say that $\mathcal{M}$(A) is a *refinement ideal of partitions.*

Lemma A.4 *If A is a locally finite refinement ideal, then with S the projective limit of $\mathcal{M}(A)$ in (27.5), For every $E \in \mathcal{M}(A)$, $[\![E]\!]$ is a partition of S, and if $E \preceq F$ then $[\![E]\!] \sqsubseteq [\![F]\!]$. Hence, $[\![\mathcal{M}(A)]\!] := \{[\![E]\!] | E \in \mathcal{M}(A)\}$ is a refinement ideal of partitions. Moreover, the mapping $\phi : x \mapsto [x]$ defines an isomorphism of test spaces from $\mathcal{M}(A)$ to $[\![\mathcal{M}(A)]\!]$.*

Proof That $[\![E]\!]$ is a partition of S is clear from the definitions. Let $y \in F$, and set $a = f_{E,F}^{-1}(y) \subseteq E$. Thus, $[\![a]\!] \subseteq [\![E]\!]$. Now

$$\bigcup [\![a]\!] = f_E^{-1}(a) = f_E^{-1} \circ f_{E,F}^{-1}(x) = (f_{E,F} \circ f_E)^{-1}(x) = f_F^{-1}(y) = [y].$$

Thus, every cell in the partition $[\![F]\!]$ is a union of cells of $[\![E]\!]$, i.e., $E \sqsubseteq F$. Since every pair of tests in $\mathcal{M}(A)$ have a common refinement with respect to $\Omega(A)$, it follows that $[\![\mathcal{M}(A)]\!]$ is a refinement ideal of partitions. It is clear that $\phi : x \mapsto [x]$ is an outcome-preserving, positive interpretation from $\mathcal{M}(A)$ onto $[\![\mathcal{M}(A)]\!]$. It remains to show it's injective. Let $x, y \in X(A)$ with $[x] = [y]$. Supposing that $x \in E \in \mathcal{M}(A)$ and $y \in F \in \mathcal{M}(A)$, let $G \preceq E, F$. Then $f_{G,E}^{-1}(x) = f_{G,F}^{-1}(y)$, so for all $\alpha \in \Omega(A)$, we have

$$\alpha(x) = \alpha(f_{G,E}^{-1}(x)) = \alpha(f_{G,F}^{-1}(y)) = \alpha(y).$$

By our standing assumption that states in $\Omega(A)$ separate outcomes in $X(A)$, $x = y$. □

Now let $\mathcal{M}$ be a test space of partitions of a set S, and let $\mathcal{M}$ be a refinement ideal with respect to ordinary refinement of partitions.

Define $\phi : \mathcal{E}v(\mathcal{M}) \to \mathcal{P}(S)$ by $\widehat{a} = \cup a$ for all $a \in \mathcal{E}v(\mathcal{M})$. Note here that an event a is a subset of a finite partition of S, that is, a finite, pairwise disjoint set of non-empty subsets of S. Now let

$$\Sigma := \{\, \widehat{a} \mid a \in \mathcal{E}v(\mathcal{M}) \,\}.$$

Lemma A.5 Σ *is an algebra of sets on* S.

Proof It is easy to see that if $a, b \subseteq E \in \mathcal{M}$, $\widehat{a} \cap \widehat{b} = \widehat{a \cap b}$, $\widehat{a \cup b} = \widehat{a} \cup \widehat{b}$, and $\widehat{a}^c = \widehat{E \setminus a}$. Thus, the set $\Sigma_E := \{\widehat{a} | a \subseteq E\}$ is a subalgebra of $\mathcal{P}(S)$. If $G \sqsubseteq E$, then the definition of refinement tells us that $\Sigma_E \subseteq \Sigma_G$. Hence, if $\mathcal{M}$ is a refinement ideal, $\{\Sigma_E | E \in \mathcal{M}\}$ is a directed family of subalgebras of $\mathcal{P}(S)$ under inclusion, whence, its union is also a subalgebra. □

Now let A be a refinement ideal in the sense of probabilistic models, i.e., $\mathcal{M}(A)$ is a refinement ideal with respect to $\Omega(A)$. For every $\alpha \in \Omega(A)$, define $\forall \widehat{a} \in \Sigma$,

$$\widehat{\alpha}(\widehat{a}) = \alpha(a).$$

Lemma A.6 $\widehat{\alpha}$ *is well-defined, and a finitely-additive probability measure on* Σ.

Proof Let $a \subseteq E \in \mathcal{M}$ and $b \subseteq F \in \mathcal{M}$. We want to show that if $\widehat{a} = \widehat{b}$, then $\alpha(a) = \alpha(b)$. Let G refine both E and F. There are canonical surjections $e : G \to E$ and $f : G \to F$, namely $e(z) = x$ where x is the unique cell of E containing $z \in G$, and similarly for f. Let $a_1 = e^{-1}(a)$ and $b_1 = f^{-1}(b)$. Then $\widehat{a_1} = \widehat{a} = \widehat{b} = \widehat{b_1}$. Since $\widehat{\ }$ is injective on $\mathcal{P}(G)$, $a_1 = b_1$. Since A is a refinement ideal with respect to $\Omega(A)$, $\alpha(a) = \alpha(a_1) = \alpha(b_1) = \alpha(b)$.

This shows that $\widehat{\alpha}$ is well defined. To see that it's additive, let $a \subseteq E \in \mathcal{M}(A)$, $b \subseteq F \in \mathcal{M}(A)$, with $\widehat{a} \cap \widehat{b} = \emptyset$. Choosing G a common refinement of E and F, we have $a_1, b_1 \subseteq G$ with $\widehat{a_1} = \widehat{a}$ and $\widehat{b_1} = \widehat{b}$, so $\widehat{a_1} \cap \widehat{b_1} = \emptyset$, whence, $a_1 \cap b_1 = \emptyset$. It follows that

$$\begin{aligned}\widehat{\alpha}(\widehat{a} \cup \widehat{b}) = \widehat{\alpha}(\widehat{a_1 \cup a_2}) = \alpha(a_1 \cup a_2) &= \alpha(a_1) + \alpha(a_2) \\ &= \widehat{\alpha}(\widehat{a_1}) + \widehat{\alpha}(\widehat{b_1}) = \widehat{\alpha}(\widehat{a}) + \widehat{\alpha}(\widehat{b}).\end{aligned}$$

□

Proposition A.7 *If* A *is a locally finite refinement ideal and* $\Omega(A)$ *separates compatible events, then there is a measurable space* (S, Σ) *and an embedding* $\mathcal{M}(A) \to \mathcal{D}(S, \Sigma)$ *such that every probability weight in* Ω *extends to a finitely-additive probability measure on* (S, Σ)*. Conversely, if A is unital and locally finite and admits such an embedding, then A embeds in a refinement ideal without loss of states.*

Proof The forward implication follows from Lemmas A.4, A.5 and A.6, while the reverse implication is more or less obvious. □

To this extent, then, the existence of common refinements is the key classical postulate.

Appendix B Semiclassical Test Spaces

Recall that $\mathcal{M}$ is *semiclassical* iff distinct tests in $\mathcal{M}$ are disjoint. Evidently, such a test space has a wealth of dispersion-free states, as we can simply choose an element x_E from each test $E \in \mathcal{M}$ and set $\delta(x) = 1$ if $x = x_E$ for the unique test containing x, and 0 otherwise. In fact, the dispersion-free states are exactly the pure states:

Lemma B.1 *Let K_i be an indexed family of convex sets, and let $\alpha = (\alpha_i) \in K = \Pi_{i \in I} K_i$. Then α is pure iff each α_i is pure.*

Proof Suppose that for some $j \in I$, α_j is not pure. Then there exist distinct points $\beta_j, \gamma_j \in K_j$ with $\alpha_j = t\beta_j + (1-t)\gamma_j$ for some $t \in (0, 1)$. Define $\widehat{\beta}, \widehat{\gamma} \in K$ by setting

$$\widehat{\beta}_i = \begin{cases} \alpha_i & i \neq j \\ \beta_j & i = j \end{cases} \text{ and } \widehat{\gamma}_i = \begin{cases} \alpha_i & i \neq j \\ \gamma_j & i = j \end{cases}$$

Then we have $t\widehat{\beta} + (1-t)\widehat{\gamma} = \alpha$, so α is not pure either. The converse is clear. □

Corollary B.2 *If $\mathcal{M}$ is semiclassical, then* $\mathbf{Pr}(\mathcal{M})_{ext} = \mathbf{Pr}(\mathcal{M})_{df}$.

Proof Since $\mathcal{M}$ is semiclassical, $\mathbf{Pr}(\mathcal{M}) \simeq \Pi_{E \in \mathcal{M}} \Delta(E)$. □

Let $S(\mathcal{M})$ denote the set of dispersion-free states on $\mathcal{M}$. It is easy to see that $S(\mathcal{M})$ is closed, and hence compact, as a subset of $[0, 1]^X$. By $\Delta(S)$, I mean the simplex of Borel probability measures on S with respect to this topology.

Appendix C Constructing Fully G-Symmetric Models

This appendix collects more technical material on the construction of fully symmetric models. In particular, we prove Theorem 4.7, which for convenience we restate below as Proposition C.2.

As discussed in Sect. 27.4.2, we are given a set E (which we wish to regard as the outcome-set of an experiment), and a group H acting transitively on E. We are also given a set Ω of physical states (those of the system to which the experiment pertains), and a function

$$p : \Omega \times E \to [0, 1]$$

assigning a probability weight $\widehat{\alpha} = p(\alpha, \cdot)$ on E to each state $\alpha \in \Omega$. We are also given a group G of "physical symmetries" acting on Ω on the right, plus a subgroup H of G acting on E (on the left), in such a way that

$$p(\alpha h, x) = p(\alpha, hx)$$

for all $h \in H$. For each $\alpha \in \Omega$, we write $\widehat{\alpha}$ both for the mapping $G \mapsto \Delta(E)$ given by $\widehat{\alpha}(g)(x) = p(\alpha g, x)$, and for the mapping $G \times E \to [0, 1]$ given by $\widehat{\alpha}(g, x) = \widehat{\alpha}(g)(x)$, whichever is more convenient. Thus, $\widehat{\alpha g_1}(g_2) = \widehat{\alpha}(g_1 g_2)$ for all $g_1, g_2 \in G$, and $\widehat{\alpha}(gh)(x) = \widehat{\alpha}(g)(hx)$ for all $g \in G, h \in H$ and $x \in E$. We assume that the mapping $\alpha \mapsto \widehat{\alpha}$ is injective, so that α is determined by the function $\widehat{\alpha}$.

Given this data, choose and fix an outcome $x_o \in E$, and let

$$\mathbb{K} := \{k \in G | \forall \alpha \in \Omega \ p(\alpha k, x_o) = p(\alpha, x_o)\}. \tag{27.6}$$

Equivalently, $k \in \mathbb{K}$ iff $\widehat{\alpha}(gk)(x_o) = \widehat{\alpha}(g)(x_o)$. Notice that the stabilizer H_o of x_o in H is a subset of $\mathbb{K}$; in particular, $\mathbb{K}$ is nonempty.

Lemma C.1 *$\mathbb{K} \leq G$, and $\mathbb{K} \cap H = H_o$.*

Proof Let $k, k' \in \mathbb{K}$. Then for all $g \in G$ and $\alpha \in \Omega$, we have $\alpha(gkk', x_o) = \alpha(gk, x_o) = \alpha(g, x_o)$ so $kk' \in \mathbb{K}$; also

$$\widehat{\alpha}(gk^{-1}, x_o) = \widehat{\alpha}((gk^{-1})k, x_o) = \widehat{\alpha}(g, x_o)$$

so $k^{-1} \in \mathbb{K}$. Thus, $\mathbb{K}$ is a subgroup of G. For the second statement, $h \in \mathbb{K} \cap H$ implies that, for all α and all g,

$$\widehat{\alpha}(g, x_o) = \widehat{\alpha}(gh, x_o) = \widehat{\alpha}(g, hx_o).$$

Since the functions $\widehat{\alpha}(g)$ separate points of E, $hx_o = x_o$, so $h \in H_o$ On the other hand, if $h \in H_o$, then $\alpha(gh, x_o) = \alpha(g, hx_o) = \alpha(g, x_o)$, so $h \in \mathbb{K}$. □

We now prove Theorem 4.5, showing that any choice of a subgroup K between H_o and $\mathbb{K}$ generates a fully G-symmetric test space, having Ω as its state space, in a canonical manner. For convenience, we restate this result here:

Proposition C.2 *With notation as above, suppose H acts transitively on E. Let K be any subgroup of G with $H_o \leq K \leq \mathbb{K}$. Set $X := G/K$, and let $x_g := gK \in X$ for all $g \in G$. Then there is a well-defined H-equivariant injection $\phi : E \to X$ given by $\phi(hx_o) = x_h$ for all $h \in H$. Moreover, identifying E with $\phi(E) \subseteq X$,*

(a) $\mathcal{M} := \{g\phi(E) | g \in G\}$ is a G-symmetric test space;
(b) For every $\alpha \in \Omega$, $\underset{\sim}{\alpha}(x_g) := \widehat{\alpha}(g, x_o)$ is a well-defined probability weight on $\mathcal{M}$;
(c) The mapping $\alpha \mapsto \underset{\sim}{\alpha}$ is a G-equivariant affine injection, where G acts on $\Delta(E)^G$ on the right.

Proof For convenience, let us write x_g for the left coset gK in $X = G/K$. Then the mapping $\phi : E \to X$ above is given by

$$\phi : hx_o \mapsto x_h.$$

To see this is well-defined, let $h, h' \in H$ with $hx_o = h'x_o$: then $h^{-1}h' \in H_o \leq K$, so that $x_h = x_{h'}$. Conversely, if $x_h = x_{h'}$, we have $h^{-1}h' \in S_o$ whence $hx_o = h'x_o$, and ϕ is injective. If $x = hx_o$, then $\phi(h'x) = h'hK = h'x_h = h\phi(x)$, so ϕ is H-equivariant. This disposes of (a). Now let $\mathcal{M} = \{ g\widetilde{E} \mid g \in G \}$. Since X is a G-set, this gives us a G test space—a symmetric one, as G acts transitively on $\mathcal{M}$ and H acts transitively on $\phi(E) \in \mathcal{M}$. For $\alpha \in \Omega$, set $\underset{\sim}{\alpha}(x_g) = \alpha(g, x_o)$. This is well-defined, since if $x_g = x_{g'}$, then $g' = gk$ for some $k \in K$, and thus

$$\underset{\sim}{\alpha}(x_{g'}) = \underset{\sim}{\alpha}(x_{gk}) = \widehat{\alpha}(gk, x_o) = \widehat{\alpha}(g, x_o) = \underset{\sim}{\alpha}(x_g).$$

To see that $\underset{\sim}{\alpha}$ is a probability weight on $\mathcal{M}$, for each $x \in E$, choose $h_x \in H$ with $h_x x_o = x$ (recalling here that H acts transitively on E). Then

$$\underset{\sim}{\alpha}(g\phi(x)) = \widehat{\alpha}(gx_{h_x}) = \widehat{\alpha}(x_{gh_x}) = \widehat{\alpha}(gh_x, x_o) = \widehat{\alpha}(g, h_x x_o) = \widehat{\alpha}(g, x).$$

Thus,

$$\sum_{x \in E} \underset{\sim}{\alpha}(g\phi(x)) = \sum_{x \in E} \widehat{\alpha}(g, x) = 1.$$

The mapping $\alpha \mapsto \underset{\sim}{\alpha}$ is obviously affine, and is injective by our assumption that $\alpha \mapsto \widehat{\alpha}$ is injective. To see that it is equivariant, note that for all $g, l \in G$ and $\alpha \in \Omega$,

$$(\underset{\sim}{\alpha g})(x_l) = (\widehat{\alpha g})(l, x_o) = \widehat{\alpha}(gl, x_o) = \underset{\sim}{\alpha}(glx_o) = \underset{\sim}{\alpha}(g, x_l) = \underset{\sim}{\alpha}g(x_l).$$

□

Thus, $(\mathcal{M}, \underset{\sim}{\Omega})$, where $\underset{\sim}{\Omega} = \{\underset{\sim}{\alpha} | \alpha \in \Omega\}$, is a G-symmetric model with state space isomorphic to Ω.

Remarks

(1) If $H_o \leq K \leq K' \leq \mathbb{K}$, we obtain a G-equivariant, surjective outcome-preserving interpretation

$$\phi_{K',K} : \mathcal{M}_K \to \mathcal{M}_{K'}$$

given by $\phi_{K',K}(gK) = gK'$ for all $g \in G$. So taking $K = H_o$ gives in this sense the least constrained symmetric G-test space containing E (in such a way as to extend the action of H on E), while $K = \mathbb{K}$ gives the most constrained. Note all such test spaces have the same rank, namely $|E|$, provided that the set of probability weights $\widehat{\alpha}(g)$, with α ranging over Ω and g ranging over G, is large enough to separate points of E.

(2) Accordingly, all probability weights on $\mathcal{M}_{K'}$ pull back to probability weights on $\mathcal{M}_K$ via $\phi^*_{K,K'}$, which is an injective G-equivariant affine mapping. Writing $\Omega_K := \mathbf{Pr}(\mathcal{M}_K)$ for the convex set of all probability weights on $\mathcal{M}_K$, we have $\Omega \subseteq \Omega_K$ (where we identify Ω with its image $\underset{\sim}{\Omega}$). Thus, in particular, we can replace Ω with $\mathrm{Pr}(\mathcal{M}_K)$ to obtain a larger state space with respect to which the same construction works.

Given the H-set E, we can obtain the initial data for this construction as follows. Let G be any group containing H as a subgroup, and let

$$G \times_H E = (G \times E)/\sim$$

where $\sim$ is the equivalence relation defined by $(g, x) \sim (h, y)$ iff there exists some $s \in H$ with $h = gs$ and $sy = x$. Letting $[g, x]$ denote the equivalence class of (g, x), we have

$$[g, x] = [gs, s^{-1}x] \text{ i.e., } [gs, x] = [g, sx]$$

for all $g \in G, x \in E$ and $s \in H$. Note that $G \times_H E$ is a G-set, with action defined by

$$g[h, x] = [gh, x]$$

which is well-defined because $[ghs, s^{-1}x] = [gh, x]$ for all $s \in H$. Now let

$$[g, E] := \{[g, x] | x \in E\} :$$

then if $[h, x] \in [g, E]$, we have $[h, x] = [g, y]$ for some $y \in E$, whence, there is some $s \in H$ with $h = gs$ and $sy = x$. Then for all $z \in E$, say $z = s'x$, we have $[h, z] = [gs, s'x] = [g, ss'x] \in [g, E]$. That is, $[h, E] \subseteq [g, E]$. By the same token, $[g, E] \subseteq [h, E]$. In other words, the sets $[g, E]$ partition $G \times_H E$. With this observation, it is easy to prove the following

Lemma C.3 *With notation as in Proposition C.2, $X_{H_o} \simeq G \times_{H_o} E$, and $\mathcal{M}_{H_o} \simeq \{[g, E] | g \in E\}$, a semiclassical test space, independent of Ω.*

Proof Let $\phi : G \times E \to G/H_o = X_{H_o}$ be given by

$$\phi(g, sx_o) = gsH_o.$$

for all $g \in G, s \in H$.

To see that this is well-defined, note that if $s_1x_o = s_2x_o$, then $s_2^{-1}s_1 := s \in H_o$, and $gs_1H_o = gs_2sH_o = gs_2H_o$. Next, observe that $\phi(g_1, s_1x_o) = \phi(g_2, s_2x_o)$ iff $g_1s_1H_o = g_2s_2H_o$ iff $g_2s_2 = g_1s_1s$ for some $s \in H_o$, whence, $[g_2, s_2x_o] = [g_2s_2, x_o] = [g_1s_1s, x_o] = [g_1, s_1x_o]$. Thus, passing to the quotient set $G \times_{H_o} E$ gives us a bijection. Since $\{[g, E] | g \in G\}$ is a partition of $G \times_{H_o} E$, we have a semiclassical test space, as advertised. □

Notice that the stabilizer of $[e, E]$ in G is exactly H_0. Now choosing any G-invariant set of probability weights on $\mathcal{M}_{H_o}$—say, the orbit of any given probability weight—we have a natural G-equivariant injection $\Omega \to \Delta(E)^G$ given by

$$\alpha \mapsto \widehat{\alpha} : \widehat{\alpha}(g)(x) = \alpha[g, x]$$

for all $\alpha \in \Omega$. The construction now proceeds as above for any choice of subgroup K of G with $H_o \leq K \leq \mathbb{K}$, where $\mathbb{K}$ is determined by Ω as in Equation (27.6). We can regard K as parameterizing the possible fully G-symmetric models containing E as a test, and having dynamical group G. If $K = H_o$, the stabilizer of x_o in H, then $\mathcal{M}$ is semiclassical. Larger choices of K further constrain the structure of $\mathcal{M}$, enforcing outcome-identifications between tests that, in turn, constrain any further enlargement of the state space that we might contemplate.

References

Alfsen, E. (1971). *Convex sets and boundary integrals*. Heidelberg/Berlin: Springer.

Bacciagaluppi, G. (2004). Classical extensions, classical representations, and Bayesian updating in quantum mechanics. arXiv:quant-ph/0403055v1.

Baez, J. (2012). Division algebras and quantum theory. *Foundations of Physics, 42*, 819–855. arXiv:1101.5690.

Barnum, H., & Wilce, A. (2016). Post-classical probability theory. In G. Chiribella & R. Spekkens (Eds.), *Quantum theory: Informational foundations and foils*. Dordrecht: Springer.

Beltrametti, E., & Bugajski, S. (1995). A classical extension of quantum mechanics. *Journal of Physics A, 28*, 3329.

Bub, J., & Pitowsky, I. (2012). Two dogmas about quantum mechanics. In S. Saunders, J. Barrett, A. Kent, & D. Wallace (Eds.), *Many worlds? Everett, quantum theory and reality* (chapter 14, pp. 433–459). Oxford. arXiv:0712.4258.

Dakič, B., & Brukner, C. (2011). Quantum theory and beyond: Is entanglement special? In H. Halvorson (Ed.), *Deep beauty: Understanding the quantum world through mathematical innovation*. Cambridge University Press. arXiv:0911.0695.

D'Ariano, M., Chiribella, G., & Perinotti, P. (2011). Informational derivation of quantum theory. *Physical Review A, 84*, 012311.

Foulis, D. J., & Randall, C. H. (1978). Manuals, morphisms and quantum mechanics. In A. R. Marlowe (Ed.), *Mathematical foundations of quantum theory*. New York: Academic Press.

Foulis, D. J., & Randall, C. H. (1981). Empirical logic and tensor products. In H. Neumann (Ed.), *Interpretations and foundations of quantum mechanics*. Mannheim: B.I.-Wissenshaftsverlag.

Hardy, L. (2001). Quantum theory from five reasonable axioms. arXiv:quant-ph/0101012.

Hellwig and Singer. (1991). "classical" in terms of general statistical models. In V. Dodonov & V. Man'ko (Eds.), *Group theoretical methods in physics* (Lecture notes in physics, Vol. 382). Berlin/Heidelberg: Springer.

Hellwig, K.-E., Bush, S., & Stulpe, W. (1993). On classical representations of finite-dimensional quantum mechanics. *International Journal of Theoretical Physics, 32*, 399.

Holevo, A. (1982). *Probabilistic and statistical aspects of quantum theory*. Amsterdam: North Holland.

Holland, S. (1995). Orthomodularity in infinite dimensions: a theorem of m. solèr. *Bulletin of the American Mathematical Society, 32*, 205–234. arXiv:math/9504224.

Kläy, M. (1988). Einstein-Podolsky-Rosen experiments: the structure of the probability space I, II. *Foundations of Physics Letters, 1*, 205–244.

Lindsay, B. (1995). *Mixture models: Theory, geometry and applications* (NSF-CBMS regional conference series in probability and statistics, Vol. 5). Hayward: Institute of Mathematical Statistics/American Statistical Association.

Masanes, L., & Müller, M. (2011). A derivation of quantum theory from physical requirements. *New Journal of Physics, 13*. arXiv:1004.1483.

Mueller, M., Barnum, H., & Ududec, C. (2014). Higher-order interference and single-system postulates characterizing quantum theory. *New Journal of Physics, 16*. arXiv:1403.4147.

Namioka, I., & Phelps, R. (1969). Tensor products of compact convex sets. *Pacific Journal of Mathematics, 31*, 469–480.

Pitowsky, I. (2005). Quantum mechanics as a theory of probability. In W. Demopolous (Ed.), *Physical theory and its interpretation: Essays in honor of Jeffrey Bub* (Western Ontario series in philosophy of science). Kluwer. arXiv:0510095.

Popescu, S., & Rohrlich, D. (1994). Quantum nonlocality as an axiom. *Foundations of Physics, 24*, 379–285.

Spekkens, R. (2005). Contextuality for preparations, transformations and unsharp measurements. *Physical Review A, 71*. arXiv:quant-ph/0406166.

Varadarajan, V. S. (1985). *The geometry of quantum theory* (2nd ed.). New York: Springer.

Werner, R. (1989). Quantum states with Einstein-Podolsky-Rosen correlations admitting a hidden-variable model. *Phys. Rev. A, 40*, 4277.

Wilce, A. (2005). Symmetry and topology in quantum logic. *International Journal of Theoretical Physics, 44*, 2303–2316.

Wilce, A. (2009). Symmetry and composition in generalized probabilistic theories. arXiv:0910.1527.

Wilce, A. (2012). Conjugates, filters and quantum mechanics. arXiv:1206.2897.

Wilce, A. (2017). Quantum logic and probability theory. In *Stanford encyclopedia of philosophy*. https://plato.stanford.edu/archives/spr2017/entries/qt-quantlog/.

Correction to: Everettian Probabilities, The Deutsch-Wallace Theorem and the Principal Principle

Harvey R. Brown and Gal Ben Porath

Correction to:
Chapter 7 in: M. Hemmo, O. Shenker (eds.), *Quantum, Probability, Logic*, Jerusalem Studies in Philosophy and History of Science, https://doi.org/10.1007/978-3-030-34316-3_7

In the original version of the book, there was a spelling error in the chapter title wherein "Everettian" was incorrectly spelt as "Eerettian". This is now corrected.

The updated version of this chapter can be found at
https://doi.org/10.1007/978-3-030-34316-3_7

M. Hemmo, O. Shenker (eds.), *Quantum, Probability, Logic*,
Jerusalem Studies in Philosophy and History of Science,
https://doi.org/10.1007/978-3-030-34316-3_28

MIX
Papier aus verantwortungsvollen Quellen
Paper from responsible sources
FSC® C105338

If you have any concerns about our products,
you can contact us on
ProductSafety@springernature.com

In case Publisher is established outside the EU,
the EU authorized representative is:
Springer Nature Customer Service Center GmbH
Europaplatz 3, 69115 Heidelberg, Germany

Printed by Libri Plureos GmbH
in Hamburg, Germany